Edited by
Jan-Erling Bäckvall

Modern Oxidation Methods

Related Titles

Mizuno, N. (ed.)

Modern Heterogeneous Oxidation Catalysis

Design, Reactions and Characterization

356 pages with 134 figures and 38 tables
2009
Hardcover
ISBN: 978-3-527-31859-9

Andersson, P. G., Munslow, I. J. (eds.)

Modern Reduction Methods

522 pages with 593 figures and 66 tables
2008
Hardcover
ISBN: 978-3-527-31862-9

Schmid, R. D., Urlacher, V. (eds.)

Modern Biooxidation

Enzymes, Reactions and Applications

318 pages with approx. 70 figures and approx. 20 tables
2007
Hardcover
ISBN: 978-3-527-31507-9

Hynes, J. T., Klinman, J. P., Limbach, H.-H., Schowen, R. L. (eds.)

Hydrogen-Transfer Reactions

1603 pages in 4 volumes with 539 figures and 69 tables
2007
Hardcover
ISBN: 978-3-527-30777-7

de Vries, J. G., Elsevier, C. J. (eds.)

The Handbook of Homogeneous Hydrogenation

1641 pages in 3 volumes with 422 figures and 254 tables
2007
Hardcover
ISBN: 978-3-527-31161-3

Edited by
Jan-Erling Bäckvall

Modern Oxidation Methods

2nd completely revised and enlarged edition

WILEY-VCH Verlag GmbH & Co. KGaA

The Editor

Prof. Dr. Jan-Erling Bäckvall
Stockholm University
Department of Organic Chem.
Arrhenius Lab.
106 91 Stockholm
Schweden

All books published by **Wiley-VCH** are carefully produced. Nevertheless, authors, editors, and publisher do not warrant the information contained in these books, including this book, to be free of errors. Readers are advised to keep in mind that statements, data, illustrations, procedural details or other items may inadvertently be inaccurate.

Library of Congress Card No.: applied for

British Library Cataloguing-in-Publication Data
A catalogue record for this book is available from the British Library.

Bibliographic information published by the Deutsche Nationalbibliothek
The Deutsche Nationalbibliothek lists this publication in the Deutsche Nationalbibliografie; detailed bibliographic data are available on the Internet at http://dnb.d-nb.de.

© 2010 Wiley-VCH Verlag & Co. KGaA, Boschstr. 12, 69469 Weinheim, Germany

All rights reserved (including those of translation into other languages). No part of this book may be reproduced in any form – by photoprinting, microfilm, or any other means – nor transmitted or translated into a machine language without written permission from the publishers. Registered names, trademarks, etc. used in this book, even when not specifically marked as such, are not to be considered unprotected by law.

Cover Design   Adam Design, Weinheim
Typesetting   Thomson Digital, Noida, India
Printing and Binding   betz-druck GmbH, Darmstadt

Printed in the Federal Republic of Germany
Printed on acid-free paper

ISBN: 978-3-527-32320-3

Contents

Preface *XI*
List of Contributors *XIII*

1   **Recent Developments in Metal-catalyzed Dihydroxylation of Alkenes** *1*
      Man Kin Tse, Kristin Schröder, and Matthias Beller
1.1   Introduction *1*
1.2   Environmentally Friendly Terminal Oxidants *3*
1.2.1   Hydrogen Peroxide *3*
1.2.2   Hypochlorite *5*
1.2.3   Chlorite *8*
1.2.4   Oxygen or Air *9*
1.3   Supported Osmium Catalyst *16*
1.3.1   Nitrogen-group Donating Support *16*
1.3.2   Microencapsulated OsO_4 *17*
1.3.3   Supports Bearing Alkenes *19*
1.3.4   Immobilization by Ionic Interaction *21*
1.4   Ionic Liquid *22*
1.5   Ruthenium Catalysts *23*
1.6   Iron Catalysts *26*
1.7   Conclusions *32*
      References *32*

2   **Transition Metal-Catalyzed Epoxidation of Alkenes** *37*
      Hans Adolfsson
2.1   Introduction *37*
2.2   Choice of Oxidant for Selective Epoxidation *38*
2.3   Epoxidations of Alkenes Catalyzed by Early Transition Metals *39*
2.4   Molybdenum and Tungsten-Catalyzed Epoxidations *42*
2.4.1   Homogeneous Catalysts – Hydrogen Peroxide as the Terminal Oxidant *42*
2.4.2   Heterogeneous Catalysts *46*
2.5   Manganese-Catalyzed Epoxidations *47*

Modern Oxidation Methods. Edited by Jan-Erling Bäckvall
Copyright © 2010 WILEY-VCH Verlag GmbH & Co. KGaA, Weinheim
ISBN: 978-3-527-32320-3

2.6	Rhenium-Catalyzed Epoxidations 52
2.6.1	MTO as Epoxidation Catalyst – Original Findings 54
2.6.2	The Influence of Heterocyclic Additives 55
2.6.3	The Role of the Additive 58
2.6.4	Other Oxidants 59
2.6.5	Solvents/Media 61
2.6.6	Solid Support 63
2.6.7	Asymmetric Epoxidations Using MTO 64
2.7	Iron-Catalyzed Epoxidations 64
2.7.1	Iron-Catalyzed Asymmetric Epoxidations 72
2.8	Ruthenium-Catalyzed Epoxidations 74
2.9	Epoxidations Using Late Transition Metals 76
2.10	Concluding Remarks 79
	References 80

3	**Organocatalytic Oxidation. Ketone-Catalyzed Asymmetric Epoxidation of Alkenes and Synthetic Applications** 85
	Yian Shi
3.1	Introduction 85
3.2	Catalyst Development 86
3.3	Synthetic Applications 98
3.4	Conclusion 109
	References 109

4	**Catalytic Oxidations with Hydrogen Peroxide in Fluorinated Alcohol Solvents** 117
	Albrecht Berkessel
4.1	Introduction 117
4.2	Properties of Fluorinated Alcohols 118
4.2.1	A Detailed Look at the Hydrogen Bond Donor Features of HFIP 120
4.3	Epoxidation of Alkenes in Fluorinated Alcohol Solvents 123
4.3.1	Alkene Epoxidation with Hydrogen Peroxide – in the Absence of Further Catalysts 123
4.3.1.1	On the Mechanism of Epoxidation Catalysis by Fluorinated Alcohols 123
4.3.2	Alkene Epoxidation with Hydrogen Peroxide – in the Presence of Further Catalysts 129
4.3.2.1	Arsines and Arsine Oxides as Catalysts 129
4.3.2.2	Arsonic Acids as Catalysts 130
4.3.2.3	Diselenides/Seleninic Acids as Catalysts 132
4.3.2.4	Rhenium Compounds as Catalysts 133
4.3.2.5	Fluoroketones as Catalysts 135
4.4	Sulfoxidation of Thioethers in Fluorinated Alcohol Solvents 136
4.5	Baeyer-Villiger Oxidation of Ketones in Fluorinated Alcohol Solvents 136

4.5.1	Acid-Catalyzed Baeyer-Villiger Oxidation of Ketones in Fluorinated Alcohol Solvents – Mechanism *139*	
4.5.2	Baeyer-Villiger Oxidation of Ketones in Fluorinated Alcohol Solvents – Catalysis by Arsonic and Seleninic Acids *141*	
4.6	Epilog *142*	
	References *143*	

5	**Modern Oxidation of Alcohols using Environmentally Benign Oxidants** *147*	
	Isabel W.C.E. Arends and Roger A. Sheldon	
5.1	Introduction *147*	
5.2	Oxoammonium based Oxidation of Alcohols – TEMPO as Catalyst *147*	
5.3	Metal-Mediated Oxidation of Alcohols – Mechanism *151*	
5.4	Ruthenium-Catalyzed Oxidations with O_2 *153*	
5.5	Palladium-Catalyzed Oxidations with O_2 *163*	
5.5.1	Gold Nanoparticles as Catalysts *169*	
5.6	Copper-Catalyzed Oxidations with O_2 *170*	
5.7	Other Metals as Catalysts for Oxidation with O_2 *174*	
5.8	Catalytic Oxidation of Alcohols with Hydrogen Peroxide *176*	
5.8.1	Biocatalytic Oxidation of Alcohols *179*	
5.9	Concluding Remarks *180*	
	References *180*	

6	**Aerobic Oxidations and Related Reactions Catalyzed by N-Hydroxyphthalimide** *187*	
	Yasutaka Ishii, Satoshi Sakaguchi, and Yasushi Obora	
6.1	Introduction *187*	
6.2	NHPI-Catalyzed Aerobic Oxidation *188*	
6.2.1	Alkane Oxidations with Dioxygen *188*	
6.2.2	Oxidation of Alkylarenes *193*	
6.2.2.1	Synthesis of Terephthalic Acid *196*	
6.2.2.2	Oxidation of Methylpyridines and Methylquinolines *199*	
6.2.2.3	Oxidation of Hydroaromatic and Benzylic Compounds *201*	
6.2.3	Preparation of Acetylenic Ketones by Direct Oxidation of Alkynes *203*	
6.2.4	Oxidation of Alcohols *205*	
6.2.5	Epoxidation of Alkenes using Dioxygen as Terminal Oxidant *208*	
6.2.6	Baeyer-Villiger Oxidation of KA Oil *209*	
6.2.7	Preparation of ε-Caprolactam Precursor from KA Oil *210*	
6.3	Functionalization of Alkanes Catalyzed by NHPI *211*	
6.3.1	Carboxylation of Alkanes with CO and O_2 *211*	
6.3.2	First Catalytic Nitration of Alkanes using NO_2 *212*	
6.3.3	Sulfoxidation of Alkanes Catalyzed by Vanadium *214*	
6.3.4	Reaction of NO with Organic Compounds *217*	
6.3.5	Nitrosation of Cycloalkanes with *t*-BuONO *219*	
6.3.6	Ritter-type Reaction with Cerium Ammonium Nitrate (CAN) *220*	

6.4	Carbon-Carbon Bond-Forming Reaction via Catalytic Carbon Radicals Generation Assisted by NHPI 222
6.4.1	Oxyalkylation of Alkenes with Alkanes and Dioxygen 222
6.4.2	Synthesis of α-Hydroxy-γ-lactones by Addition of α-Hydroxy Carbon Radicals to Unsaturated Esters 223
6.4.3	Hydroxyacylation of Alkenes using 1,3-Dioxolanes and Dioxygen 224
6.4.4	Hydroacylation of Alkenes Using NHPI as a Polarity Reversal Catalyst 226
6.4.5	Chiral NHPI Derivatives as Enantioselective Catalysts: Kinetic Resolution of Oxazolidines 228
6.5	Conclusions 229
	References 230
7	**Ruthenium-Catalyzed Oxidation for Organic Synthesis** 241
	Shun-Ichi Murahashi and Naruyoshi Komiya
7.1	Introduction 241
7.2	RuO_4-Promoted Oxidation 241
7.3	Oxidation with Low-Valent Ruthenium Catalysts and Oxidants 245
7.3.1	Oxidation of Alkenes 245
7.3.2	Oxidation of Alcohols 249
7.3.3	Oxidation of Amines 255
7.3.4	Oxidation of Amides and β-Lactams 260
7.3.5	Oxidation of Phenols 262
7.3.6	Oxidation of Hydrocarbons 265
	References 268
8	**Selective Oxidation of Amines and Sulfides** 277
	Jan-E. Bäckvall
8.1	Introduction 277
8.2	Oxidation of Sulfides to Sulfoxides 277
8.2.1	Stoichiometric Reactions 278
8.2.1.1	Peracids 278
8.2.1.2	Dioxiranes 278
8.2.1.3	Oxone and Derivatives 279
8.2.1.4	H_2O_2 in "Fluorous Phase" and Related Reactions 279
8.2.2	Chemocatalytic Reactions 280
8.2.2.1	H_2O_2 as Terminal Oxidant 280
8.2.2.2	Molecular Oxygen as Terminal Oxidant 293
8.2.2.3	Alkyl Hydroperoxides as Terminal Oxidant 295
8.2.2.4	Other Oxidants in Catalytic Reactions 297
8.2.3	Biocatalytic Reactions 297
8.2.3.1	Peroxidases 298
8.2.3.2	Ketone Monooxygenases 299
8.3	Oxidation of Tertiary Amines to *N*-Oxides 300
8.3.1	Stoichiometric Reactions 300
8.3.2	Chemocatalytic Oxidations 302

8.3.3	Biocatalytic Oxidation	*306*
8.3.4	Applications of Amine N-Oxidation in Coupled Catalytic Processes	*306*
8.4	Concluding Remarks	*308*
	References	*309*

9 Liquid Phase Oxidation Reactions Catalyzed by Polyoxometalates *315*
Ronny Neumann

9.1	Introduction	*315*
9.2	Polyoxometalates (POMs)	*316*
9.3	Oxidation with Mono-Oxygen Donors	*317*
9.4	Oxidation with Peroxygen Compounds	*323*
9.5	Oxidation with Molecular Oxygen	*331*
9.6	Heterogenization of Homogeneous Reactions – Solid-Liquid, Liquid-Liquid, and Alternative Reaction Systems	*341*
9.6.1	Solid-Liquid Reactions	*341*
9.6.2	Liquid-Liquid Reactions and Reactions in 'Alternative' Media	*343*
9.7	Conclusion	*346*
	References	*346*

10 Oxidation of Carbonyl Compounds *353*
Eric V. Johnston and Jan-E. Bäckvall

10.1	Introduction	*353*
10.2	Oxidation of Aldehydes to Carboxylic Acids	*353*
10.2.1	Metal-Free Oxidation of Aldehydes to Carboxylic Acids	*354*
10.2.2	Metal-Catalyzed Oxidation of Aldehydes to Carboxylic Acids	*355*
10.3	Oxidation of Ketones	*356*
10.3.1	Baeyer-Villiger Reactions	*356*
10.3.2	Catalytic Asymmetric Baeyer-Villiger Reactions	*356*
10.3.2.1	Chemocatalytic Versions	*357*
10.3.2.2	Biocatalytic Versions	*358*
	References	*365*

11 Manganese-Catalyzed Oxidation with Hydrogen Peroxide *371*
Wesley R. Browne, Johannes W. de Boer, Dirk Pijper, Jelle Brinksma, Ronald Hage, and Ben L. Feringa

11.1	Introduction	*371*
11.2	Bio-inspired Manganese Oxidation Catalysts	*372*
11.3	Manganese-Catalyzed Bleaching	*375*
11.4	Epoxidation and *cis*-Dihydroxylation of Alkenes	*375*
11.4.1	Manganese Salts	*376*
11.4.2	Porphyrin-Based Catalysis	*378*
11.4.3	Salen-Based Systems	*381*
11.4.4	Tri- and Tetra-azacycloalkane Derivatives	*385*
11.4.4.1	Tetra-azacycloalkane Derivatives	*386*
11.4.4.2	Triazacyclononane Derivatives	*387*

11.4.4.3	Manganese Complexes for Alkene Oxidation Based on Pyridyl Ligands 403
11.5	Manganese Catalysts for the Oxidation of Alkanes, Alcohols, and Aldehydes 406
11.5.1	Oxidation of Alkanes 406
11.5.2	Oxidation of Alcohols and Aldehydes 407
11.5.3	Sulfides, Sulfoxides, and Sulfones 408
11.6	Conclusions 411
	References 412
12	**Biooxidation with Cytochrome P450 Monooxygenases** 421
	Marco Girhard and Vlada B. Urlacher
12.1	Introduction 421
12.2	Properties of Cytochrome P450 Monooxygenases 422
12.2.1	Structure 422
12.2.2	Enzymology 423
12.2.3	Reactions Catalyzed by P450s 425
12.2.4	P450s as Industrial Biocatalysts 429
12.2.4.1	Advantages 429
12.2.4.2	Challenges in the Development of Technical P450 Applications 429
12.2.4.3	General Aspects of Industrial Application and Engineering of P450s 430
12.3	Application and Engineering of P450s for the Pharmaceutical Industry 430
12.3.1	Microbial Oxidations with P450s for Synthesis of Pharmaceuticals 431
12.3.2	Application of Mammalian P450s for Drug Development 434
12.3.2.1	Enhancement of Recombinant Expression in *E. coli* 435
12.3.2.2	Enhancement of Activity and Selectivity and Engineering of Novel Activities 436
12.3.2.3	Construction of Artificial Self-Sufficient Fusion Proteins 436
12.4	Application of P450s for Synthesis of Fine Chemicals 437
12.5	Engineering of P450s for Biocatalysis 438
12.5.1	Cofactor Substitution and Regeneration 438
12.5.1.1	Cofactor Substitution *In Vitro* 438
12.5.1.2	Cofactor Regeneration *In Vitro* 439
12.5.1.3	Cofactor Regeneration in Whole-Cells 439
12.5.2	Construction of Artificial Fusion Proteins 440
12.5.3	Engineering of New Substrate Specificities 440
12.5.3.1	P450$_{cam}$ from *Pseudomonas putida* 440
12.5.3.2	P450$_{BM3}$ from *Bacillus megaterium* 442
12.6	Future Trends 443
	References 444

Index 451

Preface

Oxidation reactions continue to play an important role in organic chemistry, and the increasing demand for selective and mild oxidation methods in modern organic synthesis has led to rich developments in the field during recent decades. Significant progress has been achieved within the area of catalytic oxidations, and this has led to a range of selective and mild processes. These reactions can be based on metal catalysis, organocatalysis, or biocatalysis, enantioselective catalytic oxidation reactions being of particular interest.

The First Edition of the multi-authored book 'Modern Oxidation Methods' was published in 2004 with the aim of fulfilling the need for an overview of the latest developments in the field. In particular, some general and synthetically useful oxidation methods that are frequently used by organic chemists were covered, including catalytic as well as noncatalytic oxidation reactions, the emphasis being on catalytic methods that employ environmentally friendly ('green') oxidants such as molecular oxygen and hydrogen peroxide. These oxidants are atom economic and lead to a minimum amount of waste.

This Second Edition has in total twelve chapters, each covering an area of contemporary interest, and now includes two additional chapters on topics that were not covered in the first book, the other chapters having been updated. One of the added chapters (Chapter 12) reviews biooxidation with cytochrome P450 monooxygenases, an area of increasing interest, and the other (Chapter 4) covers oxidations with hydrogen peroxide in fluorinated alcohol solvents. Topics that are reviewed in the updated chapters involve olefin oxidations and include osmium-catalyzed dihydroxylation, metal-catalyzed epoxidation, and organocatalytic epoxidation. In subsequent chapters, catalytic alcohol oxidation with environmentally benign oxidants and aerobic oxidations catalyzed by *N*-hydroxyphthalimides (NHPI), with a special focus on the oxidation of hydrocarbons via C–H activation, are reviewed. Other chapters include recent advances in ruthenium-catalyzed oxidations in organic synthesis, selective oxidation of amines and sulfides, oxidations catalyzed by polyoxymetalates, oxidation of carbonyl compounds, and manganese-catalyzed H_2O_2 oxidations.

Modern Oxidation Methods. Edited by Jan-Erling Bäckvall

ISBN: 978-3-527-32320-3

I hope that the Second Edition of 'Modern Oxidation Methods' will be of value to chemists involved in oxidation reactions in both academic and industrial research and that it will stimulate further development in this important field. Finally, I would like to warmly thank all the authors for their excellent contributions.

Stockholm, June 2010

Jan-E. Bäckvall

List of Contributors

Hans Adolfsson
Stockholm University
Arrhenius Laboratory
Department of Organic Chemistry
SE-106 91 Stockholm
Sweden

Isabel W.C.E. Arends
Delft University of Technology
Department of Biotechnology
Laboratory for Biocatalysis and Organic Chemistry
Julianalaan 136
2628 BL Delft
The Netherlands

Jan-Erling Bäckvall
Stockholm University
Arrhenius Laboratory
Department of Organic Chemistry
SE-106 91 Stockholm
Sweden

Matthias Beller
Leibniz-Institut für Katalyse e.V. an der Universität Rostock
Albert-Einstein-Str. 29a
D-18059 Rostock
Germany

and

University of Rostock
Center for Life Science Automation (CELISCA)
Friedrich-Barnewitz-Str. 8
D-18119 Rostock-Warnemünde
Germany

Albrecht Berkessel
University of Cologne
Chemistry Department
Greinstraße 4
D-50939 Cologne
Germany

Jelle Brinksma
University of Groningen
Stratingh Institute for Chemistry
Center for Systems Chemistry
Nijenborgh 4
9747 AG Groningen
The Netherlands

Modern Oxidation Methods. Edited by Jan-Erling Bäckvall
Copyright © 2010 WILEY-VCH Verlag GmbH & Co. KGaA, Weinheim
ISBN: 978-3-527-32320-3

List of Contributors

Wesley R. Browne
University of Groningen
Stratingh Institute for Chemistry
Center for Systems Chemistry
Nijenborgh 4
9747 AG Groningen
The Netherlands

Johannes W. de Boer
University of Groningen
Stratingh Institute for Chemistry
Center for Systems Chemistry
Nijenborgh 4
9747 AG Groningen
The Netherlands

and

Rahu Catalytics
Biopartner Center Leiden
Wassenaarseweg 72
2333 AL Leiden
The Netherlands

Ben L. Feringa
University of Groningen
Stratingh Institute for Chemistry
Center for Systems Chemistry
Nijenborgh 4
9747 AG Groningen
The Netherlands

Marco Girhard
Heinrich-Heine-Universität Düsseldorf
Institut für Biochemie
Universitätsstr. 1, Geb. 26.02
40225 Düsseldorf
Germany

Ronald Hage
Rahu Catalytics
Biopartner Center Leiden
Wassenaarseweg 72
2333 AL Leiden
The Netherlands

Yasutaka Ishii
Kansai University
Faculty of Chemistry, Materials and Bioengineering
Department of Chemistry and Materials Engineering
Suita, Osaka 564-8680
Japan

Eric V. Johnston
Stockholm University
Arrhenius Laboratory
Department of Organic Chemistry
SE-106 91 Stockholm
Sweden

Naruyoshi Komiya
Osaka University
Graduate School of Engineering Science
Department of Chemistry
1-3, Machikaneyama, Toyonaka
Osaka 560-8531
Japan

Shun-Ichi Murahashi
Okayama University of Science
Department of Applied Chemistry
1-1, Ridai-cho
Okayama 700-0005
Japan

and

Osaka University
Graduate School of Engineering Science
Department of Chemistry
1-3, Machikaneyama, Toyonaka
Osaka 560-8531
Japan

Ronny Neumann
Weizmann Institute of Science
Department of Organic Chemistry
Rehovot 76100
Israel

Yasushi Obora
Kansai University
Faculty of Chemistry, Materials and
Bioengineering
Department of Chemistry and Materials
Engineering
Suita, Osaka 564-8680
Japan

Dirk Pijper
University of Groningen
Stratingh Institute for Chemistry
Center for Systems Chemistry
Nijenborgh 4
9747 AG Groningen
The Netherlands

Satoshi Sakaguchi
Kansai University
Faculty of Chemistry, Materials and
Bioengineering
Department of Chemistry and Materials
Engineering
Suita, Osaka 564-8680
Japan

Kristin Schröder
Leibniz-Institut für Katalyse e.V. an der
Universität Rostock
Albert-Einstein-Str. 29a
D-18059 Rostock
Germany

Roger A. Sheldon
Delft University of Technology
Department of Biotechnology
Laboratory for Biocatalysis and Organic
Chemistry
Julianalaan 136
2628 BL Delft
The Netherlands

Yian Shi
Colorado State University
Department of Chemistry
Fort Collins, CO 80523
USA

Man Kin Tse
Leibniz-Institut für Katalyse e.V. an der
Universität Rostock
Albert-Einstein-Str. 29a
D-18059 Rostock
Germany

and

University of Rostock
Center for Life Science Automation
(CELISCA)
Friedrich-Barnewitz-Str. 8
D-18119 Rostock-Warnemünde
Germany

Vlada B. Urlacher
Heinrich-Heine-Universität Düsseldorf
Institut für Biochemie
Universitätsstr. 1, Geb. 26.02
40225 Düsseldorf
Germany

1
Recent Developments in Metal-catalyzed Dihydroxylation of Alkenes

Man Kin Tse, Kristin Schröder, and Matthias Beller

1.1
Introduction

The oxidative functionalization of alkenes is of major importance in the chemical industry, both in organic synthesis and in the industrial production of bulk and fine chemicals [1]. Among the various oxidation products of alkenes, 1,2-diols have numerous applications. Ethylene and propylene glycols are produced annually on a multi-million tons scale as polyester monomers and anti-freeze agents [2]. A number of 1,2-diols such as 2,3-dimethyl-2,3-butanediol, 1,2-octanediol, 1,2-hexanediol, 1,2-pentanediol, and 1,2- and 2,3-butanediol are important starting materials for the fine chemical industry. In addition, enantiomerically enriched 1,2-diols are employed as intermediates in the production of pharmaceuticals and agrochemicals. Nowadays 1,2-diols are mainly manufactured by a two-step sequence consisting of epoxidation of an alkene with a hydroperoxide, a peracid, or oxygen followed by hydrolysis of the resulting epoxide [3]. Compared to the epoxidation-hydrolysis process, dihydroxylation of C=C double bonds comprises a more atom-efficient and shorter route to 1,2-diols. In general dihydroxylation of alkenes is catalyzed by osmium, ruthenium, iron, or manganese oxo species. Though considerable advances in biomimetic non-heme complexes have been achieved in recent years, the osmium-catalyzed variant is still the most reliable and efficient method for the synthesis of *cis*-1,2-diols [4]. Using osmium as a catalyst with stoichiometric amounts of a secondary oxidant, various alkenes, including mono-, di-, and tri-substituted unfunctionalized as well as many functionalized alkenes, can be converted to the corresponding diols. Electrophilic OsO_4 reacts only slowly with electron-deficient alkenes; hence, it is necessary to employ higher amounts of catalyst and ligand for these alkenes. Recent studies have revealed that these substrates react much more efficiently when the reaction medium is maintained in an acidic state [5]. Citric acid appears to be superior for maintaining the pH in the desired range. However, it acts also as a ligand in this reaction but does not provide any asymmetric information transfer to the alkene. In contrast, it was found in another

Modern Oxidation Methods. Edited by Jan-Erling Bäckvall
Copyright © 2010 WILEY-VCH Verlag GmbH & Co. KGaA, Weinheim
ISBN: 978-3-527-32320-3

study that nonreactive internal alkenes, especially tetra-substituted ones, react faster at a constant pH value of 12.0 [6].

Since its discovery by Sharpless and coworkers, catalytic asymmetric dihydroxylation (AD) has significantly enhanced the utility of osmium-catalyzed dihydroxylation (Scheme 1.1) [7]. Numerous applications in organic synthesis have appeared in recent years [8].

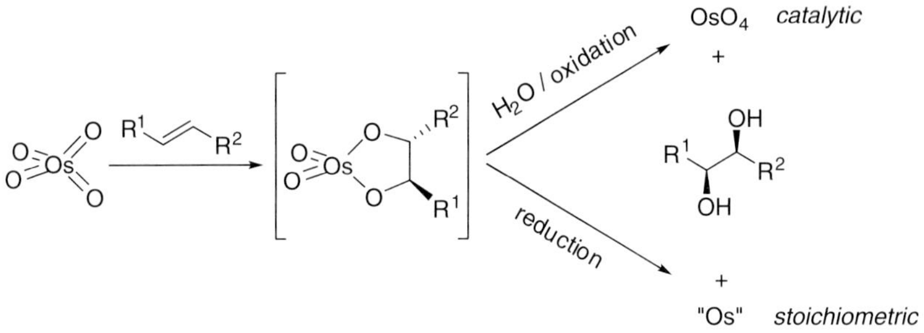

Scheme 1.1 Osmylation of alkenes.

While the enantioselectivity of the reaction has largely been advanced through extensive synthesis and screening of cinchona alkaloid ligands by the Sharpless group, larger scale applications of this method remain problematic. The minimization of the use of expensive osmium catalyst and the efficient recycling of the metal should be a primary focus for the development. Coming in second is the replacement of the costly reoxidants, which generate overstoichiometric amounts of waste in the form of Os(VI) species.

Several reoxidation processes for osmium(VI) glycolates or other osmium(VI) species have been developed. Historically, chlorates [9] and hydrogen peroxide [10] were first applied as stoichiometric oxidants; however, in both cases, the dihydroxylation often proceeds with low chemoselectivity. Other reoxidants for osmium(VI) are *tert*-butyl hydroperoxide in the presence of Et_4NOH [11] and a range of N-oxides such as N-methylmorpholine N-oxide (NMO) [12] (Upjohn process), and trimethylamine N-oxide. $K_3[Fe(CN)_6]$ gave a substantial improvement in the enantioselectivities in asymmetric dihydroxylations when it was introduced as a reoxidant for osmium(VI) species in 1990 [13]. However, $K_3[Fe(CN)_6]$ was already described as an oxidant for other Os-catalyzed oxidation reactions as early as in 1975 [14]. Today the 'AD-mix', a combination of the catalyst precursor $K_2[OsO_2(OH)_4]$, the co-oxidant $K_3[Fe(CN)_6]$, the base K_2CO_3, and the chiral ligand, is commercially available, and the dihydroxylation reaction is easy to carry out. However, the production of overstoichiometric amounts of waste continues to be a significant disadvantage of the reaction protocol.

This article updates an earlier version in the first edition of this book and summarizes the recent developments in the area of osmium-catalyzed dihydroxylations which bring this transformation closer to a 'green reaction'. Special emphasis is placed on the use of new reoxidants and recycling of the osmium catalyst. Moreover, less toxic metal catalysts such as ruthenium and iron are also discussed.

1.2
Environmentally Friendly Terminal Oxidants

1.2.1
Hydrogen Peroxide

Since the publication of the Upjohn procedure in 1976, the use of N-methylmorpholine N-oxide (NMO) based oxidants has become one of the standard methods for osmium-catalyzed dihydroxylations. However, NMO has not been fully appreciated in the asymmetric dihydroxylation for a long time since it was difficult to obtain high enantiomeric excess (ee). This drawback was significantly improved by slow addition of the alkene to the aqueous tert-BuOH reaction mixture, in which 97% ee was achieved with styrene [15].

Although hydrogen peroxide was one of the first stoichiometric oxidants used in osmium-catalyzed dihydroxylation [10a], it was not employed efficiently until recently. When hydrogen peroxide is used as a reoxidant for transition metal catalysts, a very common big disadvantage is that a large excess of H_2O_2 is required to compensate for the major unproductive peroxide decomposition to O_2.

Recently, Bäckvall and coworkers were able to improve the H_2O_2 reoxidation process significantly by using N-methylmorpholine together with an electron transfer mediator (ETM) as co-catalysts in the presence of hydrogen peroxide (Figure 1.1) [16]. Thus, a renaissance of both NMO and H_2O_2 was brought about. The mechanism of the triply catalyzed H_2O_2 oxidation is shown in Scheme 1.2.

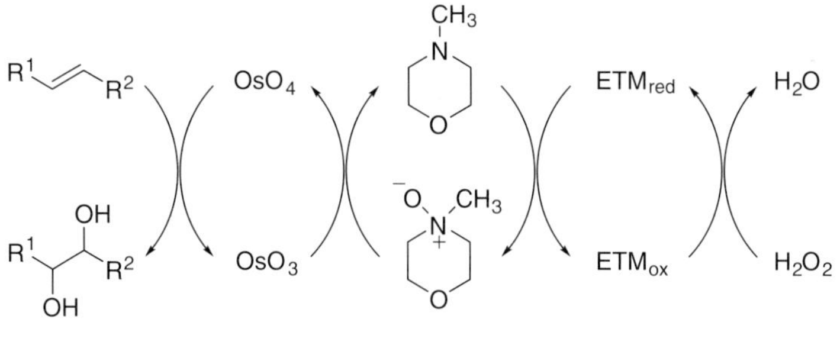

Scheme 1.2 Osmium-catalyzed dihydroxylation of alkenes using H_2O_2 as the terminal oxidant.

The oxidized electron transfer mediator (ETM_{ox}), namely the peroxo complexes of methyltrioxorhenium (MTO) and vanadyl acetylacetonate [VO(acac)$_2$] and flavin hydroperoxide, generated from its reduced form (Figure 1.1) and H_2O_2, recycles the N-methylmorpholine (NMM) to N-methylmorpholine N-oxide (NMO), which in turn reoxidizes the Os(VI) to Os(VIII). While the use of hydrogen peroxide as oxidant without any electron transfer mediators is inefficient and nonselective, various alkenes were oxidized to diols in good to excellent yields employing this mild triple catalytic system (Scheme 1.2).

By using a chiral Sharpless ligand, high enantioselectivities were obtained. In the flavin system, an increase in the addition time for alkene and H_2O_2 has a positive

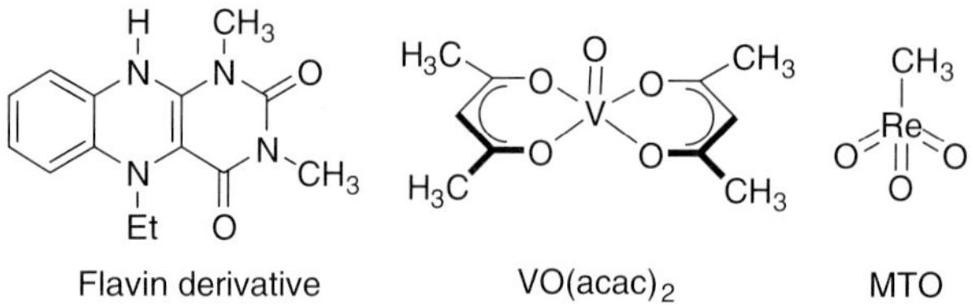

Figure 1.1 Electron transfer mediators (ETMs).

effect on the enantioselectivity. For example, α-methylstyrene was oxidized with the aid of flavin as the ETM to its corresponding vicinal diols in good yield (88%) and excellent enantiomeric excess (99% ee) (Scheme 1.3).

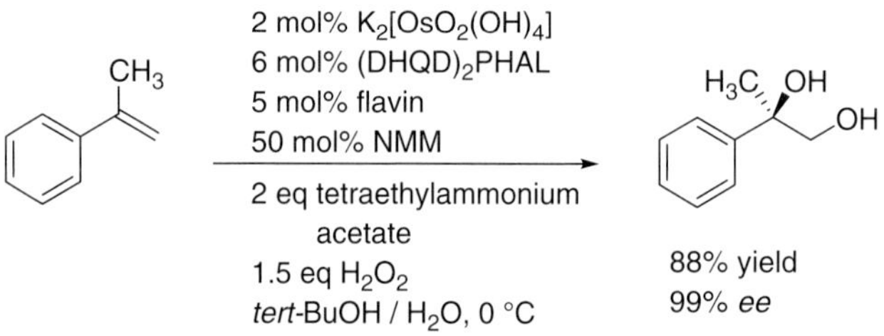

Scheme 1.3 Osmium-catalyzed dihydroxylation of α-methylstyrene using H_2O_2.

Bäckvall and coworkers have shown that other tertiary amines can assume the role of the N-methylmorpholine [16c]. They reported the first example of an enantioselective catalytic redox process where the chiral tertiary amine ligand has two different modes of operation. The primary function is to provide stereocontrol in the addition of the substrate, and the secondary function is to reoxidize the metal through the N-oxide [16c]. The results obtained with hydroquinidine 1,4-phthalazinediyl diether $(DHQD)_2PHAL$ both as electron transfer mediator and chiral ligand in the osmium-catalyzed dihydroxylation are comparable to those obtained employing NMM together with $(DHQD)_2PHAL$. The proposed catalytic cycle for the reaction is depicted in Scheme 1.4 [16c,e].

Flavin is an efficient electron transfer mediator, but is rather unstable. Several transition metal complexes, such as vanadyl acetylacetonate, can also activate hydrogen peroxide and are capable of replacing flavin in the dihydroxylation reaction [16d,g]. The co-catalyst loading can hence be reduced from 5 mol% (flavin) to 2 mol% [VO(acac)$_2$ or MTO] with comparable yield and ee. The introduction of ETM for the oxidation of tertiary amine N-oxides significantly enhanced the efficiency of H_2O_2. With 1.5 equivalents of commercially available aqueous 30% H_2O_2, good to excellent yield can be achieved [16e]. Interestingly, an ETM, MTO, catalyzes oxidation of the chiral ligand to its mono-N-oxide, which in turn can be used as the oxidant to

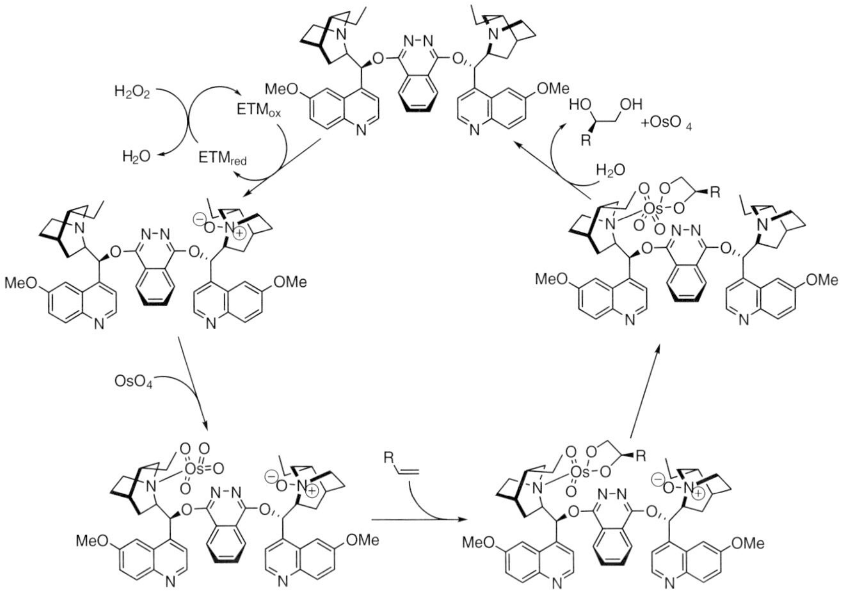

Scheme 1.4 Catalytic cycle for the enantioselective dihydroxylation of alkenes using (DHQD)$_2$PHAL for oxygen transfer and as source of chirality.

generate Os^{VIII} from Os^{VI}. The system gives phenylethanediol diols in 71% yield with 98% *ee*. This clearly confirms the role of the ETM in the regeneration of *N*-oxides during the dihydroxylation process.

1.2.2
Hypochlorite

Apart from oxygen and hydrogen peroxide, bleach is the simplest and most economical oxidant, and is widely used in industry without problems. This oxidant has only been applied in the presence of osmium complexes in two patents in the early 1970s for the oxidation of fatty acids [17]. In 2003 the first general dihydroxylation procedure of various alkenes in the presence of sodium hypochlorite as the reoxidant was described by our group [18]. Using α-methylstyrene as a model compound, 100% conversion and 98% yield of the desired 1,2-diol were demonstrated (Scheme 1.5).

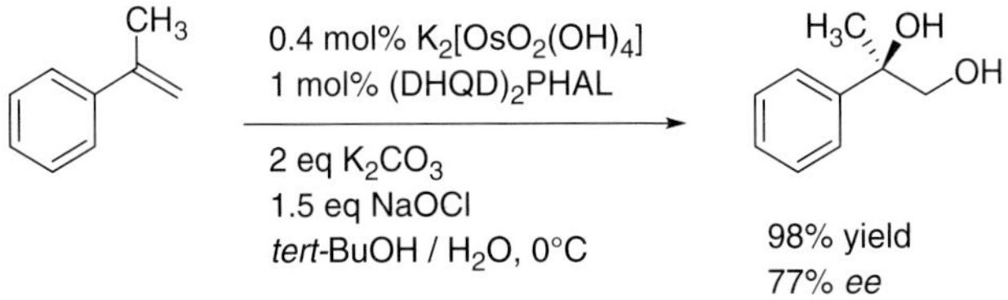

Scheme 1.5 Osmium-catalyzed dihydroxylation of α-methylstyrene using sodium hypochlorite.

The efficacy of hypochlorite is significantly higher than that of other conventional oxidants. The yield of 2-phenyl-1,2-propanediol reached 98% after only 1 h, while literature protocols using NMO [19] or $K_3[Fe(CN)_6]$ [7b] gave both in 90% yield at 0 °C. The turnover frequency of the hypochlorite system was $242\,h^{-1}$, which is a reasonable level for further industrial application [20]. Under the conditions shown in Scheme 1.5 an enantioselectivity of only 77% ee was obtained, while 94% ee was reported using $K_3[Fe(CN)_6]$ as reoxidant [7b]. The lower enantioselectivity can be explained by some involvement of the so-called second catalytic cycle with the intermediate Os(VI) glycolate being oxidized to an Os(VIII) species prior to hydrolysis (Scheme 1.6) [21].

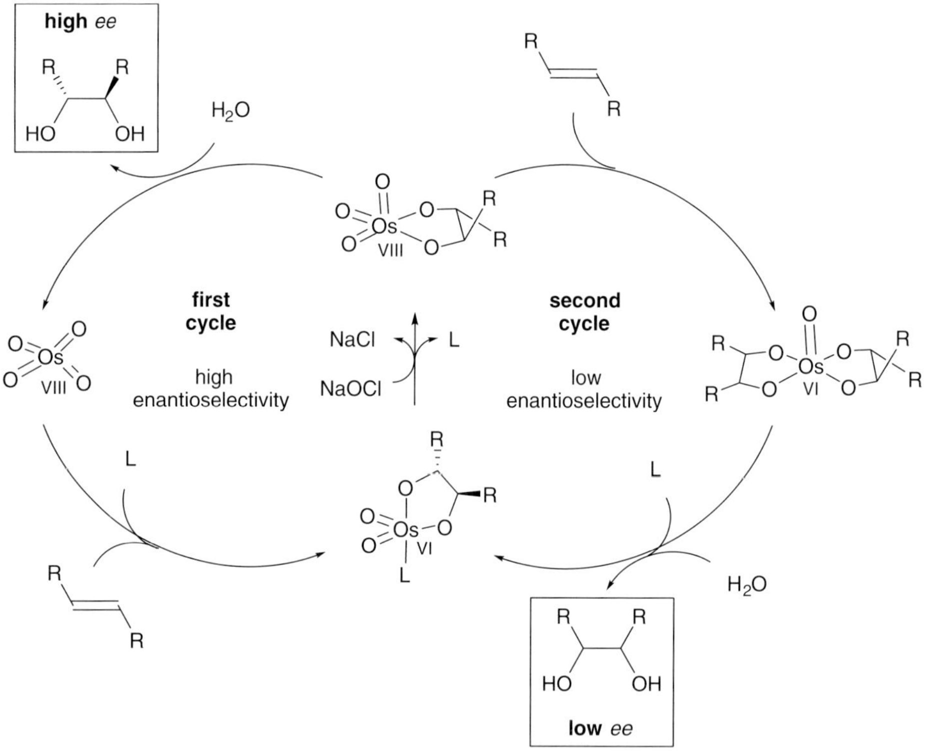

Scheme 1.6 The two catalytic cycles in asymmetric dihydroxylation.

The enantioselectivity can be improved by applying a higher ligand concentration. In the presence of 5 mol% (DHQD)$_2$PHAL a good enantioselectivity (91% ee) was observed for α-methylstyrene. Using *tert*-butylmethylether as organic co-solvent instead of *tert*-butanol, 99% yield and 89% ee with only 1 mol% (DHQD)$_2$PHAL were reported for the same substrate. An increase in the concentration of the chiral ligand in the organic phase could be an explanation of this increase in enantioselectivity. Increasing the polarity of the water phase by using a 10% aqueous NaCl

1.2 Environmentally Friendly Terminal Oxidants

Table 1.1 Asymmetric dihydroxylation of different alkenes using NaOCl as the terminal oxidant.[a]

Entry	Alkene	Time (h)	Yield (%)	Selectivity (%)	ee (%)	ee (%) Ref. [4b][b]	ee (%) Ref. [15][c]
1	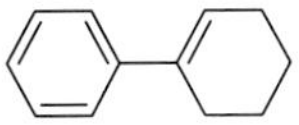 cyclohexyl-CH=CH-cyclohexyl	1	88	88	95	99	—
2	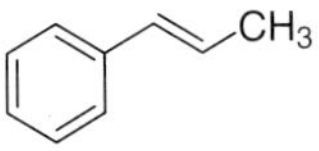 PhO-CH=CH-CH$_3$	2	93	99	95	97	98
3	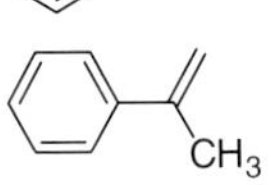 Ph-CH=C(CH$_3$)-	1	99	99	91	95	—
4	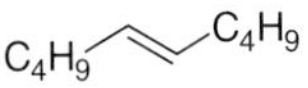 C$_4$H$_9$-CH=CH-C$_4$H$_9$	1	92	94	93	97	—
5	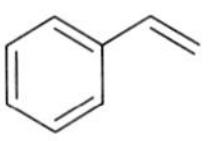 Ph-CH=CH-	1	84	84	91	97	97
6	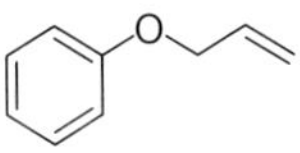 PhO-CH$_2$-O-CH=CH-	2	88	94	73	88	—
7	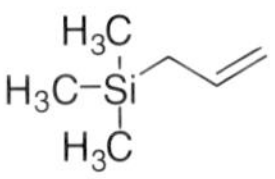 (H$_3$C)$_3$Si-CH=CH-	2	87	93	80[d]	—	—
8	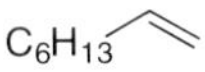 C$_6$H$_{13}$-CH=CH-	2	97	97	73	—	—
9	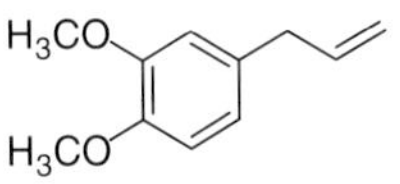 (H$_3$CO)$_2$-Ar-CH=CH-	2	94	96	34[d]	—	—
10	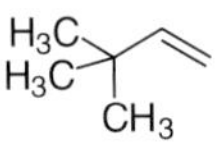 (H$_3$C)$_2$CH-CH=CH-CH$_3$	2	97	>97	80[d]	92	46

a) Reaction conditions: 2 mmol alkene, 0.4 mol% K$_2$[OsO$_2$(OH)$_4$], 5 mol% (DHQD)$_2$PHAL, 10 mL H$_2$O, 10 mL tert-BuOH, 1.5 equiv. NaOCl, 2 equiv. K$_2$CO$_3$, 0 °C.
b) K$_3$[Fe(CN)$_6$] as oxidant.
c) NMO as oxidant.
d) 5 mol% (DHQD)$_2$PYR instead of (DHQD)$_2$PHAL.

solution showed a similar positive effect. Table 1.1 shows the results of the asymmetric dihydroxylation of various alkenes with NaOCl as terminal oxidant.

Despite the slow hydrolysis of the sterically hindered Os(VI) glycolate, trans-5-decene reacted smoothly to give the corresponding diol. This result is especially impressive since addition of stoichiometric amounts of hydrolysis aids is usually necessary in the dihydroxylation of most internal alkenes in the presence of other oxidants.

Thus this 'hypochlorite-procedure' is economical, productive, and easy to manage for asymmetric dihydroxylation.

1.2.3
Chlorite

The pH of the system is of vital importance in osmium-catalyzed dihydroxylation reactions [5, 6]. An additional base, which aids the hydrolysis of the osmium glycolate, is usually present in the recipe of the general procedure. In 2004 Hormi and Junttila introduced sodium chlorite as the new reoxidant in asymmetric dihydroxylation [22]. $NaClO_2$ acts as both an oxidant and a hydroxyl ion pump in the system (Scheme 1.7). The pH value was maintained since each $NaClO_2$ provides the reaction with the necessary stoichiometric number of electrons and hydroxide ions during the reaction profile. Various alkenes can be dihydroxylated to the corresponding diols in good yield (63–80%) with good enantioselectivity (41–>99.5% ee) (Scheme 1.8).

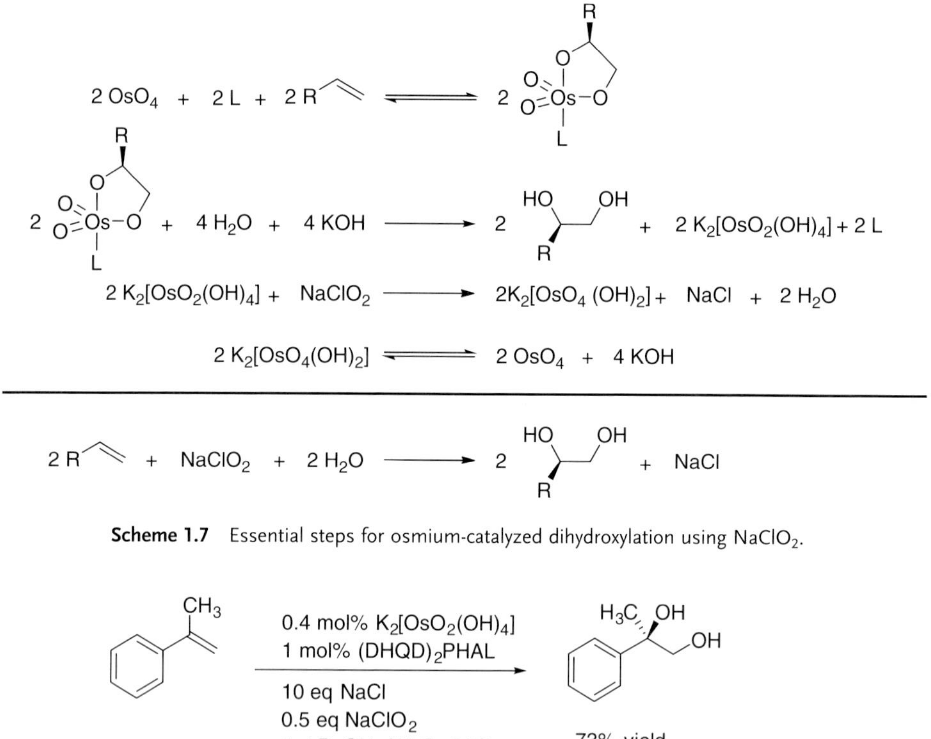

$$2\ OsO_4 + 2\ L + 2\ R\!\!-\!\!\!=\!\!\!-\ \rightleftharpoons\ 2\ [Os\ glycolate]$$

$$2\ [Os\ glycolate] + 4\ H_2O + 4\ KOH\ \longrightarrow\ 2\ \underset{R}{HO\diagdown\!\!\diagup OH} + 2\ K_2[OsO_2(OH)_4] + 2\ L$$

$$2\ K_2[OsO_2(OH)_4] + NaClO_2\ \longrightarrow\ 2K_2[OsO_4(OH)_2] + NaCl + 2\ H_2O$$

$$2\ K_2[OsO_4(OH)_2]\ \rightleftharpoons\ 2\ OsO_4 + 4\ KOH$$

$$2\ R\!\!-\!\!\!=\!\!\!- + NaClO_2 + 2\ H_2O\ \longrightarrow\ 2\ \underset{R}{HO\diagdown\!\!\diagup OH} + NaCl$$

Scheme 1.7 Essential steps for osmium-catalyzed dihydroxylation using $NaClO_2$.

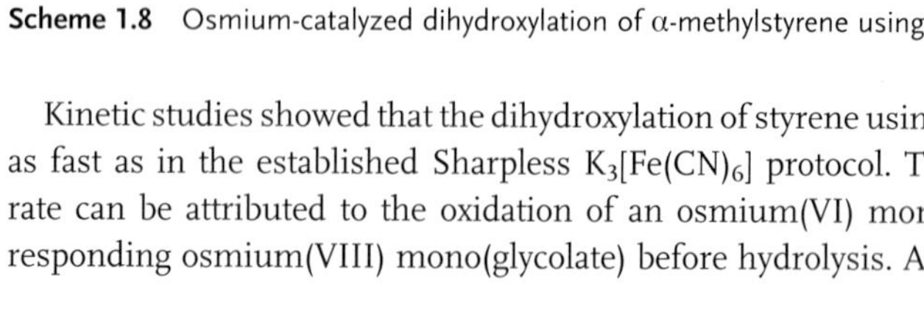

PhC(CH$_3$)=CH$_2$ $\xrightarrow[\substack{10\ \text{eq NaCl}\\ 0.5\ \text{eq NaClO}_2\\ \text{tert-BuOH / H}_2\text{O, 0 °C}}]{\substack{0.4\ \text{mol\%}\ K_2[OsO_2(OH)_4]\\ 1\ \text{mol\%}\ (DHQD)_2PHAL}}$ PhC(CH$_3$)(OH)CH$_2$OH

73% yield
93% ee

Scheme 1.8 Osmium-catalyzed dihydroxylation of α-methylstyrene using $NaClO_2$.

Kinetic studies showed that the dihydroxylation of styrene using $NaClO_2$ was twice as fast as in the established Sharpless $K_3[Fe(CN)_6]$ protocol. This higher reaction rate can be attributed to the oxidation of an osmium(VI) mono(glycolate) to corresponding osmium(VIII) mono(glycolate) before hydrolysis. As the concentration of the more electrophilic osmium(VIII) mono(glycolate) is increased, the rate-limiting hydrolysis step is accelerated [23].

1.2.4
Oxygen or Air

Several groups have reported the oxidation of alkenes in the presence of OsO_4 and oxygen. However mainly nonselective oxidation reactions take place in these systems [24]. The breakthrough came in 1999 when Krief et al. published a reaction system consisting of oxygen, catalytic amounts of OsO_4, and selenides for the asymmetric dihydroxylation of α-methylstyrene under irradiation of visible light in the presence of a sensitizer (Scheme 1.9) [25]. In this system, the selenides are oxidized to their oxides by singlet oxygen and the selenium oxides are then able to re-oxidize osmium (VI) to osmium (VIII). The reaction works with yields and ees similar to those in Sharpless AD. Potassium carbonate is used in only one tenth of the amount present in the AD mix, and air can be used instead of pure oxygen.

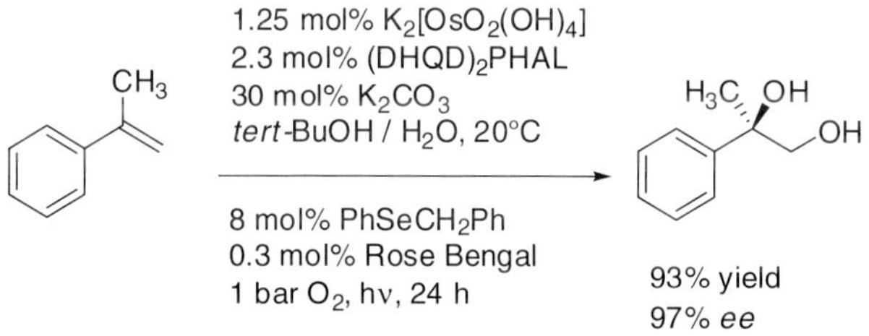

Scheme 1.9 Osmium-catalyzed dihydroxylation using $^1O_2$ and benzyl phenyl selenide.

A wide range of aromatic and aliphatic alkenes were demonstrated in this system [26]. It was shown that both yield and enantioselectivity are influenced by the pH of the reaction medium. The procedure was also applied to practical syntheses of natural product derivatives [27]. This version of the AD reaction not only uses a more ecological co-oxidant, but also requires much less material: 87 mg of material (catalyst, ligand, base, reoxidant) is required to oxidize 1 mmol of the same alkene instead of 1400 mg when AD mix is used.

In 1999 the first AD reaction using molecular oxygen without any secondary electron transfer mediator was published [28]. Osmium(VI) was readily reoxidized to osmium(VIII) in this system. We demonstrated that the osmium-catalyzed dihydroxylation of aliphatic and aromatic alkenes proceeds efficiently in the presence of dioxygen under ambient conditions. This dihydroxylation procedure constitutes a significant advancement compared to other re-oxidation procedures (Table 1.2, entry 7).

For a better comparison, a model reaction of the dihydroxylation of α-methylstyrene was examined using different stoichiometric oxidants. The yield of the 1,2-diol remained good to very good (73–99%), independently of the oxidant used.

Table 1.2 Comparison of the dihydroxylation of α-methylstyrene in the presence of different oxidants.

Entry	Oxidant	Yield (%)	Reaction conditions	ee (%)	TON	Waste (oxidant) (kg/kg diol)	Ref.
1	$K_3[Fe(CN)_6]$	90	0 °C $K_2[OsO_2(OH)_4]$ $^tBuOH/H_2O$	94[a]	450	8.1[c]	[7b]
2	NMO	90	0 °C OsO_4 Acetone/H_2O	33[b]	225	0.88[d]	[19]
3	$PhSeCH_2Ph/O_2$ $PhSeCH_2Ph/air$	89 87	12 °C $K_2[OsO_2(OH)_4]$ $^tBuOH/H_2O$	96[a] 93[a]	222 48	0.16[e] 0.16[e]	[25a]
4	NMM/Flavin/H_2O_2	88	RT OsO_4 Acetone/H_2O	99[a]	44	0.33[f]	[16a]
5	NMM/VO(acac)$_2$/ H_2O_2	86	RT	—	43	0.25[g]	[16d]
6	MTO/H_2O_2	85	RT OsO_4 Acetone/H_2O	64[a]	43	0.041[h]	[16e]
7	O_2	96	50 °C $K_2[OsO_2(OH)_4]$ ^tBuOH/aq. buffer	80[a]	192	—	[28a]
8	NaOCl	99	0 °C $K_2[OsO_2(OH)_4]$ $^tBuOH/H_2O$	91[a]	247	0.58[i]	[18]
9	$NaClO_2$	73	0 °C $K_2[OsO_2(OH)_4]$ $^tBuOH/H_2O$	93[a]	183	0.26[i]	[22]

a) Ligand: Hydroquinidine 1,4-phtalazinediyl diether.
b) Hydroquinidine p-chlorobenzoate.
c) $K_4[Fe(CN)_6]$.
d) N-Methylmorpholine (NMM).
e) PhSe(O)CH$_2$Ph.
f) NMO/Flavin-OOH.
g) NMO/VO$_2$(acac)$_2$.
h) MTO(O).
i) NaCl.

The best enantioselectivities (94–99% ee) were obtained with hydroquinidine 1,4-phthalazinediyl diether ((DHQD)$_2$PHAL) as the ligand at 0–12 °C (Table 1.2, entries 1, 3, and 4).

The dihydroxylation process with oxygen is clearly the most ecologically favorable procedure (Table 1.2, entry 7) if the production of waste from a stoichiometric reoxidant is considered. With the use of K$_3$[Fe(CN)$_6$] as oxidant, approximately 8.1 kg of iron salts per kg of product are formed. However, in the case of the Krief (Table 1.2, entry 3) and Bäckvall procedures (Table 1.2, entry 4) as well as in the presence of NaOCl (Table 1.2, entry 8) some by-products also arise because of the use of co-catalysts and co-oxidants. It should be noted that only salts and by-products formed from the oxidant were included in the calculation. Other waste products were not considered. Nevertheless, the calculations presented in Table 1.2 give a rough estimation of the environmental impact of the reaction.

Since the use of pure molecular oxygen on a larger scale might lead to safety concerns, it is even more advantageous to use air as the oxidizing agent. In fact, all current bulk oxidation processes, such as the oxidation of BTX (benzene, toluene, xylene) aromatics or alkanes to carboxylic acids, and the conversion of ethylene to ethylene oxide, use air but not pure oxygen as the oxidant [29]. The results of using air and pure oxygen have been compared in the dihydroxylation of α-methylstyrene as a model reaction (Scheme 1.10 and Table 1.3) [30].

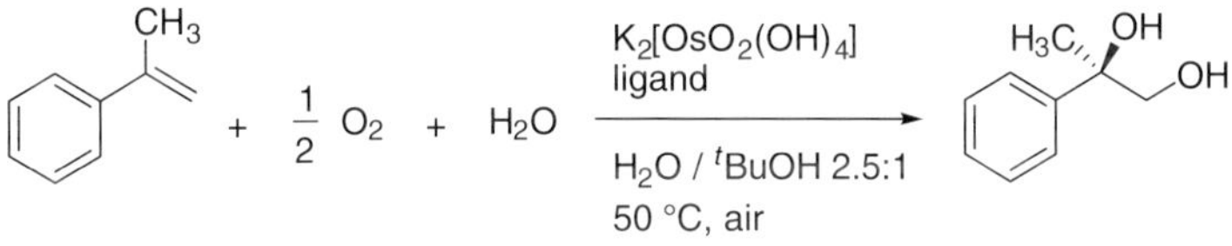

Scheme 1.10 Osmium-catalyzed dihydroxylation of α-methylstyrene.

Table 1.3 Dihydroxylation of α-methylstyrene with air.[a]

Entry	Pressure (atm)[c]	Cat. (mol%)	Ligand	L/Os	[L] (mmol/l)	Time (h)	Yield (%)	Selectivity (%)	ee (%)
1	1 (pure O$_2$)	0.5	DABCO[d]	3:1	3.0	16	97	97	—
2	1 (pure O$_2$)	0.5	(DHQD)$_2$PHAL[e]	3:1	3.0	20	96	96	80
3	1	0.5	DABCO	3:1	3.0	24	24	85	—
4	1	0.5	DABCO	3:1	3.0	68	58	83	—
5	5	0.1	DABCO	3:1	0.6	24	41	93	—
6	9	0.1	DABCO	3:1	0.6	24	76	92	—
7	20	0.5	(DHQD)$_2$PHAL	3:1	3.0	17	96	96	82
8	20	0.1	(DHQD)$_2$PHAL	3:1	0.6	24	95	95	62
9	20	0.1	(DHQD)$_2$PHAL	15:1	3.0	24	95	95	83
10[b]	20	0.1	(DHQD)$_2$PHAL	3:1	1.5	24	94	94	67
11[b]	20	0.1	(DHQD)$_2$PHAL	6:1	3.0	24	94	94	78
12[b]	20	0.1	(DHQD)$_2$PHAL	15:1	7.5	24	60	95	82

a) Reaction conditions: K$_2$[OsO$_2$(OH)$_4$], 50 °C, 2 mmol alkene, 25 mL buffer solution (pH 10.4), 10 mL tert-BuOH.
b) 10 mmol alkene, 50 mL buffer solution (pH 10.4), 20 mL tert-BuOH.
c) The autoclave was purged with air and then pressurized to the given value.
d) 1,4-Diazabicyclo[2.2.2.]octane.
e) Hydroquinidine 1,4-phthalazinediyl diether.

The dihydroxylation of α-methylstyrene in the presence of 1 atm of pure dioxygen proceeded smoothly (Table 1.3, entries 1–2), with the best results being obtained at pH 10.4. In the presence of 0.5 mol% $K_2[OsO_2(OH)_4]$/1.5 mol% DABCO or 1.5 mol% $(DHQD)_2PHAL$ at pH 10.4 and 50 °C, full conversion was achieved after 16 h or 20 h depending on the ligand. Though the total yield and selectivity of the reaction are excellent (97% and 96% respectively), the total turnover frequency of the catalyst is comparatively low (TOF = 10–12 h^{-1}). In the presence of the chiral cinchona ligand $(DHQD)_2PHAL$, an ee of 80% was observed. Sharpless et al. reported an enantioselectivity of 94% for the dihydroxylation of α-methylstyrene with $(DHQD)_2PHAL$ as the ligand using $K_3[Fe(CN)_6]$ as the reoxidant at 0 °C [31]. Studies of the ceiling ee at 50 °C (88% ee) showed that the main difference in the enantioselectivity stems from the higher reaction temperature. Using air instead of pure dioxygen gas gave only 24% of the corresponding diol after 24 h (TOF = 1 h^{-1}; Table 1.3, entry 3). Although the reaction is slow, it is important to note that the catalyst stayed active as the product continuously formed up to 58% yield after 68 h (Table 1.3, entry 4). It is noteworthy that the chemoselectivity of the dihydroxylation does not significantly decrease after prolonged reaction time. At 5–20 atm air pressure, the turnover frequency of the catalyst improved (Table 1.3, entries 5–11).

Full conversion of α-methylstyrene was achieved at an air pressure of 20 atm in the presence of 0.1 mol% of osmium, which corresponds to a turnover frequency of 40 h^{-1} (Table 1.3, entries 8–11). It is apparent that by increasing the oxygen pressure it is possible to reduce the osmium catalyst loading by a factor of 5. A decrease in the amount of osmium catalyst and ligand led to a decrease in the enantioselectivity from 82% to 62% ee. This can easily be explained by the participation of the nonstereoselective osmium glycolate as the active catalyst. The enantioselectivity can be resumed when higher concentration of the chiral ligand is applied (Table 1.3, entry 7 and 9). While the reaction at higher substrate concentration (10 mmol instead of 2 mmol) proceeded only sluggishly at 1 atm of pure oxygen; full conversion was achieved after 24 h at 20 atm of air (Table 1.3, entries 10, 11 and Table 1.4, entries 17, 18). It is interesting that under air atmosphere the chemoselectivity of the dihydroxylation remained excellent (92–96%).

As depicted in Table 1.4, various alkenes gave the corresponding diols in moderate to good yields (55–97%) with air. The enantioselectivities varied from 63–98% ee depending on the substrate. As the main side reaction is the oxidative cleavage of the C=C double bond and the yield decreases with respect to time, the chemoselectivity of the reaction patently relates to the sensitivity of the produced diol toward further oxidation. Thus, the oxidation of trans-stilbene in the biphasic mixture water/tert-butanol at pH 10.4, 50 °C, and 20 atm air pressure gave no hydrobenzoin, but gave benzaldehyde in 84% yield (Table 1.4, entry 9). Interestingly, changing the solvent to isobutyl methyl ketone (Table 1.4, entry 12) made it possible to obtain hydrobenzoin in high yield (89%) and enantioselectivity (98% ee) at pH 10.4.

Table 1.4 Dihydroxylation of various alkenes with air.[a]

Entry	Alkene	Cat. (mol%)	Ligand (L)	L/Os	[L] (mmol/L)	Time (h)	Yield (%)[b]	Selectivity (%)[b]	ee (%)
1	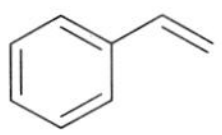 styrene	0.5	(DHQD)$_2$PHAL	3:1	3.0	24	42	42	87
2		0.5	(DHQD)$_2$PHAL	3:1	3.0	16	66	66	86
3		0.5	(DHQD)$_2$PHAL	3:1	3.0	14	76	76	87
4	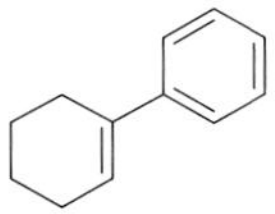 phenylcyclohexene	0.5	(DHQD)$_2$PHAL	3:1	3.0	24	88	88	89
5	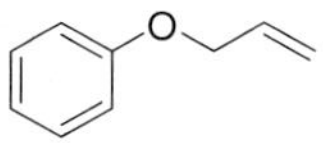 allyl phenyl ether	0.5	(DHQD)$_2$PHAL	3:1	3.0	24	63	63	67
6		0.5	(DHQD)$_2$PHAL	3:1	3.0	18	68	68	68
7		0.5	(DHQD)$_2$PHAL	3:1	3.0	14	67	67	66
8		0.5	(DHQD)$_2$PHAL	3:1	3.0	9	77	77	68
9	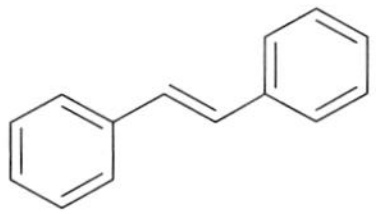 stilbene	0.5	—	—	—	24	0 (84)	0 (84)	—
10[c]		1.0	DABCO	3:1	1.5	24	4 (77)	5 (87)	—
11[c],[d]		1.0	(DHQD)$_2$PHAL	3:1	1.5	24	40 (35)	48 (42)	86
12[c],[e]		1.0	(DHQD)$_2$PHAL	3:1	1.5	24	89 (7)	89(7)	98
13[d]	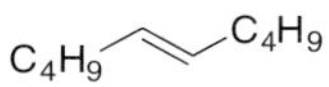 C$_4$H$_9$—CH=CH—C$_4$H$_9$	1.0	(DHQD)$_2$PHAL	3:1	6.0	24	85	85	82

(Continued)

Table 1.4 (Continued)

Entry	Alkene	Cat. (mol%)	Ligand (L)	L/Os	[L] (mmol/L)	Time (h)	Yield (%)[b]	Selectivity (%)[b]	ee (%)
14	C_6H_{13}⌒	0.5	(DHQD)$_2$PHAL	3:1	3.0	18	96	96	63
15		0.1	(DHQD)$_2$PHAL	3:1	0.6	24	95	95	44
16		0.1	(DHQD)$_2$PHAL	15:1	3.0	24	97	97	62
17[f]		0.1	(DHQD)$_2$PHAL	3:1	1.5	24	94	94	47
18[f]		0.1	(DHQD)$_2$PHAL	6:1	3.0	24	95	95	62
19	C_6F_{13}⌒	2.0	(DHQD)$_2$PYR[g]	3:1	12.0	24	55	—	68

a) Reaction conditions: K$_2$[OsO$_2$(OH)$_4$], 50 °C, 2 mmol alkene, 20 atm air, pH = 10.4, 25 mL buffer solution, 10 mL tert-BuOH, entries 9–12: 15 mL buffer solution, 20 mL tert-BuOH, entries 17–18: 50 mL buffer solution, 20 mL tert-BuOH.
b) Values in brackets are for benzaldehyde.
c) 1 mmol alkene.
d) pH = 12.
e) Isobutyl methyl ketone instead of tert-BuOH.
f) 10 mmol alkene.
g) Hydroquinidine 2,5-diphenyl-4,6-pyrimidinediyl diether.

The mechanism of the dihydroxylation reaction with oxygen or air is thought to be similar to the catalytic cycle presented by Sharpless et al. for the osmium-catalyzed dihydroxylation with $K_3[Fe(CN)_6]$ as the reoxidant (Scheme 1.11). The addition of alkene to a ligated Os(VIII) species proceeds mainly in the organic phase. Depending on the hydrolytic stability of the resulting Os(VI) glycolate complex, the rate-determining step of the reaction is either hydrolysis of the Os(VI) glycolate or the reoxidation of Os(VI) hydroxy species. There must be a minor involvement of a second catalytic cycle, as suggested for the dihydroxylation with NMO [21]. Such a second cycle would lead to significantly lower enantioselectivities, as the attack of a second alkene molecule on the Os(VIII) glycolate would occur in the absence of the chiral ligand. The observed enantioselectivities for the dihydroxylation with air are only slightly lower than those in the data previously published by the Sharpless group, despite the higher reaction temperature (50 °C vs 0 °C). Therefore the direct oxidation of the Os(VI) glycolate to an Os(VIII) glycolate does not represent a major reaction pathway.

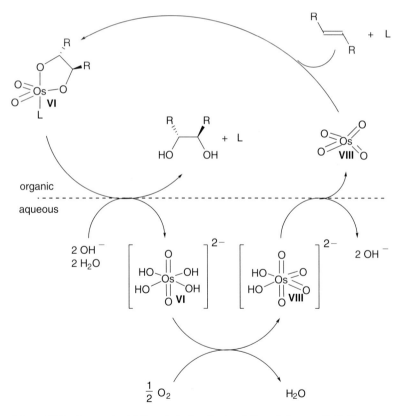

Scheme 1.11 Proposed catalytic cycle for the dihydroxylation of alkenes with OsO_4 and oxygen as the terminal oxidant.

1.3
Supported Osmium Catalyst

Hazardous toxicity and high costs are the chief drawbacks in reactions using osmium tetroxide. However, these disadvantages can be overcome by the use of stable and nonvolatile adducts of immobilized osmium tetroxide together with the advancement of a regeneration procedure for osmium tetroxide from a stoichiometric secondary oxidant, significantly improving the usability of osmium-catalyzed dihydroxylation [32]. These immobilized catalysts offer the advantages of easy and safe handling, simple separation from the reaction medium, and the possibility to reuse the expensive transition metal. Unfortunately, the instability of the support and leaching of the metal continue to be the main drawbacks.

1.3.1
Nitrogen-group Donating Support

In this context Cainelli and coworkers reported in 1989 the preparation of polymer-supported catalysts in which OsO_4 was immobilized on several amine-type polymers [33]. Such catalysts have structures of the type $OsO_4 \cdot L$ with the N-group of the polymer (=L) being coordinated to the Lewis acidic osmium center. Based upon this concept, catalytic enantioselective dihydroxylation was established by using polymers containing cinchona alkaloid derivatives [34]. However, since the amine ligands coordinate to osmium under equilibrium conditions, recovery of the osmium using polymer-supported ligands was difficult. In some cases, additional OsO_4 is needed to maintain the reactivity. Os-diolate hydrolysis seems to require detachment from the polymeric ligand, and hence causes leaching.

Herrmann and coworkers reported on the preparation of immobilized OsO_4 on poly(4-vinyl pyridine) and its use in the dihydroxylation of alkenes by means of hydrogen peroxide [35]. However, the problems of gradual polymer decomposition and osmium leaching were not solved.

OsO_4 can also be immobilized in polyaniline through nitrogen coordination as shown by Cloudary's group (Scheme 1.12) [36]. This polyaniline was applied to other Lewis acid-type transition metals like Sc, Pd, Re, and In. An electron transfer

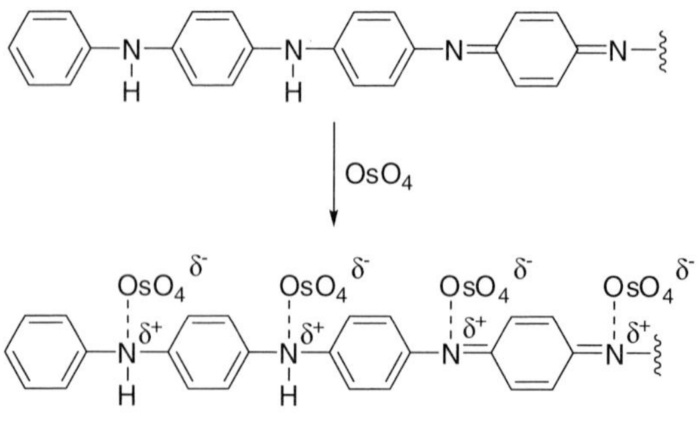

Scheme 1.12 Structure of PANI-doped with OsO_4.

mediator of MTO can be incorporated together with OsO$_4$ in this polyaniline to give a multifunction supported catalyst, which catalyzes the asymmetric dihydroxylation of *trans*-stilbene with 30% H$_2$O$_2$ as the terminal oxidant (Scheme 1.13). This system gave good isolated yields (81–85%) with excellent enantioselectivity (98% *ee*) for 5-times recycling.

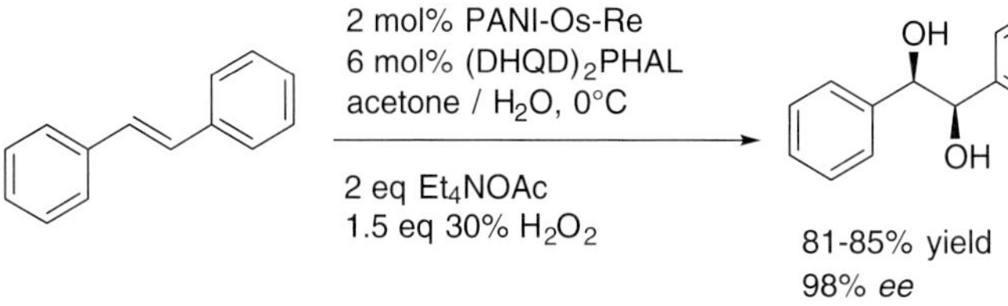

Scheme 1.13 Osmium-catalyzed dihydroxylation using polyaniline-supported OsO$_4$/MTO.

In another approach, Song and coworkers took advantage of the double bond in naturally occurring quinine [37]. The dimeric ligand (QN)$_2$PHAL is dihydroxylated *in situ*. With hydrogen bonding interaction with sucrose in the aqueous phase, the dihydroxylated ligand with coordinated OsO$_4$ stays in the aqueous phase in the reaction mixture. Hence, dihydroxylation of *trans*-stilbene can be performed 4 times in 82–92% yield and 81–99% *ee* with 0.1 mol% of OsO$_4$ and 2.5 mol% of (QN)$_2$PHAL (Scheme 1.14).

1.3.2
Microencapsulated OsO$_4$

A new strategy (at the time) of microencapsulated osmium tetroxide was published by Kobayashi and coworkers in 1998 [38]. The metal is immobilized onto a polymer on the basis of physical envelopment by the polymer and on electron interactions between the π electrons of the benzene rings of the polystyrene-based polymer and a vacant orbital of the Lewis acid. Using cyclohexene as a model compound it was shown that this microencapsulated osmium tetroxide (MC OsO$_4$) can be used as a catalyst in the dihydroxylation with NMO as stoichiometric oxidant (Scheme 1.15).

In contrast to other typical OsO$_4$-catalyzed dihydroxylations, where H$_2$O-^{t}BuOH is used as solvent system, the best yields were obtained in H$_2$O-acetone-CH$_3$CN. While the reaction was successfully carried out using NMO, moderate yields were obtained using trimethylamine N-oxide, and much lower yields were observed using hydrogen peroxide or potassium ferricyanide. The catalyst was recovered quantitatively by simple filtration and reused five times without any loss in activity.

A study on the rate of conversion of the starting material showed that the reaction proceeds faster using OsO$_4$ than using the microencapsulated catalyst. This is ascribed to the slower reoxidation of the microencapsulated osmium ester with NMO, compared to simple OsO$_4$.

Later on, acrylonitrile-butadiene-polystyrene (ABS) polymer was used as the support based on the same microencapsulation technique, and several alkenes,

1 Recent Developments in Metal-catalyzed Dihydroxylation of Alkenes

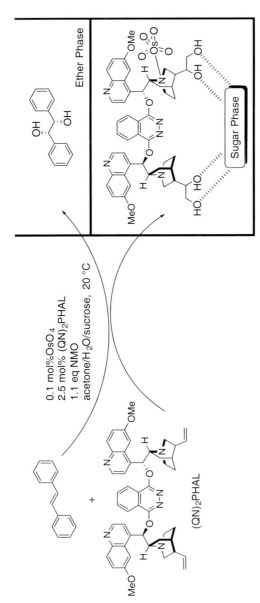

Scheme 1.14 An osmium-(QN)$_2$PHAL-sugar recyclable dihydroxylation system.

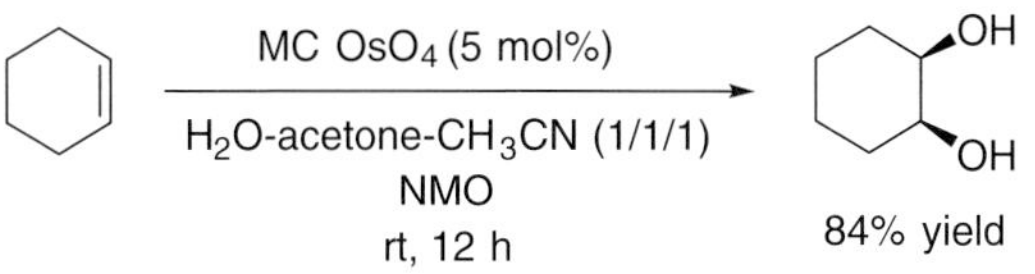

Scheme 1.15 Dihydroxylation of cyclohexene using microencapsulated osmium tetroxide (MC OsO$_4$).

including cyclic and acyclic, terminal, mono-, di-, tri-, and tetrasubstituted ones, gave the corresponding diols in high yields [39]. Enantioselectivities up to 95% ee were achieved with this type of microencapsulated OsO$_4$ when (DHQD)$_2$PHAL was introduced as a chiral ligand. However, this system requires slow addition of the alkene. After running a 100 mmol-scale experiment, more than 95% of the ABS-MC OsO$_4$ and the chiral ligand were recovered.

Later, Kobayashi and coworkers reported on a new type of microencapsulated osmium tetroxide using phenoxyethoxymethyl-polystyrene as support [40]. With this catalyst, asymmetric dihydroxylation of alkenes has been successfully performed using (DHQD)$_2$PHAL as a chiral ligand and K$_3$[Fe(CN)$_6$] as a cooxidant in H$_2$O/acetone (Scheme 1.16). This dihydroxylation does not require slow addition of the alkene, and the catalyst can be recovered quantitatively by simple filtration and reused without loss of activity. With a divinylbenzene-cross-linked polystyrene microencapsulated OsO$_4$ and a nonionic phase transfer catalyst (Triton® X-405), this system can be run in an aqueous system [41].

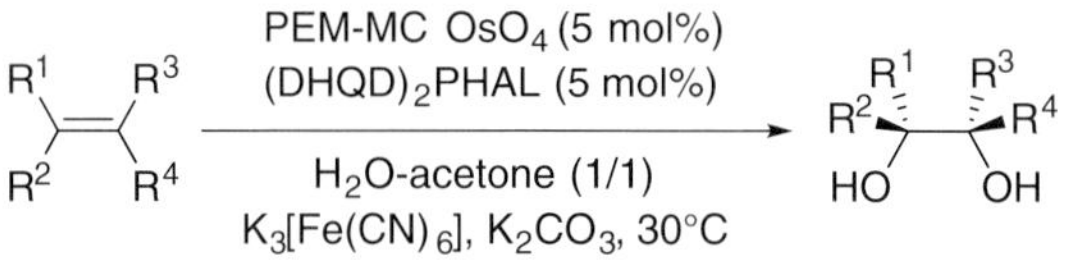

Scheme 1.16 Asymmetric dihydroxylation of alkenes using PEM-MC OsO$_4$.

A polyurea microencapsulated OsO$_4$, so-called Os EnCat™, is now commercially available. Ley et al. demonstrated that this type of microencapsulated OsO$_4$ catalyzed dihydroxylation of various alkenes using NMO as the oxidant [42]. Five times recycling was achieved with good yield (74–88%). With NaIO$_4$ as the oxidant, oxidation of alkenes give the corresponding oxidative cleavage products of aldehydes and ketones.

Microencapsulated OsO$_4$ can function as a reservoir and an efficient scavenger of homogeneous catalytic species [42, 43]. Hence, it gives results comparable to those achieved by its homogeneous analog.

1.3.3
Supports Bearing Alkenes

Jacobs and coworkers published a completely different strategy to immobilize osmium catalysts. Their approach is based on two details of the mechanism of the cis-dihydroxylation: (1) Tetrasubstituted alkenes react with OsO$_4$ smoothly to form osmate(VI) esters which are not hydrolyzed under mild conditions, and (2) an Os(VI) monodiolate complex can be reoxidized to a cis-dioxo Os(VIII) species without release of the diol. These two properties make it possible to immobilize a catalytically active osmium compound by the addition of OsO$_4$ to a tetrasubstituted alkene which is covalently linked to a silica support. Hence, subsequent addition of a second alkene results in an Os bisdiolate complex. The tetrasubstituted diolate ester which is formed at one side of the Os atom is stable, and keeps the catalyst fixed on the support material. The catalytic reaction can take place at the free coordination sites of Os (Scheme 1.17) [44].

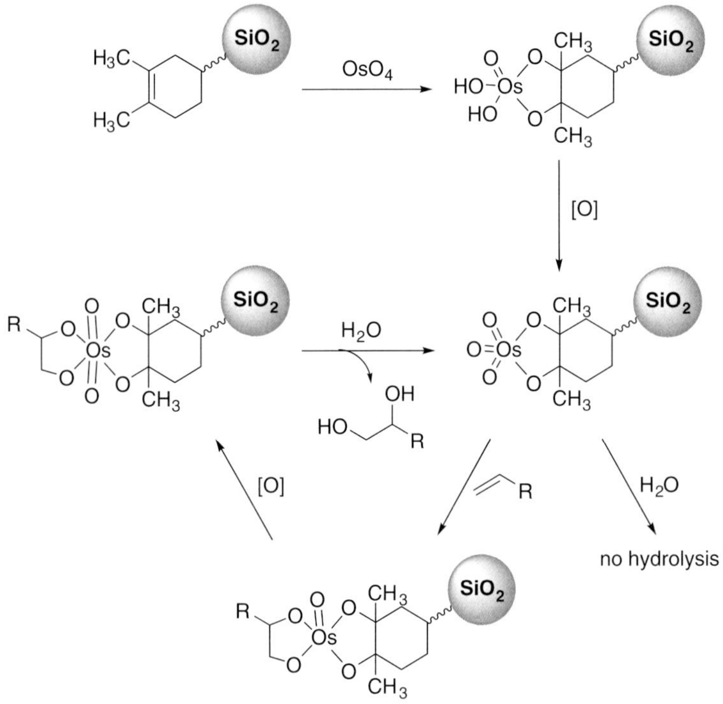

Scheme 1.17 Immobilization of Os in a tertiary diolate complex, and proposed catalytic cycle for cis-dihydroxylation.

The dihydroxylation of mono- and disubstituted aliphatic alkenes and cyclic alkenes is successfully performed using this heterogeneous catalyst and NMO as cooxidant. The amount of osmium needed is 0.25 mol% Os with respect to the alkene. This system gives excellent chemoselectivity, being equal to that in the homogeneous reaction with NMO. However, a somewhat longer reaction time is necessary. The development of an asymmetric variant of this process by addition of the typical chiral alkaloid ligands of the asymmetric dihydroxylation would be difficult, since the reactions performed with these heterogeneous catalysts take place in the so-called 'second cycle'. With cinchona alkaloid ligands, high ees are only achieved in dihydroxylations occurring in the first cycle. However, recent findings by the groups of Sharpless and Adolfsson show that even second-cycle dihydroxylations may give substantial ees [45]. Although this process needs to be optimized, further development of the concept of an enantioselective second-cycle process offers the prospect of a future heterogeneous asymmetric catalyst.

1.3.4
Immobilization by Ionic Interaction

Choudary and his group reported in 2001 the design of an ion-exchange technique for the development of recoverable and reusable osmium catalysts immobilized on layered double hydroxides (LDH), modified silica, and organic resin for asymmetric dihydroxylation [46]. An activity profile of the dihydroxylation of *trans*-stilbene with various exchanger-OsO_4 catalysts revealed that LDH-OsO_4 displays the highest activity, and the heterogenized catalysts in general have higher reactivity than $K_2[OsO_2(OH)_4]$. When *trans*-stilbene was added to a mixture of LDH-OsO_4, chiral ligand (DHQD)$_2$PHAL (1 mol% each), and NMO in H_2O-^{t}BuOH, the desired diol was obtained in 96% yield with 99% ee. Similarly, excellent ees were obtained with resin-OsO_4 and SiO_2-OsO_4 in the same reaction. All of the prepared catalysts can be recovered quantitatively by simple filtration and reused for five cycles with consistent activity. With this procedure, various alkenes ranging from mono- to trisubstituted, activated to simple, were transformed into their diols. In most cases, the desired diols are formed in higher yields, albeit with ees similar to those reported in homogeneous systems. Slow addition of the alkene to the reaction mixture is sure to achieve higher ee. This LDH-OsO_4 system presented by Choudary and coworkers is superior in terms of activity, enantioselectivity, and scope of the reaction to that of Kobayashi.

Although the LDH-OsO_4 shows excellent activity with NMO, it is deactivated when $K_3[Fe(CN)_6]$ or molecular oxygen is used as co-oxidant [47]. This deactivation is attributed to the displacement of OsO_4^{2-} by the competing anions, which include ferricyanide, ferrocyanide, and phosphate ions (from the aqueous buffer solution). To solve this problem, resin-OsO_4 and SiO_2-OsO_4 were designed and prepared by the ion exchange process on the quaternary ammonium-anchored resin and silica, respectively, as these ion exchangers are expected to prefer bivalent anions over trivalent anions. These new heterogeneous catalysts show consistent performance in the dihydroxylation of α-methylstyrene for a number of recycles using NMO, $K_3[Fe(CN)_6]$ or O_2 as reoxidant. The resin-OsO_4 catalyst, however, displays higher activity than that of the SiO_2-OsO_4 catalyst. In the presence of Sharpless ligands, various alkenes can be oxidized enantioselectively using these heterogeneous systems, and very good ees can be obtained with any of the three cooxidants. Equimolar ratios of ligand to osmium are sufficient for achieving excellent ees. This is in contrast to the homogeneous reaction, wherein a 2–3 molar excess of the expensive chiral ligand to osmium is usually employed. These studies indicate that the binding ability of these heterogeneous osmium catalysts with the chiral ligand is greater than that of the homogeneous analog.

Incidentally, this is the first reported heterogeneous osmium-catalyst-mediated AD reaction of alkenes using molecular oxygen as the cooxidant. Under identical conditions, the turnover numbers of the heterogeneous catalyst are similar to those of the homogeneous system.

Furthermore, Choudary and coworkers presented a procedure for the application of a heterogeneous catalytic system for the AD reaction in combination with hydrogen peroxide as cooxidant [48]. Here, a triple catalytic system composed of NMM and two heterogeneous catalysts was designed. Titanium silicalite acts as electron transfer mediator to perform oxidation of NMM, being used in catalytic amounts with hydrogen peroxide to provide *in situ* NMO continuously for AD of alkenes, which is catalyzed by another heterogeneous catalyst, namely silica gel-supported cinchona alkaloid [SGS-(DHQD)$_2$PHAL]-OsO$_4$. Good yields were observed for various alkenes. Again, very good *ee*s are achieved with an equimolar ratio of ligand to osmium, but slow addition of alkene and H$_2$O$_2$ is advisable. Unfortunately, recovery and reuse of the [SGS-(DHQD)$_2$PHAL]-OsO$_4$/TS-1 revealed that about 30% of the osmium became leached out during the reaction. This amount has to be replenished in each additional run.

1.4
Ionic Liquid

Recently, ionic liquids have become popular new solvents in organic synthesis [[49, 50]]. They can dissolve a wide range of organometallic compounds and are miscible with organic compounds. They are highly polar but noncoordinating. In general, ionic liquids exhibit excellent chemical and thermal stability with ease of reuse. It is possible to vary their miscibility with water and organic solvents simply by changing the counter anion. Their essentially negligible vapor pressure gives them an advantage over volatile organic solvents from the viewpoint of green chemistry.

In 2002, a process for alkene dihydroxylation by recoverable and reusable OsO$_4$ in ionic liquids was published for the first time [51]. Yanada and coworkers described the immobilization of OsO$_4$ in 1-ethyl-3-methylimidazolium tetrafluoroborate [47a]. They chose 1,1-diphenylethylene as a model compound and found that the use of 5 mol% OsO$_4$ in [emim]BF$_4$, 1.2 equiv. of NMO·H$_2$O, and room temperature were the best reaction conditions for a good yield. After 18 h, a 100% yield of the corresponding diol was obtained. OsO$_4$-catalyzed reactions with other co-oxidants such as hydrogen peroxide, sodium percarbonate, and *tert*-butyl hydroperoxide gave poor results. With anhydrous NMO the yield of diol was only 6%. After the reaction, the 1,2-diol can be extracted with ethyl acetate, and the ionic liquid containing the catalyst can be reused for further catalytic oxidation reaction. It was shown that even in the fifth run the obtained yield did not change. This new method using immobilized OsO$_4$ in an ionic liquid was applied to several substrates, including mono-, di-, and trisubstituted aliphatic alkenes, as well as to aromatic alkenes. In all cases, the desired diols were obtained in high yields.

The group of Yao developed a slightly different procedure: They used 1-butyl-3-methylimidazolium hexafluorophosphate/water/^{t}BuOH (1:1:2) as the solvent system and NMO (1.2 equiv.) as reoxidant for the osmium catalyst [47b]. Here, 2 mol% osmium is needed for efficient dihydroxylation of various alkenes. After the reaction, all volatiles were removed under reduced pressure and the product was extracted from the ionic liquid layer using ether. The ionic liquid layer containing the catalyst can be used several times with only a slight drop in catalytic activity. In order to prevent osmium leaching, 1.2 equiv. of DMAP relative to OsO_4 have to be added to the reaction mixture. This amine forms stable complexes with OsO_4, and this strong binding to a polar amine enhances its partitioning in the more polar ionic liquid layer. The mild catalyst composed of the triple components OsO_4, NMM and flavin was successfully immobilized in a [bmim][PF_6] layer [52]. With additional base, tetraethyl ammonium acetate and acetone as co-solvent α-methylstyrene was oxidized in 85% with hydrogen peroxide as terminal oxidant. It was shown that the system can be reused 5 times without any loss in activity. Song and coworkers reported on the Os-catalyzed dihydroxylation using NMO in mixtures of ionic liquids (1-butyl-3-methylimidazolium hexafluorophosphate or hexafluoroantimonate) with acetone/H_2O [53]. They used 1,4-bis(9-O-quininyl)phthalazine [$(QN)_2$PHAL] as chiral ligand. $(QN)_2$PHAL is converted to a new ligand bearing highly polar residues (four hydroxy groups in the 10,11-positions of the quinine parts) during AD reactions of alkenes. The use of $(QN)_2$PHAL instead of $(DHQD)_2$PHAL afforded the same yields and *ee*s, and, moreover, resulted in drastic improvement in the recyclability of both catalytic components. In another report Afonso and coworkers described the $K_2[OsO_2(OH)_4]/K_3[Fe(CN)_6]/(DHQD)_2$PHAL or $(DHQD)_2$PYR system for the asymmetric dihydroxylation using two different ionic liquids [54]. Both the systems used, [bmim][PF_6]/water and [bmim][PF_6]/water/^{t}BuOH (bmim = 1-*n*-butyl-3-methylimidazole), are effective for a considerable number of runs (e.g., run 1: yield 88%, *ee* 90%; run 9: yield 83%, *ee* 89%). Only after 11 or 12 cycles was a significant drop in the chemical yield and optical purity observed. More recently, in combination with ionic liquids, supercritical carbon dioxide was used in AD reactions for separation purposes [55].

In summary, it was demonstrated that the application of an ionic liquid provides a simple approach to the immobilization of the osmium catalyst for alkene dihydroxylation [56]. It is important to note that the volatility and toxicity of OsO_4 are greatly suppressed when ionic liquids are used.

1.5
Ruthenium Catalysts

As stated in the previous sections, the chief drawbacks in reactions using osmium tetroxide are its volatility, toxicity, and high cost. The less expensive and hazardous isoelectronic ruthenium tetroxide has long been appreciated for its powerful oxidation chemistry, especially oxidative cleavage of alkenes to aldehydes, ketones, and carboxylic acids [57]. Sharpless and Akashi showed for the first time that RuO_4 can *cis*-dihydroxylate *cis*- and *trans*-cyclododecene stereospecifically in only ~20% yield, in each case at −78 °C [11]. In 1994, Shing reported the first practical dihydroxylation protocol using ruthenium as the catalyst [58]. RuO_4 is generated *in situ* from $RuCl_3 \cdot 3H_2O$ with $NaIO_4$, and α-methylstyrene can thus be oxidized to the corresponding diol in 70% yield with the same oxidant (Scheme 1.18). Though the catalyst loading is relatively high, the reaction in general furnishes the diol in minutes with a wide variety of substrates [59]. The solvent mixture has a crucial effect on the reaction selectivity [60]. On the one hand the well-balanced biphasic system limits the partition of alkene substrate and diol product in the aqueous phase; hence, overoxidation to the fission product can be prevented. On the other hand it aids the hydrolysis of the Ru(VIII) glycolate, a purpose that can, incidentally, be served by benzonitrile and methylsulfonamide instead [59, 60].

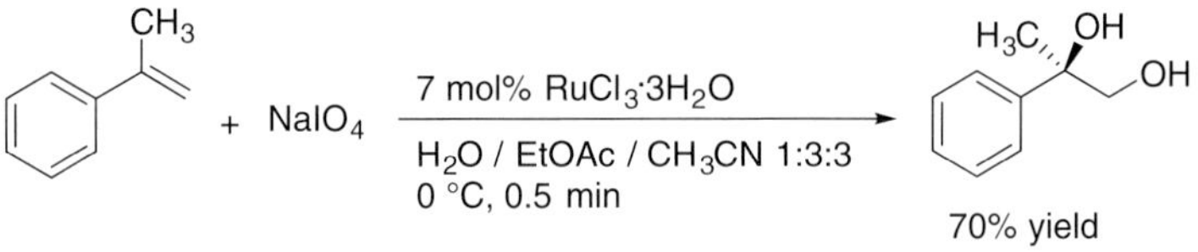

Scheme 1.18 Ruthenium-catalyzed dihydroxylation of α-methylstyrene.

Plietker and coworkers improved the system by using Brønsted or Lewis acids to increase the electrophilicity of the Ru(VIII) glycolate (Scheme 1.19). As a result, the rate of the reaction can be enhanced by an order of magnitude [61]. With this protocol, the catalyst loading can be reduced to as low as 0.5 mol% and the short reaction time (seconds or minutes) is maintained (Scheme 1.20). In the course of mechanistic studies, a novel method for ketohydroxylation was also developed [62].

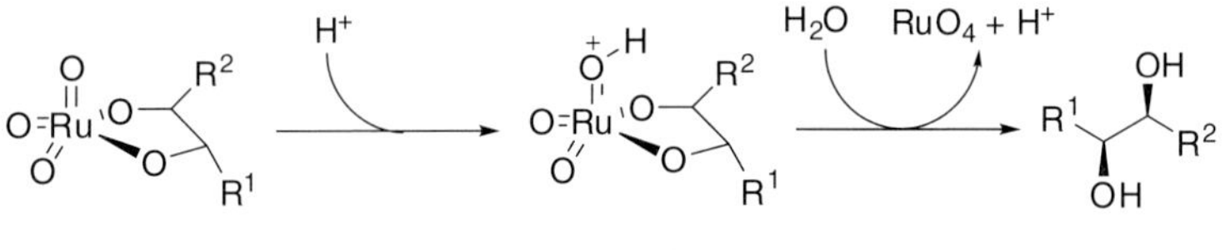

Scheme 1.19 Proton-accelerated hydrolysis of ruthenium glycolate.

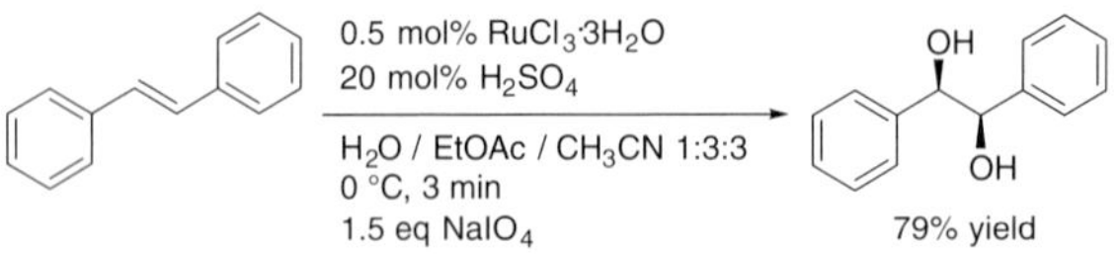

Scheme 1.20 Acid-accelerated ruthenium-catalyzed dihydroxylation of *trans*-stilbene.

Immobilization of ruthenium catalyst for dihydroxylation of alkenes was demonstrated by Yu and Che [63]. A colloidal ruthenium species was suspended with calcium hydroxyapatite in water in their grafting process. A 5 wt% Ru loading was thereby achieved on the resulting nano-ruthenium hydroxyapatite (nano-RuHAP). In a 2 mmol scale reaction, with 40 mg of nano-RuHAP catalyst (which contained 1 mol% of ruthenium) and 20 mol% H_2SO_4, the reaction was completed in 30 min (Scheme 1.21). The activity of the supported catalyst dropped slightly after recycling 4 times because of the slow dissolution of the support in acidic medium.

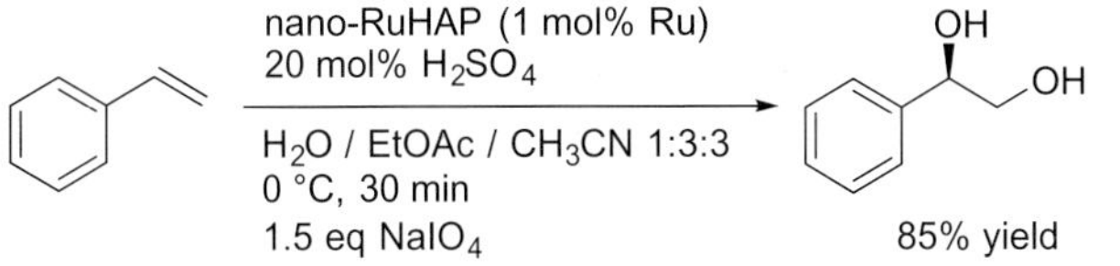

Scheme 1.21 Immobilized nano-ruthenium-catalyzed dihydroxylation of styrene.

Relatively few ruthenium complexes bearing cis-dioxo ligands have been reported. If this type of complex could transfer both oxygen atoms to an alkene simultaneously, it could provide promising applications for dihydroxylation. In 2005, Che and coworkers reported that a ruthenium complex, cis-[(Me$_3$tacn)-(CF$_3$CO$_2$)Ru^{VI}O$_2$]ClO$_4$ (Me$_3$tacn = N,N′,N″-trimethyl-1,4,7-triaza-cyclononane), reacted stoichiometrically with alkenes to give 1,2-diols in moderate to good yield [64]. They also showed in the same article that this type of complex is able to catalyze dihydroxylation reaction. For example cis-[(Me$_3$tacn)-(CF$_3$CO$_2$)Ru^{VI}O$_2$]CF$_3$CO$_2$ oxidizes cyclooctene to cis-1,2-cyclooctanediol and cyclooctene oxide in 50% and 42% yield respectively by using H_2O_2 as a terminal oxidant. An improved procedure was developed with the aid of basic Al_2O_3 and NaCl [65]. In this system, various alkenes can be dihydroxylated with a catalytic amount of fac-[(Me$_3$tacn)-Ru^{III}Cl$_3$] utilizing H_2O_2 as the terminal oxidant in moderate to excellent yield (Scheme 1.22). The presence of Al_2O_3 and NaCl significantly reduces the formation of epoxides, which in consequence produce trans-1,2-diols as the by-products.

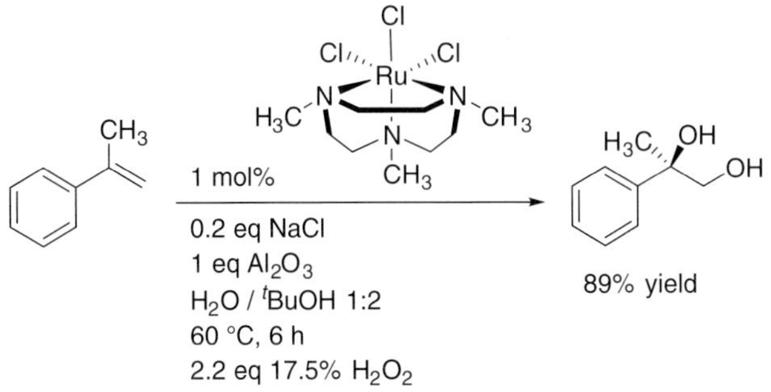

Scheme 1.22 fac-[(Me$_3$tacn)-Ru^{III}Cl$_3$]-catalyzed dihydroxylation of α-methylstyrene.

Ruthenium tetroxide has also shown its potential as a practical catalyst in the dihydroxylation of alkenes. Unlike the isoelectronic osmium tetroxide, which catalyzes dihydroxylation via a ligand accelerated reaction, alkenes can be oxidized by $NaIO_4$ efficiently with a catalytic amount of ligand-free RuO_4 [4b]. As a result, the background reaction significantly diminishes the productivity of the chiral ligand-coordinated ruthenium complex. Hence, up to now, an efficient asymmetric ruthenium dihydroxylation catalyst has not yet been reported. As demonstrated by Che and coworkers, structurally well-defined cis-[(Me$_3$tacn)-(CF$_3$CO$_2$)Ru^{VI}O$_2$]CF$_3$CO$_2$ and its precursor fac-[(Me$_3$tacn)-Ru^{III}Cl$_3$] catalyze cis-dihydroxylation of alkenes with aqueous H_2O_2. Enantioselective oxidation of alkenes to 1,2-diols is a strong possibility and is eagerly awaited by the scientific community.

1.6
Iron Catalysts

Iron is the most abundant element by mass on the earth's crust [66]. Owing to its availability, benign effect on the environment, and biological relevance, there is increasing interest in the use of iron complexes as catalysts for a wide range of reactions [67]. Although 16 iron oxides have been isolated and characterized, FeO_4, which is isolobal to RuO_4 and OsO_4, has never been observed, possibly because of its superhalogen character [68]. The Os(VI)-containing $K_2[OsO_2(OH)_4]$ is commonly used as a nonvolatile substitute for OsO_4. Its Fe(VI) alternative exists as $K_2[FeO_4]$, which clearly indicates that the hydroxyl groups on Fe are far more acidic than those on Os. In fact, $K_2[FeO_4]$ is a strong oxidant for alkane, alcohol and aniline oxidations in organic chemistry [69] and is a disinfectant for water treatment [70]. Because of its high oxidizing power, the use of $K_2[FeO_4]$ as a dihydroxylation catalyst is not very likely.

Nature takes advantage of the abundance of iron, which hence is present in nearly all living organisms [71]. Many biological systems such as hemoglobin, myoglobin, cytochrome oxygenases, and non-heme oxygenases as well as [Fe–Fe] hydrogenases are iron-containing enzymes or co-enzymes [72, 73]. In the wake of nature, biomimetic or bio-inspired approaches to the development of new synthetically useful dihydroxylation protocols with iron catalysts should be feasible [74].

Particularly interesting systems are the non-heme oxygenase mimics developed mainly by Que and coworkers [75]. In 1999, this group reported a functional model [76] for Rieske dioxygenases, which catalyze the dihydroxylation of arenes [77]. In this model system, cyclooctene was oxidized with aqueous 35% H_2O_2 in a catalytic amount of an iron complex to give the corresponding epoxide and cis-diol in 7% and 49% yield, respectively (Table 1.5, entry 1). The efficacy of the catalyst was maintained, 4 times its weight of H_2O_2 being used, and the yield of the cis-diol stayed at 55% [76]. As stated by Que, a basic requirement for performing dihydroxylation with iron catalysts is to provide two labile sites cis to one another on the complex, since this is favorable to the formation of the η^2-peroxo intermediate, which is responsible for the diol formation. In the presence of acetic acid, the selectivity towards cis-diol becomes lower (Table 1.5, entries 2 and 3) and the reaction gives more epoxide [78]. In-situ generation of peracetic acid was thus suggested to be favorable to the formation of epoxides.

Table 1.5 Epoxidation of cyclooctene with iron catalysts and H$_2$O$_2$.

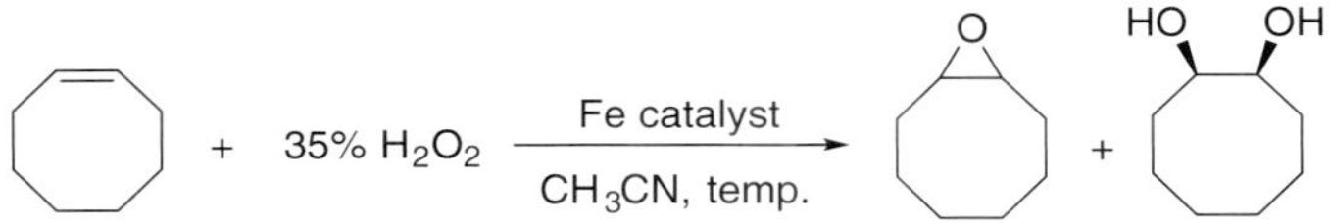

Entry	Catalyst	Alkene:H$_2$O$_2$	Temp. (°C)	Epoxide[a]	cis-diol[a]	Additive	Ref.
1	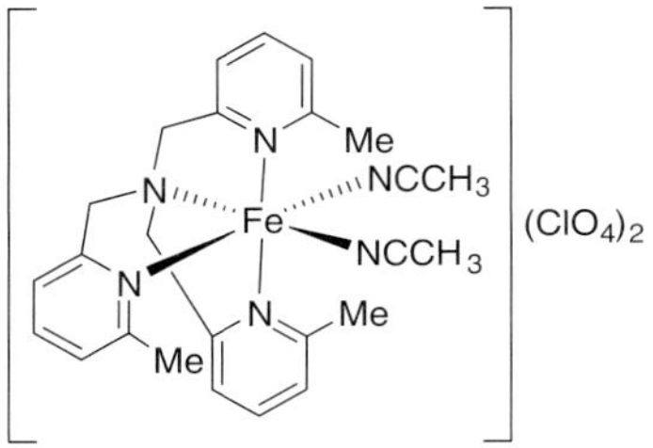 [Fe complex with Me-pyridyl ligand](ClO$_4$)$_2$, 10 mol%	100:1	25	7%	49%	nil	[76]
2	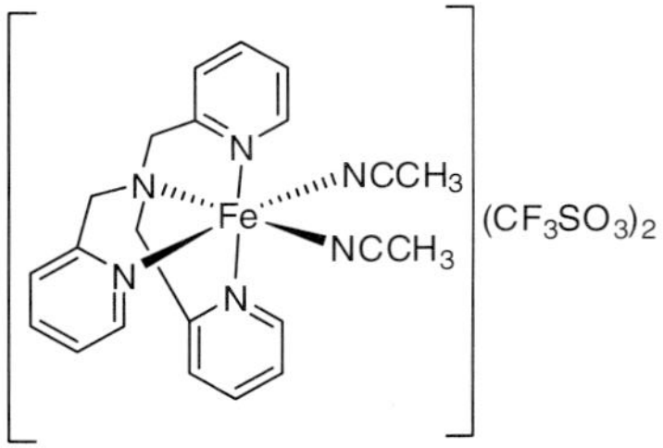 [Fe complex with pyridyl ligand](CF$_3$SO$_3$)$_2$, 7 mol%	100:2.9	r.t.	30%	41%	nil	[78a]

(Continued)

Table 1.5 (Continued)

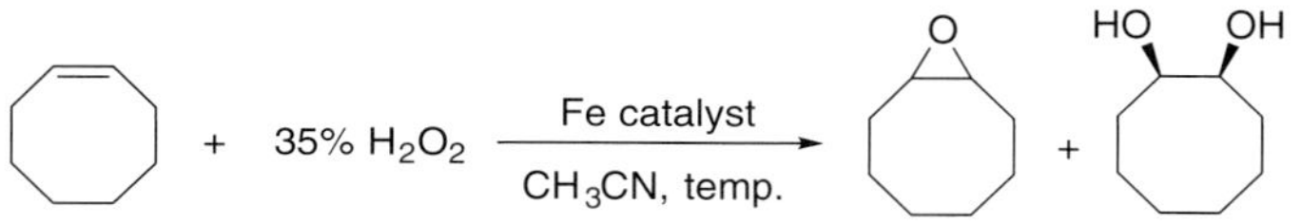

Entry	Catalyst	Alkene:H$_2$O$_2$	Temp. (°C)	Epoxide[a]	cis-diol[a]	Additive	Ref.
3	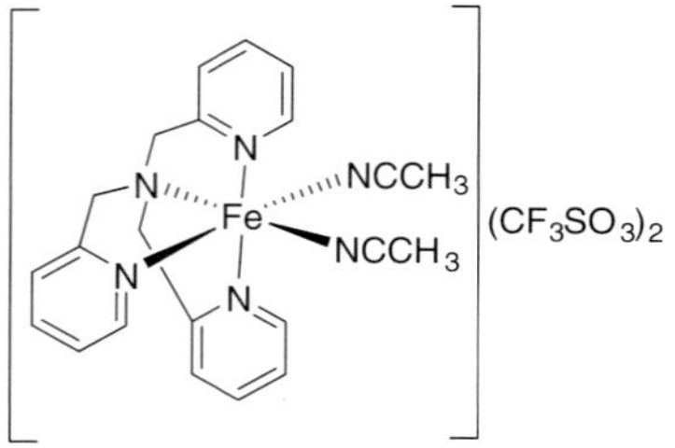 [Fe complex](CF$_3$SO$_3$)$_2$ 7 mol%	100 : 2.9	r.t.	88%	14%	200 mol% HOAc	[78a]
4	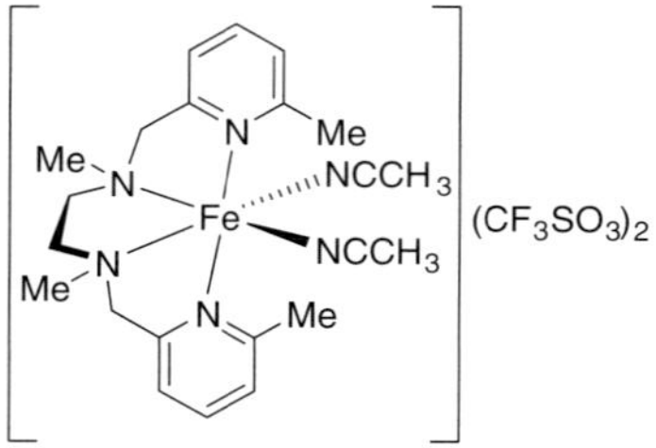 [Fe complex](CF$_3$SO$_3$)$_2$ 10 mol%	100 : 1[b]	30	15%	64%	nil	[79b]

5	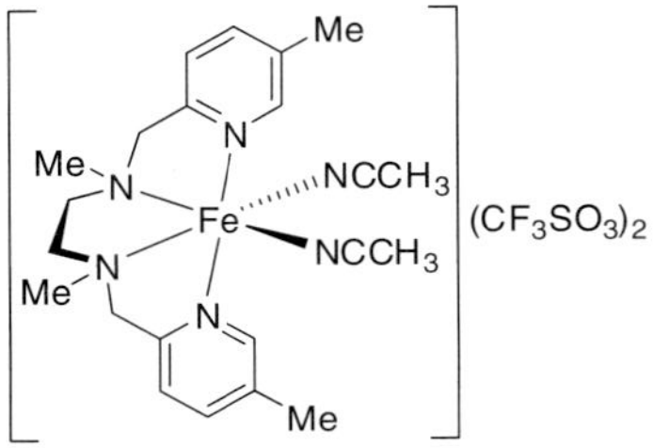 [Fe complex with methylpyridine ligands and NCCH₃](CF₃SO₃)₂ 10 mol%	100:1	25	78%	13%	nil	[79c]
6	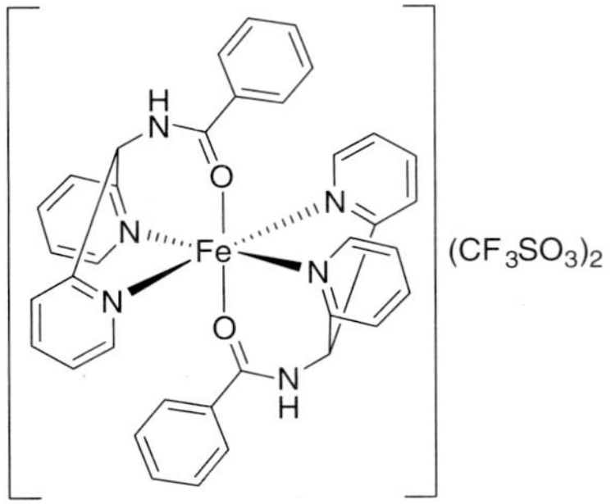 [Fe complex with pyridine/benzamide ligands](CF₃SO₃)₂ 10 mol%	100:1	r.t.	5%	70%	nil	[81a]

a) Yield based on the limiting reagent.
b) 50% H_2O_2 was used.

Steric effect is very crucial in the selectivity of *cis*-diol [79]. While the (6-Me$_2$-BPMEN)Fe(OTf)$_2$ produces *cis*-1,2-cylcooctanediol as the major product (Table 1.5, entry 4), the 5-methyl analog performs as an epoxidation catalyst (Table 1.5, entry 5). In the presence of acetic acid, the nonsubstituted (BPMEN)Fe(SbF$_6$)$_2$, also referred to as (MEP)Fe(SbF$_6$)$_2$, self-assembled to a dimer as a structural mimic of methane monooxygenase (MMO) [80]. It catalyzes the epoxidation of a range of aliphatic alkenes. Even the relatively nonreactive substrate, 1-decene, can be oxidized to the corresponding epoxide in 85% yield in 5 min.

More recently a new ligand scaffold of [di-(2-pyridyl)methyl]benzamide has been introduced by Que [81]. This novel facial *N,N,O*-ligand arrangement mimics that found for the Rieske dioxygenase. The iron(II) complex from this ligand efficiently catalyzes the dihydroxylation of various alkenes, including aliphatic and aromatic ones (Table 1.5, entry 6). It should be noted that the *cis*-diol to epoxide ratio is significantly increased. With α,β-unsaturated alkenes, even no epoxide was reported.

Based on these excellent studies, Que and coworkers developed an asymmetric dihydroxylation system with chiral iron complexes as catalysts originated from the BPMEN-type ligands (Table 1.6, entry 1) [79b]. *trans*-1,2-Cyclohexanediamine is introduced as the bridge of the two (2-pyridyl)methyl groups to provide the chiral information. Other than its nonsubstituted parent BPMEN and BPMCN ligands which give *cis*-α topology in the iron complexes, the BPMCN ligand with methyl groups at the 6-positions of the pyridines adopts a *cis*-β topology for the iron complex. This iron complex catalyzes the oxidation of *trans*-2-heptene to its corresponding *cis*-diol in 49% yield with 79% *ee* at room temperature (Table 1.6, entry 1). It gives even higher *ee* (88%) when the temperature is increased to 50 °C. Hence, participation of more than one active species, possibly different conformers, was suggested.

In 2008, Que and coworkers reported a new type of ligand bearing chiral bipyrrolidine as the chiral backbone [82]. The corresponding iron(II) complex provides general reactivity of dihydroxylation of various alkenes using H$_2$O$_2$. Both aliphatic and aromatic alkenes work nicely. For example, dihydroxylation of styrene gave the corresponding styrene oxide and 1-phenylethane-1,2-diol in <1% and 65% yield, respectively. The most striking result is the oxidation of 2-heptene. In this system, the *cis*-diol was obtained in 55% yield with 97% *ee* (Table 1.6, entry 2). The *ee*s are even comparable with those obtained from AD mixes. A closely related complex, [Fe^{II}{(S,S)-BPBP}(NCCH$_3$)$_2$](SbF$_6$)$_2$, was also reported as a hydroxylation catalyst for tertiary C—H bonds using H$_2$O$_2$ with predictable selectivity [83].

This biomimetic approach to the dihydroxylation of alkenes has given many encouraging results to the scientific community. However, vigorous decomposition of H$_2$O$_2$ by a simple Fe source from the degradation of the complex obstructs the efficient use of the oxidant [84]. Highly reactive free hydroxyl radical generated by Fenton or Gif chemistry induces the decomposition of the ligand and substrate as well as the product [85]. Hence, a high ratio of substrate to hydrogen peroxide is usually employed to solve these problems. This reduces the synthetic utility of these iron catalysts.

Table 1.6 Asymmetric epoxidation of trans-2-heptene with chiral iron complexes and H_2O_2.

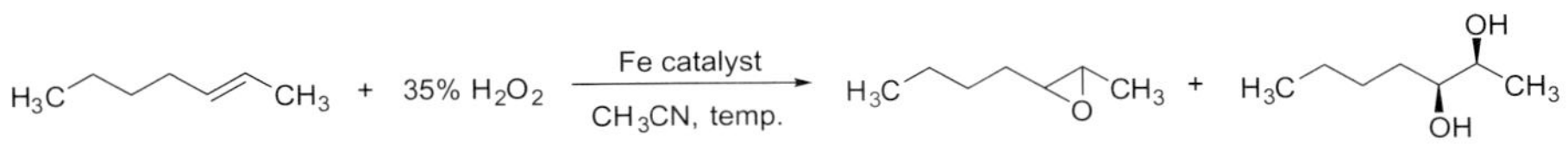

$$H_3C\diagup\diagup CH_3 + 35\% H_2O_2 \xrightarrow[CH_3CN, temp.]{Fe\ catalyst} H_3C\diagup\diagup\overset{O}{\triangle}CH_3 + H_3C\diagup\diagup\underset{OH}{\overset{OH}{|}}CH_3$$

Entry	Catalyst	Alkene:H_2O_2	Temp. (°C)	Epoxide[a]	cis-diol[a]	ee (%)[b]	Ref.
1	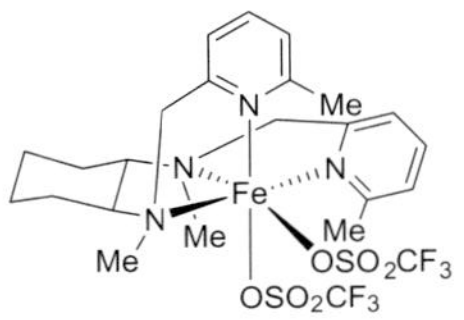 (structure) 10 mol%	50 : 1[c]	30	7%	49%	79	[79b]
2	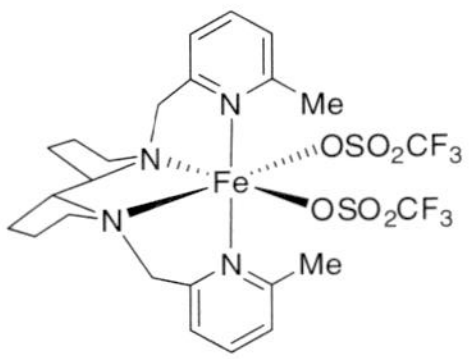 (structure) 10 mol%	50 : 1[d]	r.t.	2%	52%	97	[82]

a) Yield based on the limiting reagent.
b) Enantiomeric excess of cis-2,3-heptanediol.
c) 50% H_2O_2 was used.
d) Concentration of H_2O_2 was not mentioned.

1.7 Conclusions

Excellent techniques for the asymmetric and nonasymmetric dihydroxylation of alkenes are now available. Osmium catalysts still represent the state of the art for this purpose. Since the amount of waste generated from the process determines the usability of the method, environmentally benign oxidants such as O_2 and H_2O_2 were introduced during the last decade. Various techniques, including the use of polymer support, solid support, and ionic liquids, are now established to recycle expensive and toxic OsO_4. Great achievements have advanced the practical application of this method, and kilogram scale processes in the pharmaceutical industry have already been realized. On the other hand, isoelectronic RuO_4 has shown its potential applications in the nonasymmetric dihydroxylation of alkenes. Together with Brønsted or Lewis acids, low-catalyst loading systems have also been demonstrated. However, 'greener' oxidants like O_2 and H_2O_2 as well as the asymmetric version of these systems are still awaited. Recent insights into related biological systems have advanced iron-catalyzed dihydroxylation of alkenes into a new era. From these excellent results, it is expected that a practical protocol with iron catalysts will be available in the near future.

References

1 Beller, M. and Bolm, C. (eds) (2004) *Transition Metals for Organic Synthesis*, 2nd edn, Wiley-VCH, Weinheim.

2 Worldwide production capacities for ethylene glycol in 2000: 13.6 Mio to/a; worldwide production of 1,2-propylene glycol in 1996: 1.4 Mio to/a; Weissermel, K. and Arpe, H.J. (2003) *Industrial Organic Chemistry*, 4th edn, Wiley-VCH, Weinheim, p. 152 and 277.

3 (a) Szmant, H.H. (1989) *Organic Building Blocks of the Chemical Industry*, Wiley, New York, p. 347; (b) Werle, P. (2002) in *Ullmann's Encyclopedia of Industrial Chemistry*, 6th edn, Wiley-VCH, Weinheim.

4 Reviews: (a) Schröder, M. (1980) *Chem. Rev.*, **80**, 187; (b) Kolb, H.C., Van Nieuwenhze, M.S., and Sharpless, K.B. (1994) *Chem. Rev.*, **94**, 2483; (c) Beller, M. and Sharpless, K.B. (1996) in *Applied Homogeneous Catalysis* (eds B. Cornils and W.A. Herrmann), VCH, Weinheim, p. 1009; (d) Marko, I.E. and Svendsen, J.S. (1999) in *Comprehensive Asymmetric Catalysis II* (eds E.N. Jacobsen, A. Pfaltz, and H. Yamamoto), Springer, Berlin, p. 713; (e) Kolb, H.C. and Sharpless, K.B. (2004) in *Transition Metals for Organic Synthesis*, 2nd edn, vol. 2 (eds M. Beller and C. Bolm), Wiley-VCH, Weinheim, p. 275; (f) Zaitsev, A.B. and Adolfsson, H. (2006) *Synthesis*, 1725.

5 Dupau, P., Epple, R., Thomas, A.A., Fokin, V.V., and Sharpless, K.B. (2002) *Adv. Synth. Catal.*, **344**, 421.

6 Mehltretter, G.M., Döbler, C., Sundermeier, U., and Beller, M. (2000) *Tetrahedron Lett.*, **41**, 8083.

7 (a) Hentges, S.G. and Sharpless, K.B. (1980) *J. Am. Chem. Soc*, **102**, 4263; (b) Sharpless, K.B., Amberg, W., Bennani, Y.L., Crispino, G.A., Hartung, J., Jeong, K.-S., Kwong, H.-L., Morikawa, K., Wang, Z.-M., Xu, D., and Zhang, X.-L. (1992) *J. Org. Chem.*, **57**, 2768.

8 (a) Aladro, F.J., Guerra, F.M., Moreno-Dorado, F.J., Bustamante, J.M., Jorge, Z.D., and Massanet, G.M. (2000) *Tetrahedron Lett.*, **41**, 3209; (b) Liang, J., Moher, E.D., Moore, R.E., and Hoard, D.W. (2000) *J. Org. Chem.*, **65**, 3143;

(c) Zhou, X.D., Cai, F., and Zhou, W.S. (2001) *Tetrahedron Lett.*, **42**, 2537; (d) Hamon, D.P.G., Tuck, K.L., and Christie, H.S. (2001) *Tetrahedron*, **57**, 9499; (e) Choudary, B.M., Chowdari, N.S., Jyothi, K., Kumar, N.S., and Kantam, M.L. (2002) *Chem. Commun.*, 586; (f) Hayes, P.Y. and Kitching, W. (2002) *J. Am. Chem. Soc.*, **124**, 9718; (g) Chandrasekhar, S., Ramachandar, T., and Reddy, M.V. (2002) *Synthesis*, 1867; (h) Andreana, P.R., McLellan, J.S., Chen, Y., and Wang, P.G. (2002) *Org. Lett.*, **4**, 3875; (i) Harried, S.S., Lee, C.P., Yang, G., Lee, T.I.H., and Myles, D.C. (2003) *J. Org. Chem.*, **68**, 6646; (j) Jiang, Z.-X., Qin, Y.-Y. and Qing, F.-L. (2003) *J. Org. Chem.*, **68**, 7544; (k) Gupta, P., Fernandes, R.A., and Kumar, P. (2003) *Tetrahedron Lett.*, **44**, 4231; (l) Sayyed, I.A. and Sudalai, A. (2004) *Tetrahedron: Asymmetry*, **15**, 3111; (m) Hövelmann, C.H. and Muñiz, K. (2005) *Chem. Eur. J.*, **11**, 3951; (n) Donohoe, T.J. and Butterworth, S. (2005) *Angew. Chem. Int. Chem.*, **44**, 4766; (o) Robinson, T.V., Taylor, D.K., and Tiekink, E.R.T. (2006) *J. Org. Chem.*, **71**, 7236; (p) Donohoe, T.J., Harris, R.M., Burrows, J., and Parker, J. (2006) *J. Am. Chem. Soc.*, **128**, 13704; (q) Ahmed, M.M. and O'Doherty, G.A. (2006) *Carbohydrate Res.*, **341**, 1505; (r) Xiao, N., Jiang, Z.-X., and Yu, Y.B. (2007) *Biopolymers*, **88**, 781.
9 Hofmann, K.A. (1912) *Chem. Ber.*, **45**, 3329.
10 (a) Milas, N.A. and Sussman, S. (1936) *J. Am. Chem. Soc.*, **58**, 1302; (b) Milas, N.A., Trepagnier, J.-H., Nolan, J.T., and Iliopulos, M.I. (1959) *J. Am. Chem. Soc.*, **81**, 4730.
11 Sharpless, K.B. and Akashi, K. (1976) *J. Am. Chem. Soc.*, **98**, 1986.
12 (a) Schneider, W.P. and McIntosh, A.V. (1956) (Upjohn) US-2.769.824.(1957) *Chem. Abstr.*, **51**, 8822e; (b) VanRheenen, V., Kelly, R.C., and Cha, D.Y. (1976) *Tetrahedron Lett.*, **17**, 1973; (c) Ray, R. and Matteson, D.S. (1980) *Tetrahedron Lett.*, **21**, 449.
13 Minato, M., Yamamoto, K., and Tsuji, J. (1990) *J. Org. Chem.*, **55**, 766.
14 Singh, M.P., Singh, H.S., Arya, A.K., Singh, A.K., and Sisodia, A.K. (1975) *Indian J. Chem.*, **13**, 112.
15 Ahrgren, L. and Sutin, L. (1997) *Org. Process Res. Dev.*, **1**, 425.
16 (a) Bergstad, K., Jonsson, S.Y., and Bäckvall, J.-E. (1999) *J. Am. Chem. Soc.*, **121**, 10424; (b) Jonsson, S.Y., Fränegardh, K., and Bäckvall, J.-E. (2001) *J. Am. Chem. Soc.*, **123**, 1365; (c) Jonsson, S.Y., Adolfsson, H., and Bäckvall, J.-E. (2001) *Org. Lett.*, **3**, 3463; (d) Ell, A.H., Jonsson, S.Y., Börje, A., Adolfsson, H., and Bäckvall, J.-E. (2001) *Tetrahedron Lett.*, **42**, 2569; (e) Jonsson, S.Y., Adolfsson, H., and Bäckvall, J.-E. (2003) *Chem. Eur. J.*, **9**, 2783; (f) Éll, A.H., Closson, A., Adolfsson, H., and Bäckvall, J.-E. (2003) *Adv. Synth. Catal.*, **345**, 1012; (g) Johansson, M., Lindén, A.A., and Bäckvall, J.-E. (2005) *J. Organomet. Chem.*, **690**, 3614.
17 (a) Cummins, R.W. (1970) FMC-Corporation New York, US-Patent 3488394. (b) Cummins, R.W. (1974) FMC-Corporation New York, US-Patent 3846478.
18 Mehltretter, G.M., Bhor, S., Klawonn, M., Döbler, C., Sundermeier, U., Eckert, M., Militzer, H.-C., and Beller, M. (2003) *Synthesis*, **2**, 295.
19 Jacobsen, E.N., Markó, I., Mungall, W.S., Schröder, G., and Sharpless, K.B. (1988) *J. Am. Chem. Soc.*, **110**, 1968.
20 Beller, M., Zapf, A., and Mägerlein, W. (2001) *Chem. Eng. Techn.*, **24**, 575.
21 Wai, J.S.M., Markó, I., Svendsen, J.S., Finn, M.G., Jacobsen, E.N., and Sharpless, K.B. (1989) *J. Am. Chem. Soc.*, **111**, 1123.
22 Junttila, M.H. and Hormi, O.E.O. (2004) *J. Org. Chem.*, **69**, 4816.
23 Junttila, M.H. and Hormi, O.E.O. (2007) *J. Org. Chem.*, **72**, 2956.
24 (a) Cairns, J.F. and Roberts, H.L. (1968) *J. Chem. Soc. (C)*, 640; (b) Celanese Corp. (1966) GB-1,028,940. (1966) *Chem. Abstr.*, **65**, 3064f; (c) Myers, R.S., Michaelson, R.C., and Austin, R.G. (1984) (Exxon Corp.) US-4496779.(1984) *Chem. Abstr.*, **101**, P191362.
25 (a) Krief, A. and Colaux-Castillo, C. (1999) *Tetrahedron Lett.*, **40**, 4189; (b) Krief, A., Delmotte, C., and Colaux-Castillo, C. (2000) *Pure Appl. Chem.*, **72**, 1709.
26 (a) Krief, A. and Colaux-Castillo, C. (2001) *Synlett*, 501; (b) Krief, A. and Colaux-

Castillo, C. (2002) *Pure Appl. Chem.*, **74**, 107.
27 Krief, A., Destree, A., Durisotti, V., Moreau, N., Smal, C., and Colaux-Castillo, C. (2002) *Chem. Commun.*, 558.
28 (a) Döbler, C., Mehltretter, G.M., and Beller, M. (1999) *Angew. Chem. Int. Ed.*, **38**, 3026; (b) Döbler, C., Mehltretter, G.M., Sundermeier, U., and Beller, M. (2000) *J. Am. Chem. Soc.*, **122**, 10289.
29 Weissermel, K. and Arpe, H.J. (2003) *Industrial Organic Chemistry*, 4th edn, Wiley-VCH, Weinheim.
30 Döbler, C., Mehltretter, G.M., Sundermeier, U., and Beller, M. (2001) *J. Organomet. Chem.*, **621**, 70.
31 Bennani, Y.L., Vanhessche, K.P.M., and Sharpless, K.B. (1994) *Tetrahedron Asymm.*, **5**, 1473.
32 (a) Severeyns, A., De Vos, D.E., and Jacobs, P.A. (2002) *Top. Catal.*, **19**, 125; (b) Kobayashi, S. and Sugiura, M. (2006) *Adv. Synth. Catal.*, **348**, 1496.
33 Cainelli, C., Contento, M., Manescalchi, F., and Plessi, L. (1989) *Synthesis*, 45.
34 (a) Kim, B.M. and Sharpless, K.B. (1990) *Tetrahedron Lett.*, **31**, 3003; (b) Lohray, B.B., Thomas, A., Chittari, P., Ahuja, J.R., and Dhal, P.K. (1992) *Tetrahedron Lett.*, **33**, 5453; (c) Han, H. and Janda, K.D. (1996) *J. Am. Chem. Soc.*, **118**, 7632; (d) Gravert, D.J. and Janda, K.D. (1997) *Chem. Rev.*, **97**, 489; (e) Bolm, C. and Gerlach, A. (1997) *Angew. Chem. Int. Ed.*, **36**, 741; (f) Bolm, C. and Gerlach, A. (1998) *Eur. J. Org. Chem.*, 21; (g) Salvadori, P., Pini, D., and Petri, A. (1999) *Synlett*, 1181; (h) Petri, A., Pini, D., Rapaccini, S., and Salvadori, P. (1999) *Chirality*, **11**, 745; (i) Mandoli, A., Pini, D., Agostini, A., and Salvadori, P. (2000) *Tetrahedron: Asymmetry*, **11**, 4039; (j) Bolm, C. and Maischak, A. (2001) *Synlett*, 93; (k) Motorina, I. and Crudden, C.M. (2001) *Org. Lett.*, **3**, 2325; (l) Lee, H.M., Kim, S.-W., Hyeon, T., and Kim, B.M. (2001) *Tetrahedron: Asymmetry*, **12**, 1537; (m) Kuang, Y.-Q., Zhang, S.-Y., Jiang, R., and Wei, L.-L. (2002) *Tetrahedron Lett.*, **43**, 3669; (n) Cheng, S.K., Zhang, S.Y., Wang, P.A., Kuang, Y.Q., and Sun, X.L. (2005) *Appl. Organomet. Chem.*, **19**, 975.
35 (a) Herrmann, W.A. and Weichselbaumer, G. (1991) DE 3920917. CAN 114:121466; (b) Herrmann, W.A., Kratzer, R.M., Blümel, J., Friedrich, H.B., Fischer, R.W., Apperley, D.C., Mink, J., and Berkesi, O. (1997) *J. Mol. Catal. A: Chem.*, **120**, 197.
36 Choudary, B.M., Roy, M., Roy, S., Kantam, M.L., Sreedhar, B., and Kumar, K.V. (2006) *Adv. Synth. Catal.*, **348**, 1734.
37 Kwueon, E.K., Choi, D.S., Choi, H.Y., Lee, Y.J., Jo, C.H., Hwang, S.H., Park, Y.S., and Song, C.E. (2005) *Bull. Korean Chem. Soc.*, **26**, 1839.
38 Nagayama, S., Endo, M., and Kobayashi, S. (1998) *J. Org. Chem.*, **63**, 6094.
39 Kobayashi, S., Endo, M., and Nagayama, S. (1999) *J. Am. Chem. Soc.*, **121**, 11229.
40 Kobayashi, S., Ishida, T., and Akiyama, R. (2001) *Org. Lett.*, **3**, 2649.
41 Ishida, T., Akiyama, R., and Kobayashi, S. (2005) *Adv. Synth. Catal.*, **347**, 1189.
42 Ley, S.V., Ramarao, C., Lee, A.-L., Østergaard, N., Smith, S.C., and Shirley, I.M. (2003) *Org. Lett.*, **5**, 185.
43 Davies, I.W., Matty, L., Hughes, D.L., and Reider, P.J. (2001) *J. Am. Chem. Soc.*, **123**, 10139.
44 Severeyns, A., De Vos, D.E., Fiermans, L., Verpoort, F., Grobet, P.J., and Jacobs, P.A. (2001) *Angew. Chem. Int. Ed.*, **40**, 586.
45 (a) Andersson, M.A., Epple, R., Fokin, V.V., and Sharpless, K.B. (2002) *Angew. Chem. Int. Ed.*, **41**, 472; (b) Adolfsson, H. and Balan, D. (2001) 221st ACS National Meeting, San Diego, CA, United States.
46 Choudary, B.M., Chowdari, N.S., Kantam, M.L., and Raghavan, K.V. (2001) *J. Am. Chem. Soc.*, **123**, 9220.
47 Choudary, B.M., Chowdari, N.S., Jyothi, K., and Kantam, M.L. (2002) *J. Am. Chem. Soc.*, **124**, 5341.
48 Choudary, B.M., Chowdari, N.S., Jyothi, K., Madhi, S., and Kantam, M.L. (2002) *Adv. Synth. Catal.*, **344**, 503.
49 (a) Reviews: Welton, T. (1999) *Chem. Rev.*, **99**, 2071; (b) Wasserscheid, P., and Keim, W. (2000) *Angew. Chem. Int. Ed.*, **39**, 3772; (c) Sheldon, R. (2001) *Chem. Commun.*, 2399.
50 (a) Reynolds, J.L., Erdner, K.R., and Jones, P.B. (2002) *Org. Lett.*, **4**, 917; (b) Ansari, I.A. and Gree, R. (2002) *Org. Lett.*, **4**, 1507; (c) Mayo, K.G., Nearhoof, E.H., and Kiddle, J.J. (2002) *Org. Lett.*, **4**, 1567; (d) Fukuyama, T., Shinmen, M.,

Nishitani, S., Sato, M., and Ryu, I. (2002) *Org. Lett.*, **4**, 1691; (e) Sémeril, D., Olivier-Bourbigou, H., Bruneau, C., and Dixneuf, P.H. (2002) *Chem. Commun.*, 146; (f) Consorti, C.S., Ebeling, G., and Dupont, J. (2002) *Tetrahedron Lett.*, **43**, 753; (g) Nara, S.J., Harjani, J.R., and Salunkhe, M.M. (2002) *Tetrahedron Lett.*, **43**, 2979.

51 (a) Yanada, R. and Takemoto, Y. (2002) *Tetrahedron Lett.*, **43**, 6849; (b) Yao, Q. (2002) *Org. Lett.*, **4**, 2197.

52 Closson, A., Johansson, M., and Bäckvall, J.-E. (2004) *Chem. Commun.*, 1494.

53 Song, C.E., Jung, D.-u., Roh, E.J., Lee, S.-g., and Chi, D.Y. (2002) *Chem. Commun.*, 3038.

54 (a) Branco, L.C. and Afonso, C.A.M. (2002) *Chem. Commun.*, 3036; (b) Branco, L.C. and Afonso, C.A.M. (2004) *J. Org. Chem.*, **69**, 4381.

55 (a) Branco, L.C., Serbanovic, A., Nunes da Ponte, M., and Afonso, C.A.M. (2005) *Chem. Commun.*, 107; (b) Serbanovic, A., Branco, L.C., Nunes da ponte, M., and Afonso, C.A.M. (2005) *J. Organomet. Chem.*, **690**, 3600.

56 Feature article: Afonso, C.A.M., Branco, L.C., Candeias, N.R., Gois, P.M., Lourenco, N.M.T., Mateus, N.M.M., and Rosa, J.N. (2007) *Chem. Commun.*, 2669.

57 Carlsen, P.H.J., Katsuki, T., Martin, V.S., and Sharpless, K.B. (1981) *J. Org. Chem.*, **46**, 3936.

58 Shing, T.K.M., Tai, V.W.-F., and Tam, E.K.W. (1994) *Angew. Chem. Int. Ed. Engl.*, **33**, 2312.

59 Shing, T.K.M., Tam, E.K.W., Tai, V.W.-F., Chung, I.H.F., and Jiang, Q. (1996) *Chem. Eur. J.*, **2**, 50.

60 Shing, T.K.M. and Tam, E.K.W. (1999) *Tetrahedron Lett.*, **40**, 2179.

61 (a) Plietker, B. and Niggemann, M. (2003) *Org. Lett.*, **5**, 3353; (b) Plietker, B., Niggemann, M., and Pollrich, A. (2004) *Org. Biomol. Chem.*, **2**, 1116; (c) Plietker, B. and Niggemann, M. (2005) *J. Org. Chem.*, **70**, 2402.

62 Plietker, B. and Niggemann, M. (2004) *Org. Biomol. Chem.*, **2**, 2403.

63 Ho, C.-M., Yu, W.-Y., and Che, C.-M. (2004) *Angew. Chem. Int. Ed.*, **43**, 3303.

64 Yip, W.-P., Yu, W.-Y., Zhu, N., and Che, C.-M. (2005) *J. Am. Chem. Soc.*, **127**, 14239.

65 Yip, W.-P., Ho, C.-M., Zhu, N., Lau, T.-C., and Che, C.-M. (2008) *Chem. Asian J.*, **3**, 70.

66 (a) Mielczarek, E.V. and McGrayne, S.B. (2000) *Iron, Nature's Universal Element: Why People Need Iron & Animals Make Magnets*, Rutgers University Press, New Brunswick, N.J.; (b) Cornell, R.M. and Schwertmann, U. (2003) *The Iron Oxides: Structures, Properties, Occurrences and Uses*, Wiley-VCH, Weinheim.

67 (a) Fontecave, M., Ménage, S., and Duboc-Toia, C. (1998) *Coord. Chem. Rev.*, **178–180**, 1555; (b) Costas, M., Chen, K., and Que, L. Jr. (2000) *Coord. Chem. Rev.*, **200–202**, 517; (c) Bolm, C., Legros, J., Le Paih, J., and Zani, L. (2004) *Chem. Rev.*, **104**, 6217; (d) Vieira Soares, A.P., Portela, M.F., and Kiennemann, A. (2005) *Catal. Rev.*, **47**, 125; (e) Fürstner, A. and Martin, R. (2005) *Chem. Lett.*, **34**, 624; (f) Enthaler, S., Junge, K., and Beller, M. (2008) *Angew. Chem. Int. Ed.*, **47**, 3317.

68 (a) Wu, H., Desai, S.R., and Wang, L.-S. (1996) *J. Am. Chem. Soc.*, **118**, 5296; (b) for corrections see: Wu, H., Desai, S.R., and Wang, L.-S. (1996) *J. Am. Chem. Soc.*, **118**, 7434; (c) Gutsev, G.L., Khanna, S.N., Rao, B.K., and Jena, P. (1999) *Phy. Rev. A.*, **59**, 3681.

69 (a) Delaude, L. and Laszlo, P. (1996) *J. Org. Chem.*, **61**, 6360; (b) Caddick, S., Murtagh, L., and Weaving, R. (1999) *Tetrahedron Lett.*, **40**, 3655; (c) Huang, H., Sommerfeld, D., Dunn, B.C., Lloyd, C.R., and Eyring, E.M. (2001) *J. Chem. Soc., Dalton Trans.*, 1301.

70 (a) Jiang, J.-Q. and Lloyd, B. (2002) *Water Res.*, **36**, 1397; (b) Virender, K.S., Futaba, K., Hu, J., and Ajay, K.R. (2005) *J. Water Health*, **3**, 45.

71 Lippard, S.J. and Berg, J.M. (1994) *Principles of Bioinorganic Chemistry*, University Science Books, Mill Valley, CA.

72 Sheldon, R.A. (ed.) (1994) *Metalloporphyrins in Catalytic Oxidations*, Marcel Dekker Ltd., New York.

73 (a) Ponka, P., Schulman, H.M., and Woodworth, R.C. (1990) *Iron Transport and Storage*, CRC Press. Inc., Boca Raton, Florida, (b) Trautheim, A.X. (ed.) (1997) *Bioinorganic Chemistry: Transition Metals in Biology and their Coordination Chemistry*, Wiley-VCH, Weinheim.

74 (a) Meunier, B. (ed.) (2000) *Biomimetic Oxidations Catalyzed by Transition Metals*, Imperial College Press, London, (b) van Eldik, R. (ed.) (2006) Advances in Inorganic Chemistry, in *Homogeneous Biomimetic Oxidation Catalysis*, vol. 58, Academic Press, London.

75 (a) Que, L. Jr. and Ho, R.Y.N. (1996) *Chem. Rev.*, 96, 2607; (b) Costas, M., Mehn, M.P., Jensen, M.P., and Que, L. Jr. (2004) *Chem. Rev.*, 104, 939; (c) Oldenburg, P.D., and Que, L. Jr. (2006) *Catal. Today*, 117, 15.

76 Chen, K. and Que, L. Jr. (1999) *Angew. Chem. Int. Ed.*, 38, 2227.

77 (a) Gibson, D.T. and Parales, R.E. (2000) *Curr. Opin. Biotechnol.*, 11, 236; (b) Karlsson, A., Parales, J.V., Parales, R.E., Gibson, D.T., Eklund, H., and Ramaswamy, S. (2003) *Science*, 299, 1039.

78 Fujita, M. and Que, L. Jr. (2004) *Adv. Synth. Catal.*, 346, 190.

79 (a) Chen, K. and Que, L. Jr. (1999) *Chem. Commun.*, 1375; (b) Costas, M., Tipton, A.K., Chen, K., Jo, D.-H., and Que, L. Jr. (2001) *J. Am. Chem. Soc.*, 123, 6722; (c) Chen, K., Costas, M., Kim, J., Tipton, A.T., and Que, L. Jr. (2002) *J. Am. Chem. Soc.*, 124, 3026.

80 White, M.C., Doyle, A.G., and Jacobsen, E.N. (2001) *J. Am. Chem. Soc.*, 123, 7194.

81 (a) Oldenburg, P.D., Shteinman, A.A., and Que, L., Jr. (2005) *J. Am. Chem. Soc.*, 127, 15672; (b) Oldenburg, P.D., Ke, C.-Y., Tipton, A.A., Shteinman, A.A., and Que, L. Jr. (2006) *Angew. Chem. Int. Ed.*, 45, 7975.

82 Suzuki, K., Oldenburg, P.D., and Que, L. Jr. (2008) *Angew. Chem. Int. Ed.*, 47, 1887.

83 Chen, M.S. and White, M.C. (2007) *Science*, 318, 783.

84 De Laat, J. and Le, T.G. (2006) *Appl. Catal. B: Environmental*, 66, 137.

85 (a) Barton, D.H.R. and Taylor, D.K. (1996) *Pure Appl. Chem.*, 68, 497; (b) Barton, D.H.R. and Hu, B. (1997) *Pure Appl. Chem.*, 69, 1941.

2
Transition Metal-Catalyzed Epoxidation of Alkenes
Hans Adolfsson

2.1
Introduction

The formation of epoxides via metal-catalyzed oxidation of alkenes in the presence of molecular oxygen or hydrogen peroxide represents the most elegant and environmentally friendly route for the production of this compound class [1, 2]. This is of particular importance considering that the conservation and management of resources should be the main focus of interest when novel chemical processes are developed. Thus, the introduction and improvement of catalytic epoxidation methods where molecular oxygen or hydrogen peroxide are employed as terminal oxidants are highly desirable. However, one of today's industrial routes for the formation of simple epoxides (e.g., propylene oxide) is the *chlorohydrin* process, where alkenes are reacted with chlorine in the presence of sodium hydroxide (Scheme 2.1) [3]. This process produces 2.01 tonne NaCl and 0.102 tonne 1,2-dichloropropane as by-products per tonne of propylene oxide. These significant amounts of waste are certainly not acceptable in the long run, and efforts to replace such chemical processes with 'greener' epoxidation processes are under way. When it comes to the production of fine chemicals, non-catalyzed processes with traditional oxidants (e.g., peroxyacetic acid and *meta*-chloroperoxybenzoic acid) are often used. In these cases, though, transition metal-based systems using hydrogen peroxide as the terminal oxidant have several advantages. This chapter covers the development of efficient and selective metal-catalyzed epoxidation methods starting with early transition metals (Ti, V) and ending with group X metals (Pt). The scope and focus will be to highlight some novel approaches to alkene oxidation using environmentally benign oxidants.

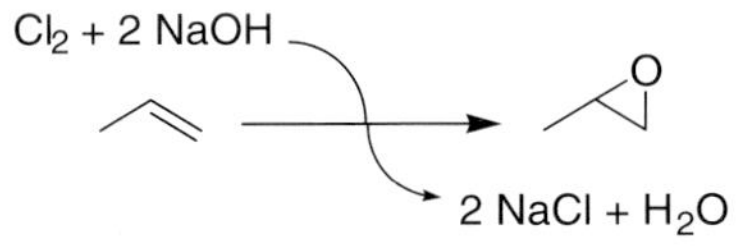

Scheme 2.1

Modern Oxidation Methods. Edited by Jan-Erling Bäckvall
Copyright © 2010 WILEY-VCH Verlag GmbH & Co. KGaA, Weinheim
ISBN: 978-3-527-32320-3

2.2
Choice of Oxidant for Selective Epoxidation

Several terminal oxidants are available for the transition metal-catalyzed epoxidation of alkenes (Table 2.1). Typical oxidants compatible with a majority of metal-based epoxidation systems include various alkyl hydroperoxides, hypochlorite, or iodosylbenzene. A problem associated with these oxidants is their low active oxygen content (Table 2.1). Considering the nature of the waste produced, there are further drawbacks using these oxidants. From an environmental and economical point of view, molecular oxygen should be the preferred oxidant in view of its high active oxygen content and the fact that that no waste product or only water is formed. One of the major limitations, however, in using molecular oxygen as terminal oxidant for the formation of epoxides is the poor product selectivity obtained in these processes [4]. In combination with the limited number of catalysts available for direct activation of molecular oxygen, this seriously limits the use of this oxidant. On the other hand, hydrogen peroxide displays much better properties as terminal oxidant. The active oxygen content of H_2O_2 is about as high as that in typical applications of molecular oxygen in epoxidations (since a reductor is required in almost all cases), and the waste produced by employing this oxidant is plain water. As in the case of molecular oxygen, the epoxide selectivity using H_2O_2 can sometimes be relatively poor, although recent developments have led to transition metal-based protocols where excellent reactivity and epoxide selectivity can be obtained [5]. The various oxidation systems available for the selective epoxidation of alkenes using transition metal catalysts and hydrogen peroxide will be covered in the following sections.

Table 2.1 Oxidants used in transition metal-catalyzed epoxidations, and their active oxygen content.

Oxidant	Active oxygen content (wt. %)	Waste product
Oxygen (O_2)	100	Nothing or H_2O
Oxygen (O_2)/reductor	50	H_2O
H_2O_2	47	H_2O
NaOCl	21.6	NaCl
CH_3CO_3H	21.1	CH_3CO_2H
t-BuOOH (TBHP)	17.8	t-BuOH
$KHSO_5$	10.5	$KHSO_4$
BTSP[a]	9	Hexamethyldisiloxane
PhIO	7.3	PhI

a) Bistrimethylsilyl peroxide.

2.3
Epoxidations of Alkenes Catalyzed by Early Transition Metals

High-valent early transition metals like titanium(IV) and vanadium(V) have been shown to efficiently catalyze the epoxidation of alkenes. The preferred oxidants using these catalysts are various alkyl hydroperoxides, typically *tert*-butyl hydroperoxide (TBHP) or ethylbenzene hydroperoxide (EBHP). One of the routes for the industrial production of propylene oxide is based on a heterogeneous Ti^{IV}/SiO_2 catalyst, which employs EBHP as the terminal oxidant [6].

The Sharpless-Katsuki asymmetric epoxidation (AE) protocol for the enantioselective formation of epoxides from allylic alcohols is a milestone in asymmetric catalysis [7]. This historically highly important and classical asymmetric transformation is catalyzed by a dimeric complex formed from titanium isopropoxide in combination with a tartrate ester (**1**). TBHP is the preferred terminal oxidant, and the reaction has been widely used in various synthetic applications. There are several excellent reviews covering the scope and utility of the AE reaction [8]. As a complement to the Sharpless-Katsuki AE-reaction of allylic alcohols, Yamamoto recently presented some highly interesting results using novel vanadium catalysts (**2** and **3**) based on C_2-symmetric bis-hydroxamic acid ligands [9]. The newly designed catalysts were indeed superior to previously reported vanadium systems, and excellent enantioselectivities were achieved in the epoxidation of a series of differently substituted allylic alcohols. In particular, the AE reaction on *cis*-substituted alkenes, which normally tend to give lower enantioselectivities, resulted in excellent *ee*s employing catalyst **3**. In addition, this epoxidation protocol tolerates the use of aqueous TBHP, and does not suffer from the ligand deceleration effects normally observed for vanadium catalysts.

On the other hand, the use of hydrogen peroxide as an oxidant in combination with early transition metal catalysts (Ti and V) is rather limited. The reason for the poor reactivity can be traced to severe inhibition of the metal complexes by strongly coordinating ligands like alcohols, in particular, water. The development of the heterogeneous titanium(IV)-silicate catalyst (TS-1) by chemists at Enichem represented a breakthrough for reactions performed with hydrogen peroxide [10]. This hydrophobic molecular sieve demonstrated excellent properties (i.e., high catalytic activity and selectivity) for the epoxidation of small linear alkenes in methanol. The substrates are adsorbed into the micropores of the TS-1 catalyst, which efficiently prevents the inhibition by water observed using the Ti^{IV}/SiO_2 catalyst. After the epoxidation reaction, the TS-1 catalyst can easily be separated and reused. To enlarge the scope of this epoxidation method and thereby allow for the oxidation of a wider range of substrates, several different titanium-containing silicate zeolites have been prepared. Consequently, the scope has been somewhat improved, but the best epoxidation results using titanium silicates as catalysts are obtained with smaller, nonbranched substrates.

The problems encountered with high oxophilicity of early transition metal catalyst systems have severely limited the use of aqueous hydrogen peroxide as terminal oxidant. However, recently the group of Katsuki reported on a significant breakthrough in the field of titanium-catalyzed asymmetric epoxidation of nonfunctionalized alkenes [11–13]. The successful catalysts used in this system are based on a combination of titanium and reduced salen-type ligands (also known as salenan and salan, respectively). The oxidation of terminal alkenes such as styrene and 1-octene using catalyst 4 resulted in good yields of the epoxides (90 and 70% respectively) in high enantioselectivity (93 and 87% ee respectively). Most strikingly, only 1 mol% of the catalyst and 1.05 equivalents of 30% aqueous hydrogen peroxide were required in the epoxidations. When this catalytic system was employed for the epoxidation of 1,2-dihydronaphthalene, the corresponding epoxide was obtained in 99% yield and 99% ee. The dimeric µ-oxo-bridged catalyst 4 with its half-reduced salen ligands has a rather elaborate structure and a somewhat complicated preparation procedure. Nevertheless, its stability is remarkable since the homochiral complex keeps its dimeric structure in methanol for more than 24 h. For comparison, the corresponding salen-type complex, which is completely inactive as an epoxidation catalyst under these conditions, dissociates immediately into monomeric titanium-salen species in methanol. The difference in catalytic activity observed between 4 and the corresponding [Ti(salen)] complex is believed to be due to the presence of an intramolecular hydrogen bonding between the amine N–H and a peroxo-group on the metal, which activates the complex for the transfer of oxygen to the substrate. A synthetically less challenging and more flexible analog of 4 has also been evaluated as an epoxidation catalyst in the presence of aqueous hydrogen peroxide. Complex 5a (Ar = Ph), easily prepared directly from titanium isopropoxide and the salan ligand, is, however, less active and selective than 4.

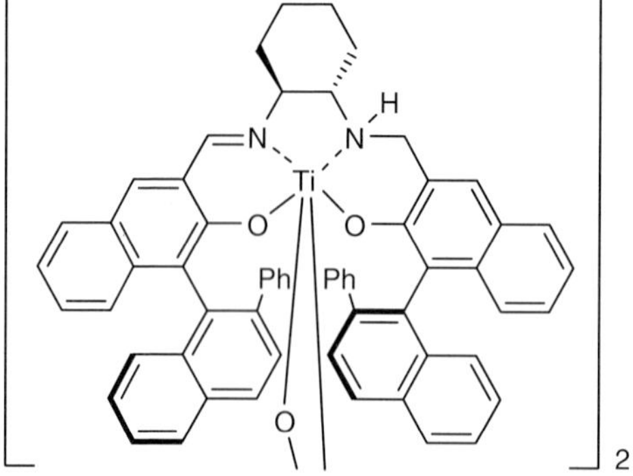

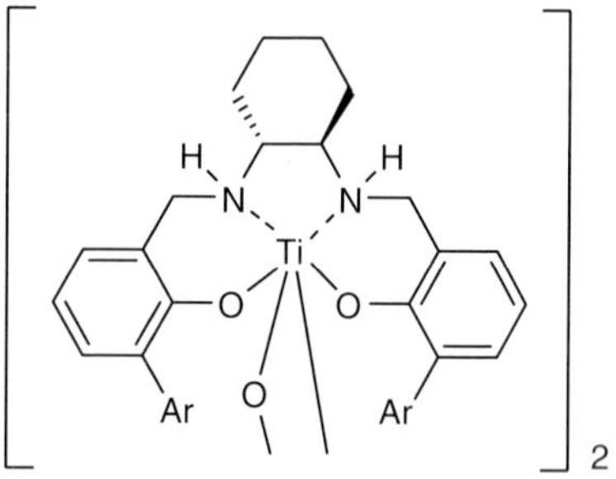

5a Ar = Ph
5b Ar = o-MeOC$_6$H$_4$
5c Ar = o-CF$_3$C$_6$H$_4$

Table 2.2 Asymmetric titanium-catalyzed epoxidation.

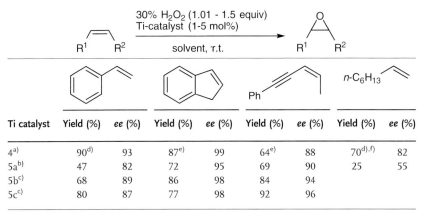

Ti catalyst	Yield (%)	ee (%)	Yield (%)	ee (%)	Yield (%)	ee (%)	Yield (%)	ee (%)
4[a)]	90[d)]	93	87[e)]	99	64[e)]	88	70[d),f)]	82
5a[b)]	47	82	72	95	69	90	25	55
5b[c)]	68	89	86	98	84	94		
5c[c)]	80	87	77	98	92	96		

a) 1 mol% catalyst, reaction time 24 h.
b) 5 mol% catalyst, reaction time 24 h.
c) Catalyst prepared *in situ* from 5 mol% titanium isopropoxide and 6 mol% of the corresponding ligand, reaction time 9 h.
d) Reaction performed in dichloromethane.
e) Reaction performed in ethyl acetate.
f) Reaction time 48 h.

Moreover, a structural modification of the initial salan ligand used in the study led to the development of catalysts which are able to form epoxides with a high degree of stereocontrol. As shown in Table 2.2, catalysts based on ligands **5b** and **5c** (Ar = o-MeOC$_6$H$_4$ and o-CF$_3$C$_6$H$_4$, respectively) are able to epoxidize a range of alkenes in good yields and with excellent selectivity. As in the case of the corresponding manganese-salen systems, *cis*-substituted alkenes are the preferred substrates. The epoxidation reaction is reported to be stereospecific, since no *trans*-epoxides are formed starting from *cis*-alkenes.

The oxidation of simple alkenes like cyclohexene using monovanadium complexes and hydrogen peroxide as terminal oxidant predominantly results in allylic oxidations instead of epoxide formation. However, when vanadium is incorporated into polyoxometalates the results are significantly improved. Mizuno and coworkers recently reported that the tetra-*n*-butylammonium salt of a polyoxometalate with the composition [γ-1,2-H$_2$SiV$_2$W$_{10}$O$_{40}$]$^{4-}$ acts as a highly active and selective epoxidation catalyst employing aqueous H$_2$O$_2$ as terminal oxidant [14, 15]. This polyoxometalate, which has a bridging di-VO-core, VO-(μ-OH)$_2$-VO, catalyzes the stereospecific epoxidation of a wide range of unfunctionalized alkenes, dienes, and hydroxyl-functionalized alkenes, in the presence of only one equivalent of oxidant per unsaturation (Scheme 2.2). The limitation of this system appears to be trans-disubstituted alkenes, which showed poor reactivity (*trans*-2-octene gave only 6% yield of the epoxide under the conditions presented in Scheme 2.4). Mechanistic studies and DFT calculations indicate severe steric interactions between the substituents of such substrates and the polyoxometalate framework [16].

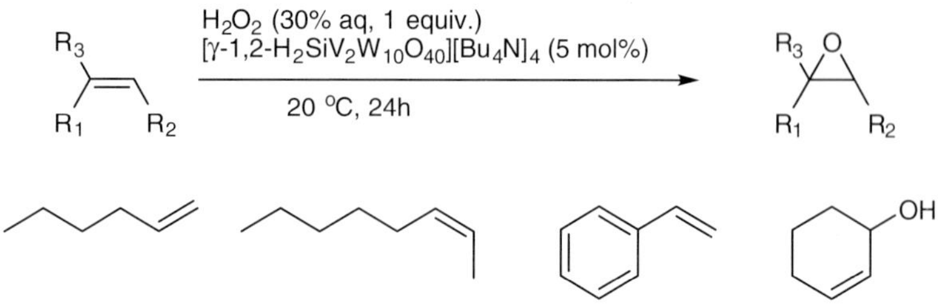

Scheme 2.2

Reaction conditions: H$_2$O$_2$ (30% aq, 1 equiv.), [γ-1,2-H$_2$SiV$_2$W$_{10}$O$_{40}$][Bu$_4$N]$_4$ (5 mol%), 20 °C, 24h

Substrates and results:
- 1-octene: yield 92%, selectivity 99%
- cis-internal alkene: yield 90%, selectivity 99%
- styrene: yield 88%, selectivity 99%
- 2-cyclohexen-1-ol: yield 87%, 95% (syn:anti 12:88)

2.4
Molybdenum and Tungsten-Catalyzed Epoxidations

Epoxidation systems based on molybdenum and tungsten catalysts have been extensively studied for more than 40 years. The typical catalysts, Mo^{VI}-oxo or W^{VI}-oxo species do, however, behave quite differently depending on whether anionic or neutral complexes are employed. Whereas the former catalysts, especially tungstates under phase-transfer conditions, are able to efficiently activate aqueous hydrogen peroxide for the formation of epoxides. Neutral molybdenum or tungsten complexes do react with hydrogen peroxide, but better selectivities are often achieved using organic hydroperoxides (e.g., *t*-butyl hydroperoxide) as terminal oxidants [17, 18].

2.4.1
Homogeneous Catalysts – Hydrogen Peroxide as the Terminal Oxidant

Payne and Williams reported in 1959 on the selective epoxidation of maleic, fumaric, and crotonic acids using a catalytic amount of sodium tungstate (2 mol%) in combination with aqueous hydrogen peroxide as the terminal oxidant [19]. The key to success was careful control of the pH (4–5.5) in the reaction media. These electron-deficient substrates were notoriously difficult to selectively oxidize using the standard techniques (peroxy acid reagents) available at the time. Previous attempts to use sodium tungstate and hydrogen peroxide led to the isolation of the corresponding diols because of rapid hydrolysis of the intermediate epoxides. Significant improvements of this catalytic system were introduced by Venturello and coworkers [20, 21]. They found that the addition of phosphoric acid and the introduction of quaternary ammonium salts as PTC-reagents considerably increased the scope of the reaction. The active tungstate catalysts are often generated *in situ*, although catalytically active peroxo-complexes like (*n*-hexyl$_4$N)$_3$[PO$_4$(W(O)(O$_2$)$_2$)$_4$] have been isolated and characterized (Figure 2.1) [22].

In more recent work, Noyori and coworkers found conditions for the selective epoxidation of aliphatic terminal alkenes either in toluene, or using a completely solvent-free reaction setup [23, 24]. One of the disadvantages with the previous

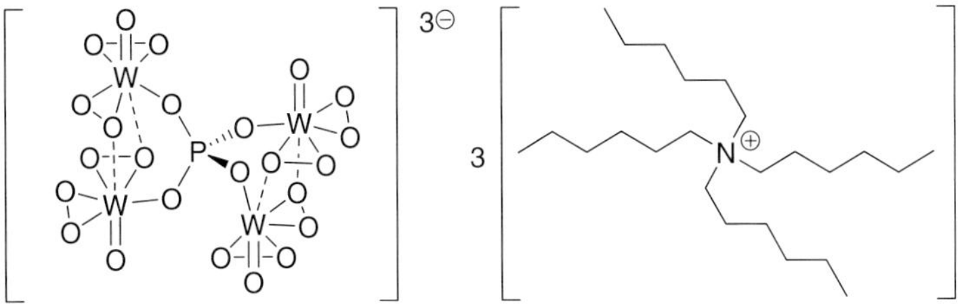

Figure 2.1 The Venturello $(n\text{-hexyl}_4N)_3[PO_4(W(O)(O_2)_2)_4]$ catalyst.

systems was the use of chlorinated solvents [20, 21]. The conditions established by Noyori, however, provided an overall 'greener' epoxidation process since the reactions efficiently were performed in nonchlorinated solvents. In this reaction, sodium tungstate (2 mol%), (aminomethyl)phosphonic acid and methyltri-n-octylammonium bisulfate (1 mol% of each) were employed as catalysts for the epoxidation using aqueous hydrogen peroxide (30%) as the terminal oxidant. The epoxidation of various terminal alkenes using the above-mentioned conditions (90 °C, no solvent added) gave high yields for a number of substrates (Table 2.3). The work-up procedure was exceptionally simple, since the product epoxides could be distilled directly from the reaction mixture. The use of proper additives turned out to be crucial for a successful outcome of these epoxidation reactions.

When the (aminomethyl)phosphonic acid was replaced by other phosphonic acids or simply by phosphoric acid, significantly lower conversions were obtained. The nature of the phase-transfer reagent was further established as an important parameter. The use of ammonium bisulfate (HSO_4^-) was superior to the corresponding chloride or hydroxide salts. The size, and hence the lipophilicity of the ammonium ion was important, since tetra-n-butyl- and tetra-n-hexylammonium bisulfate were inferior to phase-transfer agents containing larger alkyl-groups. The epoxidation system was later extended to encompass other substrates, like simple

Table 2.3 Epoxidation of terminal alkenes using the Noyori system.

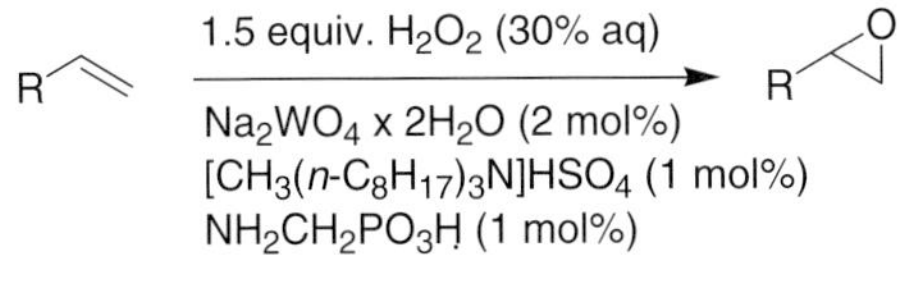

Entry	Alkene	Time (h)	Conversion (%)	Yield (%)
1	1-octene	2	89	86
2	1-decene	2	94	93
3[a]	1-decene	4	99	99
4[a]	allyl octyl ether	2	81	64
5[a]	styrene	3	70	2

a) 20 mmol alkene in 4 mL toluene.

alkenes with different substitution patterns and alkenes containing various functionalities (alcohols, ethers, ketones, and esters).

A major limitation of this method is the low pH at which the reactions are performed. This led to substantially lower yields in reactions with substrates leading to acid-sensitive epoxides, where competing ring-opening processes effectively reduced the usefulness of the protocol. As an example, the oxidation of styrene led to 70% conversion after 3 h at 70 °C, although the observed yield for styrene oxide was only 2% (Table 2.3, entry 5).

The epoxidation method developed by Noyori has subsequently been applied to the direct formation of dicarboxylic acids from alkenes [25]. Cyclohexene was oxidized to adipic acid in 93% yield using the tungstate/ammonium bisulfate system and 4 equivalents of hydrogen peroxide. The selectivity problem associated with the Noyori protocol was to a certain degree circumvented by the improvements introduced by Jacobs and coworkers [26]. To the standard catalytic mixture were added additional amounts of (aminomethyl)phosphonic acid and Na_2WO_4, and the pH of the reaction medium was adjusted to 4.2–5 with aqueous NaOH. These changes allowed for the formation of epoxides from α-pinene, 1-phenyl-1-cyclohexene, and indene, in high conversions and with good selectivity (Scheme 2.3).

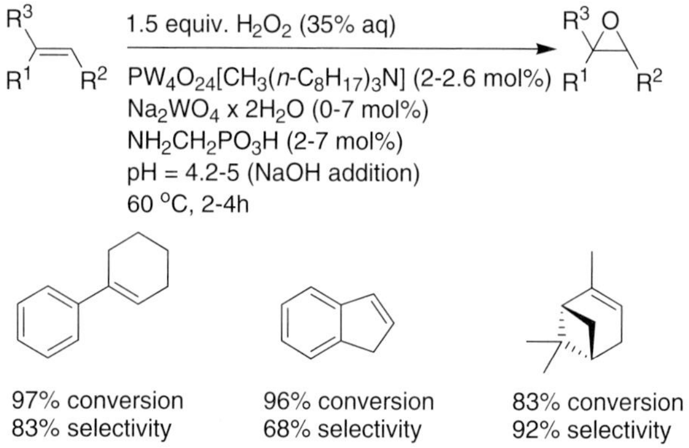

Scheme 2.3

Another highly efficient tungsten-based system for the epoxidation of alkenes was introduced by Mizuno and coworkers [27, 28]. The tetrabutylammonium salt of a Keggin-type silicodecatungstate $[\gamma\text{-}SiW_{10}O_{34}(H_2O)_2]^{4-}$ (Scheme 2.3) was found to catalyze the epoxidation of various alkene substrates using aqueous hydrogen peroxide as the terminal oxidant. The characteristics of this system are very high epoxide selectivity (99%) and excellent efficiency in the use of the terminal oxidant (99%). Terminal as well as di- and tri-substituted alkenes were all epoxidized in high yields within reasonably short reaction times using 0.16 mol% catalyst (1.6 mol% in tungsten, Scheme 2.4). The X-ray structure of the catalyst precursor revealed 10 tungsten atoms connected to a central SiO_4 unit. In situ infrared spectroscopy of the reaction mixture during the epoxidation reaction indicated high structural stability of

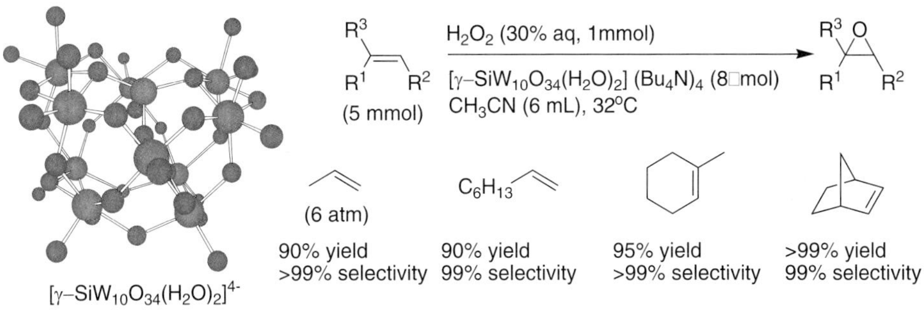

Scheme 2.4

[γ–SiW$_{10}$O$_{34}$(H$_2$O)$_2$]$^{4-}$

R^{1}R^{2}C=CR^3 (5 mmol) + H$_2$O$_2$ (30% aq, 1mmol) $\xrightarrow{[\gamma\text{–SiW}_{10}\text{O}_{34}(\text{H}_2\text{O})_2] (\text{Bu}_4\text{N})_4 (8\ \text{mol}) \\ \text{CH}_3\text{CN (6 mL), 32°C}}$ epoxide

- propene (6 atm): 90% yield, >99% selectivity
- C$_6$H$_{13}$–CH=CH$_2$: 90% yield, 99% selectivity
- cyclohexene: 95% yield, >99% selectivity
- norbornene: >99% yield, 99% selectivity

the catalyst. Furthermore, it was demonstrated that the catalyst can be recovered and reused up to 5 times without loss of activity or selectivity (epoxidation of cyclooctene). Interestingly, the often encountered problem with hydrogen peroxide decomposition was negligible using this catalyst. The efficient use of hydrogen peroxide (99%) combined with the high selectivity and productivity in propylene epoxidation may well open up industrial applications.

The use of molybdenum catalysts in combination with hydrogen peroxide is not as common as that of tungsten catalysts. There are, however, a number of examples where molybdates have been employed for the activation of hydrogen peroxide. A catalytic amount of sodium molybdate in combination with monodentate ligands (e.g., hexa-alkyl phosphorus triamides or pyridine-N-oxides) and sulfuric acid allowed for the epoxidation of simple linear or cyclic alkenes [29]. The selectivity obtained using this method was quite low, and significant amounts of diols were formed, even though highly concentrated hydrogen peroxide (>70%) was employed.

More recently, Sundermeyer and coworkers reported on the use of long-chain trialkylamine oxides, trialkylphosphane oxides, or trialkylarsane oxides as monodentate ligands for neutral molybdenum peroxo complexes [30]. These compounds were employed as catalysts for the epoxidation of 1-octene and cyclooctene with aqueous hydrogen peroxide (30%), under biphasic conditions (CHCl$_3$). The epoxide products were obtained in high yields with good selectivity. The high selectivity achieved using this method was ascribed to high solubility of the product in the organic phase, thus protecting the epoxide from hydrolysis. This protocol has not been employed for the formation of hydrolytically sensitive epoxides, and the generality of the method can thus be questioned. A recent example of a highly efficient molybdenum-based protocol for alkene epoxidation using aqueous hydrogen peroxide was developed by Bhattacharyya and coworkers [31]. The use of oxodiperoxomolybdenum(VI) complexes ligated with 8-quinolinol (QOH) allowed for epoxidation of a series of various alkenes. With the most active catalyst, [PPh$_4$][Mo(O)(O$_2$)$_2$(QO)], a highly impressive turnover number (TON) of 14 800 was observed for the epoxidation of cyclohexene, using 3–5 equivalents of aqueous hydrogen peroxide in acetonitrile at room temperature. This catalytic system requires sodium bicarbonate (25 mol%) as co-catalyst, and since this additive renders the solution basic, rather sensitive epoxides like styrene oxide can also be produced in excellent yields.

2.4.2
Heterogeneous Catalysts

One problem associated with the above-described peroxotungstate-catalyzed epoxidation system is the separation of the catalyst after completion of the reaction. To overcome this obstacle, efforts to prepare heterogeneous tungstate catalysts have been conducted. De Vos and coworkers employed tungsten catalysts derived from sodium tungstate and layered double hydroxides (LDH – coprecipitated $MgCl_2$, $AlCl_3$, and NaOH) for the epoxidation of simple alkenes and allyl alcohols with aqueous hydrogen peroxide [32]. They found that, depending on the nature of the catalyst (either hydrophilic or hydrophobic catalysts were used), different reactivities and selectivities were obtained for nonpolar and polar alkenes respectively. The hydrophilic LDH-WO_4 catalyst was particularly effective for the epoxidation of allyl and homo-allyl alcohols, whereas the hydrophobic catalyst (containing p-toluensulfonate) showed better reactivity with nonfunctionalized substrates.

Gelbard and coworkers have reported on the immobilization of tungsten catalysts using polymer-supported phosphine oxide, phosphonamide, phosphoramide, and phosphotriamide ligands [33]. Employing these heterogeneous catalysts together with hydrogen peroxide for the epoxidation of cyclohexene resulted in moderate to good conversion of the substrate, although in most cases low epoxide selectivity was observed. A significantly more selective heterogeneous catalyst was obtained by Jacobs and coworkers upon treatment of the macroreticular ion-exchange resin Amberlite IRA-900 with an ammonium salt of the Venturello anion $\{PO_4[WO(O_2)_2]_4\}^{3-}$ [26, 34]. The catalyst formed was used for the epoxidation of a number of terpenes, and high yields and good selectivity of the corresponding epoxides were achieved.

In a different strategy, siliceous mesoporous MCM-41-based catalysts were prepared [34]. Quaternary ammonium salts and alkyl phosphoramides, respectively, were grafted onto MCM-41, and the material obtained was treated with tungstic acid for the preparation of heterogeneous tungstate catalysts. The catalysts were employed in the epoxidation of simple cyclic alkenes with aqueous hydrogen peroxide (35%) as terminal oxidant, but conversion and selectivity for epoxide formation was rather low. In the case of cyclohexene, the selectivity could be improved by the addition of pyridine. The low tungsten leaching ($<2\%$) using these catalysts is certainly advantageous.

A particularly interesting system for the epoxidation of propylene to propylene oxide, working under pseudo-heterogeneous conditions, was reported by Zuwei and coworkers [35]. The catalyst, which was based on the Venturello anion combined with long-chain alkylpyridinium cations, showed unique solubility properties. In the presence of hydrogen peroxide the catalyst was fully soluble in the solvent (a 4 : 3 mixture of toluene and tributylphosphate), but when no more oxidant was left, the tungsten catalyst precipitated and could simply be removed from the reaction mixture (Scheme 2.5). Furthermore, this epoxidation system was combined with the 2-ethylanthraquinone (EAQ)/2-ethylanthrahydroquinone (EAHQ) process for hydrogen peroxide formation (Scheme 2.6), and good conversion and selectivity were

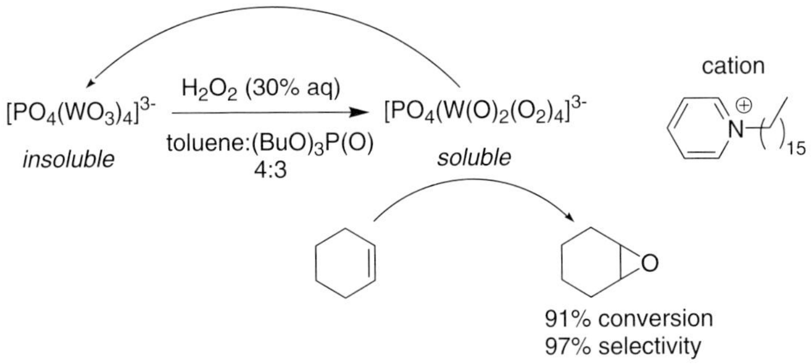

Scheme 2.5

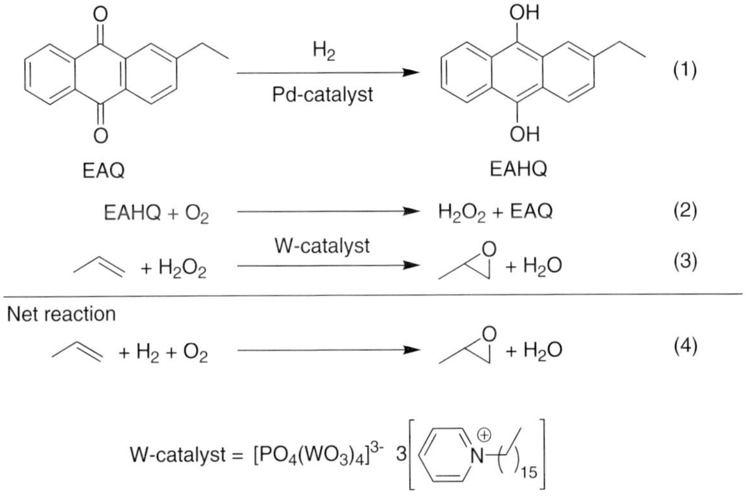

Scheme 2.6

obtained for propylene oxide in three consecutive cycles. The catalyst was recovered by centrifugation in between every cycle and used directly in the next reaction.

2.5
Manganese-Catalyzed Epoxidations

Historically, the interest of using manganese complexes as catalysts for the epoxidation of alkenes comes from biologically relevant oxidative manganese porphyrins. The terminal oxidants compatible with manganese porphyrins were initially restricted to iodosylbenzene, sodium hypochlorite, alkyl peroxides and hydroperoxides, N-oxides, $KHSO_5$, and oxaziridines. Molecular oxygen can also be used in the

Table 2.4 Manganese-porphyrin catalyzed epoxidation of cis-cyclooctene using aqueous H_2O_2 (30%).

Entry	Catalyst	Additive	Temp. (°C)	Time	Yield (%)
1	**6** 2.5 mol%	Imidazole (0.6 equiv.)	20	45 min	90%
2	**6** 0.5 mol%	N-Hexyl-imidazole (0.5 mol%) Benzoic acid (0.5 mol%)	0	15 min	100%
3	**7** 0.1 mol%	—	0	3 min	100%

presence of an electron source. The use of hydrogen peroxide often results in oxidative decomposition of the catalyst due to the potency of this oxidant. However, the introduction of chlorinated porphyrins (**6**) allowed hydrogen peroxide to be used as terminal oxidant [36]. These catalysts, discovered by Mansuy and coworkers, were demonstrated to resist decomposition, and when used together with imidazole or imidazolium carboxylates as additives, efficient epoxidation of alkenes were achieved (Table 2.4, entries 1 and 2).

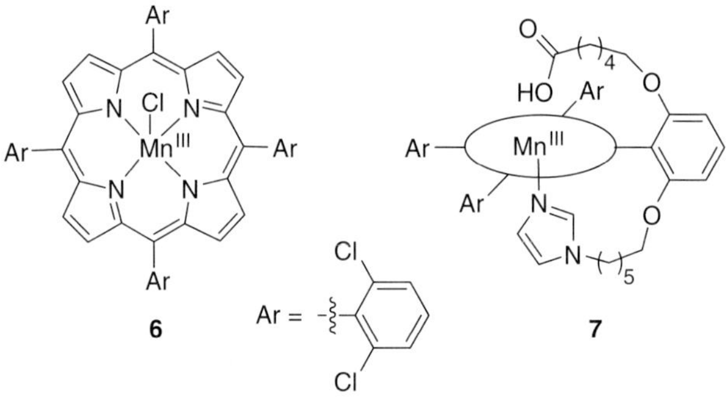

The observation that imidazoles and carboxylic acids significantly improved the epoxidation reaction led to the development of Mn-porphyrin complexes containing these groups covalently linked to the porphyrin platform as attached pendant arms (**7**) [37]. When these catalysts were employed in the epoxidation of simple alkenes with hydrogen peroxide, enhanced oxidation rates in combination with perfect product selectivity was obtained (Table 2.4, entry 3). In contrast to epoxidations catalyzed by other metals, the Mn-porphyrin system yields products with scrambled stereochemistry. For example, the epoxidation of cis-stilbene using Mn(TPP)Cl (TPP = tetraphenylporphyrin) and iodosylbenzene generated cis- and trans-stilbene oxide in a ratio of 35:65. The low stereospecificity was improved using heterocyclic additives, like pyridines or imidazoles. The epoxidation system using hydrogen peroxide as the terminal oxidant was reported to be stereospecific for cis-alkenes, whereas trans-alkenes are poor substrates with these catalysts.

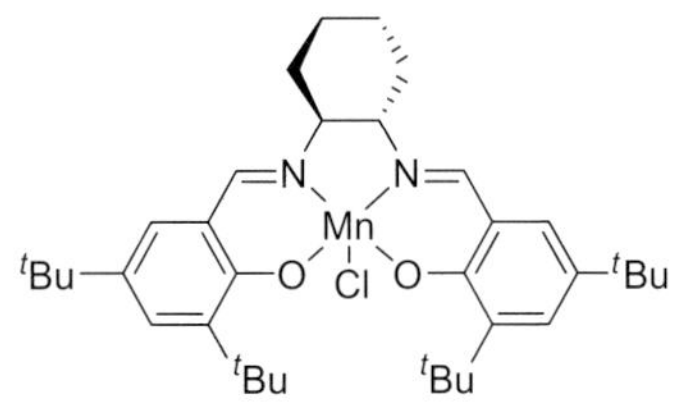

A breakthrough for manganese epoxidation catalysts came in the early 1990s when the groups of Jacobsen and Katsuki independently discovered that chiral Mn-salen complexes (**8**) catalyzed the enantioselective formation of epoxides [38–40]. The discovery that simple nonchiral Mn-salen complexes could be used as catalysts for alkene epoxidation had already been established about 5 years earlier, and the typical terminal oxidants used with these catalysts closely resemble those of the porphyrin systems [41]. In contrast to the titanium-catalyzed asymmetric epoxidation discovered by Sharpless, the Mn-salen system does not require pre-coordination of the alkene substrate to the catalyst, and hence unfunctionalized alkenes could be efficiently and selectively oxidized. The enantioselectivity was shown to be highly sensitive toward the substitution pattern of the alkene substrate. Excellent selectivity (>90% *ee*) was obtained for aryl- or alkynyl-substituted terminally substituted, *cis*-di-substituted, and tri-substituted alkenes, whereas trans-di-substituted alkenes were epoxidized with low rates and low *ee* (<40%). The typical oxidant used in Mn-salen asymmetric epoxidations is NaOCl. However, more recent work by the groups of Berkessel and Katsuki have opened up ways for hydrogen peroxide to be employed [42, 43]. Berkessel found that imidazole additives were crucial for the formation of the active oxo-manganese intermediates, and a manganese catalyst (**9**) based on a salen ligand incorporating a pendant imidazole was used for the asymmetric epoxidation using aqueous H_2O_2. Yields and enantioselectivity, however, did not reach the levels obtained when other oxidants were used. In the work of Katsuki, *N*-methylimidazole was present as an additive in the epoxidation of a chromene derivative using the sterically hindered Mn-salen catalyst (**10**) (Scheme 2.7). At low substrate concentration (0.1 M) the yield was only 17%, but

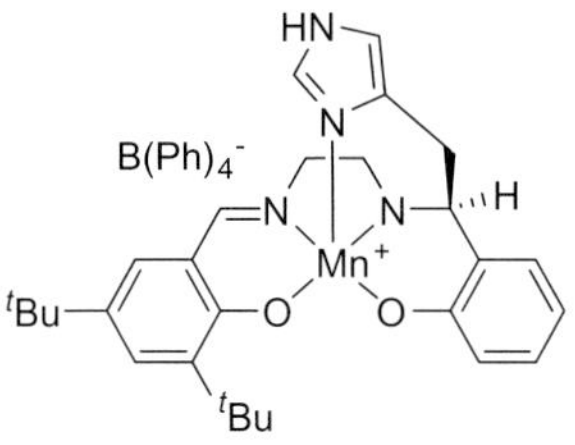

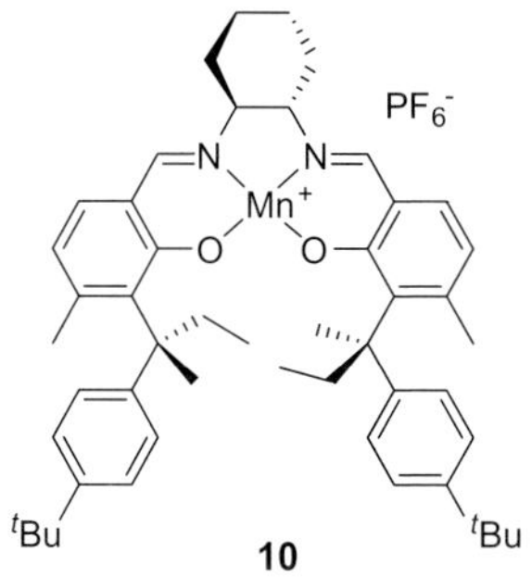

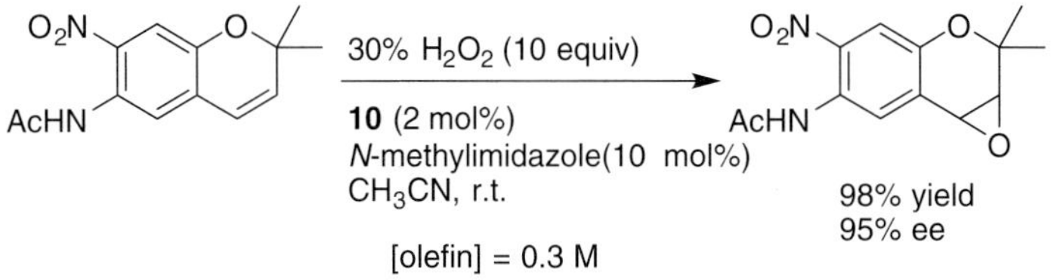

Scheme 2.7

performing the reaction at higher concentration gave substantially more product. Unfortunately, other substrates were not oxidized with the same yield and degree of enantioselectivity.

Pietikäinen reported on the use of ammonium acetate (20 mol%) as an additive together with the Jacobsen-catalyst (**8**) for the epoxidation of a number of alkenes with aqueous hydrogen peroxide as the terminal oxidant [44]. In general, this protocol generated epoxides in 50–90% yield with an enantioselectivity up to 96%; however, only a narrow range of substrates has been examined. A major problem with aqueous hydrogen peroxide in the Mn-salen-catalyzed reactions is associated with Mn-catalyzed oxidant decomposition and catalyst deactivation due to the presence of water. Anhydrous hydrogen peroxide, either in the form of the urea/hydrogen peroxide adduct (UHP) or the triphenylphosphine oxide/H_2O_2 adduct, has been employed to circumvent this problem [45–47]. Although epoxide yield and enantioselectivity are in the range of what can be obtained using NaOCl, the catalyst loading is often significantly higher, and the removal of urea or Ph_3PO constitutes an additional problem.

11 **12** **13**

Apart from porphyrin and salen catalysts, manganese complexes of N-alkylated 1,4,7-triazacyclononane (e.g., TMTACN, **11**) have been found to efficiently catalyze the epoxidation of alkenes in the presence of acid additives (typically oxalic, ascorbic, or squaric acid) and hydrogen peroxide [48–50]. Reactions performed without acid present required a huge excess (about 100 equivalents) of hydrogen peroxide for efficient epoxidation. Manganese complexes based on the TACN ligands have also found use as catalysts for cis-dihydroxylation of alkenes [51, 52]. The use of the dimeric manganese complex $[Mn_2O_3(TMTACN)]^{2+}$ in combination with trichloroacetic acid (1–25 mol%) allowed for efficient oxidation of cyclooctene to a mixture of cyclooctene oxide and cis-cyclooctane-1,2-diol. The reaction required slow addition of hydrogen peroxide (6 h), but only 1.3 equivalents of the oxidant was required. Interestingly, when salicylic acid (1 mol%) was employed as a co-catalyst, the formation of cyclooctene oxide was the major product observed in the reaction mixture. Hence, the chemoselectivity could to a certain point be controlled by the choice of additive. Ribas and Costas have studied manganese triflate complexes based on the 2-pyridyl derivative of TACN (**12**), and they found that complex **13** was a robust and active catalyst for the epoxidation of alkenes using peracetic acid as the terminal oxidant [53]. More recently, this catalyst was found to be compatible with aqueous hydrogen peroxide if the epoxidation reactions were performed with acetic acid (14 equiv.) as an additive (Scheme 2.8) [54]. Using only 0.1 mol% of catalyst **13**, a series of

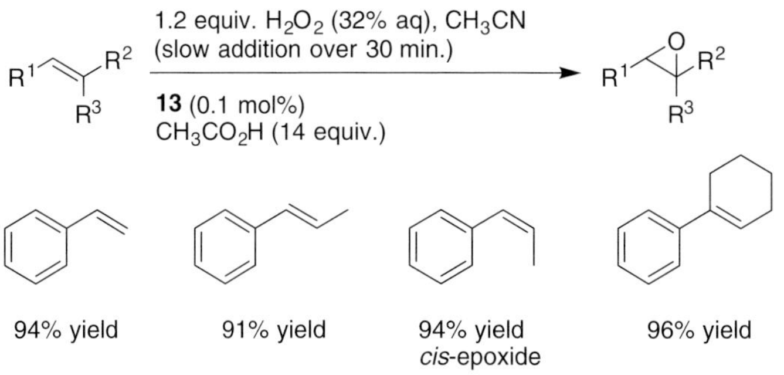

2.5 Manganese-Catalyzed Epoxidations | 51

Scheme 2.8

Reaction conditions: 1.2 equiv. H_2O_2 (32% aq), CH_3CN (slow addition over 30 min.), **13** (0.1 mol%), CH_3CO_2H (14 equiv.)

Products:
- styrene oxide: 94% yield
- β-methylstyrene oxide: 91% yield
- cis-β-methylstyrene oxide: 94% yield, cis-epoxide
- 1-phenylcyclohexene oxide: 96% yield

differently functionalized alkenes were effectively and selectively converted into their corresponding epoxides in rather short reaction times (35–90 min).

The rather difficult preparation of the TACN ligands has led to an increased activity in order to find alternative ligands with similar coordinating properties. In this respect, pyridyl-amine ligands represent an interesting alternative. Feringa and coworkers found that the dinuclear manganese complex **15**, prepared from the tetra-pyridyl ligand **14**, was an efficient catalyst for the epoxidation of simple alkenes [55]. Only 0.1 mol% of the catalyst (**15**) was required for high conversion (87%) of cyclohexene into its corresponding epoxide. An excess of aqueous hydrogen peroxide (8 equivalents) was used because of the usual problem of peroxide decomposition in the presence of manganese complexes.

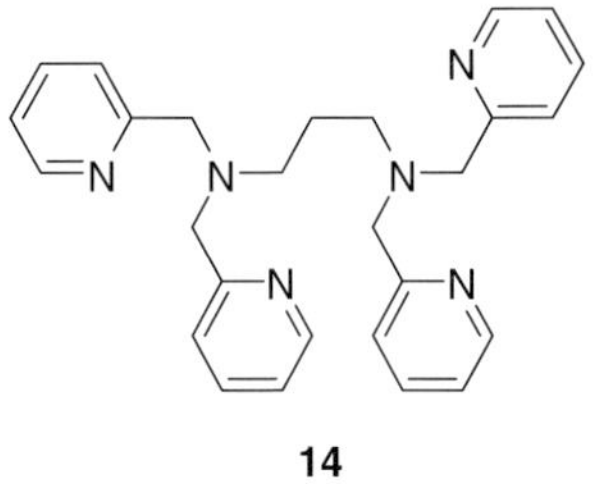

14, 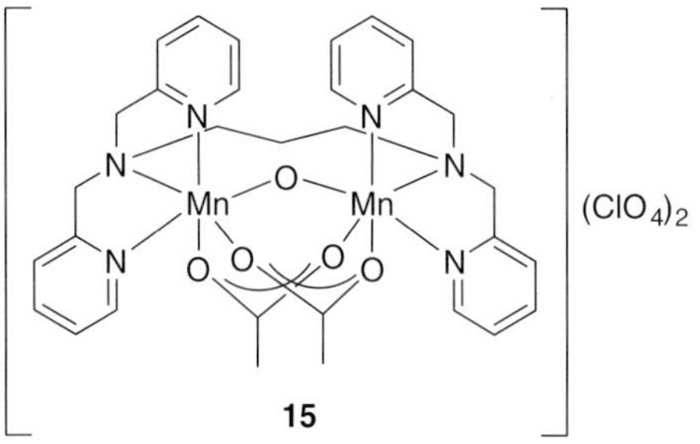 **15** $(ClO_4)_2$

A catalytic process of particular simplicity uses the manganese sulfate/hydrogen peroxide system. On screening different metal salts, Lane and Burgess found that simple manganese(II) and (III) salts catalyzed the formation of epoxides in DMF or t-BuOH, using aqueous hydrogen peroxide (Scheme 2.9) [56]. It was further established that the addition of bicarbonate was of importance for the epoxidation reaction.

Using spectroscopic methods, it was established that peroxymonocarbonate (HCO_4^-) is formed on mixing hydrogen peroxide and bicarbonate [57]. In the absence of the metal catalyst, the oxidizing power of the peroxymonocarbonate formed in situ with respect to its reaction with alkenes was demonstrated to be moderate. In the initial reaction setup, this $MnSO_4$-catalyzed epoxidation required a considerable excess of hydrogen peroxide (10 equiv.) for efficient epoxide formation.

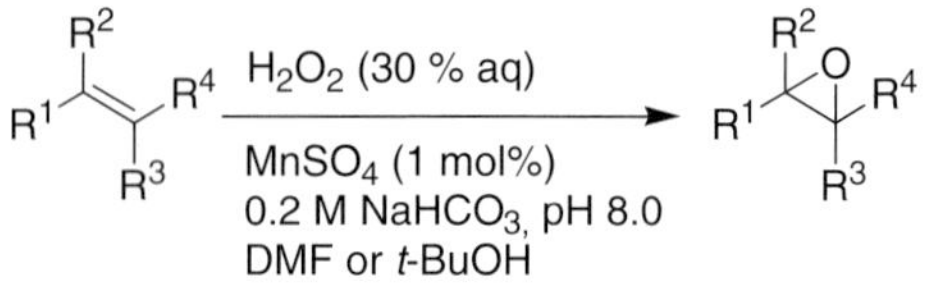

Scheme 2.9

$$\underset{R^3}{\overset{R^2}{\underset{R^1}{\diagup}}}\!\!=\!\!\underset{}{\overset{}{\diagdown}}\!R^4 \quad \xrightarrow[\substack{\text{MnSO}_4\ (1\ \text{mol\%}) \\ 0.2\ \text{M NaHCO}_3,\ \text{pH 8.0} \\ \text{DMF or }t\text{-BuOH}}]{\text{H}_2\text{O}_2\ (30\ \%\ \text{aq})} \quad \underset{R^3}{\overset{R^2\ \ \ O}{\underset{R^1}{\diagup\!\!\!\diagdown}}}\!R^4$$

Regarding the scope of the reaction, it was found that electron-rich substrates like di-, tri- and tetra-substituted alkenes were giving moderate to good yields of their corresponding epoxides. Styrene and styrene derivatives were also demonstrated to react smoothly, whereas mono-alkyl-substituted substrates were completely unreactive under these conditions. The basic reaction medium used was very beneficial for product protection, and hence acid sensitive epoxides were formed in good yields. Different additives were screened in order to improve this epoxidation system, and it was found that the addition of sodium acetate was beneficial for reactions performed in t-BuOH. Similarly, the addition of salicylic acid improved the outcome of the reaction performed in DMF. The use of these additives efficiently reduced the number of hydrogen peroxide equivalents necessary for a productive epoxidation (Table 2.5). The reaction is not completely stereospecific, since the epoxidation of cis-4-octene yielded a cis/trans mixture of the product (1 : 1.45 without additive and 1 : 1.1 in the presence of 4 mol% salicylic acid).

The use of the ionic liquid [bmim][BF$_4$] further improved the Burgess epoxidation system [58]. Chan and coworkers found that replacing sodium bicarbonate with tetramethylammoniun bicarbonate and performing the reaction in [bmim][BF$_4$] allowed for efficient epoxidation of a number of different alkenes, including substrates leading to acid labile epoxides (e.g., di-hydronaphthalene (99% yield) and 1-phenylcyclohexene (80% yield)).

2.6
Rhenium-Catalyzed Epoxidations

The use of rhenium-based systems for the epoxidation of alkenes has increased considerably during the last 20 years [59, 60]. In 1989, Jørgensen stated that 'the catalytic activity of rhenium in epoxidation reactions is low' [2]. The very same year, a number of patents were released describing the use of porphyrin complexes containing rhenium as catalysts for the production of epoxides. The first major breakthrough, however, came in 1991 when Herrmann introduced methyltrioxorhenium (MTO, **16**) as a powerful catalyst for alkene epoxidation, using hydrogen peroxide as the terminal oxidant [61]. This commercially available organometallic rhenium compound was initially formed in tiny amounts in the reaction between $(CH_3)_4ReO$ and air, and was first detected by Beattie and Jones in 1979 [62].

Today, there is a whole range of organorhenium oxides available, and these can be considered as one of the best examined classes of organometallic compounds [63, 64]. From a catalytic point of view, though, MTO is one of few organorhenium oxides that

Table 2.5 Manganese sulfate-catalyzed epoxidation of alkenes using aqueous H_2O_2 (30%)[a].

Alkene	No additive		Salicylic acid (4 mol%)	
	Equiv. H_2O_2	Yield	Equiv. H_2O_2	Yield
 cyclohexene	10	99	2.8	96
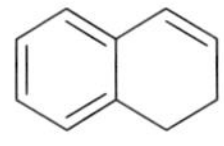 dihydronaphthalene	10	87	5	97[b]
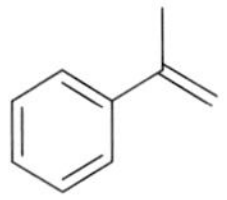 α-methylstyrene derivative	10	96	5	95[b]
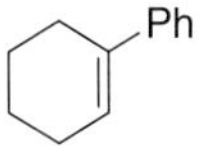 1-phenylcyclohexene	10	95	5	95[b]
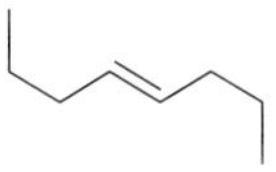 branched alkene	25	60	25	75
 internal alkene	25	54	25	75
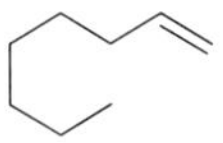 terminal alkene	25	0	25	0

a) Conditions according to Scheme 2.5.
b) Isolated yields.

have been shown to effectively act as a catalyst in epoxidation reactions. Regarding the physical properties of organorhenium oxides, MTO shows the greatest thermal stability (decomposing at >300 °C), apart from the catalytically inert 18-electron complex (η^5-C$_5$Me$_5$)ReO$_3$. Furthermore, the high solubility of MTO in virtually any solvent from pentane to water makes this compound particularly attractive.

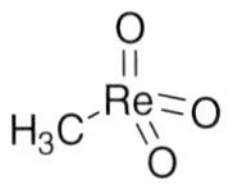

16

For the catalytic epoxidation application, perhaps the most important feature of MTO is its ability to activate hydrogen peroxide (5–85%) without decomposing the oxidant. The half-life of hydrogen peroxide in the presence of MTO is 20 000 times longer than it is in the presence of RuCl$_3$ and 50 times longer than in the presence of MnO$_2$ [65]. Upon treatment of MTO with hydrogen peroxide a rapid equilibrium takes place according to Scheme 2.10.

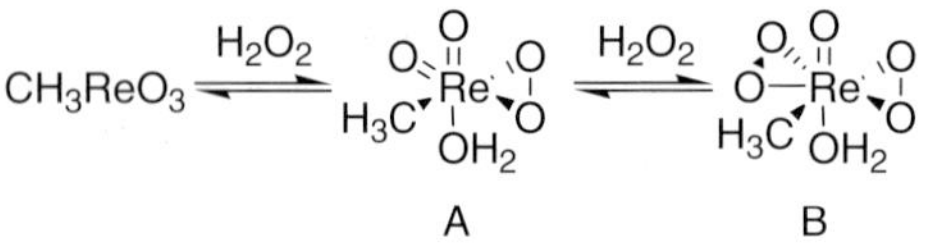

$$CH_3ReO_3 \xrightleftharpoons{H_2O_2} \underset{A}{\begin{array}{c}O\diagdown\underset{\underset{OH_2}{|}}{Re}\diagup O\\ H_3C\diagup\diagdown O\end{array}} \xrightleftharpoons{H_2O_2} \underset{B}{\begin{array}{c}O\diagdown\underset{\underset{OH_2}{|}}{Re}\diagup O\\ O-\diagdown O\\ H_3C\end{array}}$$

Scheme 2.10

The reaction with one equivalent of hydrogen peroxide generates a mono-peroxo complex (**A**) which undergoes further reaction to yield a bis-peroxorhenium complex (**B**). The formation of the peroxo complexes is evident from the appearance of an intense yellow color of the solution. Both peroxo complexes (**A** and **B**) have been detected by their methyl resonances using ^1H and ^{13}C NMR spectroscopy [66]. Furthermore, the structure of the bis-peroxo complex (**B**) has been determined by X-ray crystallography [67]. In solution, **B** is the most abundant species in the equilibrium, suggesting that this is the thermodynamically most stable peroxo complex. The coordination of a water molecule to **B** has been established by NMR-spectroscopy, however no such coordination have been observed for **A**, indicating either no coordinated water or high lability of such a ligand. The protons of the coordinated water molecule in **B** are highly acidic, and this has important implications for the epoxidation reaction (see below). Regarding catalytic activity, however, it has been demonstrated that both complexes are active as oxygen-transfer species. Whereas decomposition of the MTO catalyst under basic conditions is often negligible, the presence of hydrogen peroxide completely changes the situation. The combination of basic media and H_2O_2 rapidly induces an irreversible decomposition of MTO according to Scheme 2.11, and this deleterious side reaction is usually a great problem in the catalytic system [66].

$$CH_3ReO_3 \xrightarrow[\text{Pyridine}]{H_2O_2} HOReO_3 \cdot 2py + CH_3OH$$

Scheme 2.11

In this oxidative degradation, MTO decomposes into catalytically inert perrhenate and methanol. The decomposition reaction is accelerated at higher pH, presumably through the reaction between the more potent nucleophile HO_2^- and MTO. The decomposition of MTO, occurring under basic conditions, is rather problematic, since the selectivity for epoxide formation certainly benefits from the use of non-acidic conditions.

2.6.1
MTO as Epoxidation Catalyst – Original Findings

The rapid formation of peroxo complexes in the reaction between MTO and hydrogen peroxide makes this organometallic compound useful as an oxidation catalyst. In the original report on alkene epoxidation using MTO, Herrmann and coworkers employed a preformed solution of hydrogen peroxide in *tert*-butanol as the terminal oxidant [61]. This solution was prepared by mixing *tert*-butanol and aqueous hydrogen peroxide followed by the addition of anhydrous $MgSO_4$. After filtration, this essentially water-free solution of hydrogen peroxide was used in the epoxidation reactions. It was further reported that MTO, or rather its peroxo-complexes, was stable for weeks in this solution if kept at low temperatures (below 0 °C). However, later studies by Espenson revealed the instability of MTO in hydrogen peroxide solutions [66]. Epoxidation of various alkenes using 0.1–1 mol% of MTO and the H_2O_2/*t*-BuOH solution resulted generally in high conversion to epoxide, but a significant amount of *trans*-1,2-diol was often formed via ring opening of the epoxide. The reason for using 'anhydrous' hydrogen peroxide was of course to attempt to avoid the latter side-reaction; however, since hydrogen peroxide generates water upon reaction with MTO it was impossible to work under strictly water-free conditions. The ring-opening process can either be catalyzed directly by MTO, using the intrinsic metal Lewis acidity, or simply by protonation of the epoxide. To overcome this problem, Herrmann used an excess of amines (e.g., 4,4'-dimethyl-2,2'-bipyridine, quinine, and cinchonine), which would coordinate to the metal and thus suppress the ring-opening process [68]. This resulted in better selectivity for the epoxide at the expense of decreased, or in some cases completely inhibited, catalytic activity. In an attempt to overcome the problems of low selectivity for epoxide formation and decreased catalytic activity obtained using amine additives, Adam introduced the urea/hydrogen peroxide adduct (UHP) as the terminal oxidant for the MTO-catalyzed system [69]. This resulted in substantially better selectivity for several alkenes, although substrates leading to highly acid-sensitive epoxides still suffered from deleterious ring opening reactions.

2.6.2
The Influence of Heterocyclic Additives

The second major discovery concerning the use of MTO as an epoxidation catalyst came in 1997, when Sharpless and coworkers reported on the use of sub-stoichiometric amounts of pyridine as co-catalysts in the system [70]. The switch of solvent from *tert*-butanol to dichloromethane and the introduction of 12 mol% of pyridine allowed for the synthesis of even very sensitive epoxides using aqueous hydrogen peroxide as the terminal oxidant. A significant rate acceleration was also observed for the epoxidation reaction performed in the presence of pyridine. This discovery was the first example of an efficient MTO-based system for epoxidation under neutral-to-basic conditions. Under these conditions the detrimental acid-induced decomposition of the epoxide is effectively avoided. Employing the novel system, a variety of alkene-substrates were converted into their corresponding epoxides in high yield and with high epoxide selectivity (Scheme 2.12 and Table 2.6).

The increased rate of epoxidation observed in the presence of added pyridine has been studied by Espenson and Wang and was to a certain degree explained as an accelerated formation of peroxorhenium species in the presence of pyridine [71]. Stabilization of the rhenium catalyst through pyridine coordination was also

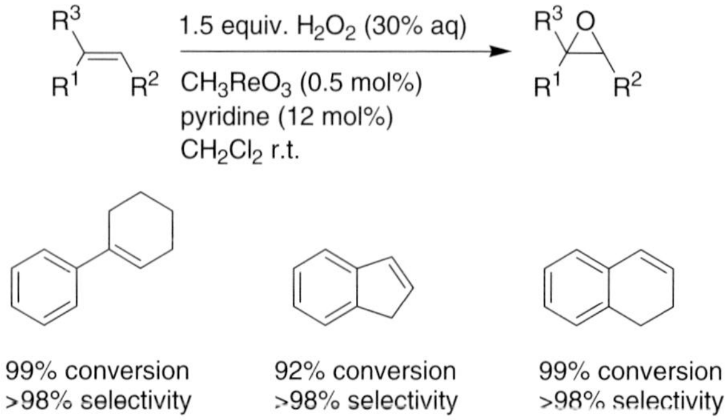

Scheme 2.12

$$R^1R^2C=CR^3H \xrightarrow[\substack{CH_3ReO_3 \text{ (0.5 mol\%)} \\ \text{pyridine (12 mol\%)} \\ CH_2Cl_2 \text{ r.t.}}]{1.5 \text{ equiv. } H_2O_2 \text{ (30\% aq)}} R^1R^2C\text{-}CR^3H \text{ (epoxide)}$$

- 1-phenylcyclohexene: 99% conversion, >98% selectivity
- indene: 92% conversion, >98% selectivity
- dihydronaphthalene: 99% conversion, >98% selectivity

Table 2.6 MTO-catalyzed epoxidation of alkenes using H_2O_2.

Alkene	No additive[a]	Pyridine[b]	3-Cyanopyridine[b]	Pyrazole[b]	3-Methylpyrazole[b]
 cyclohexene	90 (5)	96 (6)			96 (5)[d]
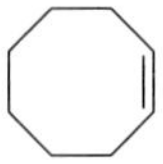 cyclooctene	100 (2)[b]	99 (2)		89 (0.02)	99 (4)[d]
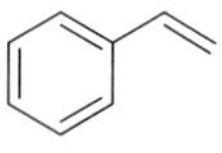 styrene		84 (16)	96 (5)[c]	96 (5)	92 (5)[e]
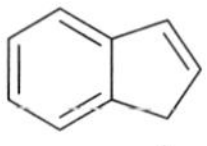 indene	48 (37)	96 (5)			87 (1)[e]
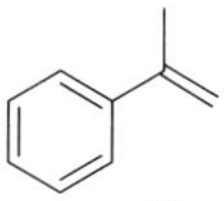 α-methylstyrene		82 (6)	74 (1.5)[c]	93 (1.5)	92 (3)[f]
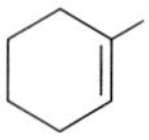 1-phenylcyclohexene		98 (1)	96 (1)[c]	95 (1)	95 (5)[f]
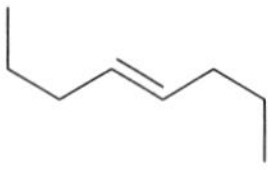 (Z)-alkene	95 (2)	91 (24)	97 (12)		98 (8)[f]
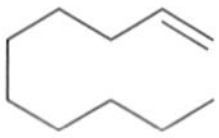 1-decene	75 (72)	82 (48)	99 (14)	99 (14)	91 (8)

The catalytic loading is 0.5 mol% MTO unless otherwise stated. The figures in the table refer to yields (%) obtained in the epoxidation. Figures within parentheses are reaction times (h).
a) Anhydrous H_2O_2 in t-BuOH.
b) Aqueous H_2O_2 (30%).
c) Pyridine and 3-cyanopyridine (6 mol% of each)
d) 0.05 mol% MTO.
e) 0.2 mol% MTO.
f) 0.1 mol% MTO.

detected, although the excess of pyridine needed in the protocol unfortunately led to increased catalyst deactivation. As can be seen above, MTO is stable under acidic conditions, but at high pH an accelerated decomposition of the catalyst into perrhenate and methanol occurs. The Brønsted basicity of pyridine leads to increased amounts of HO_2^-, which speeds up the formation of the peroxo-complexes and the decomposition of the catalyst. Hence, the addition of pyridine to the epoxidation system led to certain improvements regarding rate and selectivity for epoxide formation at the expense of catalyst lifetime. This turned out to be a minor problem for highly reactive substrates such as tetra-, tri- and cis-di-substituted alkenes, since these compounds are converted into epoxides at a rate significantly higher than the rate of catalyst decomposition. Less electron-rich substrates such as terminal alkenes, however, react more slowly with electrophilic oxygen-transfer agents, and require longer reaction times to reach acceptable conversions. When the reaction was performed in the presence of added pyridine (12 mol%), neither 1-decene nor styrene was fully converted, even after prolonged reaction times.

A major improvement regarding epoxidation of terminal alkenes was achieved upon replacing pyridine ($pK_a = 5.4$) with its less basic analog 3-cyanopyridine ($pK_a = 1.9$) [72]. This improvement turned out to be general for a number of different terminal alkenes, regardless of the existence of steric hindrance in the α-position of the alkene or whether other functional groups were present in the substrate (Scheme 2.13).

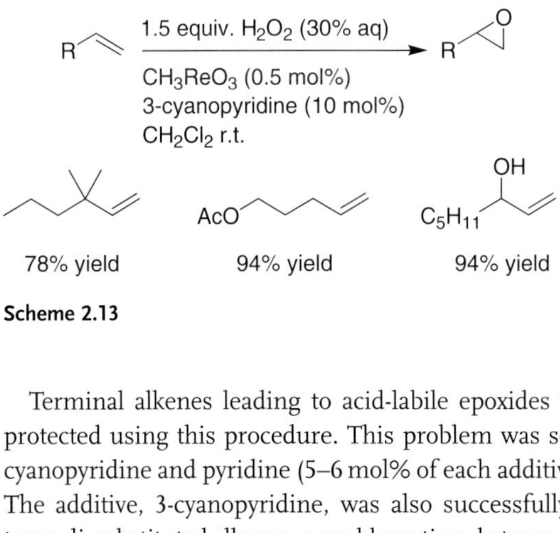

Scheme 2.13

Terminal alkenes leading to acid-labile epoxides were, however, not efficiently protected using this procedure. This problem was solved by using a cocktail of 3-cyanopyridine and pyridine (5–6 mol% of each additive) in the epoxidation reaction. The additive, 3-cyanopyridine, was also successfully employed in epoxidation of trans-di-substituted alkenes, a problematic substance class using the parent pyridine system [73]. In these reactions, the amount of the MTO catalyst could be reduced to 0.2–0.3 mol% with only 1–2 mol% of 3-cyanopyridine added. Again, acid-sensitive epoxides were obtained using a mixture of 3-cyanopyridine and the parent pyridine. It should be pointed out that the pyridine additives do undergo oxidation reactions, forming the corresponding pyridine-N-oxides [74]. This will of course effectively decrease the amount of additive present in the reaction mixture. In fact, as pointed out by Espenson, the use of a pyridinium salt (mixture of pyridine and, for example, acetic acid) can be more effective in protecting the additive from N-oxidation [71] (Adolfsson, H. and Sharpless, K. B. unpublished results.). This can be beneficial for slow-reacting substrates, where N-oxidation would compete with alkene epoxidation. The Herrmann group introduced an improvement to the Sharpless system by employing pyrazole as an additive [75]. Compared to pyridine, pyrazole is a less basic heterocycle ($pK_a = 2.5$) and does not undergo N-oxidation by the MTO/H_2O_2 system. Furthermore, employing pyrazole as the additive allowed for the formation of certain acid-sensitive epoxides. Recently, Yamazaki presented results using 3-methylpyrazole (10 mol%) as an additive in the MTO-catalyzed epoxidation of alkenes [76]. A huge number of substrates were screened with excellent results (Table 2.6). The major improvement found using 3-methylpyrazole as the additive, instead of any of the previously used heterocyclic compounds, is the low catalytic loading of MTO which can be used in the epoxidations. Typically, 0.5 mol% was used in the protocols containing pyridine derivatives or pyrazole. However, in the presence of 3-methylpyrazole, the catalyst loading can be decreased to 0.05–0.2 mol%. Regarding the choice of additive, 3-methylpyrazole is perhaps the most effective for the majority of alkenes, although for certain acid-labile compounds, pyridine would be the preferred additive (Table 2.6) [77].

2.6.3
The Role of the Additive

The use of various heterocyclic additives in the MTO-catalyzed epoxidation has been demonstrated to be of great importance for substrate conversion as well as for product selectivity. Regarding the selectivity, the role of the additive is obviously to protect the product epoxides from deleterious, acid-catalyzed (Brønsted or Lewis acid) ring-opening reactions. This is achieved partly by direct coordination of the heterocyclic additive to the rhenium metal, thereby significantly decreasing the Lewis acidity of the metal, and partly by increasing the pH of the reaction medium, the additives being basic in nature.

Concerning the accelerating effects observed when pyridine or pyrazole is added to the MTO-system, there are a number of different suggestions available. One likely explanation is that the additives do serve as phase-transfer agents. Hence, when MTO is added to an aqueous H_2O_2 solution, an immediate formation of the peroxo-complexes **A** and **B** (cf. Scheme 2.10) occurs, which is visualized by the intense bright yellow color of the solution. If a non-miscible organic solvent is added, the yellow color is still present in the aqueous layer, but addition of pyridine to this mixture results in an instantaneous transfer of the peroxo-complexes into the organic phase. The transportation of the active oxidants into the organic layer would thus favor the epoxidation reaction, since the alkene concentration is significantly higher in this phase (Scheme 2.14). Additionally, the rate with which MTO is converted into **A** and **B** is accelerated when basic heterocycles are added. This has been attributed to the Brønsted-basicity of the additives, which increases the amount of peroxide anion present in the reaction mixture. A higher concentration of HO_2^- is, however, detrimental to the MTO-catalyst, but the coordination of a Lewis base to the metal seems to have a positive effect in protecting the catalyst from decomposition.

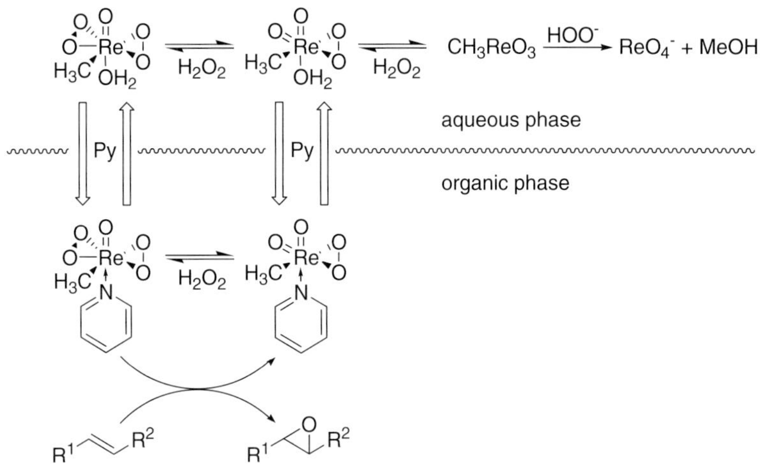

Scheme 2.14

2.6.4
Other Oxidants

While aqueous hydrogen peroxide certainly is the most practical oxidant for MTO-catalyzed epoxidations, the use of other terminal oxidants can sometimes be advantageous. As mentioned above, the urea-hydrogen peroxide adduct has been employed in alkene epoxidations. The anhydrous conditions obtained using UHP improved the system by decreasing the amount of diol formed in the reaction. The absence of significant amounts of water further helped in preserving the active catalyst from decomposition. A disadvantage, however, is the poor solubility of UHP in many organic solvents, which makes these reactions heterogeneous.

Another interesting terminal oxidant which has been applied in MTO-catalyzed epoxidations is sodium percarbonate (SPC) [78]. The fundamental structure of SPC consists of hydrogen peroxide encapsulated via hydrogen bonding in a matrix of sodium carbonate [79]. It slowly decomposes in water and in organic solvents to release hydrogen peroxide. This process is intrinsically safe, as is borne out by its common use as an additive in household washing detergents and toothpaste. When this 'solid form' of hydrogen peroxide was employed in MTO-catalyzed (1 mol%) oxidation of a wide range alkenes, good yields of the corresponding epoxides were obtained. An essential requirement for a successful outcome of the reaction was the addition of an equimolar amount (with respect to the oxidant) of trifluoroacetic acid. In the absence of this acid or with acetic acid added, little or no reactivity was observed. The role of the acid in this heterogeneous system is to facilitate the slow release of hydrogen peroxide. Despite the presence of acid, even hydrolytically sensitive epoxides were formed in high yields. This can be explained by an efficient buffering of the system by $NaHCO_3$ and CO_2, formed in the reaction between trifluoroacetic acid and SPC. The initial pH was measured to be 2.5, but after 15 min a constant pH of 10.5 was established, ensuring protection of acid-sensitive products.

Bis-trimethylsilyl peroxide (BTSP) represents another form of 'anhydrous' hydrogen peroxide [80]. The use of strictly anhydrous conditions in MTO-catalyzed alkene epoxidations would efficiently eliminate problems with catalyst deactivation and product decomposition due to ring opening reactions. BTSP, which is the di-silylated form of hydrogen peroxide, has been used in various organic transformations [81]. Upon reaction, BTSP is converted to hexamethyldisiloxane, thereby assuring anhydrous conditions. In initial experiments, MTO showed little or no reactivity toward BTSP under stoichiometric conditions [82]. This was very surprising, considering the high reactivity observed for BTSP compared to hydrogen peroxide in oxidation of sulfides to sulfoxides [83]. The addition of one equivalent of water to the MTO/BTSP mixture, however, rapidly facilitated the generation of the active peroxo-complexes. This was explained by hydrolytic formation of H_2O_2 from BTSP in the presence of MTO (Scheme 2.15). In fact, other proton sources proved to be equally effective in promoting this hydrolysis. Thus, under strictly water-free conditions no epoxidation occurred when the MTO/BTSP system was used. The addition of trace amounts of a proton source triggered the activation of BTSP, and the formation of epoxides was observed.

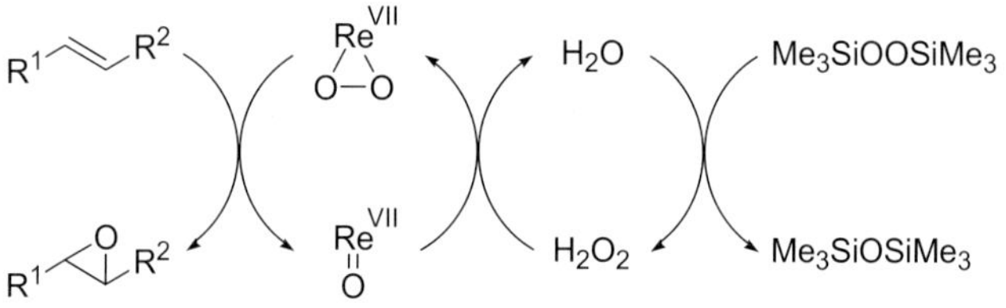

Scheme 2.15

Under optimal conditions, MTO (0.5 mol%), water (5 mol%) and 1.5 equiv. of BTSP were used for efficient epoxide formation. The discovery of these essentially water-free epoxidation conditions led to another interesting breakthrough, namely the use of inorganic oxorhenium compounds as catalyst precursors [82, 84]. The catalytic activity of rhenium compounds like Re_2O_7, $ReO_3(OH)$, and ReO_3 in oxidation reactions with aqueous hydrogen peroxide as the terminal oxidant is typically very poor. Attempts to form epoxides using catalytic Re_2O_7 in 1,4-dioxane with H_2O_2 (60%) at elevated temperatures (90 °C) mainly yielded 1,2-diols [85]. However, when hydrogen peroxide was replaced by BTSP in the presence of a catalytic amount of a proton source, any of the inorganic rhenium oxides Re_2O_7, $ReO_3(OH)$, or ReO_3 was just as effective as MTO in alkene epoxidations. In fact, the use of ReO_3 proved to be highly practical, since this compound is hydrolytically stable, in contrast to Re_2O_7. There are several benefits associated with these epoxidation conditions. The amount of BTSP used in the reaction can easily be monitored using gas chromatography. Furthermore, the simple workup procedure associated with this protocol is very appealing, since evaporation of solvent (typically dichloromethane) and formed hexamethyldisiloxane yields the epoxide. For a comparison on the efficiency of different oxidants used together with MTO, see Table 2.7.

2.6.5
Solvents/Media

The high solubility of the MTO catalyst in almost any solvent opens up a broad spectrum of reaction media from which to choose when performing epoxidations. The most commonly used solvent, however, is still dichloromethane. From an environmental point of view this is certainly not the most appropriate solvent in large scale epoxidations. Interesting solvent effects for the MTO-catalyzed epoxidation were reported by Sheldon and coworkers, who performed the reaction in trifluoroethanol [86]. The change from dichoromethane to the fluorinated alcohol allowed for a further reduction of the catalyst loading down to 0.1 mol%, even for terminal alkene substrates. It should be pointed out that this protocol does require 60% aqueous hydrogen peroxide for efficient epoxidations.

Bégué and coworkers more recently reported an improvement of this method by performing the epoxidation reaction in hexafluoro-2-propanol [87]. They found that the activity of hydrogen peroxide was significantly increased in this fluorous alcohol, as compared to trifluoroethanol, which allowed for the use of 30% aqueous H_2O_2. Interestingly, the nature of the substrate and the choice of additive turned out to have important consequences for the lifetime of the catalyst. Cyclic di-substituted alkenes were efficiently epoxidized using 0.1 mol% MTO and 10 mol% pyrazole as the catalytic mixture for tri-substituted substrates, although the use of the additive 2,2'-bipyridine turned out to be crucial for high conversion (Scheme 2.16). The use of pyrazole in the latter case proved to be highly deleterious for the catalyst, as indicated by the loss of the yellow color of the reaction solution. This observation is certainly contradictory, since more basic additives normally decrease the catalyst lifetime. The fact that full conversion of long-chain terminal alkenes was obtained after 24 h using pyrazole as the additive and the observation that the catalyst was still active after this period of time are very surprising considering the outcome with more functionalized substrates. To increase conversion for substrates which showed poor solubility in hexafluoro-2-propanol, trifluoromethylbenzene was added as a co-solvent. In this way, 1-dodecene was converted to its corresponding epoxide in high yield.

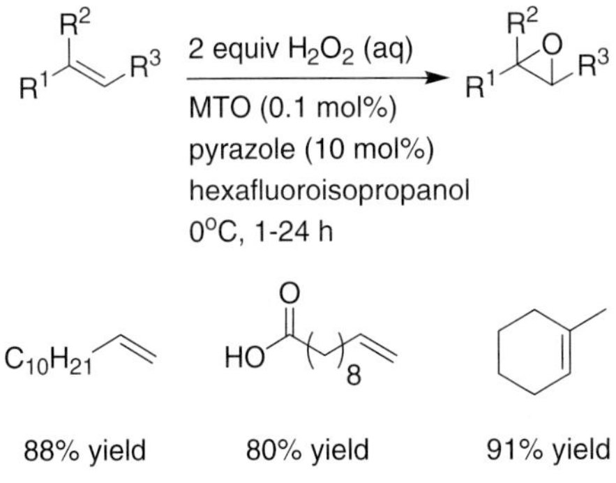

$R^1R^2C=CR^3$ → (with 2 equiv H_2O_2 (aq), MTO (0.1 mol%), pyrazole (10 mol%), hexafluoroisopropanol, 0°C, 1-24 h) → epoxide

$C_{10}H_{21}$—CH=CH₂
88% yield

HO-C(O)-(CH₂)₈-CH=CH₂
80% yield

1-methylcyclohexene
91% yield

Scheme 2.16

The use of nonvolatile ionic liquids as environmentally benign solvents has received significant attention in recent years. Abu-Omar and coworkers developed an efficient MTO-catalyzed epoxidation protocol using 1-ethyl-3-methylimidazolium tetrafluoroborate, [emim]BF$_4$, as solvent and urea-hydrogen peroxide (UHP) as the terminal oxidant [88, 89]. A major advantage of this system is the high solubility of UHP, MTO, and its peroxo-complexes, making the reaction medium completely homogeneous. Employing these essentially water-free conditions, high conversions and good epoxide selectivity were obtained for the epoxidation of variously substituted alkenes. Replacing UHP with aqueous hydrogen peroxide for the epoxidation of 1-phenylcyclohexene resulted in a poor yield of this acid-sensitive epoxide, and the corresponding diol was formed instead. A disadvantage of this system as compared to other MTO protocols is the high catalyst loading (2 mol%) required for efficient epoxide formation. Recently, a solvent-free protocol for epoxidation using the MTO catalyst was introduced by Yamazaki [90]. The combination of MTO with 10 mol% 3-methylpyrazole allowed for a series of alkenes to efficiently be converted to their corresponding epoxides in good to excellent yields without addition of any organic solvent. Somewhat longer reaction times were required for simple alkenes, while alkenols reacted faster. A summary of results obtained using different solvent system is presented in Table 2.7.

Table 2.7 MTO-catalyzed epoxidation of alkenes with H$_2$O$_2$, anhydrous or in fluorous solvents[a].

Alkene	UHP[b]	SPC[c]	BTSP[d]	UHP[e] Ionic liquid	H$_2$O$_2$[f] CF$_3$CH$_2$OH	H$_2$O$_2$[g] (CF$_3$)$_2$CHOH
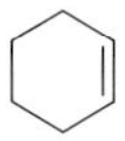 cyclohexene	97 (18)			99 (8)	99 (0.5)	
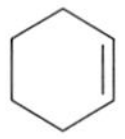 cyclohexene		94 (2)		95 (8)	99 (1)	93 (1)
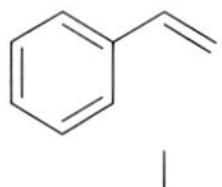 styrene derivative	44 (19)	96 (12)		95 (8)	82 (2)	
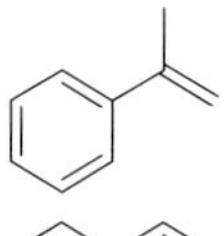 α-methylstyrene	55 (21)[h]	91 (3)				
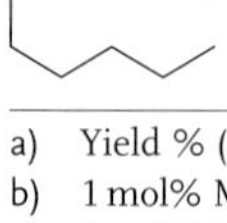 1-octene		94 (15)	94 (14)	46 (72)	97 (21)	88 (24)[i]

a) Yield % (reaction time h).
b) 1 mol% MTO.
c) 1 mol% MTO, 12 mol% pyrazole.
d) 0.5 mol% Re$_2$O$_7$.
e) 2 mol% MTO.
f) 0.1 mol% MTO, 10 mol% pyrazole, 60% H$_2$O$_2$.
g) 0.1 mol% MTO, 10 mol% pyrazole, 30% H$_2$O$_2$.
h) Additional 26% of the diol was formed.
i) 1-Dodecene was used as substrate.

2.6.6
Solid Support

The immobilization of catalysts or catalyst precursors on solid supports in order to simplify reaction procedures and to increase the stability of the catalyst is a common technique to render homogeneous systems heterogeneous. The MTO catalyst can be transferred onto polymeric material in a number of different ways. When an aqueous solution of MTO is heated for several hours (about 70 °C), a golden-colored polymeric material is formed. The composition of this organometallic polymer is $[H_{0.5}(CH_3)_{0.92}ReO_3]$ [61]. This polymeric form of MTO is nonvolatile, stable to air and moisture, and insoluble in all non-coordinating solvents. It can be used as a catalyst precursor for epoxidation of alkenes, since it is soluble in hydrogen peroxide, where it reacts to form the peroxo-rhenium species. Of course, the 'heterogeneous' property of this material is lost upon usage, but from a storage perspective, the polymeric MTO offers some advantages. MTO can, however, easily be immobilized by the addition of a polymeric material containing Lewis basic groups with the ability to coordinate to the rhenium center. A number of different approaches have been reported. Herrmann and coworkers described the use of polyvinylpyridines as the organic support, but the resulting MTO-polymer complex showed low catalytic activity [61]. A serious drawback with this supported catalyst was the oxidation of the polymeric backbone, leading to loss of the rhenium catalyst.

In a recent improvement to this approach, poly(4-vinylpyridine) and poly(4-vinylpyridine) *N*-oxides were used as the catalyst carrier [91]. The MTO catalyst obtained from 25% cross-linked poly(4-vinylpyridine) proved to efficiently catalyze the formation of even hydrolytically sensitive epoxides in the presence of aqueous hydrogen peroxide (30%). This catalyst could be recycled up to 5 times without any significant loss of activity. Attempts have been made to immobilize MTO with the use of either microencapsulation techniques, including sol-gel techniques, to form silica-bound rhenium compounds, or by the attachment of MTO to silica tethered with polyethers. These approaches have provided catalysts with good activity using aqueous hydrogen peroxide as the terminal oxidant [91–93]. In the latter case, high selectivity for epoxide formation was also obtained for very sensitive substrates (e.g., indene).

An alternative approach to immobilization of the catalyst on a solid support is to perform the MTO-catalyzed epoxidation reactions in the presence of NaY zeolites. This technique has been employed by Malek and Ozin, and later by Bein *et al.*, who used highly activated zeolites for the preparation of NaY/MTO using vacuum sublimation [94, 95]. More recently, Adam and coworkers found a significantly simpler approach toward this catalyst. The active catalyst was formed by mixing unactivated NaY zeolite with hydrogen peroxide (85%) in the presence of MTO and the substrate alkene [96]. Using this catalytic mixture, various alkenes were transformed into their corresponding epoxides without the formation of diols (typical diol formation was <5%). The MTO catalyst is positioned inside the 12 Å supercages of the NaY zeolite; hence, the role of the zeolite is to act as an absorbent for the catalyst and to provide heterogeneous microscopic reaction vessels for the reaction. The supernatant liquid was demonstrated to be catalytically inactive, even if Lewis bases (pyridine) were present. The high selectivity for epoxide formation was attributed to inhibition of the Lewis acid-mediated hydrolysis of the product by means of steric hindrance.

Recently, Omar Bouh and Espenson reported that MTO supported on niobia catalyzed the epoxidation of various fatty oils using UHP as the terminal oxidant [97]. Oleic acid, elaidic acid, linoleic acid, and linolenic acid were all epoxidized in high yields (80–100%) in less than two hours. Furthermore, it was demonstrated that the catalyst could be recovered and reused without loss of activity.

2.6.7
Asymmetric Epoxidations Using MTO

The MTO-based epoxidation system offers a particularly effective and practical route for the formation of racemic epoxides. Very few attempts to prepare chiral MTO complexes and to employ them in catalytic asymmetric epoxidation have so far been made, and the few existing reports are unfortunately quite discouraging. In the epoxidation of *cis*-β-methylstyrene with MTO and hydrogen peroxide in the presence of the additive (S)-(N,N-dimethyl)-1-phenylethylamine, a product with an enantiomeric excess of 86% has been claimed [98]. The epoxides from other substrates like styrene and 1-octene were obtained in significantly lower enantioselectivity (13% *ee*). Furthermore, the MTO-catalyzed epoxidation of 1-methylcyclohexene with L-prolineamide, (+)-2-aminomethylpyrrolidine or (R)-1-phenylethylamine as additives was reported to yield the product in low yield and enantioselectivity (up to 20% *ee*) [99]. A significant amount of the diol was formed in these reactions. More recently, Herrmann and coworkers reported on the use of chiral pyrazole derivatives as ligands in the MTO-catalyzed epoxidation of *cis*-β-methylstyrene [100]. Conversion and enantiomeric excess were unfortunately rather poor, with a maximum of 27% *ee*. They also reported on the use of chiral diols as ligands for the epoxidation of the same substrate (*cis*-β-methylstyrene), the best enantioselectivity obtained being 41% *ee*, although in this particular case the reaction was only completed to 5% conversion. Hence, a general protocol for the enantioselective formation of epoxides using rhenium catalysts is still lacking. There would certainly be a breakthrough if such a system could be developed, considering the efficiency of the MTO-catalyzed epoxidation reactions using hydrogen peroxide as the terminal oxidant.

2.7
Iron-Catalyzed Epoxidations

The use of iron salts and complexes for alkene epoxidation is in many respects similar to that of manganese catalysts. Thus, iron porphyrins can be used as epoxidation catalysts, but often conversion and selectivity is inferior to that obtained with its

2.7 Iron-Catalyzed Epoxidations

manganese counterpart. The possibilities to efficiently use hydrogen peroxide as the terminal oxidant are limited because of its rapid decomposition, catalyzed by iron. However, a recent breakthrough by Beller and coworkers [101] has resulted in a series of catalytic systems which use only 2–3 equivalents of aqueous hydrogen peroxide for the epoxidation of predominantly aromatic alkenes (see below). The traditional iron catalysts used for epoxidation of alkenes are either of the porphyrin type or based on various non-heme biomimetics. Traylor and coworkers found conditions where a polyfluorinated Fe(TPP)-catalyst (**17**) was employed in the epoxidation of cyclooctene to yield the corresponding epoxide in high yield (Scheme 2.17) [102]. High catalyst loading (5 mol%) and slow addition of the oxidant was required, which certainly limits the usefulness of this procedure.

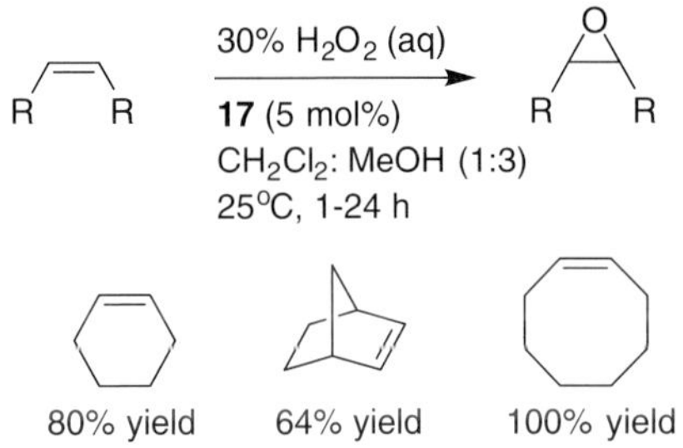

Scheme 2.17

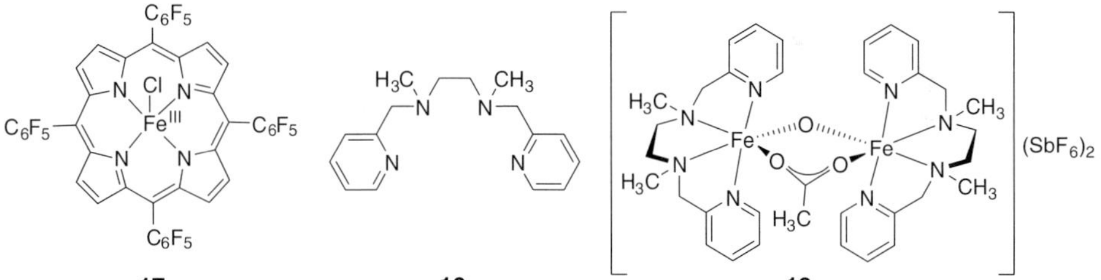

Recently, a number of iron complexes with biomimetic non-heme ligands were introduced as catalysts for alkane hydroxylation, alkene epoxidation and dihydroxylation. These complexes were demonstrated to activate hydrogen peroxide without the formation of free hydroxyl radicals, a feature commonly observed in iron oxidation chemistry. A particularly efficient catalytic system for selective epoxidation of alkenes was developed by Jacobsen and coworkers [103]. In this protocol, a tetradentate ligand (BPMEN = N,N'-dimethyl-N,N'-bis(2-pyridylmethyl)-diaminoethane, **18**) was combined with an iron(II) precursor and acetic acid to yield a self-assembled μ-oxo, carboxylate-bridged diiron(III) complex (**19**). This dimeric iron-complex, resembling the active site found in the hydroxylase methane monooxygenase (MMO), was demonstrated to efficiently epoxidize alkenes in the presence of aqueous hydrogen peroxide (50%). This catalyst turned out to be particularly active for the epoxidation of terminal alkenes, which normally are the most difficult substrates to oxidize. Thus, 1-dodecene was transformed into its corresponding epoxide in 90% yield within 5 min using 3 mol% of the catalyst. This system was also effective for the epoxidation of other simple non-terminal alkenes, like cyclooctene and trans-5-decene (Scheme 2.18).

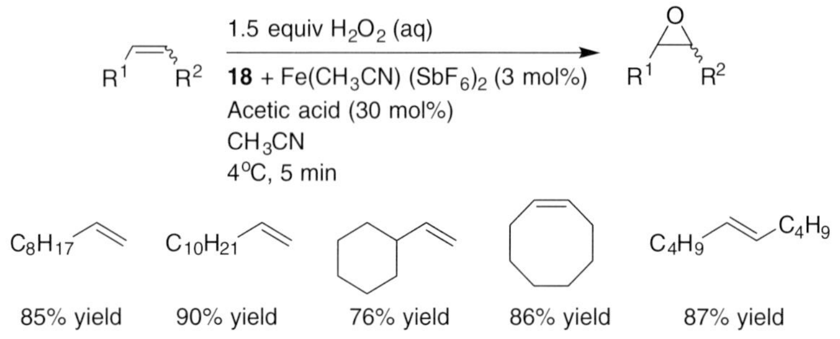

Scheme 2.18

Que and coworkers reported on a similar monomeric iron complex, formed with the BPMEN ligand but excluding acetic acid [104]. This complex was able to epoxidize cyclooctene in reasonably good yield (75%), but at the same time a small amount of the cis-diol (9%) was formed. The latter feature, observed with this class of complexes, has been further studied, and more selective catalysts have been prepared. Even though poor conversion is often obtained with the current catalysts, this method represents an interesting alternative to other cis-dihydroxylation systems [105, 106]. Using similar chiral ligands based on 1,2-diaminocyclohexane resulted in complexes which were able to catalyze the formation of epoxides in low yields and in low enantioselectivity (0–12% ee). The simultaneous formation of cis-diols was occurring with significantly better enantioselectivity (up to 82% ee), but these products were also obtained in low yields.

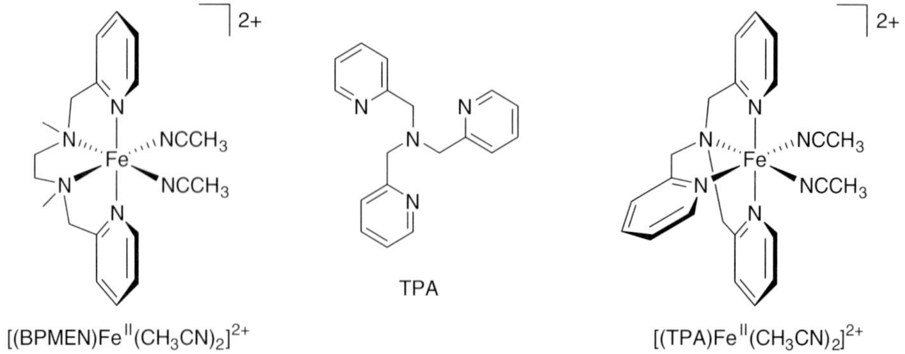

Upon closer inspection of the Fe(II)-complexes formed with either the BPMEN or the tetradentate trispicolylamine (TPA) ligands, respectively, Que and coworkers found that the presence of acetic acid in the catalytic mixture had a strong influence on the selectivity of the oxidation process [107, 108]. Hence, the use of either [(BPMEN)Fe(CH$_3$CN)$_2$](OTf)$_2$ or [(TPA)Fe(CH$_3$CN)$_2$](OTf)$_2$ as catalysts for the oxidation of cyclooctene or cis-2-heptene in a mixture of acetonitrile and acetic acid (1:2), with hydrogen peroxide as terminal oxidant, led predominantly to the formation of the epoxide (Scheme 2.19). It should be noted that slow addition of hydrogen peroxide (60 min) was required for successful alkene oxidation. When the corresponding reactions were performed without addition of acetic acid, the oxidation of cyclooctene using [(BPMEN)Fe(CH$_3$CN)$_2$](OTf)$_2$ as catalyst precursor resulted in the formation of 63% cyclooctene oxide and 6% of the corresponding diol. Using the TPA-containing catalyst gave merely 32% epoxide and 37% diol.

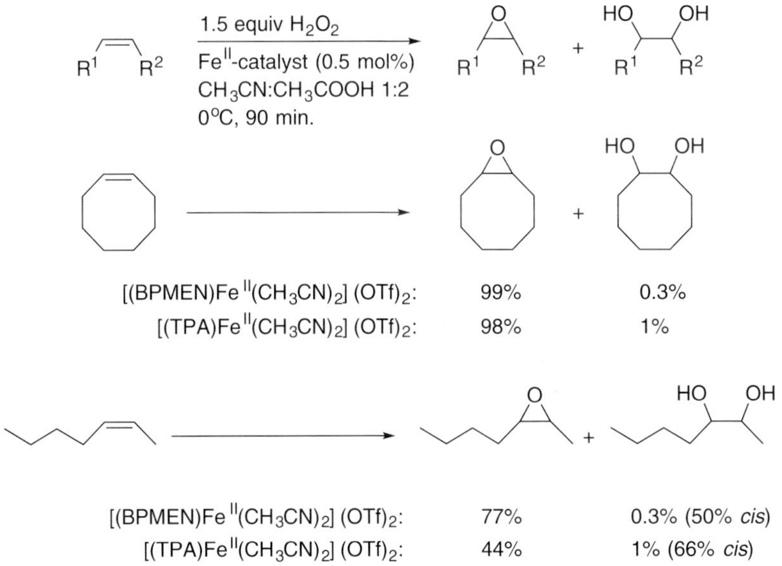

Scheme 2.19

An additional contribution to the field of biomimetic non-heme iron complexes for alkene oxidation was recently reported by Klein Gebbink and coworkers [109]. They found that iron(II) complexes formed with the neutral ligand propyl 3,3-bis(1-methylimidazol-1-yl)propionate (**20**) were active as catalysts for the oxidation of various simple alkenes. When complex **21** was employed as the catalyst for the oxidation of cyclooctene in the presence of 10 equivalents of hydrogen peroxide, a mixture of epoxide and diol in a ratio of 2.5 : 1 was obtained. However, the conversion was rather poor (39%).

20

21

The major breakthrough in the field of iron-catalyzed epoxidation using aqueous hydrogen peroxide as the terminal oxidant was recently made by Beller and coworkers. The combination of iron(III) chloride hexahydrate with a pyridine carboxylic acid ligand and an organic base allowed for alkene epoxidation under seemingly neutral conditions. After screening several different catalytic systems containing a variety of pyridine ligands and organic or inorganic bases, the combination of pyridine-2,6-dicarboxylic acid (H_2pydic) and pyrrolidine turned out to be the most successful one [101, 110]. Using slow addition of hydrogen peroxide and the above-mentioned catalyst combination allowed for the efficient epoxidation of a series of aryl substituted alkenes (Table 2.8). Interestingly, this catalyst system seems to be most efficient and selective for the epoxidation of mono-substituted or trans-1,2-disubstituted aromatic alkenes. The epoxidation of cis-1,2-disubstituted or tri-substituted substrates led in all cases to lower yields and poor chemoselectivity. Furthermore, aliphatic alkenes were less favorable substrates, leading in all cases to lower conversions and selectivity.

In further optimizations, Beller and coworkers examined various benzyl amines as replacement for pyrrolidine in the $FeCl_3$-H_2pydic catalyst system. They found that the use of different benzyl amines resulted in higher yields and better selectivity for the formation of epoxides, predominantly from aliphatic alkenes [111]. As seen in Table 2.8, the epoxidation of trans-1,2-disubstituted alkenes such as *trans*-2-octene and *trans*-5-decene resulted in high yields, whereas aliphatic terminal alkenes appear to be problematic substrates. The epoxidation of aromatic alkenes using this catalytic system gave similar results to those obtained using pyrrolidine as base.

In a recent study, the above-mentioned iron(III) chloride catalyst system was further simplified [112, 113]. Beller and coworkers examined the use of a series of simple imidazoles as ligands for iron, and the *in-situ*-formed complexes were used as epoxidation catalysts in combination with aqueous hydrogen peroxide in *tert*-amyl alcohol. Initially the best catalyst combination consisted of a mixture of $FeCl_3 \times 6H_2O$ (5 mol%) and 5-chloro-1-methylimidazole (15 mol%). Upon further optimization, it was found that the use of 12 mol% of 2,6-diisopropyl-*N*-phenylimidazole (IPrPIm) together with the iron(III) chloride salt resulted in a superior catalyst. The latter catalyst system displays reactivity features similar to those of the H_2pydic/pyrrolidine system, in that mono-substituted or trans-1,2-disubstituted aromatic alkenes are easily converted to their corresponding epoxides, whereas other substrates behave more erratically.

Table 2.8 Iron-catalyzed epoxidations.

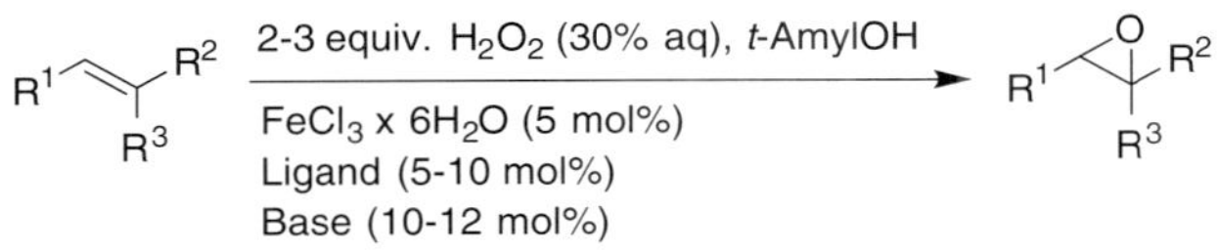

$$R^1\text{-CH=CR}^2R^3 \xrightarrow[\text{Base (10-12 mol\%)}]{\text{2-3 equiv. H}_2O_2 \text{ (30\% aq), }t\text{-AmylOH}}_{\substack{\text{FeCl}_3 \times 6H_2O \text{ (5 mol\%)} \\ \text{Ligand (5-10 mol\%)}}} R^1\text{-CH(O)CR}^2R^3$$

Ligands:

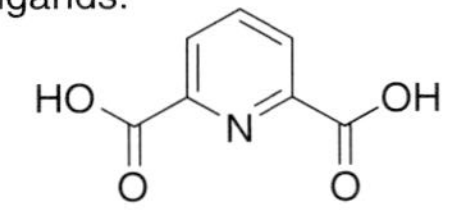

pyridine 2,6-dicarboxylic acid (H₂pydic) (5 mol%)

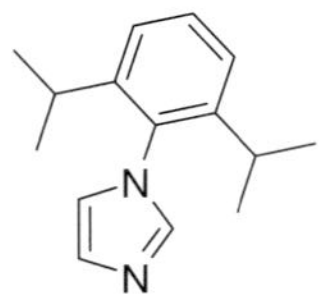

2,6-diisopropyl-N-phenyl imidazole (IPrPIm) (10 mol%)

Bases:

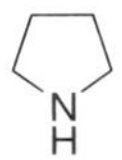

pyrrolidine (10 mol%)

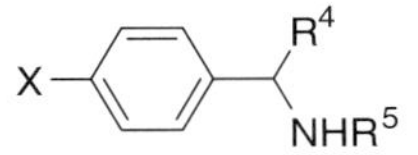

22a) X = Cl, R⁴ = Me, R⁵ = H
22b) X = H, R⁴ = Et, R⁵ = H
(12 mol%)

Entry	Substrate	Catalyst combinations: FeCl₃ × 6H₂O (5 mol%) and:		
		H₂pydic/pyrrolidine[a] Yield (%)	H₂pydic/benzylamine[b] Yield (%)	IPrPIm[c] Yield (%)
1	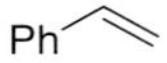 Ph-CH=CH₂	93	91[d]	87
2	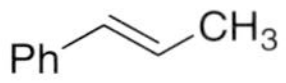 Ph-CH=CH-CH₃	95		94
				(Continued)

2 Transition Metal-Catalyzed Epoxidation of Alkenes

Table 2.8 (Continued)

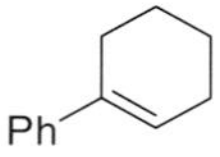

Entry	Substrate	Catalyst combinations: FeCl$_3$ × 6H$_2$O (5 mol%) and:		
		H$_2$pydic/pyrrolidine[a] Yield (%)	H$_2$pydic/benzylamine[b] Yield (%)	IPrPIm[c] Yield (%)
3	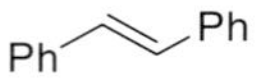 Ph–CH=CH–Ph	56		
4	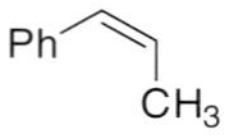 Ph–CH=CH–CH$_3$	84	84[d]	66
5	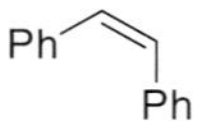 Ph–CH=CH–Ph	8	24[d]	23
6	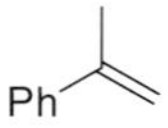 CH$_2$=C(Ph)(CH$_3$)	64		
7	Ph-cyclohexenyl	21		
8	 terminal alkene		32[d]	18
9	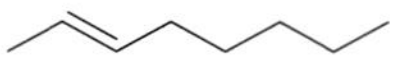 internal alkene	36	87[d]	

| 10 |  [alkene structure] | 32 | 96[e] | 37 |
| 11 | 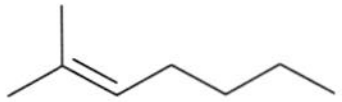 [alkene structure] | | 58[e] | |

a) H₂pydic (5 mol%) and pyrrolidine (10 mol%). Slow addition of 2 equiv. H₂O₂ over 60 min at r.t.
b) H₂pydic (5 mol%) and benzylamine (12 mol%). Slow addition of 2 equiv. H₂O₂ over 60 min at r.t.
c) IPrPIm (10 mol%). Slow addition of 3 equiv. H₂O₂ over 60 min at r.t.
d) Benzylamine 22a was employed.
e) Benzylamine 22b was employed.

2.7.1
Iron-Catalyzed Asymmetric Epoxidations

The use of iron catalysts in combination with aqueous hydrogen peroxide as the terminal oxidant for the asymmetric epoxidation of alkenes has been significantly less studied. Using high-throughput screening techniques, Francis and Jacobsen discovered a novel iron-based protocol for the preparation of enantiomerically enriched epoxides [114]. In this system, chiral complexes prepared from polymer-supported peptide-like ligands and iron(II) chloride were evaluated as catalysts for the epoxidation of trans-β-methylstyrene employing aqueous hydrogen peroxide (30%) as the terminal oxidant. The best polymer-supported catalysts yielded the corresponding epoxide in up to 78% conversion with enantioselectivities ranging from 15 to 20% ee. Employing a homogeneous catalyst derived from this combinatorial study, trans-β-methylstyrene was epoxidized in 48% ee after 1 h (100% conversion, 5 mol% catalyst, 1.25 equiv. 50% hydrogen peroxide in t-BuOH) (Scheme 2.20) [115].

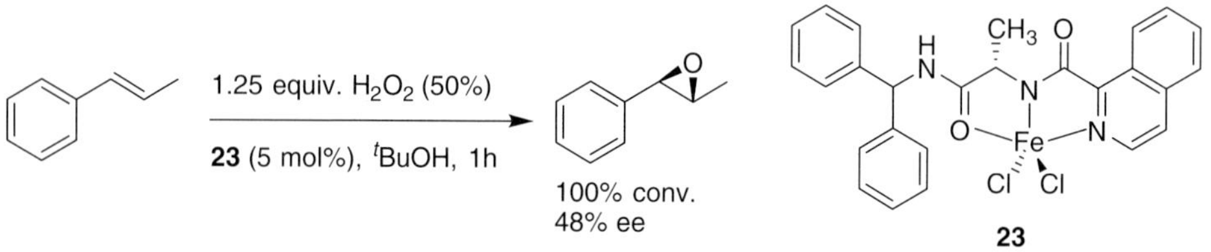

Scheme 2.20

Recently, Gelalcha, Beller, and coworkers reported that the combination of the FeCl$_3$-H$_2$pydic system with chiral monosulfonated diamine ligands (e.g., TsDPEN) resulted in complexes which catalyzed the formation of enantiomerically enriched epoxides using aqueous hydrogen peroxide as the terminal oxidant (Scheme 2.21)

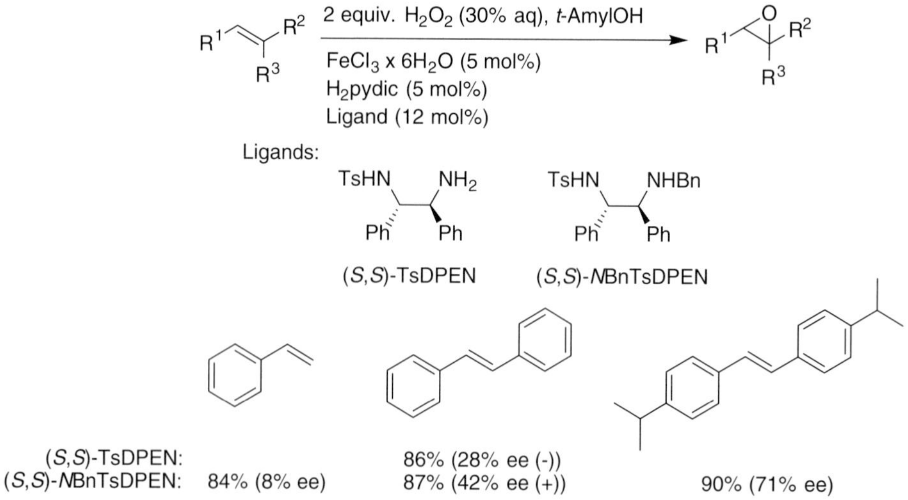

Scheme 2.21

2.7 Iron-Catalyzed Epoxidations

[116, 117]. As seen in Scheme 2.21, a reasonably high enantioselectivity can be obtained with sterically hindered substrates, whereas simple alkenes gave significantly lower *ees*. Interestingly, products of opposite stereochemistry were obtained in the epoxidation of *trans*-stilbene using the two different (*S,S*)-diamine ligands shown in Scheme 2.21.

An interesting chiral bimetallic iron complex was recently presented by Kwong and coworkers [118]. Starting from the enantiomerically pure sexipyridine ligand **24** and iron(II) chloride, the bimetallic iron complex [Fe$_2$O(**24**)Cl$_4$] was obtained (Scheme 2.22).

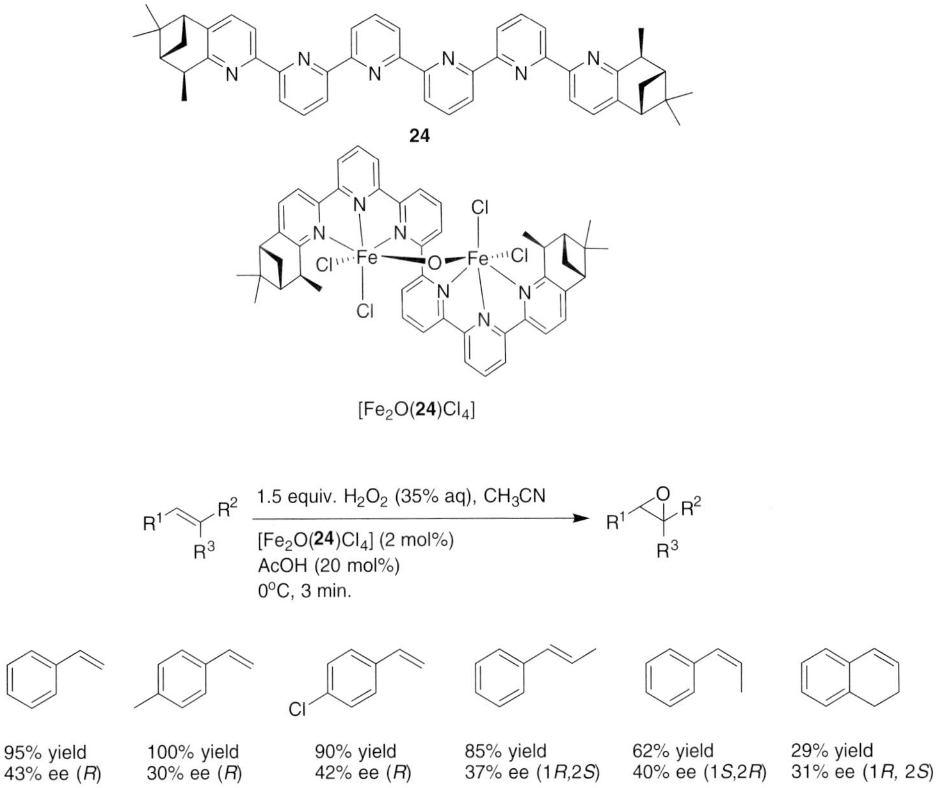

Scheme 2.22

| 95% yield | 100% yield | 90% yield | 85% yield | 62% yield | 29% yield |
| 43% ee (R) | 30% ee (R) | 42% ee (R) | 37% ee (1R,2S) | 40% ee (1S,2R) | 31% ee (1R, 2S) |

Using 2 mol% of complex [Fe$_2$O(**24**)Cl$_4$] as the catalyst for the epoxidation of a series of aromatic alkenes resulted in the formation of the corresponding epoxides in good yields and in moderate enantioselectivity.

Even though only moderate enantiomeric excesses were obtained using the chiral iron-containing catalysts described above, these examples are of particular interest, since further elaboration along these lines could produce more selective catalysts.

2.8
Ruthenium-Catalyzed Epoxidations

High-valent ruthenium oxides (e.g., RuO_4) are powerful oxidants and react readily with alkenes, mostly resulting in cleavage of the double bond [119]. If reactions are performed with very short reaction times (0.5 min) at 0 °C it is possible to obtain better control of the reactivity and thereby obtain cis-diols. On the other hand, the use of less reactive low-valency ruthenium complexes in combination with various terminal oxidants has been described for the preparation of epoxides from simple alkenes [120]. In the more successful earlier cases, ruthenium porphyrins were used as catalysts, especially in combination with N-oxides as terminal oxidants [121–123]. Two examples are shown in Scheme 2.23, where terminal alkenes are oxidized in the presence of catalytic amounts of Ru porphyrins **25** and **26** employing the sterically hindered 2,6-dichloropyridine N-oxide (2,6-DCPNO) as the oxidant. The use of this particular oxidant was crucial for high epoxide yield, since simple pyridine, generated from reaction with pyridine N-oxide, significantly lowered the epoxidation rate by coordination to the ruthenium catalyst. Interestingly, aerobic epoxidation of alkenes under 1 bar of molecular oxygen without any additional reducing agent has been reported with catalyst **26** [124].

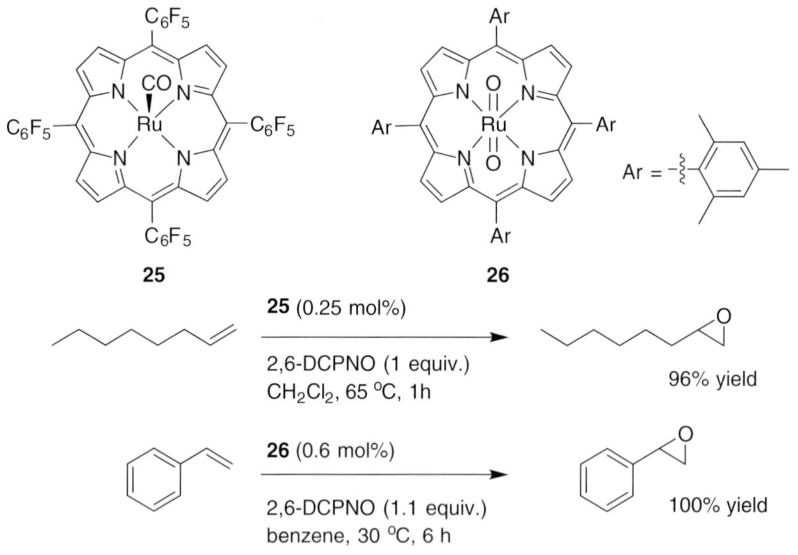

Scheme 2.23

Nishiyama and coworkers reported that the combination of pyridine-2,6-dicarboxylic acid (H_2pydic) and terpyridine ligated to ruthenium(II) resulted in a complex with reasonably high catalytic activity for the epoxidation of *trans*-stilbene in the presence of various high-valent iodine compounds or TBHP as terminal oxidants [125]. Based on these findings, Beller and coworkers initially reported that RuCl$_3$ and H$_2$pydic acted as an efficient catalyst for the epoxidation of alkenes using 30% aqueous hydrogen peroxide as the terminal oxidant [126]. With RuCl$_3$ loadings ranging from 0.01–5 mol% and slow addition of the oxidant to avoid ruthenium-catalyzed decomposition of H$_2$O$_2$, a number of different alkenes were successfully converted into epoxides in yields up to 99%. Further developments using the Nishiyama ligand combination (H$_2$pydic and terpyridine) allowed for an even more efficient catalytic process [127]. Selected examples using the latter protocol are exemplified in Scheme 2.24.

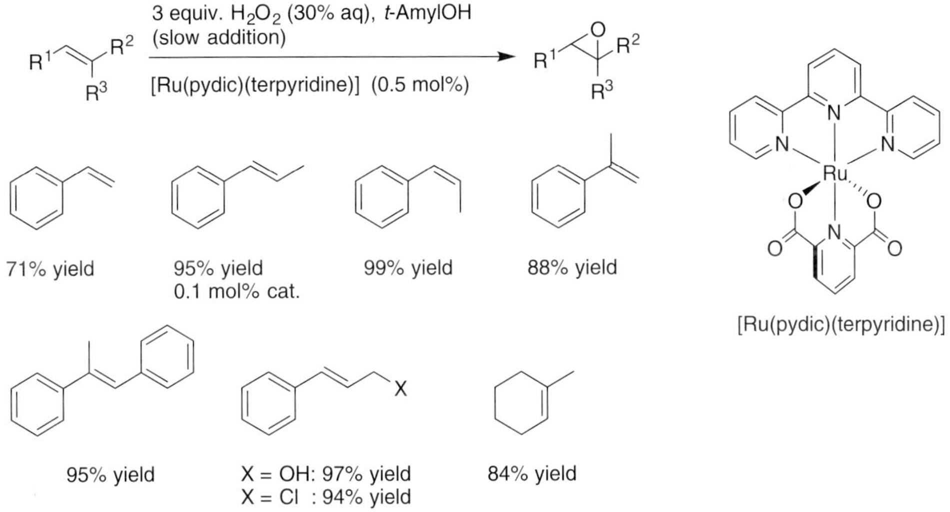

Scheme 2.24

Asymmetric epoxidation of alkenes has been reported using ruthenium catalysts based on either chiral porphyrins or on pyridine-2,6-bisoxazoline (pybox) ligands. Berkessel et al. reported that catalysts **27** and **28** were efficient catalysts for the enantioselective epoxidation of aryl-substituted alkenes [128]. Enantioselectivity up to 83% was obtained in the epoxidation of 1,2-dihydronaphthalene using catalyst **28** and 2,6-DCPNO. Simple alkenes like 1-octene reacted poorly and gave epoxides of low enantioselectivity. The mixed-ligand ruthenium complex **29**, containing a chiral tridentate Schiff base ligand derived from D-glucose and two triphenylphosphines ligands, was recently prepared, and the catalytic properties of this complex were evaluated in the epoxidation of styrene and styrene derivatives [129]. The catalytic activity was rather poor, giving epoxides in modest yields, although very promising enantioselectivities were obtained. In the epoxidation of 4-chlorostyrene with 1 mol% catalyst loading, the corresponding epoxide was formed in 45% yield and in 94% ee using one equivalent of aqueous TBHP as the terminal oxidant. A certain degree of over-oxidation was observed using this catalytic system, and significant amounts of aldehydes (8–18%) were formed along with the epoxides.

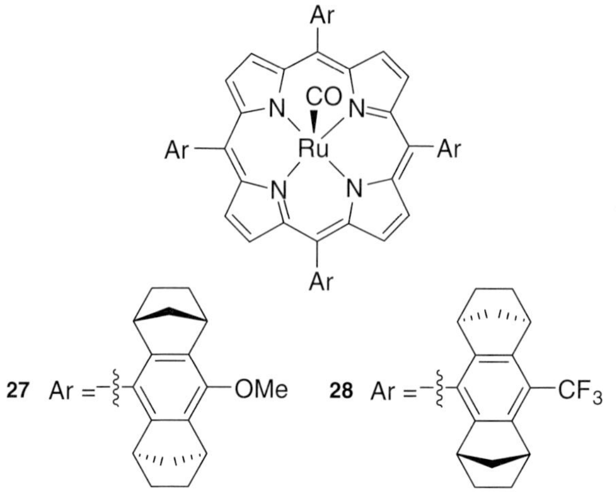

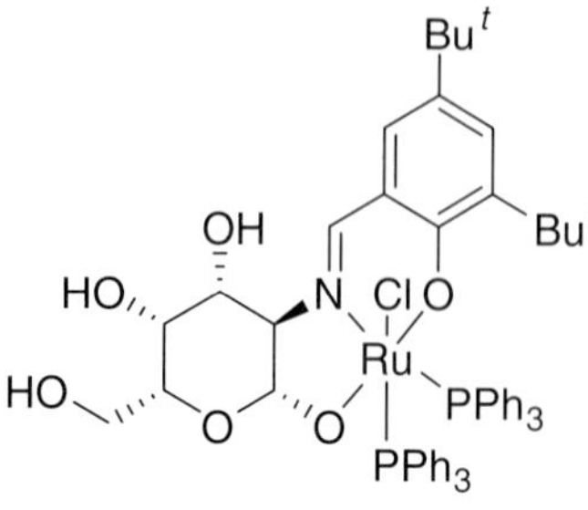

27 Ar = -OMe   **28** Ar = -CF$_3$ **29**

The use of pybox ligands in ruthenium-catalyzed asymmetric epoxidations was first reported by Nishiyama et al., who used catalyst **31** in combination with either iodosyl benzene, bisacetoxyiodo benzene [PhI(OAc)$_2$], or TBHP for the oxidation of *trans*-stilbene [130]. In the best result, using PhI(OAc)$_2$ as oxidant, they obtained *trans*-stilbene oxide in 80% yield and 63% *ee*. More recently, Beller and coworkers have reexamined this catalytic system and found that asymmetric epoxidations could be performed using ruthenium catalysts **30** and **31** and 30% aqueous hydrogen peroxide [131–133]. A development of the pybox ligand led to ruthenium complex **32**, which turned out to be the most efficient catalyst for asymmetric alkene epoxidation. Thus, using 5 mol% of **32** and slow addition of hydrogen peroxide, a number of aryl substituted alkenes were epoxidized in yields up >99% and enantioselectivity up to 84% (Scheme 2.25).

2.9
Epoxidations Using Late Transition Metals

The use of transition metal complexes from group IX as catalysts or reagents for alkene epoxidation is so far rather limited. Cobalt is the main metal to have been investigated, since a number of cobalt complexes are known to directly activate molecular oxygen, which thereafter can be used in further oxidative processes. Mukaiyama and coworkers reported that cobalt(II)-Schiff base complexes catalyzed the epoxidation of various alkenes using molecular oxygen as the terminal oxidant and ketones or aldehydes as reductants.[134, 135] Isolated yields up to 98% were reported for some substrates, but typically these protocols resulted in lower epoxide yields. When it comes to epoxidation processes mediated by group X metals, the situation is similar to that for group IX metals. There are a number of reports which discuss the use of nickel complexes as catalysts for the epoxidation of alkenes. From a synthetic point of view, no efficient and selective nickel-based systems have been

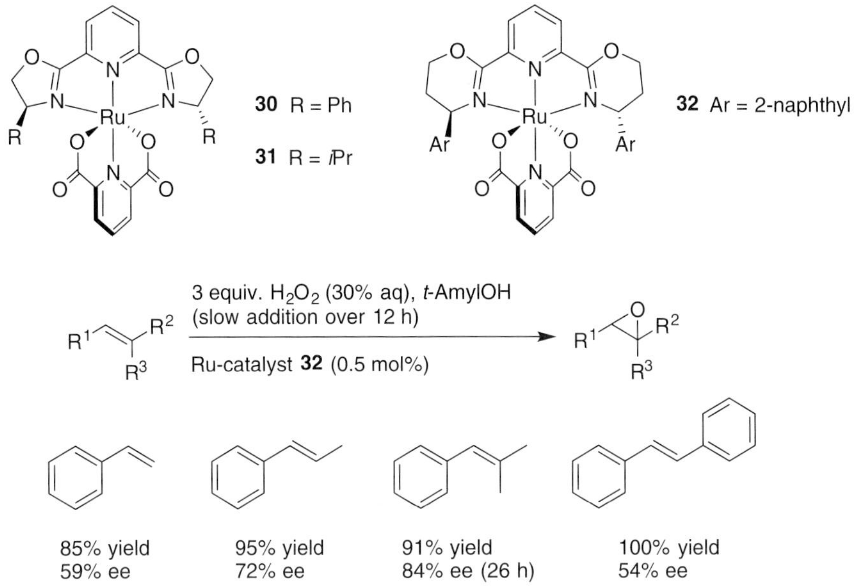

2.9 Epoxidations Using Late Transition Metals | 77

30 R = Ph
31 R = iPr

32 Ar = 2-naphthyl

$$R^1\text{-CH=CR}^2\text{R}^3 \xrightarrow[\text{Ru-catalyst 32 (0.5 mol\%)}]{\text{3 equiv. H}_2\text{O}_2 \text{ (30\% aq), }t\text{-AmylOH} \\ \text{(slow addition over 12 h)}} R^1\text{-epoxide-R}^2\text{R}^3$$

85% yield	95% yield	91% yield	100% yield
59% ee	72% ee	84% ee (26 h)	54% ee

Scheme 2.25

discovered [136–139]. On the contrary, Strukul and coworkers recently reported on a highly interesting epoxidation system based on cationic electron-deficient platinum (II) complexes, which in the presence of aqueous hydrogen peroxide catalyzed the formation of epoxides from simple terminal alkenes [140]. The catalyst precursor in the Strukul epoxidation system is the organometallic Pt(II) complex **33**, containing a pentafluorophenyl ligand along with a bidentate diphosphine and a coordinated water molecule. The catalytic activity using a number of different diphosphine ligands was investigated, and reasonable yields were obtained in the epoxidation of terminal alkenes using 2 mol% catalyst.

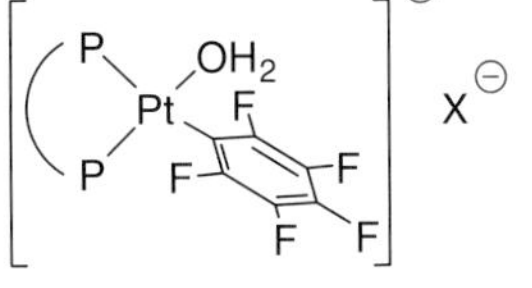

33

The introduction of chiral bidentate diphosphines allowed for asymmetric epoxidations of terminal alkenes. Screening a number of different chiral diphosphines in the epoxidation of 4-methylpentene using 2 mol% of catalyst and 1 equivalent of aqueous hydrogen peroxide as the terminal oxidant revealed that the catalyst containing S,S-Chiraphos (**34**) allowed for an efficient chirality transfer from the catalyst to the substrate [141]. With this particular catalyst combination, the R-isomer of 4-methylpentene oxide was obtained in 60% yield and 75% enantiomeric excess. The activity and selectivity of the platinum complex **34** turned out to be quite general, and a number of different terminal alkenes were efficiently and selectively converted to their corresponding epoxides (Scheme 2.26).

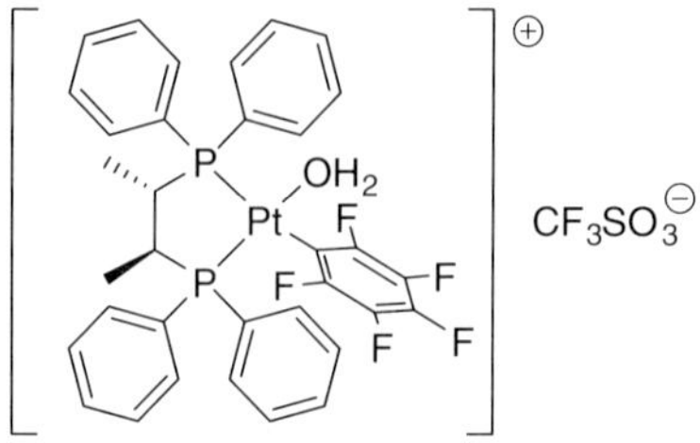

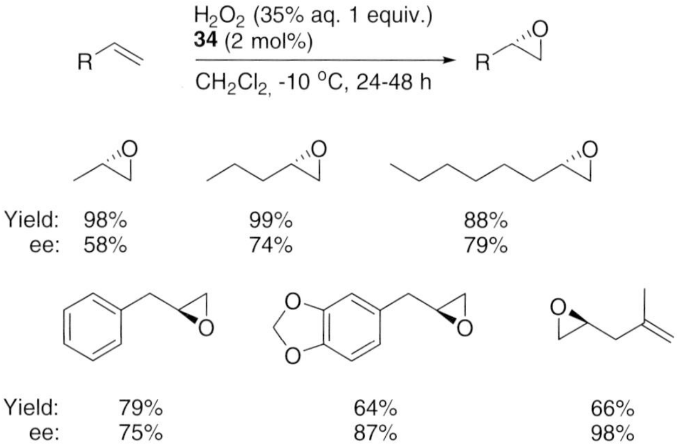

Scheme 2.26

In addition to the outstanding selectivities obtained in the epoxidation of terminal alkenes, a particularly interesting feature of this catalytic system is the inherent reactivity of the catalyst. Most epoxidation systems known favor the oxidation of the most electron-rich double bond; however, when catalyst **34** was employed in the epoxidation of dienes containing both internal and terminal alkenes, only the terminal position reacted [142]. Hence, this catalyst shows unique selectivity properties. In further studies of this catalytic system, Strukul and coworkers developed an environmentally benign enantioselective epoxidation protocol in which reactions were performed in a water-surfactant medium [143]. The best conditions were obtained using 1 mol% of **34** and one equivalent of aqueous hydrogen peroxide in a mixture of water and the surfactant Triton-X100. Under these reaction conditions, a number of terminal alkenes were converted into their corresponding epoxides in good yields and with moderate to good enantioselectivity within a short reaction time. This protocol is still in its immature stage, and we will surely see further developments along these lines in the near future.

Regarding the epoxidation activity of group XI metals, the industrial formation of ethylene oxide is a classic example of a silver-catalyzed epoxidation process [144]. In this oxidation reaction using molecular oxygen as the terminal oxidant, the active catalyst consists of silver particles supported on α-alumina. In order to moderate the activity of the catalyst and thereby further increase epoxide selectivity, the catalyst system is treated with ppm levels of gaseous chlorocarbons and controlled amounts of alkali. Nevertheless, a problem when using these silver-based catalytic systems is the high activity of the heterogeneous catalyst, which often leads to ethylene combustion instead of formation of ethylene oxide. For higher alkenes, for example, propene, this is the major reaction, and therefore Ag catalysts are poor mediators for alkene epoxidation of substrates other than ethylene. Since propene oxide is such an important commodity chemical, great effort is being put into finding new and more selective catalysts for the epoxidation of propene. Some of the latest developments even involve supported gold catalysts [145, 146].

2.10
Concluding Remarks

The epoxidation of alkenes using transition metal-based catalysts is certainly a well-studied reaction. There are, however, only a few really good and general systems that work with environmentally benign oxidants (e.g., aqueous hydrogen peroxide). A comparison of the efficiency figures obtained with the catalysts described in this chapter is presented in Table 2.9.

A striking feature of the various catalytic epoxidation methods presented in this chapter in general and in Table 2.9 in particular is the poor substrate-to-catalyst ratio observed for most systems. In the best cases, the catalyst loading is as low as 0.01 mol %, but most often 0.5–1 mol% of the catalyst needs to be added to ensure full conversion of the starting alkene. In comparison to the corresponding highly efficient catalysts existing for the reduction of alkenes using molecular hydrogen, there is a long way to go before oxidation catalysts reach the same level of efficiency. When it comes to substrate scope, it is evident from the content of this chapter that there are advantages and limitations with almost all available epoxidation systems. The environmentally attractive TS-1 system is highly efficient, but is restricted to linear substrates. The various tungsten systems available efficiently produce epoxides from simple substrates, but acid-sensitive products undergo further reactions, thus effectively reducing the selectivity of the process. MTO is a highly active epoxidation catalyst, and when it is combined with the most appropriate heterocyclic additives, even hydrolytically sensitive products are obtained in good yield and selectivity. Most of the MTO-catalyzed reactions are, however, performed in chlorinated or fluorinated solvents. Regarding asymmetric processes using hydrogen peroxide as the terminal oxidant, there are only a few reported systems that produce epoxides with respectable enantioselectivity values. The ruthenium- and iron-based catalysts developed by Beller and coworkers are promising candidates, and further developments along these lines may produce more selective systems.

Table 2.9 Transition metal-catalyzed epoxidation of alkenes using H_2O_2 as the terminal oxidant.

Catalyst	S/C	Solvent	Temp. (°C)	1-Alkene[h] yield (%)/ TOF (h^{-1})	Cyclooctene yield (%)/ TOF (h^{-1})	Ref.
Ti[a]	100	MeOH	25	74/108[i]	—	10
W[b]	50/500	Toluene	90	91/12	98/122	24
Mo[c]	4000/10000	CH_3CN	25	94/3760	80/24000	31
Mn[d]	1000	CH_3CN/CH_3CO_2H	0	90/600[i]	95/1600	54
Re[e]	200	CH_2Cl_2	25	99/14	89/8900	76
Re[e]	1000	CF_3CH_2OH	25	97/48	99/990	87
Fe[f]	33	CH_3CN	4	85/337	86/341	104
Pt[g]	50	DCE[j]	0	81/10	—	143

a) TS-1.
b) Na_2WO_4, $NH_2CH_2PO_3H_2$, $(C_8H_{17})_3NCH_3^+HSO_4^-$.
c) $[PPh_4][Mo(O)(O_2)_2(QO)]$.
d) $[Mn(OTf)_2(pyTACN)]$.
e) MTO, pyrazole.
f) 19.
g) 34.
h) 1-Decene.
i) 1-Octene.
j) DCE = dichloroethane.

In conclusion, the above summary of oxidation methods shows that there is still room for further improvement in the field of selective alkene epoxidation. The development of active and selective catalysts which are capable of oxidizing a broad range of alkene substrates employing aqueous hydrogen peroxide as terminal oxidant in inexpensive environmentally benign solvents remains a continuing challenge.

References

1 Sheldon, R.A. (2002) *Applied Homogeneous Catalysis with Organometallic Compounds*, 2nd edn, 1 (eds B. Cornils and W.A. Herrmann), Wiley-VCH, Weinheim, pp. 412–427.
2 Jørgensen, K.A. (1989) *Chem. Rev.*, **89**, 431.
3 Kahlich, D., Wiechern, K., and Lindner, J. (1993) *Ullmann's Encyclopedia of Industrial Chemistry*, 5th edn, vol. **A22** (eds B. Elvers, S. Hawkins, W. Russey, and G. Schultz), VCH, Weinheim, pp. 239–260.
4 Monnier, J.R. (2001) *Appl. Catal. A-Gen.*, **221**, 73.
5 Lane, B.S. and Burgess, K. (2003) *Chem. Rev.*, **103**, 2457.
6 Sheldon, R.A. (1981) *Aspects of Homogeneous Catalysis*, vol. 4 (ed. R. Ugo), Reidel, Dordrecht, pp. 3–70.
7 Katsuki, T. and Sharpless, K.B. (1980) *J. Am. Chem. Soc.*, **102**, 5974.
8 For a comprehensive review, see: Katsuki, T. (1999) *Comprehensive Asymmetric Catalysis*, **II** (eds E.N.

Jacobsen, A. Pfaltz, and H. Yamamoto), Springer, Heidelberg, pp. 621–648.
9 Zhang, W., Basak, A., Kosugi, Y., Hoshino, Y., and Yamamoto, H. (2005) *Angew. Chem. Int. Ed.*, **44**, 4389.
10 Notari, B. (1993) *Catal. Today*, **18**, 163.
11 Matsumoto, K., Sawada, Y., Saito, B., Sakai, K., and Katsuki, T. (2005) *Angew. Chem. Int. Ed.*, **44**, 4935.
12 Sawada, Y., Matsumoto, K., Kondo, S., Watanabe, H., Ozawa, T., Suzuki, K., Saito, B., and Katsuki, T. (2006) *Angew. Chem. Int. Ed.*, **45**, 3478.
13 Matsumoto, K., Sawada, Y., and Katsuki, T. (2006) *Synlett*, 3545.
14 Nakagawa, Y., Kamata, K., Kotani, M., Yamaguchi, K., and Mizuno, N. (2005) *Angew. Chem. Int. Ed.*, **44**, 5136.
15 Mizuno, N., Nakagawa, Y., and Yamaguchi, K. (2006) *J. Mol. Catal. A*, **251**, 286.
16 Nakagawa, Y. and Mizuno, N. (2007) *Inorg. Chem.*, **46**, 1727.
17 Sharpless, K.B. and Verhoeven, T.R. (1979) *Aldrichmica. Acta*, **12**, 63.
18 Thiel, W.R. (1998) *Transition Metals for Organic Synthesis*, **2** (eds M. Beller and C. Bolm), Wiley-VCH, Weinheim, pp. 290–300.
19 Payne, G.B. and Williams, P.H. (1959) *J. Org. Chem.*, **24**, 54.
20 Venturello, C., Alneri, E., and Ricci, M. (1983) *J. Org. Chem.*, **48**, 3831.
21 Venturello, C. and D'Aloisio, R. (1988) *J. Org. Chem.*, **53**, 1553.
22 Venturello, C., D'Aloisio, R., Bart, J.C.J., and Ricci, M. (1985) *J. Mol. Catal.*, **32**, 107.
23 Sato, K., Aoki, M., Ogawa, M., Hashimoto, T., and Noyori, R. (1996) *J. Org. Chem.*, **61**, 8310.
24 Sato, K., Aoki, M., Ogawa, M., Hashimoto, T., Paynella, D., and Noyori, R. (1997) *Bull. Chem. Soc. Jpn.*, **70**, 905.
25 Sato, K., Aoki, M., and Noyori, R. (1998) *Science*, **281**, 1646.
26 Villa de P., A.L., Sels, B.F., De Vos, D.E., and Jacobs, P.A. (1999) *J. Org. Chem.*, **64**, 7267.
27 Kamata, K., Yonehara, K., Sumida, Y., Yamaguchi, K., Hikichi, S., and Mizuno, N. (2003) *Science*, **300**, 964.
28 Kamata, K., Kotani, M., Yamaguchi, K., Hikichi, S., and Mizuno, M. (2007) *Chem. Eur. J.*, **13**, 639.
29 Bortolini, O., Di Furia, F., Modena, G., and Seraglia, R. (1985) *J. Org. Chem.*, **50**, 2688.
30 Wahl, G., Kleinhenz, D., Schorm, A., Sundermeyer, J., Stowasser, R., Rummey, C., Bringmann, G., Fickert, C., and Kiefer, W. (1999) *Chem. Eur. J.*, **5**, 3237.
31 Maiti, S.K., Dinda, S., and Bhattacharyya, R. (2008) *Tetrahedron Lett.*, **49**, 6205.
32 Sels, B.F., De Vos, D.E., and Jacobs, P.A. (1996) *Tetrahedron Lett.*, **37**, 8557.
33 For a comprehensive summary, see Gelbard, G. (2000) *C. R. Chim.*, **3**, 757.
34 Hoegaerts, D., Sels, B.F., De Vos, D.E., Verpoort, F., and Jacobs, P.A. (2000) *Catal. Today*, **60**, 209.
35 Zuwei, X., Ning, Z., Yu, S., and Kunlan, L. (2001) *Science*, **292**, 1139.
36 Battioni, P., Renaud, J.-P., Bartoli, J.F., Reina-Artiles, M., Fort, M., and Mansuy, D. (1988) *J. Am. Chem. Soc.*, **110**, 8462.
37 Anelli, P.L., Banfi, L., Legramandi, F., Montanari, F., Pozzi, G., and Quici, S. (1993) *J. Chem. Soc., Perkin Trans. 1*, 1345.
38 Zhang, W., Loebach, J.L., Wilson, S.R., and Jacobsen, E.N. (1990) *J. Am. Chem. Soc.*, **112**, 2801.
39 Irie, R., Noda, K., Ito, Y., Matsumoto, N., and Katsuki, T. (1990) *Tetrahedron Lett.*, **31**, 7345.
40 Jacobsen, E.N. and Wu, M.H. (1999) *Comprehensive Asymmetric Catalysis*, **II** (eds E.N. Jacobsen, A. Pfaltz, and H. Yamamoto), Springer, Heidelberg, pp. 649–677.
41 Srinivasan, K., Michaud, P., and Kochi, J.K. (1986) *J. Am. Chem. Soc.*, **108**, 2309.
42 Berkessel, A., Frauenkron, M., Schwenkreis, T., Steinmetz, A., Baum, G., and Fenske, D. (1996) *J. Mol. Catal. A: Chem.*, **113**, 321.
43 Irie, R., Hosoya, N., and Katsuki, T. (1994) *Synlett*, 255.
44 Pietikäinen, P. (1998) *Tetrahedron*, **54**, 4319.
45 Pietikäinen, P. (2001) *J. Mol. Catal. A: Chem.*, **165**, 73.

46 Kureshy, R.I., Khan, N.H., Abdi, S.H.R., Patel, S.T., and Jasra, R.V. (2001) *Tetrahedron: Asymmetr.*, **12**, 433.

47 Kureshy, R.I., Kahn, N.H., Abdi, S.H.R., Singh, S., Ahmed, I., Shukla, R.S., and Jasra, R.V. (2003) *J. Catal.*, **219**, 1.

48 Hage, R., Iburg, J.E., Kerschner, J., Koek, J.H., Lempers, E.L.M., Martens, R.J., Racheria, U.S., Russell, S.W., Swarthoff, T., van Vliet, M.R.P., Warnaar, J.B., van der Wolf, L., and Krijnen, B. (1994) *Nature*, **369**, 637.

49 De Vos, D.E., Sels, B.F., Reynaers, M., Rao, Y.V.S., and Jacobs, P.A. (1998) *Tetrahedron Lett.*, **39**, 3221.

50 Berkessel, A. and Sklorz, C.A. (1999) *Tetrahedron Lett.*, **40**, 7965.

51 de Boer, J.W., Brinksma, J., Browne, W.R., Meetsma, A., Alsters, P.L., Hage, R., and Feringa, B.L. (2005) *J. Am. Chem. Soc.*, **127**, 7990.

52 de Boer, J.W., Browne, W.R., Brinksma, J., Alsters, P.L., Hage, R., and Feringa, B.L. (2007) *Inorg. Chem.*, **46**, 6353.

53 Garcia-Bosch, I., Company, A., Fontrodona, X., Ribas, X., and Costas, M. (2008) *Org. Lett.*, **10**, 2095.

54 Garcia-Bosch, I., Ribas, X., and Costas, M. (2009) *Adv. Synth. Catal.*, **351**, 348.

55 Brinksma, J., Hage, R., Kerschner, J., and Feringa, B.L. (2000) *Chem. Commun.*, 537.

56 Lane, B.S. and Burgess, K. (2001) *J. Am. Chem. Soc.*, **123**, 2933.

57 Lane, B.S., Vogt, M., DeRose, V.J., and Burgess, K. (2002) *J. Am. Chem. Soc.*, **124**, 11946.

58 Tong, K.-H., Wong, K.-Y., and Chan, T.H. (2003) *Org. Lett.*, **5**, 3423.

59 Kühn, F.E. and Herrmann, W.A. (2001) *Chemtracts – Org. Chem.*, **14**, 59.

60 Owens, G.S., Arias, A., and Abu-Omar, M.M. (2000) *Catal. Today*, **55**, 317, Please observe; this review contains several serious errors in the chapter dealing with MTO-catalyzed epoxidations (i.e. Tables 2.5 and 2.6).

61 A: Herrmann, W., Fischer, R.W., and Marz, D.W. (1991) *Angew. Chem. Int. Ed. Engl.*, **30**, 1638.

62 Beattie, I.R. and Jones, P.J. (1979) *Inorg. Chem.*, **18**, 2318.

63 Herrmann, W.A. and Kühn, F.E. (1997) *Acc. Chem. Res.*, **30**, 169.

64 Romão, C.C., Kühn, F.E., and Herrmann, W.A. (1997) *Chem. Rev.*, **97**, 3197.

65 Kühn, F.E., Scherbaum, A., and Herrmann, W.A. (2004) *J. Organomet. Chem.*, **689**, 4149.

66 Abu-Omar, M.M., Hansen, P.J., and Espenson, J.H. (1996) *J. Am. Chem. Soc.*, **118**, 4966.

67 Herrmann, W.A., Fischer, R.W., Scherer, W., and Rauch, M.U. (1993) *Angew. Chem. Int. Ed. Engl.*, **32**, 1157.

68 Herrmann, W.A., Fischer, R.W., Rauch, M.U., and Scherer, W. (1994) *J. Mol. Catal.*, **86**, 243.

69 Adam, W. and Mitchell, C.M. (1996) *Angew. Chem. Int. Ed. Engl.*, **35**, 533.

70 Rudolph, J., Reddy, K.L., Chiang, J.P., and Sharpless, K.B. (1997) *J. Am. Chem. Soc.*, **119**, 6189.

71 Wang, W.-D. and Espenson, J.H. (1998) *J. Am. Chem. Soc.*, **120**, 11335.

72 Copéret, C., Adolfsson, H., and Sharpless, K.B. (1997) *Chem. Commun.*, 1565.

73 Adolfsson, H., Copéret, C., Chiang, J.P., and Yudin, A.K. (2000) *J. Org. Chem.*, **65**, 8651.

74 Copéret, C., Adolfsson, H., Khuong, T.-A.V., Yudin, A.K., and Sharpless, K.B. (1998) *J. Org. Chem.*, **63**, 1740.

75 Herrmann, W.A., Kratzer, R.M., Ding, H., Thiel, W.R., and Glas, H. (1998) *J. Organomet. Chem.*, **555**, 293.

76 Yamazaki, S. (2007) *Org. Biomol. Chem.*, **5**, 2109.

77 Adolfsson, H., Converso, A., and Sharpless, K.B. (1999) *Tetrahedron Lett.*, **40**, 3991.

78 Vaino, A.R. (2000) *J. Org. Chem.*, **65**, 4210.

79 McKillop, A. and Sanderson, W.R. (1995) *Tetrahedron*, **51**, 6145.

80 Jackson, W.P. (1990) *Synlett*, 536.

81 Jost, C., Wahl, G., Kleinhenz, D., and Sundermeyer, J. (2000) *Peroxide Chemistry* (ed. W. Adam), Wiley-VCH, Weinheim, pp. 341–364.

82 Yudin, A.K. and Sharpless, K.B. (1997) *J. Am. Chem. Soc.*, **119**, 11536.

83 Curci, R., Mello, R., and Troisi, L. (1986) *Tetrahedron*, **42**, 877.

84 Yudin, A.K., Chiang, J.P., Adolfsson, H., and Copéret, C. (2001) *J. Org. Chem.*, **66**, 4713.
85 Warwel, S., Rüsch den Klaas, M., and Sojka, M. (1991) *J. Chem. Soc., Chem. Commun.*, 1578.
86 van Vliet, M.C.A., Arends, I.W.C.E., and Sheldon, R.A. (1999) *Chem. Commun.*, 821.
87 Iskra, J., Bonnet-Delpon, D., and Bégué, J.-P. (2002) *Tetrahedron Lett.*, **43**, 1001.
88 Owens, G.S. and Abu-Omar, M.M. (2000) *Chem. Commun.*, 1165.
89 Owens, G.S., Durazo, A., and Abu-Omar, M.M. (2002) *Chem. Eur. J.*, **8**, 3053.
90 Yamazaki, S. (2008) *Tetrahedron*, **64**, 9253.
91 Saladino, R., Neri, V., Pelliccia, A.R., Caminiti, R., and Sadun, C. (2002) *J. Org. Chem.*, **67**, 1323.
92 Dallmann, K. and Buffon, R. (2000) *Catal. Commun.*, **1**, 9.
93 Neumann, R. and Wang, T.-J. (1997) *Chem. Commun.*, 1915.
94 Malek, A. and Ozin, G. (1995) *Adv. Mater.*, **7**, 160.
95 Bein, T., Huber, C., Moller, K., Wu, C.-G., and Xu, L. (1997) *Chem. Mater.*, **9**, 2252.
96 Adam, W., Saha-Möller, C.R., and Weichold, O. (2000) *J. Org. Chem.*, **65**, 2897.
97 Omar Bouh, A. and Espenson, J.H. (2003) *J. Mol. Catal. A: Chem.*, **200**, 43.
98 Tucker, C.E. and Davenport, K.G. (1997) Hoechst Celanese Corporation US Patent 5618958.
99 Sabater, M.J., Domine, M.E., and Corma, A. (2002) *J. Catal.*, **210**, 192.
100 Haider, J.J., Kratzer, R.M., Herrmann, W.A., Zhao, J., and Kühn, F.E. (2004) *J. Organomet. Chem.*, **689**, 3735.
101 Anilkumar, G., Bitterlich, B., Gelalcha, F.G., Tse, M.K., and Beller, M. (2007) *Chem. Commun.*, 289.
102 Traylor, T.G., Tsuchiya, S., Byun, Y.-S., and Kim, C. (1993) *J. Am. Chem. Soc.*, **115**, 2775.
103 White, M.C., Doyle, A.G., and Jacobsen, E.N. (2001) *J. Am. Chem. Soc.*, **123**, 7194.
104 Chen, K. and Que, L. Jr. (1999) *Chem. Commun.*, 1375.
105 Costas, M., Tipton, A.K., Chen, K., Jo, D.-H., and Que, L. Jr. (2001) *J. Am. Chem. Soc.*, **123**, 6722.
106 Chen, K., Costas, M., Kim, J., Tipton, A.K., and Que, L. Jr. (2002) *J. Am. Chem. Soc.*, **124**, 3026.
107 Fujita, M. and Que, L. Jr. (2004) *Adv. Synth. Catal.*, **346**, 190.
108 Mas-Ballesté, R. and Que, L. Jr. (2007) *J. Am. Chem. Soc.*, **129**, 15964.
109 Bruijnincx, P.C.A., Buurmans, I.L.C., Gosiewska, S., Moelands, M.A.H., Lutz, M., Spek, A.S., van Koten, G., and Klein Gebbink, R.J.M. (2008) *Chem. Eur. J.*, **14**, 1228.
110 Bitterlich, B., Anilkumar, G., Gelalcha, F.G., Spilker, B., Grotevendt, A., Jackstell, R., Tse, M.K., and Beller, M. (2007) *Chem. Asian J.*, **2**, 521.
111 Bitterlich, B., Schröder, K., Tse, M.K., and Beller, M. (2008) *Eur. J. Org. Chem.*, 4867.
112 Schröder, K., Tong, X., Bitterlich, B., Tse, M.K., Gelalcha, F.G., Brückner, A., and Beller, M. (2007) *Tetrahedron Lett.*, **48**, 6339.
113 Schröder, K., Enthaler, S., Bitterlich, B., Schulz, T., Spannenberg, A., Tse, M.K., Junge, K., and Beller, M. (2009) *Chem. Eur. J.*, **15**, 5471.
114 Francis, M.B. and Jacobsen, E.N. (1999) *Angew. Chem. Int. Ed. Engl.*, **38**, 937.
115 Jacobsen, E.N. (2007) Abstr. Pap.-Am. Chem. Soc. 219th 2007, 7-INORG.
116 Gelalcha, F.G., Bitterlich, B., Anilkumar, G., Tse, M.K., and Beller, M. (2007) *Angew. Chem. Int. Ed.*, **46**, 7293.
117 G: Gelalcha, F., Anilkumar, G., Tse, M.K., Brückner, A., and Beller, M. (2008) *Chem. Eur. J.*, **14**, 7687.
118 Yeung, H.-L., Sham, K.-C., Tsang, C.-S., Lau, T.-C., and Kwong, H.-L. (2008) *Chem. Commun.*, 3801.
119 Murahashi, S.-I. and Komiya, N. (2004) *Ruthenium in Organic Synthesis* (ed S-.I. Murahashi), Wiley-VCH, Weinheim, pp. 53–87.
120 Barf, G.A. and Sheldon, R.A. (1995) *J. Mol. Catal, A: Chem.*, **102**, 23.
121 Groves, J.T., Bonchio, M., Carofiglio, T., and Shalyaev, K. (1996) *J. Am. Chem. Soc.*, **118**, 8961.
122 Higuchi, T., Othake, H., and Hirobe, M. (1989) *Tetrahedron Lett.*, **30**, 6545.
123 Ohtake, H., Higuchi, T., and Hirobe, M. (1992) *Tetrahedron Lett.*, **33**, 2521.

124 Groves, J.T.and Quinn, R. (1985) *J. Am. Chem. Soc.*, **107**, 5790.
125 Nishiyama, H., Shimida, T., Itoh, H., Sugiyama, H., and Motoyama, Y. (1997) *Chem. Commun.*, 1863.
126 Klawonn, M., Tse, M.K., Bhor, S., Döbler, C., and Beller, M. (2004) *J. Mol. Catal. A: Chem.*, **218**, 13.
127 Tse, M.K., Klawonn, M., Bohr, S., Döbler, C., Anilkumar, G., Hugl, H., Mägerlein, W., and Beller, M. (2005) *Org. Lett.*, **7**, 987.
128 Berkessel, A., Kaiser, P., and Lex, J. (2003) *Chem. J. Eur.*, **9**, 4746.
129 Chatterjee, D., Basak, S., Riahi, A., and Muzart, J. (2006) *J. Mol. Catal. A. Chem.*, **255**, 283.
130 Nishiyama, H., Shimida, T., Itoh, H., Sugiyama, H., and Motoyama, Y. (1997) *Chem. Commun.*, 1863.
131 Tse, M.K., Döbler, C., Bhor, S., Klawonn, M., Mägerlein, W., Hugl, H., and Beller, M. (2004) *Angew. Chem. Int. Ed.*, **43**, 5255.
132 Tse, M.K., Bhor, S., Klawonn, M., Anilkumar, G., Jiao, H., Döbler, C., Spanneberg, A., Mägerlein, W., Hugl, H., and Beller, M. (2006) *Chem. Eur. J.*, **12**, 1855.
133 Tse, M.K., Bhor, S., Klawonn, M., Anilkumar, G., Jiao, H., Döbler, C., Spanneberg, A., Mägerlein, W., Hugl, H., and Beller, M. (2006) *Chem. Eur. J.*, **12**, 1875.
134 Takai, T., Hata, E., Yorozu, K., and Mukaiyama, T. (2077) *Chem Lett.*, **1992**.
135 Mukaiyama, T., Yorozu, K., Takai, T., and Yamada, T. (1993) *Chem Lett.*, 439.
136 Yoon, H., Wagler, T.R., O'Connor, K.J., and Burrows, C.J. (1990) *J. Am. Chem. Soc.*, **112**, 4568.
137 Wentzel, B.B., Gosling, P.A., Feiters, M.C., and Nolte, R.J.M. (1998) *J. Chem. Soc. Dalton Trans.*, 2241.
138 Wentzel, B.B., Leinonen, S.M., Thomson, S., Sherington, D.C., Feiters, M.C., and Nolte, R.J.M. (2000) *J. Chem. Soc. Perkin Trans. 1.*, 3428.
139 Ben-Daniel, R., Khenkin, A.M., and Neumann, R. (2000) *Chem. Eur. J.*, **6**, 3722.
140 Pizzo, E., Sgarbossa, P., Scarso, A., Michelin, R.A., and Strukul, G. (2006) *Organometallics*, **25**, 3056.
141 Colladon, M., Scarso, A., Sgarbossa, P., Michelin, R.A., and Strukul, G. (2006) *J. Am. Chem. Soc.*, **128**, 14006.
142 Colladon, M., Scarso, A., Sgarbossa, P., Michelin, R.A., and Strukul, G. (2007) *J. Am. Chem. Soc.*, **129**, 7680.
143 Colladon, M., Scarso, A., and Strukul, G. (2007) *Adv. Synth. Catal.*, **349**, 797.
144 Lambert, R.M., Williams, F.J., Cropley, R.L., and Palermo, A. (2005) *J. Mol. Catal. A. Chem.*, **228**, 27.
145 Sinha, A.K., Seelan, S., Tsubota, S., and Haruta, M. (2004) *Angew. Chem. Int. Ed.*, **43**, 1546.
146 Chowdhury, B., Bravo-Suárez, J.J., Daté, M., Tsubota, S., and Haruta, M. (2006) *Angew. Chem. Int. Ed.*, **45**, 412.

3
Organocatalytic Oxidation. Ketone-Catalyzed Asymmetric Epoxidation of Alkenes and Synthetic Applications

Yian Shi

3.1
Introduction

The asymmetric epoxidation of alkenes constitutes a powerful approach to enantiomerically enriched epoxides, a class of highly versatile intermediates in organic synthesis [1]. Various effective epoxidation systems have been developed, including epoxidation of allylic [2, 3] and homoallylic [4] alcohols, metal-catalyzed epoxidation of unfunctionalized alkenes [5–7], and the nucleophilic epoxidation of electron-deficient alkenes [8]. During the past 10–15 years, much effort has been devoted to chiral ketone-catalyzed asymmetric epoxidation (Scheme 3.1). The subject has been described in great detail in the first edition [9] and other reviews [10]. This chapter provides an update on progress in this area since the first edition [9].

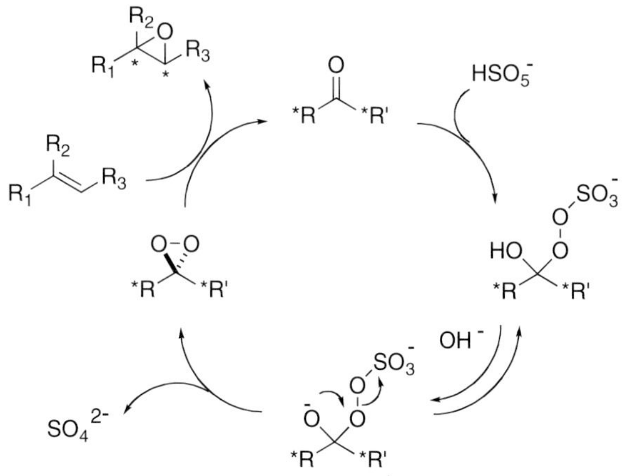

Scheme 3.1 Chiral ketone-catalyzed asymmetric epoxidation of alkenes.

Modern Oxidation Methods. Edited by Jan-Erling Bäckvall
Copyright © 2010 WILEY-VCH Verlag GmbH & Co. KGaA, Weinheim
ISBN: 978-3-527-32320-3

3.2 Catalyst Development

Dioxiranes are generated from an oxidant such as Oxone ($2KHSO_5{\cdot}KHSO_4{\cdot}K_2SO_4$) and ketones, and can be used *in situ* (Scheme 3.1) [11, 12]. While one may expect that asymmetric epoxidation could be possible using a chiral ketone as catalyst, it has been quite challenging to develop an effective catalyst with a broad substrate scope. As summarized in Schemes 3.2–3.7 , many types of chiral ketones of varying structures have been investigated and reported by various laboratories [13–44].

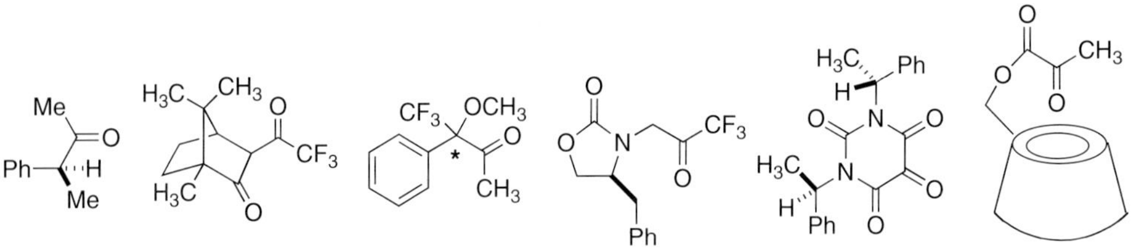

1 (ref. 13a) **2** (ref. 13b) **3** (ref. 13b) **4** (ref. 35b) **5** (ref. 36) **6** (ref. 43a)

Scheme 3.2 Selected examples of ketones with an attached chiral moiety.

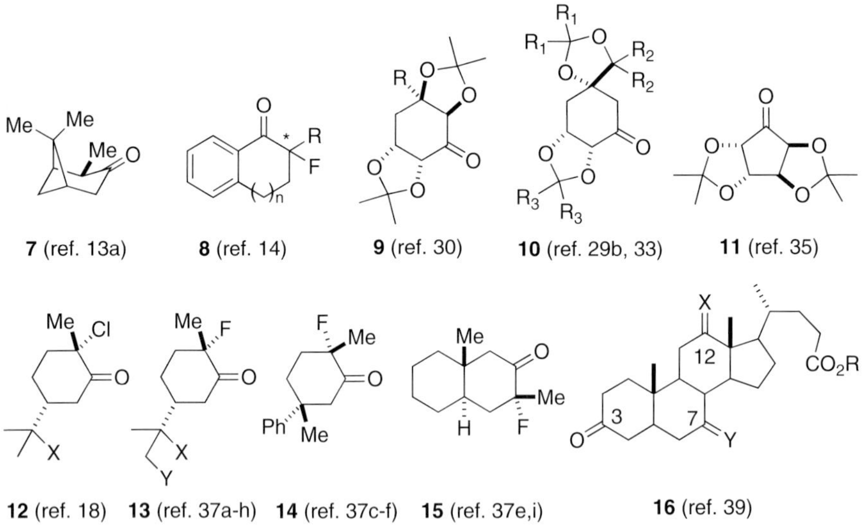

7 (ref. 13a) **8** (ref. 14) **9** (ref. 30) **10** (ref. 29b, 33) **11** (ref. 35)

12 (ref. 18) **13** (ref. 37a-h) **14** (ref. 37c-f) **15** (ref. 37e,i) **16** (ref. 39)

Scheme 3.3 Selected examples of carbocyclic ketones.

Trans and trisubstituted alkenes conjugated with aromatic groups appear to be more effective substrates, likely because of the large size of aromatic substituents, thus providing more efficient stereodifferentiation. A number of chiral ketones can provide good to high *ees* for this class of alkenes (Table 3.1). However, trans and

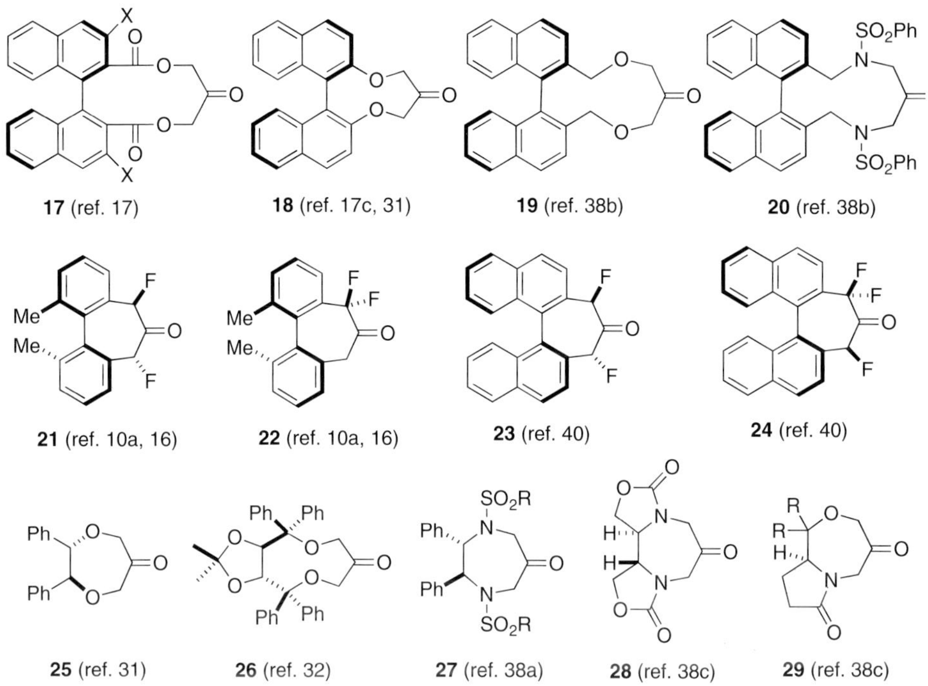

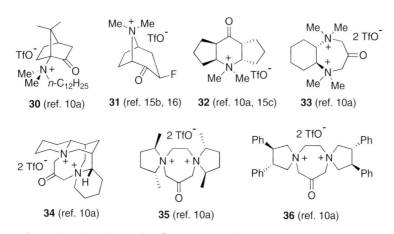

Scheme 3.4 Selected examples of C₂-symmetric binaphthyl-based and related ketones.

Scheme 3.5 Selected examples of ammonium and bis(ammonium) ketones.

trisubstituted alkenes without conjugated aromatic substituents as well as cis and terminal alkenes have generally proven to be extremely challenging (Scheme 3.8), and very few chiral ketones are effective for these alkenes.

Fructose-derived ketone **42** has been shown to be a highly general ketone catalyst (typically 20–30 mol% used) for epoxidation of a wide variety of trans and trisubsti-

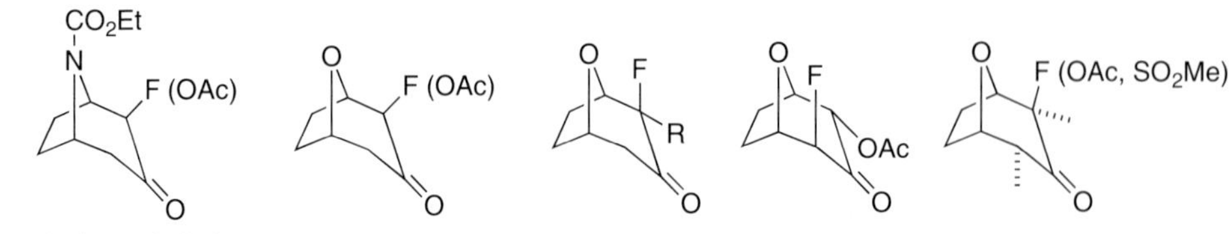

88 | *3 Organocatalytic Oxidation. Ketone-Catalyzed Asymmetric Epoxidation of Alkenes*

37 (ref. 34a,b,d,e)   **38** (ref. 34b,c,g)   **39** (ref. 34g)   **40** (ref. 34g)   **41** (ref. 41)

Scheme 3.6 Selected examples of bicyclo[3.2.1]octan-3-ones and related ketones.

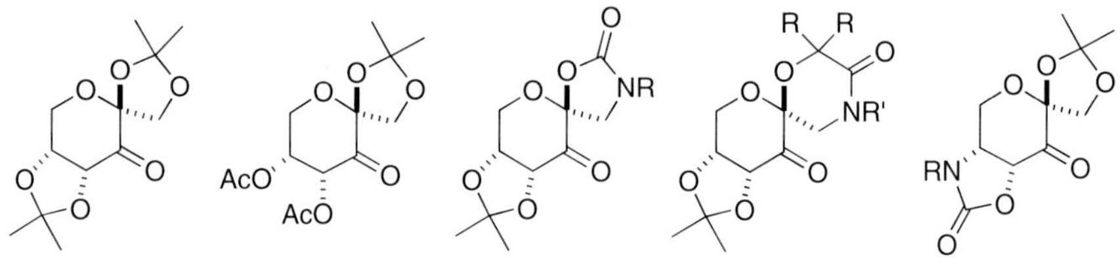

42 (ref. 20, 21)   **43** (ref. 22)   **44** (ref. 24)   **45** (ref. 27)   **46** (ref. 28)

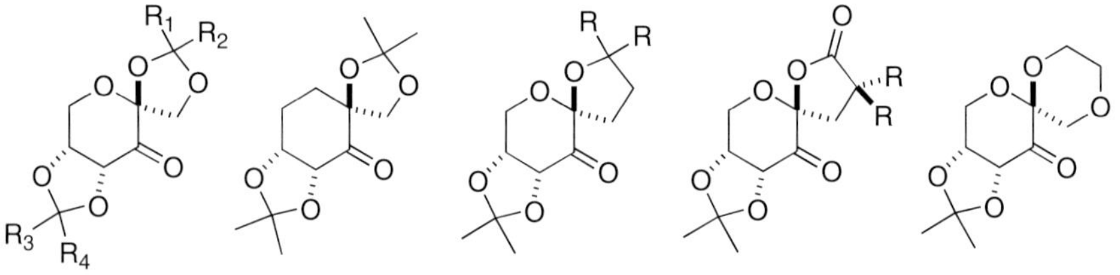

47 (ref. 29a)   **48** (ref. 29b)   **49** (ref. 29c)   **50** (ref. 29c)   **51** (ref. 29a)

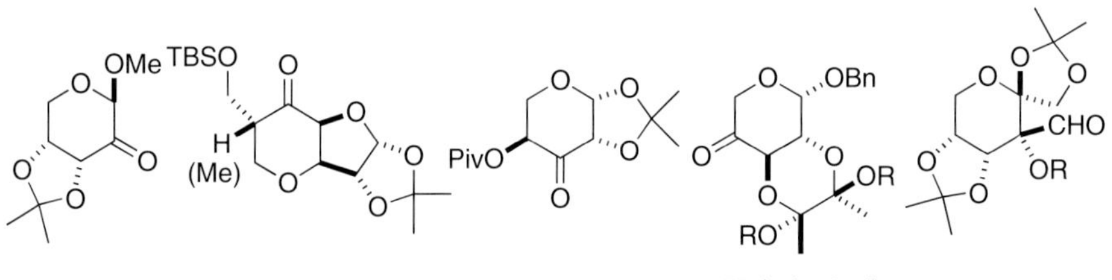

52 (ref. 29a)   **53** (ref. 42a)   **54** (ref. 42b)   **55** (ref. 42c-f)   **56** (ref. 44)

Scheme 3.7 Selected examples of carbohydrate-based and related ketones.

tuted alkenes with or without aromatic substituents [20c], including vinyl silanes [20i], hydroxyalkenes [20d], enol silyl ethers and esters [20g,j,l,45], conjugated dienes [20e], and enynes (Scheme 3.9) [20f,h]. A reliable transition state model has been established for epoxidation with ketone **42**. The epoxidation of trans and trisubstituted alkenes proceeds mainly through spiro **A**, with planar **B** competing (Figure 3.1) [20]. The competition between spiro **A** and planar **B** is dependent on the substituents on the alkenes [10d,h,20,25]. In general, transition state spiro **A** is favored by decreasing the size of R_1 and/or increasing the size of R_3, thus giving high

Table 3.1 Representative Asymmetric Epoxidation of Aryl Alkenes with Chiral Ketones.

Entry	Ketone	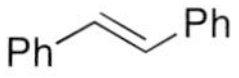 Ph–Ph	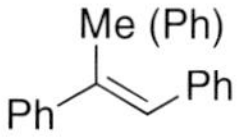 Me(Ph)/Ph,Ph	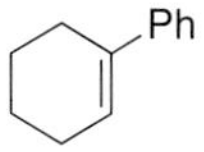 Ph-cyclohexene	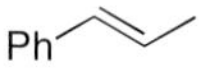 Ph~	Ref.
1	1–3				9.5–18% ee	13a,b
2	4	25% ee	31–33% ee	34% ee		35b
3	6	36% ee	35% ee		26% ee	43a
4	7				12.5% ee	13a
5	9 (R=CMe$_2$OH)	96% ee	92% ee	85% ee	80% ee	30
6	10	32–85% ee	25–85% ee		38–46% ee	29b,33
7	12	42–87% ee				18
8	13, 14	60–90% ee				37a–h
9	15	86% ee			70% ee	37e,i
10	16	43–98% ee	20–85% ee		15–78% ee	39
11[a]	17	32–84% ee	67–81% ee	64–71% ee		17c
12	18	12–26% ee			29% ee	17c,31a
13	21, 22	94% ee		59% ee	85–88% ee	10a,16
14	23, 24				83–86% ee	40
15	25	59% ee			20% ee	31a
16	26	65% ee	81% ee			32
17	27	20–30% ee				38a
18	28, 29	6–64% ee	70–73% ee	78–82% ee	57–62% ee	38c
19	31	58% ee		7% ee	35% ee	16
20	32			58% ee	34% ee	10a
21	33–36				8–40% ee	10a
22	37 (F)	76% ee	73–83% ee	69% ee	56% ee	34d
23[b]	37 (OAc), 38	83–93% ee$_{max}$	98% ee$_{max}$	82% ee$_{max}$	70% ee$_{max}$	34b–d,g
24	41 (F)	68% ee	66% ee		34% ee	41
25	42	98% ee	96–97% ee	98% ee	95% ee	20c
26	53	11–71% ee	47% ee		29% ee	42a
27	54	67% ee	43–68% ee		40% ee	42b
28	55	67–83% ee	80–90% ee	60–85% ee	69–72% ee	42d
29	56	64–94% ee	53–92% ee	48–67% ee		44

a) Up to 95% ee obtained for 4,4′-substituted stilbenes.
b) ee$_{max}$ (100 × product ee/ketone ee).

enantioselectivity for the epoxidation [20c]. Spiro **A** is also favored by conjugating groups such as phenyl, alkenyl, alkynyl, and so on. These groups enhance the stabilizing secondary orbital interaction in the spiro transition state by lowering the energy of the π^* orbital of the reacting alkene [46, 47].

Ketones **43** and **46** (Scheme 3.7), obtained by replacing the fused ketal of **42** with more electron-withdrawing diacetates and an oxazolidinone respectively, show significantly enhanced reactivity. Only 1–5 mol% of ketone **46** is required to achieve good conversions for the epoxidation of a variety of trans and trisubstituted alkenes with high ees [28]. A variety of α,β-unsaturated esters can be epoxidized with ketone **43**

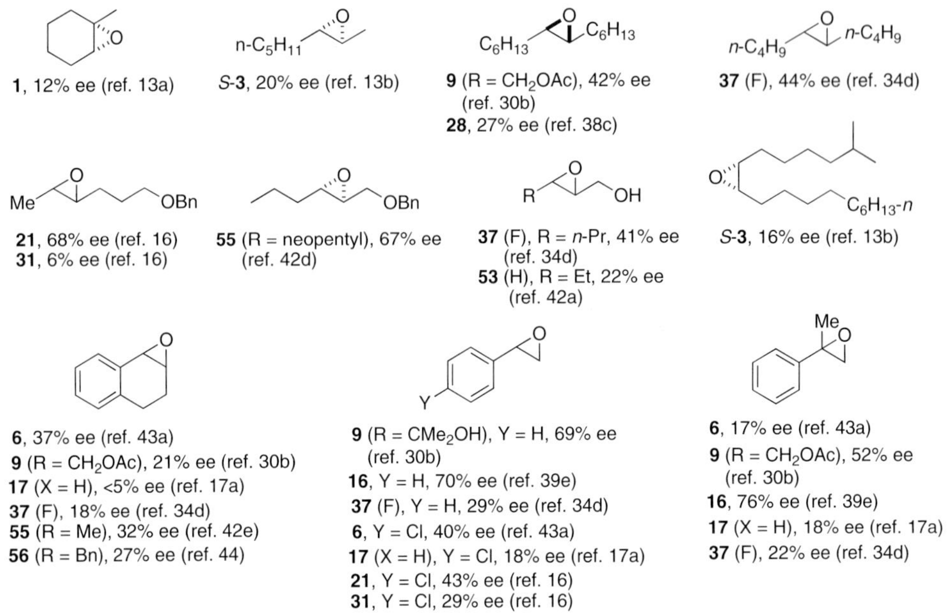

Scheme 3.8 Examples of epoxidation of alkenes with chiral ketones.

in good yields and high *ees* (Scheme 3.10). Interestingly, high *ees* have also been obtained for some *cis*-alkenes with **43** (Scheme 3.10) [22,23c].

In efforts to extend the asymmetric epoxidation to other types of alkenes, glucose-derived oxazolidinone ketone **44** was developed. In initial studies [24a–d], Boc-ketone **44a** was found to be highly enantioselective for various *cis*-alkenes, with no isomerization being observed in the epoxidation of acyclic systems (Scheme 3.11) [24a,c]. Certain terminal alkenes were also epoxidized in encouragingly high *ees* (Scheme 3.11) [24b,c]. It appears that the stereodifferentiation for cis and terminal alkenes with **44a** probably results from an apparent attractive interaction between the R_π group of the alkene and the oxazolidinone moiety of the ketone catalyst in spiro transition state **C** (Figure 3.2) [24a–c]. Further studies show that a carbocyclic analog of ketone **44a** provides higher *ees* for styrenes (89–93% *ee*) than **44a** [25]. The replacement of the pyranose oxygen with a less electronegative carbon raises the energy of the nonbonding orbital of the dioxirane and consequently enhances the stabilizing interaction of the nonbonding orbital of the dioxirane with the π^* orbital of the alkene, thus further favoring the spiro transition state over the planar one [25].

Encouraged by the enantioselectivity obtained with ketone **44a**, a series of N-aryl-substituted ketones (examples shown in Scheme 3.12) were investigated to further understand the N-substituent effect on the epoxidation and to search for more practical ketone catalysts [24e,i]. Ketones such as **44b,c** can be readily prepared in four steps in large quantities from inexpensive glucose and anilines [24l].

Like ketone **44a**, these N-aryl-substituted ketones have also been found to be highly effective for a variety of alkenes, including *cis*-β-methylstyrenes [24f], chromenes

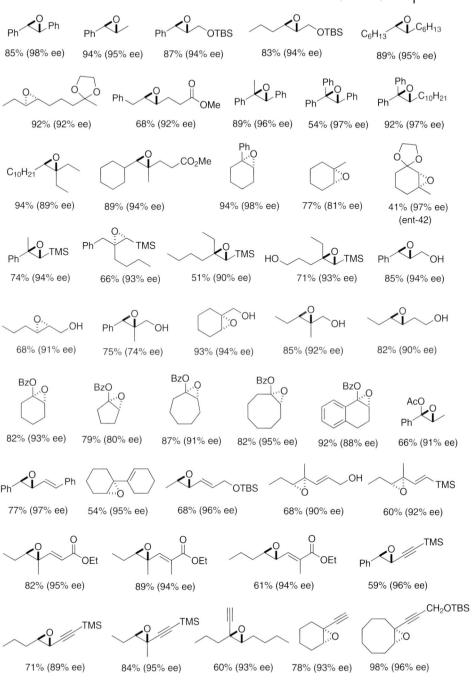

Scheme 3.9 Selected examples of epoxidation of trans and trisubstituted alkenes with ketone **42**.

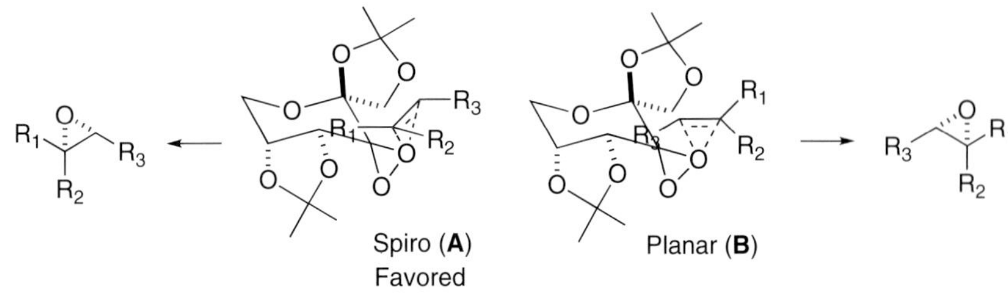

Figure 3.1 The spiro and planar transition states for epoxidation with ketone **42**.

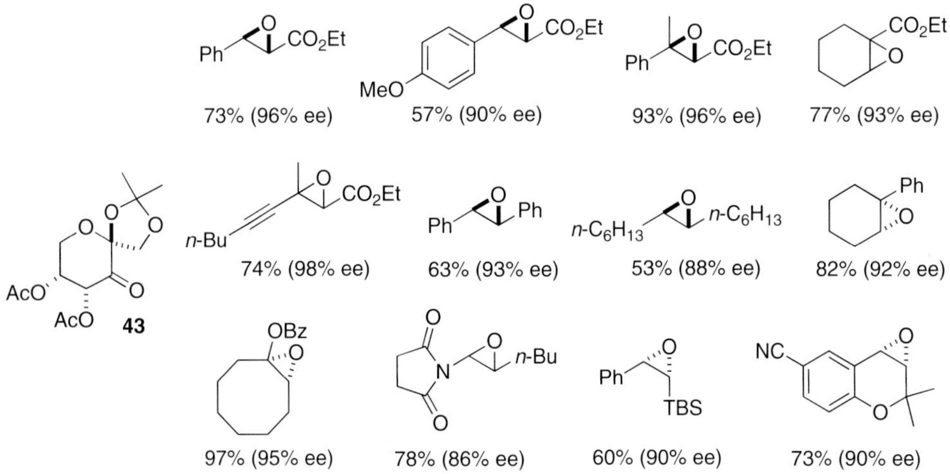

Ph─epoxide─CO₂Et: 73% (96% ee)

MeO-C₆H₄─epoxide─CO₂Et: 57% (90% ee)

Ph─epoxide─CO₂Et (trisubstituted): 93% (96% ee)

cyclohexene oxide─CO₂Et: 77% (93% ee)

Ketone **43** (with AcO, AcO, acetonide groups)

n-Bu─alkyne epoxide─CO₂Et: 74% (98% ee)

Ph─epoxide─Ph: 63% (93% ee)

n-C₆H₁₃─epoxide─n-C₆H₁₃ (trans): 53% (88% ee)

Ph-fused cyclohexene oxide: 82% (92% ee)

cyclooctene oxide with OBz: 97% (95% ee)

succinimide-N-CH₂-epoxide-n-Bu: 78% (86% ee)

Ph─epoxide─TBS: 60% (90% ee)

NC-chromene oxide: 73% (90% ee)

Scheme 3.10 Epoxidation of alkenes with ketone **43** (0.1–0.3 equiv.).

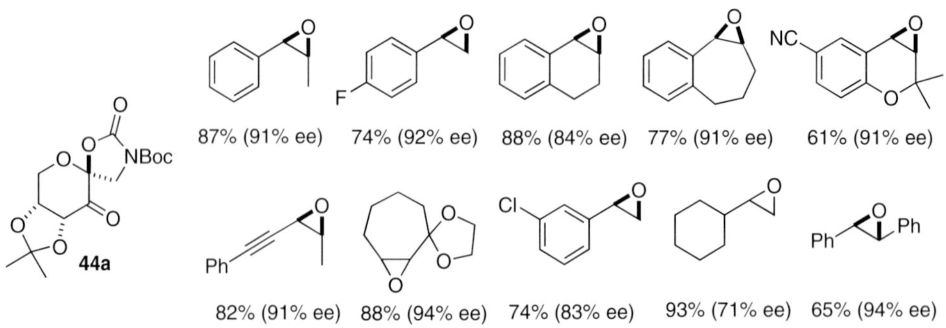

Ketone **44a** (with NBoc group)

styrene oxide: 87% (91% ee)

4-F-styrene oxide: 74% (92% ee)

dihydronaphthalene oxide: 88% (84% ee)

benzocycloheptene oxide: 77% (91% ee)

NC-chromene oxide: 61% (91% ee)

Ph─alkyne epoxide: 82% (91% ee)

spiro dioxolane epoxide: 88% (94% ee)

Cl-phenyl epoxide: 74% (83% ee)

3-Cl-phenyl epoxide: 93% (71% ee)

cyclohexyl epoxide: 65% (94% ee)

Ph─epoxide─Ph (cis): (shown)

Scheme 3.11 Epoxidation of alkenes with ketone **44a** (0.15–0.30 equiv.).

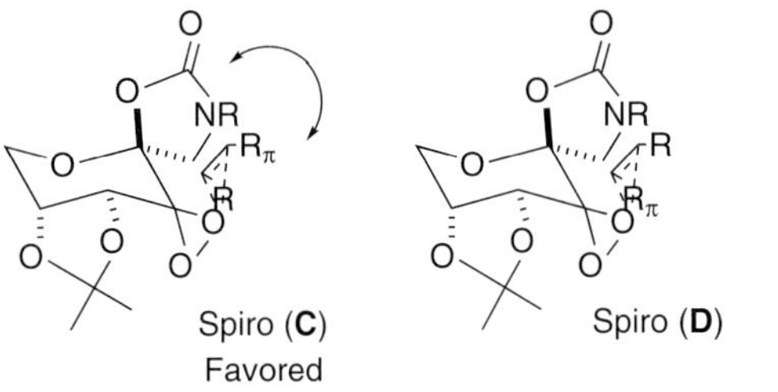

Figure 3.2 Competing transition states for the epoxidation with ketone 44.

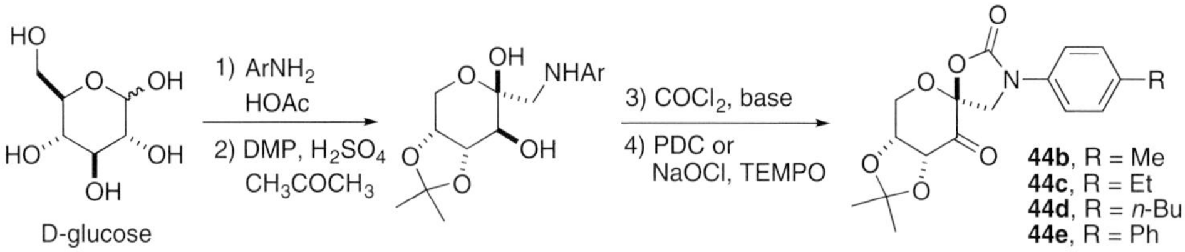

Scheme 3.12 Synthesis of N-aryl-substituted ketones.

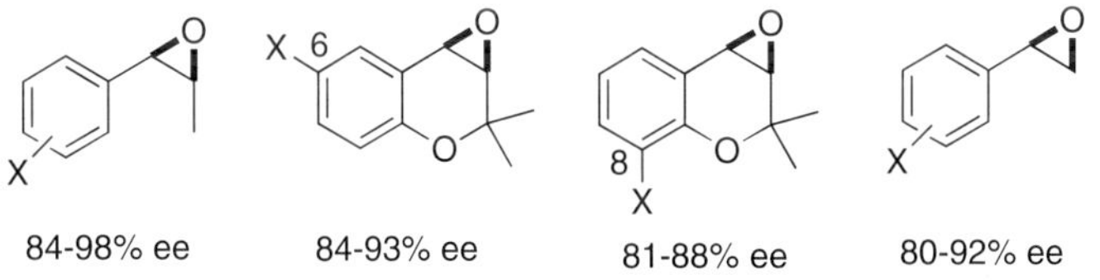

44b, R = Me
44c, R = Et
44d, R = n-Bu
44e, R = Ph

84–98% ee   84–93% ee   81–88% ee   80–92% ee

Scheme 3.13 Epoxidation of cis-β-methylstyrenes, chromenes, and styrenes with ketones **44b,c**.

[24j], and styrenes [24i] (Scheme 3.13). For cis-β-methylstyrenes, the substituents on the phenyl group of the alkene (either electron-donating or electron-withdrawing group) increased the ee of the epoxidation, which suggests that these substituents have additional beneficial non-bonding interactions with the phenyl group of the ketone catalyst (van der Waals forces and/or hydrophobic interactions). As a result, the spiro transition state **E** is further favored and the enantioselectivity of the epoxidation is increased (Figure 3.3) [24f]. Such interactions are also supported by

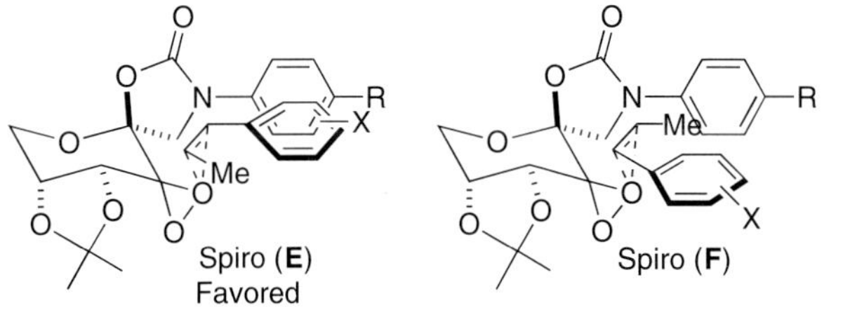

Figure 3.3 Two reacting approaches for the epoxidation of cis-beta-methylstyrene with ketone 44.

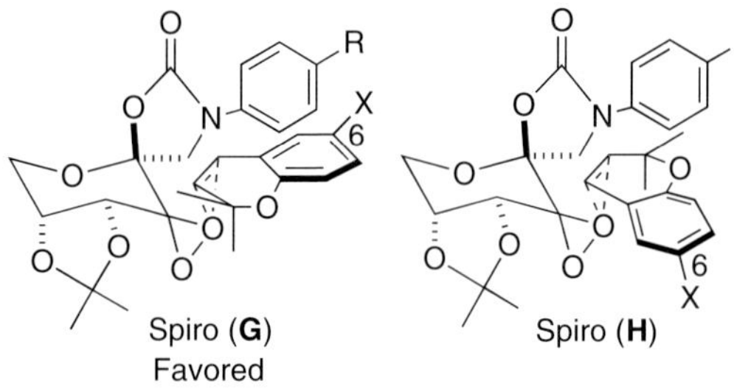

Figure 3.4 Two reacting approaches for the epoxidation of 6-subsituted chromenes with ketone 44.

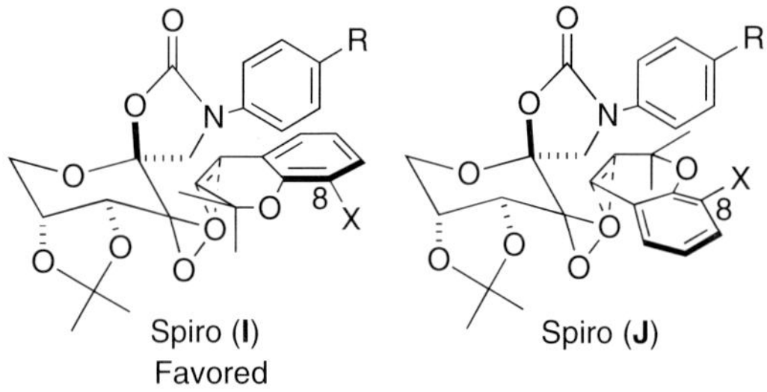

Figure 3.5 Two reacting approaches for the epoxidation of 8-subsituted chromenes with ketone 44.

the observation that 6-substituted 2,2-dimethylchromenes give higher *ee*s than 8-substituted ones (Scheme 3.13). The substituent at the 6-position of the chromenes is in the proximity of the phenyl group of the catalyst in spiro transition state **G** and can cause the aforementioned nonbonding interactions, thus further favoring this transition state (Figure 3.4). However, such an interaction is not possible for 8-substituted chromenes since the substituents are distal to the phenyl group of the catalyst in spiro transition state **I** (Figure 3.5) [24j].

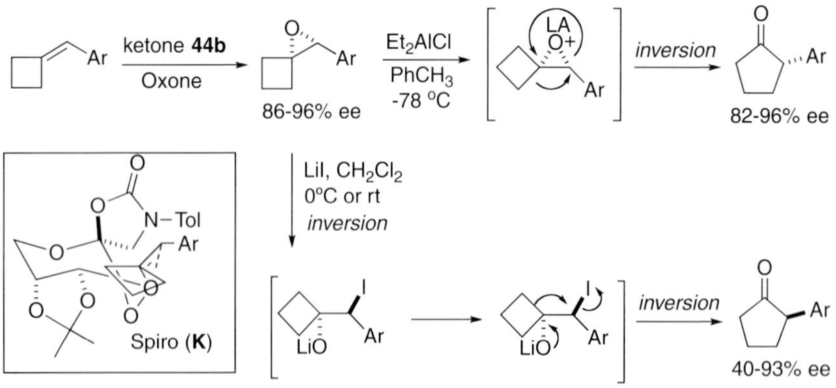

Scheme 3.14 Epoxidation with ketone **44b** and subsequent epoxide rearrangement.

3.2 Catalyst Development

The attractive interaction between the aryl group of the alkene and the oxazolidinone of the ketone catalyst makes trisubstituted benzylidenecyclobutanes effective substrates for the epoxidation mainly via spiro transition state **K** (Scheme 3.14) [24g]. The epoxides are obtained in high *ee*s and can be rearranged to optically active 2-aryl cyclopentanones using Et$_2$AlCl or LiI with either inversion or net retention of configuration, respectively. Optically active 2-alkyl-2-aryl cyclopentanones are also obtained in 70–90% *ee* from tetrasubstituted benzylidenecyclobutanes after epoxide rearrangement (Scheme 3.15) [24h]. When benzylidenecyclopropanes are used for the epoxidation, optically active γ-aryl-γ-butyrolactones are obtained in 71–91% *ee* and reasonable yields via sequential epoxidation, epoxide rearrangement, and Baeyer-Villiger oxidation (Scheme 3.16) [24k,48]. If more ketone catalyst and less Oxone are used to suppress the Baeyer-Villiger oxidation, chiral cyclobutanones can also be isolated [24k,48].

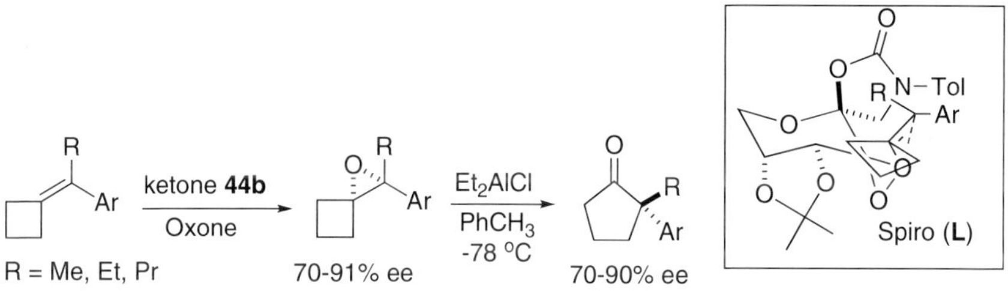

Scheme 3.15 Epoxidation with ketone **44b** and subsequent epoxide rearrangement.

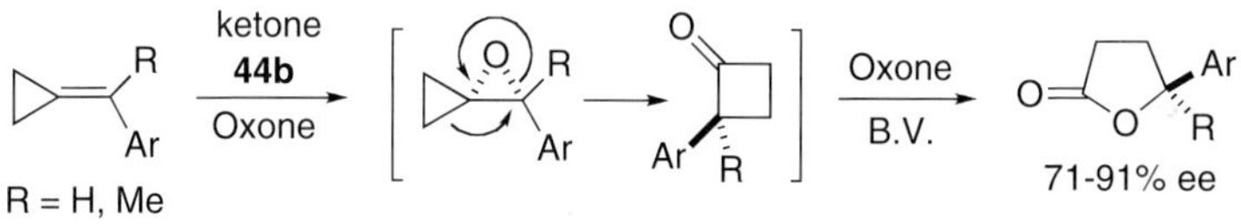

Scheme 3.16 Epoxidation with ketone **44b**, subsequent epoxide rearrangement, and Baeyer-Villiger oxidation.

Conjugated *cis*-dienes [24m] and *cis*-enynes [24n] are found to be effective substrates, giving *cis*-epoxides in high *ee*s with no isomerization observed during the epoxidation (Scheme 3.17). Alkenes and alkynes prefer to be in the proximity of the oxazolidinone of the catalyst in the transition state, likely due to attractive interactions with the oxazolidinone (Figures 3.6 and 3.7). It appears that hydrophobic interactions between the substituents on the diene and enyne and the oxazolidinone moiety of the ketone catalyst (possibly N-aryl group) also have significant influence on the enantioselectivity. High enantioselectivity can be obtained for nonconjugated *cis*-alkenes if the difference in hydrophilicity between the two alkene substituents is large. For example, the corresponding lactone is obtained in 91% *ee* when *cis*-dec-4-enoic acid is subjected to the epoxidation conditions (Scheme 3.18) [49]. As with ketone **42** [26a,b], epoxidation with ketone **44c** can also employ H$_2$O$_2$ as the primary oxidant instead of Oxone [26c].

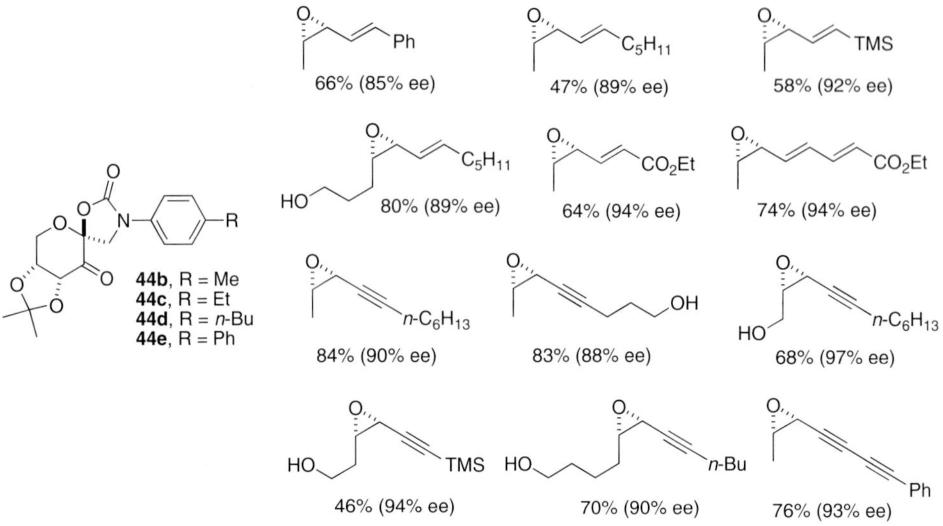

Scheme 3.17 Epoxidation of cis-dienes and cis-enynes with ketone **44b–e** (0.1–0.3 equiv.).

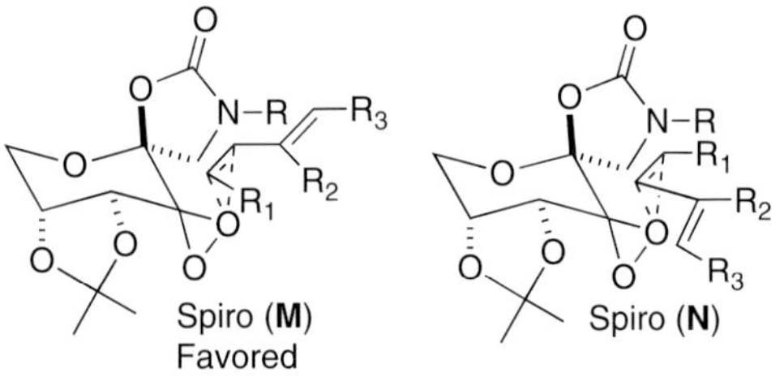

Figure 3.6 Two reacting approaches for the epoxidation of cis-dienes with ketone 44.

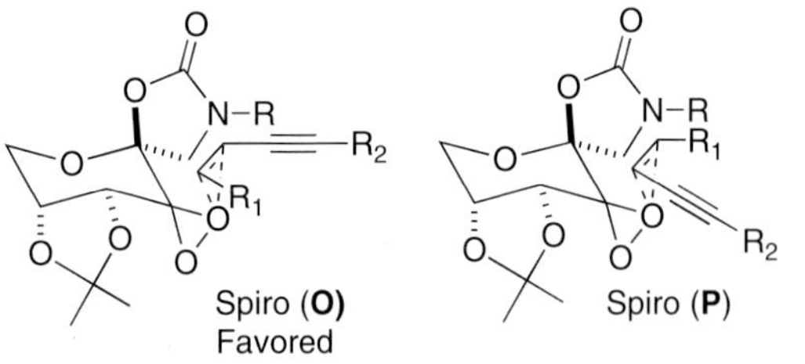

Figure 3.7 Two reacting approaches for the epoxidation of cis-enyes with ketone 44.

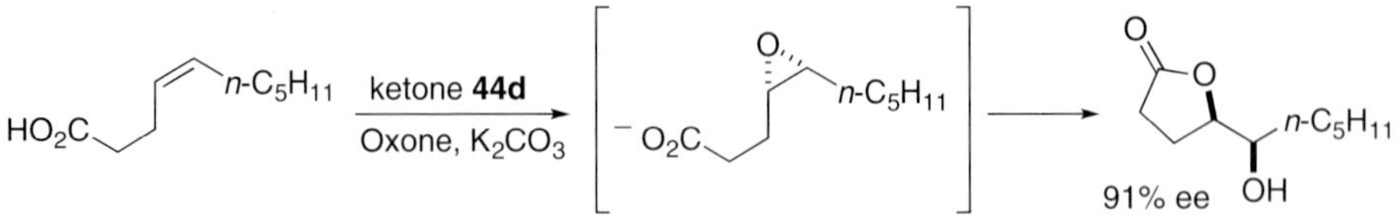

Scheme 3.18 Epoxidation of cis-dec-4-enoic acid with ketone **44d**.

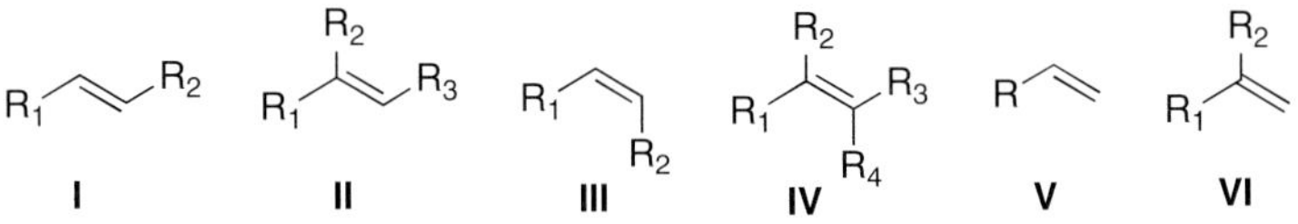

3.2 Catalyst Development

Figure 3.8 Six classes of alkenes for the assymmetric epoxidation.

In the search for ketone catalysts for asymmetric epoxidation of 1,1-disubstituted terminal alkenes (**VI**) (Figure 3.8), morpholinone ketone **45a** was found to be a promising catalyst with up to 88% *ee* obtained (Scheme 3.19) [27a]. It appears that the epoxidation proceeds mainly via planar transition state **Q** because of the attraction between the phenyl group of the alkene and the six-membered morpholinone moiety (Figure 3.9). Good *ee*s are also obtained for some cis and trisubstituted alkenes with **45a** (Scheme 3.19) [27a]. In an effort to search for ketone catalysts with broader substrate scope, ketone **45b** was designed to combine the steric features of ketone **42** and electronic features of ketone **45a** [27b]. Ketone **45b** is indeed found to be a highly effective catalyst for trans and trisubstituted alkenes (Scheme 3.20). However, ketone **45b** gives lower *ee*s than **45a** for 1,1-disubstituted and cis alkenes (Scheme 3.20). The methyl groups introduced onto the morpholinone moiety significantly influence the competition among various spiro and planar transition states. These results provide useful insights for designing catalysts with broader scope in the future. As illustrated in Schemes 3.9–3.11 and 3.13–3.20, carbohydrate-derived ketones such as **42–45** are effective for a wide range of alkenes including trans, trisubstituted, cis, tetrasubstituted, and terminal alkenes (**I–VI**) (Figure 3.8). The substrate scope will certainly be further expanded with the development of new catalysts.

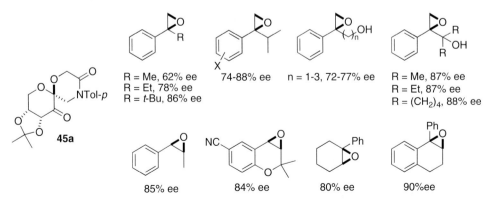

Scheme 3.19 Epoxidation of alkenes with ketone **45a**.

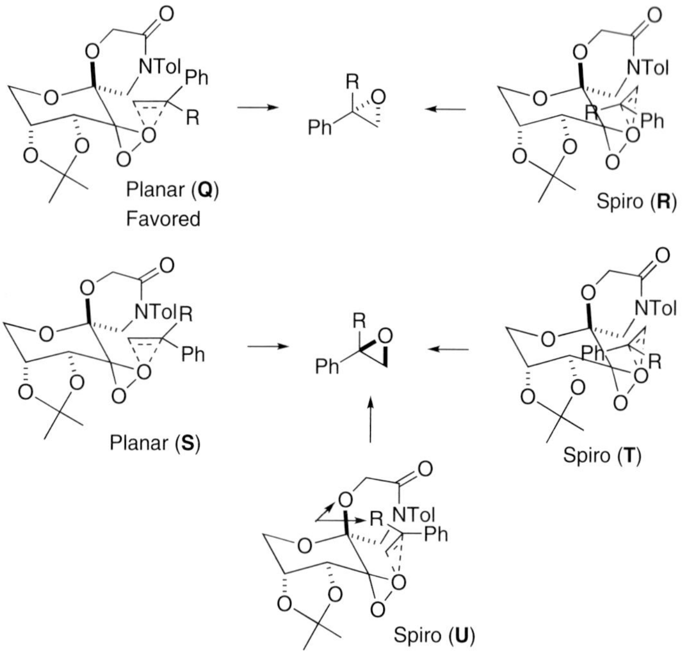
NTol
Ph
R
Planar (Q)
Favored
Spiro (R)
Planar (S)
Spiro (T)
Spiro (U)

Figure 3.9 Possible competing transition states for epoxidation with ketone **45a**.

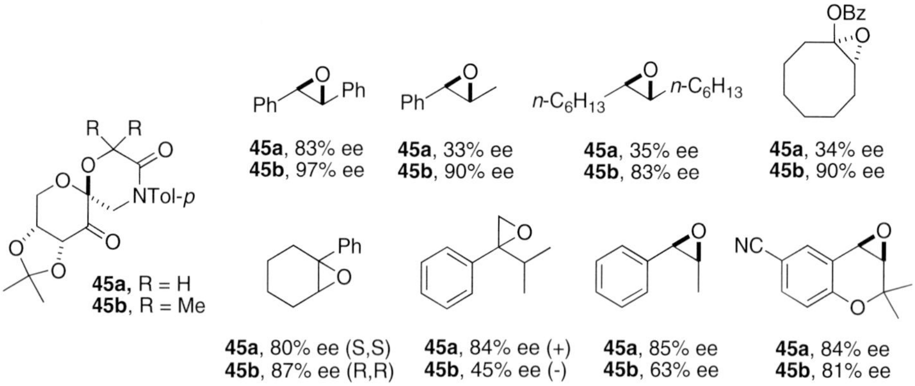
NTol-p
45a, R = H
45b, R = Me
45a, 83% ee
45b, 97% ee
45a, 33% ee
45b, 90% ee
45a, 35% ee
45b, 83% ee
45a, 34% ee
45b, 90% ee
45a, 80% ee (S,S)
45b, 87% ee (R,R)
45a, 84% ee (+)
45b, 45% ee (-)
45a, 85% ee
45b, 63% ee
45a, 84% ee
45b, 81% ee

Scheme 3.20 Comparison of epoxidation with ketones **45a** and **45b**.

3.3 Synthetic Applications

Ketone **42** is readily available from fructose and is a highly effective epoxidation catalyst. Various applications of this ketone have been reported, and some of them are highlighted hereafter.

Epoxides are frequently contained in biologically active and medicinally significant compounds. For example, cryptophycin 52 (**60**) possesses potent antitumor activities and has undergone advanced clinical studies for tumor treatment (Scheme 3.21) [50]. One of the synthetic transformations involves a stereoselective epoxidation. Among various epoxidation systems examined, Moher and coworkers at Eli Lilly reported that good diastereoselectivity can be obtained by the epoxidation with ketone **42** (Scheme 3.21) [50].

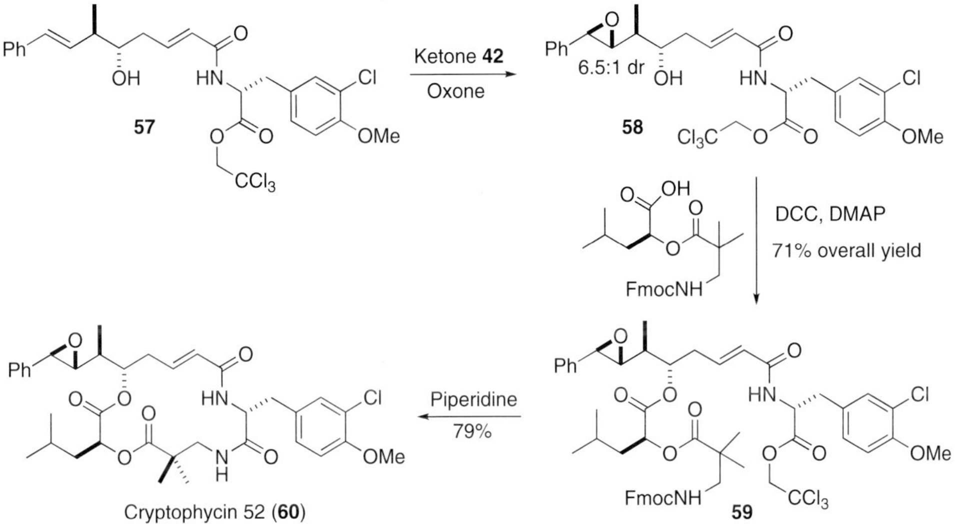

Scheme 3.21 Synthesis of cryptophycin 52 (**60**).

Epothilone is a highly potent anticancer agent. In their synthesis of highly potent epothilone analogs [51], Altmann and coworkers reported that alkene precursor **61** was stereoselectively epoxidized with ketone ***ent*-42** to give compounds **62a,b** (Scheme 3.22). It appears that the epoxidation is compatible with several functional groups such as the benzimidazole in **61**, and the existing ketone in the macrocycle did not appear to interfere with the epoxidation [51, 52].

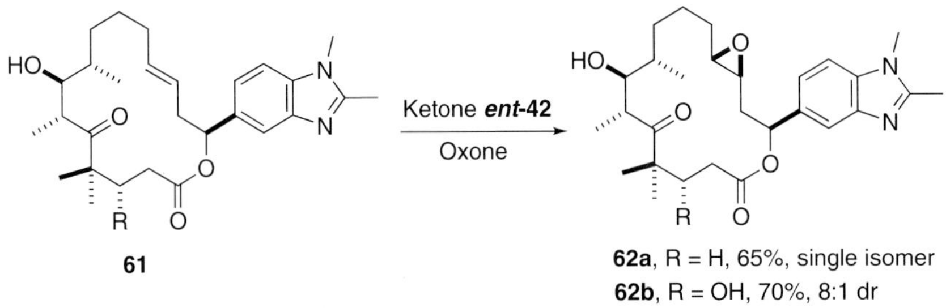

Scheme 3.22 Synthesis of epothilone analogs **62a,b**.

In their synthesis of pladienolide B (**65**) (potent antitumor agent) (Scheme 3.23), Kotake and coworkers recently reported that sulfone **63** was epoxidized with ketone **42** in 71% yield and >99% *de* after recrystallization, and the resulting epoxide was then attached to the macrocyclic ring via Julia-Kocienski alkenation [53].

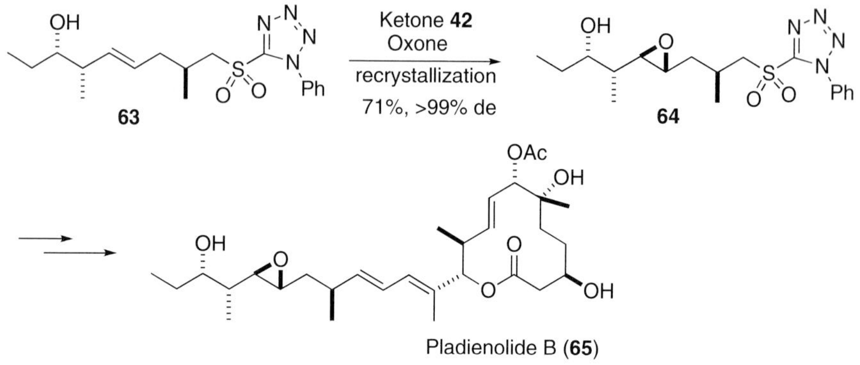

Scheme 3.23 Synthesis of pladienolide B (**65**).

Epoxides can be opened by various nucleophiles, and epoxidation with ketone **42** has been utilized to synthesize various molecules using epoxides as intermediates. For example, McDonald and coworkers reported that epoxide **67**, resulting from the epoxidation of **66** with ketone **42** using hydrogen peroxide, was converted into 1-deoxy-5-hydroxy-sphingolipid analog **68** by stereo- and regioselective opening of the epoxide and subsequent transformations (Scheme 3.24) [54].

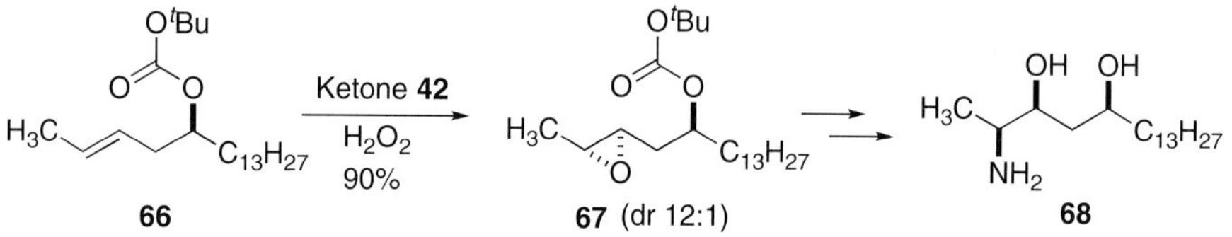

Scheme 3.24 Synthesis of aminodiol **68**.

Marshall and coworkers reported that the internal trisubstituted alkene of **69** was stereoselectively epoxidized with ketone **42** in 80% yield, and the resulting epoxide (**70**) was transformed into the bistetrahydrofuran C17–C32 segment (**72**) of antibiotic ionomycin (Scheme 3.25) [55].

Taber and coworkers reported that crude epoxide **74**, obtained from the epoxidation of **73** with ketone **42**, was regioselectively opened with allylmagnesium chloride to give alcohol **75** in 73% overall yield and 96% *ee*. Alcohol **75** was subsequently converted into (−)-mesembrine (**76**) in five steps (Scheme 3.26) [56].

In their two-directional synthesis of (−)-longilene peroxide, Morimoto and coworkers reported that diene **77** was epoxidized with ketone ***ent*-42** to give tricyclic compound **78** after cyclization. Compound **78** was then transformed into (−)-long-

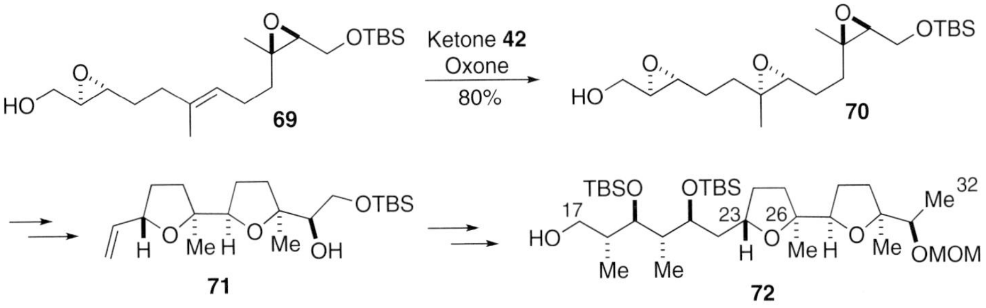

Scheme 3.25 Synthesis of bistetrahydrofuran C17-C32 fragment (**72**).

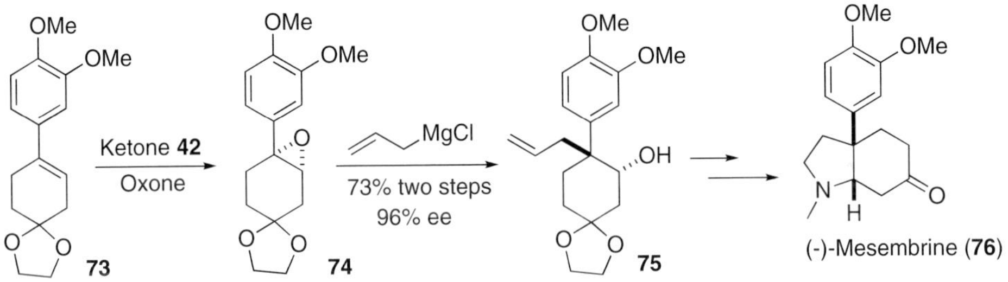

Scheme 3.26 Synthesis of (−)-mesembrine (**76**).

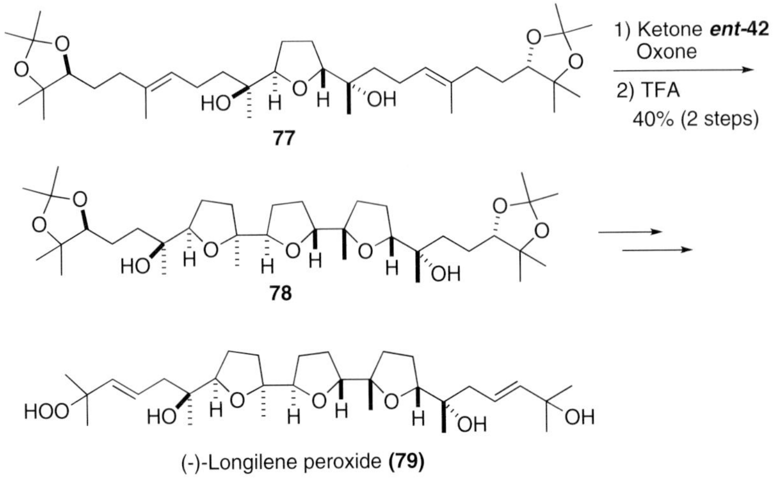

(−)-Longilene peroxide (**79**)

Scheme 3.27 Synthesis of (−)-longilene peroxide (**79**).

ilene peroxide in several steps, with determination of the absolute configuration as well (Scheme 3.27) [57].

In the first total synthesis of cytotoxic bromotriterpene polyether (+)-aurilol (**85**) (Scheme 3.28) [58], Morimoto and coworkers reported that epoxide **81**, resulting from

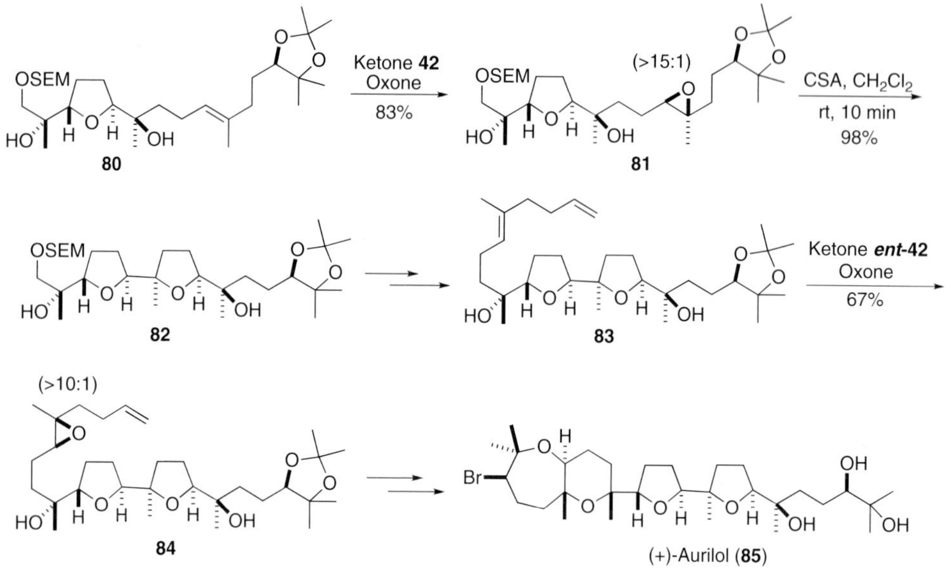

Scheme 3.28 Synthesis of (+)-aurilol (**85**).

the epoxidation of **80** with ketone **42**, underwent acid-catalyzed 5-*exo* cyclization to produce bistetrahydrofuran **82** with the desired stereochemistry. Compound **82** was transformed into diene **83** over several steps. Diene **83** was selectively epoxidized only at the trisubstituted alkene with ***ent*-42** to give epoxide **84**, which was eventually transformed into (+)-aurilol (**85**) [58]. The epoxidation with ketones **42** and ***ent*-42** has also been used in the synthesis and stereochemical assignment of (+)-intricatetraol [59], (+)-enshuol [60], (+)-omaezakianol [61], and other compounds [62, 63].

Ready and coworkers reported that compound **86** was epoxidized with ketone **42** to give (+)-nigellamine A$_2$ (**87**) after acylation with nicotinic acid in 51% yield over two steps (Scheme 3.29) [64]. Among three double bonds present in **86**, the desired C$_7$-C$_8$ double bond was selectively epoxidized with the desired stereochemistry.

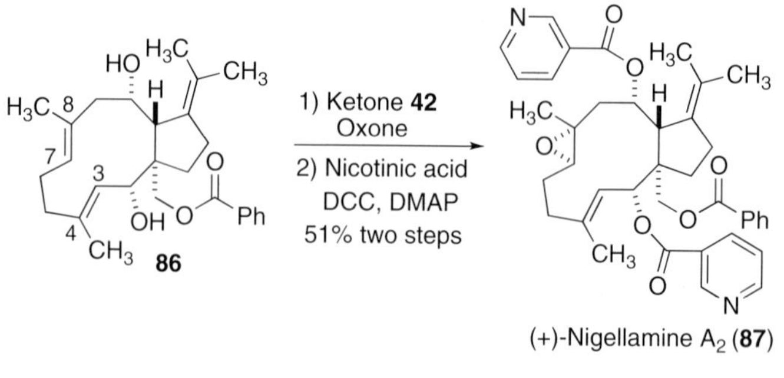

Scheme 3.29 Synthesis of (+)-nigellamine A$_2$ (**87**).

In their synthesis of the proposed glabrescol (**90**) (Scheme 3.30) [65], Kodama and coworkers reported that compound **88** was regio- and stereoselectively epoxidized at the central alkene with ketone **42** to give compound **89** after ring closure. The alkene inductively deactivated with the allylic acetate is less reactive.

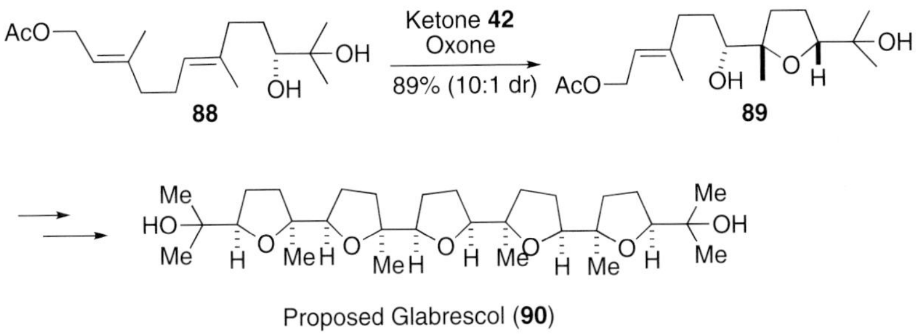

Scheme 3.30 Synthesis of proposed glabrescol (**90**).

In their synthesis of various isomers of annonaceous bis-THF acetogenins, Sinha and coworkers reported that compound **91** containing two trans double bonds can be selectively epoxidized at alkene **a** with ketone **42** to provide mono-THF lactone **93** in 54% overall yield after cyclization with CSA (Scheme 3.31) [66].

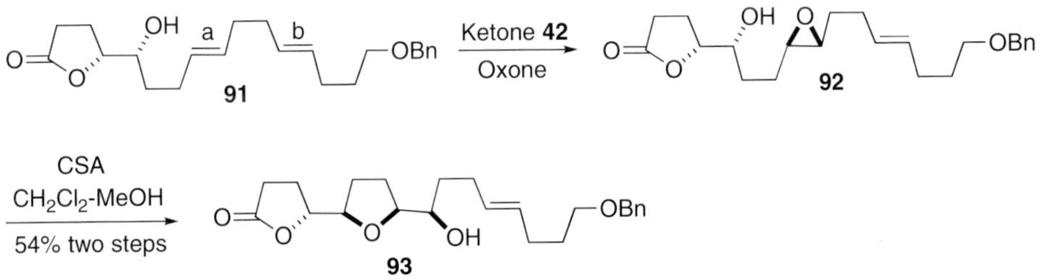

Scheme 3.31 Synthesis of tetrahydrofuran lactone **93**.

In their total synthesis of *ent*-abudinol B (**97**) and *ent*-nakorone (**98**), McDonald and coworkers reported that epoxidation of triene-yne **94** with ketone **42** selectively occurred on the two more electron-rich double bonds, leaving the alkene inductively deactivated by the allylic sulfone group intact (Scheme 3.32). Resulting bis-epoxide **95** was further transformed into *ent*-abudinol B (**97**) and *ent*-nakorone (**98**) [67].

In their second generation synthesis of *ent*-abudinol B (**97**) (Scheme 3.33) [68], farnesyl acetate **99** was selectively epoxidized at the two alkenes with ketone **42** and transformed into compound **101** via bromide **100**. The cyclization of compound **101** with TMSOTf gave tricyclic compound **102**, which was alkenated and preferentially epoxidized at the two trisubstituted alkenes with ketone **42** by careful control of the reaction conditions, leaving the 1,1-disubstituted alkene intact. The cyclization of diepoxide **103** with TMSOTf gave *ent*-abudinol (**97**) in 15% yield along with the alkene isomer and partially cyclized product.

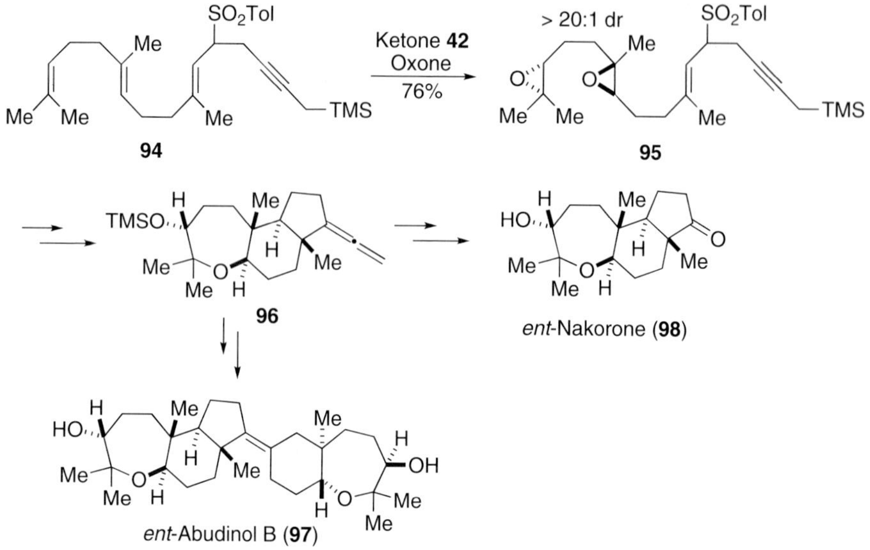

3 Organocatalytic Oxidation. Ketone-Catalyzed Asymmetric Epoxidation of Alkenes

Scheme 3.32 Synthesis of ent-abudinol B (**97**) and ent-nakorone (**98**).

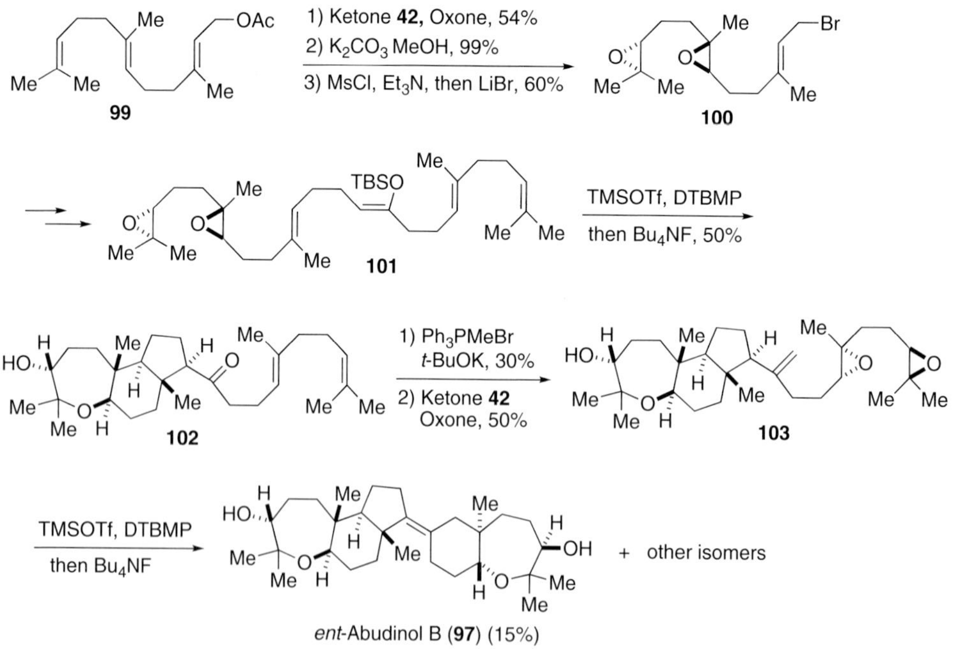

Scheme 3.33 Second generation synthesis of ent-abudinol B (**97**).

Corey and coworkers reported that (R)-2,3-dihydroxy-2,3-dihydrosqualene (**104**) was rapidly converted into proposed glabrescol **90** in 31% overall yield by enantio-selective pentaepoxidation with ketone **42** and subsequent cascade epoxide opening (Scheme 3.34) [69]. In their efforts to determine the correct structure of glabrescol, tetraene **106** was epoxidized with ketone **42**. The resulting epoxide **107** was transformed into a chiral C_2-symmetric pentacyclic oxasqualenoid **109** (Scheme 3.35), which matches the reported isolated natural product glabrescol [70, 71].

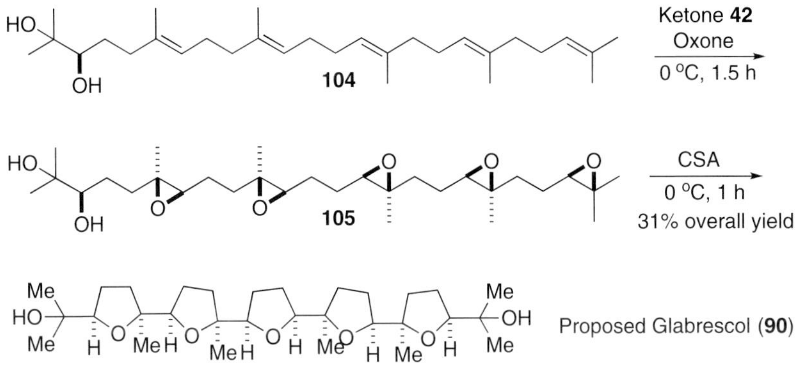

Scheme 3.34 Synthesis of proposed glabrescol (**90**).

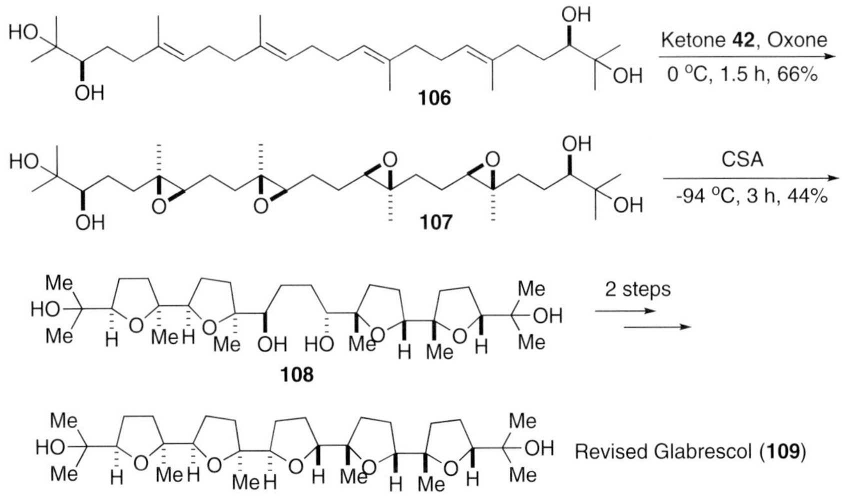

Scheme 3.35 Synthesis of revised glabrescol (**109**).

Kishi and coworkers reported that triene **110** was stereoselectively epoxidized with ketone **42** to give compound **112** in 52% overall yield after epoxide opening with LiAlH$_4$. (+)-Glisoprenin A (**113**) was obtained from **112** in five steps (Scheme 3.36) [72].

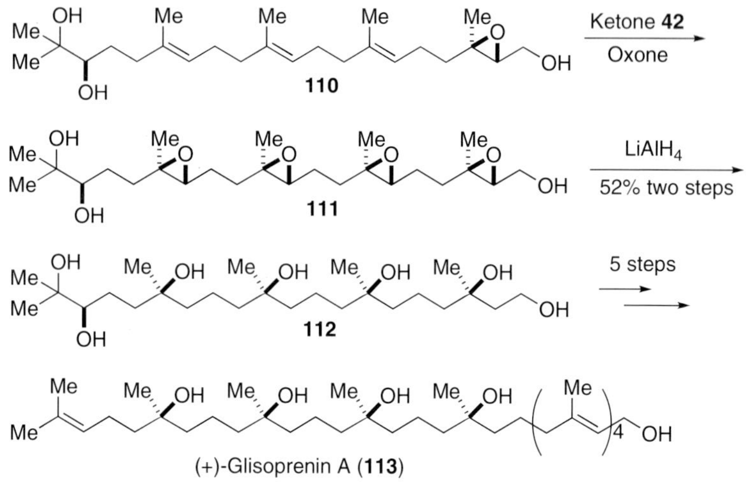

Scheme 3.36 Synthesis of (+)-glisoprenin A (**113**).

McDonald and coworkers reported a series of studies on biomimetic syntheses of fused polycyclic ethers [73, 74]. For example, polyepoxide **115**, resulting from the epoxidation of triene **114** with ketone **42**, was transformed into fused polycyclic ether **116** via the BF_3-Et_2O-promoted cascade *endo*-oxacyclization (Scheme 3.37) [73d].

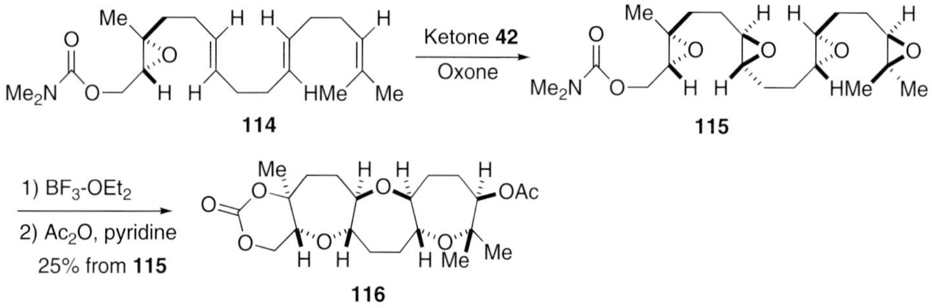

Scheme 3.37 Polyepoxide cyclization.

Jamison and coworkers reported that triepoxide **118**, obtained from the epoxidation of vinylsilane **117** with ketone **42**, cyclized with Cs_2CO_3/CsF in MeOH to give fused tetracyclic tetrahydropyran **119** in 20% overall yield after acetylation (Scheme 3.38) [75]. The $SiMe_3$ group directed the *endo* cyclization and disappeared at the end [76].

More recently, Jamison and coworkers reported that water can act as the optimal reaction medium. Polyepoxide **121** selectively cyclized to form the fused tetrahydropyran rings when the epoxide-opening reactions were carried out in water (Scheme 3.39) [77, 78].

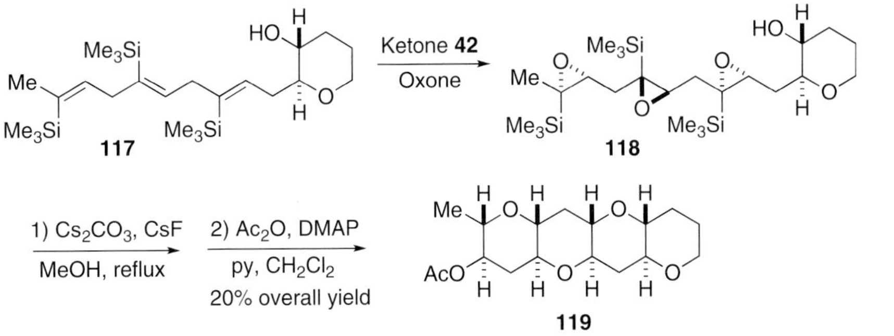

Scheme 3.38 Polyether synthesis via cascade epoxide cyclization.

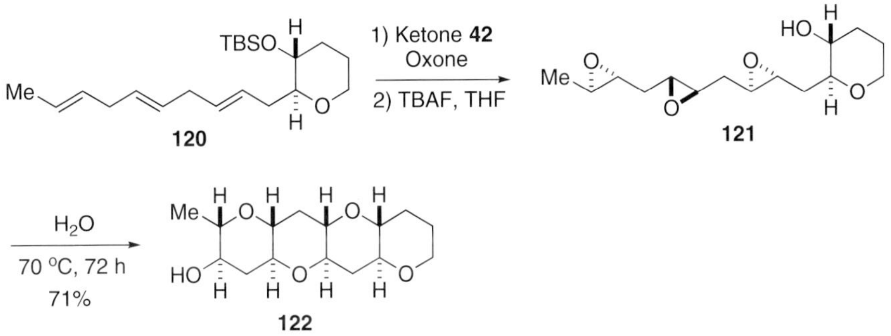

Scheme 3.39 Water-promoted cascade epoxide cyclization.

Many naturally occurring polyethers such as brevitoxin B (**123**) have been proposed to biosynthetically derive from the polyepoxidation of polyenes and subsequent cyclization of the resulting polyepoxides (Scheme 3.40) [79, 80]. Such a biomimetic polyene-polyepoxide-polycyclization approach would provide a potentially powerful and versatile strategy for the synthesis of polyethers, with quick assembly of stereochemically complex molecules from relatively simple achiral polyalkene precursors. The epoxidation with ketone **42** should provide a valuable method to investigate this biosynthetic hypothesis and possible application of the proposed pathway in the synthesis of polyethers.

Epoxidation with ketone **42** was also carried out on pilot-plant scale at DSM Pharma Chemicals. Crude alkene **127**, prepared from 4-pentynoic acid by hydroboration and Suzuki coupling with 3-fluorobenzyl chloride, was epoxidized with ketone **42** to give lactone **128** (59 lb) in around 63% overall yield and 88% *ee* after cyclization, without need for isolation of any intermediates during the entire process (Scheme 3.41) [81].

As illustrated by the above examples and others [82], the epoxidation with ketone **42** has been utilized in various syntheses. The reaction can be applied to rather complex substrates with various functional groups. When more than one alkene is present, certain alkenes can often be selectively epoxidized with carefully controlled reaction

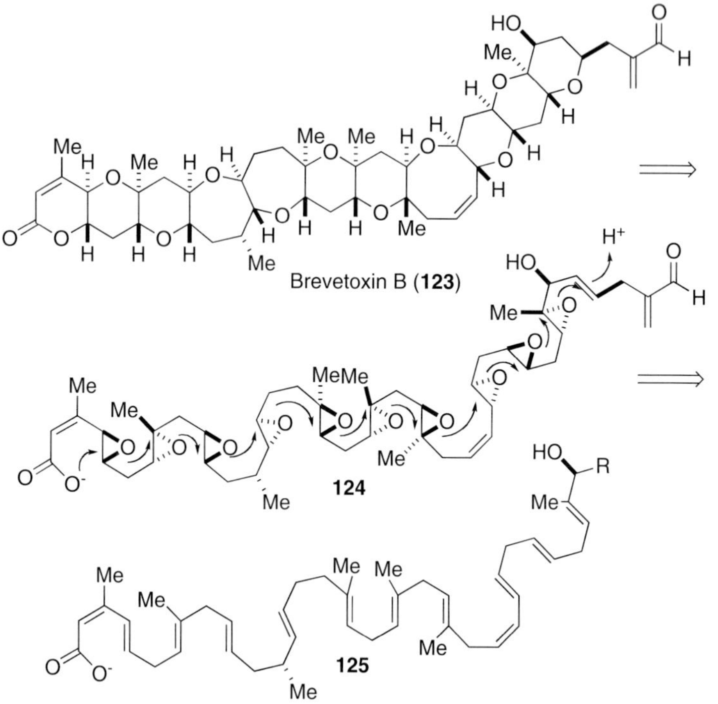

Scheme 3.40 Proposed biosynthetic pathway of brevitoxin B (**123**).

conditions. On the other hand, when a substrate contains several reaction sites, multiple epoxidations can be simultaneously achieved to build molecular complexity rather quickly after the epoxide transformations. In addition, the epoxidation proceeds in a predictable manner, and can be used to establish the stereochemical structure of natural products. Furthermore, the epoxidation is carried out under mild and environmentally friendly conditions requiring easy workup, and has proven to be a viable process on a large scale. The effectiveness and simplicity of this epoxidation make it extremely useful in organic synthesis.

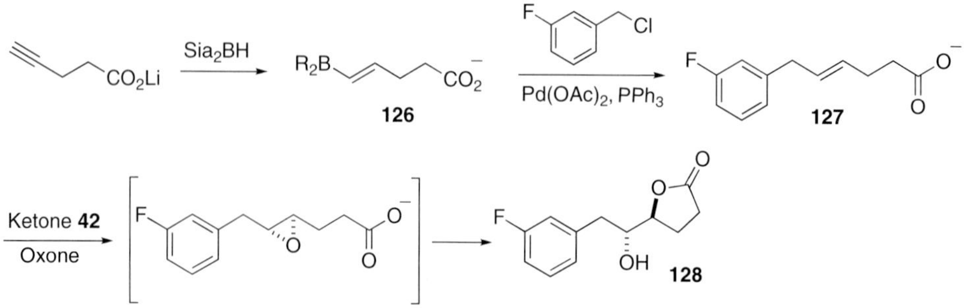

Scheme 3.41 Synthesis of lactone **128** on pilot plant scale.

3.4
Conclusion

Chiral ketone-catalyzed asymmetric epoxidation has received extensive interest, and ketones with a wide variety of structures have been investigated in a number of laboratories. However, the epoxidation reactivity and selectivity are extremely sensitive to the steric and electronic nature of the ketone catalyst, which creates a great challenge to discovering and developing effective catalysts. Very few ketone catalysts have a good a range of substrate scope and are practically useful. The practicality, generality, predictability, and scalability of carbohydrate-derived ketones such as ketone **42** make them an invaluable class of organic catalysts for asymmetric epoxidation.

In contrast to metal-based chiral catalysis, asymmetric catalysis with small, organic molecules did not gain significant growth until the mid-1990s [83]. With the development of the field, various effective systems catalyzed by organic molecules have been discovered and developed. Organic catalysis [84a,b], organocatalysis [84b], or organocatalytic [84b] are phrases commonly used to describe these non-metal-catalyzed reaction processes. Fructose-derived ketone **42** represents an early example of a general and practical organic catalyst. It can effectively epoxidize a wide variety of trans and trisubstituted alkenes without requiring a directing functional group, which has traditionally been challenging for metal catalysts and is so even today. The generality and practicality displayed by fructose-derived ketone **42** was unprecedented for non-metal systems at the time, which opened up a new perspective on simple, small, chiral, organic molecules as viable catalysts for practical asymmetric reactions.

Acknowledgment

The author is grateful to Richard Cornwall, Sunliang Cui, Haifeng Du, Renzhong Fu, Brian Nettles, Thomas Ramirez, Yuehong Wen, O. Andrea Wong (in particular), Baoguo Zhao, and other research group members for their assistance during the preparation of this manuscript.

References

1. For leading reviews, see: (a) Besse, P. and Veschambre, H. (1994) *Tetrahedron*, **50**, 8885; (b) Bonini, C. and Righi, G. (2002) *Tetrahedron*, **58**, 4981.
2. For leading reviews on titanium-catalyzed asymmetric epoxidation of allylic alcohols, see: (a) Johnson, R.A. and Sharpless, K.B. (1993) *Catalytic Asymmetric Synthesis* (ed. I. Ojima), VCH, New York, p. 103; (b) Katsuki, T. and Martin, V.S. (1996) *Org. React.*, **48**, 1; (c) Johnson, R.A. and Sharpless, K.B. (2000) Chapter 6A, in *Catalytic Asymmetric Synthesis* (ed. I. Ojima), VCH, New York.
3. For leading references on vanadium-catalyzed asymmetric epoxidation of allylic alcohols, see: (a) Murase, N., Hoshino, Y., Oishi, M., and Yamamoto, H. (1999) *J. Org. Chem.*, **64**, 338; (b) Hoshino, Y. and Yamamoto, H. (2000) *J. Am. Chem.*

Soc., **122**, 10452; (c) Zhang, W., Basak, A., Kosugi, Y., Hoshino, Y., and Yamamoto, H. (2005) *Angew. Chem. Int. Ed.*, **44**, 4389; (d) Bourhani, Z. and Malkov, A.V. (2005) *Chem. Commun.*, 4592; (e) Malkov, A.V., Bourhani, Z., and Kočovský, P. (2005) *Org. Biomol. Chem.*, **3**, 3194.

4 For leading references on vanadium-catalyzed asymmetric epoxidation of homoallylic alcohols, see: (a) Makita, N., Hoshino, Y., and Yamamoto, H. (2003) *Angew. Chem. Int. Ed.*, **42**, 941; (b) Zhang, W. and Yamamoto, H. (2007) *J. Am. Chem. Soc.*, **129**, 286.

5 For leading reviews on metal-catalyzed unfunctionalized alkenes, see: (a) Jacobsen, E.N. (1993) Chapter 4.2, in *Catalytic Asymmetric Synthesis* (ed. I. Ojima), VCH, New York; (b) Collman, J.P., Zhang, X., Lee, V.J., Uffelman, E.S., and Brauman, J.I. (1993) *Science*, **261**, 1404; (c) Mukaiyama, T. (1996) *Aldrichimica Acta*, **29**, 59; (d) Katsuki, T. (2000) Chapter 6B in *Catalytic Asymmetric Synthesis* (ed. I. Ojima), VCH, New York; (e) McGarrigle, E.M. and Gilheany, D.G. (2005) *Chem. Rev.*, **105**, 1563; (f) Xia, Q.-H., Ge, H.-Q., Ye, C.-P., Liu, Z.-M., and Su, K.-X. (2005) *Chem. Rev.*, **105**, 1603.

6 For leading references on titanium-catalyzed asymmetric epoxidation of unfunctionalized alkenes with H_2O_2, see: (a) Matsumoto, K., Sawada, Y., Saito, B., Sakai, K., and Katsuki, T. (2005) *Angew. Chem. Int. Ed.*, **44**, 4935; (b) Sawada, Y., Matsumoto, K., Kondo, S., Watanabe, H., Ozawa, T., Suzuki, K., Saito, B., and Katsuki, T. (2006) *Angew. Chem. Int. Ed.*, **45**, 3478; (c) Matsumoto, K., Sawada, Y., and Katsuki, T. (2006) *Synlett*, 3545; (d) Sawada, Y., Matsumoto, K., and Katsuki, T. (2007) *Angew. Chem. Int. Ed.*, **46**, 4559.

7 For a recent report on chiral molybdenum-catalyzed asymmetric epoxidation of unfunctionalized alkenes, see: Barlan, A.U., Basak, A., and Yamamoto, H. (2006) *Angew. Chem. Int. Ed.*, **45**, 5849.

8 For leading reviews, see: (a) Porter, M.J. and Skidmore, J. (2000) *Chem. Commun.*, 1215; (b) Lauret, C. and Roberts, S.M. (2002) *Aldrichimica Acta*, **35**, 47; (c) Nemoto, T., Ohshima, T., and Shibasaki, M. (2002) *J. Synth. Org. Chem. Jpn.*, **60**, 94; (d) Kelly, D.R. and Roberts, S.M. (2006) *Biopolymers*, **84**, 74; (e) Shibasaki, M., Kanai, M., and Matsunaga, S. (2006) *Aldrichimica Acta*, **39**, 31.

9 Shi, Y. (2004) Chapter 3, in *Modern Oxidation Methods* (ed. J.-E. Bäckvall), Wiley-VCH, Weinheim.

10 For leading reviews on asymmetric epoxidation by chiral ketones, see: (a) Denmark, S.E. and Wu, Z. (1999) *Synlett*, 847; (b) Frohn, M. and Shi, Y. (2000) *Synthesis*, 1979; (c) Shi, Y. (2002) *J. Synth. Org. Chem. Jpn.*, **60**, 342; (d) Shi, Y. (2004) *Acc. Chem. Res.*, **37**, 488; (e) Yang, D. (2004) *Acc. Chem. Res.*, **37**, 497;(f) Shi, Y. (2006) Chapter 10, in *Handbook of Chiral Chemicals* (ed. D. Ager), CRC Press, Taylor & Francis Group, Boca Raton; (g) Goeddel, D. and Shi, Y. (2008) *Science of Synthesis*, vol. **37** (ed. C.J. Forsyth), Georg Thieme Verlag KG, Stuttgart, Germany, p. 277. (h) Wong, O.A. and Shi, Y. (2008) *Chem. Rev.*, **108**, 3958; (i) Wong, O.A. and Shi, Y. (2010) *Top. Curr. Chem.*, vol. **291** (ed. B. List), Springer, New York, pp. 201–232.

11 For general leading references on dioxiranes: (a) Murray, R.W. (1989) *Chem. Rev.*, **89**, 1187; (b) Adam, W., Curci, R., and Edwards, J.O. (1989) *Acc. Chem. Res.*, **22**, 205; (c) Curci, R., Dinoi, A., and Rubino, M.F. (1995) *Pure Appl. Chem.*, **67**, 811; (d) Adam, W. and Smerz, A.K. (1996) *Bull. Soc. Chim. Belg.*, **105**, 581; (e) Adam, W., Saha-Möller, C.R., and Ganeshpure, P.A. (2001) *Chem. Rev.*, **101**, 3499; (f) Adam, W., Saha-Möller, C.R., and Zhao, C.-G. (2002) *Org. React.*, **61**, 219.

12 For examples of *in situ* generation of dioxiranes see: (a) Edwards, J.O., Pater, R.H., Curci, R., and Di Furia, F. (1979) *Photochem. Photobiol.*, **30**, 63; (b) Curci, R., Fiorentino, M., Troisi, L., Edwards, J.O., and Pater, R.H. (1980) *J. Org. Chem.*, **45**, 4758; (c) Gallopo, A.R. and Edwards, J.O. (1981) *J. Org. Chem.*, **46**, 1684; (d) Cicala, G., Curci, R., Fiorentino, M., and Laricchiuta, O. (1982) *J. Org. Chem.*, **47**, 2670; (e) Corey, P.F. and Ward, F.E. (1986) *J. Org. Chem.*, **51**, 1925; (f) Adam, W., Hadjiarapoglou, L., and Smerz, A. (1991) *Chem. Ber.*, **124**, 227; (g) Kurihara, M., Ito, S., Tsutsumi, N., and Miyata, N. (1994)

Tetrahedron Lett., **35**, 1577; (h) Denmark, S.E., Forbes, D.C., Hays, D.S., DePue, J.S., and Wilde, R.G. (1995) *J. Org. Chem.*, **60**, 1391; (i) Yang, D., Wong, M.-K., and Yip, Y.-C. (1995) *J. Org. Chem.*, **60**, 3887; (j) Denmark, S.E. and Wu, Z. (1997) *J. Org. Chem.*, **62**, 8964; (k) Boehlow, T.R., Buxton, P.C., Grocock, E.L., Marples, B.A., and Waddington, V.L. (1998) *Tetrahedron Lett.*, **39**, 1839; (l) Frohn, M., Wang, Z.-X., and Shi, Y. (1998) *J. Org. Chem.*, **63**, 6425; (m) Yang, D., Yip, Y.-C., Jiao, G.-S., and Wong, M.-K. (1998) *J. Org. Chem.*, **63**, 8952; (n) Yang, D., Yip, Y.-C., Tang, M.-W., Wong, M.-K., and Cheung, K.-K. (1998) *J. Org. Chem.*, **63**, 9888; (o) Shu, L. and Shi, Y. (2000) *J. Org. Chem.*, **65**, 8807.

13 For ketones 1–3 and 7, see: (a) Curci, R., Fiorentino, M., and Serio, M.R. (1984) *Chem. Commun.*, 155; (b) Curci, R., D'Accolti, L., Fiorentino, M., and Rosa, A. (1995) *Tetrahedron Lett.*, **36**, 5831.

14 For ketone 8, see: Brown, D.S., Marples, B.A., Smith, P., and Walton, L. (1995) *Tetrahedron*, **51**, 3587.

15 For ketones 30–36 see: (a) Ref. 12h. (b) Denmark, S.E., Wu, Z., Crudden, C.M., and Matsuhashi, H. (1997) *J. Org. Chem.*, **62**, 8288; (c) Denmark, S.E. and Wu, Z. (1998) *J. Org. Chem.*, **63**, 2810; (d) Ref. 10a.

16 For ketones 21 and 22, see: Denmark, S.E. and Matsuhashi, H. (2002) *J. Org. Chem.*, **67**, 3479.

17 For ketones 17 and 18, see: (a) Yang, D., Yip, Y.-C., Tang, M.-W., Wong, M.-K., Zheng, J.-H., and Cheung, K.-K. (1996) *J. Am. Chem. Soc.*, **118**, 491; (b) Yang, D., Wang, X.-C., Wong, M.-K., Yip, Y.-C., and Tang, M.-W. (1996) *J. Am. Chem. Soc.*, **118**, 11311; (c) Yang, D., Wong, M.-K., Yip, Y.-C., Wang, X.-C., Tang, M.-W., Zheng, J.-H., and Cheung, K.-K. (1998) *J. Am. Chem. Soc.*, **120**, 5943.

18 For ketone 12, see: Yang, D., Yip, Y.-C., Chen, J., and Cheung, K.-K. (1998) *J. Am. Chem. Soc.*, **120**, 7659.

19 For additional studies on ketone 17, see: (a) Furutani, T., Hatsuda, M., Imashiro, R., and Seki, M. (1999) *Tetrahedron: Asymmetry*, **10**, 4763; (b) Seki, M., Furutani, T., Hatsuda, M., and Imashiro, R. (2000) *Tetrahedron Lett.*, **41**, 2149; (c) Kuroda, T., Imashiro, R., and Seki, M. (2000) *J. Org. Chem.*, **65**, 4213; (d) Seki, M., Yamada, S.-i., Kuroda, T., Imashiro, R., and Shimizu, T. (2000) *Synthesis*, 1677; (e) Hatsuda, M., Hiramatsu, H., Yamada, S.-i., Shimizu, T., and Seki, M. (2001) *J. Org. Chem.*, **66**, 4437; (f) Furutani, T., Hatsuda, M., Shimizu, T., and Seki, M. (2001) *Biosci. Biotechnol. Biochem.*, **65**, 180; (g) Seki, M., Furutani, T., Imashiro, R., Kuroda, T., Yamanaka, T., Harada, N., Arakawa, H., Kusama, M., and Hashiyama, T. (2001) *Tetrahedron Lett.*, **42**, 8201; (h) Furutani, T., Imashiro, R., Hatsuda, M., and Seki, M. (2002) *J. Org. Chem.*, **67**, 4599; (i) Imashiro, R. and Seki, M. (2004) *J. Org. Chem.*, **69**, 4216.

20 For leading references on ketone 42, see: (a) Tu, Y., Wang, Z.-X., and Shi, Y. (1996) *J. Am. Chem. Soc.*, **118**, 9806; (b) Wang, Z.-X., Tu, Y., Frohn, M., and Shi, Y. (1997) *J. Org. Chem.*, **62**, 2328; (c) Wang, Z.-X., Tu, Y., Frohn, M., Zhang, J.-R., and Shi, Y. (1997) *J. Am. Chem. Soc.*, **119**, 11224; (d) Wang, Z.-X., and Shi, Y. (1998) *J. Org. Chem.*, **63**, 3099; (e) Frohn, M., Dalkiewicz, M., Tu, Y., Wang, Z.-X., and Shi, Y. (1998) *J. Org. Chem.*, **63**, 2948; (f) Cao, G.-A., Wang, Z.-X., Tu, Y., and Shi, Y. (1998) *Tetrahedron Lett.*, **39**, 4425; (g) Zhu, Y., Tu, Y., Yu, H., and Shi, Y. (1998) *Tetrahedron Lett.*, **39**, 7819; (h) Wang, Z.-X., Cao, G.-A., and Shi, Y. (1999) *J. Org. Chem.*, **64**, 7646; (i) Warren, J.D. and Shi, Y. (1999) *J. Org. Chem.*, **64**, 7675; (j) Zhu, Y., Manske, K.J., and Shi, Y. (1999) *J. Am. Chem. Soc.*, **121**, 4080; (k) Frohn, M., Zhou, X., Zhang, J.-R., Tang, Y., and Shi, Y. (1999) *J. Am. Chem. Soc.*, **121**, 7718; (l) Zhu, Y., Shu, L., Tu, Y., and Shi, Y. (2001) *J. Org. Chem.*, **66**, 1818; (m) Tu, Y., Frohn, M., Wang, Z.-X., and Shi, Y. (2003) *Org. Synth.*, **80**, 1; (n) Wang, Z.-X., Shu, L., Frohn, M., Tu, Y., and Shi, Y. (2003) *Org. Synth.*, **80**, 9; (o) Lorenz, J.C., Frohn, M., Zhou, X., Zhang, J.-R., Tang, Y., Burke, C., and Shi, Y. (2005) *J. Org. Chem.*, **70**, 2904.

21 For ketone **ent**-42, see: (a) Ref. 20c. (b) Zhao, M.-X. and Shi, Y. (2006) *J. Org. Chem.*, **71**, 5377.

22 For ketone 43, see: (a) Wu, X.-Y., She, X., and Shi, Y. (2002) *J. Am. Chem. Soc.*, **124**, 8792; (b) Wang, B., Wu, X.-Y., Wong, O.A.,

Nettles, B., Zhao, M.-X., Chen, D., and Shi, Y. (2009) *J. Org. Chem.*, **74**, 3986.

23 For additional studies on ketone 43, also see: (a) Nieto, N., Molas, P., Benet-Buchholz, J., and Vidal-Ferran, A. (2005) *J. Org. Chem.*, **70**, 10143; (b) Nieto, N., Munslow, I.J., Fernández-Pérez, H., and Vida-Ferran, A. (2008) *Synlett*, 2856; (c) Hirooka, Y., Nitta, M., Furuta, T., and Kan, T. (2008) *Synlett*, 3234.

24 For ketone 44, see: (a) Tian, H., She, X., Shu, L., Yu, H., and Shi, Y. (2000) *J. Am. Chem. Soc.*, **122**, 11551; (b) Tian, H., She, X., Xu, J., and Shi, Y. (2001) *Org. Lett.*, **3**, 1929; (c) Tian, H., She, X., Yu, H., Shu, L., and Shi, Y. (2002) *J. Org. Chem.*, **67**, 2435; (d) Shu, L., Shen, Y.-M., Burke, C., Goeddel, D., and Shi, Y. (2003) *J. Org. Chem.*, **68**, 4963; (e) Shu, L., Wang, P., Gan, Y., and Shi, Y. (2003) *Org. Lett.*, **5**, 293; (f) Shu, L., and Shi, Y. (2004) *Tetrahedron Lett.*, **45**, 8115; (g) Shen, Y.-M., Wang, B., and Shi, Y. (2006) *Angew. Chem. Int. Ed.*, **45**, 1429; (h) Shen, Y.-M., Wang, B., and Shi, Y. (2006) *Tetrahedron Lett.*, **47**, 5455; (i) Goeddel, D., Shu, L., Yuan, Y., Wong, O.A., Wang, B., and Shi, Y. (2006) *J. Org. Chem.*, **71**, 1715; (j) Wong, O.A. and Shi, Y. (2006) *J. Org. Chem.*, **71**, 3973; (k) Wang, B., Shen, Y.-M., and Shi, Y. (2006) *J. Org. Chem.*, **71**, 9519; (l) Zhao, M.-X., Goeddel, D., Li, K., and Shi, Y. (2006) *Tetrahedron*, **62**, 8064; (m) Burke, C.P. and Shi, Y. (2006) *Angew. Chem. Int. Ed.*, **45**, 4475; (n) Burke, C.P. and Shi, Y. (2007) *J. Org. Chem.*, **72**, 4093.

25 For a carbocyclic analog of ketone 44, see: Hickey, M., Goeddel, D., Crane, Z., and Shi, Y. (2004) *Proc. Natl. Acad. Sci. USA*, **101**, 5794.

26 For epoxidation with ketones 42 and 44 using H_2O_2, see: (a) Shu, L. and Shi, Y. (1999) *Tetrahedron Lett.*, **40**, 8721; (b) Shu, L. and Shi, Y. (2001) *Tetrahedron*, **57**, 5231; (c) Burke, C.P., Shu, L., and Shi, Y. (2007) *J. Org. Chem.*, **72**, 6320.

27 For ketone 45, see: (a) Wang, B., Wong, O.A., Zhao, M.-X., and Shi, Y. (2008) *J. Org. Chem.*, **73**, 9539; (b) Wong, O.A., Wang, B., Zhao, M.-X., and Shi, Y. (2009) *J. Org. Chem.*, **74**, 6335–6338.

28 For ketone 46, see: Tian, H., She, X., and Shi, Y. (2001) *Org. Lett.*, **3**, 715.

29 For ketones 10, 47–52, see: (a) Tu, Y., Wang, Z.-X., Frohn, M., He, M., Yu, H., Tang, Y., and Shi, Y. (1998) *J. Org. Chem.*, **63**, 8475; (b) Wang, Z.-X., Miller, S.M., Anderson, O.P., and Shi, Y. (2001) *J. Org. Chem.*, **66**, 521; (c) Crane, Z., Goeddel, D., Gan, Y., and Shi, Y. (2005) *Tetrahedron*, **61**, 6409.

30 For ketone 9, see: (a) Wang, Z.-X. and Shi, Y. (1997) *J. Org. Chem.*, **62**, 8622; (b) Wang, Z.-X., Miller, S.M., Anderson, O.P., and Shi, Y. (1999) *J. Org. Chem.*, **64**, 6443.

31 For ketones 18 and 25, see: (a) Song, C.E., Kim, Y.H., Lee, K.C., Lee, S-g., and Jin, B.W. (1997) *Tetrahedron: Asymmetry*, **8**, 2921; (b) Kim, Y.H., Lee, K.C., Chi, D.Y., Lee, S-g., and Song, C.E. (1999) *Bull. Korean. Chem. Soc.*, **20**, 831.

32 For ketone 26, see: Adam, W. and Zhao, C.-G. (1997) *Tetrahedron: Asymmetry*, **8**, 3995.

33 For ketone 10, see: Adam, W., Saha-Möller, C.R., and Zhao, C.-G. (1999) *Tetrahedron: Asymmetry*, **10**, 2749.

34 For ketones 37, 38, and related ketones, see: (a) Armstrong, A. and Hayter, B.R. (1998) *Chem. Commun.*, 621; (b) Armstrong, A., Hayter, B.R., Moss, W.O., Reeves, J.R., and Wailes, J.S. (2000) *Tetrahedron: Asymmetry*, **11**, 2057; (c) Armstrong, A., Moss, W.O., and Reeves, J.R. (2001) *Tetrahedron: Asymmetry*, **12**, 2779; (d) Armstrong, A., Ahmed, G., Dominguez-Fernandez, B., Hayter, B.R., and Wailes, J.S. (2002) *J. Org. Chem.*, **67**, 8610; (e) Sartori, G., Armstrong, A., Maggi, R., Mazzacani, A., Sartorio, R., Bigi, F., and Dominguez-Fernandez, B. (2003) *J. Org. Chem.*, **68**, 3232; (f) Armstrong, A. and Tsuchiya, T. (2006) *Tetrahedron*, **62**, 257; (g) Armstrong, A., Dominguez-Fernandez, B., and Tsuchiya, T. (2006) *Tetrahedron*, **62**, 6614.

35 For ketones 4 and 11, see: (a) Armstrong, A. and Hayter, B.R. (1997) *Tetrahedron: Asymmetry*, **8**, 1677; (b) Armstrong, A. and Hayter, B.R. (1999) *Tetrahedron*, **55**, 11119.

36 For ketone 5, see: Carnell, A.J., Johnstone, R.A.W., Parsy, C.C., and Sanderson, W.R. (1999) *Tetrahedron Lett.*, **40**, 8029.

37 For ketones 13–15 and related ketones, see: (a) Solladié-Cavallo, A. and Bouérat, L. (2000) *Tetrahedron: Asymmetry*, **11**, 935; (b) Solladié-Cavallo, A. and Bouérat, L. (2000) *Org. Lett.*, **2**, 3531; (c) Solladié-Cavallo, A.,

Jierry, L., Bouérat, L., and Taillasson, P. (2001) *Tetrahedron: Asymmetry*, **12**, 883; (d) Solladié-Cavallo, A., Bouérat, L., and Jierry, L. (2001) *Eur. J. Org. Chem.*, 4557; (e) Solladié-Cavallo, A., Jierry, L., and Klein, A. (2003) *C. R. Chimie*, **6**, 603; (f) Freedman, T.B., Cao, X., Nafie, L.A., Solladié-Cavallo, A., Jierry, L., and Bouérat, L. (2004) *Chirality*, **16**, 467; (g) Solladié-Cavallo, A., Jierry, L., Norouzi-Arasi, H., and Tahmassebi, D. (2004) *J. Fluorine Chem.*, **125**, 1371; (h) Solladié-Cavallo, A., Jierry, L., Lupattelli, P., Bovicelli, P., and Antonioletti, R. (2004) *Tetrahedron*, **60**, 11375; (i) Solladié-Cavallo, A., Jierry, L., Klein, A., Schmitt, M., and Welter, R. (2004) *Tetrahedron: Asymmetry*, **15**, 3891.

38 For ketones 19, 20, 27–29, and related ketones, see: (a) Matsumoto, K. and Tomioka, K. (2001) *Heterocycles*, **54**, 615; (b) Matsumoto, K. and Tomioka, K. (2001) *Chem. Pharm. Bull.*, **49**, 1653; (c) Matsumoto, K. and Tomioka, K. (2002) *Tetrahedron Lett.*, **43**, 631.

39 For ketone 16 and related ketones, see: (a) Bortolini, O., Fogagnolo, M., Fantin, G., Maietti, S., and Medici, A. (2001) *Tetrahedron: Asymmetry*, **12**, 1113; (b) Bortolini, O., Fantin, G., Fogagnolo, M., Forlani, R., Maietti, S., and Pedrini, P. (2002) *J. Org. Chem.*, **67**, 5802; (c) Bortolini, O., Fantin, G., Fogagnolo, M., and Mari, L. (2004) *Tetrahedron: Asymmetry*, **15**, 3831; (d) Devlin, F.J., Stephens, P.J., and Bortolini, O. (2005) *Tetrahedron: Asymmetry*, **16**, 2653; (e) Bortolini, O., Fantin, G., Fogagnolo, M., and Mari, L. (2006) *Tetrahedron*, **62**, 4482.

40 For ketones 23, 24, and related ketones, see: Stearman, C.J. and Behar, V. (2002) *Tetrahedron Lett.*, **43**, 1943.

41 For ketone 41, see: Klein, S. and Roberts, S.M. (2002) *J. Chem. Soc., Perkin Trans. 1*, 2686.

42 For ketones 53–55, and related ketones, see: (a) Shing, T.K.M. and Leung, G.Y.C. (2002) *Tetrahedron*, **58**, 7545; (b) Shing, T.K.M., Leung, Y.C., and Yeung, K.W. (2003) *Tetrahedron*, **59**, 2159; (c) Shing, T.K.M., Leung, G.Y.C., and Yeung, K.W. (2003) *Tetrahedron Lett.*, **44**, 9225; (d) Shing, T.K.M., Leung, G.Y.C., and Luk, T. (2005) *J. Org. Chem.*, **70**, 7279; (e) Shing, T.K.M., Luk, T., and Lee, C.M. (2006) *Tetrahedron*, **62**, 6621; (f) Shing, T.K.M. and Luk, T. (2009) *Tetrahedron: Asymmetry*, **20**, 883.

43 For ketone 6 and related ketones, see: (a) Chan, W.-K., Yu, W.-Y., Che, C.-M., and Wong, M.-K. (2003) *J. Org. Chem.*, **68**, 6576; (b) Rousseau, C., Christensen, B., Petersen, T.E., and Bols, M. (2004) *Org. Biomol. Chem.*, **2**, 3476.

44 For aldehyde 56 and related ketones, see: Bez, G. and Zhao, C.-G. (2003) *Tetrahedron Lett.*, **44**, 7403.

45 For a recent report on highly enantioselective epoxidation of silyl enol ethers with ketone 42 and subsequent regioselective epoxide opening, see: Lim, S.M., Hill, N., and Myers, A.G. (2009) *J. Am. Chem. Soc.*, **131**, 5763.

46 Spiro transition state is proposed to be the major transition state for the epoxidation with dimethyldioxirane based on the observation that *cis*-hexene is epoxidized 7–9 times faster than *trans*-hexene: (a) Baumstark, A.L. and McCloskey, C.J. (1987) *Tetrahedron Lett.*, **28**, 3311; (b) Baumstark, A.L. and Vasquez, P.C. (1988) *J. Org. Chem.*, **53**, 3437.

47 For leading references on computational studies on transition states for the dioxirane epoxidation, see: (a) Bach, R.D., Andrés, J.L., Owensby, A.L., Schlegel, H.B., and McDouall, J.J.W. (1992) *J. Am. Chem. Soc.*, **114**, 7207; (b) Houk, K.N., Liu, J., DeMello, N.C., and Condroski, K.R. (1997) *J. Am. Chem. Soc.*, **119**, 10147; (c) Jenson, C., Liu, J., Houk, K.N., and Jorgensen, W.L. (1997) *J. Am. Chem. Soc.*, **119**, 12982; (d) Deubel, D.V. (2001) *J. Org. Chem.*, **66**, 3790; (e) Singleton, D.A., and Wang, Z. (2005) *J. Am. Chem. Soc.*, **127**, 6679.

48 For a synthesis of chiral γ-aryl-γ-butyrolactones using ketone 42, see: (a) Yoshida, M., Ismail, M.A.-H., Nemoto, H., and Ihara, M. (1999) *Heterocycles*, **50**, 673; (b) Yoshida, M., Ismail, M.A.-H., Nemoto, H., and Ihara, M. (2000) *J. Chem. Soc., Perkin Trans. 1*, 2629.

49 Burke, C.P. and Shi, Y. (2009) *Organic Letters*, **11**, 5150–5153.

50 Hoard, D.W., Moher, E.D., Martinelli, M.J., and Norman, B.H. (2002) *Org. Lett.*, **4**, 1813, and references cited therein.

51 (a) Cachoux, F., Isarno, T., Wartmann, M., and Altmann, K.-H. (2005) *Angew. Chem. Int. Ed.*, **44**, 7469; (b) Cachoux, F., Isarno, T., Wartmann, M., and Altmann, K.-H. (2006) *ChemBioChem*, **7**, 54.

52 For a review, see: Feyen, F., Cachoux, F., Gertsch, J., Wartmann, M., and Altmann, K.-H. (2008) *Acc. Chem. Res.*, **41**, 21.

53 Kanada, R.M., Itoh, D., Nagai, M., Niijima, J., Asai, N., Mizui, Y., Abe, S., and Kotake, Y. (2007) *Angew. Chem. Int. Ed.*, **46**, 4350.

54 Wiseman, J.M., McDonald, F.E., and Liotta, D.C. (2005) *Org. Lett.*, **7**, 3155.

55 (a) Marshall, J.A. and Mikowski, A.M. (2006) *Org. Lett.*, **8**, 4375; (b) Marshall, J.A. and Hann, R.K. (2008) *J. Org. Chem.*, **73**, 6753.

56 Taber, D.F. and He, Y. (2005) *J. Org. Chem.*, **70**, 7711.

57 Morimoto, Y., Iwai, T., and Kinoshita, T. (2001) *Tetrahedron Lett.*, **42**, 6307.

58 Morimoto, Y., Nishikawa, Y., and Takaishi, M. (2005) *J. Am. Chem. Soc.*, **127**, 5806.

59 (a) Morimoto, Y., Takaishi, M., Adachi, N., Okita, T., and Yata, H. (2006) *Org. Biomol. Chem.*, **4**, 3220; (b) Morimoto, Y., Okita, T., Takaishi, M., and Tanaka, T. (2007) *Angew. Chem. Int. Ed.*, **46**, 1132.

60 Morimoto, Y., Yata, H., and Nishikawa, Y. (2007) *Angew. Chem. Int. Ed.*, **46**, 6481.

61 Morimoto, Y., Okita, T., and Kambara, H. (2009) *Angew. Chem. Int. Ed.*, **48**, 2538.

62 Morimoto, Y., Takaishi, M., Iwai, T., Kinoshita, T., and Jacobs, H. (2002) *Tetrahedron Lett.*, **43**, 5849.

63 For a review, see: Morimoto, Y. (2008) *Org. Biomol. Chem.*, **6**, 1709.

64 Bian, J., Van Wingerden, M., and Ready, J.M. (2006) *J. Am. Chem. Soc.*, **128**, 7428.

65 Hioki, H., Kanehara, C., Ohnishi, Y., Umemori, Y., Sakai, H., Yoshio, S., Matsushita, M., and Kodama, M. (2000) *Angew. Chem. Int. Ed.*, **39**, 2552.

66 Das, S., Li, L.-S., Abraham, S., Chen, Z., and Sinha, S.C. (2005) *J. Org. Chem.*, **70**, 5922.

67 Tong, R., Valentine, J.C., McDonald, F.E., Cao, R., Fang, X., and Hardcastle, K.I. (2007) *J. Am. Chem. Soc.*, **129**, 1050.

68 Tong, R. and McDonald, F.E. (2008) *Angew. Chem. Int. Ed.*, **47**, 4377.

69 Xiong, Z. and Corey, E.J. (2000) *J. Am. Chem. Soc.*, **122**, 4831.

70 Xiong, Z. and Corey, E.J. (2000) *J. Am. Chem. Soc.*, **122**, 9328.

71 For synthesis and structure revision of glabrescol, also see: Morimoto, Y., Iwai, T., and Kinoshita, T. (2000) *J. Am. Chem. Soc.*, **122**, 7124.

72 Adams, C.M., Ghosh, I., and Kishi, Y. (2004) *Org. Lett.*, **6**, 4723.

73 (a) McDonald, F.E., Wang, X., Do, B., and Hardcastle, K.I. (2000) *Org. Lett.*, **2**, 2917; (b) McDonald, F.E., Bravo, F., Wang, X., Wei, X., Toganoh, M., Rodríguez, J.R., Do, B., Neiwert, W.A., and Hardcastle, K.I. (2002) *J. Org. Chem.*, **67**, 2515; (c) Bravo, F., McDonald, F.E., Neiwert, W.A., Do, B., and Hardcastle, K.I. (2003) *Org. Lett.*, **5**, 2123; (d) Valentine, J.C., McDonald, F.E., Neiwert, W.A., and Hardcastle, K.I. (2005) *J. Am. Chem. Soc.*, **127**, 4586; (e) Tong, R., McDonald, F.E., Fang, X., and Hardcastle, K.I. (2007) *Synthesis*, 2337.

74 For a review, see: Valentine, J.C. and McDonald, F.E. (2006) *Synlett*, 1816.

75 Simpson, G.L., Heffron, T.P., Merino, E., and Jamison, T.F. (2006) *J. Am. Chem. Soc.*, **128**, 1056.

76 For SiMe$_3$-based strategy for polyether synthesis, also see: Heffron, T.P. and Jamison, T.F. (2003) *Org. Lett.*, **5**, 2339.

77 Vilotijevic, I. and Jamison, T.F. (2007) *Science*, **317**, 1189.

78 For additional examples, see: (a) Morten, C.J. and Jamison, T.F. (2009) *J. Am. Chem. Soc.*, **131**, 6678; (b) Van Dyke, A.R. and Jamison, T.F. (2009) *Angew. Chem. Int. Ed.*, **48**, 4430.

79 For leading references, see: (a) Lin, Y.-Y., Risk, M., Ray, S.M., Van Engen, D., Clardy, J., Golik, J., James, J.C., and Nakanishi, K. (1981) *J. Am. Chem. Soc.*, **103**, 6773; (b) Shimizu, Y., Chou, H.-N., Bando, H., Van Duyne, G., and Clardy, J.C. (1986) *J. Am. Chem. Soc.*, **108**, 514; (c) Pawlak, J., Tempesta, M.S., Golik, J., Zagorski, M.G., Lee, M.S., Nakanishi, K., Iwashita, T., Gross, M.L., and Tomer, K.B. (1987) *J. Am. Chem. Soc.*, **109**, 1144; (d) Nakanishi, K. (1985) *Toxicon*, **23**, 473.

80 For a leading reference, see: Nicolaou, K.C. (1996) *Angew. Chem. Int. Ed. Engl.*, **35**, 588.

81 Ager, D.J., Anderson, K., Oblinger, E., Shi, Y., and VanderRoest, J. (2007) *Org. Proc. Res. Devel.*, **11**, 44.

82 For other synthetic applications of ketone 42, see: (a) Tokiwano, T., Fujiwara, K., and Murai, A. (2000) *Synlett*, 335; (b) Bluet, G. and Campagne, J.-M. (2000) *Synlett*, 221; (c) Shen, K.-H., Lush, S.-F., Chen, T.-L., and Liu, R.-S. (2001) *J. Org. Chem.*, **66**, 8106; (d) Guz, N.R., Lorenz, P., and Stermitz, F.R. (2001) *Tetrahedron Lett.*, **42**, 6491; (e) Olofsson, B. and Somfai, P. (2002) *J. Org. Chem.*, **67**, 8574; (f) McDonald, F.E. and Wei, X. (2002) *Org. Lett.*, **4**, 593; (g) Kumar, V.S., Aubele, D.L., and Floreancig, P.E. (2002) *Org. Lett.*, **4**, 2489; (h) Olofsson, B. and Somfai, P. (2003) *J. Org. Chem.*, **68**, 2514; (i) Madhushaw, R.J., Li, C.-L., Su, H.-L., Hu, C.-C., Lush, S.-F., and Liu, R.-S. (2003) *J. Org. Chem.*, **68**, 1872; (j) Smith, A.B. III and Fox, R.J. (2004) *Org. Lett.*, **6**, 1477; (k) Zhang, Q., Lu, H., Richard, C., and Curran, D.P. (2004) *J. Am. Chem. Soc.*, **126**, 36; (l) Halim, R., Brimble, M.A., and Merten, J. (2005) *Org. Lett.*, **7**, 2659; (m) Kumar, V.S., Wan, S., Aubele, D.L., and Floreancig, P.F. (2005) *Tetrahedron: Asymmetry*, **16**, 3570; (n) Lecornué, F., Paugam, R., and Ollivier, J. (2005) *Eur. J. Org. Chem.*, 2589; (o) Iwasaki, J., Ito, H., Nakamura, M., and Iguchi, K. (2006) *Tetrahedron Lett.*, **47**, 1483; (p) Curran, D.P., Zhang, Q., Richard, C., Lu, H., Gudipati, V., and Wilcox, C.S. (2006) *J. Am. Chem. Soc.*, **128**, 9561; (q) Wan, S., Gunaydin, H., Houk, K.N., and Floreancig, P.E. (2007) *J. Am. Chem. Soc.*, **129**, 7915; (r) Neighbors, J.D., Mente, N.R., Boss, K.D., Zehnder, D.W. II, and Wiemer, D.F. (2008) *Tetrahedron Lett.*, **49**, 516; (s) Mente, N.R., Neighbors, J.D., and Wiemer, D.F. (2008) *J. Org. Chem.*, **73**, 7963; (t) Chapelat, J., Buss, A., Chougnet, A., and Woggon, W.-D. (2008) *Org. Lett.*, **10**, 5123; (u) Emmanuvel, L. and Sudalai, A. (2008) *Tetrahedron Lett.*, **49**, 5736; (v) Shichijo, Y., Migita, A., Oguri, H., Watanabe, M., Tokiwano, T., Watanabe, K., and Oikawa, H. (2008) *J. Am. Chem. Soc.*, **130**, 12230; (w) Yu, M. and Snider, B.B. (2009) *Org. Lett.*, **11**, 1031.

83 For an early leading review, see: Dalko, P.I. and Moisan, L. (2001) *Angew. Chem. Int. Ed.*, **40**, 3726.

84 For leading references, see: (a) Kagan, H.B. (1999) Chapter 2, in *Comprehensive Asymmetric Catalysis* (eds E.N. Jacobsen, A. Pfaltz, and H. Yamamoto), Springer-Verlag, Berlin, Heidelberg, Germany; (b) Ahrendt, K.A., Borths, C.J., and MacMillan, D.W.C (2000) *J. Am. Chem. Soc.*, **122**, 4243.

4
Catalytic Oxidations with Hydrogen Peroxide in Fluorinated Alcohol Solvents

Albrecht Berkessel

4.1
Introduction

In organic chemistry, oxidations constitute a particularly important and difficult class of transformations. This is because, in principle, all organic matter (be it the substrate, the solvent, or the catalyst), in the presence of an oxidant such as hydrogen peroxide, is only kinetically stable. Thermodynamically, full degradation to carbon dioxide and water should occur. In other words, for an efficient and selective oxidation process, two requirements need to be met: (i) substrate oxidation should proceed efficiently, but only to the point wanted (e.g., thioether to sulfoxide, but not further to sulfone); and (ii) both the solvent and the catalyst(s) employed should be as stable as possible in the presence of the oxidant. Fluorinated alcohol solvents [1] are particularly advantageous in this respect as they are able to activate oxidants such as hydrogen peroxide toward electrophilic oxidation by (multiple) hydrogen bonding. Furthermore, and once again by hydrogen bonding, they can alter the reactivity of the substrates and the products in such a way that multiple oxidations can be selectively arrested, for example, once the first oxygen atom transfer has occurred. This type of activity *and* selectivity enhancement is found, for example, when the sulfoxidation of thioethers is performed in fluoroalcohol solvents. Fluoroalcohols are highly stable toward oxidation, in accordance with (ii) above [2, 3]. Furthermore, most substrates are soluble in or miscible with fluorinated alcohols – another important point with regard to process efficiency. Scheme 4.1 summarizes a number of readily (i.e. commercially) available fluorinated alcohols and diols, together with their common acronyms. Of the nine compounds shown, TFIP and PhTFE are chiral, the other seven being achiral. With regard to application as solvents in oxidations, trifluoroethanol (TFE) and 1,1,1,3,3,3-hexafluoro-2-propanol (HFIP) are employed most frequently.

This review focuses on the use of hydrogen peroxide as the terminal oxidant. The virtues of hydrogen peroxide as a readily available and 'green' carrier of active oxygen have been praised numerous times [4–6]. In the current context, it is particularly gratifying that hydrogen peroxide is activated toward electrophilic oxidations by

Modern Oxidation Methods. Edited by Jan-Erling Bäckvall

ISBN: 978-3-527-32320-3

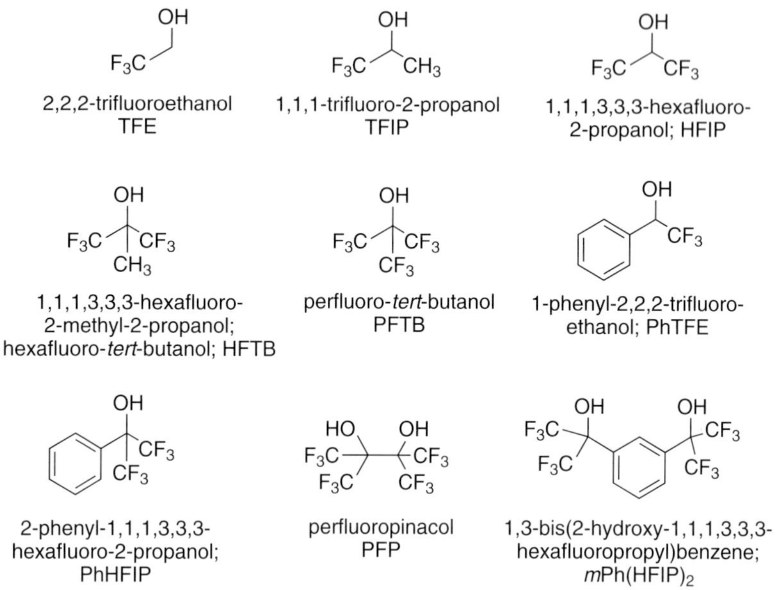

Scheme 4.1

- 2,2,2-trifluoroethanol; TFE — F_3C-CH_2-OH
- 1,1,1-trifluoro-2-propanol; TFIP — $F_3C-CH(OH)-CH_3$
- 1,1,1,3,3,3-hexafluoro-2-propanol; HFIP — $(F_3C)_2CH-OH$
- 1,1,1,3,3,3-hexafluoro-2-methyl-2-propanol; hexafluoro-tert-butanol; HFTB — $(F_3C)_2C(OH)CH_3$
- perfluoro-tert-butanol; PFTB — $(F_3C)_3C-OH$
- 1-phenyl-2,2,2-trifluoroethanol; PhTFE — $Ph-CH(OH)-CF_3$
- 2-phenyl-1,1,1,3,3,3-hexafluoro-2-propanol; PhHFIP — $Ph-C(OH)(CF_3)_2$
- perfluoropinacol; PFP — $(F_3C)_2C(OH)-C(OH)(CF_3)_2$
- 1,3-bis(2-hydroxy-1,1,1,3,3,3-hexafluoropropyl)benzene; mPh(HFIP)$_2$

fluorinated alcohols through multiple hydrogen bonding – a highly synergistic combination of solvent and oxidant [7, 8]. Nevertheless, it should be noted that fluorinated alcohols have also proven beneficial as solvents for alkene epoxidations proceeding via a dioxirane as the active oxidant [9, 10]. The latter is typically generated from a fluoroketone by treatment with persulfate, in the form of Oxone. In HFIP/water mixtures, and using fluoroketones as catalysts, the amounts of Oxone and bicarbonate buffer required could be significantly reduced. Thus, fluorinated alcohols once again (and significantly) enhance the overall efficiency of the oxidation process.

4.2
Properties of Fluorinated Alcohols

The presence of C–F instead of C–H next to the C–OH group confers specific properties to fluorinated alcohols (Scheme 4.2):

1) Fluorinated alcohols are (mildly) acidic: as exemplified by the pK_a values of 12.4 for TFE and 9.3 for HFIP, fluorinated alcohols are significantly more acidic than their nonfluorinated counterparts [11].
2) Fluorinated alcohols are strong hydrogen bond donors: the specific α-values for TFE and HFIP are 1.51 and 1.96, respectively [12a]. The remarkable strength of H-bond donation by, for example, HFIP is nicely illustrated by the existence of its 1 : 1-complex with THF, which can be distilled without decomposition [12b]. See Section 4.2.1 for a more detailed look at the hydrogen bond donor features

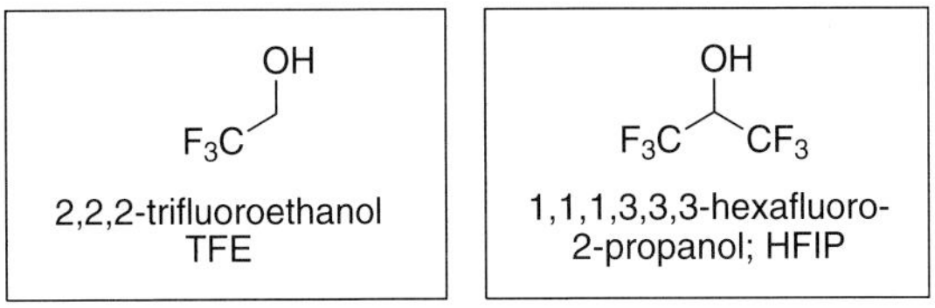

Scheme 4.2

	2,2,2-trifluoroethanol TFE (F$_3$C-CH$_2$-OH)	1,1,1,3,3,3-hexafluoro-2-propanol; HFIP ((F$_3$C)$_2$CH-OH)
boiling point [°C]	73.8	58.6
melting point [°C]	-43.5	-5
pK_a	12.4	9.3
dielectric constant ε	26.7	16.7
polarity, Reichert constant E^N_T	0.898	1.068
ionizing power Y	1.8	3.82
hydrogen-bond donor ability α	1.51	1.96
hydrogen-bond acceptor ability β	0	0
nucleophilicity constant N	-2.78	-4.23
auto-association constant [l·mol^{-1}]	0.65	0.13

of fluorinated alcohols and Section 4.3.1.1 for the effects of multiple hydrogen bonding in epoxidation catalysis.

3) **Fluorinated alcohols are poor hydrogen bond acceptors**: the specific β-values for TFE and HFIP are both 0 [13]. For comparison, the hydrogen bond acceptor constant β of ethanol is 0.77. As a consequence of properties 2 and 3 in the present list, auto-association of the solvent molecules is less pronounced in fluorinated alcohols than in nonfluorinated analogs. For example, the dimerization constant for ethanol is 0.89 L mol^{-1}, whereas that for TFE is 0.65 L mol^{-1}, and for HFIP 0.13 L mol^{-1} [14]. However, as discussed in more detail in Section 4.2.1, two important additional features need to be taken into account when nucleophilic/hydrogen bond-accepting substrates are dissolved in fluorinated alcohols: First, the presence of hydrogen bond acceptors triggers self-association/aggregation of the solvent (a 'sergeant and soldiers' effect); second, the hydrogen bond donor ability of fluorinated alcohol aggregates is enhanced compared with that of the monomers ('H-bond enhanced H-bonding').

4) **Fluorinated alcohols are highly polar solvents**: With regard to the Reichardt E^N_T-value, HFIP ($E^N_T = 1.068$) even exceeds the scale defined by tetramethylsilane ($E^N_T = 0$) and water ($E^N_T = 1$) [15].

5) **Fluorinated alcohols have high ionizing power**: This important property of fluorinated alcohols is reflected by their Y-values of 1.80 (TFE) and even 3.82 for HFIP [13].
6) **Fluorinated alcohols are non-nucleophilic**: Compared to, for example, ethanol with an N-constant of 0, N equals −2.78 for TFE and even −4.23 for HFIP [13].

As a consequence of the above summarized properties of fluorinated alcohols, they are ideal solvents for the generation of cationic or radical-cationic species or reaction intermediates. This effect has been exploited numerous times, for example, in the investigation of organic radical cations [11]. At the same time, it lies at the heart of the catalysis of alkene epoxidation with hydrogen peroxide and of Baeyer-Villiger-type oxidations of carbonyl compounds. The mechanisms of the latter two reaction types are discussed in more detail in Sections 4.3.1.1 (epoxidation) and Section 4.5.1 (Baeyer-Villiger).

4.2.1
A Detailed Look at the Hydrogen Bond Donor Features of HFIP

For understanding the catalytic properties of fluorinated alcohols, it is necessary to take a closer look at the hydrogen bond donor properties of HFIP and the factors by which they are influenced [16]. The hydrogen bond donor ability of fluorinated

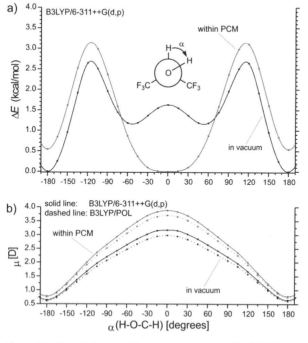

Figure 4.1 Potential energy (a) and dipole moment (b) of HFIP versus HOCH-dihedral angle in vacuum (black) and within a PCM (gray).

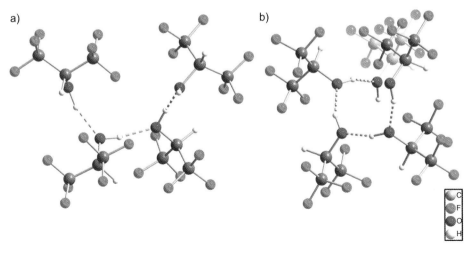

Figure 4.2 Single-crystal X-ray structures of HFIP: (a) view perpendicular to the helix axis, and (b) view along the helix axis.

alcohols, and in particular HFIP, is mainly dependent on two parameters: (i) the conformation of the alcohol monomer along the C−O bond [16, 17] and (ii) the cooperative aggregation to hydrogen-bonded alcohol clusters [16].

In a polarizable environment, the absolute minimum structure of HFIP carries the OH synclinal (*sc*) or almost synperiplanar (*sp*) to the adjacent CH (Figure 4.1) [16]. On the basis of quantum-chemical considerations as well as single-crystal X-ray structures in which HFIP acts as hydrogen bond donor, HFIP always takes on such a *sc* or even *sp* conformation. In this conformation, the hydrogen bond donor ability of HFIP is significantly increased (Figures 4.2 and 4.3) [16].

Furthermore, the hydrogen bond donor ability of an HFIP hydroxyl group is greatly enhanced upon coordination of a second or even third molecule of HFIP

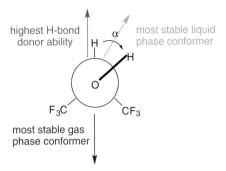

Figure 4.3 Dependence of the properties of monomeric HFIP on the conformation along the CO bond.

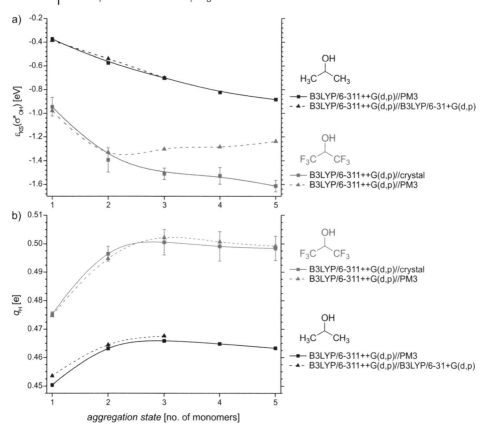

Figure 4.4 LUMO energy (σ^*OH) (a) and natural charge qH of the hydroxyl proton (b) versus aggregation state of HFIP.

(Figures 4.4 and 4.5). Aggregation beyond the trimer has no significant additional effect [16, 18].

Therefore, our mechanistic investigation of the epoxidation of alkenes with hydrogen peroxide, as summarized in Section 4.3.1.1, will be constrained to reaction pathways which (i) involve HFIP in an *sc* or even *sp* conformation and (ii) to hydrogen-bonded HFIP aggregates comprising up to four alcohol monomers.

Figure 4.5 Aggregation-induced hydrogen bonding enhancement of HFIP.

4.3
Epoxidation of Alkenes in Fluorinated Alcohol Solvents

The first reports on the use of fluorinated alcohols, and in particular of HFIP in oxidations with hydrogen peroxide, can be found in the patent literature of the late 1970s and early 1980s [19, 20]. Typically, 60% aqueous hydrogen peroxide was used in the presence of metal catalysts. A number of reports on alkene epoxidations in fluorinated alcohols, both in the absence and in the presence of additional catalysts, have followed.

4.3.1
Alkene Epoxidation with Hydrogen Peroxide – in the Absence of Further Catalysts

In 2000, *Neimann and Neumann* reported on alkene epoxidation by H_2O_2 in fluorinated alcohol solvents without the addition of further catalysts [21]. Shortly thereafter, in 2001, *Sheldon et al.* reported about their results, also on alkene epoxidation in fluorinated alcohol solvents [22]. In the latter study, it became clear that buffering the reaction mixtures, preferably by addition of Na_2HPO_4 improves the overall efficiency of the process, presumably by suppressing acid-catalyzed degradation of the product epoxides. Scheme 4.3 summarizes the results obtained using TFE as solvent, whereas the results for HFIP are summarized in Scheme 4.4.

Inspection of Schemes 4.3 and 4.4 reveals that, in the absence of further catalysts, epoxidation with hydrogen peroxide in TFE or HFIP is especially effective for relatively electron-rich alkenes such as *cis*-cyclooctene, 1-methylcyclohexene, or 3-carene. For less electron-rich alkenes such as 1-octene, the epoxidation is typically slow, and low conversions result even after longer reaction times and at reflux temperature. Generally speaking, the noncatalyzed epoxidation has three parameters that can be adjusted to the individual alkene: (i) HFIP has a stronger activating effect than TFE, (ii) hydrogen peroxide can be used in higher concentrations, if necessary, and (iii) the reaction temperature can be varied up to reflux of the solvent. As evidenced by the work of *Neimann and Neumann*, styrenes appear not to provide epoxide in useful yields. In their study, a mixture of products, resulting from ring-opening of the epoxide and from C–C-bond cleavage, was obtained [21].

4.3.1.1 On the Mechanism of Epoxidation Catalysis by Fluorinated Alcohols
Kinetic investigations of the epoxidation of Z-cyclooctene by aqueous H_2O_2 in HFIP by *Berkessel et al.* showed that the reaction follows a first-order dependence with respect to the substrate alkene as well as to the oxidant, suggesting a monomolecular participation of these components in the rate-determining step [16, 18, 23]. On the other hand, a kinetic order of 2–3 with respect to the concentration of HFIP was observed for several co-solvents. The large negative $\Delta S^\dagger$ of -39 cal mol^{-1} K^{-1} points to a highly ordered *TS* of the rate-determining reaction step: typical $\Delta S^\dagger$-values for alkene epoxidations by peracids range from -18 to -30 cal mol^{-1} K^{-1} [24]. These

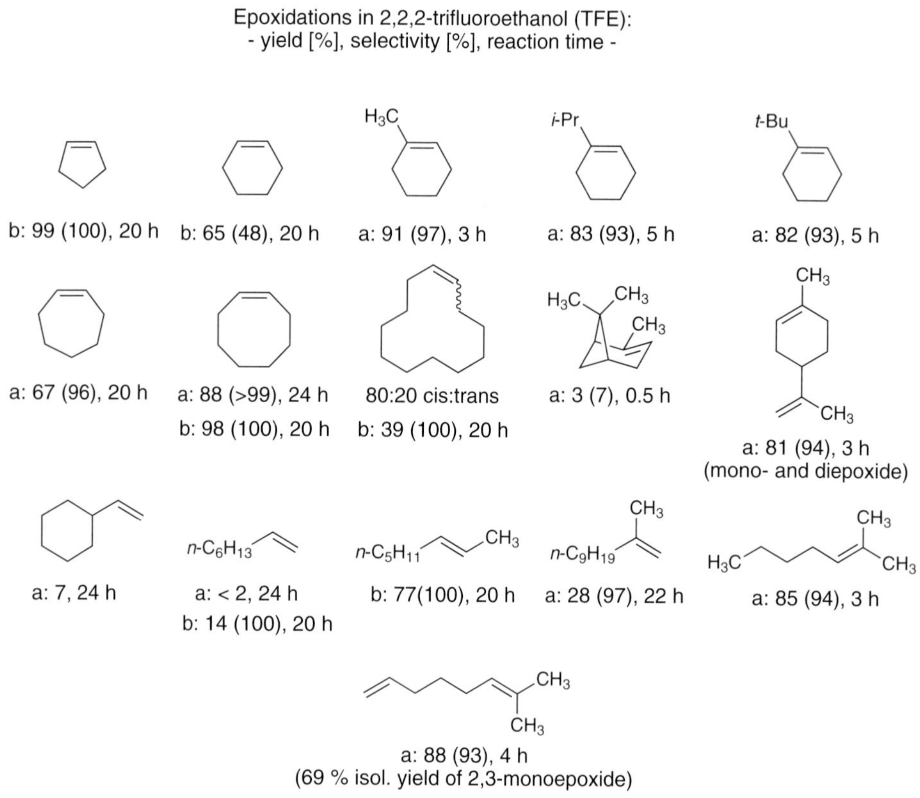

Scheme 4.3

Epoxidations in 2,2,2-trifluoroethanol (TFE):
- yield [%], selectivity [%], reaction time -

- cyclopentene: b: 99 (100), 20 h
- cyclohexene: b: 65 (48), 20 h
- 1-methylcyclohexene (H₃C): a: 91 (97), 3 h
- i-Pr-cyclohexene: a: 83 (93), 5 h
- t-Bu-cyclohexene: a: 82 (93), 5 h
- cycloheptene: a: 67 (96), 20 h
- cyclooctene: a: 88 (>99), 24 h; b: 98 (100), 20 h
- cyclododecene: 80:20 cis:trans; b: 39 (100), 20 h
- α-pinene (H₃C, CH₃, CH₃): a: 3 (7), 0.5 h
- limonene (CH₃, CH₃): a: 81 (94), 3 h (mono- and diepoxide)
- vinylcyclohexane: a: 7, 24 h
- n-C₆H₁₃-CH=CH₂: a: < 2, 24 h; b: 14 (100), 20 h
- n-C₅H₁₁-CH=CH-CH₃: b: 77(100), 20 h
- n-C₉H₁₉-C(CH₃)=CH₂: a: 28 (97), 22 h
- H₃C-CH=CH-CH₃ (internal): a: 85 (94), 3 h
- 2,6-dimethyl-2,6-octadiene: a: 88 (93), 4 h (69 % isol. yield of 2,3-monoepoxide)

a: Sheldon et al., ref. 22. Reaction conditions: 5 mmol alkene, 10 mmol 60 % H_2O_2, 5 mol-% Na_2HPO_4, 5 mL trifluoroethanol, reflux (ca. 80 °C).
b: Neumann and Neimann, ref. 21. Reaction conditions: 1.2 mmol alkene, 2 mmol 60 % H_2O_2, 1 mL trifluoroethanol, 60 °C.

experimental results provide the basis for our calculations in which one to four molecules of HFIP were added to the transition state of the reaction.

The first quantum-chemical investigation of the mechanism of alkene epoxidation in fluoroalcohols was carried out by *Shaik et al.* [25]. In the absence of kinetic data, a monomolecular mode of activation by the fluorinated alcohols for all reaction pathways was assumed [25]. In our work [18], we first compared the transition state, which does not involve HFIP-participation [**TS(e,0)**], with single-HFIP involvement [**TS(e,1)** and **TS(e,1)'**] (Figure 4.6). Particular emphasis was then put on the twofold HFIP-activated complex (Figure 4.7) for a detailed inspection of the hydrogen-bond-assisted epoxidation. All relevant characteristics of higher-order activation (as shown e.g., in Figure 4.8) are already present in the transition states **TS(e,2)** and **TS(e,2)'** (Figure 4.7).

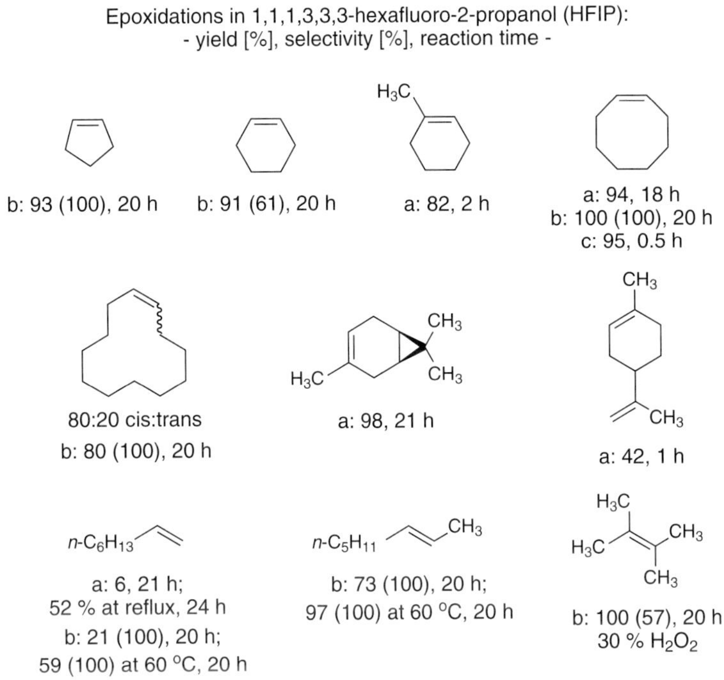

4.3 Epoxidation of Alkenes in Fluorinated Alcohol Solvents

Epoxidations in 1,1,1,3,3,3-hexafluoro-2-propanol (HFIP):
- yield [%], selectivity [%], reaction time -

b: 93 (100), 20 h

b: 91 (61), 20 h

a: 82, 2 h

a: 94, 18 h
b: 100 (100), 20 h
c: 95, 0.5 h

80:20 cis:trans
b: 80 (100), 20 h

a: 98, 21 h

a: 42, 1 h

$n\text{-}C_6H_{13}$
a: 6, 21 h;
52 % at reflux, 24 h
b: 21 (100), 20 h;
59 (100) at 60 °C, 20 h

$n\text{-}C_5H_{11}$
b: 73 (100), 20 h;
97 (100) at 60 °C, 20 h

b: 100 (57), 20 h
30 % H_2O_2

a: Sheldon et al., ref. 22. Reaction conditions: alkene 1M in HFIP, 2 eq. 60 % H_2O_2, 5 mol-% Na_2HPO_4, room temperature.
b: Neumann and Neimann, ref. 21. Reaction conditions: 1.2 mmol alkene, 2 mmol 60 % H_2O_2, 1 mL HFIP, 22 °C.
c: Berkessel and Adrio, ref. 23. Reaction conditions: cyclooctene 31 μmol, 42 % H_2O_2, 23 eq., 1.2 mL HFIP, 50 °C.

Scheme 4.4

In the 2:1 pre-complex **C2** composed of HFIP and H_2O_2, hydrogen peroxide is coordinated by the two alcoholic hydroxyl groups in a cyclic fashion, one HFIP acting as a hydrogen bond donor toward the leaving OH (hydrogen bond length 1.767 Å), and the other one as a hydrogen bond acceptor (hydrogen bond length 1.906 Å), deprotonating the hydroxyl group which is transferred to be the epoxide oxygen atom (**C2**, Figure 4.7) [18]. The 'internal' hydrogen bond between the two fluorinated alcohols (hydrogen bond length 1.823 Å) cooperatively increases the hydrogen bond donor ability of the alcohol molecule which activates the leaving OH group. By this hydrogen bond pattern, a polarization of the O–O bond is achieved (the donated and accepted hydrogen bonds are not equal in length and angle), and an electron-deficient oxygen atom is generated, ready for electrophilic attack on the alkene double bond. In the corresponding transition state (**TS(e,2)**, Figure 4.7) the shorter hydrogen bond (in which HFIP acts as H-bond donor) is extremely contracted to 1.409 Å (−0.358 Å), whereas the longer hydrogen bond (in which HFIP acts as acceptor) is slightly decreased in length to 1.864 Å (−0.042 Å). Thus, the acidity of the donor

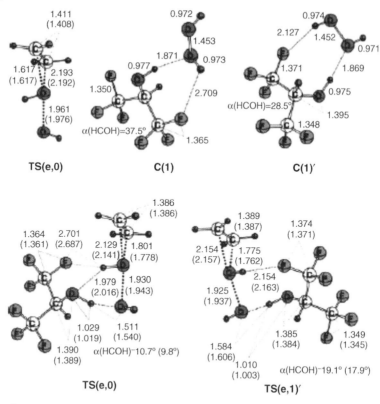

Figure 4.6 Stationary-point structures for the epoxidation of ethene with hydrogen peroxide in the absence and in the presence of one molecule of HFIP, optimized at RB3LYP/6-31+G(d,p) (selected bond lengths in Å; RB3LYP/6-311++G(d,p) results in parentheses).

HFIP molecule is cooperatively increased by shortening of the HFIP internal hydrogen bond from 1.823 Å to 1.692 Å (−0.131 Å).

A second potential reaction path (**C(2)′**, **TS(e,2)′**, Figure 4.7) for twofold HFIP activation was calculated, which differs from **TS(e,2)** with regard to the hydrogen bond from H_2O_2 to HFIP. Here, a fluorine atom of the trifluoromethyl group serves as hydrogen bond acceptor, and not a second hydroxy function. Both transition states are similar in energy, but the corresponding pre-complex **C(2)′**, consisting of H_2O_2 and two HFIP molecules, lies 18.4 kJ mol^{-1} above **C(2)**.

An analysis of the hydrogen bonding parameters shows that, in all cases where HFIP donates a hydrogen bond to the oxidant, this hydrogen bond is significantly contracted in the transition state, usually by more than 0.3 Å. The result of this significant contraction is the formation of a 'low-barrier hydrogen bond' [26], characterized by an increase in covalency which effectively exerts the pronounced stabilization of the highly polar transition states through charge transfer. Hydrogen bonds between two HFIP molecules show the same trend, being regularly shortened by about 0.1 Å. This effect clearly indicates a cooperative enhancement of

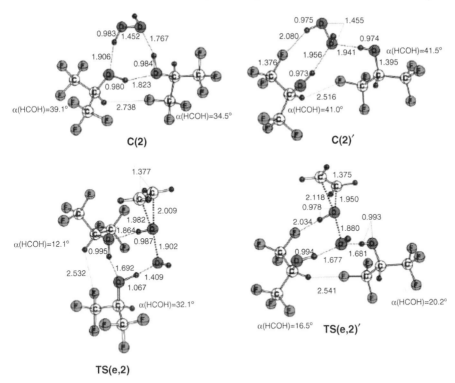

Figure 4.7 Stationary-point structures for the epoxidation of ethene with hydrogen peroxide in the presence of two molecules of HFIP, optimized at RB3LYP/6- 31 + G(d,p).

hydrogen bonding. Additionally, we find a reduction of the HOCH-dihedral angles in 14 of the 16 HFIP molecules within the seven calculated reaction pathways [18]. This result is in agreement with the analysis of the hydrogen-bonding properties of HFIP, as the hydrogen bond donor ability is maximized toward the *sp* conformation of the alcohol.

Proceeding from the transition states to the resulting products, IRC analysis demonstrates that, along this reaction path, a subsequent and barrier-free, cascade-like proton transfer toward the formation of the epoxide and water takes place. Figure 4.9 shows the overall dependence of the activation parameters on the number of HFIP molecules involved. Interestingly, the activation enthalpy of the epoxidation decreases steadily from zero to fourth order in HFIP. As expected, the activation entropy $-T\Delta S^{\dagger}$ shows a continuous increase with increasing numbers of specifically coordinated HFIP molecules. Because of the increasing entropic contribution, the value of $\Delta G^{\dagger}$ approaches saturation when three or four HFIP molecules are involved. For methanol, however, no influence of explicit coordination of the solvent on the activation parameters of oxygen transfer could be found, so it seems solely to act as a polar reaction medium. In line with this result, no significant epoxidation catalysis results from the use of methanol as solvent.

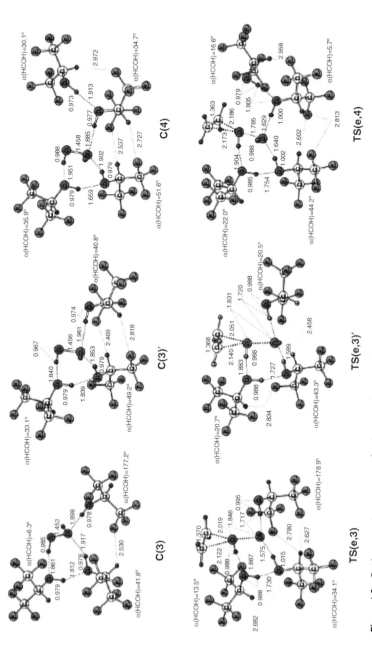

Figure 4.8 Stationary-point structures for the epoxidation of ethene with hydrogen peroxide in the presence of three to four molecules of HFIP, optimized at RB3LYP/6-31+G(d,p).

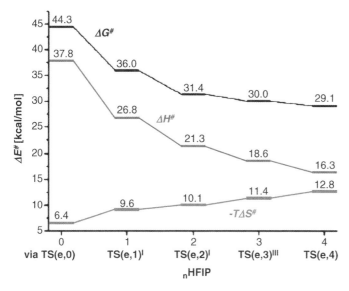

Figure 4.9 Activation parameters versus number of HFIP molecules for the epoxidation of ethane within a solution model at 298 K (RB3LYP/6-311++G(d,p)//RB3LYP/6-31+G(d,p)).

4.3.2
Alkene Epoxidation with Hydrogen Peroxide – in the Presence of Further Catalysts

The catalysts applied to alkene epoxidation in fluorinated alcohol solvents can be subdivided into those which are metal/chalconide-based and those which are purely organic in nature (Scheme 4.5). The former comprise arsanes/arsane oxides [27, 28], arsonic acids [29, 30], seleninic acids/diselenides [31–35], and rhenium compounds such as Re_2O_7 and MTO (methylrhenium trioxide) [36, 37]. As shown in Scheme 4.5, their catalytic activity is ascribed to the intermediate formation of, for example, perseleninic/perarsonic acids or bisperoxorhenium complexes. In other words, their catalytic effect is due to the equilibrium transformation of hydrogen peroxide to kinetically more active peroxidic species.

A second class of catalysts comprises perfluorinated ketones such as hexafluoroacetone [38] and perfluoro-2-octanone [39]. This class of catalyst is assumed to reversibly form peroxyhemihydrates when exposed to hydrogen peroxide [40]; the latter are the active oxygen-transfer agents. Oxygen transfer to the substrate is accompanied by formation of the ketone's hydrate, which, in equilibrium, regenerates the fluoroketone catalyst.

4.3.2.1 Arsines and Arsine Oxides as Catalysts
Inspired by a claim in the patent literature [27], Sheldon et al. identified di-n-butylphenylarsine as a highly active catalyst for the epoxidation of cyclohexene as the substrate, and with 60% hydrogen peroxide in TFE as the solvent [28]. Among a series of purely aromatic arsines, electron-rich ones (such as Ph_3As) were found to be

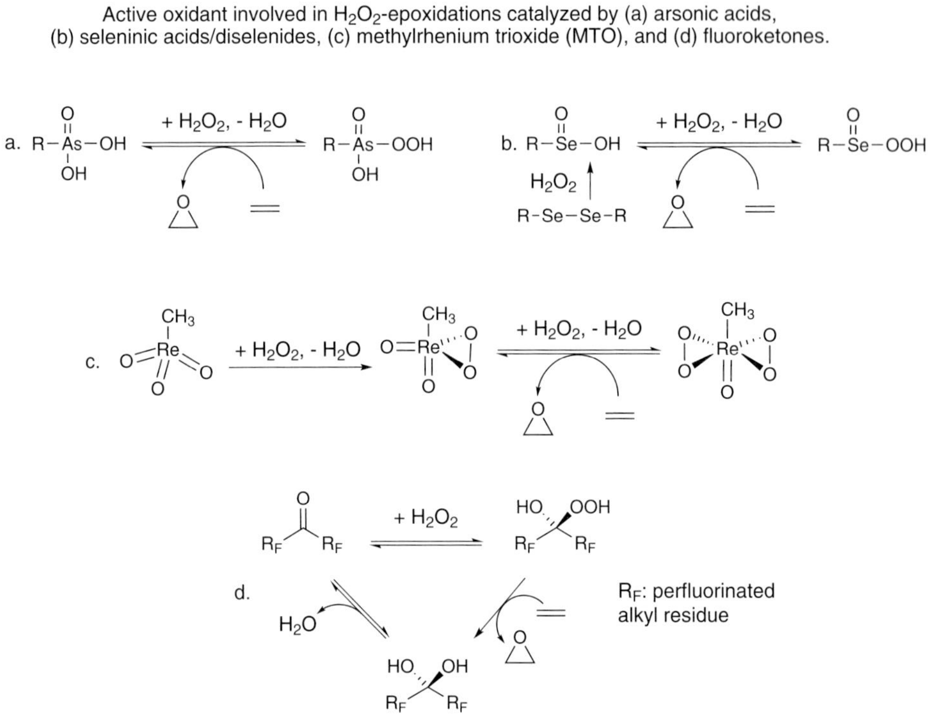

Active oxidant involved in H_2O_2-epoxidations catalyzed by (a) arsonic acids, (b) seleninic acids/diselenides, (c) methylrhenium trioxide (MTO), and (d) fluoroketones.

Scheme 4.5

more active than electron-deficient ones [such as $(C_6F_5)_3As$]. In fact, the arsines employed are pre-catalysts, as under the reaction conditions they are promptly oxidized to the corresponding arsine oxides. The latter are presumed to be the active catalysts. No mechanistic information regarding the further fate of the arsine oxide appears to exist. It may be assumed that a species analogous to the peroxyhemihydrates generated from perfluoroketones may be formed (Scheme 4.5). However, at this point in time, this assumption is purely speculative. Scheme 4.6 summarizes the results obtained in epoxidations with di-n-butylphenylarsine as catalyst. The latter arsine was identified as the most advantageous catalyst in the preceding screening.

By modifying the n-butyl substituents in di-n-butylphenylarsine to $-(CH_2)_2$-C_8F_{17}, Sheldon et al. obtained a catalyst system that, in principle, can be recycled by extraction with prefluorinated solvents [28]. Thus far, the partitioning of the corresponding arsine oxide proved unsatisfactory. Nevertheless, at least partial recovery could be achieved by crystallization from hexane or acetone.

4.3.2.2 Arsonic Acids as Catalysts

Epoxidation catalysis by arsonic acids (both free and polymer-bound) had been reported as early as 1979 by Jacobson et al. [29]. However, these early experiments were typically carried out in 1,4-dioxane as solvent. The latter had been identified as an optimal solvent, as it is miscible both with the aqueous oxidant and the organic

4.3 Epoxidation of Alkenes in Fluorinated Alcohol Solvents

Epoxidations with 60 % H_2O_2 in 2,2,2-trifluoroethanol (TFE), catalyzed by di-n-butylphenyl arsine
- yield [%], catalyst loading [mol-%], reaction time -

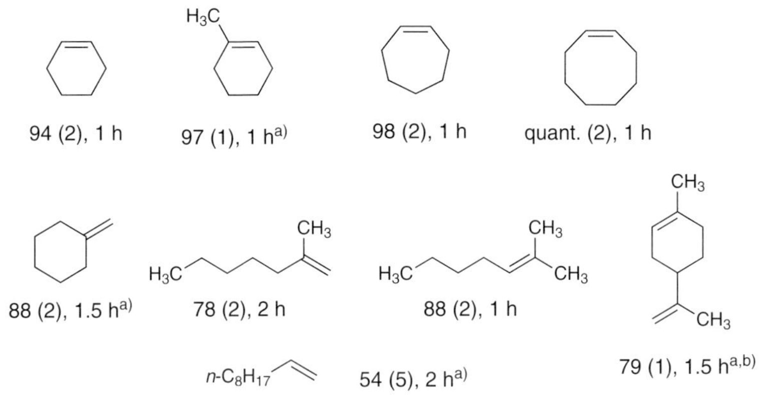

cyclohexene	methylcyclohexene	cycloheptene	cyclooctene
94 (2), 1 h	97 (1), 1 h[a]	98 (2), 1 h	quant. (2), 1 h

methylenecyclohexane	2-methyl-2-hexene	2-methyl-2-heptene	limonene
88 (2), 1.5 h[a]	78 (2), 2 h	88 (2), 1 h	79 (1), 1.5 h[a,b]

n-C$_8$H$_{17}$-CH=CH$_2$: 54 (5), 2 h[a]

Reaction conditions: 5 mmol alkene, 10 mmol 60 % H_2O_2, and amount of di-n-butylphenyl arsine indicated in 5 mL trifluoroethanol, reflux under N_2 (ca. 80 °C).
[a] 5 mol-% Na_2HPO_4 added as buffer. [b] preferential epoxidation of trisubstituted double bond, 5 % of the diepoxide were formed as well.

Scheme 4.6

substrate, and because it promotes the swelling of the polystyrene supports used to immobilize the arsonic acid catalyst [29]. In 2001, we observed that the catalytic effect of benzenearsonic acid in the homogeneous-phase epoxidation of 1-octene is potentiated enormously when switching the solvent from 1,4-dioxane to fluorinated alcohols, preferably HFIP [30]. Our results are summarized in Table 4.1. Inspection

Table 4.1

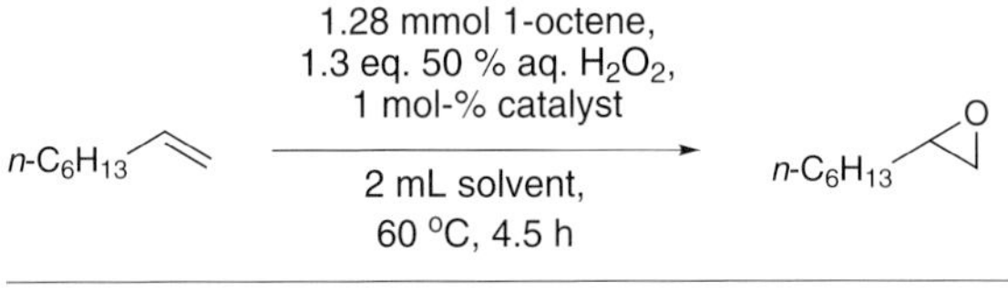

n-C$_6$H$_{13}$-CH=CH$_2$ →(1.28 mmol 1-octene, 1.3 eq. 50 % aq. H_2O_2, 1 mol-% catalyst, 2 mL solvent, 60 °C, 4.5 h)→ n-C$_6$H$_{13}$-epoxide

Solvent	Catalyst	Conversion of 1-octene [%]	Yield of epoxide [%]
HFIP	None	9	9
HFIP	Ph-AsO$_3$H$_2$	Quant.	95
HFIP	Ph-PO$_3$H$_2$	7	7
HFIP	HCl	8	7
HFIP	CF$_3$CO$_2$H	9	8
HFIP	p-TsOH	8	7
1,4-Dioxane	Ph-AsO$_3$H$_2$	<1	<1

of Table 4.1 furthermore reveals that epoxidation catalysis is not just an effect of acidity, but is specific for the arsonic acid: the closely related benzenephosphonic acid, hydrochloric acid, and toluenesulfonic acid do not accelerate the epoxidation beyond the background reaction.

With regard to the mechanism, *Jacobson et al.* proposed that the arsonic acid catalyst is transformed to the perarsonic acid by hydrogen peroxide, the perarsonic acid being the active oxidant (Scheme 4.5) [29]. This may well be the case for arsonic acid-catalyzed epoxidations performed in 1,4-dioxane as solvent. When using fluorinated alcohols as solvent, we believe that reversible ester formation is involved in the mechanism (Scheme 4.7). This assumption is based on ESI-MS studies of the reaction mixtures [23].

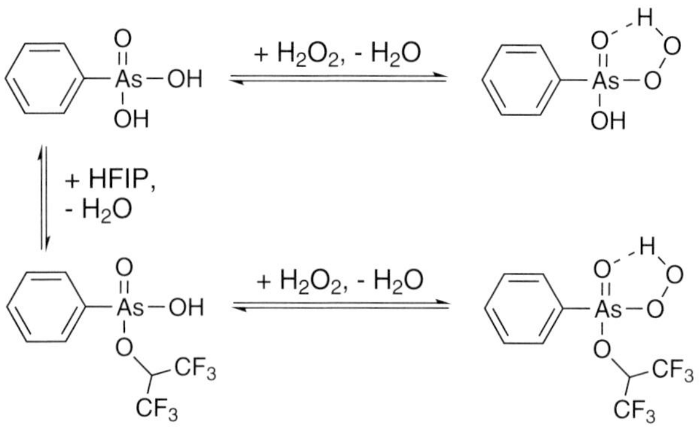

Scheme 4.7

4.3.2.3 Diselenides/Seleninic Acids as Catalysts

Under the reaction conditions, diselenides are converted to seleninic acids, which are the true catalytically active species (Scheme 4.5) [31–34]. The first report on epoxidation catalysis by a seleninic acid (benzeneseleninic acid, employed in stoichiometric amount together with hydrogen peroxide) appears to be by *Grieco et al.*, published in 1977 [33]. The transition to a truly catalytic and highly efficient system was achieved by *Sheldon et al.* in 2001 [34]. A systematic solvent screening once again identified TFE as affording the highest turnover frequencies, at high selectivity. Buffering with sodium acetate proved crucial to prevent ring-opening of acid-sensitive epoxides. With regard to the substitution pattern on the diselenide precatalyst, 3,5-bis(trifluoromethyl) disubstitution was found to be the best (diselenide shown in Scheme 4.8). Epoxidation results obtained under optimized conditions are summarized in Scheme 4.8 [34]. With as little as 0.25 mol% of the diselenide catalyst precursor, impressive epoxide yields and selectivities were obtained for a number of di- and trisubstituted alkenes carrying aliphatic substituents. The hydroxyl group of an allylic alcohol is tolerated, just as a methyl and TMS ether or an acetoxy substituent. Styrene turned out to undergo epoxidation with low yield and selectivity. 1-Octene, as a prototypical mono-substituted terminal alkene, gave the epoxide with

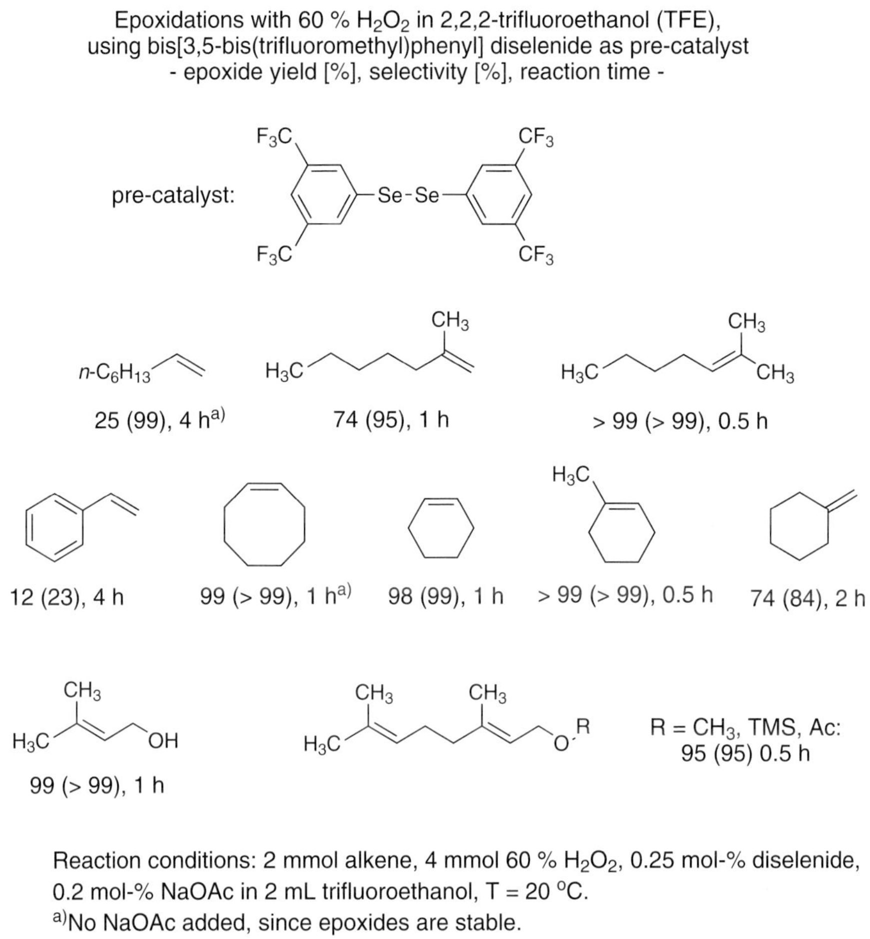

4.3 Epoxidation of Alkenes in Fluorinated Alcohol Solvents

Epoxidations with 60 % H_2O_2 in 2,2,2-trifluoroethanol (TFE), using bis[3,5-bis(trifluoromethyl)phenyl] diselenide as pre-catalyst
- epoxide yield [%], selectivity [%], reaction time -

pre-catalyst: (3,5-(CF$_3$)$_2$C$_6$H$_3$)-Se-Se-(3,5-(CF$_3$)$_2$C$_6$H$_3$)

n-C$_6$H$_{13}$-CH=CH$_2$: 25 (99), 4 h[a]

2-methyl-2-heptene: 74 (95), 1 h

2,3-dimethyl-2-butene type: > 99 (> 99), 0.5 h

styrene: 12 (23), 4 h

cyclooctene: 99 (> 99), 1 h[a]

cyclohexene: 98 (99), 1 h

1-methylcyclohexene: > 99 (> 99), 0.5 h

methylenecyclohexane: 74 (84), 2 h

2-methyl-2-buten-1-ol: 99 (> 99), 1 h

geranyl-OR: R = CH$_3$, TMS, Ac: 95 (95) 0.5 h

Reaction conditions: 2 mmol alkene, 4 mmol 60 % H_2O_2, 0.25 mol-% diselenide, 0.2 mol-% NaOAc in 2 mL trifluoroethanol, T = 20 °C.
[a] No NaOAc added, since epoxides are stable.

Scheme 4.8

very high selectivity, but relatively slowly, and a higher catalyst loading (about 5 mol%) is required for full conversion.

4.3.2.4 Rhenium Compounds as Catalysts

Rhenium compounds, and in particular methylrhenium trioxide (MTO), have attracted considerable attention in oxidation catalysis [36, 37]. Unlike many other transition metal compounds, their activity in promoting nonproductive decomposition of hydrogen peroxide is low. The epoxidizing activity of MTO was originally discovered by Herrmann et al. [41, 42]. In the initial studies, selectivity for epoxide formation was often relatively low because of (Lewis-) acid-catalyzed ring-opening. The addition of bases proved beneficial with regard to epoxide yield, but reduced the activity of the catalyst. Typically, hydrogen peroxide in high concentration had to be used. As an alternative, the urea-hydrogen peroxide clathrate could be used as the terminal oxidant. Sharpless et al. later on found that the addition of an excess of pyridine, and in particular 3-cyanopyridine (relative to the catalyst MTO) afforded both higher reaction rates and selectivities, and provided the possibility of using aqueous hydrogen peroxide [43, 44]. Shortly after this discovery, pyrazole was shown to be even more effective [45, 46].

The synthetic effort necessary for the preparation of MTO makes commercially available inorganic rhenium compounds such as perrhenic acid ($HReO_4$) an attractive alternative. Unfortunately, the catalytically active perrhenic acid is to an even larger extent plagued by subsequent ring opening of the epoxide products – an effect of its even higher acidity (relative to MTO). Simple addition of bases is not a viable solution, as the resulting perrhenate salts are catalytically inactive. Sheldon et al. discovered that a combination of perrhenic acid with tertiary arsines forms a catalytic system which is able to efficiently epoxidize a variety of alkenes with aqueous hydrogen peroxide. With regard to the solvent, TFE once again proved the best and gave high epoxide yields at high reaction rates (Scheme 4.9) [47]. As little as 0.1 mol% catalyst loading is sufficient to epoxidize, for example, cyclooctene in excellent yield. At somewhat higher catalyst loading (0.5 mol%), terminal alkenes such as 1-octene or 1-decene can be epoxidized in good yield as well. Less satisfactory results were obtained with very electron-rich alkenes such as phenyl allyl ether (26%).

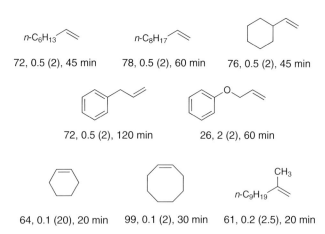

Scheme 4.9

4.3.2.5 Fluoroketones as Catalysts

Sheldon et al. observed that the addition of catalytic amounts of hexafluoroacetone to HFIP as solvent quite significantly improved the uncatalyzed epoxidation of alkenes observed in this solvent [38] (the epoxidation of alkenes with H_2O_2 in HFIP and in the absence of further catalysts is presented in Section 4.3.1, Scheme 4.4). As already mentioned, perfluoroketones have gained in importance as epoxidation catalysts, as they react reversibly with hydrogen peroxide to give a perhydrate, which is the active oxygen transfer agent (Scheme 4.5) [40]. The results obtained in HFIP, using 5 mol% of hexafluoroacetone as catalyst, are summarized in Scheme 4.10 [38].

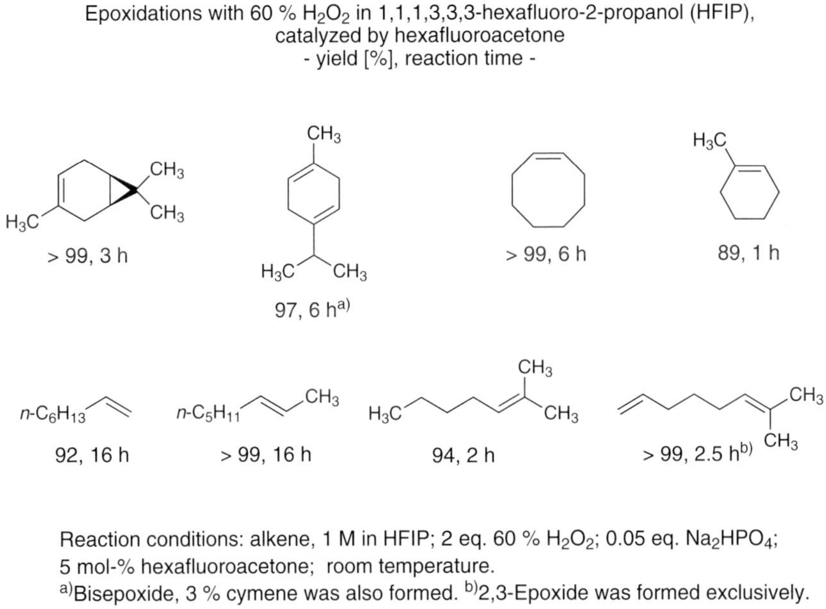

Epoxidations with 60 % H_2O_2 in 1,1,1,3,3,3-hexafluoro-2-propanol (HFIP), catalyzed by hexafluoroacetone
- yield [%], reaction time -

> 99, 3 h

97, 6 h^{a)}

> 99, 6 h

89, 1 h

92, 16 h

> 99, 16 h

94, 2 h

> 99, 2.5 h^{b)}

Reaction conditions: alkene, 1 M in HFIP; 2 eq. 60 % H_2O_2; 0.05 eq. Na_2HPO_4; 5 mol-% hexafluoroacetone; room temperature.
a) Bisepoxide, 3 % cymene was also formed. b) 2,3-Epoxide was formed exclusively.

Scheme 4.10

At ambient temperature and pressure, hexafluoroacetone is a (toxic) gas, and catalyst recovery is impractical. As a nonvolatile alternative, Sheldon et al. employed perfluoroheptadecan-9-one. After completion of the epoxidation, this catalyst can be recovered from halogenated solvents such as dichloroethane or TFE by simple cooling of the reaction mixture [39]. Furthermore, this long-chain perfluorinated ketone has the potential for immobilization in fluorous phases. When applied to epoxidation with perfluoroheptadecan-9-one as catalyst, TFE gave yields of epoxide comparable to those obtained in dichloroethane. However, to achieve good yields of acid-sensitive epoxides, buffering by Na_2HPO_4 was necessary.

Fluorinated alcohols, and in particular HFIP, have also proven beneficial for epoxidations of alkenes using persulfate (Oxone) as the terminal oxidant and fluoroketones as catalysts [9, 10]. Under these conditions, the fluorinated ketones are converted to dioxiranes, which are the active epoxidizing species. Typical catalysts are fluorinated derivatives of 2-nonanone or 2-decanone. However, as a terminal oxidant different from hydrogen peroxide is required, a detailed discussion of these catalytic epoxidations is not within the scope of this review.

4.4
Sulfoxidation of Thioethers in Fluorinated Alcohol Solvents

In 1998, Bégué et al. reported that the oxidation of thioethers with 1–2 equiv. of 30% aqueous hydrogen peroxide in HFIP at room temperature affords the corresponding sulfoxides in high yields and in high selectivity (Scheme 4.11) [9, 10, 48, 49]. Most remarkably, no sulfones were formed under these conditions – even the re-submission of sulfoxides to the reaction conditions did not result in further oxidation to the sulfone. Once again, hydrogen bond donation from the fluorinated alcohol to hydrogen peroxide most likely facilitates electrophilic oxygen transfer to the thioether. On the other hand, hydrogen bonding to the product sulfoxide renders the latter inert toward further oxidation (Scheme 4.11). Double bonds (C=C) remain unaffected under the reaction conditions, and no N-oxidation of, for example, pyridines occurred. The latter can once again be ascribed to involvement of the pyridine's lone pair in hydrogen bonding to the solvent. Overall, the sulfoxidation proceeds under neutral conditions, and acid-sensitive functional groups (such as glycosides, Scheme 4.11) are typically not affected. The sulfoxidation procedure can be combined with a fluoroalcohol-promoted ring-opening of epoxides by thiophenols, providing β-hydroxy sulfoxides in a one-pot procedure [48].

Under the same reaction conditions (i.e., fluorinated alcohol solvent, room temperature, 2 equiv. aqueous hydrogen peroxide), thiols are cleanly oxidized to disulfides (Scheme 4.12) [50].

4.5
Baeyer-Villiger Oxidation of Ketones in Fluorinated Alcohol Solvents

In 2000, Neumann and Neimann reported that the treatment of certain ketones with 60% aqueous hydrogen peroxide in HFIP as solvent, and in the absence of further catalysts, results in lactone formation (Scheme 4.13) [21]. After 20 h reaction time at 60 °C, good lactone yields were obtained from cycloalkanones. 2-Octanone, as a prototypical acyclic alkanone, reacted sluggishly, whereas no lactone formation was observed for acetophenone (Scheme 4.13).

Our independent study of ε-caprolactone formation from cyclohexanone revealed that this process is Brønsted-acid catalyzed [30]. As summarized in Table 4.2, basically no conversion of the starting cyclohexanone occurred within 4 h in acid-free HFIP. Unlike alkene epoxidation (see Section 4.3.2.2), the catalytic activity was not specific for the arsonic acid, but all Brønsted acids tested were catalytically active in HFIP as solvent. Para-toluenesulfonic acid was identified as the most practical catalyst and

4.5 Baeyer-Villiger Oxidation of Ketones in Fluorinated Alcohol Solvents | 137

Sulfoxidations in 1,1,1,3,3,3-hexafluoro-2-propanol (HFIP)
or 2,2,2-trifluoroethanol (TFE):
- yield of sulfoxide [%], reaction time -

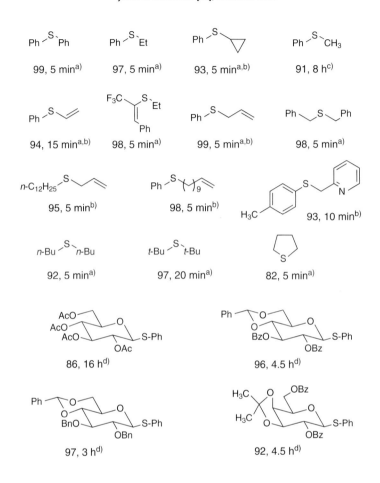

[a)] Bégué et al., ref. 48. Reaction conditions: sulfide 2 mmol, 30 % H_2O_2 4 mmol, 2. 5 mL HFIP, room temperature.
[b)] Bégué et al., ref. 49. Reaction conditions: sulfide 2 mmol, 30 % H_2O_2 4 mmol, 2. 5 mL HFIP, room temperature.
[c)] Bégué et al., ref. 51. Reaction conditions: sulfide 100 mmol, 30 % H_2O_2 110 mmol, 50 mL TFE, 0 °C - room temperature.
[d)] Bégué et al., ref. 49. Reaction conditions: sulfide 0.5 mmol, 30 % H_2O_2 1 mmol, 2. 5 mL HFIP, 25 °C; product obtained as 1:1-mixture of diastereomers.

Scheme 4.11

Oxidation of thiols and thiophenols to disulfides in 1,1,1,3,3,3-hexafluoro-2-propanol (HFIP) or 2,2,2-trifluoroethanol (TFE):
- yield of disulfide [%], reaction time -

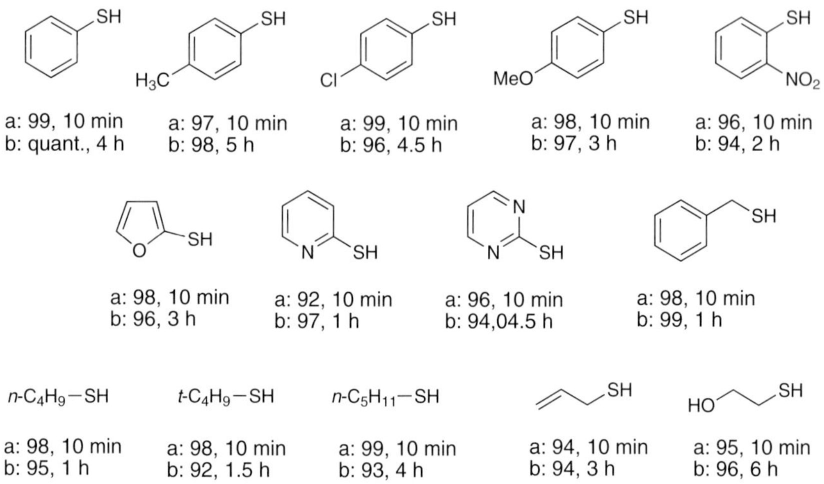

PhSH
a: 99, 10 min
b: quant., 4 h

4-MeC₆H₄-SH
a: 97, 10 min
b: 98, 5 h

4-ClC₆H₄-SH
a: 99, 10 min
b: 96, 4.5 h

4-MeOC₆H₄-SH
a: 98, 10 min
b: 97, 3 h

2-NO₂C₆H₄-SH
a: 96, 10 min
b: 94, 2 h

furfuryl-SH (2-furyl-SH)
a: 98, 10 min
b: 96, 3 h

2-pyridyl-SH
a: 92, 10 min
b: 97, 1 h

2-pyrimidyl-SH
a: 96, 10 min
b: 94, 04.5 h

BnSH
a: 98, 10 min
b: 99, 1 h

n-C₄H₉–SH
a: 98, 10 min
b: 95, 1 h

t-C₄H₉–SH
a: 98, 10 min
b: 92, 1.5 h

n-C₅H₁₁–SH
a: 99, 10 min
b: 93, 4 h

allyl-SH (CH₂=CHCH₂SH with extra CH₂)
a: 94, 10 min
b: 94, 3 h

HO-CH₂CH₂-SH
a: 95, 10 min
b: 96, 6 h

a: Reaction conditions: thiol/thiophenol 1 mmol, 30 % H₂O₂ 1.1 mmol, 1 mL HFIP, room temperature.
b: Reaction conditions: thiol/thiophenol 1 mmol, 30 % H₂O₂ 1.1 mmol, 1 mL TFE, room temperature.

Scheme 4.12

Oxidation of ketones to lactones and esters in 1,1,1,3,3,3-hexafluoro-2-propanol (HFIP):
- yield of lactone/ester [%], reaction time -

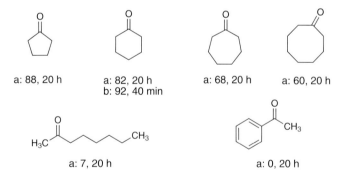

cyclopentanone: a: 88, 20 h
cyclohexanone: a: 82, 20 h; b: 92, 40 min
cycloheptanone: a: 68, 20 h
cyclooctanone: a: 60, 20 h

H₃C-CO-(CH₂)ₙ-O-CH₃ ester: a: 7, 20 h
PhC(O)OCH₃: a: 0, 20 h

a: Neumann and Neimann, ref. 21. Reaction conditions: ketone 1.2 mmol, 60 % H₂O₂ 2 mmol, 1 mL HFIP, 60 °C.
b: Berkessel and Andreae, ref. 30. Reaction conditions: ketone 19.3 mmol, 50 % H₂O₂ 24.7 mmol, 1 mol-% p-toluenesulfonic acid, 24 mL HFIP, 60 °C.

Scheme 4.13

Table 4.2

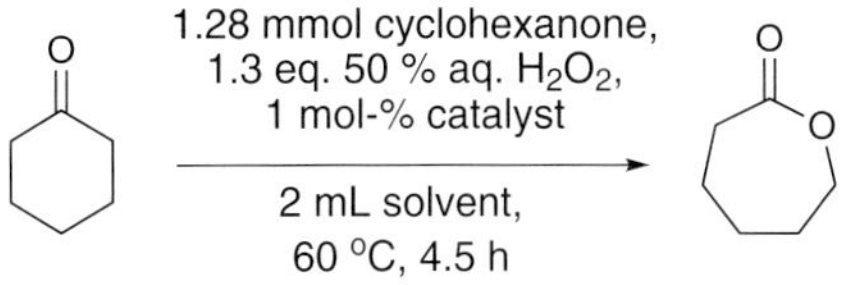

Solvent	Catalyst	Reaction time [min]	Conversion of ketone [%]	Yield of lactone [%]
HFIP	None	240	<1	<1
HFIP	Ph-AsO$_3$H$_2$	240	Quant.	85
HFIP	Ph-PO$_3$H$_2$	75	86	84
HFIP	HCl	65	90	65
HFIP	CF$_3$CO$_2$H	65	65	90
HFIP	p-TsOH	40	Quant.	92
1,4-Dioxane	Ph-AsO$_3$H$_2$	240	<1	<1

Reaction conditions: 1.28 mmol cyclohexanone, 1.3 eq. 50 % aq. H$_2$O$_2$, 1 mol-% catalyst, 2 mL solvent, 60 °C, 4.5 h.

afforded 92% of the lactone product after only 40 min reaction time at 60 °C. Note that only 1 mol% of the catalyst was used, at a very moderate excess of hydrogen peroxide (1.3 equiv. relative to ketone, 50% aqueous H$_2$O$_2$). The mechanism of acid catalysis under these conditions is discussed below in Section 4.5.1.

4.5.1
Acid-Catalyzed Baeyer-Villiger Oxidation of Ketones in Fluorinated Alcohol Solvents – Mechanism

Our detailed study of the mechanism of cyclohexanone conversion to ε-caprolactone revealed that it differs from that of the 'classical' Baeyer-Villiger reaction [53]. Unstrained ketones such as cyclohexanone are well known to form various peroxidic adducts with hydrogen peroxide [54]. In HFIP, only the formation of the spiro-bisperoxide 7,8,15,16-tetraoxadispiro[5.2.5.2]hexadecane could be observed. This spiro-bisperoxide is a compound that has been known since the 1940s [55]. In fact, it is an isomer of two equivalents of caprolactone, but numerous earlier attempts to isomerize it to ε-caprolactone in standard organic solvents under Brønsted- or Lewis-acid catalysis failed to produce reasonable yields of lactone. In HFIP, however, Brønsted-acids induce the spontaneous and quantitative rearrangement to two equivalents of ε-caprolactone. In summary, the oxidation of cyclohexanone with hydrogen peroxide is a 'Baeyer-Villiger-like' transformation, as it produces the same lactone product, but proceeds via a different mechanism involving the spiro-bisperoxide as an intermediate (Scheme 4.14) [53].

Typically, ω-hydroxycaproic acid is formed as a by-product (which may be overlooked, as its GC-detection requires prior silylation). The time-course of the reaction clearly indicated that this hydroxyacid does not result from subsequent hydrolysis of

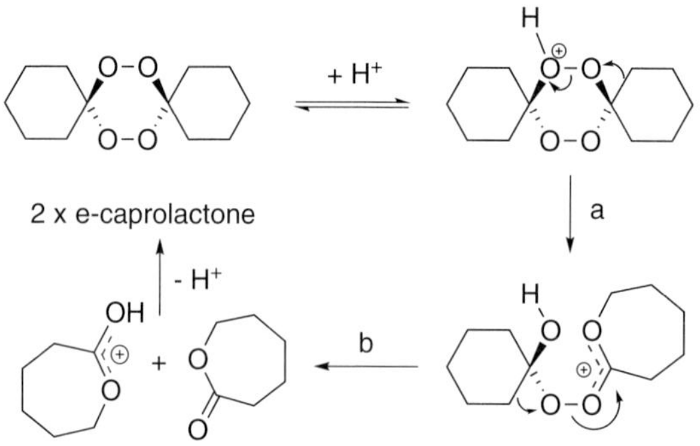

Scheme 4.14

the lactone product, but that it is formed by a separate pathway. Most likely, the lactonium cation shown as an intermediate in lactone formation (Scheme 4.14) is the branching point: attack of water results in ring opening and hydroxy acid formation (Scheme 4.15). We could show by ^{18}O-isotopic labeling (H$_2$^{18}O) and mass-spectrometric analysis that the attack of the water molecule occurs at the acyl carbon atom (path B, Scheme 4.15) and not in the sense of a nucleophilic attack at the ring carbon atoms (path A, Scheme 4.15) [53]. In the former case, the oxygen label should end up in the hydroxy function of the by-product, whereas incorporation into the carboxyl group should result from pathway B. In fact, exclusive incorporation into the carboxyl group was observed experimentally.

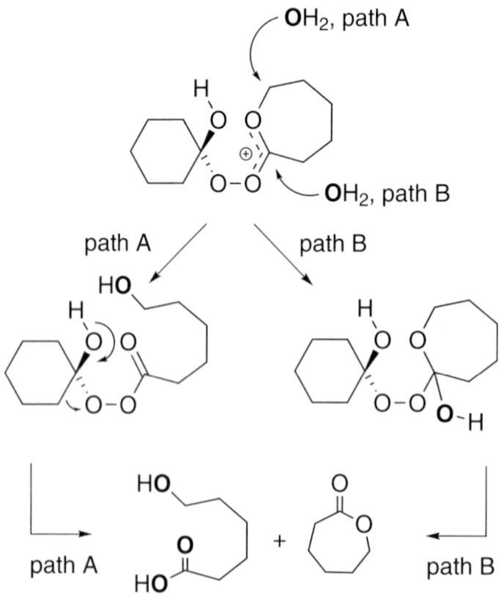

Scheme 4.15

4.5.2
Baeyer-Villiger Oxidation of Ketones in Fluorinated Alcohol Solvents – Catalysis by Arsonic and Seleninic Acids

It was noted in Sections 4.3.2.2 and 4.3.2.3 that arsonic acids and seleninic acids are efficient catalysts for the epoxidation of alkenes. For both types of catalyst, significant enhancement of catalyst activity and selectivity was observed in fluorinated alcohol solvents compared to, for example, 1,4-dioxane.

Arsonic acids: Arene arsonic acids, together with polymer-bound variants, have also been applied to the catalysis of the Baeyer-Villiger oxidation of ketones by Jacobson et al. as early as 1979 [56]. Typically, 1,4-dioxane was used as the solvent, together with high-concentration hydrogen peroxide (90%). Using an up to fivefold excess of hydrogen peroxide, the Baeyer-Villiger oxidation of a number of substrates can be performed efficiently. For example, with a catalyst loading of about 10 mol%, 2-methylcyclohexanone yields 80% of methyl ε-caprolactone after 7 h at 80 °C. Apparently, no fluorinated alcohol solvents were tested in this study [56]. The relative behavior of the various substrates used led to the conclusion that under these conditions (1,4-dioxane, arsonic acid catalysts, high-concentration H_2O_2), a 'normal' Baeyer-Villiger oxidation, initiated by the attack of the persarsonic acid on the ketone, is operating. As mentioned above (Sections 4.5 and 4.5.1), changing to HFIP as solvent, in combination with simple Brønsted-acid catalysts such as *p*-TsOH, leads to a highly efficient catalytic system that operates by a different mechanism, that is, via intermediate formation of a spiro-bisperoxide and its subsequent rearrangement.

Seleninic acids: Seleninic acids of the type used for alkene epoxidation (cf. Section 4.3.2.3) have also been employed as catalysts for the Baeyer-Villiger oxidation of ketones with hydrogen peroxide, mainly by Syper [57] and by Sheldon et al. [58, 59]. In most cases, halogenated solvents such as dichloromethane or 1,2-dichloroethane were used. In a study of solvent effects, Sheldon et al. observed that, once again, TFE and in particular 1,1,1,3,3,3-hexafluoro-2-propanol are superior to dichloromethane with regard to selectivity and rate [58]. However, the effects are not as pronounced as in the case of alkene epoxidation (e.g., a factor of about 2 in rate between dichloromethane and HFIP, and 1.3 for TFE). The Baeyer-Villiger oxidation of a series of ketones and aldehydes was studied in TFE, and the results are summarized in Table 4.3 [58].

Once again, the relative rates and product distributions follow the pattern typical for 'classical' Baeyer-Villiger oxidations. Together with the relatively low accelerating effect of the fluorinated alcohol solvent, it may be concluded that, also under these conditions of selenium catalysis, a 'classical' mechanism based on perseleninic acid is operating. Finally, it should be mentioned that diselenides (as pre-catalysts, see Section 4.3.2.3) with long-chain perfluoroalkyl substituents have been synthesized and successfully applied to Baeyer-Villiger oxidations in fluorous bi- and triphasic systems [59]. Several chiral diselenides for the *in-situ* formation of chiral (per)seleninic acids have been synthesized by Uemura et al. and tested in the asymmetric Baeyer-Villiger oxidation of a number of ketones [60]. High chemical yields and enantiomeric excesses up to 19% *ee* were observed. However, 1,4-dioxane, DME, and THF were used as solvents in this study, and no fluorinated alcohol solvents appear to have been tested.

Table 4.3

pre-catalyst: 3,5-bis(trifluoromethyl)phenyl diselenide [(3,5-(CF$_3$)$_2$C$_6$H$_3$)Se-Se(3,5-(CF$_3$)$_2$C$_6$H$_3$)]

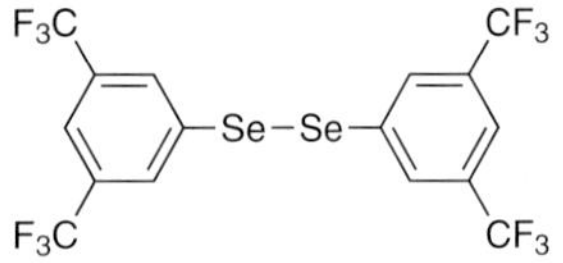

Substrate	Product	Reaction time [h]	Conversion [%]	Selectivity [%]
Cyclobutanone	γ-Butyrolactone	1	99	90
Cyclopentanone	δ-Valerolactone	8	95	94
Cyclohexanone	ε-Caprolactone	4	95	99
Adamantanone	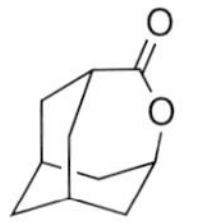 (lactone)	1	100	99
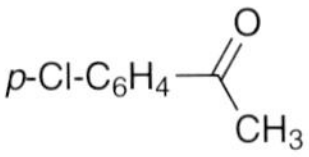 p-Cl-C$_6$H$_4$-CO-CH$_3$	p-Cl-C$_6$H$_4$O-Ac	24	25	50
	p-Cl-C$_6$H$_4$OH			50
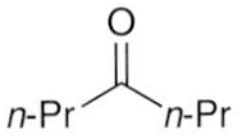 n-Pr-CO-n-Pr	n-PrCO$_2$H	24	25	50
	n-PrCO$_2$-n-Pr			50
3,4,5-(MeO)$_3$-C$_4$H$_2$-CHO	3,4,5-(MeO)$_3$-C$_4$H$_2$-OH	0.25	99	99[a]
p-Me-C$_4$H$_4$-CHO	p-Me-C$_4$H$_4$-OH	2	99	55[a]
	p-Me-C$_4$H$_4$-CO$_2$H			45
p-NO$_2$-C$_4$H$_4$-CHO	p-NO$_2$-C$_4$H$_4$-CO$_2$H	2	98	98[b]
n-C$_7$H$_{15}$-CHO	n-C$_7$H$_{15}$-CO$_2$H	3	88	96[b]
Ph-C$_2$H$_4$-CHO	Ph-C$_2$H$_4$-CO$_2$H	3	>90	99[b]
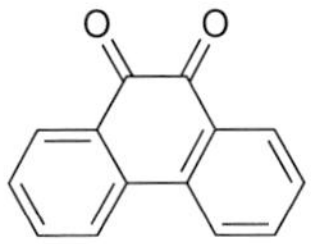 phenanthrenequinone	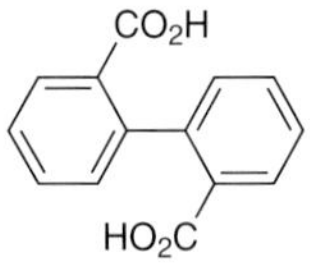 diphenic acid (2,2'-(HO$_2$C)$_2$-biphenyl)	16	>98	95

Reaction conditions: 2 mmol substrate, 1.0 mol% diselenide, 4 mmol 60% H$_2$O$_2$, in 2 mL 2,2,2-trifluoroethanol (TFE), $T = 20\,°C$.
a) Selectivity to substituted phenol after base hydrolysis.
b) 60 °C.

4.6
Epilog

Over recent years, it has become obvious that fluorinated alcohol solvents are excellent media for effecting various types of oxidations with hydrogen peroxide as the terminal oxidant. The stability and easy recovery of these solvents, their excellent solubilizing properties and miscibility, together with the high selectivities of the transformations and the advantageous properties of hydrogen peroxide make these systems almost 'ideal'. Mechanistic studies have revealed that the catalytic effects of fluorinated alcohol solvents hinge mostly on their pronounced hydrogen bond donor ability, together with low nucleophilicity. In some cases, a detailed picture of how these supramolecular solvent-substrate interactions operate could be obtained. As an appealing next step, it may be envisaged that synthetic chemistry may provide 'second generation' fluorinated solvents/catalysts that incorporate these features and do not depend on the statistics of supramolecular chemistry for the formation of active aggregates.

References

1 Earlier review on fluorinated alcohols as reaction medium: Bégué, J.-P., Bonnet-Delpon, D., and Crousse, B. (2004) *Synlett*, 18–29.
2 Imperiali, B. and Abeles, R.H. (1986) *Tetrahedron Lett.*, **27**, 135–138.
3 (a) Thaisrivongs, S., Pals, D.T., Kati, W.M., Turner, S.R., Thomasco, L.M., and Watt, W. (1986) *J. Med. Chem.*, **29**, 2080–2087; (b) Fearon, K., Spaltenstein, A., Hopkins, P.B., and Gelb, M.H. (1987) *J. Med. Chem.*, **30**, 1617–1622.
4 (a) Arends, I.W.C.E. and Sheldon, R.A. (2004) *Modern Oxidation Methods* (ed. J.-E. Bäckvall), Wiley-VCH, Weinheim, pp. 83–118; (b) Brinksma, J., de Boer, J.W., Hage, R., and Feringa, B.L., *ibid*. 295–326.
5 Arends, I.W.C.E. and Sheldon, R.A. (2002) *Top. Catal.*, **19**, 133–141.
6 Strukul, G. (2003) *Catalytic Oxidations with Hydrogen Peroxide as Oxidant*, Kluwer.
7 Berkessel, A. and Etzenbach-Effers, K. (2009) *Hydrogen Bonding in Organic Synthesis* (ed. P. Pihko), Wiley-VCH, Weinheim, pp. 15–38.
8 Berkessel, A. and Etzenbach-Effers, K. (2010) *Top. Curr. Chem.*, **291**, 1–27.
9 Legros, J., Crousse, B., Bourdon, J., Bonnet-Delpon, D., and Bégué, J.-P. (2001) *Tetrahedron Lett.*, **42**, 4463–4466.
10 Legros, J., Crousse, B., Bonnet-Delpon, D., and Bégué, J.-P. (2002) *Tetrahedron*, **58**, 3993–3998.
11 Eberson, L., Hartshorn, M.P., and Persson, O. (1995) *J. Chem. Soc., Perkin Trans 2*, 1735–1744.

12 (a) Kamlet, M.J., Abboud, J.M.L., Abraham, M.H., and Taft, R.W. (1983) *J. Org. Chem.*, **48**, 2877–2887; (b) Middleton, W.J. and Lindsey, R.V. (1964) *J. Am. Chem. Soc.*, **86**, 4948–4952.
13 (a) Schadt, F.L., Bentley, T.W., and Schleyer, P.v.R. (1976) *J. Am. Chem. Soc.*, **98**, 7667–7674; (b) Allard, B., Casadevall, A., Casadevall, E., and Largeau, C. (1979) *Nouv. J. Chim.*, **3**, 335–342; (c) Allard, B., Casadevall, A., Casadevall, E., and Largeau, C. (1980) *Nouv. J. Chim.*, **4**, 539–545; (d) Leonard, N.J. and Neelima, A. (1995) *Tetrahedron Lett.*, **36**, 7833–7836.
14 (a) Murto, J., Kivinen, A., and Strandman, L. (1971) *Suomen Kemistilehti B*, **44**, 308; (b) Schall, H., Häber, T., and Suhm, M.A. (2000) *J. Phys. Chem. A*, **104**, 265–274.
15 (a) Reichard, C. (1994) *Chem. Rev.*, **94**, 2319–2358; (b) Reichardt, C. (2002) *Solvents and Solvent Effects in Organic Chemistry*, 3rd edn, Wiley-VCH, Weinheim.
16 Berkessel, A., Adrio, J.A., Hüttenhain, D., and Neudörfl, J.M. (2006) *J. Am. Chem. Soc.*, **128**, 8421–8426.
17 Maiti, N.C., Zhu, Y., Carmichael, I., Serianni, A.S., and Anderson, V.E. (2006) *J. Org. Chem.*, **71**, 2878–2880.
18 Berkessel, A. and Adrio, J.A. (2006) *J. Am. Chem. Soc.*, **128**, 13412–13420.
19 Shrine, T.M. and Kim, L. (1977) US 4024165 (Patent to Shell Oil Company).

20 Romanelli, M.G. (1983) Exxon EP 0096130.
21 Neimann, K. and Neumann, R. (2000) *Org. Lett.*, **2**, 2861–2863.
22 van Vliet, M.C.A., Arends, I.W.C.E., and Sheldon, R.A. (2001) *Synlett*, 248–250.
23 Berkessel, A. and Adrio, J.A. (2004) *Adv. Synth. Catal.*, **346**, 275–280.
24 Dryuk, V.G. (1976) *Tetrahedron*, **32**, 2855–2866.
25 de Visser, S.P., Kaneti, J., Neumann, R., and Shaik, S. (2003) *J. Org. Chem.*, **68**, 2903–2912.
26 Cleland, W.W., Frey, P.A., and Gerlt, J.A. (1998) *J. Biol. Chem.*, **273**, 25529–25532.
27 Johnstone en Francsicsné-Czinege, R.A.W.,WO 98/17640, 30 April 1998.
28 van Vliet, M.C.A., Arends, I.W.C.E., and Sheldon, R.A. (1999) *Tetrahedron Lett.*, **40**, 5239–5242.
29 Jacobson, S.E., Mares, F., and Zambri, P.M. (1979) *J. Am. Chem. Soc.*, **101**, 6946–6950.
30 Berkessel, A. and Andreae, M.R.M. (2001) *Tetrahedron Lett.*, **42**, 2293–2295.
31 For a review on selenium-promoted oxidations of organic compounds, see: Mlochowski, J., Brzaszcz, M., Giurg, M., Palus, J., and Wojtowicz, H. (2003) *Eur. J. Org. Chem.*, 4329–4339.
32 Syper, L. and Mlochowski, J. (1987) *Tetrahedron*, **43**, 207–213.
33 Grieco, P.A., Yokoyama, Y., Gilman, S., and Nishizawa, M. (1977) *J. Org. Chem.*, **42**, 2034–2036.
34 ten Brink, G.-J., Fernandes, B.C.M., van Vliet, M.C., Arends, I.W.C.E., and Sheldon, R.A. (2001) *J. Chem. Soc., Perkin Trans. 1*, 224–228.
35 For a report on selenium-catalyzed epoxidations with hydrogen peroxide in perfluroalkanes (not fluorinated alcohols), see: Betzemeier, B., Lhermitte, F., and Knochel, P. (1999) *Synlett*, 489–491.
36 Adolfsson, H. (2004) *Modern Oxidation Methods* (ed. J-.E. Bäckvall), Wiley-VCH, Weinheim, pp. 21–49.
37 Romao, C.C., Kühn, F.E., and Herrmann, W.A. (1997) *Chem. Rev.*, **97**, 3197–3246.
38 van Vliet, M.C.A., Arends, I.W.C.E., and Sheldon, R.A. (2001) *Synlett*, 1305–1307.
39 van Vliet, M.C.A., Arends, I.W.C.E., and Sheldon, R.A. (1999) *Chem. Commun.*, 263–264.
40 Adam, W., Saha-Möller, C.R., and Ganeshpure, P.A. (2001) *Chem. Rev.*, **101**, 3499–3548.
41 Herrmann, W.A., Fischer, R.W., and Marz, D.W. (1991) *Angew. Chem. Int. Ed.*, **30**, 1638–1641.
42 Herrmann, W.A., Fischer, R.W., Scherer, W., and Rauch, M.U. (1993) *Angew. Chem. Int. Ed.*, **32**, 1157–1160.
43 Rudolph, J., Reddy, K.L., Chiang, J.P., and Sharpless, K.B. (1997) *J. Am. Chem. Soc.*, **119**, 6189–6190.
44 Copéret, C., Adolfsson, H., and Sharpless, K.B. (1997) *Chem. Commun.*, 1565–1566.
45 Herrmann, W.A., Kühn, F.E., Fischer, R.W., Thiel, W.R., and Ramao, C.C. (1992) *Inorg. Chem.*, **31**, 4431–4432.
46 Herrmann, W.A., Kratzer, R.M., and Fischer, R.W. (1997) *Angew. Chem., Int. Ed. Engl.*, **36**, 2652–2654.
47 van Vliet, M.C.A., Arends, I.W.C.E., and Sheldon, R.A. (2000) *J. Chem. Soc., Perkin Trans. 1*, 377–380.
48 Ravikumar, K.S., Bégué, J.P., and Bonnet-Delpon, D. (1998) *Tetrahedron Lett.*, **39**, 3141–3144.
49 Ravikumar, K.S., Zhang, Y.M., Bégué, J.P., and Bonnet-Delpon, D. (1998) *Eur. J. Org. Chem.*, 2937–2940.
50 Kesavan, V., Bonnet-Delpon, D., and Bégué, J.-P. (2000) *Tetrahedron Lett.*, **41**, 2895–2898.
51 Ravikumar, K.S., Kesavan, V., Crousse, B., Bonnet-Delpon, D., and Bégué, J.P. (2003) *Org. Synth., Coll.*, **80**, 184–186.
52 Kesavan, V., Bonnet-Delpon, D., and Bégué, J.P. (2000) *Synthesis*, 223–225.
53 Berkessel, A., Andreae, M.R.M., Schmickler, H., and Lex, J. (2002) *Angew. Chem. Int. Ed.*, **41**, 4481–4484.
54 Berkessel, A. and Vogl, N. (2006) *The Chemistry of Peroxides*, vol. 2 (ed. Z. Rappoport), Wiley, Chichester, pp. 307–596.
55 Dilthey, W., Inckel, M., and Stephan, H. (1940) *J. Prakt. Chem.*, **154**, 219–237.

56 Jacobson, S.E., Mares, F., and Zambri, P.M. (1979) *J. Am. Chem. Soc.*, **101**, 6938–6946.
57 Syper, L. (1989) *Synthesis*, 167–172.
58 ten Brink, G.-J., Vis, J.-M., Arends, I.W.C.E., and Sheldon, R.A. (2001) *J. Org. Chem.*, **66**, 2429–2433.
59 ten Brink, G.-J., Vis, J.-M., Arends, I.W.C.E., and Sheldon, R.A. (2002) *Tetrahedron*, **58**, 3977–3983.
60 Miyake, Y., Nishibayashi, Y., and Uemura, S. (2002) *Bull. Chem. Soc. Jpn.*, **75**, 2233–2237.

5
Modern Oxidation of Alcohols using Environmentally Benign Oxidants
Isabel W.C.E. Arends and Roger A. Sheldon

5.1
Introduction

The oxidation of primary and secondary alcohols to the corresponding carbonyl compounds plays a central role in organic synthesis [1]. However, standard organic textbooks [2] still recommend classical oxidation methods using stoichiometric quantities of inorganic oxidants, notably chromium(VI) reagents [3] or ruthenium or manganese salts [4], which are highly toxic and environmentally polluting. Other classical non-green methods are based on the use of high-valent iodine compounds (notably the Dess Martin reagent) or involve the stoichiometric use of DMSO (Swern oxidation) [4]. However, the state of the art in alcohol oxidation nowadays is far better. Numerous catalytic methods are now known which can be used to oxidize alcohols using either O_2 or H_2O_2 as the oxidant. These oxidants are to be preferred because they are inexpensive and produce water as the sole by-product. In this review we focus on the use of metal catalysts to mediate the selective oxidation of alcohols using O_2 or H_2O_2 as the primary oxidant. Mainly homogeneous catalysts are described, but, where relevant, heterogeneous catalysts, in particular metal nanoparticles, are covered as well. For an excellent review on heterogeneous oxidation of alcohols and carbohydrates we refer to the publications by Gallezot *et al.* [5]. Before turning to metal-mediated oxidation of alcohols, we first describe the recent developments in catalytic oxoammonium-mediated oxidation of alcohols.

5.2
Oxoammonium based Oxidation of Alcohols – TEMPO as Catalyst

A very useful and frequently applied method in the fine chemical industry to convert alcohols into the corresponding carbonyl compounds is the use of oxoammonium salts as oxidants, as shown in Figure 5.1 [6]. These are very selective oxidants for alcohols, operate under mild conditions, and tolerate a large variety of functional groups. The oxidation proceeds in acidic as well as alkaline media.

Modern Oxidation Methods. Edited by Jan-Erling Bäckvall
Copyright © 2010 WILEY-VCH Verlag GmbH & Co. KGaA, Weinheim
ISBN: 978-3-527-32320-3

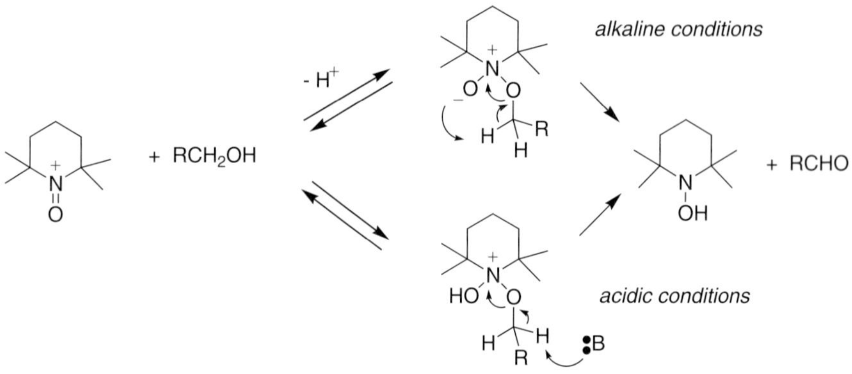

Figure 5.1 Mechanism for the oxoammonium-catalyzed oxidation of alcohols.

The oxoammonium is generated *in situ* from its precursor, 2,2′,6,6′-tetramethylpiperidine-*N*-oxyl (TEMPO), or derivatives thereof, which is used in catalytic quantities (see Figure 5.2). Various oxidants can be applied as the final oxidant [7–12]. In particular, the TEMPO-bleach protocol using bromide as co-catalyst introduced by Anelli *et al.* is finding wide application in organic synthesis [7]. TEMPO is used in amounts as low as 1 mol% relative to the substrate, and full conversion of substrates can commonly be achieved within 30 min.

The major drawbacks of this method are the use of NaOCl as the oxidant, the need for addition of 10 mol% bromide as a cocatalyst, and the necessity to use chlorinated solvents. Recently a great deal of effort has been devoted toward a greener oxoammonium-based method, for example, by replacing TEMPO by heterogeneous variations or replacing NaOCl by a combination of metal as cocatalyst and molecular oxygen as oxidant. Examples of heterogeneous variants of TEMPO are anchoring TEMPO to solid supports such as silica [13, 14] and the mesoporous silica,

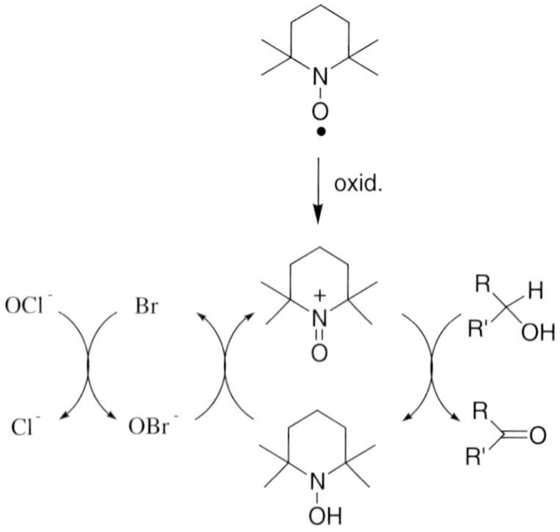

Figure 5.2 TEMPO-catalyzed oxidation of alcohols using hypochlorite as the oxidant.

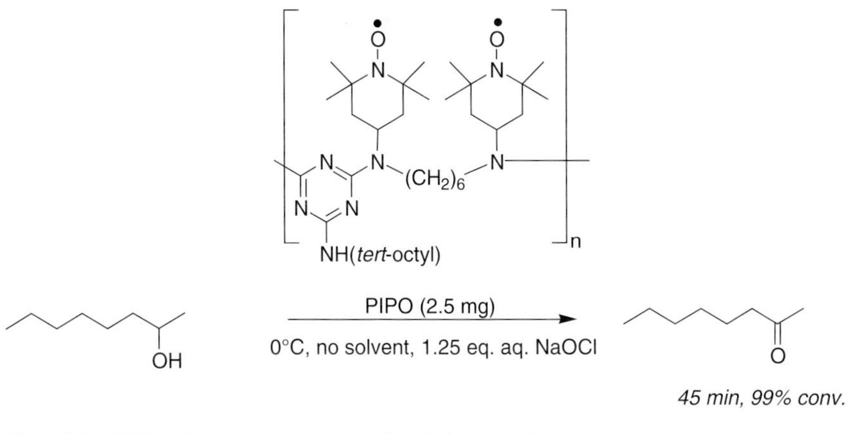

5.2 Oxoammonium based Oxidation of Alcohols – TEMPO as Catalyst

Figure 5.3 PIPO as heterogeneous catalyst for alcohol oxidation.

MCM-41, [15] or by entrapping TEMPO in sol-gel [16]. Alternatively, in our group we developed an oligomeric TEMPO (Figure 5.3) derived from Chimassorb 944 [17].

This new polymer-immobilized TEMPO, which we refer to as PIPO (polymer-immobilized piperidinyl oxyl), proved to be a very effective catalyst for the oxidation of alcohols with hypochlorite (Figure 5.3) [17]. Under the standard conditions, PIPO is dissolved in the dichloromethane layer. In contrast, in the absence of solvent, PIPO was a very effective recyclable heterogeneous catalyst. Furthermore, the enhanced activity of PIPO compared to TEMPO made the use of a bromide cocatalyst redundant. Hence, the use of PIPO, in an amount equivalent to 1 mol% of nitroxyl radical, provided an effective (heterogeneous) catalytic method for the oxidation of a variety of alcohols with 1.25 equiv. of 0.35 M NaOCl (pH 9.1) in a bromide-free and chlorinated hydrocarbon solvent-free medium (Figure 5.4). Under these environmentally benign conditions, PIPO was superior to the already mentioned heterogeneous TEMPO systems and homogeneous TEMPO [18]. In the solvent-free system, primary alcohols such as octan-1-ol gave low selectivities to the corresponding aldehyde owing to overoxidation to the carboxylic acid. This problem was circumvented by using n-hexane as the solvent, in which PIPO is insoluble, affording an increase in aldehyde selectivity from 50 to 94%.

Various approaches have been used to replace the hypochlorite oxidant with the greener molecular oxygen or air. For example, Neumann and coworkers showed that a heteropolyacid, a known redox catalyst, was able to generate oxoammonium ions in-situ with 2 atm of molecular oxygen at 100 °C [19]. In another approach, a combination of manganese and cobalt (5 mol%) was able to generate oxoammonium ions under acidic conditions at 40 °C [20]. Results for these two methods are compared in Table 5.1. Although these conditions are still susceptible to improvement, both processes use molecular oxygen as the ultimate oxidant and are chlorine free, and therefore constitute valuable examples of progress in this area. Later, we discuss examples where the use of TEMPO in combination with a Ru or Cu catalyst results in even more active catalytic systems. However, the mechanism in these cases is metal-based instead of oxoammonium-based and is therefore listed in the

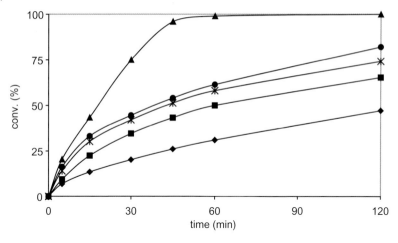

Figure 5.4 Bleach oxidation of octan-2-ol under chlorinated hydrocarbon solvent- and bromide-free conditions using 1 mol% of nitroxyl catalyst: (▲) PIPO (3.19 mmol/g; amine linker) [17]; (●) MCM-41 TEMPO (0.60 mmol/g; ether linker) [15]; (∗) SiO$_2$ TEMPO (0.87 mmol/g, amine linker) [13]; (■) SiO$_2$ TEMPO (0.40 mmol/g, amide linker) [14]; (◆) TEMPO.

Table 5.1 Aerobic oxoammonium-based oxidation of alcohols.

Substrate	Aldehyde or ketone yield[a]		
	2 mol% Mn(NO$_3$)$_2$ 2 mol% Co(NO$_3$)$_2$ 10 mol% TEMPO acetic acid, 40 °C 1 atm O$_2$[b]	1 mol% H$_5$[PMo$_{10}$V$_2$O$_{40}$] 3 mol% TEMPO acetone, 100 °C 2 atm O$_2$[c]	Laccase (5.9 U mL^{-1}); Water pH 4.5; 15 mol% TEMPO; 25 °C, 1 atm air[d]
n-C$_6$H$_{13}$CH$_2$OH	97% (6 h)		
n-C$_7$H$_{15}$CH$_2$OH		98% (18 h)	
n-C$_9$H$_{19}$CH$_2$OH			
n-C$_7$H$_{15}$CH(CH$_3$)OH	100% (5 h)		
n-C$_6$H$_{13}$CH(CH$_3$)OH		96% (18 h)	
PhCH$_2$OH	98% (10 h)[e]	100% (6 h)	87% (7 h)
PhCH(CH$_3$)OH	98% (6 h)[e]		67% (5.5 h)
cis-C$_3$H$_7$CH=CHCH$_2$OH		100% (10 h)	
PhCH=CHCH$_2$OH	99% (3 h)		94% (24 h)

a) GLC Yields.
b) Cecchetto et al. [20].
c) Daniel et al. [19].
d) Arends et al. [22].
e) Reaction performed at 20 °C with air.

appropriate section. Another approach to the generation of oxoammonium ions *in situ* is an enzymatic one. Laccase, an abundant highly potent redox enzyme, is capable of oxidizing TEMPO to the oxoammonium ion [21]. Although this method requires large amounts of TEMPO (up to 30 mol%) and long reaction times (24 h), it demonstrates that a combination of laccase and TEMPO is able to catalyze the aerobic oxidation of alcohols. In a more recent study [22] this method was optimized, and 10 mol% TEMPO was sufficient to give good conversion and excellent selectivity (see Table 5.1). The combinations of TEMPO with $NaNO_2$ and 1,3-dibromo-5,5-dimethylhydantoin [23] or Fe(III) salts [24] as cocatalysts have also been shown to catalyze the aerobic oxidation of alcohols in water at 80 °C.

5.3
Metal-Mediated Oxidation of Alcohols – Mechanism

Noble metal salts, for example, of Pd(II) or Pt(II), undergo reduction by primary and secondary alcohols in homogeneous solution. Indeed, the ability of alcohols to reduce Pd(II) was described as early as 1828 by Berzelius, who showed that K_2PdCl_4 was reduced to palladium metal in an aqueous ethanolic solution [25]. The reaction involves a β-hydride elimination from an alkoxymetal intermediate and is a commonly used method for the preparation of noble metal hydrides (Eq. (5.1)). In the presence of dioxygen this leads to catalytic oxidative dehydrogenation of the alcohol, for example, with palladium salts [26–30].

$$M-O-CHR_2 \longrightarrow MH + R_2C=O \qquad (5.1)$$

The aerobic oxidation of alcohols catalyzed by low-valent late transition metal ions, particularly those of Group VIII elements, involves an oxidative dehydrogenation mechanism. In the catalytic cycle (see Figure 5.5) a hydridometal species, formed by β-hydride elimination from an alkoxymetal intermediate, is reoxidized by dioxygen, presumably via insertion of O_2 into the M–H bond with formation of H_2O_2. Alternatively, an alkoxymetal species can decompose to a proton and the reduced form of the catalyst (see Figure 5.5), either directly or via the intermediacy of a hydridometal intermediate. These reactions are promoted by bases as cocatalysts, which presumably facilitate the formation of an alkoxymetal intermediate and/or β-hydride elimination. Examples of metal ions that operate via this pathway are Pd (II), Ru(II) and Rh(III).

Metal-catalyzed oxidations of alcohols with peroxide reagents can be conveniently divided into two categories involving peroxometal and oxometal species, respectively, as the active oxidant (Figure 5.6). In the peroxometal pathway the metal ion remains in the same oxidation state throughout the catalytic cycle, and no stoichiometric oxidation is observed in the absence of the peroxide. In contrast, oxometal pathways involve a two-electron change in the oxidation state of the metal ion, and a stoichiometric oxidation is observed, with the oxidized form of the catalyst, in the

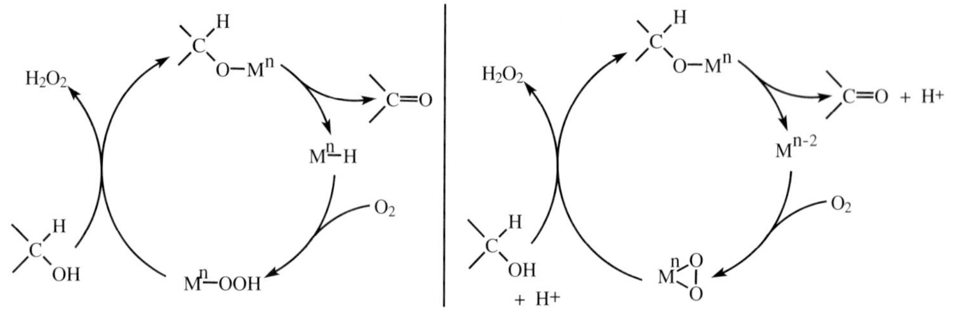

Figure 5.5 Hydridometal pathways for alcohol oxidation.

absence of, for example, H_2O_2. Indeed, this is a test for distinguishing between the two pathways.

Peroxometal pathways are typically observed with early transition metal ions with a d^0 configuration, for example, Mo(VI), W(VI), Ti(IV), and Re(VII), which are relatively weak oxidants. Oxometal pathways are characteristic of late transition elements and first row transition elements, for example, Cr(VI), Mn(V), Os(VIII), Ru(VI), and Ru(VIII), which are strong oxidants in high oxidation states. Some metals can operate via both pathways depending, among other things, on the substrate; for example, vanadium(V) operates via a peroxometal pathway in alkene epoxidations, but an oxometal pathway is involved in alcohol oxidations [31].

In some cases, notably ruthenium, the aerobic oxidation of alcohols is catalyzed by both low- and high-valent forms of the metal (see later). In the former case the reaction involves (see Figure 5.5) the formation of a hydridometal species (or its equivalent), while the latter involves an oxometal intermediate (see Figure 5.6), which is regenerated by reaction of the reduced form of the catalyst with dioxygen instead of

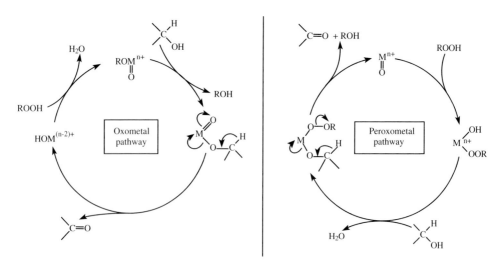

Figure 5.6 Oxometal versus peroxometal pathways in metal-catalyzed alcohol oxidations.

a peroxide. It is difficult to distinguish between the two, and one should bear in mind, therefore, that aerobic oxidations with high-valent oxometal catalysts could involve the formation of low-valent species, even metal nanoparticles, as the actual catalyst.

5.4 Ruthenium-Catalyzed Oxidations with O_2

Ruthenium compounds are widely used as catalysts in organic synthesis [32, 33] and have been extensively studied as catalysts for the aerobic oxidation of alcohols [34]. In 1978, Mares and coworkers [35] reported that $RuCl_3 \cdot nH_2O$ catalyzes the aerobic oxidation of secondary alcohols into the corresponding ketones, albeit in modest yields. In 1981, Ito and Matsumoto showed that $RuCl_3$ and $RuCl_2(Ph_3P)_3$ catalyze the aerobic oxidation of activated allylic and benzylic alcohols under mild conditions [36]; for example, the oxidation of retinol to retinal could be performed at 25 °C (57% yield was obtained after 48 h). Aliphatic primary and secondary alcohols were more efficiently oxidized using trinuclear ruthenium carboxylates, $Ru_3O(O_2CR)_6L_n$ (L = H_2O, Ph_3P) as the catalysts [37]. With lower aliphatic alcohols, for example, 1-propanol, 2-propanol, and 1-butanol, activities were about 10 times higher than with $RuCl_3$ and $RuCl_2(PPh_3)_3$. Somewhat higher activities were reached using $RuCl_2(PPh_3)_3$ as the catalyst with ionic liquids as solvents (Figure 5.7). The latter have been tested as environmentally friendly solvents for a large variety of reactions [38]. In this particular case tetramethylammoniumhydroxide and aliquat® 336 (tricaprylylmethylammonium chloride) were used as the solvent, and rapid conversion of benzyl alcohol was observed [39]. Moreover the tetramethylammonium hydroxide/$RuCl_2(PPh_3)_3$ could be reused after extraction of the product.

Ruthenium compounds are widely used as catalysts for hydrogen transfer reactions. These systems can be readily adapted to the aerobic oxidation of alcohols by employing dioxygen, in combination with a hydrogen acceptor as a cocatalyst, in a multistep process. For example, Bäckvall and coworkers [40] used low-valent ruthenium complexes in combination with a benzoquinone and a cobalt-Schiff's base complex. The coupled catalytic cycle is shown in Figure 5.8. A low-valent ruthenium complex reacts with the alcohol to afford the aldehyde or ketone product and a ruthenium dihydride. The latter undergoes hydrogen transfer to the benzoquinone

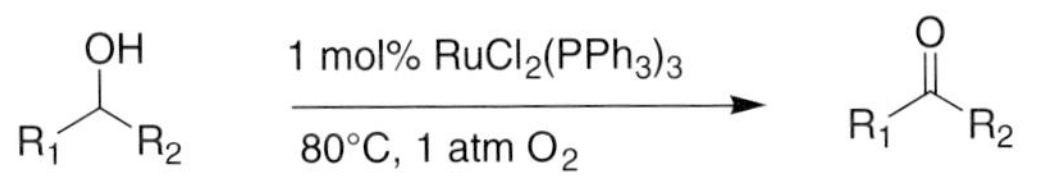

$$R_1R_2CHOH \xrightarrow[80°C,\ 1\ atm\ O_2]{1\ mol\%\ RuCl_2(PPh_3)_3} R_1R_2C=O$$

solvent:	tetramethylammonium hydroxide	aliquat
R_1=$PhCH_2$, R_2=H	91% conv. (5h)	58% conv. (5h)
R_1, R_2, = c-C_7H_{14}	61% conv. (11h)	92% conv. (11h)
R_1=C_6H_{13}, R_2=CH_3	43% conv. (25h)	81% conv. (25h)

Figure 5.7 Aerobic Ru-catalyzed oxidation in ionic liquids.

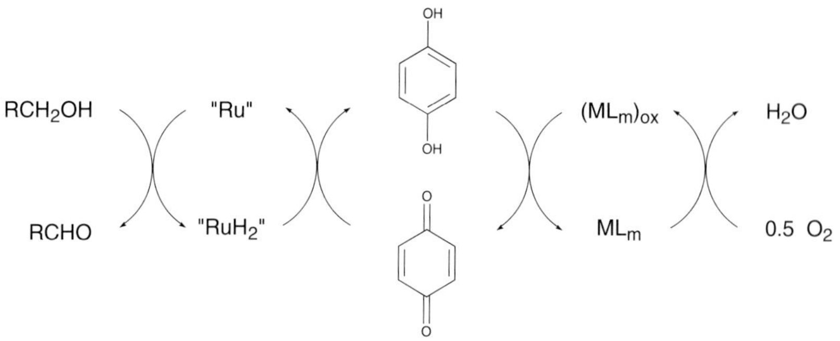

Figure 5.8 Ruthenium catalyst in combination with a hydrogen acceptor for aerobic oxidation.

to give hydroquinone with concomitant regeneration of the ruthenium catalyst. The cobalt-Schiff's base complex catalyzes the subsequent aerobic oxidation of the hydroquinone to benzoquinone to complete the catalytic cycle. Optimization of the electron-rich quinone, combined with the so-called 'Shvo' Ru-catalyst, led to one of the fastest catalytic systems reported for the oxidation of secondary alcohols [40]. The reaction conditions and results for selected alcohols are reported in Table 5.2.

The regeneration of the benzoquinone can also be achieved with dioxygen in the absence of the cobalt cocatalyst. Thus, Ishii and coworkers [41] showed that a combination of $RuCl_2(Ph_3P)_3$, hydroquinone, and dioxygen, in $PhCF_3$ as solvent, oxidized primary aliphatic, allylic, and benzylic alcohols to the corresponding aldehydes in quantitative yields (Eq. (5.2)).

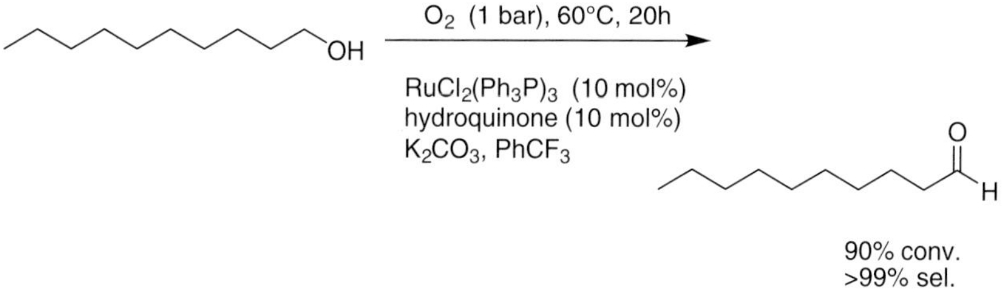

90% conv.
>99% sel.

(5.2)

Table 5.2 Ruthenium/quinone/Co-salen-catalyzed aerobic oxidation of secondary alcohols[a].

Substrate	Time (h)	Isolated yield
n-$C_6H_{13}CH(CH_3)OH$	1	92%
Cyclohexanol	1	92%
Cyclododecanol	1.5	86%
$PhCH(CH_3)OH$	1	89%
L-menthol	2	80%

a) According to Ref. [40] (2002). Reaction conditions: 1 mmol substrate, 1 mL toluene, 100 °C, 1 atm air; employing 0.5 mol% [($C_4Ph_4COHOCC_4Ph_4$)(µ-H)(CO)$_4$Ru$_2$], 20 mol% 2,6-dimethoxy-1,4-benzoquinone, and 2 mol% bis(salicylideniminato-3-propyl)methylamino-cobalt(II) as catalysts.

5.4 Ruthenium-Catalyzed Oxidations with O_2

A combination of $RuCl_2(Ph_3P)_3$ and the stable nitroxyl radical, 2,2′,6,6′-tetramethylpiperidine-N-oxyl (TEMPO), is a remarkably effective catalyst for the aerobic oxidation of a variety of primary and secondary alcohols, giving the corresponding aldehydes and ketones, respectively, in >99% selectivity [42]. The best results were obtained using 1 mol% of $RuCl_2(Ph_3P)_3$ and 3 mol% of TEMPO (Eq. (5.3)).

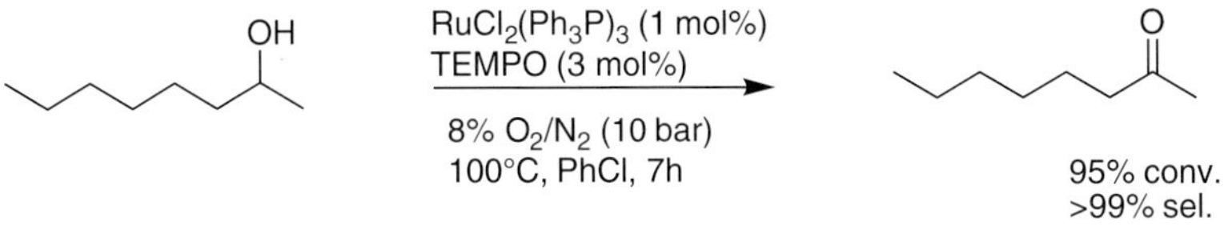

$$\text{2-octanol} \xrightarrow[\substack{8\% \, O_2/N_2 \, (10 \, \text{bar}) \\ 100°C, \, PhCl, \, 7h}]{\substack{RuCl_2(Ph_3P)_3 \, (1 \, \text{mol\%}) \\ TEMPO \, (3 \, \text{mol\%})}} \text{2-octanone} \quad \substack{95\% \, \text{conv.} \\ >99\% \, \text{sel.}}$$

(5.3)

The results obtained in the oxidation of representative primary and secondary aliphatic alcohols and allylic and benzylic alcohols using this system are shown in Table 5.3.

Primary alcohols give the corresponding aldehydes in high selectivity; for example, 1-octanol affords 1-octanal in >99% selectivity. Over-oxidation to the corresponding carboxylic acid, normally a rather facile process, is completely suppressed in the presence of a catalytic amount of TEMPO. For example, attempted oxidation of octanal under the reaction conditions, in the presence of 3 mol% TEMPO, gave no reaction in one week. In contrast, in the absence of TEMPO, octanal was completely converted to octanoic acid within 1 h under the same conditions. These results are consistent with over-oxidation of aldehydes occurring via a free radical autoxidation mechanism. TEMPO suppresses this reaction by efficiently scavenging free radical intermediates, resulting in the termination of free radical chains; that is, it acts as an antioxidant. Allylic alcohols were selectively converted to the corresponding unsat-

Table 5.3 Ruthenium-TEMPO-catalyzed oxidation of primary and secondary alcohols to the corresponding aldehyde using molecular oxygen[a].

Substrate	S/C ratio[b]	Time (h)	Conv.(%)[c]
n-$C_7H_{15}CH_2OH$	50	7	85
n-$C_6H_{13}CH(CH_3)OH$	100	7	98
Adamantan-2-ol	100	7	92
Cyclooctanol	100	7	92
$(CH_3)_2C=CHCH_2OH$	67	7	96
$(CH_3)_2C=CH(CH_2)_2CH(CH_3)=CHCH_2OH$[d]	67	7	91
$PhCH_2OH$[e]	200	2.5	>99
$(4-NO_2)PhCH_2OH$[e]	200	6	97
$PhCH(CH_3)$-OH	100	4	>99

a) 15 mmol substrate, 30 mL chlorobenzene, $RuCl_2(PPh_3)_3$/TEMPO ratio of 1/3, 10 mL min^{-1} O_2/N_2 (8/92; v/v), $P = 10$ atm, $T = 100\,°C$.
b) Substrate/Ru ratio.
c) Conversion of substrate, selectivity to aldehyde or ketone >99%.
d) Geraniol.
e) 1 atm O_2.

urated aldehydes in high yields. No formation of the isomeric saturated ketones via intramolecular hydrogen transfer, which is known to be promoted by ruthenium-phosphine complexes [43], was observed.

Although, in separate experiments, secondary alcohols are oxidized faster than primary ones, in competition experiments the Ru/TEMPO system displayed a preference for primary over secondary alcohols. This can be explained by assuming that initial complex formation between the alcohol and the ruthenium precedes the rate-limiting hydrogen transfer and determines substrate specificity; that is, complex formation with a primary, not a secondary, alcohol is favored.

An oxidative hydrogenation mechanism, analogous to that proposed by Bäckvall for the Ru/quinone system (see above), can be envisaged for the Ru/TEMPO system (see Figure 5.9).

The intermediate hydridoruthenium species is most probably $RuH_2(Ph_3P)_3$, as was observed in $RuCl_2(Ph_3P)_3$-catalyzed hydrogen transfer reactions [44]. The observation that $RuH_2(Ph_3P)_4$ exhibits the same activity as $RuCl_2(Ph_3P)_3$ in the Ru/TEMPO-catalyzed aerobic oxidation of 2-octanol is consistent with this notion. The TEMPO acts as a hydrogen transfer mediator by promoting the regeneration of the ruthenium catalyst via oxidation of the ruthenium hydride, resulting in the concomitant formation of the corresponding hydroxylamine, TEMPOH. The latter then undergoes rapid re-oxidation to TEMPO, by molecular oxygen, to complete the catalytic cycle (see Figure 5.9).

A linear increase in the rate of 2-octanol oxidation was observed with increasing TEMPO concentration in the range 0–4 mol%, but above 4 mol% further addition of TEMPO had a negligible effect on the rate. Analogous results were observed by Bäckvall and coworkers [45] in the Ru/benzoquinone system and were attributed to a change in the rate-limiting step. Hence, by analogy, we propose that at relatively low TEMPO/Ru ratios (up to 4 : 1), reoxidation of the ruthenium hydride species is the slowest step, while at high ratios dehydrogenation of the alcohol becomes rate-limiting.

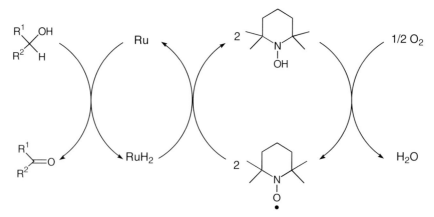

Figure 5.9 Ruthenium/TEMPO-catalyzed aerobic oxidation of alcohols.

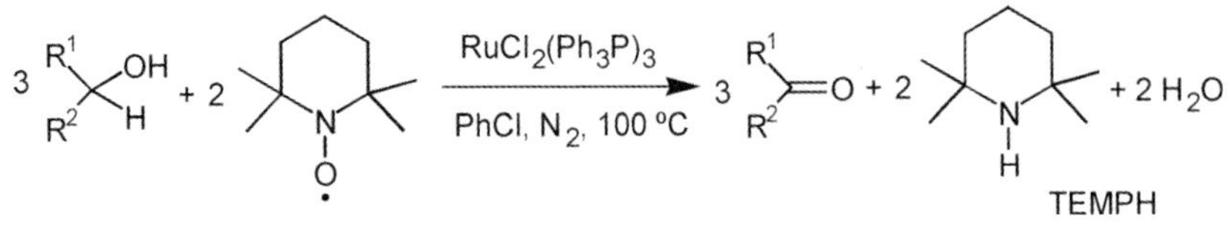

5.4 Ruthenium-Catalyzed Oxidations with O_2

$$3 \; R^1\text{-CH(OH)-}R^2 + 2 \; \text{TEMPO} \xrightarrow[\text{PhCl, N}_2, 100\,°C]{\text{RuCl}_2(\text{Ph}_3\text{P})_3} 3 \; R^1\text{-C(=O)-}R^2 + 2 \; \text{TEMPH} + 2 \; H_2O \quad (5.4)$$

Under an inert atmosphere, $RuCl_2(Ph_3P)_3$ catalyzes the stoichiometric oxidation of 2-octanol by TEMPO to give 2-octanone and the corresponding piperidine, TEMPH, in a stoichiometry of 3 : 2, as represented in (Eq. (5.4)) [39].

This result can be explained by assuming that the initially formed TEMPOH (see above) undergoes disproportionation to TEMPH and the oxoammonium cation (Eq. (5.5)). Reduction of the latter by the alcohol affords another molecule of TEMPOH, and this ultimately leads to the formation of the ketone and TEMPH in the observed stoichiometry of 3 : 2. The observation that attempts to prepare TEMPOH [46] under an inert atmosphere always resulted in the formation of TEMPH is consistent with this hypothesis.

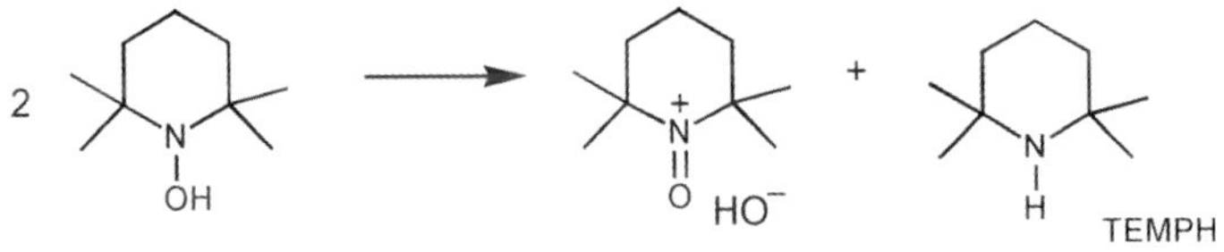

$$2 \; \text{TEMPOH} \longrightarrow \text{TEMPO}^+ + \text{HO}^- + \text{TEMPH} \quad (5.5)$$

Based on the results discussed above, the detailed catalytic cycle depicted in Figure 5.10 is proposed for the Ru/TEMPO-catalyzed aerobic oxidation of alcohols.

The alcohol oxidations discussed above involve as a key step the oxidative dehydrogenation of the alcohol to form low-valent hydridoruthenium intermediates. On the other hand, high-valent oxoruthenium species are also able to dehydrogenate alcohols via an oxometal mechanism (see Figure 5.6). It has long been known that ruthenium tetroxide, generated by reaction of ruthenium dioxide with periodate, smoothly oxidizes a variety of alcohols to the corresponding carbonyl compounds [47].

Griffith and coworkers [48] reported the synthesis of the organic soluble tetra-n-butylammoniumperruthenate (TBAP), $n\text{-}Bu_4N^+ RuO_4^-$, in 1985. They later found that tetra-n-propylammoniumperruthenate (TPAP), $n\text{-}Pr_4N^+ RuO_4^-$, is even easier to prepare, from RuO_4 and $n\text{-}Pr_4NOH$ in water [49, 50]. TBAB and TPAP are air stable, nonvolatile, and soluble in a wide range of organic solvents. Griffith and Ley [51, 52] subsequently showed that TPAP is an excellent catalyst for the selective oxidation of a wide variety of alcohols using N-methylmorpholine-N-oxide (NMO) as the stoichiometric oxidant (Eq. (5.6)).

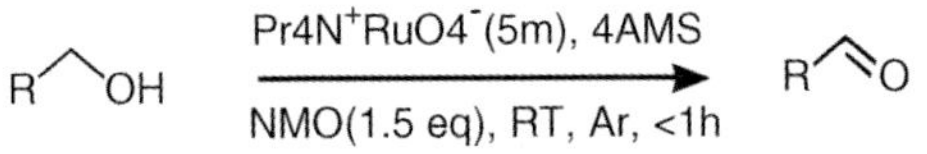

$$R\text{-CH}_2\text{OH} \xrightarrow[\text{NMO(1.5 eq), RT, Ar, <1h}]{\text{Pr}_4N^+RuO_4^- (5m), 4AMS} R\text{-CHO} \quad (5.6)$$

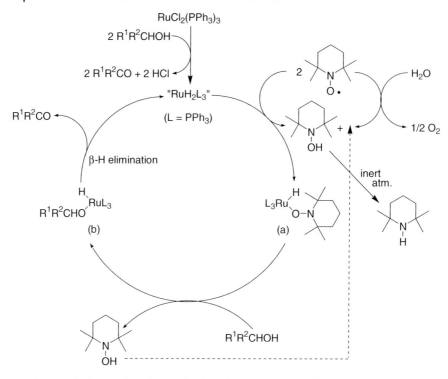

Figure 5.10 Proposed mechanism for the ruthenium/TEMPO-catalyzed oxidation of alcohols.

The groups of Ley [53] and Marko [54] independently showed that TPAP is able to catalyze the oxidation of alcohols using dioxygen as the stoichiometric oxidant. In particular, polymer-supported perruthenate (PSP), prepared by anion exchange of $KRuO_4$ with a basic anion exchange resin (Amberlyst A-26), has emerged as a versatile catalyst for the aerobic oxidation (Eq. (5.7)) of alcohols [55]. However the activity was about 4 times lower than that of homogeneous TPAP, and this catalyst could not be recycled, which was attributed to oxidative degradation of the polystyrene support. PSP displays a marked preference for primary versus secondary alcohol functionalities [55]. The problem of deactivation was also prominent for the homogeneous TPAP oxidation, which explains the high (10 mol%) loading of catalyst required.

$$R\text{—}OH \xrightarrow[\text{toluene, 0.5-0.8 h}]{\bullet\text{—}NMe_3\ RuO_4^-\ (10m\%)\atop O_2,\ 75\text{-}85°C} R\text{—}O \qquad (5.7)$$

yield 56-95%

Examples illustrating the scope of TPAP-catalyzed aerobic oxidation of primary and secondary alcohols to the corresponding aldehydes are shown in Table 5.4.

5.4 Ruthenium-Catalyzed Oxidations with O_2

Table 5.4 Perruthenate-catalyzed oxidation of primary and secondary alcohols to aldehydes using molecular oxygen.

Substrate	Carbonyl yield[a]		
	Toluene, 75–85 °C 10 mol% polymer-supported perruthenate (PSP)[b]	Toluene, 70–80 °C, 4 Å MS, 5 mol% tetrapropyl-ammoniumperruthenate (TPAP)[c]	Toluene, 75 °C, 10 mol% TPAP-doped sol-gel ormosil[d]
$C_7H_{15}CH_2OH$	91% (8 h)		70% (7 h)
$C_9H_{19}CH_2OH$		73% (0.5 h)[e]	
$C_9H_{19}CH(CH_3)OH$		88% (0.5 h)	
$(H_3C)_2N(CH_2)_2CH_2OH$	>95% (8 h)		
$PhCH_2OH$	>95% (0.5 h)		100% (0.75 h)
$(4\text{-}Cl)PhCH_2OH$		81% (0.5 h)	
$Ph\text{-}CH=CHCH_2OH$	>95% (1 h)	70% (0.5 h)	90% (5 h)

a) Yields at 100% conversion.
b) Ley et al. [55].
c) Marko et al. [54].
d) Ciriminna et al. [57].
e) 94% conversion, no molecular sieves were added.

A heterogeneous TPAP-catalyst was developed, which could be recycled successfully and displayed no leaching, by tethering the tetraalkylammonium perruthenate to the internal surface of mesoporous silica (MCM-41). It was shown [56] to catalyze the selective aerobic oxidation of primary and secondary allylic and benzylic alcohols (Figure 5.11). Surprisingly, both cyclohexanol and cyclohexenol were unreactive,

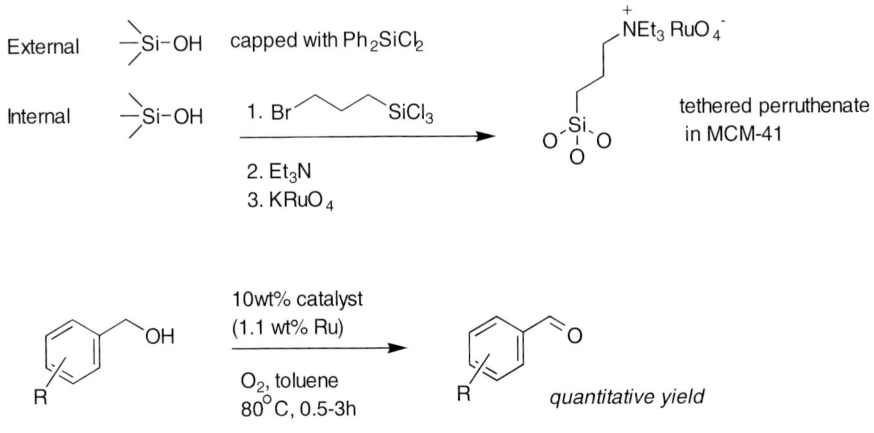

Figure 5.11 Aerobic alcohol oxidation catalyzed by perruthenate tethered to the internal surface of MCM-41.

although these substrates can easily be accommodated in the pores of MCM-41. No mechanistic interpretation for this surprising observation was offered by the authors.

Another variation on this theme involves straightforward doping of methyl-modified silica, referred to as ormosil, with tetrapropylammonium perruthenate via the sol-gel process [57] (see Table 5.4). A serious disadvantage of this system is the low turnover frequency (1.0 and $1.8\,h^{-1}$) observed for primary aliphatic alcohol and allylic alcohol respectively.

Little attention has been paid to the mechanism of perruthenate-catalyzed alcohol oxidations [58]. Although TPAP can act as a three-electron oxidant ($Ru^{VII} \rightarrow Ru^{IV}$), the fact that it selectively oxidizes cyclobutanol to cyclobutanone and tert-butyl phenylmethanol to the corresponding ketone, militates against free radical intermediates and is consistent with a heterolytic, two-electron oxidation [58, 59]. Presumably, the key step involves β-hydride elimination from a high-valent, for example, alkoxyruthenium(VII) intermediate followed by reoxidation of the lower-valent ruthenium by dioxygen. However, as shown in Figure 5.12, if this involved the Ru(VII)/Ru(V) couple, the reoxidation would require the close proximity of two ruthenium centres, which would seem unlikely in a polymer-supported catalyst. A plausible alternative, which can occur at an isolated ruthenium center, involves the oxidation of a second molecule of alcohol, resulting in the reduction of ruthenium(V) to ruthenium(III), followed by reoxidation of the latter to ruthenium(VII) by dioxygen (see Figure 5.12).

More detailed mechanistic studies are obviously necessary in order to elucidate the details of this fascinating reaction. It is worth noting, in this context, that the reaction of TPAP with 2-propanol was found to be autocatalytic, possibly because of the formation of colloidal RuO_2 [60]. Another possible alternative is one involving the initial formation of oxoruthenium(VI), followed by cycling between ruthenium(VI), ruthenium(IV), and possibly ruthenium(II).

We note, in this context, that James and coworkers [61] showed that a trans-dioxoruthenium(VI) complex of meso-tetrakismesitylporphyrin dianion (tmp) oxidizes isopropanol, in a stoichiometric reaction, with concomitant formation of a dialkoxyruthenium(IV)-tmp complex (Eq. (5.8)).

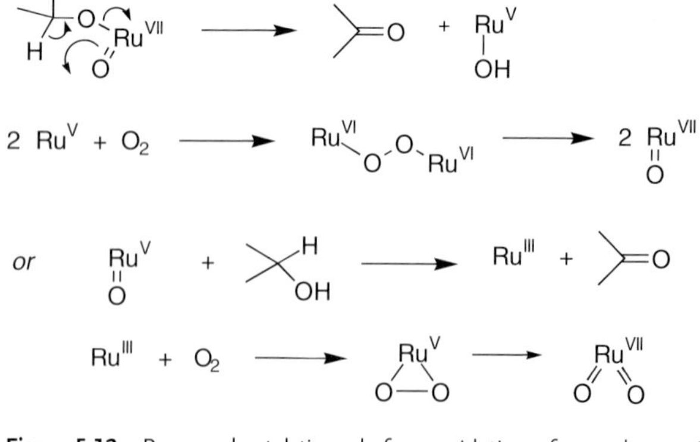

Figure 5.12 Proposed catalytic cycle for reoxidation of perruthenate in the oxidation of alcohols.

$$Ru^{VI}(tmp)O_2 + 3\,Me_2CHOH \longrightarrow Ru^{IV}(tmp)(OCHMe_2)_2 + Me_2CO + 2\,H_2O \quad (5.8)$$

The oxoruthenium(VI) complex was prepared by exposing a benzene solution of *trans*-Ru^{II} (tmp)(MeCN)$_2$ to air at 20 °C. Addition of isopropanol to the resulting solution, in the absence of air, afforded the dialkoxyruthenium(IV) complex, in quantitative yield, within 24 h. In the presence of air, benzene solutions of the dioxoruthenium(VI) or the dialkoxyruthenium(IV) complex effected catalytic oxidation of isopropanol at room temperature, albeit at a modest rate (1.5 catalytic turnovers per day). Interestingly, with the dialkoxyruthenium(IV) complex, catalytic oxidation was observed with air but not with dry oxygen, suggesting that hydrolysis to an oxoruthenium(IV) complex is necessary for a catalytic cycle.

Mizuno and Yamaguchi [62] reported ruthenium on alumina to be a powerful and recyclable catalyst for selective alcohol oxidation. This method was effective with a large range of substrates (see Eq. (5.9) and Table 5.5) and tolerates the presence of sulfur and nitrogen groups. Only primary aliphatic alcohols required the addition of hydroquinone. Turnover frequencies ranging from 4 h^{-1} (for secondary allylic alcohols) to 18 h^{-1} (for 2-octanol) were obtained in trifluorotoluene, while, in the solvent-free oxidation at 150 °C, a TOF of 300 h^{-1} was observed for 2-octanol.

$$R_1R_2CH(OH) + 0.5\,O_2 \xrightarrow[\text{PhCF}_3,\,83°C,\,1\,\text{atm}\,O_2]{2.5\,\text{mol\%}\,Ru(OH)_3\,\text{on}\,\gamma\text{-Al}_2O_3} R_1R_2C{=}O + H_2O \quad (5.9)$$

The catalyst consists of highly dispersed Ru(OH)$_3$ on the surface of γ-Al$_2$O$_3$. Based, among other things, on the fact that this catalyst is also capable of performing a transfer hydrogenation using 2-propanol as the hydrogen donor, it was concluded that the mechanism of this reaction proceeds via a hydridometal pathway.

Table 5.5 Ru(OH)$_3$-Al$_2$O$_3$ catalyzed oxidation of primary and secondary alcohols to the corresponding aldehydes and ketones using O$_2$[a].

Substrate	Time (h)	Conv. (%)	Select. (%)
n-C$_6$H$_{13}$CH(CH$_3$)OH	2	91	>99
cyclooctanol	6	81	>99
n-C$_7$H$_{15}$CH$_2$OH[b]	4	87	98
PhCH(CH$_3$)OH	1	>99	>99
(CH$_3$)$_2$C=CH(CH$_2$)$_2$CH(CH$_3$)=CHCH$_2$OH[c]	6	89	97
PhCH$_2$OH	1	>99	>99
(4-NO$_2$)PhCH$_2$OH	3	97	>99

a) According to Ref. [62]; 2.5 mol% Ru/Al$_2$O$_3$, PhCF$_3$ as solvent, 83 °C, 1 atm O$_2$; conversion and yields determined by GLC.
b) 5 mol% Ru/Al$_2$O$_3$ and 5 mol% hydroquinone (to suppress overoxidation) were used.
c) Geraniol.

Ruthenium-exchanged hydrotalcites (Ru-HT) were shown by Kaneda and co-workers [63] to be heterogeneous catalysts for the aerobic oxidation of reactive allylic and benzylic alcohols. Hydrotalcites are layered anionic clays consisting of a cationic brucite layer with anions (hydroxide or carbonate) situated in the interlayer region. Various cations can be introduced into the brucite layer by ion exchange. For example, Ru-HT with the formula $Mg_6Al_2Ru_{0.5}(OH)_{16}CO_3$, was prepared by treating an aqueous solution of $RuCl_3 \cdot 3H_2O$, $MgCl_2 \cdot 6H_2O$ and $AlCl_3 \cdot H_2O$ with a solution of NaOH and Na_2CO_3 followed by heating at 60 °C for 18 h [63]. The resulting slurry was cooled to room temperature, filtered, washed with water and dried at 110 °C for 12 h. The resulting Ru-HT showed the highest activity among a series of hydrotalcites exchanged with, for example, Fe, Ni, Mn, V, and Cr.

Subsequently, the same group showed that the activity of the Ru-HT was significantly enhanced by the introduction of cobalt(II), in addition to ruthenium(III), into the brucite layer [64]. For example, cinnamyl alcohol underwent complete conversion in 40 min in toluene at 60 °C in the presence of Ru/Co-HT, compared with 31% conversion under the same conditions with Ru-HT. A secondary aliphatic alcohol, 2-octanol, was smoothly converted into the corresponding ketone, but primary aliphatic alcohols, for example, 1-octanol, exhibited extremely low activity. The authors suggested that the introduction of cobalt induced the formation of higher oxidation states of ruthenium, for example, Ru(IV) to Ru(VI), leading to a more active oxidation catalyst. However, on the basis of the reported results it is not possible to rule out low-valent ruthenium species as the active catalyst in a hydridometal pathway. The results obtained in the oxidation of representative alcohols with Ru-HT and Ru-Co-HT are compared in Table 5.6.

Table 5.6 Oxidation of various alcohols to their corresponding aldehydes or ketones with Ru-hydrotalcites using molecular oxygen.[a]

Substrate	Ru-Mg-Al-CO$_3$-HT[b]		Ru-Co-Al-CO$_3$-HT[c]	
	Time	Yield (%)	Time	Yield (%)
PhCH=CHCH$_2$OH	8 h	95[d]	40 min	94
PhCH$_2$OH	8 h	95[d]	1 h	96
4-ClPhCH$_2$OH	8 h	61[e]	1.5 h	95
PhCH(CH$_3$)OH	18 h	100	1.5 h	100
n-C$_6$H$_{13}$CH(CH$_3$)OH	—	—	2 h	97
(CH$_3$)$_2$C=CH(CH$_2$)$_2$CH(CH$_3$)=CHCH$_2$OH[f]	—	—	12 h	71[g]

a) 2 mmol substrate, 0.3 g hydrotalcite (≈14 mol%), in toluene, 60 °C, 1 atm O$_2$. Conversion 100%.
b) See Ref. [63].
c) See Ref. [64].
d) Conv. 98%.
e) Conv. 64%.
f) Geraniol.
g) Conversion 89%.

5.5
Palladium-Catalyzed Oxidations with O_2

Palladium(II) is also capable of mediating the oxidation of alcohols via the hydridometal pathway shown in Figure 5.5. Excellent reviews on the activation of oxygen by palladium related to alcohol oxidation have been written by Stahl [65] and Muzart [66]. Blackburn and Schwarz first reported [67] the $PdCl_2$-NaOAc-catalyzed aerobic oxidation of alcohols in 1977. However, activities were very low, with turnover frequencies of the order of $1\,h^{-1}$. Subsequently, much effort has been devoted to finding synthetically useful methods for the palladium-catalyzed aerobic oxidation of alcohols. For example, the giant palladium cluster, $Pd_{561}phen_{60}(OAc)_{180}$ [68], was shown to catalyze the aerobic oxidation of primary allylic alcohols to the corresponding α,β-unsaturated aldehydes (Eq. (5.10)) [69].

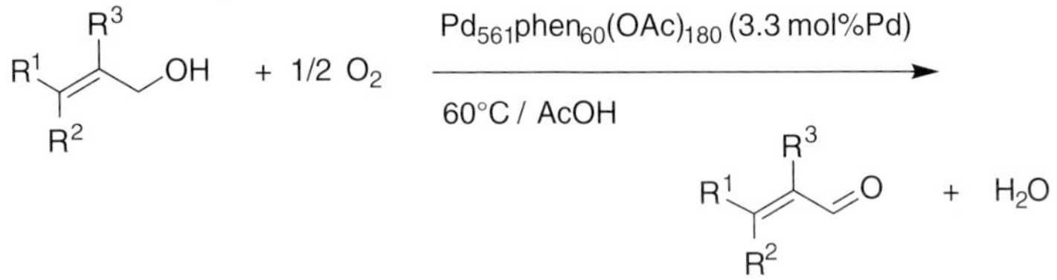

$$R^1R^2C=C(R^3)CH_2OH + 1/2\,O_2 \xrightarrow[60\,^\circ C\,/\,AcOH]{Pd_{561}phen_{60}(OAc)_{180}\,(3.3\,mol\%Pd)} R^1R^2C=C(R^3)CHO + H_2O \quad (5.10)$$

In 1998, Peterson and Larock showed that $Pd(OAc)_2$ in combination with $NaHCO_3$ as a base in DMSO as solvent catalyzed the aerobic oxidation of primary and secondary allylic and benzylic alcohols to the corresponding aldehydes and ketones, respectively, in fairly good yields [70]. In both cases, ethylene carbonate and DMSO acted both as the solvent and as the ligand necessary for a smooth reoxidation [71]. Similarly, $PdCl_2$, in combination with sodium carbonate and a tetraalkylammonium salt, Adogen 464, as a phase transfer catalyst, catalyzed the aerobic oxidation of alcohols; for example, 1,4- and 1,5-diols afforded the corresponding lactones (Eq. (5.11)) [72, 73].

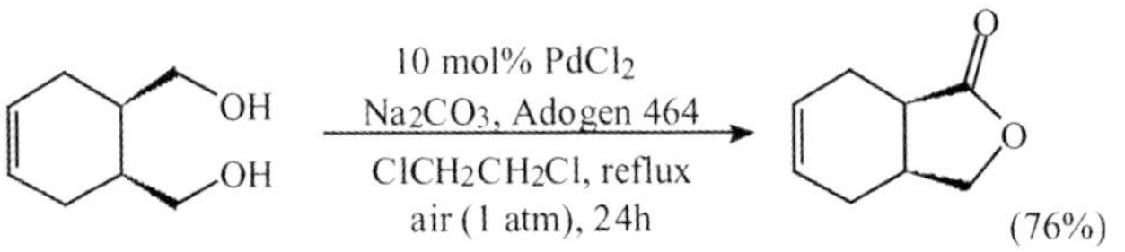

$$\text{cyclohexene-diol} \xrightarrow[\substack{ClCH_2CH_2Cl,\,reflux \\ air\,(1\,atm),\,24h}]{\substack{10\,mol\%\,PdCl_2 \\ Na_2CO_3,\,Adogen\,464}} \text{bicyclic lactone (76\%)} \quad (5.11)$$

However, these methods suffer from low activities and/or narrow scope. Uemura and coworkers [74, 75] reported an improved procedure involving the use of $Pd(OAc)_2$ (5 mol%) in combination with pyridine (20 mol%) and 3 Å molecular sieves (500 mg per mmol of substrate) in toluene at 80 °C. This system smoothly catalyzed the aerobic oxidation of primary and secondary aliphatic alcohols to the corresponding aldehydes and ketones, respectively, in addition to benzylic and allylic alcohols. Representative examples are summarized in Table 5.7. The corresponding lactones were afforded by 1,4- and 1,5-diols. This approach could also be employed under fluorous biphasic conditions [76].

Table 5.7 Pd(II)-catalyzed oxidation of various alcohols to their corresponding ketones or aldehydes in the presence of pyridine using molecular oxygen[a].

Substrate	Conversion after 2 h	Yield (aldehyde/ketone)
PhCH$_2$OH	100%	100%
4-ClPhCH$_2$OH	100%	98%
n-C$_{11}$H$_{23}$CH$_2$OH	97%	93%[b]
n-C$_{10}$H$_{21}$CH(CH$_3$)OH	98%	97%[b]
PhCH=CHCH$_2$OH	46%	35%[b]

a) Data from Ref. [75] Reaction conditions: alcohol 1.0 mmol, 5 mol% Pd(OAc)$_2$, 20 mol% pyridine, 500 mg MS3 Å, toluene 10 mL, 80 °C, 1 atm O$_2$, 2 h.
b) Isolated yield.

Stahl and coworkers conducted mechanistic studies on both the Pd/DMSO and the Pd/pyridine system [77, 78]. Kinetic studies revealed that in the Pd/pyridine system the rate exhibits no dependence on the oxygen pressure, and kinetic isotope effect studies support turnover-limiting substrate oxidation. In contrast the Pd/DMSO system features turnover-limiting oxidation of palladium(0) (see Figure 5.13). Moreover in the Pd/pyridine system, pyridine is very effective in oxidizing palladium(0) by molecular oxygen, but at the same time inhibits the rate of alcohol oxidation by palladium(II).

Although this methodology constitutes an improvement on those previously reported, turnover frequencies were still generally $<10\,\text{h}^{-1}$, and hence there is considerable room for further improvement. Attempts to replace either pyridine by triethylamine [79], or Pd(OAc)$_2$ by palladacycles [80], resulted in lower activities.

We showed that the water-soluble palladium(II) complex of sulfonated bathophenanthroline is a stable, recyclable catalyst for the aerobic oxidation of alcohols in a two-phase aqueous-organic medium, for example, in Eq. (5.12) [81, 82]. Reactions were generally complete in 5 h at 100 °C/30 atm air with as little as 0.25 mol% catalyst. No organic solvent is required (unless the substrate is a solid) and the product ketone is easily recovered by phase separation. The catalyst is stable and remains in the aqueous phase which can be recycled to the next batch.

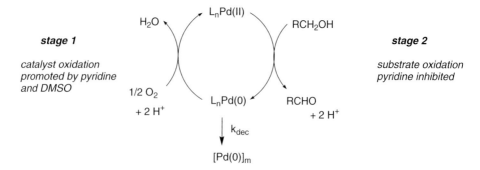

Figure 5.13 Mechanistic insights on Pd/pyridine and Pd/DMSO systems.

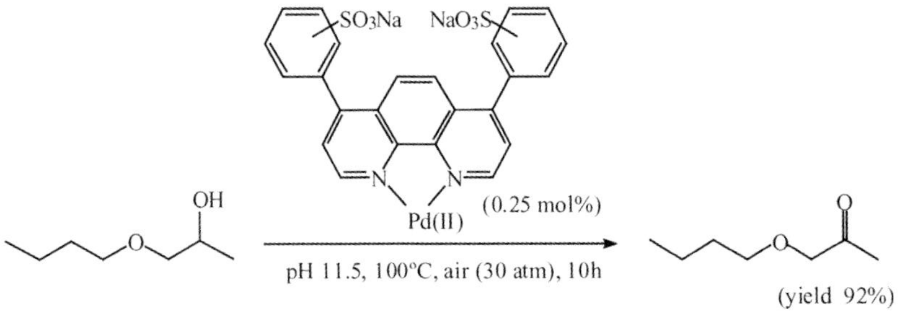

5.5 Palladium-Catalyzed Oxidations with O_2

$$\text{OH substrate} \xrightarrow[\text{pH 11.5, 100°C, air (30 atm), 10h}]{\text{Pd(II) (0.25 mol\%)}} \text{ketone product} \quad \text{(yield 92\%)} \quad (5.12)$$

A wide range of alcohols were oxidized, with TOFs ranging from 10 to 100 h^{-1}, depending on the solubility of the alcohol in water. (Since the reaction occurs in the aqueous phase the alcohol must be at least sparingly soluble in water.) Thus, in a series of straight-chain secondary alcohols, the TOF decreased from 100 to 13 h^{-1} on increasing the chain length from 1-pentanol to 1-nonanol. Representative examples of secondary alcohols that were smoothly oxidized using this system are collected in Table 5.8. The corresponding ketones were obtained in >99% selectivity in virtually all cases.

Primary alcohols afforded the corresponding carboxylic acids via further oxidation of the aldehyde intermediate; for example, 1-hexanol afforded 1-hexanoic acid in 95% yield. It is important to note, however, that this was achieved without the requirement of one equivalent of base to neutralize the carboxylic acid product (which is the case with supported noble metal catalysts [5]). In contrast, when 1 mol% TEMPO (4 equivalents per Pd) was added the aldehyde was obtained in high yield; for example, 1-hexanol afforded 1-hexanal in 97% yield. Some representative examples of primary alcohol oxidations using this system are shown in Table 5.9. The TEMPO was previously shown to suppress the autoxidation of aldehydes to the carboxylic acids (see earlier).

Table 5.8 Conversion of secondary alcohols to their corresponding ketones using a Pd(II)-bathophenanthroline-complex in a two-phase system[a].

Substrate	Time	Conversion (%)	Selectivity[b] (%)	Isolated yield (%)
n-C$_3$H$_7$CH(CH$_3$)OH	5 h	100	100	90
n-C$_4$H$_9$CH(CH$_3$)OH	10 h	100	100	90
Cyclopentanol	5 h	100	100	90
PhCH(CH$_3$)OH	10 h	90	100	85
CH$_3$CH=CHCH(CH$_3$)OH	10 h	95	83[c]	79
n-C$_4$H$_9$OCH$_2$CH(CH$_3$)OH	10 h	100	100	92

a) Reaction conditions 20 mmol alcohol, 0.05 mmol PhenS*Pd(OAc)$_2$, 1 mmol NaOAc, 100 °C, 30 atm air.
b) Selectivity to ketone, determined by gas chromatography with an external standard.
c) Ether (17%) was formed.

Table 5.9 Conversion of primary alcohols to their corresponding aldehydes or acids using a Pd(II)-bathophenanthroline-complex in a two-phase system[a].

Substrate	Product	Time	Conv. (%)	Select.[b] (%)	Isolated yield (%)
n-C$_4$H$_9$CH$_2$OH[c]	n-C$_4$H$_9$CHO	15 h	98	97[d]	90
n-C$_5$H$_9$CH$_2$OH	n-C$_5$H$_9$COOH	12 h	95	90[e]	80
PhCH$_2$OH	PhCHO	10 h	100	99.8[d]	93
(CH$_3$)$_2$CH=CHCH$_2$OH	(CH$_3$)$_2$CH=CHCHO	10 h	95	83[d]	79

a) Reaction conditions 10 mmol alcohol, 0.05 mmol PhenS∗Pd(OAc)$_2$, 1 mmol NaOAc, 100 °C, 30 atm air.
b) Selectivity to product, determined by gas chromatography with an external standard.
c) TEMPO(4 eq. to Pd) was added.
d) Acid was formed as the major by-product.
e) Hexanal and hexanoate were formed.

Compared to existing systems for the aerobic oxidation of alcohols, the Pd-bathophenanthroline system is among the fastest catalytic systems reported to date. It requires no solvent, and product/catalyst isolation involves simple phase separation. The system has broad scope but is not successful with all alcohols. Some examples of unreactive alcohols are shown in Figure 5.14. Low reactivity was generally observed with alcohols containing functional groups which could strongly coordinate to the palladium.

The reaction is half-order in palladium and first-order in the alcohol substrate when measured with a water-soluble alcohol to eliminate the complication of mass transfer [82]. A possible mechanism is illustrated in Figure 5.15. The resting catalyst is a dimeric complex containing bridging hydroxyl groups. Reaction with the alcohol in the presence of a base, added as a cocatalyst (NaOAc) or free ligand, affords a monomeric alkoxy palladium(II) intermediate, which undergoes β-hydride elimi-

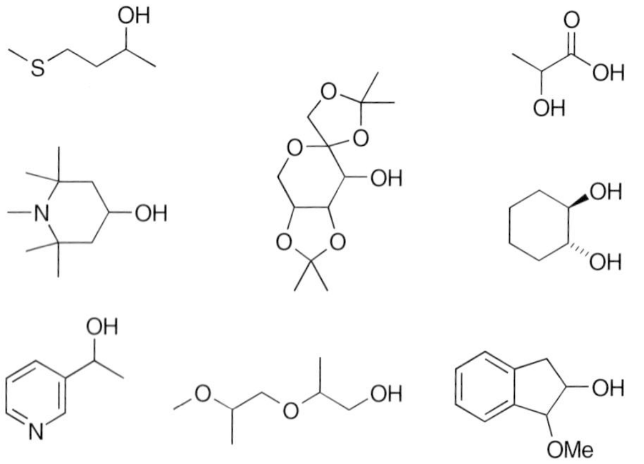

Figure 5.14 Unreactive alcohols in the Pd-bathophenanthroline-catalytic system.

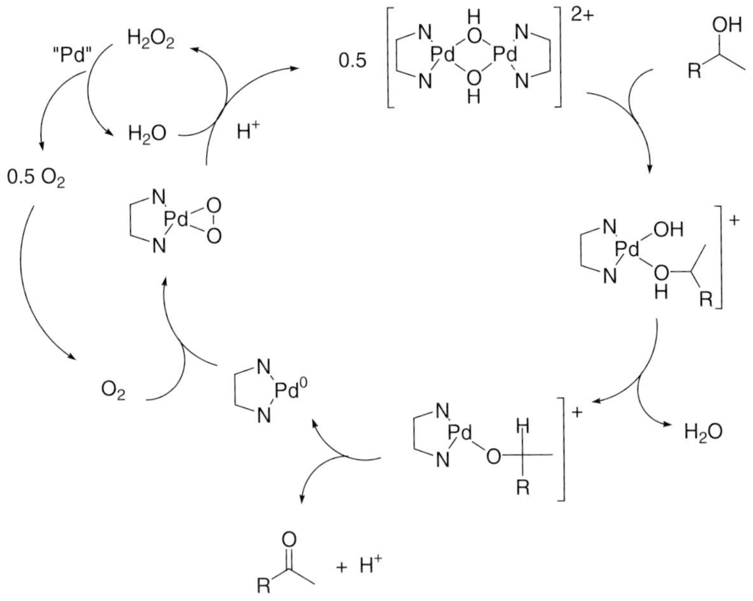

Figure 5.15 Mechanism of Pd-bathophenanthroline catalyzed oxidation of alcohols.

nation to give the carbonyl compound, water, and a palladium(0) complex. Oxidative addition of dioxygen to the latter affords a palladium(II) η-peroxo complex which can react with the alcohol substrate to regenerate the catalytic intermediate, presumably with concomitant formation of hydrogen peroxide, as was observed in analogous systems [83].

According to the proposed mechanism, the introduction of substituents at the 2 and 9 positions in the PhenS ligand would, as a result of steric hindrance (see Figure 5.15), promote dissociation of the dimer and enhance the reactivity of the catalyst. This proved to be the case: introduction of methyl groups at the 2 and 9 positions (the ligand is commercially available and is known as bathocuproin) tripled the activity in 2-hexanol oxidation [82].

Under cosolvent conditions using water/ethylene carbonate or water/DMSO, Pd-neocuproin was found to be even more active (Figure 5.16) [84]. This system is exceptional because of its activity (a TOF of $1800 \, h^{-1}$ was observed with 2-hexanol) and functional group tolerance, such as C=C bonds, C≡C bonds, halides, α-carbonyls, ethers, amines, and so on, giving it broad synthetic utility.

However, a more detailed examination of the results obtained with the Pd(II) bathophenthroline and Pd(II) neocuproin complexes revealed a remarkable difference in the oxidation of the unsaturated alcohol substrate shown in Figure 5.17. With the former, the major product was derived from oxidation of the alkene double bond, while the latter afforded >99% selective oxidation of the alcohol moiety. This suggested that we were concerned with totally different types of catalyst. Indeed, further investigation revealed that the Pd(II) neocuproin complex dissociates completely to afford Pd nanoclusters which are the actual catalyst [85].

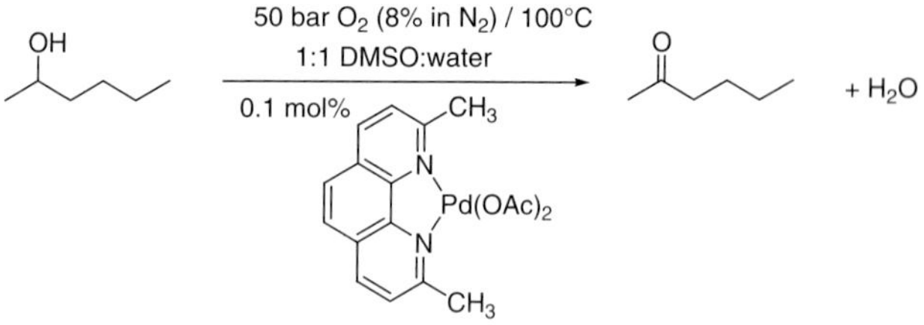

5 Modern Oxidation of Alcohols using Environmentally Benign Oxidants

Reaction scheme: 2-hexanol → 2-hexanone + H_2O

Conditions: 50 bar O_2 (8% in N_2) / 100°C, 1:1 DMSO:water, 0.1 mol% Pd(OAc)$_2$ with neocuproine (2,9-dimethyl-1,10-phenanthroline) ligand.

Substrate	t(h)	TOF$_0$ (h^{-1})	Conv.(%)	Sel.(%)
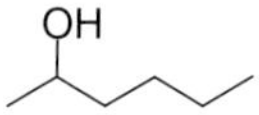 2-hexanol (OH)	2	>>500	100	100
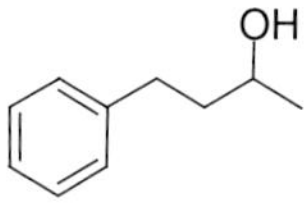 4-phenyl-2-butanol	2	>>500	99	99
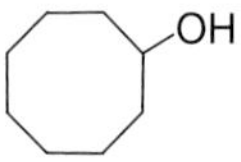 cyclooctanol	3	400	93	99+
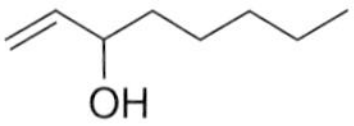 oct-7-en-2-ol (allylic alcohol)	3	400	95	96
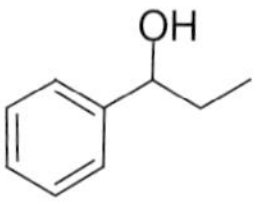 1-phenylethanol (sec-benzylic)	3	300	80	99+
 1-heptanol	2.5	200	40	75(heptanoic acid)/ 20 (heptanal)
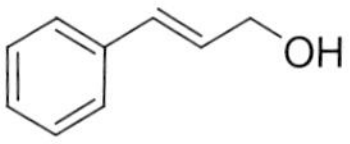 cinnamyl alcohol	10	135	88	99+
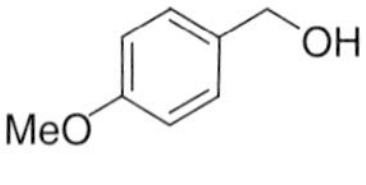 4-methoxybenzyl alcohol (MeO-C$_6$H$_4$-CH$_2$OH)	3	>>330	100	100

Figure 5.16 Pd-neocuproin as catalyst for alcohol oxidation under water/cosolvent conditions.

Moiseev and coworkers [68, 86] had previously shown that giant Pd clusters (nowadays known as Pd nanoclusters) are good catalysts for the oxidation of alcohol moieties and selectively oxidize allylic C−H bonds in alkenes. More recently, Pd nanoparticles supported on hydroxyapatite [87] were shown to be an excellent catalyst for aerobic alcohol oxidations.

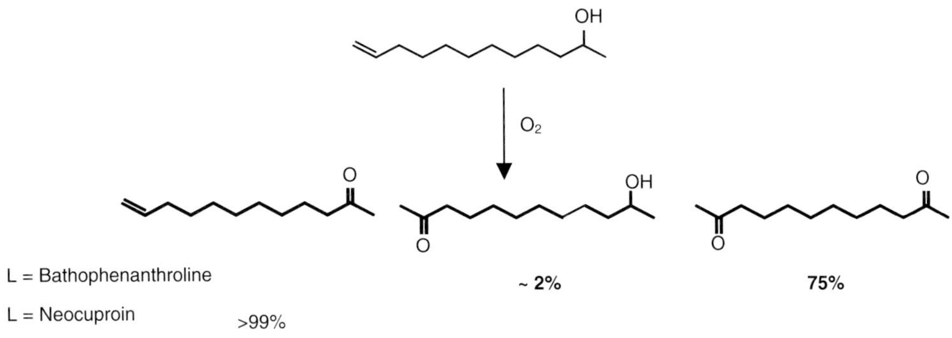

Figure 5.17 Selectivity in oxidation of unsaturated alcohol oxidation using the Pd-bathophenanthroline catalyst [82] versus the Pd-neocuproin catalyst [84]. For conditions see Eq. (5.12) and Figure 5.16.

L = Bathophenanthroline
L = Neocuproin   >99%   ~ 2%   75%

The most common heterogeneous palladium catalyst is Pd-on-C which is generally used for the aerobic oxidation of water-soluble alcohols, for example, carbohydrates [88]. Other heterogeneous palladium catalysts have also been described, such as those formed by introduction of Pd ions into the brucite layer of hydrotalcite [89] or by supporting $PdCl_2(PhCN)_2$ on hydroxyapatite [90]. However, in the light of the above discussion, some or all of these systems may involve supported palladium nanoparticles. Other recent examples include the use of palladium nanoparticles entrapped in aluminum hydroxide [91], resin-dispersed Pd nanoparticles [92], and poly(ethylene glycol)-stabilized palladium nanoparticles in $scCO_2$ [93]. Although in some cases the activities for activated alcohols obtained with these Pd-nanoparticles are impressive, the conversion of aliphatic alcohols is still rather slow.

5.5.1
Gold Nanoparticles as Catalysts

Recently, gold nanoparticles have emerged as one of the most active catalysts for aerobic alcohol oxidations and are especially selective for polyalcohols. Rossi and coworkers [94] were pioneers in the use of Au nanoclusters as catalysts for the aerobic oxidation of alcohol moieties in aqueous media. They showed that Au nanoclusters are excellent catalysts for the aerobic oxidation of carbohydrates, for example, glucose to gluconic acid [94]. Similarly, Christensen and coworkers reported the aerobic oxidation of aqueous (bio)ethanol to acetic acid over Au-on-Mg_2AlO_4 [95]. Interestingly, when the oxidation is performed in methanol, the methyl ester of the corresponding carboxylic acid is obtained; for example, the renewable raw materials furfural and hydroxymethylfurfural gave methyl furoate and the dimethyl ester of furan-1,4-dicarboxylic acid, respectively [96].

Similarly, Corma and coworkers showed that gold nanoparticles deposited on nanocrystalline ceria form an excellent recyclable catalyst for the aerobic oxidation of alcohols [97, 98] (Figure 5.18). Another example of a gold catalyst with exceptional activity is a 2.5% Au-2.5% Pd/TiO_2 [99] (Figure 5.18). As mentioned above, Au is now

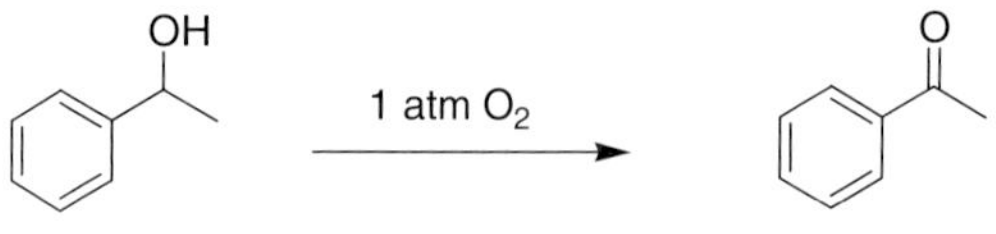

OH → (1 atm O₂) → ketone (acetophenone from 1-phenylethanol)

Catalyst	Conditions	Temp	TOF
Au on nanocrystalline CeO$_2$:	solvent-free	80 °C	TOF 74 h^{-1} (90% yld)
	solvent-free	160 °C	TOF 12500 h^{-1} (99% sel.)
2.5% Au-Pd-alloys on TiO$_2$:	solvent-free	160 °C	TOF 269000 h^{-1}

Figure 5.18 Au-nanoparticles for catalytic oxidation of alcohols.

considered as the catalyst of choice for carbohydrate oxidation. Similarly, glycerol can be oxidized to glyceric acid with 100% selectivity using either 1% Au/charcoal or 1% Au/graphite catalyst under mild reaction conditions (60 °C, 3 h, water as solvent) [100].

5.6
Copper-Catalyzed Oxidations with O$_2$

Copper would seem to be an appropriate choice of metal for the catalytic oxidation of alcohols with dioxygen since it comprises the catalytic center in a variety of enzymes, for example, galactose oxidase, which catalyze this conversion *in vivo* [101, 102]. However, despite extensive efforts [103], synthetically useful copper-based systems have generally not been forthcoming. For instance, in the absence of other metals, CuCl in combination with 2,2'-bipyridine (bipy) as base/ligand shows catalytic activity in the aerobic oxidation of alcohols. However, benzhydrol is the only suitable substrate, and at least one equivalent of bipy (relative to substrate) is required to reach complete conversion. On the other hand, with *ortho*-phenanthroline as ligand, CuCl$_2$ can catalyze the aerobic oxidation of a variety of primary and secondary alcohols to the corresponding carboxylic acids and ketones in alkaline media [103].

A special class of active copper-based aerobic oxidation systems comprises the biomimetic models of galactose oxidase, that is, Cu(II)-phenoxyl radical complexes reported by Stack and Wieghardt [104–107]. Just like the enzyme itself, these monomeric Cu(II) species are effective only with easily oxidized benzylic and allylic alcohols, simple primary and secondary aliphatic alcohols being largely unreactive. A good example of a biomimetic model of galactose oxidase is [Cu(II)BSP], in which BSP stands for a salen-ligand with a binaphthyl backbone (Figure 5.19). The rate-determining step (RDS) of this interesting system was suggested to involve inner sphere one-electron transfer from the alkoxide ligand to Cu(II) followed by hydrogen transfer to the phenoxyl radical yielding Cu(I), phenol, and the carbonyl product (Figure 5.19) [108].

Marko and coworkers [109, 110] showed that a combination of CuCl (5 mol%), phenanthroline (5 mol%), and di-*tert*-butylazodicarboxylate, DBAD (5 mol%), in the presence of 2 equivalents of K$_2$CO$_3$, catalyzes the aerobic oxidation of allylic and

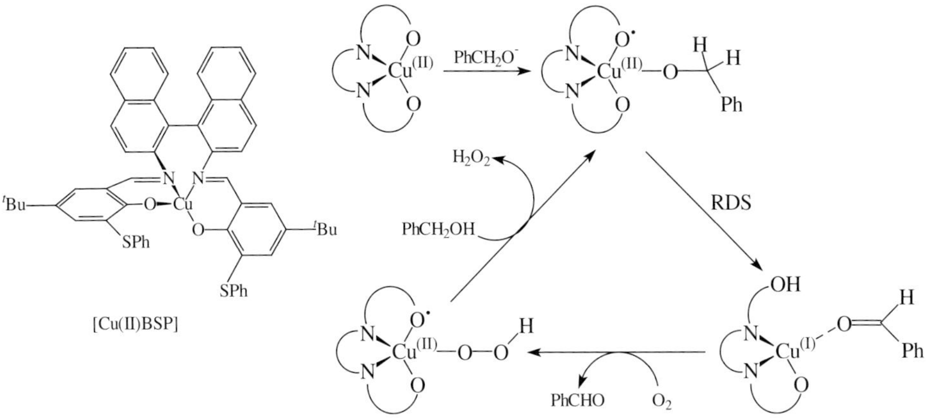

Figure 5.19 [Cu(II)BSP]-catalyzed aerobic oxidation of benzyl alcohol.

benzylic alcohols (Eq. (5.13)). Primary aliphatic alcohols, for example, 1-decanol, could be oxidized but required 10 mol% catalyst for smooth conversion.

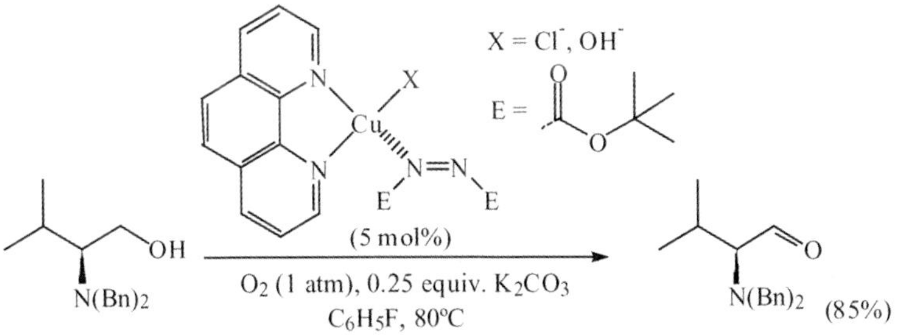

(5.13)

The nature of the copper counterion was critical, with chloride, acetate and triflate proving to be the most effective. Polar solvents such as acetonitrile inhibit the reaction, whereas smooth oxidation takes place in apolar solvents such as toluene. An advantage of the system is that it tolerates a variety of functional groups (see Table 5.10 for examples). Serious drawbacks of the system are the low activity, the need for two equivalents of K_2CO_3 (relative to substrate), and the expensive DBAD as a cocatalyst. According to a later report [111] the amount of K_2CO_3 can be reduced to 0.25 equivalents by changing the solvent to fluorobenzene.

The active catalyst is heterogeneous, being adsorbed on the insoluble K_2CO_3 (filtration gave a filtrate devoid of activity). Besides fulfilling a role as a catalyst support, the K_2CO_3 acts both as a base and as a water scavenger. The mechanism illustrated in Figure 5.20 was postulated to explain the observed results.

Semmelhack et al. [112] reported that the combination of CuCl and 4-hydroxy TEMPO catalyzes the aerobic oxidation of alcohols. However, the scope was limited to active benzylic and allylic alcohols, and activities were low (10 mol% of catalyst was needed for smooth reaction). They proposed that the copper catalyzes the reoxidation

Table 5.10 Copper-catalyzed aerobic oxidation of alcohols to the corresponding aldehyde or ketone using DBAD and K_2CO_3 [a].

Substrate	Carbonyl yield[b]
MeS-PhCH$_2$OH	81%
Ph-CH=CHCH$_2$OH	89%
$(CH_3)_2C=CH(CH_2)_2CH(CH_3)=CHCH_2OH$ [c]	71%
$C_9H_{19}CH_2OH$	65%
$C_9H_{19}CH(CH_3)OH$	88%

a) Table adapted from Ref. [110]. Conditions: 5 mol% CuCl, 5 mol% phenanthroline, 5 mol% DBAD-H$_2$ (DBAD = dibutylazodicarboxylate), 2 equiv. K_2CO_3, gentle stream of O_2, solvent is toluene, 90 °C. After 1 h reaction was complete.
b) Isolated yields at 100% conversion.
c) Geraniol.

of TEMPO to the oxoammonium cation. Based on our results with the Ru/TEMPO system we doubted the validity of this mechanism. Hence, we subjected the Cu/TEMPO to the same mechanistic studies described above for the Ru/TEMPO system [113]. The results of stoichiometric experiments under anaerobic conditions,

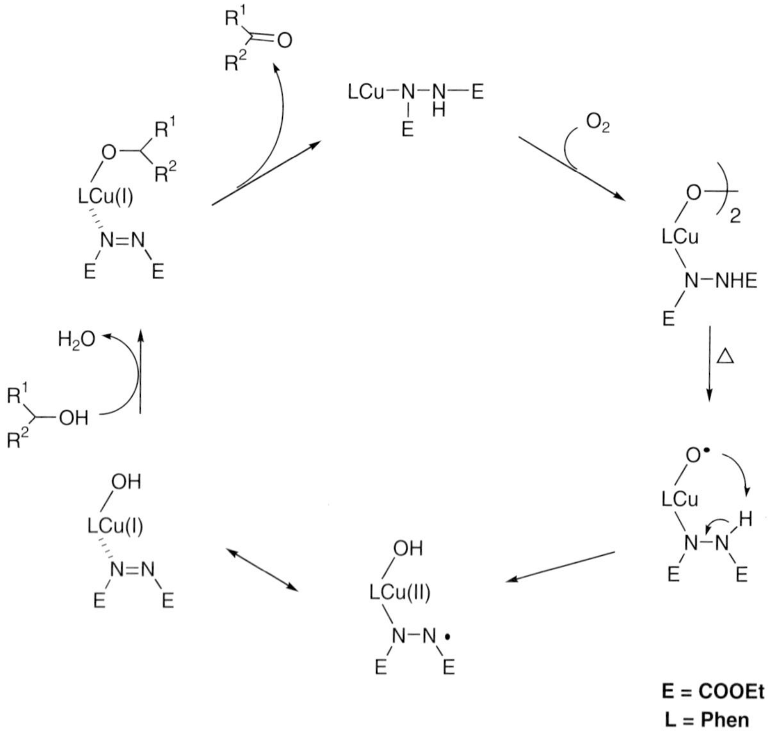

Figure 5.20 Mechanism of CuCl.phen-catalyzed oxidation of alcohols using DEAD-H$_2$ (diethylazo dicarboxylate) as an additive.

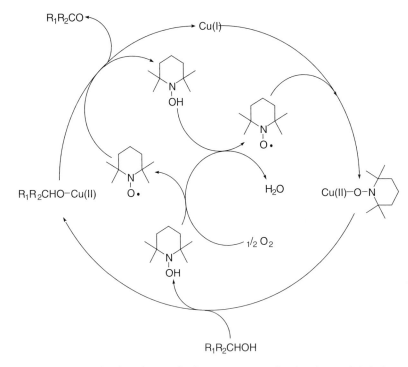

Figure 5.21 Postulated mechanism for the Cu/TEMPO-catalyzed oxidation of alcohols.

Hammett correlations, and kinetic isotope effect studies showed a similar pattern to those with the Ru/TEMPO system, that is, that they are inconsistent with a mechanism involving an oxoammonium species as the active oxidant. Hence, we propose the mechanism shown in Figure 5.21 for Cu/TEMPO-catalyzed aerobic oxidation of alcohols.

We have shown, in stoichiometric experiments, that reaction of copper(I) with TEMPO affords a piperidinyloxyl copper(II) complex. Reaction of the latter with a molecule of alcohol afforded the alkoxycopper(II) complex and TEMPOH. Reaction of the alkoxycopper(II) complex with a second molecule of TEMPO gave the carbonyl compound, copper(I), and TEMPOH. This mechanism resembles that proposed for the aerobic oxidation of alcohols catalyzed by the copper-dependent enzyme, galactose oxidase, and mimics thereof. Finally, TEMPOH is reoxidized to TEMPO by oxygen. We have also shown that copper in combination with PIPO affords an active and recyclable catalyst for alcohol oxidation [18].

In addition, many variations of the Cu/Tempo system have been published. In a first example, Knochel et al. [114] showed that CuBr.Me$_2$S with perfluoroalkyl-substituted bipyridine as the ligand was capable of oxidizing a large variety of primary and secondary alcohols in a fluorous biphasic system of chlorobenzene and perfluorooctane (see Eq. (5.14)). In the second example Ansari and Gree [115] showed that the combination of CuCl and TEMPO can be used as a catalyst in 1-butyl-3-methylimidazolium hexafluorophosphate, an ionic liquid, as the solvent. However, in this case turnover frequencies were still rather low even for benzylic alcohol (around $1.3\,h^{-1}$).

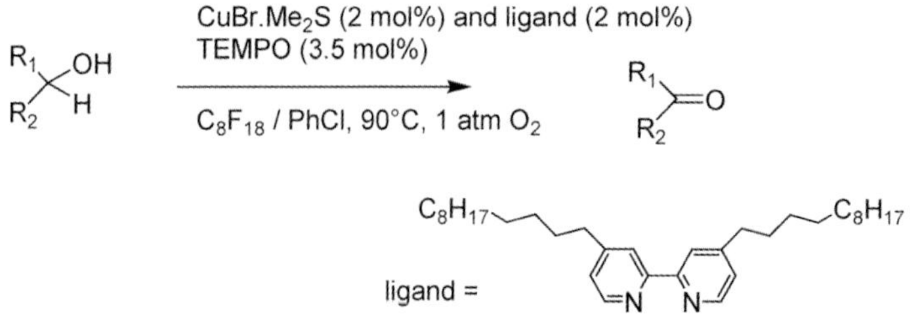

$$\underset{R_2}{\overset{R_1}{>}}\!\!\!\underset{H}{\overset{OH}{\longrightarrow}} \xrightarrow[C_8F_{18}\,/\,PhCl,\,90°C,\,1\,atm\,O_2]{CuBr.Me_2S\,(2\,mol\%)\,and\,ligand\,(2\,mol\%)\,\,TEMPO\,(3.5\,mol\%)} \underset{R_2}{\overset{R_1}{>}}\!\!=\!O \quad (5.14)$$

ligand = C_8H_{17}-bipyridine-C_8H_{17}

We developed an alternative to the above described system [116] for the selective aerobic oxidation of primary alcohols to aldehydes based on the combination $Cu^{II}Br_2(Bpy)$–TEMPO (Bpy = 2,2'-bipyridine). The reactions were carried out under air at room temperature in aqueous acetonitrile and were catalyzed by a [copper^{II}(bipyridine ligand)] complex and TEMPO and base (KOtBu) as cocatalysts (Eq. (5.15)).

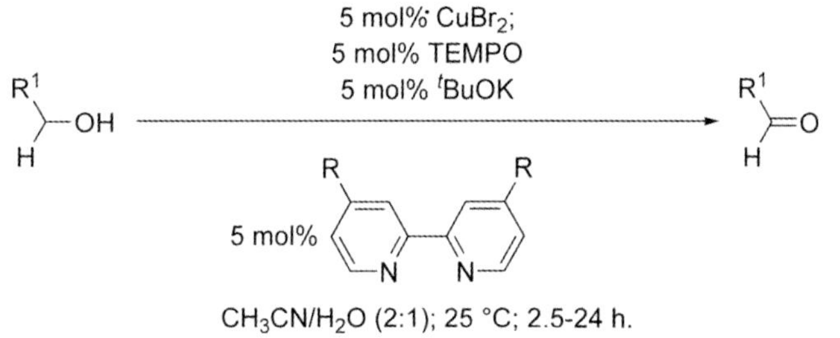

$$\underset{H}{\overset{R^1}{>}}\!\!-OH \xrightarrow[\substack{5\,mol\%\,CuBr_2;\\5\,mol\%\,TEMPO\\5\,mol\%\,^tBuOK\\5\,mol\%\,R\text{-bipy-}R\\CH_3CN/H_2O\,(2\!:\!1);\,25\,°C;\,2.5\text{-}24\,h.}]{} \underset{H}{\overset{R^1}{>}}\!\!=\!O \quad (5.15)$$

Several primary benzylic, allylic, and aliphatic alcohols were successfully oxidized, with excellent conversions (61–100%) and high selectivities (Table 5.11). The system displays a remarkable selectivity toward primary alcohols. This selectivity for the oxidation of primary alcohols resembles that encountered for certain copper-dependent enzymes. In the mechanism proposed, TEMPO coordinates side-on to the copper during the catalytic cycle.

More recently, Repo and coworkers [117] showed that the selective aerobic oxidation of benzyl alcohol to benzaldehyde could be performed with TEMPO in combination with Cu(II)/phenanthroline in aqueous alkaline with no added organic solvent. Finally, an interesting recent development in copper-catalyzed oxidation of alcohols is the use of copper nanoparticles on hydrotalcite as a heterogeneous catalyst for the liquid phase dehydrogenation of alcohols in the absence of an oxidant [118].

5.7
Other Metals as Catalysts for Oxidation with O_2

In addition to ruthenium, other late and first-row transition elements are capable of dehydrogenating alcohols via an oxometal pathway. Some are used as catalysts, in

Table 5.11 CuBr$_2$(Bpy)-TEMPO-catalyzed oxidation of alcohols to the corresponding aldehydes under air[a] [116].

Alcohol	Time (h.)	Yield (%)
PhCH$_2$OH	2.5	100
PhCH(CH$_3$)OH	5	no reaction
(CH$_3$)$_2$C=CH(CH$_2$)$_2$CH(CH$_3$)=CHCH$_2$OH[c]	5	100
C$_7$H$_{15}$CH$_2$OH	24	61[b]

a) For conditions, see Eq. (5.15).
b) 95% conversion at 40 °C.
c) Geraniol.

combination with H$_2$O$_2$ or RO$_2$H, for the oxidative dehydrogenation of alcohols (see later). By analogy with ruthenium, one might expect that regeneration of the active oxidant with dioxygen would be possible. For example, one could easily envisage alcohol oxidation by oxovanadium(V) followed by reoxidation of the resulting vanadium(III) by dioxygen. However, scant attention appears to have been paid to such possibilities. The aerobic oxidation of 1-propanol to 1-propanal (94–99% selectivity) in the gas phase at 210 °C over a V$_2$O$_5$ catalyst modified with an alkaline earth metal oxide (10 mol%) was described in 1979 [119], but subsequently only sporadic attention was paid to vanadium-catalyzed oxidations of alcohols [120]. Fiegel and Sobczak recently reported [121] that vanadium, for example, 0.5 mol% of Bu$_4$NVO$_3$ or VO(acac)$_2$Cl, in combination with 5 mol% NHPI, catalyzes the aerobic oxidation of primary and secondary alcohols in acetonitrile at 75 °C.

Similarly, Co(acac)$_3$ in combination with N-hydroxyphthalimide (NHPI) as cocatalyst mediates the aerobic oxidation of primary and secondary alcohols to the corresponding carboxylic acids and ketones, respectively (see, for example, Eq. (5.16) [122]).

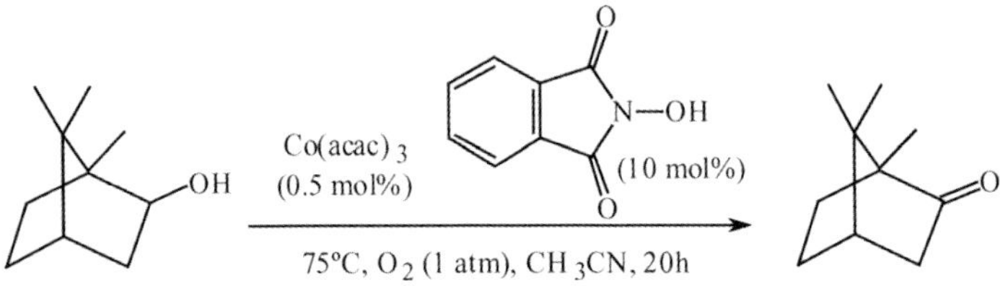

$$\text{borneol} \xrightarrow[\text{75°C, O}_2 \text{ (1 atm), CH}_3\text{CN, 20h}]{\substack{\text{Co(acac)}_3 \text{ (0.5 mol\%)} \\ \text{NHPI (10 mol\%)}}} \text{camphor} \quad (5.16)$$

By analogy with other oxidations mediated by the Co/NHPI catalyst studied by Ishii and coworkers [123, 124], Eq. (5.16) probably involves a free-radical mechanism. We attribute the promoting effect of NHPI to its ability to efficiently scavenge alkylperoxy radicals, suppressing the rate of termination by combination of these radicals. The resulting PINO radical subsequently abstracts a hydrogen atom from the α-C–H bond of the alcohol to propagate the autoxidation chain (Eqs. (5.17)–(5.19)).

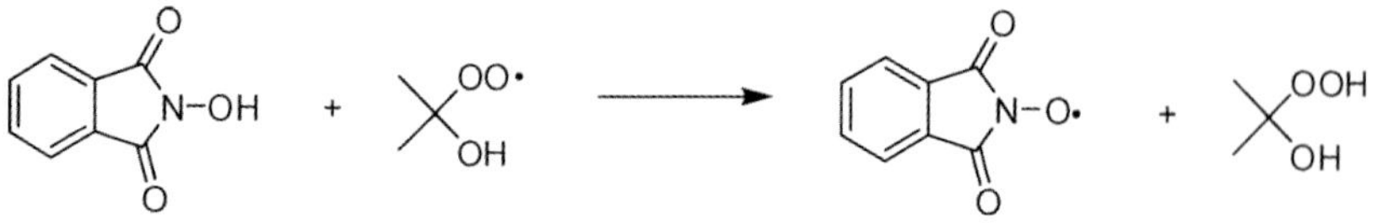

$$\text{Phth-N-OH} + \text{Me}_2\text{C(OH)OO}\cdot \longrightarrow \text{Phth-N-O}\cdot + \text{Me}_2\text{C(OH)OOH} \quad (5.17)$$

$$\text{Phth-N-O}\cdot + \text{Me}_2\text{C(H)OH} \longrightarrow \text{Phth-N-OH} + \text{Me}_2\text{C(OH)}\cdot \quad (5.18)$$

$$\text{Me}_2\text{C(OH)}\cdot + \text{O}_2 \longrightarrow \text{Me}_2\text{C(OH)OO}\cdot \quad (5.19)$$

A nickel-substituted hydrotalcite was also reported as a catalyst for the aerobic oxidation of benzylic and allylic alcohols [125]. Analogous to cobalt, nickel is expected to catalyze oxidation via a free-radical mechanism.

After their leading publication on the osmium-catalyzed dihydroxylation of alkenes in the presence of dioxygen [126], Beller et al. [127] reported that alcohol oxidations could also be performed using the same conditions (see Eq. (5.20)). The reactions were carried out in a buffered two-phase system with a constant pH of 10.4. Under these conditions a remarkable catalyst productivity (TON up to 16 600 for acetophenone) was observed. The pH value is critical in order to ensure the reoxidation of Os(VI) to Os(VIII). The scope of this system seems to be limited to benzylic and secondary alcohols.

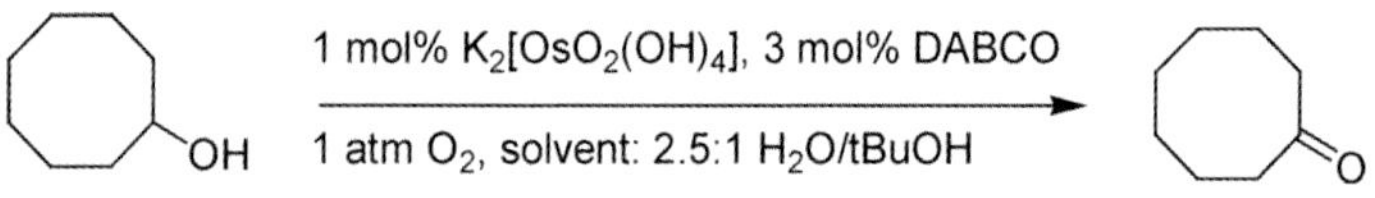

$$\text{cyclooctanol} \xrightarrow[\text{1 atm O}_2, \text{ solvent: 2.5:1 H}_2\text{O/tBuOH}]{\text{1 mol\% K}_2[\text{OsO}_2(\text{OH})_4], \text{ 3 mol\% DABCO}} \text{cyclooctanone} \quad (5.20)$$

5.8
Catalytic Oxidation of Alcohols with Hydrogen Peroxide

In the aerobic oxidations discussed in the preceding sections, the most effective catalysts tend to be late transition elements, for example, Ru and Pd, that operate via oxometal or hydridometal mechanisms. In contrast, the most effective catalysts with H_2O_2 (or RO_2H) as the oxidant tend to be early transition metal ions with a d^0 configuration, for example, Mo(VI), W(VI), and Re(VII), which operate via peroxometal pathways. Ruthenium and palladium are generally not effective with H_2O_2

because they display high catalase activity, that is, they catalyze rapid decomposition of H_2O_2. Early transition elements, on the other hand, are generally poor catalysts for H_2O_2 decomposition.

One of the few examples of ruthenium-based systems is the $RuCl_3 \cdot 3H_2O$/didecyldimethylammonium bromide combination reported by Sasson and coworkers [128]. This system catalyzes the selective oxidation of a variety of alcohols, at high (625:1) substrate:catalyst ratios, in an aqueous/organic biphasic system. However, 3–6 equivalents of H_2O_2 were required, reflecting the propensity of ruthenium for catalyzing non-productive decomposition of H_2O_2.

Jacobsen and coworkers [129] showed, in 1979, that anionic molybdenum(VI) and tungsten(VI) peroxo complexes are effective oxidants for the stoichiometric oxidation of secondary alcohols to the corresponding ketones. Subsequently, Trost and Masuyama [130] showed that ammonium molybdate, $(NH_4)_6Mo_7O_{24} \cdot 4H_2O$ (10 mol%), is able to catalyze the selective oxidation of secondary alcohols to the corresponding ketones, using hydrogen peroxide in the presence of tetrabutylammonium chloride and a stoichiometric amount of a base (K_2CO_3). It is noteworthy that a more hindered alcohol moiety was oxidized more rapidly than a less hindered one (see, for example Eq. (5.21)).

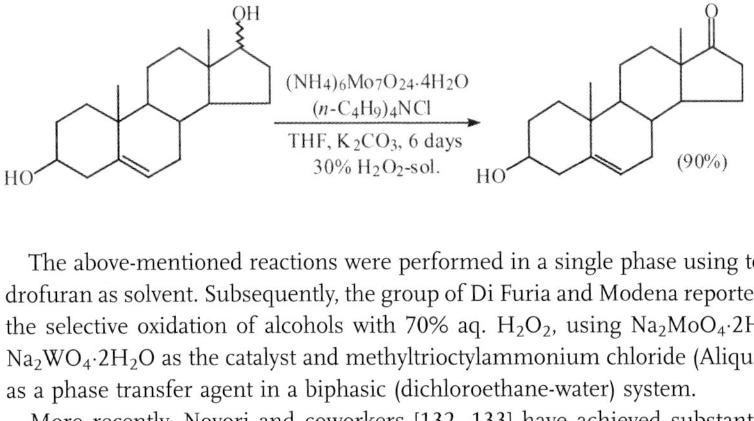

$$(5.21)$$

Reaction conditions: $(NH_4)_6Mo_7O_{24} \cdot 4H_2O$, $(n\text{-}C_4H_9)_4NCl$, THF, K_2CO_3, 6 days, 30% H_2O_2-sol. (90%)

The above-mentioned reactions were performed in a single phase using tetrahydrofuran as solvent. Subsequently, the group of Di Furia and Modena reported [131] the selective oxidation of alcohols with 70% aq. H_2O_2, using $Na_2MoO_4 \cdot 2H_2O$ or $Na_2WO_4 \cdot 2H_2O$ as the catalyst and methyltrioctylammonium chloride (Aliquat 336) as a phase transfer agent in a biphasic (dichloroethane-water) system.

More recently, Noyori and coworkers [132, 133] have achieved substantial improvements in the sodium tungstate-based biphasic system by employing a phase transfer agent containing a lipophilic cation and bisulfate as the anion, for example, $CH_3(n\text{-}C_8H_{17})_3NHSO_4$. This afforded a highly active catalytic system for the oxidation of alcohols using 1.1 equivalents of 30% aq. H_2O_2 in a solvent-free system. For example, 1-phenylethanol was converted to acetophenone with turnover numbers up to 180 000. As with all Mo- and W-based systems, the Noyori system shows a marked preference for secondary alcohols (see, for example, Eq. (5.22)).

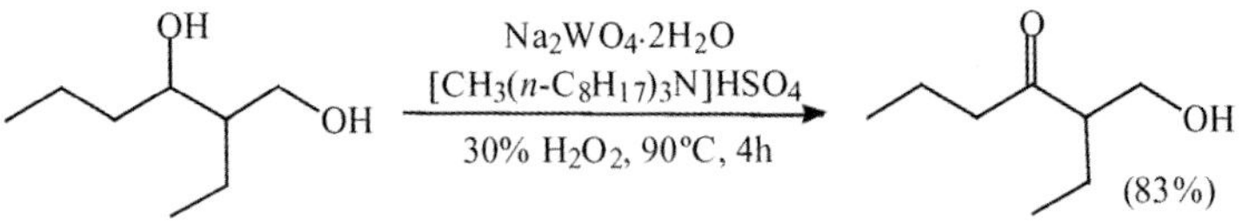

$$(5.22)$$

Reaction conditions: $Na_2WO_4 \cdot 2H_2O$, $[CH_3(n\text{-}C_8H_{17})_3N]HSO_4$, 30% H_2O_2, 90°C, 4h (83%)

Unsaturated alcohols generally undergo selective oxidation of the alcohol moiety (Eqs. (5.23) and (5.24)), but, when an allylic alcohol contained a reactive trisubstituted double bond, selective epoxidation of the double bond was observed (Eq. (5.25)).

$$\underset{(CH_2)_8}{\overset{OH}{\bigwedge}} \longrightarrow \underset{(CH_2)_8}{\overset{O}{\bigwedge}} \quad (100\% \text{ selectivity}) \tag{5.23}$$

$$\underset{OH}{\overset{(CH_2)_8CH_3}{\bigwedge}} \longrightarrow \underset{O}{\overset{(CH_2)_8CH_3}{\bigwedge}} \quad (80\% \text{ selectivity}) \tag{5.24}$$

$$\underset{(CH_2)_6CH_3}{\overset{OH}{\bigwedge}} \longrightarrow \underset{(CH_2)_6CH_3}{\overset{OH}{\bigwedge}} \quad (100\% \text{ selectivity}) \tag{5.25}$$

Molybdenum- and tungsten-containing heteropolyanions are also effective catalysts for alcohol oxidations with H_2O_2 [134–136]. For example, $H_3PMo_{12}O_{40}$ or $H_3PW_{12}O_{40}$, in combination with cetylpyridinium chloride as a phase transfer agent, were shown by Ishii and coworkers [137] to be effective catalysts for alcohol oxidations with H_2O_2 in a biphasic, chloroform/water system.

Methyltrioxorhenium (MTO) also catalyzes the oxidation of alcohols with H_2O_2 via a peroxometal pathway [137, 138]. Primary benzylic and secondary aliphatic alcohols afforded the corresponding aldehydes and ketones, respectively, albeit using two equivalents of H_2O_2. In the presence of bromide ion the rate was increased by a factor 1000 [137]. In this case the active oxidant could be hypobromite (HOBr), formed by MTO-catalyzed oxidation of bromide ion by H_2O_2.

For titanium, the titanium silicalite TS-1, an isomorphously substituted molecular sieve [139], is a truly heterogeneous catalyst for oxidations with 30% aq. H_2O_2, including the oxidation of alcohols [140].

A dinuclear manganese(IV) complex of trimethyl triazacyclononane (tmtacn) catalyzed the selective oxidation of reactive benzylic alcohols with hydrogen peroxide in acetone [141]. However, a large excess (up to 8 equivalents) of H_2O_2 was required, suggesting that there is substantial nonproductive decomposition of the oxidant. Moreover, we note that the use of acetone as a solvent for oxidations with H_2O_2 is not recommended owing to the formation of explosion-sensitive peroxides. The exact nature of the catalytically active species in this system is rather obscure; for optimum activity it was necessary to pretreat the complex with H_2O_2 in acetone. Presumably the active oxidant is a high-valent oxomanganese species, but further studies are necessary to elucidate the mechanism.

Free nano-iron has been reported as an oxidation catalyst for the oxidation of benzyl alcohol applying hydrogen peroxide as the oxidant [142]. Although the turnover number based on iron is only modest (TON around 30), the high selectivity of 97% towards benzaldehyde is striking in comparison to the use of homogeneous Fe(NO$_3$)$_3$ as a catalyst. The selectivity based on hydrogen peroxide is approximately 30%. All reactions were performed in water as solvent (Eq. (5.26)).

$$\text{PhCH}_2\text{OH} \xrightarrow[\text{30 wt\% HOOH}]{\substack{1 \text{ mol\% Fe}^{3+} \\ 75°C / 12 \text{ h}}} \text{PhCHO}$$

	TON	Selectivity
γ-Fe$_2$O$_3$	32	97%
Fe(NO$_3$)$_3$	25	35%

(5.26)

5.8.1
Biocatalytic Oxidation of Alcohols

Enzymatic methods for the oxidation of alcohols are becoming more important [143]. Besides the already mentioned oxidases (laccase, see above and, for example, glucose oxidase and galactose oxidase), the enzymes that are responsible in vivo for (asymmetric) alcohol dehydrogenation are the alcohol dehydrogenases. Commercially viable protocols are currently available for recycling the NAD(P)$^+$ cofactor [143]. Recently, a stable NAD$^+$-dependent alcohol dehydrogenase from *Rhodococcus rubber* was reported, which accepts acetone as co-substrate for NAD$^+$ regeneration and at the same time performs the desired alcohol oxidation (Figure 5.22). Alcohol concentrations up to 50% v/v could be applied [144]. With a chiral alcohol substrate, kinetic resolution is observed, that is, one of the enantiomers is selectively converted. For example, for 2-octanol, maximum 50% conversion is obtained and the residual alcohol has an *ee* of >99%. On the other hand, for 2-butanol, no chiral discrimination occurs, and 96% conversion was observed. Also steric requirements exist: while cyclohexanol is not oxidized, cyclopentanol is a good substrate. Much is to be expected in the future from alcohol oxidations by flavin-dependent oxidases. New oxidases have been discovered, albeit still exhibiting a small alcohol scope [145].

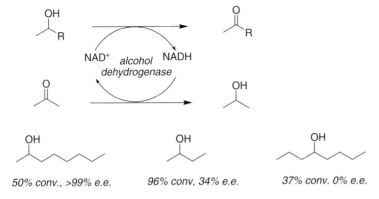

Figure 5.22 Biocatalytic oxidation of alcohol using acetone for co-factor recycling.

5.9
Concluding Remarks

The economic importance of alcohol oxidations in the fine chemical industry will, in the future, continue to stimulate the quest for effective catalysts that utilize dioxygen or hydrogen peroxide as the primary oxidant. Although much progress has been made in recent years there is still room for further improvement with regard to catalyst activity and scope in organic synthesis. A better understanding of mechanistic details regarding the nature of the active intermediate and the rate-determining step would certainly facilitate this since many of these systems are poorly understood. The enantioselective oxidation of chiral alcohols, as described for ruthenium-based systems [146] and the palladium-based systems reported by Sigman, Stolz, and coworkers [147, 148] remains of interest, and biocatalytic systems, for example, alcohol dehydrogenases, are particularly effective in this respect [143, 144].

The current trend toward the production of chemicals from feedstocks based on renewable resources rather than fossil fuels such as oil or natural gas has also led to a need for effective methods for the conversion of carbohydrates, from glycerol to polysaccharides, to commercially valuable products. In many cases this will involve selective aerobic oxidation of one or more alcohol moieties.

In short, we expect the interest in selective, environmentally benign oxidation of alcohols, using both chemo- and biocatalytic methodologies, to continue unabated in the future.

References

1 Ley, S.V., Norman, J., Griffith, W.P., and Marsden, S.P. (1994) *Synthesis*, 639.
2 See Organic Textbooks, e.g. Clayden, J., Greeves, N., Warren, S., and Wothers, P. (2001) *Organic Chemistry*, Oxford Univ. Press, New York; Bruice, P.Y. (2005) *Organic Chemistry*, 5th edn, Prentice Hall, New Jersey.
3 Cainelli, G. and Cardillo, G. (1984) *Chromium Oxidations in Organic Chemistry*, Springer, Berlin.
4 Sheldon, R.A. and Kochi, J.K. (1981) *Metal-Catalysed Oxidations of Organic Compounds*, Academic Press, New York.
5 For recent reviews see: Besson, M. and Gallezot, P. (2000) *Catal. Today*, 57, 127; Mallat, T. and Baiker, A. (1994) *Catal. Today*, 19, 247; Besson, M. and Gallezot, P. (2001) in *Fine Chemicals through Heterogeneous Catalysis* (eds R.A. Sheldon and H. van Bekkum), Wiley-VCH, Weinheim, p. 491.
6 de Nooy, A.E.J., Besemer, A.C., and van Bekkum, H. (1996) *Synthesis*, 1153, and refs. cited herein.
7 Anelli, P.L., Biffi, C., Montanari, F., and Quici, S. (1987) *J. Org. Chem.*, 52, 2559.
8 Cella, J.A., Kelley, J.A., and Kenehan, E.F. (1975) *J. Org. Chem.*, 40, 1860; Rychovsky, S.D. and Vaidyanathan, R. (1999) *J. Org. Chem.*, 64, 310.
9 Inokuchi, T., Matsumoto, S., Nishiyama, T., and Torii, S. (1990) *J. Org. Chem.*, 55, 462.
10 Zhao, M., Li, J., Mano, E., Song, Z., Tschaen, D.M., Grabowski, E.J.J., and Reider, P.J. (1999) *J. Org. Chem.*, 64, 2564.
11 Jenny, C.-J., Lohri, B., and Schlageter, M. (1997) Eur. Patent. 0775684A1.

12 Bolm, C., Magnus, A.S., and Hildebrand, J.P. (2000) *Org. Lett.*, **2**, 1173.
13 Bolm, C. and Fey, T. (1999) *Chem. Commun.*, 1795; Brunel, D., Fajula, F., Nagy, J.B., Deroide, B., Verhoef, M.J., Veum, L., Peters, J.A., and van Bekkum, H. (2001) *Appl. Catal. A: General*, **213**, 73.
14 Brunel, D., Lentz, P., Sutra, P., Deroide, B., Fajula, F., and Nagy, J.B. (1999) *Stud. Surf. Sci. Catal.*, **125**, 237.
15 Verhoef, M.J., Peters, J.A., and van Bekkum, H. (1999) *Stud. Surf. Sci. Catal.*, **125**, 465.
16 Ciriminna, R., Blum, J., Avnir, D., and Pagliaro, M. (2000) *Chem. Commun.*, 1441.
17 Dijksman, A., Arends, I.W.C.E., and Sheldon, R.A. (2000) *Chem. Commun.*, 271.
18 Dijksman, A., Arends, I.W.C.E., and Sheldon, R.A. (2001) *Synlett*, 102.
19 Daniel, R.B., Alsters, P., and Neumann, R. (2001) *J. Org. Chem.*, **66**, 8650.
20 Cecchetto, A., Fontana, F., Minisci, F., and Recupero, F. (2001) *Tetrahedron Lett.*, **42**, 6651.
21 Fabbrini, M., Galli, C., Gentilli, P., and Macchitella, D. (2001) *Tetrahedron Lett.*, **42**, 7551; d'Acunzo, F., Baiocco, P., Fabbrini, M., Galli, C., and Gentili, P. (1995) *Eur. J. Org. Chem.*, 4195; Astolfi, P., Brandi, P., Galli, C., Gentilli, P., Gerini, M.F., Greci, L., and Lanzalunga, O. (2005) *New. J. Chem.*, **29**, 1308.
22 Arends, I.W.C.E., Li, Y.X., and Sheldon, R.A. (2006) *Biocat. Biotrans.*, **24**, 443; Arends, I.W.C.E., Li, Y.X., Ausan, R., and Sheldon, R.A. (2006) *Tetrahedron*, **62**, 6659.
23 Liu, R., Dong, C., Liang, X., Wang, X., and Hu, X. (2005) *J. Org. Chem.*, **70**, 729; see also Wang, X., Liu, R.H., Jin, Y., and Liang, X. (2008) *Chem. Eur. J.*, **14**, 2679.
24 Wang, N., Liu, R., Chen, J., and Liang, X. (2005) *Chem. Commun.*, 5322; Wang, X. and Liang, X. (2008) *Chin. J. Catal.*, **29**, 935.
25 Berzelius, J.J. (1828) *Ann.*, **13**, 435.
26 Lloyd, W.G. (1967) *J. Org. Chem.*, **32**, 2816.
27 Blackburn, T.F. and Schwarz, J.J. (1977) *Chem. Soc. Chem. Commun.*, 157.
28 Aït-Mohand, S. and Muzart, J.J. (1998) *J. Mol. Catal. A: Chemical*, **129**, 135.
29 Peterson, K.P. and Larock, R.C.J. (1998) *Org. Chem.*, **63**, 3185.
30 Nishimura, T., Onoue, T., Ohe, K., and Uemura, S. (1998) *Tetrahedron Lett.*, **39**, 6011; Nishimura, T., Onoue, T., Ohe, K., and Uemura, S. (1999) *J. Org. Chem.*, **64**, 6750.
31 Sheldon, R.A. (1993) *Top. Curr. Chem.*, **164**, 21.
32 Naota, T., Takaya, H., and Murahashi, S.-I. (1998) *Chem. Rev.*, **98**, 2599.
33 Murahashi, S.-I. and Komiya, N. (2000) in *Biomimetic Oxidations Catalyzed by Transition Metal Complexes* (ed. B. Meunier), Imperial College Press, London, p. 563; see also Gore, E. (1983) *Plat. Met. Rev.*, **27**, 111.
34 Sheldon, R.A., Arends, I.W.C.E., and Dijksman, A. (2000) *Catal. Today*, **57**, 158.
35 Tang, R., Diamond, S.E., Neary, N., and Mares, F.J. (1978) *J. Chem. Soc. Chem. Commun.*, 562.
36 Matsumoto, M. and Ito, S.J. (1981) *Chem. Soc. Chem. Commun.*, 907.
37 Bilgrien, C., Davis, S., and Drago, R.S. (1987) *J. Am. Chem. Soc.*, **109**, 3786.
38 Sheldon, R.A. (2001) *Chem. Commun.*, **23**, 2399.
39 Wolfson, A., Wuyts, S., de Vos, D.E., Vancelecom, I.F.J., and Jacobs, P.A. (2002) *Tetrahedron Lett.*, **43**, 8107.
40 Bäckvall, J.E., Chowdhury, R.L., and Karlsson, U. (1991) *J. Chem. Soc. Chem. Commun.*, 473; Wang, G.-Z., Andreasson, U., and Bäckvall, J.E. (1994) *J. Chem. Soc. Chem. Commun.*, 1037; Csjernyik, G., Ell, A., Fadini, L., Pugin, B., and Bäckvall, J.E. (2002) *J. Org. Chem.*, **67**, 1657.
41 Hanyu, A., Takezawa, E., Sakaguchi, S., and Ishii, Y. (1998) *Tetrahedron Lett.*, **39**, 5557.
42 Dijksman, A., Arends, I.W.C.E., and Sheldon, R.A. (1999) *Chem. Commun.*, 1591; Dijksman, A., Marino-González, A., Mairata i Payeras, A., Arends, I.W.C.E., and Sheldon, R.A. (2001) *J. Am. Chem. Soc.*, **123**, 6826; for a related study see Inokuchi, T., Nakagawa, K., and Torii, S. (1995) *Tetrahedron Lett.*, **36**, 3223.

43 Bäckvall, J.-E. and Andreasson, U. (1993) *Tetrahedron Lett.*, **34**, 5459; Trost, B.M. and Kulawiec, R.J. (1991) *Tetrahedron Lett.*, **32**, 3039.

44 Aranyos, A., Csjernyik, G., Szabo, K.J., and Bäckvall, J.-E. (1999) *Chem. Commun.*, 351.

45 Karlson, U., Wang, G.-Z., and Bäckvall, J.-E. (1994) *J. Org. Chem.*, **59**, 1196.

46 Paleos, C.M. and Dais, P.J. (1977) *Chem. Soc. Chem. Commun.*, 345.

47 Beynon, P.J., Collins, P.M., Gardiner, D., and Overend, W.G. (1968) *Carbohydr. Res.*, **6**, 431; see also Friedrich, H.B. (1999) *Plat. Met. Rev.*, **43**, 94.

48 Dengel, A.C., Hudson, R.A., and Griffith, W.P. (1985) *Trans. Met. Chem.*, **10**, 98.

49 Griffith, W.P., Ley, S.V., Whitcombe, G.P., and White, A.D. (1987) *Chem. Commun.*, 1625.

50 Dengel, A.C., El-Hendawy, A.M., and Griffith, W.P. (1989) *Trans. Met. Chem.*, **40**, 230.

51 Griffith, W.P. and Ley, S.V. (1990) *Aldrichim. Acta*, **23**, 13.

52 Ley, S.V., Norman, J., Griffith, W.P., and Marsen, S.P. (1994) *Synthesis*, 639.

53 Lenz, R. and Ley, S.V. (1997) *J. Chem. Soc. Perkin Trans. 1*, 3291.

54 Marko, I.E., Giles, P.R., Tsukazaki, M., Chelle-Regnaut, I., Urch, C.J., and Brown, S.M. (1997) *J. Am. Chem. Soc.*, **119**, 12661.

55 Hinzen, B., Lenz, R., and Ley, S.V. (1998) *Synthesis*, 977.

56 Bleloch, A., Johnson, B.F.G., Ley, S.V., Price, A.J., Shepard, D.S., and Thomas, A.N. (1999) *Chem. Commun.*, 1907.

57 Pagliaro, M. and Ciriminna, R. (2001) *Tetrahedron Lett.*, **42**, 4511; For the use of a Fluorogel analogue for reactions in ScCO$_2$ see Ciriminna, R., Campestrini, S., and Pagliaro, M. (2006) *Org. Biomol. Chem.*, **4**, 2637.

58 Hasan, M., Musawir, M., Davey, P.N., and Kozhevnikov, I. (2002) *J. Mol. Catal. A: Chemical*, **180**, 77.

59 Rocek, J. and Ng, C.-S. (1974) *J. Am. Chem. Soc.*, **96**, 1522.

60 Lee, D.G., Wang, Z., and Chandler, W.D. (1992) *J. Org. Chem.*, **57**, 3276.

61 Cheng, S.Y.S., Rajapakse, N., Rettig, S.J., and James, B.R. (1994) *J. Chem. Soc. Chem. Commun.*, 2669; see also Rajapakse, N., James, B.R., and Dolphin, D. (1990) *Stud. Surf. Sci. Catal.*, **55**, 109.

62 Yamaguchi, K. and Mizuno, N. (2002) *Angew. Chem. Int. Ed.*, **41**, 4538.

63 Kaneda, K., Yamashita, T., Matsushita, T., and Ebitani, K. (1998) *J. Org. Chem.*, **63**, 1750.

64 Matsushita, T., Ebitani, K., and Kaneda, K. (1999) *Chem. Commun.*, 265.

65 Stahl, S.S. (2005) *Science*, **309**, 1824.

66 Muzart, J. (2003) *Tetrahedron*, **59**, 5789; Muzart, J. (2006) *Chem. Asian J.*, **1**, 508.

67 Blackburn, T.F. and Schwartz, J.J. (1977) *Chem. Soc. Chem. Commun.*, 157.

68 Vargaftik, M.N., Zagorodnikov, V.P., Storarov, I.P., and Moiseev, I.I. (1989) *J. Mol. Catal.*, **53**, 315; see also Moiseev, I.I. and Vargaftik, M.N. (1998) in *Catalysis by Di- and Polynuclear Metal Cluster Complexes* (eds R.D. Adamsand F.A. Cotton), Wiley-VCH, Weinheim, p. 395.

69 Kaneda, K., Fujii, M., and Morioka, K. (1996) *J. Org. Chem.*, **61**, 4502; Kaneda, K., Fujie, Y., and Ebitani, K. (1997) *Tetrahedron Lett.*, **38**, 9023.

70 Peterson, K.P. and Larock, R.C.J. (1998) *Org. Chem.*, **63**, 3185.

71 van Benthem, R.A.T.M., Hiemstra, H., van Leeuwen, P.W.N.M., Geus, J.W., and Speckamp, W.N. (1995) *Angew. Chem.*, **107**, 500; (1995) *Angew. Chem. Int. Ed.*, **34**, 457.

72 Ait-Mohand, S., Hénin, F., and Muzart, J. (1995) *Tetrahedron Lett.*, **36**, 2473.

73 Ait-Mohand, S. and Muzart, J.J. (1998) *J. Mol. Catal. A: Chemical*, **129**, 135.

74 Nishimura, T., Onoue, T., Ohe, K., and Uemura, S. (1998) *Tetrahedron Lett.*, **39**, 6011.

75 Nishimura, T., Onoue, T., Ohe, K., and Uemura, S.J. (1999) *J. Org. Chem.*, **64**, 6750; Nishimura, T., Ohe, K., and Uemura, S.J. (1999) *J. Am. Chem. Soc.*, **121**, 2645.

76 Nishimura, T., Maeda, Y., Kakiuchi, N., and Uemura, S. (2000) *J. Chem. Soc. Perkin Trans. 1*, 4301.

77 Steinhoff, B.A., Fix, S.R., and Stahl, S.S. (2002) *J. Am. Chem. Soc.*, **124**, 766.

78 Steinhoff, B.A. and Stahl, S.S. (2002) *Org. Lett.*, **4**, 4179–4181.
79 Schultz, M.J., Park, C.C., and Sigman, M.S. (2002) *Chem. Commun.*, 3034.
80 Hallman, K. and Moberg, C. (2001) *Adv. Synth. Catal.*, **343**, 260.
81 ten Brink, G.-J., Arends, I.W.C.E., and Sheldon, R.A. (2000) *Science*, **287**, 1636.
82 ten Brink, G.-J., Arends, I.W.C.E., and Sheldon, R.A. (2002) *Adv. Synth. Catal.*, **344**, 355.
83 Bianchi, D., Bortolo, R., D'Aloisio, R., and Ricci, M. (1999) *Angew. Chem. Int. Ed.*, **38**, 706; Bianchi, D., Bortolo, R., D'Aloisio, R., and Ricci, M. (1999) *J. Mol. Catal. A: Chemical*, **150**, 87.
84 ten Brink, G.J., Arends, I.W.C.E., Hoogenraad, M., Verspui, G., and Sheldon, R.A. (2003) *Adv. Synth. Catal.*, **345**, 1341.
85 Mifsud, M., Parkhomenko, K.V., Arends, I.W.C.E., and Sheldon, R.A. (2010) *Tetrahedron.*, **66**, 1040.
86 Kovtun, G., Kameneva, T., Hladyi, S., Starchevsky, M., Pazdersky, Y., Stolarov, I., Vargaftik, M., and Moiseev, I.I. (2002) *Adv. Synth. Catal.*, **344**, 957; Moiseev, I.I. and Vargaftik, M.N. (1998) *New J. Chem.*, **22**, 1217; Pasichnyk, P.I., Starchevsky, M.K., Pazdersky, Y.A., Zagorodnikov, V.P., Vargaftik, M.N., and Moiseev, I.I. (1994) *Mendeleev Commun.*, 1–2; Moiseev, I.I., Stromnova, T.A., and Vargaftik, M.N. (1994) *J. Mol. Catal.*, **86**, 71.
87 Mori, K., Hara, T., Mizugaki, T., Ebitani, K., and Kaneda, K. (2004) *J. Am. Chem. Soc.*, **126**, 10657.
88 For an example in toluene see Keresszegi, C., Burgi, T., Mallat, T., and Baiker, A. (2002) *J. Catal.*, **211**, 244.
89 Nishimura, T., Kakiuchi, N., Inoue, M., and Uemura, S. (2000) *Chem. Commun.*, 1245; see also Kakiuchi, N., Nishimura, T., Inoue, M., and Uemura, S. (2001) *Bull. Chem. Soc. Jpn.*, **74**, 165.
90 Mori, K., Yamaguchi, K., Hara, T., Mizugaki, T., Ebitani, K., and Kaneda, K. (2002) *J. Am. Chem. Soc.*, **124**, 11573.
91 Kwon, M.S., Kim, N., Park, C.M., Lee, J.S., Kang, K.Y., and Park, J. (2005) *Org. Lett.*, **7**, 1077.
92 Uozumi, Y. and Nakao, R. (2003) *Angew. Chem. Int. Ed.*, **42**, 194.
93 Hou, Z., Theyssen, N., Brinkmann, A., and Leitner, W. (2005) *Angew. Chem. Int. Ed.*, **44**, 1346.
94 Prati, L. and Rossi, M. (1997) *Stud. Surf. Sci. Catal.*, **110**, 509; Prati, L. and Rossi, M. (1998) *J. Catal.*, **176**, 552; Bianchi, C., Porta, F., Prati, L., and Rossi, M. (2000) *Top. Catal.*, **13**, 231; Comotti, M., Della Pina, C., Matarrese, R., Rossi, M., and Siani, A. (2005) *Appl. Catal. A: General*, **291**, 204; Prati, L. and Porta, F. (2005) *Appl. Catal. A: General*, **291**, 199.
95 Christensen, C.H., Jørgensen, B., Rass-Hansen, J., Egeblad, K., Madfsen, R., Klitgaard, S.K., Hansen, S.M., Hansen, M.R., Andersen, H.C., and Riisager, A. (2006) *Angew. Chem. Int. Ed.*, **45**, 4648.
96 Taarning, E., Nielsen, I.S., Egeblad, K., Madsen, R., and Christensen, C.H. (2008) *ChemSusChem*, **1**, 75–78; see also Taarning, E., Madsen, A.T., Marchetti, J.M., Egeblad, K., and Christensen, C.H. (2008) *Green Chem.*, **10**, 408.
97 Abad, A., Concepcion, P., Corma, A., and Garcia, H. (2005) *Angew. Chem. Int. Ed.*, **44**, 4066.
98 See also Tsunoyama, H., Sakurai, H., Negishi, Y., and Tsukuda, T. (2005) *J. Am. Chem. Soc.*, **127**, 9374.
99 Enache, D.I., Edwards, J.K., Landon, P., Solsona-Espriu, B., Carley, A.F., Herzing, A.A., Watanabe, M., Kiely, C.J., Knight, D.W., and Hutchings, G.J. (2006) *Science*, **311**, 362.
100 Carrettin, S., McMorn, P., Johnston, P., Griffin, K., and Hutchings, G.J. (2002) *Chem. Commun.*, 696.
101 Ito, N., Phillips, S.E.V., Stevens, C., Ogel, Z.B., McPherson, M.J., Keen, J.N., Yadav, K.D.S., and Knowles, P.F. (1991) *Nature*, **350**, 87.
102 Drauz, K., and Waldmann, H. (1995) Chapter 6, in *Enzyme Catalysis in Organic Synthesis*, VCH, Weinheim.
103 For example see: Skibida, I.P. and Sakharov, A.M. (1996) *Catal. Today*, **27**, 187; Sakharov, A.M. and Skibida, I.P. (1988) *J. Mol. Catal.*, **48**, 157; Feldberg, L. and Sasson, Y.L. (1994) *J. Chem. Soc. Chem. Commun.*, 1807;

Capdevielle, P., Sparfel, D., Baranne-Lafont, J., Cuong, N.K., and Maumy, D. (1993) *J. Chem. Res. (S)*, 10; Munakata, M., Nishibayashi, S., and Sakamoto, S. (1980) *J. Chem. Soc. Chem. Commun.*, 219; Bhaduri, S. and Sapre, N.Y. (1981) *J. Chem. Soc. Dalton Trans.*, 2585; Jallabert, C. and Rivière, H. (1977) *Tetrahedron Lett.*, 1215; (1980) *J. Mol. Catal.*, **7**, 127; Jallabert, C., Lapinte, C., and Rivière, H. (1986) *J. Mol. Catal.*, **14**, 75; Jallabert, C. and Rivière, H. (1980) *Tetrahedron*, **36**, 1191.

104 Wang, Y., DuBois, J.L., Hedman, B., Hodgson, K.O., and Stack, T.D.P. (1998) *Science*, **279**, 537.

105 Chauhuri, P., Hess, M., Flörke, U., and Wieghardt, K. (1998) *Angew. Chem. Int. Ed.*, **37**, 2217.

106 Chauhuri, P., Hess, M., Weyhermüller, T., and Wieghardt, K. (1998) *Angew. Chem. Int. Ed.*, **38**, 1095.

107 Mahadevan, V., Klein Gebbink, R.J.M., and Stack, T.D.P. (2000) *Curr. Opin. Chem. Biol.*, **4**, 228.

108 Whittaker, M.M., Ekberg, C.A., Peterson, J., Sendova, M.S., Day, E.P., and Whittaker, J.W. (2000) *J. Mol. Catal. B: Enzymatic*, **8**, 3.

109 Marko, E.I., Giles, P.R., Tsukazaki, M., Brown, S.M., and Urch, C.J. (1996) *Science*, **274**, 2044; Marko, I.E., Tsukazaki, M., Giles, P.R., Brown, S.M., and Urch, C.J. (1997) *Angew. Chem. Int. Ed. Engl.*, **36**, 2208.

110 Marko, I.E., Giles, P.R., Tsukazaki, M., Chellé-Regnaut, I., Gautier, A., Brown, S.M., and Urch, C.J. (1999) *J. Org. Chem.*, **64**, 2433.

111 Marko, I.E., Gautier, A., Chellé-Regnaut, I., Giles, P.R., Tsukazaki, M., Urch, C.J., and Brown, S.M. (1998) *J. Org. Chem.*, **63**, 7576.

112 Semmelhack, M.F., Schmid, C.R., Cortés, D.A., and Chou, C.S. (1984) *J. Am. Chem. Soc.*, **106**, 3374.

113 Dijksman, A. (2001) Thesis Delft University of Technology, Delft.

114 Betzemeier, B., Cavazzine, M., Quici, S., and Knochel, P. (2000) *Tetrahedron Lett.*, **41**, 4343.

115 Ansari, I.A. and Gree, R. (2002) *Org. Lett.*, **4**, 1507.

116 Gamez, P., Arends, I.W.C.E., Reedijk, J., and Sheldon, R.A. (2003) *Chem. Commun.*, 2414; Gamez, P., Arends, I.W.C.E., Sheldon, R.A., and Reedijk, J. (2004) *Adv. Synth. Catal.*, **346**, 805.

117 Fiegel, P.J., Leskela, M., and Repo, T. (2007) *Adv. Synth. Catal.*, **349**, 1173; see also Geislmeir, D., Jary, W.G., and Falk, H. (2005) *Monatsh. Chem.*, **136**, 1591.

118 Mitsudone, T., Mikami, Y., Ebata, K., Mizugaki, T., Jitsukawa, K., and Kaneda, K. (2008) *Chem. Commun.*, 4804.

119 Minachev, Kh.M., Antoshin, G.V., Klissurski, .D.G., Guin, N.K., and Abadzhijeva, N.Ts. (1979) *React. Kinet. Catal. Lett.*, **10**, 163.

120 Kirihara, M., Ochiai, Y., Takizawa, S., Takahata, H., and Nemoto, H. (1999) *Chem. Commun.*, 1387.

121 Fiegel, P.J. and Sobczak, J.M. (2007) *New J. Chem.*, **31**, 1668.

122 Iwahama, T., Sakaguchi, S., Nishiyama, Y., and Ishii, Y. (1995) *Tetrahedron Lett.*, **36**, 6923.

123 Yoshino, Y., Hanyashi, Y., Iwahama, T., Sakaguchi, S., and Ishii, Y. (1997) *J. Org. Chem.*, **62**, 6810; Kato, S., Iwahama, T., Sakaguchi, S., and Ishii, Y. (1998) *J. Org. Chem.*, **63**, 222; Sakaguchi, S., Kato, S., Iwahama, T., and Ishii, Y. (1988) *Bull. Chem. Soc. Jpn.*, **71**, 1.

124 see also Minisci, F., Punta, C., Recupero, F., Fontana, F., and Pedulli, G.F. (2002) *Chem. Commun.*, **7**, 688.

125 Choudary, B.M., Lakshmi Kantam, M., Ateeq Rahman, Ch., Reddy, V., and Rao, K.K. (2001) *Angew. Chem. Int. Ed.*, **40**, 763.

126 (a) Döbler, C., Mehltretter, G., and Beller, M. (1999) *Angew. Chem. Int. Ed.*, **38**, 3026; (b) Döbler, C., Mehltretter, G., Sundermeier, G.M., and Beller, M.J. (2000) *J. Am. Chem. Soc.*, **122**, 10289.

127 Döbler, C., Mehltretter, G.M., Sundermeier, U., Eckert, M., Militzer, H.-C., and Beller, M. (2001) *Tetrahedron Lett.*, **42**, 8447.

128 Barak, G., Dakka, J., and Sasson, Y. (1988) *J. Org. Chem.*, **53**, 3553.

129 Jacobsen, S.E., Muccigrosso, D.A., and Mares, F. (1979) *J. Org. Chem.*, **44**, 921; see also Bortolini, O., Campestrini, S., Di Furia, F., and Modena, G. (1987) *J. Org. Chem.*, **52**, 5467.

130 Trost, B.M. and Masuyama, Y. (1984) *Tetrahedron Lett.*, **25**, 173.

131 Bortolini, O., Conte, V., Di Furia, F., and Modena, G. (1986) *J. Org. Chem.*, **51**, 2661.

132 Sato, K., Aoki, M., Takagi, J., and Noyori, R. (1997) *J. Am. Chem. Soc.*, **119**, 12386.

133 Sato, K., Takagi, J., Aoki, M., and Noyori, R. (1998) *Tetrahedron Lett.*, **39**, 7549.

134 Ishii, Y., Yamawaki, K., Yoshida, T., Ura, T., and Ogawa, M. (1987) *M. J. Org. Chem.*, **52**, 1868; Ishii, Y., Yamawaki, K., Ura, T., Yamada, H., Yoshida, T., and Ogawa, M. (1988) *J. Org. Chem.*, **53**, 3587; Yamawaki, K., Nishihara, H., Yoshida, T., Ura, T., Yamada, H., Ishii, Y., and Ogawa, M. (1988) *Synth. Commun.*, **18**, 869; Yamawaki, K., Yoshida, T., Nishihara, H., Ishii, Y., and Ogawa, M. (1986) *Synth. Commun.*, **16**, 537.

135 Venturello, C. and Gambaro, M. (1991) *J. Org. Chem.*, **56**, 5924.

136 Neumann, R. and Gara, M. (1995) *J. Am. Chem. Soc.*, **117**, 5066.

137 Zauche, T.H. and Espenson, J.H. (1995) *Inorg. Chem.*, **37**, 6827.

138 Espenson, J.H., Zhu, Z., and Zauche, T.H. (1991) *J. Org. Chem.*, **64**, 1191.

139 Arends, I.W.C.E., Sheldon, R.A., Wallau, M., and Schuchardt, U. (1997) *Angew. Chem. Int. Ed. Engl.*, **36**, 1144.

140 Maspero, F. and Romano, U. (1994) *J. Catal.*, **146**, 476.

141 Zondervan, C., Hage, R., and Feringa, B.L. (1997) *Chem. Commun.*, 419; Brinksma, J., Rispens, M.T., Hage, R.H., and Feringa, B.L. (2002) *Inorg. Chem. Acta*, **337**, 75.

142 Shi, F., Tse, M.K., Pohl, M.-M., Brückner, A., Zhang, S., and Beller, M. (2007) *Angew. Chem. Int. Ed.*, **46**, 8866.

143 For recent reviews see: Kroutil, W., Mang, H., Edegger, K., and Faber, K. (2004) *Adv. Synth. Catal.*, **346**, 125; de Wildeman, S.M., Sonke, T., Schoemaker, H.E., and May, O. (2007) *Acc. Chem. Res.*, **40**, 1260–1266; Moore, J.C., Pollard, D.J., Kosiek, B., and Devine, P.N. (2007) *Acc. Chem. Res.*, **40**, 1412.

144 Stampfer, W., Kosjek, B., Moitzi, C., Kroutil, W., and Faber, K. (2002) *Angew. Chem. Int. Ed. Engl.*, **41**, 1014; Stampfer, W., Kosjek, B., Faber, K., and Kroutil, W. (2003) *J. Org. Chem.*, **68**, 402; Edegger, K., Mang, H., Faber, K., Gross, J., and Kroutil, W. (2006) *J. Mol. Catal. A: Chemical*, **251**, 66.

145 Forneris, F., Heuts, D.P.H.M., Delvecho, M., Rovida, S., Fraaije, M.W., and Maltevi, A. (2008) *Biochem.*, **47**, 978; van Hellemond, E.W., Leferink, N.G.H., Heuts, D.P.H.M., Fraaije, M.W., and van Berkel, W.J.H. (2006) *Adv. Appl. Microbiol.*, **60**, 17.

146 Masutani, K., Uchida, T., Irie, R., and Katsuki, T. (2000) *Tetrahedron Lett.*, **41**, 5119.

147 Jensen, D.R., Pugsley, J.S., and Sigman, M.S. (2001) *J. Am. Chem. Soc.*, **123**, 7475.

148 Ferreira, E.M. and Stolz, B.M. (2001) *J. Am. Chem. Soc.*, **123**, 7725.

6
Aerobic Oxidations and Related Reactions Catalyzed by N-Hydroxyphthalimide

Yasutaka Ishii, Satoshi Sakaguchi, and Yasushi Obora

6.1
Introduction

The development of the petrochemical industry has led to a situation in which over 90% of organic chemicals are derived from petroleum. These include a wide range of oxygen-containing molecules such as alcohols, aldehydes, ketones, epoxides, and carboxylic acids which are starting materials for the production of, in particular, plastics and synthetic fiber materials for polyamides, polyesters, polycarbonates, and so on. For instance, ethylene oxide, acrolein, acrylic acid, and methacrolein are produced by the vapor-phase partial oxidation of lower alkenes like ethylene, propylene, and butenes [1], while acetic acid, KA oil (a mixture of cyclohexanone and cyclohexanol), benzoic acid, terephthalic acid, and phenol accompanied with acetone are manufactured by the liquid-phase catalytic oxidation of alkanes such as butane, cyclohexane, and alkylbenzenes, for example [2]. Liquid-phase aerobic oxidation, which is generally referred to as autoxidation, is extensively practiced in industry worldwide, although the efficiency of this oxidation methodology is not necessarily as high [2a,3]. As a result, nitric acid is still widely used as a useful oxidizing agent for manufacturing carboxylic acids like adipic acid, nicotinic acid, pyromellitic acid, and so on [2a,4]. Nowadays, however, environmentally unacceptable traditional oxidation methods using high-valent metal oxo complexes, halogens, and nitric acid are being replaced by cleaner oxidation methods.

Although the partial aerobic oxidation of alkanes leading to alcohols and carbonyl compounds has considerable potential from both ecological and economical viewpoints, current oxidation technology is not always fully feasible, as it incurs extensive oxidative cleavage or concomitant combustion of alkanes. The most important liquid-phase oxidation methods include the transformation of *p*-xylene to terephthalic acid, and cyclohexane to KA oil [2a]. However, the reaction conditions are often harsh, the reagent mixture is corrosive, and the reaction is often unselective. Therefore, it is apparent that selective transformation of hydrocarbons, especially saturated hydrocarbons like alkanes, to valuable oxygenated compounds is an extremely important area of contemporary industrial chemistry. Thus, considerable research effort has been undertaken aimed at the development of selective oxidations of alkanes with

Modern Oxidation Methods. Edited by Jan-Erling Bäckvall
Copyright © 2010 WILEY-VCH Verlag GmbH & Co. KGaA, Weinheim
ISBN: 978-3-527-32320-3

molecular oxygen, leading to alcohols, ketones, and carboxylic acids [5, 6]. A number of oxidations of alkanes with dioxygen, catalyzed by transition metals, in the presence of reducing agents such as aldehydes have appeared in the literature, but these are mainly on a laboratory scale [6]. In recent years, there has been a growing demand for the development of fundamentally new and environmentally benign catalytic systems, for the oxidation of hydrocarbons, that are operative on an industrial scale under moderate conditions in the liquid phase with a high degree of selectivity.

Recently, we have developed an innovative strategy for the catalytic generation of a type of carbon radical from hydrocarbons. This is a phthalimide N-oxyl (PlNO) radical generated *in situ* from N-hydroxyphthalimide (NHPI) and molecular oxygen in the presence (or absence) of a cobalt ion under mild conditions. The carbon radicals, derived from a variety of hydrocarbons under the influence of molecular oxygen, lead to oxygenated products like alcohols, ketones, and carboxylic acids in good yields. The development of NHPI-catalyzed reactions has been reported in review papers [7].

Here we show a novel methodology for the functionalizations of hydrocarbons, including oxygenation, nitration, sulfoxidation, epoxidation, carboxylation, and oxyalkylation through generation of the catalytic carbon radical. In particular, the NHPI-catalyzed aerobic oxidations of alkanes, which are very important in industry worldwide, are described in detail.

6.2
NHPI-Catalyzed Aerobic Oxidation

NHPI (Scheme 6.1) was first used as a catalyst by Grochowski *et al.* [8a] for the addition of ethers to diethylazodicarboxylate, and as an efficient mediator by Masui *et al.* [8b,c] for the electrochemical oxidation of alcohols. A patent has been granted for a process involving the oxidation of allylic hydrogen of isoprenoid with dioxygen using NHPI in the presence of a radical initiator [8d]. In 1995, the authors of this chapter found that NHPI serves as a carbon radical-producing catalyst (CRPC) from hydrocarbons in both the presence and the absence of transition metals like Co and Mn ions under dioxygen [9].

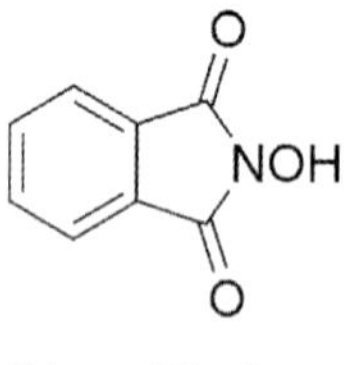

Scheme 6.1 Structure of NHPI.

6.2.1
Alkane Oxidations with Dioxygen

During the past several decades, a number of catalytic systems have been developed for the oxidation of alkanes with dioxygen in the presence of reducing agents, for example, H_2, metals, and aldehydes, among others, under mild conditions [10–14]. In 1981 Tabushi *et al.* reported the oxidation of adamantane to 1- and 2-adamantanols by the Mn(III)porphyrin/Pt/H_2 system under a dioxygen atmosphere at room temperature [11]. Barton developed a family of systems, the so-called Gif systems, for aerobic oxidation and oxidative functionalization of alkanes under mild conditions using Fe and Zn as reductants [12a–c]. Alkane oxidation using aldehydes as reducing agents was reported by Murahashi *et al.*, who attempted the aerobic oxidation of cyclohexane and adamantane by ruthenium or iron catalysts in the presence of acetaldehyde [13a–c]. There have been several reports on the photooxidations of alkanes with O_2 catalyzed by polyoxotungstates [15], heteropolyoxometalates [16], and $FeCl_3$ [17]. Shul'pin *et al.* carried out the vanadium-catalyzed oxidation of alkanes to alcohols, ketones, and hydroperoxides by O_2 in combination with H_2O_2 [18]. Lyon and Ellis reported that halogenated metalloporphyrin complexes are efficient catalysts for the oxidation of isobutane with dioxygen [19]. Mizuno *et al.* have shown that heteropolyanions containing Fe catalyze the aerobic oxidation of cyclohexane and adamantane into the corresponding alcohols and ketones [20]. An Ru(III)-EDTA system [25], Ru-substituted polyoxometalate [22], and $[Co(NCMe)_4]$ $(PF_6)_2$ [23] have been reported to catalyze the aerobic oxidation of cyclohexane and adamantane. However, effective and selective methods for the catalytic oxygenation of alkanes with dioxygen still remain a major challenge.

Adipic acid, which is used as a raw material for nylon-6,6 and polyester, is the most important of all the aliphatic dicarboxylic acids manufactured at present. The current production of adipic acid consists of a two-step oxidation process involving the aerobic oxidation of cyclohexane in the presence of a soluble Co catalyst at 150–170 °C to a KA oil, and the nitric acid oxidation of the KA oil to adipic acid [2a,24]. The drawbacks of this process are that the oxidation in the first step must be operated to give 3–6% conversion of cyclohexane to keep a high selectivity (80%) of the KA oil, and that the nitric acid oxidation evolves a large amount of undesired global-warming nitrogen oxides, in particular N_2O. Therefore, the direct conversion of cyclohexane to adipic acid with molecular oxygen has long been sought, as this would be a desirable and promising method in industrial chemistry worldwide. Tanaka *et al.* succeeded in achieving conversion of cyclohexane to adipic acid under 30 atm of O_2 by the use of a higher concentration of Co(III) acetate combined with acetaldehyde or cyclohexanone, which serves as promoter [26]. More recently, Noyori and coworkers reported the oxidation of cyclohexene to adipic acid with aqueous hydrogen peroxide by a polyoxometalate having a phase-transfer function as an alternative clean route [27].

The direct conversion of cyclohexane to adipic acid was successfully achieved by the use of a combined catalyst of NHPI with Co and Mn ions [28]. The oxidation of cyclohexane (**1**) in the presence of a catalytic amount of NHPI (10 mol%) and Mn $(acac)_2$ (1 mol%) under dioxygen atmosphere (1 atm) in acetic acid at 100 °C for 20 h gave adipic acid (**4**) in 73% selectivity at 73% conversion (Eq. (6.1)). This is the first example of a one-step oxidation of **1** to **4** under normal pressure of dioxygen at a reasonably low reaction temperature with high conversion and selectivity. The oxidation using the NHPI/Co(OAc)$_2$ system in acetonitrile, on the other hand, gave cyclohexanone (**2**) in good selectivity (Eq. (6.1)) [29]. The oxidation by the NHPI/Co system in acetonitrile provides an alternative direct route to cyclohexanone, although the autoxidation of cyclohexane leads to a KA oil mixture with cyclohexanol as a main product [2a,25]. The present catalytic system can be extended to the oxidation of large-membered cycloalkanes to the corresponding dicarboxylic acids. Cyclooctane, cyclodecane, and cyclododecane were oxidized to suberic acid, sebacic acid, and dodecanedioic acid, respectively.

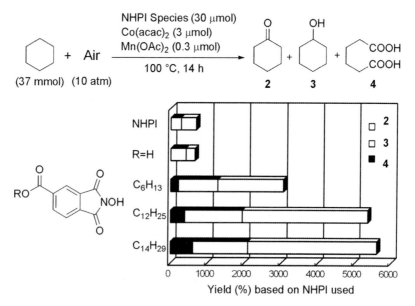

(6.1)

The aerobic oxidations of **1** must be carried out in an appropriate solvent such as acetic acid or acetonitrile because of the lower solubility of NHPI in nonpolar solvents such as hydrocarbons. It is noteworthy that the NHPI-catalyzed reaction of **1** could proceed without any solvent by the use of a lipophilic NHPI derivative. Of a series of 4-alkyloxycarbonyl N-hydroxyphthalimides examined as lipophilic NHPI catalysts, 4-lauryloxycarbonyl N-hydroxyphthalimide was found to be an efficient catalyst for the aerobic oxidation of **1** under solvent-free conditions (Figure 6.1) [30].

Figure 6.1 Oxidation of **1** catalyzed by NHPI derivatives under solvent-free conditions.

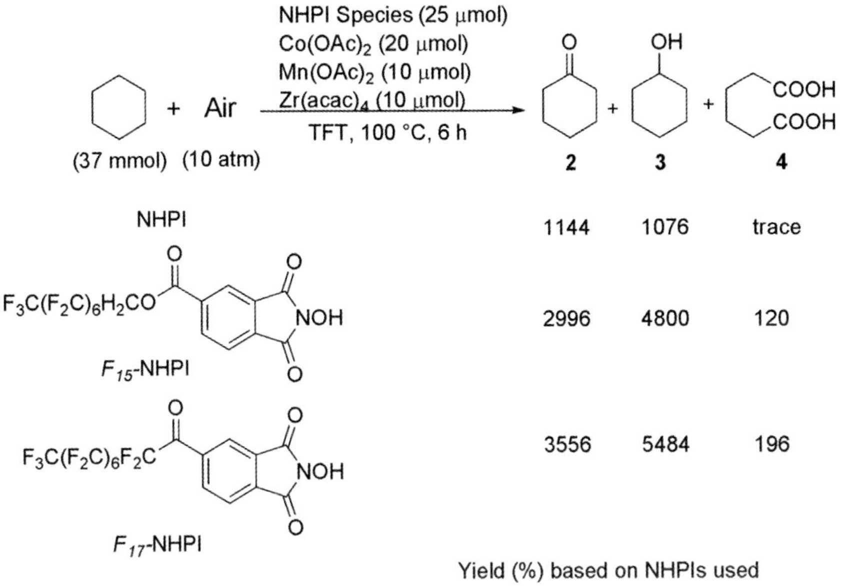

6.2 NHPI-Catalyzed Aerobic Oxidation

Catalyst	2	3	4
NHPI	1144	1076	trace
F_{15}-NHPI	2996	4800	120
F_{17}-NHPI	3556	5484	196

Yield (%) based on NHPIs used

Reaction conditions: cyclohexane (37 mmol) + Air (10 atm), NHPI Species (25 μmol), Co(OAc)$_2$ (20 μmol), Mn(OAc)$_2$ (10 μmol), Zr(acac)$_4$ (10 μmol), TFT, 100 °C, 6 h, giving products 2 (ketone), 3 (alcohol), and 4 (adipic acid).

F_{15}-NHPI: F$_3$C(F$_2$C)$_6$H$_2$CO-substituted phthalimide N-OH
F_{17}-NHPI: F$_3$C(F$_2$C)$_6$F$_2$C-substituted phthalimide N-OH

Figure 6.2 Oxidation of **1** catalyzed by F-NHPI derivatives in TFT.

The oxidation of **1** by fluorinated NHPI (F_{15}- and F_{17}-NHPI) catalysts was considerably accelerated by the addition of a small amount of [Zr(acac)$_4$] in a fluorinated solvent like trifluorotoluene (TFT) to afford selectively KA oil (**2** and **3**). The formation of adipic acid (**4**) was markedly suppressed in TFT, and the solvent used is a crucial factor to determine the product selectivity (Figure 6.2) [31].

Methane and ethane, which are the main components of natural gas, have been used for a long time as a clean fuel worldwide. Since ethane comprises 5–10% of the natural gas, a vast amount of it is obtained when methane is produced. If a new methodology for converting ethane to useful oxygen-containing compounds like ethanol and acetic acid could be developed, it would vastly contribute to the efficient use of feedstock. Previously, Sen et al. reported liquid-phase oxidation of ethane to acetic acid and formic acid by Pd/C with H$_2$O$_2$ generated in situ from H$_2$, arising from a metal-catalyzed water gas shift reaction with CO, water, and O$_2$. [32]. Ethane can be converted into N,N-dimethylpropylamine by the reaction with N,N-trimethylamine N-oxide in using Cu(OAc)$_2$ as the catalyst [33]. Ethane oxidation with H$_2$O$_2$ catalyzed homogeneously by V-containing polyphosphomolybdates has been carried out by Shul'pin et al. [34].

Catalytic aerobic oxidation of ethane to acetic acid was successfully performed through a catalytic radical process using NHPI derivatives combined with a Co(II) salt in acetonitrile or propanoic acid. Among the catalysts examined, N,N-dihydroxypyromellitimide (NDHPI) was found to be the best. For instance, when a mixture of ethane (20 atm) and air (20 atm) in acetonitrile was allowed to react in the presence of NDHPI (100 μmol) and Co(OAc)$_2$ (30 μmol) at 150 °C for 15 h, 830 μmol of acetic acid was obtained, and the turnover number (TON) of NDHPI reached 8.3 (Eq. (6.2)). In this reaction, other products such as ethanol or acetaldehyde were not detected at all.

In the oxidation of ethane using NHPI as a catalyst under these conditions, the amount of NHPI used was twice that of NDHPI, but the yield of acetic acid and the TON of the catalyst were 530 µmol and 2.7, respectively. The highest TON (15.3) was obtained when the reaction was carried out using NDPHI combined with $CoCl_2$ in propanoic acid [35].

$$C_2H_6 \text{ (20 atm)} + \text{Air (20 atm)} \xrightarrow[150\,°C]{\text{Catalyst}} CH_3COOH$$

Catalyst (µmol)	Solvent	Yield (µmol)	TON
NHPI/Co(OAc)$_2$ (200 / 30)	CH$_3$CN	530	2.7
NDHPI/Co(OAc)$_2$ (100 / 30)	CH$_3$CN	830	8.3
NDHPI/CoCl$_2$ (100 / 30)	CH$_3$CH$_2$COOH	1532	15.3

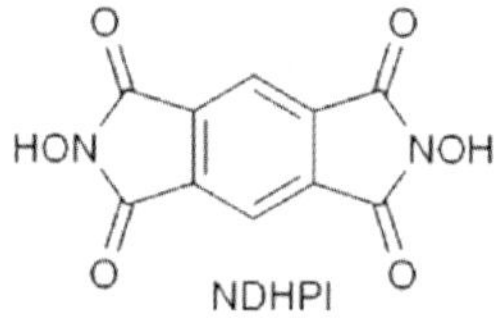

(6.2)

NDHPI

The autoxidation of isobutane is now mainly carried out to obtain *tert*-butyl hydroperoxide [36]. Halogenated metalloporphyrin complexes are reported to be efficient catalysts for the aerobic oxidation of isobutane [18, 37]. It was found that the oxidation of isobutane by air (10 atm) catalyzed by NHPI and Co(OAc)$_2$ in benzonitrile at 100 °C produced *tert*-butyl alcohol in high yield (81%) along with acetone (14%) (Eq. (6.3)) [38]. 2-Methylbutane was converted into the carbon-carbon bond-cleaved products, acetone and acetic acid, rather than the alcohols, as principal products. These cleaved products seem to be formed via β-scission of an alkoxy radical derived from the decomposition of a hydroperoxide by Co ions. The extent of the β-scission is known to depend on the stability of the radicals released from the alkoxy radicals [39]. It is thought that the β-scission of a *tert*-butoxy radical to acetone and a methyl radical occurs with more difficulty than that of a 2-methylbutoxy radical to acetone and an ethyl radical. As a result, isobutane produces *tert*-butyl alcohol as the principal product, while 2-methylbutane affords mainly acetone and acetic acid.

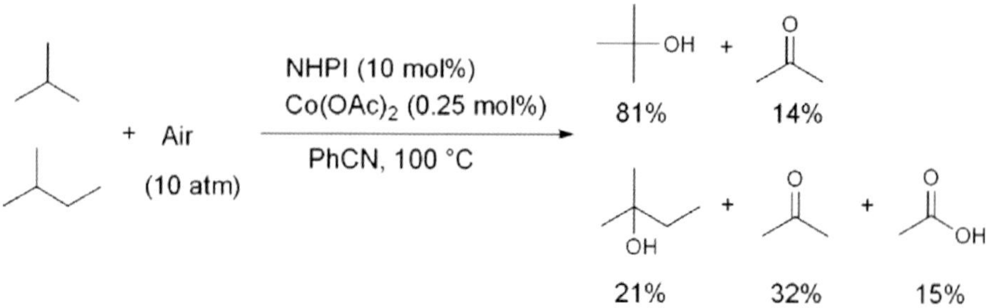

(6.3)

6.2 NHPI-Catalyzed Aerobic Oxidation

There have been a few reports on the catalytic hydroxylation of adamantane with dioxygen in the presence of aldehydes [12]. Mizuno et al. reported that the aerobic oxidation of adamantane by the PW_9-Fe_2Ni heteropolyanion without any reducing agents gives 1-adamantanol and 2-adamantanone at 29% conversion [19a]. The NHPI-catalyzed aerobic oxidation of adamantane is considerably accelerated by adding a small amount of a Co salt [28, 40, 41]. Thus, the oxidation of adamantane (**5**) in the presence of NHPI (10 mol%) and Co(acac)$_2$ (0.5 mol%) in acetic acid under dioxygen (1 atm) for 6 h produced 1-adamantanol (**6**) (43%), 1,3-adamantanediol (**8**) (40%), and 2-adamantanone (**7**) (8%) (Eq. (6.4)). The reactivity of the tertiary C−H bond relative to the secondary C−H bond in the oxidation by NHPI/Co(II) was 31.1. This value is considerably higher than that attained by the conventional autoxidation (3.8–5.4). The preferential oxidation of the tertiary C−H bond over the secondary bond may be attributed to the electron-deficient character of PINO which is a key radical species in the NHPI-catalyzed oxidation (see below).

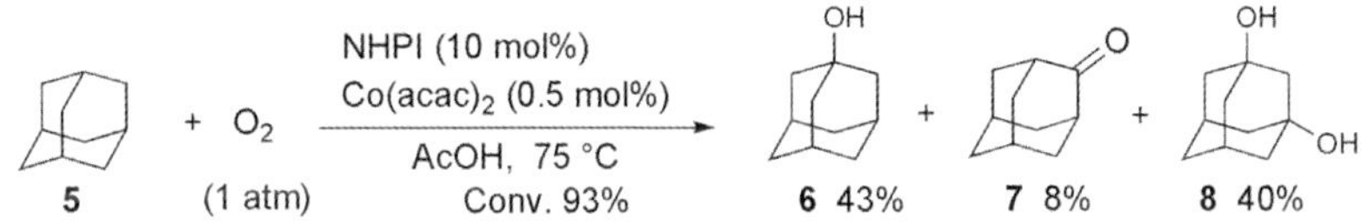

$$\mathbf{5} + O_2 \text{ (1 atm)} \xrightarrow[\text{AcOH, 75 °C} \atop \text{Conv. 93%}]{\text{NHPI (10 mol\%)} \atop \text{Co(acac)}_2 \text{ (0.5 mol\%)}} \mathbf{6}\ 43\% + \mathbf{7}\ 8\% + \mathbf{8}\ 40\% \tag{6.4}$$

It is important that the oxidation led to diol **8** in high selectivity, because **8** is rarely produced by conventional oxidation. Hirobe obtained **8** in 25% yield by the oxidation of **5** using a Ru complex with 2,6-dichloropyridine N-oxide as the oxidant [42]. In the stepwise hydroxylation of **5** by the NHPI/Co(acac)$_2$ system, the diol **8** and the triol **9** were obtained in high selectivity (Eqs. (6.5) and (6.6)). These alcohols are now manufactured as important components of photoresistant polymer materials on an industrial scale by Daicel Chemical Industry Ltd..

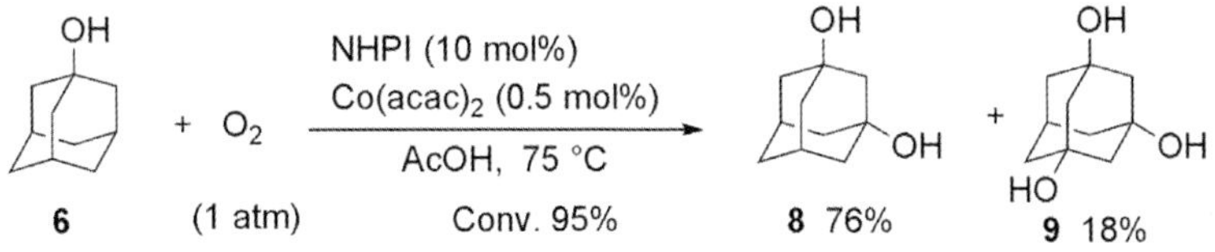

$$\mathbf{6} + O_2 \text{ (1 atm)} \xrightarrow[\text{AcOH, 75 °C} \atop \text{Conv. 95%}]{\text{NHPI (10 mol\%)} \atop \text{Co(acac)}_2 \text{ (0.5 mol\%)}} \mathbf{8}\ 76\% + \mathbf{9}\ 18\% \tag{6.5}$$

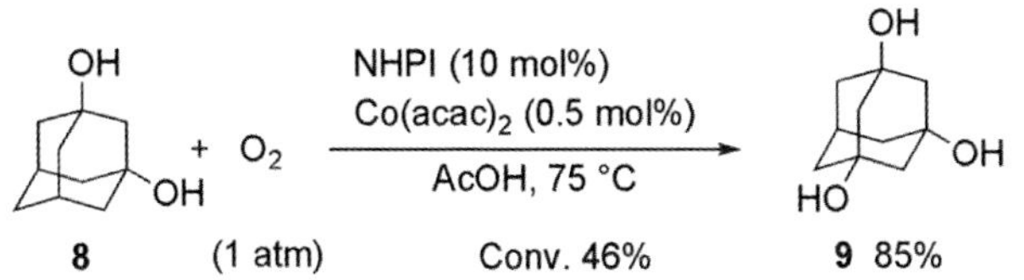

$$\mathbf{8} + O_2 \text{ (1 atm)} \xrightarrow[\text{AcOH, 75 °C} \atop \text{Conv. 46%}]{\text{NHPI (10 mol\%)} \atop \text{Co(acac)}_2 \text{ (0.5 mol\%)}} \mathbf{9}\ 85\% \tag{6.6}$$

6.2.2
Oxidation of Alkylarenes

Aerobic oxidation of alkylbenzenes is a promising subject in industrial chemistry. Many bulk chemicals such as terephthalic acid, phenol, benzoic acid, and so on are manufactured by homogeneous liquid-phase oxidations with O_2 [2, 43]. The largest-scale liquid-phase oxidation is the conversion of p-xylene to terephthalic acid, which is chiefly used as polyethylene terephthalate polymer raw material [2a]. m-Xylene is also commercially oxidized to isophthalic acid. Benzoic acid derived from the oxidation of toluene is an important raw material in the production of various pharmaceuticals and pesticides. Commercially important cumene hydroperoxide and ethylbenzene hydroperoxide are also manufactured by the aerobic oxidation of isopropylbenzene and ethylbenzene, respectively [2a,5a]. These oxidation processes are usually operated at higher temperatures and pressures of air. A great deal of effort has been made to develop the homogeneous oxidations of alkylbenzenes with better selectivity under milder conditions. The first successful oxidation of a variety of alkylbenzenes with O_2 by the use of NHPI as the catalyst under very mild conditions is achieved.

Currently, the oxidation of toluene is commercially practiced in the presence of a catalytic amount of cobalt(II) 2-ethylhexanoate under a pressure of 10 atm of air at 140–190 °C [44]. The oxidation of toluene under normal pressure of dioxygen at room temperature is achieved by the use of a combined catalyst of NHPI and a Co(II) species. The fact that the toluene was oxidized with dioxygen through the catalytic process in high yield under ambient conditions is very important from ecological and technical viewpoints as a promising strategy in oxidation chemistry. As a typical example, the oxidation of toluene in the presence of NHPI (10 mol%) and Co(OAc)$_2$ (0.5 mol%) in acetic acid under an atmosphere of O_2 at 25 °C for 20 h afforded benzoic acid and benzaldehyde in 81 and 3% yields, respectively (Eq. (6.7)) [45]. This finding suggests that an efficient cleavage of a C—H bond, having the bond dissociation energy (BDE) of 88 kcal mol^{-1} (corresponding to the BDE of toluene), is possible at room temperature by the use of NHPI catalyst. However, when Co(III) was employed in place of Co(II), no reaction took place at all at room temperature.

Representative results for the NHPI-catalyzed aerobic oxidation of various alkylbenzenes in the presence of Co(OAc)$_2$ in acetic acid under ambient conditions are listed in Table 6.1. Both p- and o-xylenes are selectively oxidized to p- and o-toluic acids without the formation of dicarboxylic acids. o-Ethyltoluene undergoes selective oxidation to form a mixture of the corresponding alcohol and ketone in which the ethyl moiety was selectively functionalized. It is of interest to examine the effect of substituents on the aromatic ring in the oxidation of substituted toluenes. p-Methoxytoluene is more rapidly oxidized than the toluene itself, while

6.2 NHPI-Catalyzed Aerobic Oxidation

Table 6.1 Aerobic oxidation of various alkylbenzenes at room temperature[a].

Run	Substrate	Time (h)	Conv. (%)	Products (Yield (%))
1	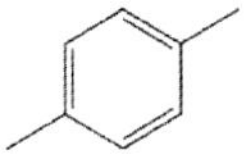 p-xylene	20	95	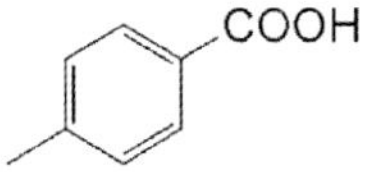 p-toluic acid (85)
2	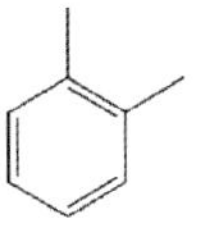 o-xylene	20	93	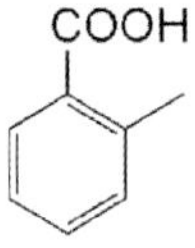 o-toluic acid (83)
3[b]	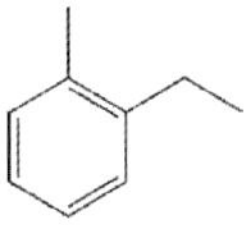 o-ethyltoluene	20	82	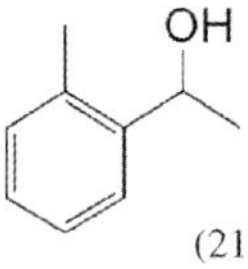 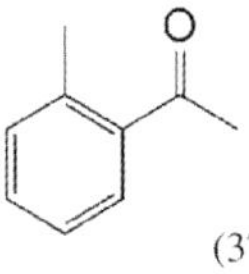 1-(o-tolyl)ethanol (21); o-methylacetophenone (37)
4	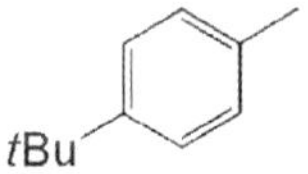 4-tBu-toluene	20	95	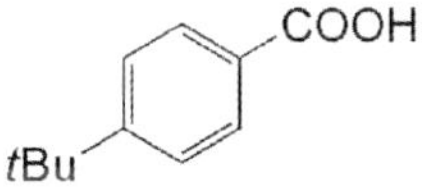 4-tBu-benzoic acid (91)
5[b]	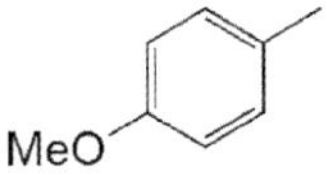 4-MeO-toluene	6	89	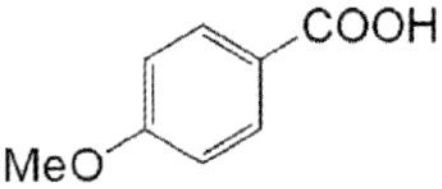 4-MeO-benzoic acid (80)
6	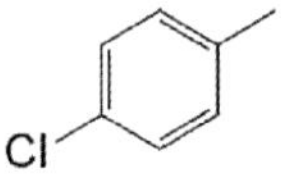 4-Cl-toluene	20	71	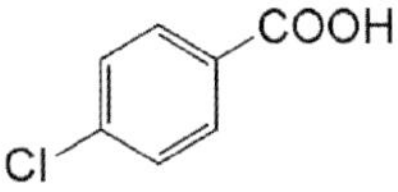 4-Cl-benzoic acid (67)
7	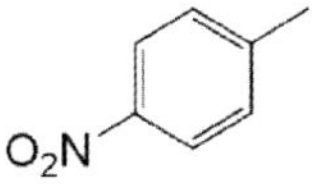 4-O$_2$N-toluene	20	No reaction	
8	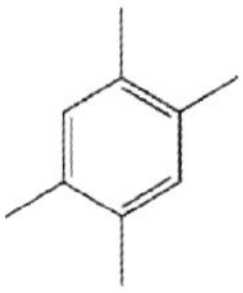 1,2,4-trimethylbenzene	12	>99	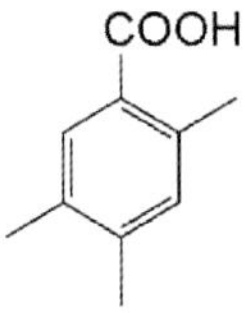 4,5-dimethylbenzoic acid (93)

a) Substrates (3 mmol) were allowed to react in the presence of NHPI (10 mol%) and Co(OAc)$_2$ (0.5 mol%) in AcOH (5 mL) under dioxygen (1 atm) at 25 °C.
b) CH$_3$CN was used as the solvent.

p-chlorotoluene is oxidized at a relatively slow rate. An electron-donating substituent anchoring to toluene stabilizes the partial positive charge on the benzylic carbon atom in the transition state for the abstraction of a benzylic hydrogen atom by PINO, possessing an electrophilic character (Scheme 6.2) (see below) [46]. Therefore, the oxidation of toluenes having electron-donating groups by the NHPI catalyst is facilitated. Indeed, p-nitrotoluene substituted by a strongly electron-withdrawing nitro group, is not oxidized at all under these conditions. Recently, various substituted NHPI derivatives were prepared and studied by Nolte et al. in the aerobic oxidation of ethylbenzene [47]. It was found that NHPI, with an electron-withdrawing fluorine substituent, increases the oxidation rate, while NHPI substituted by a methoxy group decreases the rate.

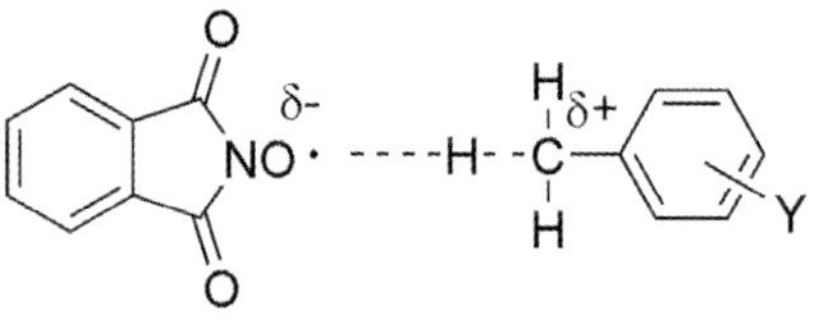

Scheme 6.2 Transition state for the reaction of PINO with substituted benzenes.

A plausible reaction pathway for the aerobic oxidation of alkanes catalyzed by NHPI and Co(II) is illustrated in Scheme 6.3. A labile dioxygen complex such as superoxocobalt(III) or peroxocobalt(III) complexes is known to be formed by the complexation of Co(II) with O_2. The *in situ* generation of PINO from NHPI by the action of the cobalt(III)-oxygen complex formed is a key step in the present oxidation. The next step involves the hydrogen atom abstraction from alkanes by PINO to form alkyl radicals. Trapping the resulting alkyl radicals by dioxygen provides peroxy radicals, which are eventually converted into oxygenated products through alkyl hydroperoxides. In fact, on exposing NHPI in benzonitrile containing a small amount of $Co(OAc)_2$ to dioxygen at $80\,°C$, an ESR signal attributed to PINO as a triplet signal having hyperfine splitting (hfs) by the nitrogen atom ($g = 2.0074$, $A_N = 4.3$ G) is observed (Figure 6.3). The g-value and hyperfine splitting constants observed here are consistent with those ($g = 2.0073$, $A_N = 4.23$ G) of PINO reported previously [48]. In addition, PINO is observed during the oxidation of toluene by the NHPI/Co(II) system under ambient conditions [45]. Quite recently, Minisci, Pedulli, and coworkers found by means of ESR spectroscopy that the BDE value of the O–H bond for NHPI is $>86\,\text{kcal mol}^{-1}$. This suggests that PINO could abstract the benzylic hydrogen atom of toluene, whose BDE is $88\,\text{kcal mol}^{-1}$ [49].

6.2.2.1 Synthesis of Terephthalic Acid
Terephthalic acid (TPA) as well as dimethyl terephthalate (DMT) have recently become very important as raw material for polyethylene terephthalate [50]. In 1999, about 17×10^6 tonnes of TPA were manufactured worldwide, and its

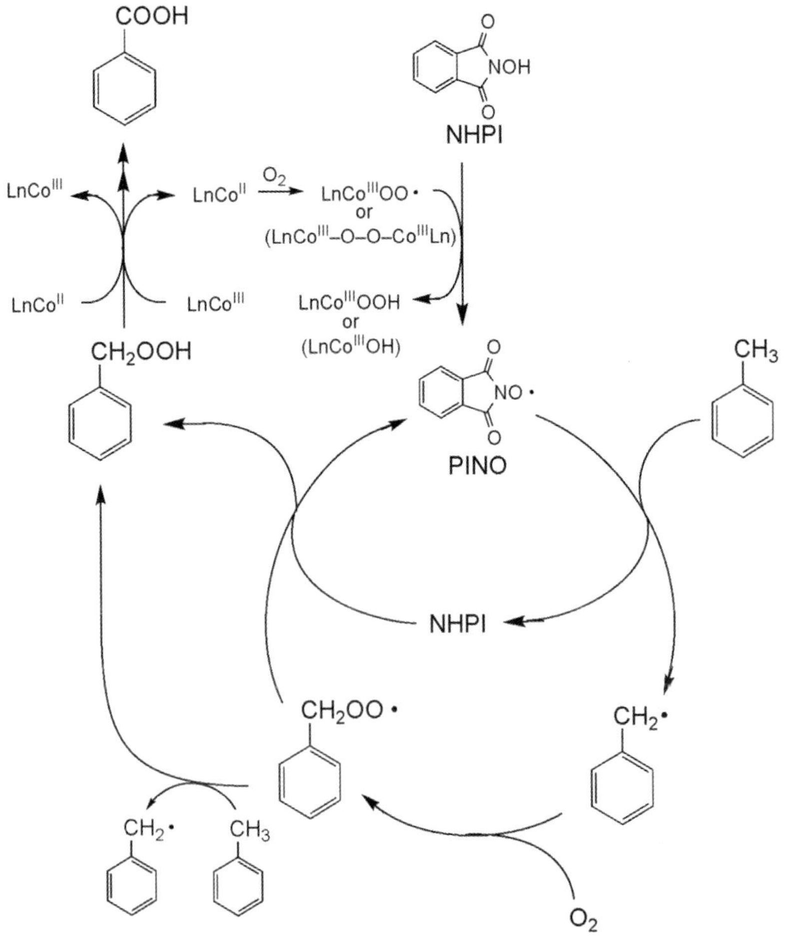

Scheme 6.3 A plausible reaction path for the aerobic oxidation of toluene catalyzed by NHPI combined with Co(II).

production has been estimated to have been increasing at a minimum growth rate of 10% annually by the year 2002. Until the 1980s, the following four-step process developed by Witten and modified by Hercules and Dynamit-Nobel (Witten-Hercules process) had been mainly operated to produce DMT [2a, 50]. The first step is the conversion of p-xylene (PX) to p-toluic acid (PTA). It then passes to an esterification step to form methyl p-toluate, which is subjected to further oxidation to monomethyl terephthalate, followed by esterification to DMT. From the 1990s, these processes were changed to the aerobic one-stage oxidation of PX to TPA by the combined use of cobalt and manganese salts in the presence of bromide as a promoter in acetic acid at 175–225 °C under 15–30 atm of air, followed by hydrogenation of the crude TPA to remove 4-carboxybenzaldehyde (4-CBA) by a Pd

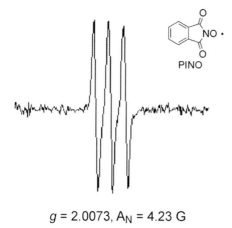

Figure 6.3 ESR spectrum of PINO.

catalyst [50–52]. This process was developed by Scientific Design and Amoco Ltd. (Amoco process). Currently, about 70% of TPA produced worldwide is based on the Amoco process, and almost all of the new plants adopt this method. However, there are several disadvantages in the Amoco process: (i) significant combustion of acetic acid used as the solvent to form CO and CO_2, (ii) use of the highly corrosive bromide ion, which calls for the use of vessels lined with expensive metals like titanium, and (iii) contamination of 4-CBA in crude TPA, which necessitates elaborate hydrogenation and recrystallization procedures in manufacturing the purified TPA required for PET. Therefore, a new oxidation system for the production of TPA is desired to overcome these disadvantages. Partenheimer recently published a review devoted to the aerobic oxidation of alkylbenzenes, especially PX, using the Co/Mn/Br system [52].

The aerobic oxidation of PX to TPA was examined by the NHPI catalyst to develop a halogen-free catalytic system [53]. The oxidation of PX with dioxygen (1 atm) in the presence of catalytic amounts of NHPI (20 mol%) and Co(OAc)$_2$ (0.5 mol%) in acetic acid at 100 °C for 14 h produced TPA in 67% yield and PTA (15%) together with small amounts of 4-CBA, 4-carboxybenzyl alcohol, 1,4-diacetoxymethylbenzene, and 4-acetoxymethylbenzoic acid, as well as several unidentified compounds in 1–2% yields, respectively, at over 99% conversion (Eq. (6.8)). The yield of TPA is improved to 82% when Mn(OAc)$_2$ (0.5 mol%) is added to the NHPI/Co(OAc)$_2$ system. The synergistic effect of Co and Mn salts in the aerobic oxidation of alkylbenzenes has been well documented [52, 54, 55]. From a practical point of view, it is important that the aerobic oxidation of PX under air (30 kg cm^{-2}) by the NHPI/Co/Mn system is completed within 3 h at 150 °C to form TPA in 84% yield (Eq. (6.9)). Both o- and m-xylenes were also successfully converted into the corresponding dicarboxylic acids, isophthalic acid and phthalic acid, respectively, in high yields.

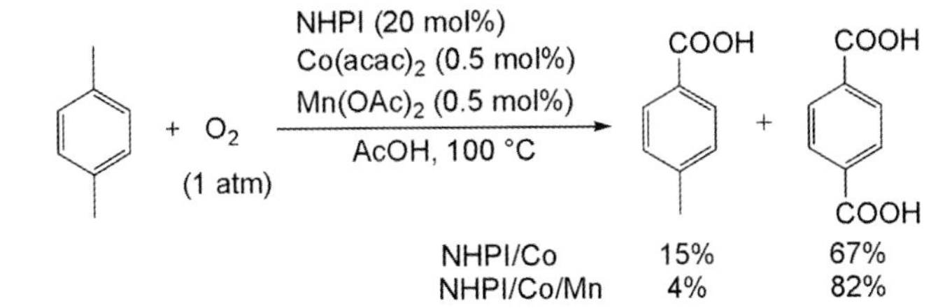

6.2 NHPI-Catalyzed Aerobic Oxidation

$$
\text{p-xylene} + O_2 \;(1\,\text{atm}) \xrightarrow[\text{AcOH, 100 °C}]{\substack{\text{NHPI (20 mol\%)} \\ \text{Co(acac)}_2 \text{ (0.5 mol\%)} \\ \text{Mn(OAc)}_2 \text{ (0.5 mol\%)}}} \text{4-CBA} + \text{TPA} \tag{6.8}
$$

	4-CBA	TPA
NHPI/Co	15%	67%
NHPI/Co/Mn	4%	82%

$$
\text{xylene} + \text{Air}\;(30\,\text{atm}) \xrightarrow[\text{AcOH, 150 °C}]{\substack{\text{NHPI (15 mol\%)} \\ \text{Co(acac)}_2 \text{ (0.5 mol\%)} \\ \text{Mn(OAc)}_2 \text{ (0.5 mol\%)}}} \text{methylbenzoic acid} + \text{phthalic acid isomer} \tag{6.9}
$$

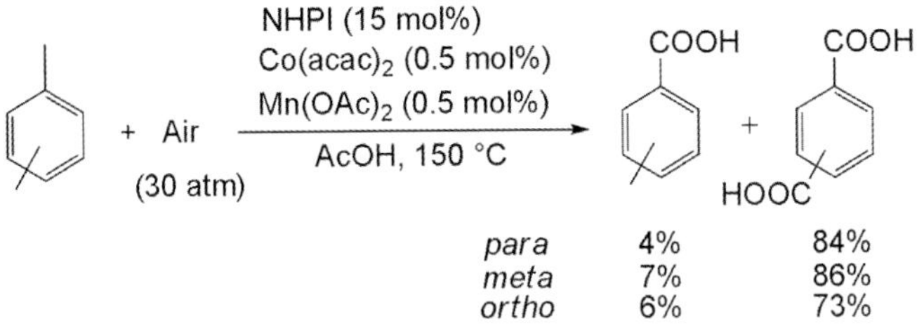

	methylbenzoic acid	diacid
para	4%	84%
meta	7%	86%
ortho	6%	73%

As shown in Eq. (6.8), about 20 mol% of NHPI must be used to obtain TPA in satisfactory yield (over 80%), because NHPI gradually decomposes to inert phthalimide and phthalic anhydride during the oxidation. If the NHPI used could be reduced by a simple modification, the present oxidation would be more desirable. Efforts to reduce the amount of the NHPI led to the discovery of an efficient catalyst, N-acetoxyphthalimide (NAPI), which can be easily prepared by the reaction of NHPI with acetic anhydride. Surprisingly, PX was oxidized to TPA in high yield (80%) even by the use of 5 mol% of NAPI, Co(OAc)$_2$ (0.5 mol%) and Mn(OAc)$_2$ (0.5 mol%) (Eq. (6.10)). The effect of NAPI is considered to be to resist the rapid decomposition to phthalimide or phthalic anhydride at the early stage of the reaction where violent chain reactions take place, since NAPI is gradually hydrolyzed to NHPI by water present in acetic acid as well as by the water resulting during the oxidation.

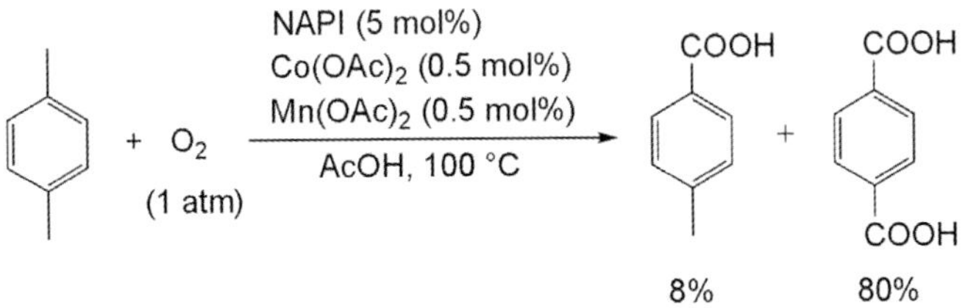

$$
\text{p-xylene} + O_2 \;(1\,\text{atm}) \xrightarrow[\text{AcOH, 100 °C}]{\substack{\text{NAPI (5 mol\%)} \\ \text{Co(OAc)}_2 \text{ (0.5 mol\%)} \\ \text{Mn(OAc)}_2 \text{ (0.5 mol\%)}}} \text{4-CBA (8\%)} + \text{TPA (80\%)} \tag{6.10}
$$

6.2.2.2 Oxidation of Methylpyridines and Methylquinolines

Pyridinecarboxylic acids are useful and important intermediates in pharmaceutical syntheses. Although the synthesis of these carboxylic acids by the aerobic oxidation of alkylpyridines is straightforward, the oxidation is usually difficult to carry out selectively owing to their low reactivities [52, 56]. Pyridinecarboxylic acids are readily prepared by the oxidation of alkylpyridines with nitric acid or by the hydrolysis of pyridinecarboxamides derived from pyridinecarbonitrile [57]. According to the recent literature, nicotinic acid is obtained in about 50% yield at 52% conversion by the oxidation of β-picoline in the presence of Co(OAc)$_2$ and Mn(OAc)$_2$ using LiCl as a promoter under air (16 atm) at 170 °C [58]. The aerobic oxidation of β-picoline to nicotinic acid catalyzed by NHPI has been examined [59]. Nicotinic acid is used as a precursor of vitamin B$_3$ and is commercially manufactured on a large scale by nitric acid oxidation of 5-ethyl-2-methylpyridine [58a]. The oxidation of β-picoline in the presence of NHPI (10 mol%) and Co(OAc)$_2$ (1.5 mol%) under dioxygen (1 atm) at 100 °C for 15 h in acetic acid affords nicotinic acid in 76% yield at 82% conversion (Eq. (6.11a)) [59]. This is the first successful oxidation of picolines with O$_2$ under mild conditions. In contrast to the oxidation of β-picoline by the NHPI/Co/Mn system where nicotinic acid was formed in good yield, γ-picoline is oxidized with some difficulty under these conditions to form 4-pyridinecarboxylic acid in low yield (22%). After optimization of the reaction conditions, 4-pyridinecarboxylic acid was obtained by the use of NHPI (20 mol%), Co(OAc)$_2$ (1 mol%) and Mn(OAc)$_2$ (1 mol%) in 60% yield at 67% conversion (Eq. (6.11b)) [59].

$$\text{β-picoline} + O_2 \text{ (1 atm)} \xrightarrow[\text{AcOH, 100 °C}]{\text{NHPI (10 mol\%)}, \text{ Co(OAc)}_2 \text{ (1.5 mol\%)}} \text{nicotinic acid} \quad (6.11a)$$

Conv. 82% → 92%

$$\text{γ-picoline} + \text{Air (20 atm)} \xrightarrow[\text{AcOH, 150 °C}]{\text{NAPI (20 mol\%)}, \text{ Co(OAc)}_2 \text{ (0.5 mol\%)}, \text{ Mn(OAc)}_2 \text{ (0.5 mol\%)}} \text{4-pyridinecarboxylic acid} \quad (6.11b)$$

Conv. 67% → 90%

Quinolines and their derivatives are common in natural products, and have attractive applications as pharmaceuticals and agrochemicals [60]. For example, 3-quinolinecarboxylic acid derivatives are reported to be potent inhibitors of bacterial DNA gyrase. So far, there have been only limited methods for the preparation of quinolinecarboxylic acids despite their potential importance [61]. The synthesis of quinolinecarboxylic acids from the corresponding methylquinolines by direct oxidation seems to be the simplest method, but the reaction has been difficult to carry out selectively because of low reactivity of the methyl group bearing the quinoline ring. Classically, the oxidation was examined by the use of a stoichiometric amount of a metal oxidant like KMnO$_4$ [62], CrO$_3$ [62], or nickel peroxide [63], or by Pd-catalyzed oxidation with H$_2$O$_2$ [64].

Treatment of 3-methylquinoline (**10**) by the NHPI-Co-Mn system under the reaction condition used for β-picoline, however, results in the recovery of the starting **10**. This is believed to be because the activation of O$_2$ by the Co(II) becomes difficult, probably because of coordination of **10** to Co(OAc)$_2$. As described below, the nitration of alkanes with NO$_2$ is enhanced in the presence of NHPI catalyst, in which the generation of PINO from NHPI is easily achieved by NO_2 without any transition metal. Hence, the oxidation of **10** by adding a small amount of NO_2 produced 3-quinolinecarboxylic acid. For instance, the oxidation of **10** with O_2 (1 atm) catalyzed by NHPI (20 mol%), $Co(OAc)_2$ (2 mol%) and $Mn(OAc)_2$ (0.1 mol%) in the presence of NO_2 (10 mol%) gave 3-quinolinecarboxylic acid in 75% yield at 90% conversion (Eq. (6.12)). Other methylquinolines are also successfully oxidized under reaction conditions similar to those used for **10** [65].

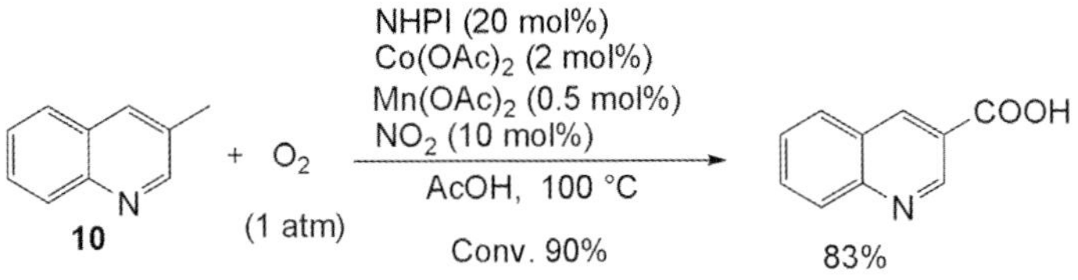

$$\text{(6.12)}$$

6.2.2.3 Oxidation of Hydroaromatic and Benzylic Compounds

Various hydroaromatic and benzylic compounds can be oxidized under a normal pressure of dioxygen catalyzed by NHPI even in the absence of a transition metal species, giving the corresponding oxygenated compounds in good yields. For example, treatment of fluorene with dioxygen in the presence of a catalytic amount of NHPI in benzonitrile at 100 °C for 20 h affords fluorenone in 80% yield. Similarly, xanthene produced xanthone in excellent yield (Eq. (6.13)) [8, 66]. After our finding of NHPI catalysis in the aerobic oxidation, Einhorn et al. reported the oxidation of these substrates with O_2 at room temperature in the presence of NHPI and acetaldehyde, and they concluded that the active species is the PINO formed by the reaction of NHPI with an acetylperoxy radical (Scheme 6.4) [67]. They prepared chiral N-hydroxyimides and used them as the catalyst in the asymmetric oxidation of indanes to give indanones in 8% ee (Figure 6.4) [68].

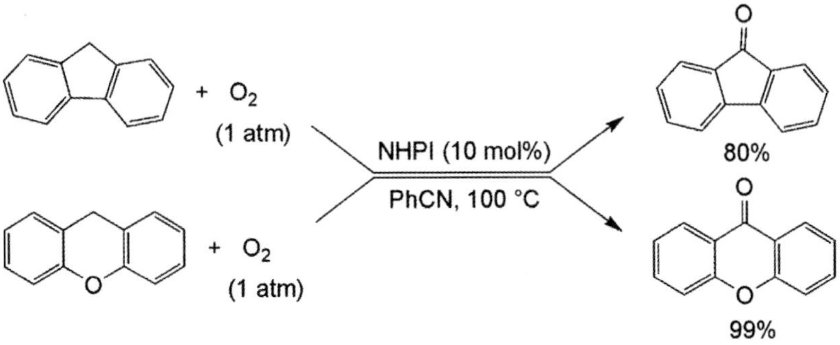

$$\text{(6.13)}$$

Hydroperoxides are used not only as oxidizing agents of alkenes but also as important precursors for the synthesis of phenols. For instance, α-hydroperoxyethylbenzene, obtained by aerobic oxidation of ethylbenzene, is used as an active oxygen

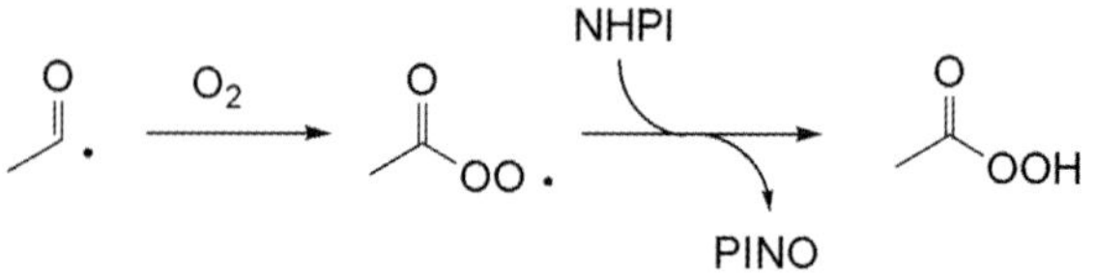

Scheme 6.4 Formation of PINO by reaction of NHPI with acetylperoxy radical.

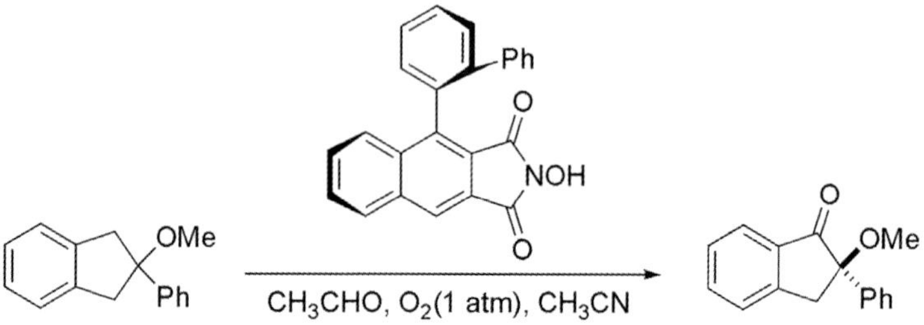

Figure 6.4 Oxidation with chiral N-hydroxyimides prepared by Einhorn et al.

carrier in the epoxidation of propylene, which is known as the Halcon process [69]. The cumene-phenol process (Hock Process) based on the decomposition of cumene hydroperoxide with sulfuric acid to phenol and acetone is the current method for phenol synthesis used worldwide [70]. An efficient approach to phenols through the formation of hydroperoxides from alkylbenzenes is successfully achieved by aerobic oxidation using NHPI as a catalyst. The oxidation of several alkylbenzenes with dioxygen by NHPI followed by treatment with an acid affords phenols in good yields. For example, the aerobic oxidation of cumene in the presence of a catalytic amount of NHPI at 75 °C and subsequent treatment with H_2SO_4 leads to phenol in 81% selectivity at 90% conversion (Eq. (6.14)) [71]. Hydroquinone (61%) and 4-isopropylphenol (33%) are obtained from 1,4-diisopropylbenzene. More recently, Sheldon et al. have reported the highly selective oxidation of cyclohexylbenzene to cyclohexylbenzene-1-hydroperoxide (CHBH). The aerobic oxidation of cyclohexylbenzene in the presence of NHPI (0.5 mol%) and some CHBH (2 mol%) as an initiator without solvent affords the desired CHBH (98% selectivity) at 32% conversion [72]. They considered that this oxidation provides an overall co-product-free route to phenol production. The acid-catalyzed decomposition of the CHBH would give a mixture of phenol and cyclohexanone, which is subsequently dehydrogenated with an appropriate catalyst to form phenol (Scheme 6.5).

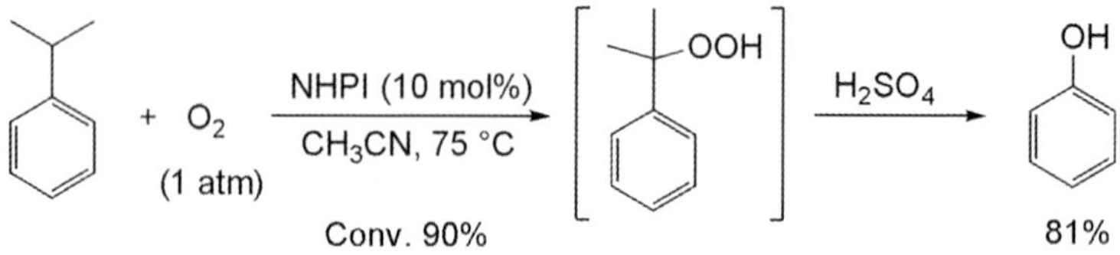

(6.14)

A one-pot synthesis of phenol and cyclohexane from cyclohexylbenzene was later achieved by using NHPI catalyst followed by treatment with sulfuric acid, which

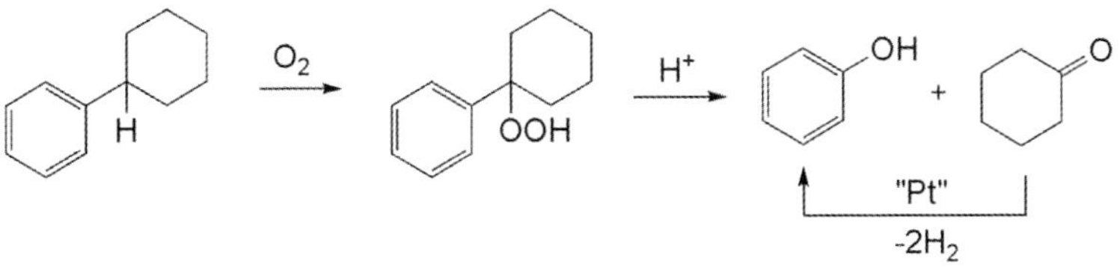

Scheme 6.5 A new route to phenol synthesis suggested by Sheldon et al.

resulted in phenol (93% selectivity) and cyclohexanone (91% selectivity) at 25% conversion (Eq. (6.15)) [73]. Similarly, the oxidation of 2,6-diisopropylnaphthalene with air (20 atm) in the presence of NHPI/AIBN led to 2,6-naphthalenediol (Eq. (6.16)) [74].

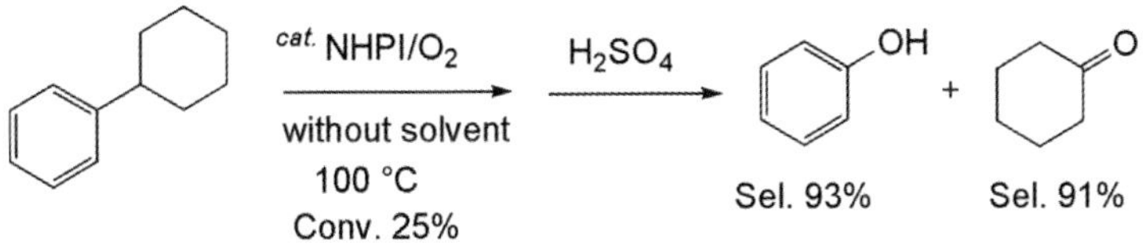

$$(6.15)$$

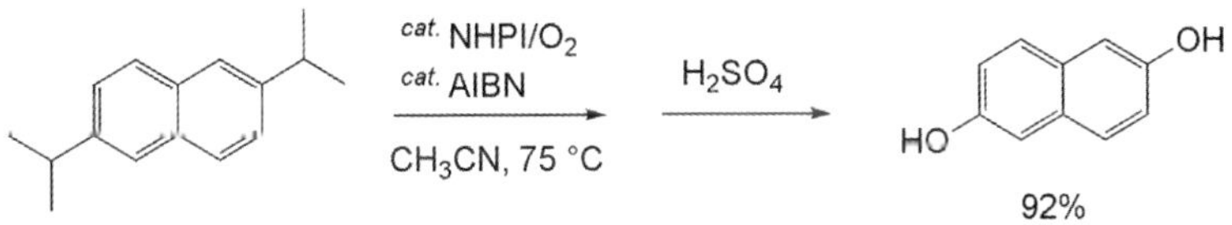

$$(6.16)$$

Various substituted phenols were selectively synthesized by a one-pot reaction through the NHPI-catalyzed aerobic oxidation of 1,1′-diarylethanes to hydroperoxides followed by treatment with dilute sulfuric acid [75]. For example, the oxidation of 1-(4-methoxyphenyl)-1-phenylethane was performed under O_2 (1 atm) in the presence of AIBN (3 mol%) and NHPI (10 mol%) in MeCN (3 mL) at 75 °C for 15 h, and treatment with sulfuric acid afforded 4-methoxyphenol and acetophenone in 61% yield (97% selectivity) at 63% conversion (Eq. (6.17)). In this reaction, the degradation of hydroperoxides was selectively induced to give more electron-rich phenols in high selectivity (Scheme 6.6).

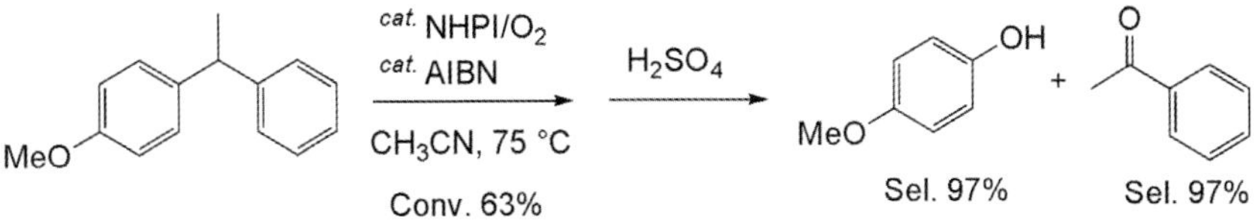

$$(6.17)$$

6.2.3
Preparation of Acetylenic Ketones by Direct Oxidation of Alkynes

α,β-Acetylenic carbonyl compounds (ynones) are important intermediates in organic synthesis, since further elaboration of ynones can lead to highly valuable compounds

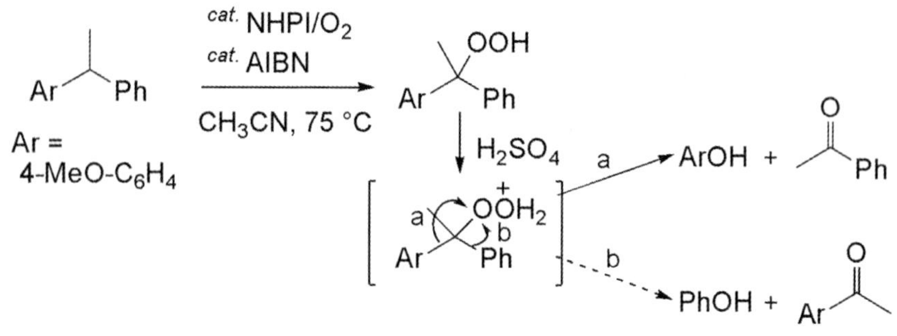

Scheme 6.6 A strategy of the selective phenol synthesis.

such as heterocyclic compounds [76], α,β-unsaturated ketones [77], cyclopentenones [78], nucleosides [79], chiral pheromones [80], and so on. Several methods have been reported for the synthesis of conjugated acetylenic ketones based on a coupling reaction of acetylenides with activated acylating reagents such as acid chloride or anhydrides [81]. Alternatively, the selective oxidation of alkynes to ynones can be carried out by the use of CrO_3/TBHP [82], CrO_3(pyridine)$_2$ [83], Na_2CrO_4/acetic anhydride [83], and SeO_2/TBHP systems [84], but these oxidations are not wholly successful [85]. An alternative approach for preparing ynones is the oxygenation of the propargylic C–H bonds of alkynes with dioxygen, since the bond dissociation energy of these bonds (87 ± 2 kcal/mol for 2-pentyne) is approximately equal to that of the benzylic C–H bonds of alkylbenzenes (88 ± 1 kcal/mol for toluene) [86]. However, the conventional oxidation of alkynes with dioxygen at higher temperatures (around 150 °C) results in undesired over-oxidation products like carboxylic acids. Since the aerobic oxidation of alkylbenzenes by the NHPI catalyst could be effected even at room temperature, the NHPI-catalyzed oxidation of alkynes at lower temperature is expected to suppress undesired side reactions.

Treatment of 4-octyne with dioxygen (1 atm) under the influence of NHPI (10 mol%) and Cu(acac)$_2$ (0.5 mol%) in acetonitrile at room temperature for 30 h produces 4-octyn-3-one (77%) and 4-octyn-3-ol (22%) at 70% conversion (Eq. (6.18)) [87]. The same reaction at 50 °C for 6 h gives the ynone in 70% yield based on 83% conversion. This oxidation would offer a facile catalytic method for the preparation of conjugated ynones from alkynes, since 1-decyne, upon treatment with TBHP in the presence of SeO$_2$, leads to the acetylenic alcohol, 1-decyn-3-ol, rather than the ynone, 1-decyn-3-one [84]. The NHPI/Cu(II) system can also promote the oxidation of acetylenic alcohols to ketones. The reaction of 1-octyn-3-ol under dioxygen (1 atm) in the presence of NHPI and Cu(acac)$_2$ affords 1-octyn-3-one in 95% yield based on 57% conversion.

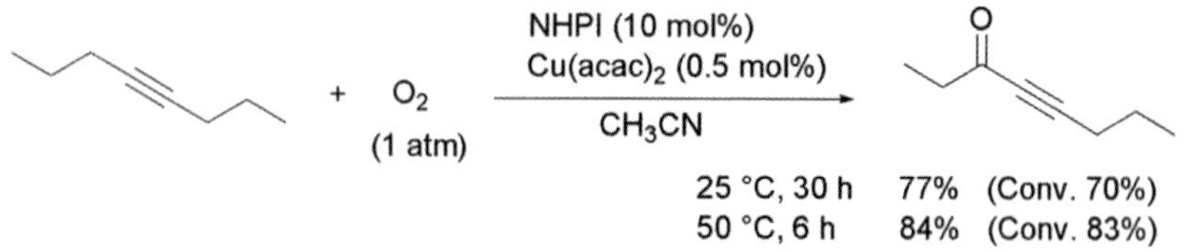

(6.18)

6.2.4
Oxidation of Alcohols

The oxidation of alcohols to the corresponding carbonyl compounds is a transformation frequently used in organic synthesis [88]. There have been many catalytic methods for the aerobic oxidation of alcohols to the corresponding carbonyl compounds [89, 90], but some of these oxidations are carried out in the presence of a reducing agent such as an aldehyde which is eventually converted into carboxylic acid, or they are severely limited to some reactive alcohols such as benzylic and allylic alcohols. Recently, a few aerobic oxidations involving nonactivated alcohols have appeared, although expensive metal catalysts such as Ru and Pd must be employed to effect the oxidation [91]. In 1996, Markó and coworkers developed an efficient aerobic oxidation system of aliphatic alcohols using an inexpensive $CuCl_2$/phenanthroline catalyst combined with azodicarboxylate [92]. Reusable heterogeneous catalysts consisting of Ru or Pd have been reported by Kaneda [93] and Uemura [94], respectively. Sheldon and coworkers have succeeded in the aerobic oxidation of alcohols by a water-soluble Pd catalyst [95].

As described in the preceding sections, alkanes are oxidized by the NHPI/Co(II) system with dioxygen under mild conditions. This catalytic system is expected to promote the aerobic oxidation of the hydroxyl functions of alcohols to carbonyl functions [96, 97]. The oxidation of 2-octanol in ethyl acetate at 70 °C in the presence of NHPI (10 mol%) and $Co(OAc)_2$ (0.5 mol%) under dioxygen (1 atm) gives rise to 2-octanone in quantitative yield. Benzoic acids such as m-chlorobenzoic acid (MCBA) enhance the oxidation of alcohols to carbonyl compounds. 2-Octanol can be converted into 2-octanone with O_2 even at room temperature by adding a catalytic amount of MCBA to the NHPI/$Co(OAc)_2$ system (Eq. (6.19)). The aerobic oxidation of aliphatic alcohols at room temperature has been reported only by Ley et al., who used $[Bu_4N]^+[RuO_4]^-$ assisted by 4-Å molecular sieves [91b].

$$C_6H_{13}\text{-CH(OH)-CH}_3 + O_2 \xrightarrow[\text{AcOEt}]{\substack{\text{NHPI (10 mol\%)} \\ \text{Co(OAc)}_2 \text{ (0.5 mol\%)} \\ \text{MCBA (5 mol\%)}}} C_6H_{13}\text{-CO-CH}_3 \quad (6.19)$$

(1 atm)   MCBA = m-chlorobenzoic acid

25 °C, 20 h   75%
70 °C, 3 h   90%

Figure 6.5 shows the oxidation of secondary and primary alcohols under ambient conditions by the NHPI/$Co(OAc)_2$/MCBA system. Aromatic and cyclic alcohols afford the corresponding ketones in good to quantitative yields. Primary alcohols are also oxidized to carboxylic acids in good yields, although MCPBA (m-chloroperbenzoic acid) is added instead of MCBA. Lauryl alcohol leads to lauric acid (66% yield),

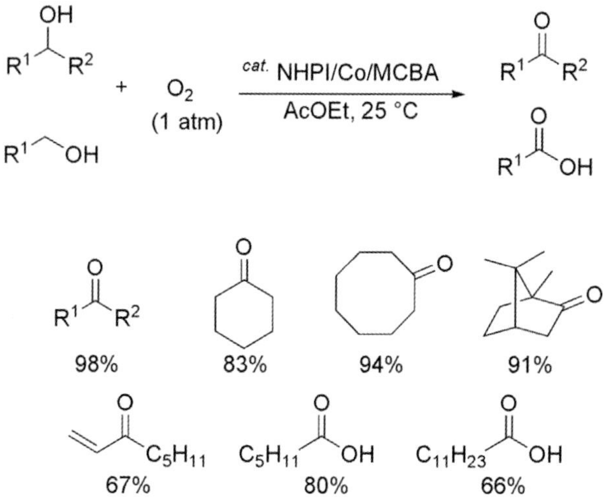

Figure 6.5 Aerobic oxidation of alcohols by NHPI/Co/MCBA system.

which is used as a surfactant raw material. In this oxidation, which proceeds through a free radical process, primary alcohols are rapidly converted into carboxylic acids without isolation of aldehydes, because the hydrogen atom abstraction from aldehydes to afford acyl radicals takes place more easily than that from alcohols to furnish α-hydroxyalkyl radicals [5a]. The oxidation of allylic alcohols is easily achieved.

In contrast to oxidations of diols with stoichiometric oxidants such as $NaIO_4$, $Pb(OAc)_4$ [98], or hydrogen peroxide [99], which are often used in organic synthesis, little work has been done so far for the oxidation of diols with dioxygen [100]. Recently, Uemura and coworkers have reported the $Pd(OAc)_2$-catalyzed lactonization of α,ω-primary diols with dioxygen in the presence of pyridine and 3-Å molecular sieves [101]. Oxidative cleavage of aliphatic and cyclic 1,2-diols with O_2 furnishes aldehydes and dialdehydes, respectively, using $Ru(PPh_3)_3Cl_2$ on active carbon [102]. The oxidation of 1,2-octanediol with dioxygen catalyzed by NHPI combined with $Co(acac)_3$ afforded heptanoic acid in 70% yield at 80% conversion (Table 6.2) [92]. A precursor to heptanoic acid is an α-ketol, since 1-hydroxy-2-octanone is obtained as a principal product at the limited stage of the reaction (Scheme 6.7). An independent oxidation of the α-ketol leads to the carboxylic acid in good yield. Woodward and coworkers have applied the NHPI-catalyzed oxidation to the carbon-carbon bond cleavage of diols to carboxylic acids [103].

Unlike 1,2-diols, internal vicinal diols such as 2,3-octanediol are selectively oxidized to diketones such as 2,3-octanedione rather than cleaved to carboxylic acids. The conversion of vicinal diols to diketones is usually performed by oxidation with metal oxidants such as $AgCO_3$ [104] and permanganate [105], by a TEMPO/NaOCl system under electrochemical conditions [106], or by a catalytic method using heteropolyoxometalates and H_2O_2 [107]. Interestingly, 1,3- and 1,4-diols are

Table 6.2 Oxidation of various diols with dioxygen[a].

Diol	Conv. (%)	Products (Yield/%)			
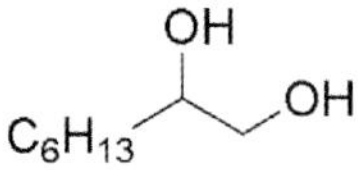 C_6H_{13}-CH(OH)-CH$_2$OH	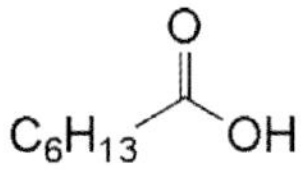 80	C_6H_{13}-C(O)-OH (70)			
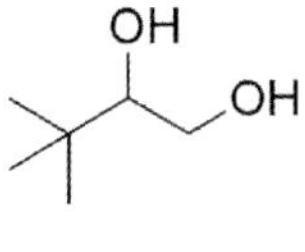 (CH$_3$)$_3$C-CH(OH)-CH$_2$OH	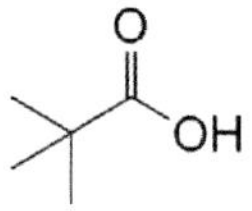 89	(CH$_3$)$_3$C-C(O)-OH (71)			
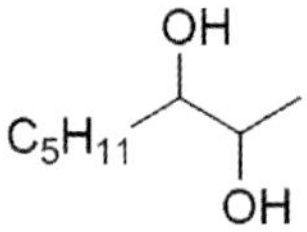 C_5H_{11}-CH(OH)-CH(OH)-CH$_3$	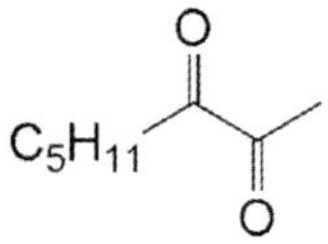 96	C_5H_{11}-C(O)-C(O)-CH$_3$ (86)			
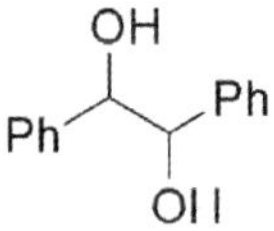 Ph-CH(OH)-CH(OH)-Ph (hydrobenzoin)	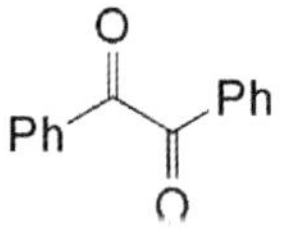 97	Ph-C(O)-C(O)-Ph (84)	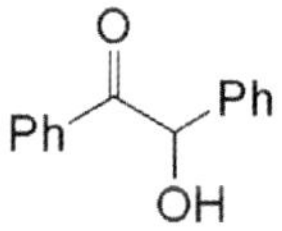	Ph-C(O)-CH(OH)-Ph (12)	
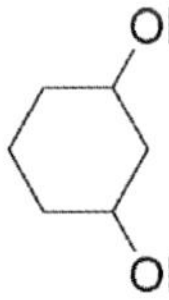 cyclohexane-1,3-diol	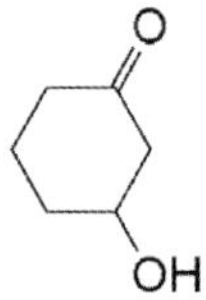 80	3-hydroxycyclohexanone (80)			
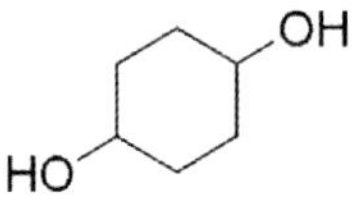 cyclohexane-1,4-diol	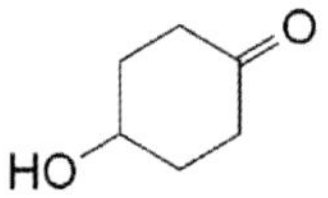 88	4-hydroxycyclohexanone (72)	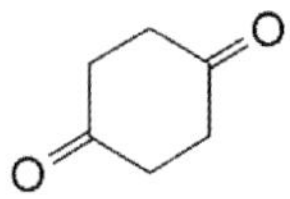	1,4-cyclohexanedione (16)	
 HO-(CH$_2$)$_5$-OH	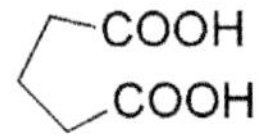 80	glutaric acid (HOOC-(CH$_2$)$_3$-COOH) (66)			

a) Diols (3 mmol) were allowed to react with molecular oxygen (1 atm) in the presence of NHPI (10 mol%) and Co(acac)$_3$ (1 mol%) in CH$_3$CN (5 mL).

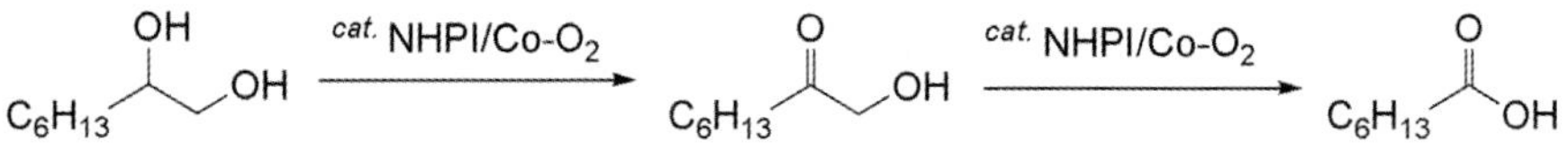

C_6H_{13}-CH(OH)-CH$_2$OH $\xrightarrow{\text{cat. NHPI/Co-O}_2}$ C_6H_{13}-C(O)-CH$_2$OH $\xrightarrow{\text{cat. NHPI/Co-O}_2}$ C_6H_{13}-C(O)-OH

Scheme 6.7 Oxidation of 1,2-octanediol to heptanoic acid.

selectively converted into the corresponding hydroxy ketones rather than to the diketones. An α,ω-diol such as 1,5-pentanediol gives rise to the dicarboxylic acid in good yield. The present reaction provides an alternative and useful route to dicarboxylic acids from diols with dioxygen.

6.2.5
Epoxidation of Alkenes using Dioxygen as Terminal Oxidant

The epoxidation of alkenes using dioxygen via a catalytic process is a challenging subject in the field of oxidation chemistry. Much effort has been devoted to the epoxidation of alkenes with dioxygen using transition metals as catalysts [5d,5e,6a,108–115]. For instance, β-diketonate complexes of Ni, V, and Fe are reported to catalyze efficiently the epoxidation of alkenes with dioxygen in the presence of an aldehyde, alcohol, or acetal as a reducing agent under mild conditions [109j]. On the other hand, Ru-porphyrin complexes [113] and Ru-substituted polyoxometalates, $\{[WZnRu_2(OH)(H_2O)](ZnW_9O_{34})_2\}^{11-}$ [114], catalyze the epoxidation of alkenes without any reducing agents.

The hexafluoroacetone (HFA)-catalyzed epoxidation of alkenes utilizing H_2O_2 obtained *in situ* by the NHPI-catalyzed aerobic oxidation of alcohols was examined (Scheme 6.8). A hydroperoxide derived from HFA and H_2O_2 has been reported to epoxidize various alkenes in fair to good yields [116, 117]. This epoxidation system seems to be an interesting industrial strategy, for it does not require the storage and transportation of explosive H_2O_2 [118]. In addition, the resulting ketones can be easily reduced to the original alcohols. 2-Octene was allowed to react under O_2 (1 atm) in the presence of 1-phenylethanol under the influence of catalytic amounts of NHPI (10 mol%) and HFA (10 mol%) in benzonitrile at 80 °C for 24 h, giving 2,3-epoxyoctane in 93% selectivity based on 93% conversion (Eq. (6.20)) [119]. This is the first successful epoxidation with H_2O_2 generated *in situ* from alcohols and O_2 without any metal catalysts. The important feature of this reaction is that the epoxidation of

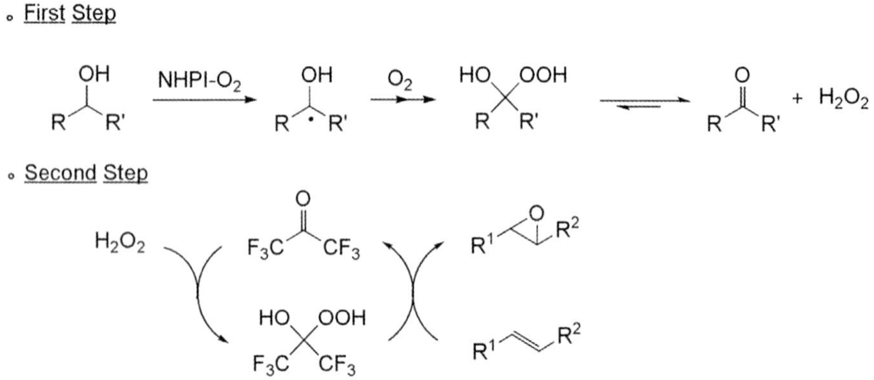

- First Step

- Second Step

Scheme 6.8 A new strategy for the epoxidation of alkenes.

cis- and trans-2-octenes proceeded in a stereospecific manner to form cis- and trans-2,3-epoxyoctanes respectively, in high yields., although O_2 is used as a terminal oxidant.

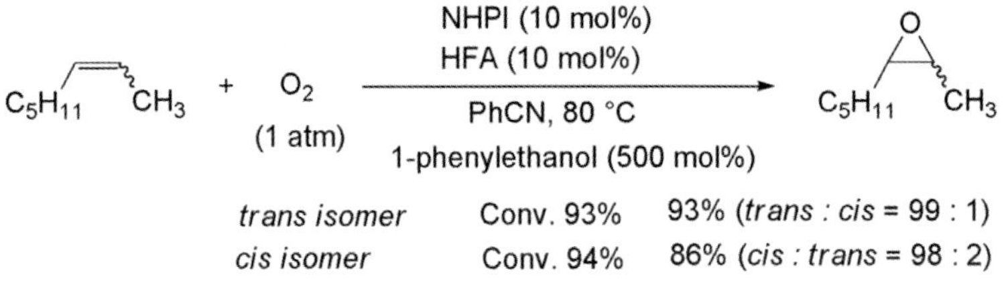

$$C_5H_{11}\text{-CH=CH-CH}_3 + O_2 \text{ (1 atm)} \xrightarrow[\text{1-phenylethanol (500 mol\%)}]{\text{NHPI (10 mol\%), HFA (10 mol\%)}} C_5H_{11}\text{-epoxide-CH}_3$$

	Conv.	Yield
trans isomer	93%	93% (trans : cis = 99 : 1)
cis isomer	94%	86% (cis : trans = 98 : 2)

(6.20)

6.2.6
Baeyer-Villiger Oxidation of KA Oil

KA oil, a mixture of cyclohexanone and cyclohexanol obtained by the aerobic oxidation of cyclohexane, is an important intermediate in petroleum industrial chemistry for the production of adipic acid and ε-caprolactam, which are key materials for manufacturing 6,6-nylon and 6-nylon, respectively [120]. Baeyer-Villiger oxidation is a frequently used synthetic tool for conversion of cycloalkanones to lactones. Usually, this transformation is carried out by the use of peracids like peracetic acid and mCPBA [121], hydrogen peroxide [122], and bis(trimethylsilyl) peroxide [123]. However, the catalytic Baeyer-Villiger oxidation using dioxygen is limited to the in situ generation of peracids using excess aldehydes and O_2 [124]. In industry, ε-caprolactone is manufactured by the reaction of cyclohexanone with peracetic acid generated by the aerobic oxidation of acetaldehyde [120]. From both a synthetic and an industrial point of view, it is very convenient that the KA oil can be used as the starting material for the production of ε-caprolactone with molecular oxygen via a catalytic process.

A new strategy for ε-caprolactone synthesis is outlined in Scheme 6.9. The aerobic oxidation of cyclohexanol (3) catalyzed by NHPI gives a mixture of cyclohexanone (2)

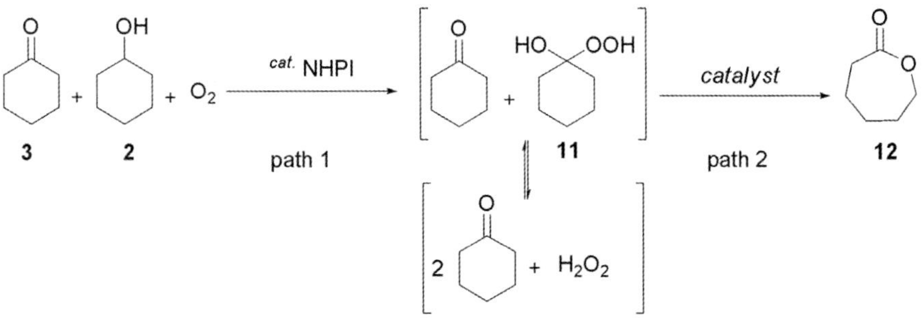

Scheme 6.9 A new strategy for the Baeyer-Villiger oxidation of KA-oil.

and hydrogen peroxide through the formation of 1-hydroxy-1-hydroperoxycyclohexane (**11**) (path 1). Treatment of the resulting reaction mixture with an appropriate catalyst would produce ε-caprolactone (**12**) (path 2). A KA oil consisting of a 1:1 mixture of **3** and **2** was employed as a model starting material. If the aerobic oxidation of the KA oil in the presence of NHPI is completed, 2 equiv. of **3** and 1 equiv. of H_2O_2 are expected to be formed. Treatment of a 1:1 mixture of **3** (6 mmol) and **2** (6 mmol) by catalytic amounts of NHPI (0.6 mmol) and 2,2′-azobisisobutyronitrile (AIBN) (0.3 mmol) under an O_2 atmosphere in CH_3CN at 75 °C for 15 h, followed by $InCl_3$ (0.45 mmol) at 25 °C for 6 h affords **12** in 57% selectivity based on the KA oil reacted, and 77% of KA oil was recovered (Eq. (6.21)). Water-stable Lewis acids such as $Sc(OTf)_3$ and $Gd(OTf)_3$ afford ε-caprolactone in somewhat lower yields [125].

$$3 + 2 + O_2 \;(1\,\text{atm}) \xrightarrow[\text{CH}_3\text{CN, 75 °C}]{\text{NHPI (10 mol\%)}} \xrightarrow[\text{75 °C}]{\text{InCl}_3\,(7.5\,\text{mol\%})} 12 \quad (6.21)$$

Conv. 23% 57%

6.2.7
Preparation of ε-Caprolactam Precursor from KA Oil

ε-Caprolactam (CL) is a very important monomer for the production of nylon-6, and about 4.2 million tons of CL were manufactured worldwide in 1998 [126]. Most current methods of CL production involve the conversion of cyclohexanone with hydroxylamine sulfate into cyclohexanone oxime followed by Beckmann rearrangement by the action of oleum and then treatment with ammonia, giving CL. A serious drawback of this process is the co-production of a large amount of ammonium sulfate waste [126, 127]. Raja and Thomas reported a method for one-step production of cyclohexanone oxime and CL by the reaction of cyclohexanone with ammonia under high-pressure air (34.5 atm) in the presence of a bifunctional molecular sieve catalyst [128]. Hydrogen peroxide oxidation of cyclohexanone in the presence of NH_3 catalyzed by titanium silicate is reported to produce CL [129]. In patent work, on the other hand, the transformation of 1,1′-peroxydicyclohexylamine (PDHA) to a 1:1 mixture of CL and cyclohexanone by LiBr has been reported [130].

It is interesting to develop a novel route to the CL precursor, PDHA, which was hitherto prepared by hydrogen peroxide oxidation of cyclohexanone (**3**) followed by treatment with ammonia [126, 130]. Because of the ease of transformation of PDHA to a 1:1 mixture of CL and **3** under the influence of an appropriate catalyst such as lithium halides, the CL production via PDHA is considered to be a superior candidate for a next-generation waste-free process for CL. The NHPI-catalyzed aerobic oxidation of KA oil was applied to the synthesis of PDHA without formation of any ammonium sulfate waste. The strategy is outlined in Scheme 6.10. The NHPI-catalyzed oxidation of KA oil (a mixture of **3** and **2**) with O_2 produces 1,1′-dihydroxydicyclohexyl peroxide, which seems to exist in equilibrium with cyclohexanone and H_2O_2 (path 1). Subsequent treatment of the resulting reaction mixture

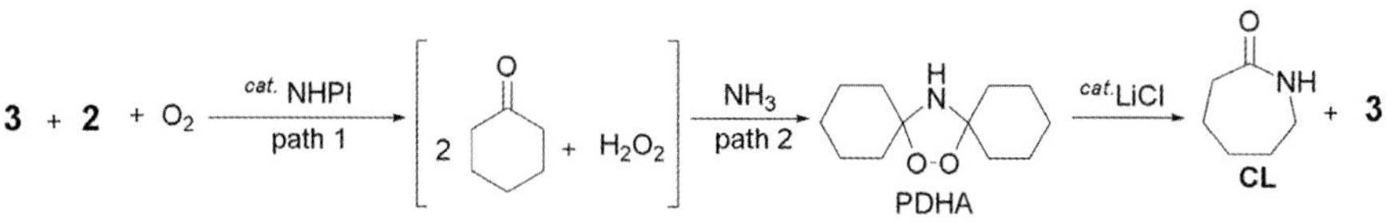

6.3 Functionalization of Alkanes Catalyzed by NHPI

Scheme 6.10 A new strategy for the synthesis of ε-caprolactam precursor, PDHA.

with NH_3 would afford PDHA (path 2). A 1 : 2 mixture of **3** and **2** was reacted under dioxygen atmosphere (1 atm) in the presence of small amounts of NHPI and AIBN in ethyl acetate at 60 °C for 20 h, followed by the reaction with an ammonia at atmospheric pressure at 70 °C for 2 h to give 84% of PDHA at 24% conversion of KA oil (Eq. (6.22)). This route provides a more economical and environmentally friendly process than that by the current method using hydroxylamine sulfate.

$$3 + 2 + O_2 \text{ (1 atm)} \xrightarrow[\text{AcOEt, 60 °C} \atop \text{Conv. 24\%}]{\text{NHPI (10 mol\%)}} \xrightarrow[\text{70 °C}]{NH_3 \text{ (gas)}} \text{PDHA} \quad 84\% \tag{6.22}$$

6.3 Functionalization of Alkanes Catalyzed by NHPI

6.3.1 Carboxylation of Alkanes with CO and O_2

Carbonylation and carboxylation of alkanes with carbon monoxide (CO) are challenging transformations in organic synthesis [131]. There have been several important discoveries including the Rh-catalyzed photocarbonylation of alkanes by Tanaka [132, 133] and the carboxylation of methane with CO/O_2 using Pd/Cu [134] or $RhCl_3$ [135] catalysts by Fujiwara and Sen et al. The carbonylation of adamantanes under the influence of Lewis acid and superacids has also been reported [136, 137]. Following the first report on the free-radical-mediated carbonylation by Coffmann et al. in 1952 [138], this type of reaction was not investigated for a long time, probably because it has to be conducted under extremely high CO pressure (200–300 atm) [139]. In 1990, Ryu performed a successful free-radical carbonylation of alkyl halides with CO mediated by tributyltin hydride [131e,140]. Sen et al. disclosed a free-radical carboxylation of methane to acetic acid by the use of peroxydisulfate as a radical source [141]. Benzophenone- [142] and polyoxotungstate-photocatalyzed [143] as well as mercury-photosensitized [144] carbonylations of cyclohexane afford cyclohexanecarbaldehyde. The trapping of alkyl radicals generated from alkanes under the influence of NHPI catalyst by CO followed by O_2 leads to carboxylic acids (Scheme 6.11) [145]. The carboxylation of adamantane under CO/air (15/1 atm) in the presence of NHPI (10 mol%) in a mixed solvent of acetic acid and 1,2-dichloroethane at 95 °C for 4 h affords 1-adamantanecarboxylic acid, 1,3-adamantanedicarboxylic

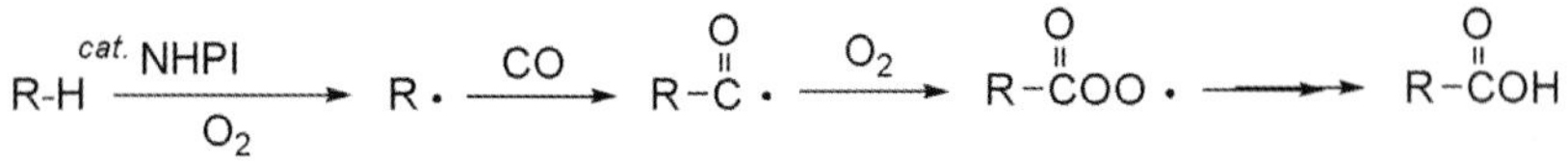

Scheme 6.11 Carbonylation of alkanes with CO and O_2 catalyzed by NHPI.

$$R-H \xrightarrow[O_2]{cat.\ NHPI} R\cdot \xrightarrow{CO} R-\overset{\overset{O}{\|}}{C}\cdot \xrightarrow{O_2} R-\overset{\overset{O}{\|}}{C}OO\cdot \longrightarrow R-\overset{\overset{O}{\|}}{C}OH$$

acid, 2-adamantanecarboxylic acid, and several oxygenated products such as 1-adamantanol and 2-adamantanone (Eq. (6.23)).

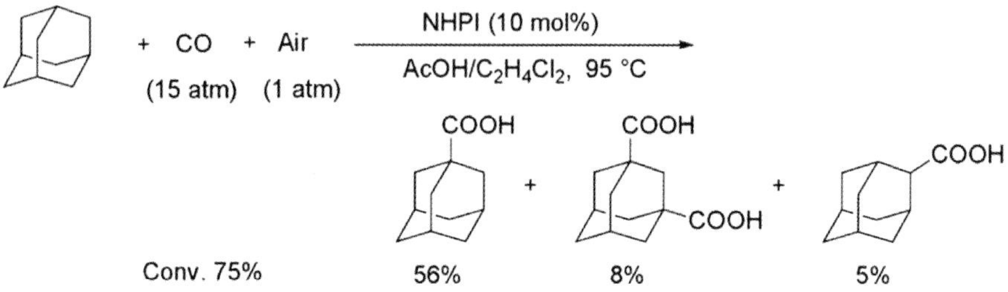

$$\text{Adamantane} + CO + \text{Air} \xrightarrow[\text{AcOH/C}_2\text{H}_4\text{Cl}_2,\ 95\ °\text{C}]{\text{NHPI (10 mol\%)}} \text{1-COOH-adamantane} + \text{1,3-(COOH)}_2\text{-adamantane} + \text{2-COOH-adamantane}$$

Conv. 75%   56%   8%   5%

(6.23)

The present strategy was successfully applied to the preparation of adamantane-dicarboxylic acid, which is an interesting monomer in polymer chemistry, through a stepwise procedure (Eq. (6.24)), although the dicarboxylic acid is difficult to obtain by conventional methods. Similarly, 1,3-dimethyladamantane and *endo*-tricyclo[5.2.1.0] decane were carboxylated to the respective mono- and dicarboxylic acids.

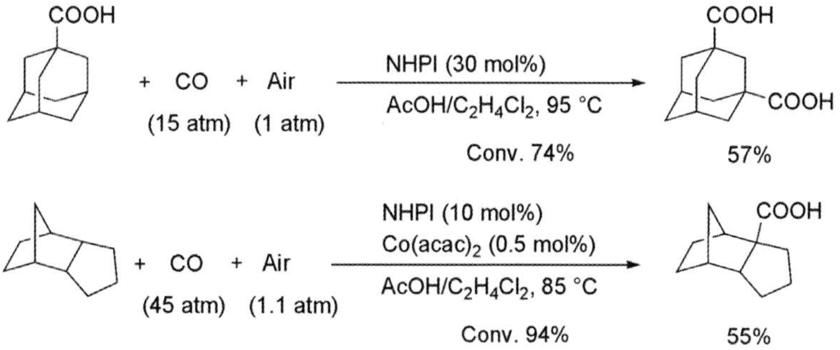

$$\text{1-adamantanecarboxylic acid} + CO + \text{Air} \xrightarrow[\text{AcOH/C}_2\text{H}_4\text{Cl}_2,\ 95\ °\text{C}]{\text{NHPI (30 mol\%)}} \text{1,3-adamantanedicarboxylic acid}$$

(15 atm) (1 atm)   Conv. 74%   57%

$$\text{endo-tricyclodecane} + CO + \text{Air} \xrightarrow[\text{AcOH/C}_2\text{H}_4\text{Cl}_2,\ 85\ °\text{C}]{\text{NHPI (10 mol\%), Co(acac)}_2\text{ (0.5 mol\%)}} \text{carboxylated product}$$

(45 atm) (1.1 atm)   Conv. 94%   55%

(6.24)

6.3.2
First Catalytic Nitration of Alkanes using NO_2

Nitration of lower alkanes such as methane and ethane with nitric acid or nitrogen dioxide is industrially practiced to produce nitroalkanes [146, 147]. However, a major problem in current industrial nitration is that the reaction must be run at fairly high temperature (250–400 °C) because of the difficulty of obtaining C–H bond homolysis by NO_2 [146]. Under such high temperatures, higher alkanes undergo not only homolysis of the C–H bonds but also cleavage of the C–C skeleton. Hence, the nitration is limited to lower alkanes [147]. The nitration of propane results in all of the possible nitroalkanes, that is, nitromethane, nitroethane, 1-nitropropane, and 2-nitropropane [146, 148]. Therefore, the catalytic nitration of alkanes under mild conditions would offer a promising and superior alternative. Since NO_2 is a paramagnetic molecule, the generation of PINO from NHPI by the action of NO_2 in analogy with O_2 is expected. Indeed, when NO_2 was added to NHPI in benzene at room temperature, an ESR signal attributable to PINO is instantly observed as a triplet. As a typical result, the nitration of cyclohexane with NO_2 by NHPI without any solvent under air (1 atm) proceeds smoothly even at 70 °C to give nitrocyclohexane (70% based on NO_2 used) and cyclohexyl nitrite (7%) along with a small amount of an oxygenated product, cyclohexanol (5%) (Eq. (6.25)) [149].

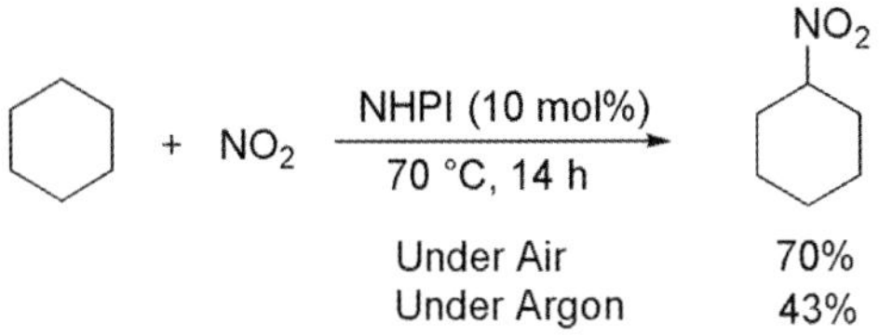

$$\text{C}_6\text{H}_{12} + NO_2 \xrightarrow[70\,°C,\,14\,h]{\text{NHPI (10 mol\%)}} \text{C}_6\text{H}_{11}NO_2 \quad (6.25)$$

Under Air 70%
Under Argon 43%

It is important that the NHPI-catalyzed nitration is conducted under air, since NO generated in the course of the reaction can be readily reoxidized to NO_2 by O_2. In the absence of air, the yield of nitrocyclohexane decreases to 43%. After the nitration, the NHPI catalyst can be separated from the reaction mixture by simple filtration and reused repeatedly. Nitrocyclohexane is easily reduced to cyclohexanone oxime. Therefore, this nitration provides an alternative practical route to cyclohexanone oxime, which is a raw material for ε-caprolactam leading to nylon-6 [150, 151].[1]

A plausible pathway is shown in Scheme 6.12. The hydrogen atom abstraction from the hydroxyimide group of NHPI is induced by NO_2 to form PINO, a key radical species. The PINO abstracts the hydrogen atom from an alkane to give an alkyl radical, which is readily trapped by NO_2 to form a nitroalkane. The HNO_2 formed is converted into HNO_3, H_2O, and NO which is easily oxidized to NO_2 under air [152]. The most promising feature of the NHPI-catalyzed nitration of alkanes by NO_2 is that

NHPI + NO_2 ⟶ PINO + HNO_2
R-H + PINO ⟶ R• + NHPI
R• + NO_2 ⟶ R-NO_2
$3HNO_2$ ⟶ HNO_3 + H_2O + 2NO
2NO + O_2 ⟶ $2NO_2$

Scheme 6.12 A possible reaction path for alkane nitration catalyzed by NHPI.

---

1) Although ε-caprolactam is currently produced by the reaction of cyclohexanone with hydroxylamine followed by a Beckmann rearrangement with sulfuric acid, the efficiency for the production of cyclohexanone by aerobic oxidation of cyclohexane is not high.

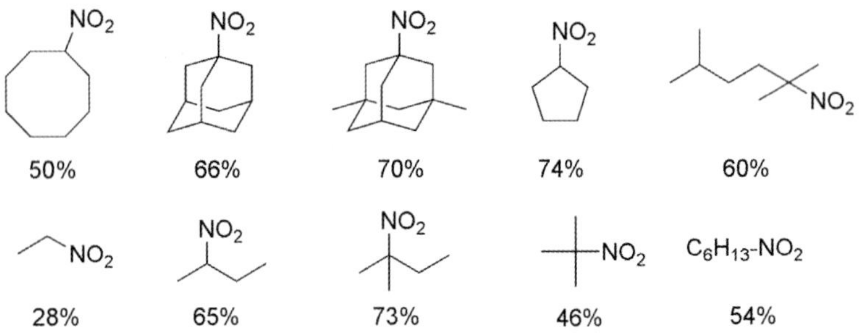

Figure 6.6 Nitration of various alkanes by NHPI/NO$_2$ system.

Structures shown with yields:
- Cyclooctane-NO$_2$: 50%
- Decalin-NO$_2$: 66%
- Adamantane-NO$_2$ (1-position): 70%
- Methyladamantane-NO$_2$: 74%
- 2,3-dimethylbutane tertiary NO$_2$: 60%
- Propyl-NO$_2$: 28%
- Isobutane t-NO$_2$: 65%
- 2-methylbutane t-NO$_2$: 73%
- Neopentyl-type -NO$_2$: 46%
- C$_6$H$_{13}$-NO$_2$: 54%

the nitration can be conducted under air at moderate temperature. Owing to the higher concentration of NO$_2$ than air, the alkyl radicals formed can react selectively with NO$_2$ rather than with O$_2$ to give nitroalkanes in preference to oxygenated products. The conventional nitration is difficult to carry out in air, because the nitration must be carried out at high temperature (250–400 °C). At these temperatures, the resulting alkyl radicals react not only with NO$_2$ but also with O$_2$ to provide a complex mixture of products [147b]. By the use of the NHPI catalyst, the highly selective nitration of higher alkanes with NO$_2$/air under mild conditions was realized for the first time.

A wide variety of alkanes were successfully nitrated by the NHPI/NO$_2$ system (Figure 6.6). In addition, nitric acid instead of NO$_2$ was found to act as an efficient nitrating reagent. For example, the reaction of adamantane with concentrated HNO$_3$ in the presence of catalytic amounts of NHPI in PhCF$_3$ at 60 °C under Ar afforded nitroadamantane and 1,3-dinitroadamantane in 64 and 3% yields, respectively (Eq. (6.26)).

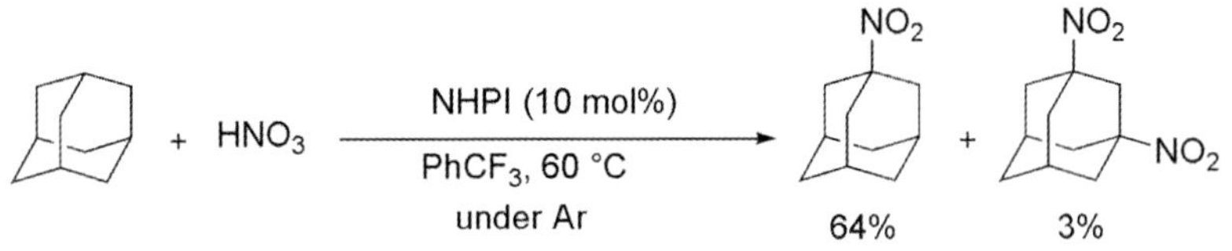

$$\text{adamantane} + \text{HNO}_3 \xrightarrow[\text{PhCF}_3,\ 60\ °C \atop \text{under Ar}]{\text{NHPI (10 mol\%)}} \text{1-nitroadamantane (64\%)} + \text{1,3-dinitroadamantane (3\%)}$$

(6.26)

6.3.3
Sulfoxidation of Alkanes Catalyzed by Vanadium

The sulfoxidation of aromatic hydrocarbons has been extensively studied, but work on the sulfoxidation of saturated hydrocarbons to alkanesulfonic acids remains at a less satisfactory level. The Strecker reaction using alkyl halides, preferably alkyl bromides, and alkali metal or ammonium sulfides, is commonly used for the synthesis of alkanesulfonic acids [153]. Another procedure, the oxidation of thiols with bromine in the presence of water or hydrogen peroxide and acetic acid, has been reported [154]. Attempts to realize the sulfoxidation of alkanes with SO_2 and O_2 have not been fully studied in spite of their importance, because of the difficulty of selective cleavage of the C—H bond in alkanes. Only a few reactions are reported for the sulfoxidation of alkanes such as cyclohexane via a radical process using a mixture of SO_2 and O_2 by means of the photo- and peroxide-initiated techniques [155]. However, the efficiency of the sulfoxidation by these methods is at an insufficient level. Therefore, if alkanes can be sulfoxidated catalytically by SO_2/O_2 without irradiation with light or initiation by a peroxide, such a method has enormous synthetic potential and provides a very attractive route to alkanesulfonic acids. The direct sulfoxidation of alkanes using SO_2 and O_2 was efficiently catalyzed by a vanadium species in the presence or absence of NHPI [156]. The reaction of adamantane with a mixture of SO_2 and O_2 (0.5/0.5 atm) in the presence of NHPI (10 mol%) and $VO(acac)_2$ (0.5 mol%) in acetic acid at 40 °C for 2 h produced 1-adamantanesulfonic acid in 95% selectivity based on 65% conversion (Eq. (6.24)). Smith obtained the same product in 15% yield by the photosulfoxidation of adamantane with SO_2/O_2 in the presence of H_2O_2 [157]. Surprisingly, 1-adamantanesulfonic acid was obtained with high selectivity and at moderate conversion even in the absence of the NHPI (Eq. (6.27)).

$$\text{Adamantane} + SO_2/O_2 \xrightarrow[\text{AcOH, 40 °C}]{\substack{\text{NHPI (10 mol\%)} \\ \text{VO(acac)}_2 \text{ (0.5 mol\%)}}} \text{1-adamantane-}SO_3H$$

(0.5/0.5 atm)

NHPI/VO(acac)$_2$ 95% (Conv. 65%)
VO(acac)$_2$ 98% (Conv. 43%)

(6.27)

In order to assess the potential of various metal ions in this sulfoxidation, a series of first-row transition metal salts, $TiO(acac)_2$, $Cr(acac)_3$, $Mn(acac)_3$, $Fe(acac)_3$, $Co(acac)_2$, $Ni(OAc)_2$, and $Cu(OAc)_2$ was tested. It is interesting to note that no sulfoxidation was induced by these metal salts [158]. From a survey of vanadium compounds, $VO(acac)_2$ and $V(acac)_3$ were found to be efficient catalysts. $VO(acac)_2$ promotes the reaction even at room temperature, affording the sulfonic acid in 81% selectivity at 64% conversion after 24 h. The addition of a small amount of hydroquinone stopped the reaction. This indicates that a radical chain process is involved in this catalytic sulfoxidation. A variety of alkanes was successfully sulfoxidized by a mixture of SO_2 and O_2, giving the corresponding alkanesulfonic acids in high selectivities (Figure 6.7). Adamantane having either an electron-withdrawing or electron-donating group was sulfoxidized in good selectivity in a range of about 60–70% conversion. The aliphatic hydrocarbon, octane, afforded a mixture of 2-, 3-, and 4-octanesulfonic acids. The sulfoxidation of alkanes seems to proceed via the reaction steps shown in Scheme 6.13.

The sulfoxidation may be initiated by one-electron transfer from an alkane to a V(V) species generated *in situ* from $VO(acac)_2$ and O_2 to form an alkyl cation radical which

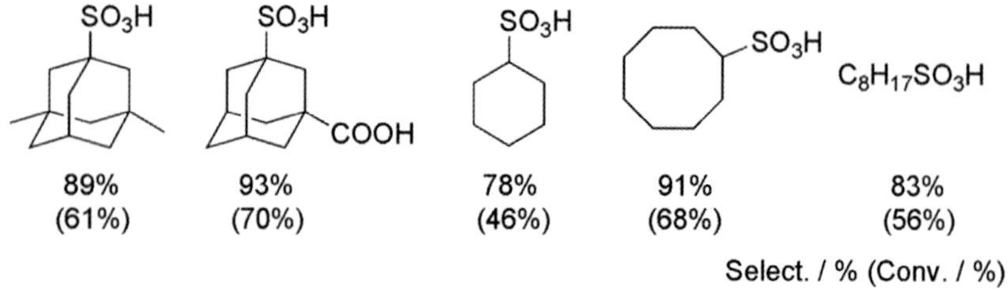

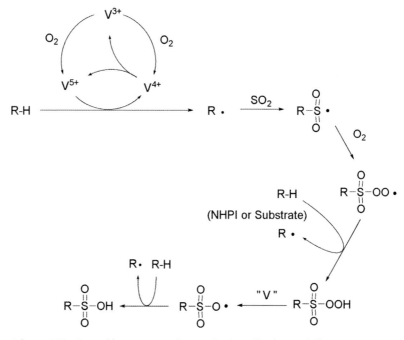

Figure 6.7 Sulfoxidation of various alkanes catalyzed by VO(acac)$_2$.

Scheme 6.13 A possible reaction mechanism for the sulfoxidation of alkanes.

readily liberates a proton to form an alkyl radical. The V(IV) species is reported to undergo disproportionation to V(V) and V(III) in the oxidative polymerization of diphenyl disulfide by the vanadium ion under a dioxygen atmosphere [159]. In addition, α-hydroxycarbonyl compounds are oxidized to α-dicarbonyl compounds by VOCl$_3$ and VO(acac)$_2$ under an oxygen atmosphere [160]. The resulting radical is trapped by SO$_2$ and then O$_2$ to generate an alkanesulfonylperoxy radical, which is finally converted into an alkanesulfonic acid through the well-known reaction path [161].

6.3.4
Reaction of NO with Organic Compounds

In recent years, much attention has been paid to nitric oxide (NO), a molecule having a free radical character, in the fields of biochemistry and medical science [162, 163]. However, its application to synthetic organic chemistry is quite limited because of the scarcity of information available on the chemical behavior of NO and the difficulty of controlling its reactivity [164–168]. Recently, Yamada et al. have reported that NO can be used as a nitrogen source for the synthesis of nitrogen-containing compounds such as 2-nitrosocarboxamides [164] and nitroalkenes [165]. A novel utilization of NO in organic synthesis with the use of NHPI has been developed [169, 170]. The reaction of adamantane with NO (1 atm) in the presence of NHPI (10 mol%) in a mixed solvent of benzonitrile and acetic acid at 100 °C for 20 h afforded 1-N-adamantylbenzamide in substantial yield along with a small amount of nitroadamantane (Eq. (6.28)) [169]. This reaction provides a novel and alternative modified Ritter-type reaction, although there are a few reports on the transformation of adamantane to the amide by means of the anodic oxidation [171] of nitronium tetrafluoroborate [172].

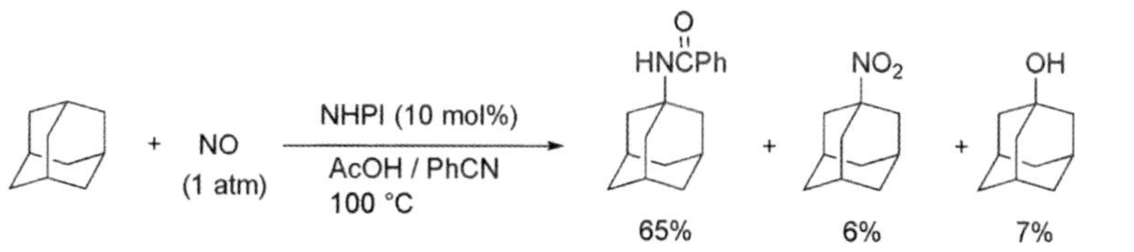

$$\text{(6.28)}$$

On the other hand, benzyl ethers react with NO in the presence of the NHPI catalyst to afford the corresponding aromatic aldehydes (Eq. (6.29)) [170]. The reaction of 4-methoxymethyltoluene catalyzed by NHPI (10 mol%) under NO (1 atm) for 5 h leads to p-tolualdehyde in 50% yield. tert-Butoxymethyltoluene and tert-butyl benzyl ethers are converted into the corresponding aldehydes in good yields.

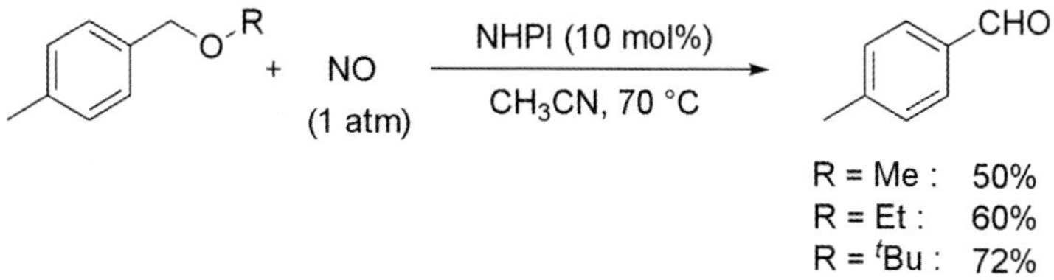

$$\text{(6.29)}$$

R = Me : 50%
R = Et : 60%
R = ^{t}Bu : 72%

The most important application of this procedure is the transformation of ethers to benzenedicarbaldehydes, which are attractive starting materials in pharmaceutical synthesis [173]. 1,3-Dihydro-2-benzofuran and 1,3-di-tert-butoxymethyl- and

1,4-dimethoxymethylbenzenes are converted into the respective dialdehydes in good yields (Eq. (6.30)) [170]. Of the various indirect procedures to obtain dialdehydes, hydrolysis of α,α,α′,α′-tetrabromoxylenes is usually used, although the preparation of bromides is troublesome [174]. Therefore, the present procedure provides a very convenient and direct route to benzenedicarbaldehydes.

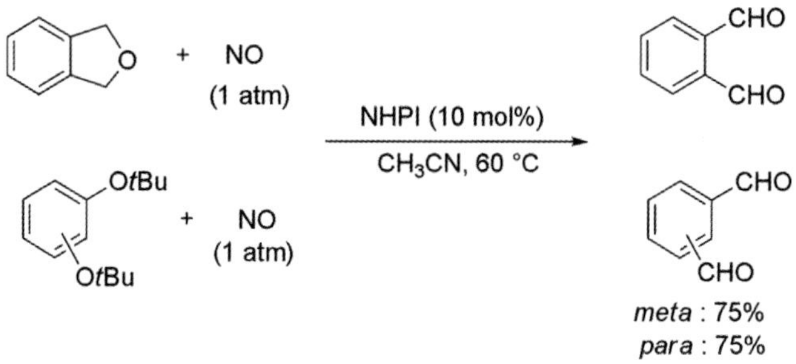

(6.30)

Mechanistically, the reactions of adamantane and ethers with NO are rationally explained by considering the formation of carbocations as transient intermediates (Scheme 6.14). The generation of PINO from NHPI in the presence of NO is confirmed by ESR measurements [170], but the formation of PINO by this method may be due to traces of NO_2 contained in the reaction system. On the other hand, Suzuki has suggested the formation of a cationic species via a diazonium nitrate in the nitration of alkenes with NO [175]. The nucleophilic attack of the benzonitrile and water on the adamantyl and benzylic cations would result in the amide and aldehyde, respectively [170].

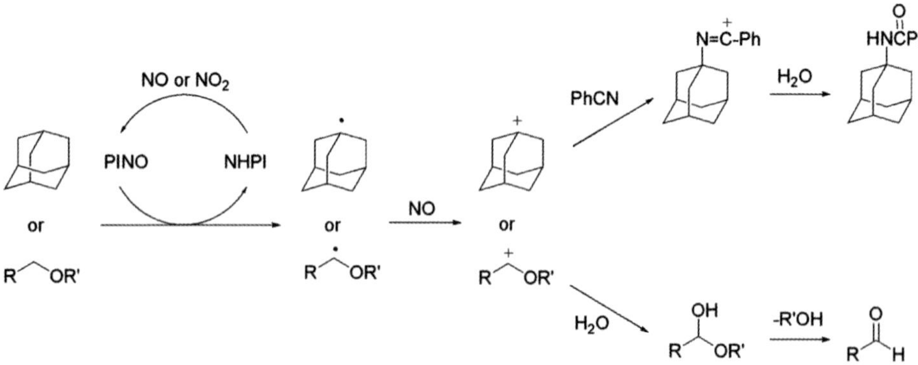

Scheme 6.14 Reaction of adamantane or ethers with NO through the formation of carbocations as transient intermediates.

6.3.5
Nitrosation of Cycloalkanes with t-BuONO

The nitrosation of cycloalkanes with *tert*-butyl nitrite (*t*-BuONO) is successfully achieved without photo-irradiation under halogen-free conditions by using NHPI as a catalyst [176]. Various cycloalkanes were allowed to react with *t*-BuONO in the presence of NHPI in acetic acid at 80 °C for 2 h to give the corresponding nitrosocycloalkanes (Eq. (6.31)). Most of the *tert*-butyl moiety of *t*-BuONO was found to be converted into *tert*-butyl alcohol. Since *tert*-butyl alcohol is known to react with NO_2 or sodium nitrite to produce *t*-BuONO [177], the nitrite may be regenerated from the *tert*-butyl alcohol formed (Scheme 6.15). This reaction is a new green route for the synthesis of lactam precursors from cycloalkanes. A plausible reaction path is outlined in Scheme 6.16. The reaction is initiated by the formation of PINO and NO from reaction of NHPI with *t*-BuONO. Since NHPI is easily oxidized with a weak oxidizing agent, *t*-BuONO may serve as an oxidizing agent for NHPI and allow the formation of PINO and NO. The PINO thus generated abstracts the hydrogen atom from cyclohexane, giving a cyclohexyl radical and NHPI. Subsequently, the cyclohexyl radical formed is trapped by NO to produce nitrosocyclohexane.

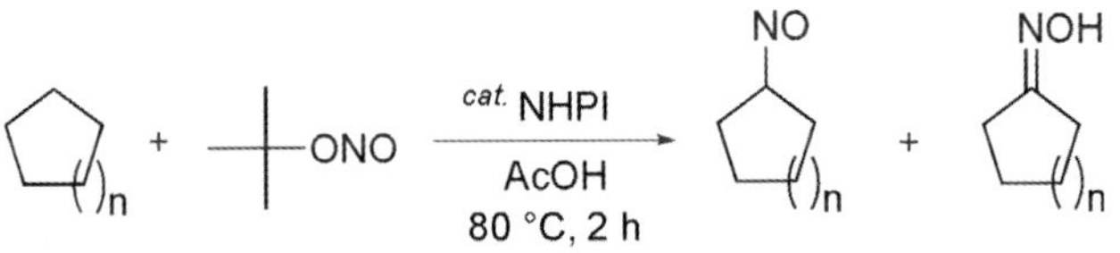

n	NO product	NOH product
n=1	49%	2%
n=2	63%	0%
n=3	28%	8%
n=4	0%	56%
n=8	5%	28%

(6.31)

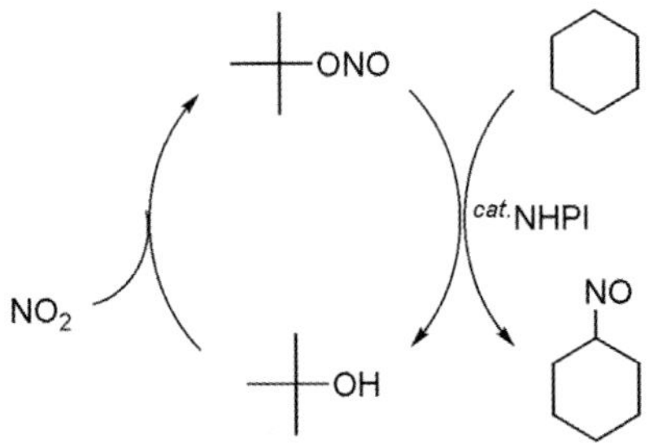

Scheme 6.15 A new route to ε-caprolactam precursor.

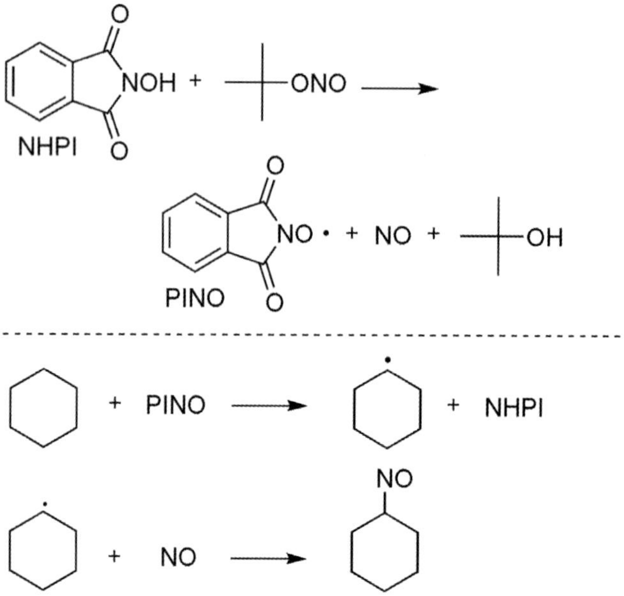

Scheme 6.16 A plausible reaction pathway for the reaction of cyclohexane with *t*-BuONO catalyzed by NHPI.

6.3.6
Ritter-type Reaction with Cerium Ammonium Nitrate (CAN)

Ritter-type reaction of alkane with nitrile forming an amide has been accomplished by the use of Br_2/H_2SO_4 [178], NO_2BF_4 [179], $AlCl_3/CH_2Cl_2$ [180], electrolysis [181], $Pb(OAc)_4$ [182] and HNO_3/CCl_4 [183], these methods, however, are limited to the reaction of adamantane or its derivatives. Hill et al. demonstrated that lower alkanes such as isobutane react with acetonitrile in the presence of a polyoxometalate under photo-assistance to form the corresponding acetoamide in high selectivity [184]. The reaction is postulated to proceed through the formation of an alkyl radical followed by one-electron oxidation by the W ion to a carbocation.

The Ritter-type reaction of adamantane is accomplished using the NHPI/NO system. In this section, we show that NHPI combined with cerium ammonium nitrate (CAN) serves as an efficient system for the generation of both PINO from NHPI and carbocations from alkyl radicals. Thus, benzylic compounds first undergo the amidation with alkyl nitrile under mild conditions to form amides in good yields. The reaction of ethylbenzene in the presence of CAN and NHPI in EtCN under argon at 80°C for 6 h produced N-(1-phenylethyl)propionamide in 84% yield at 61% conversion (Eq. (6.32)). The NHPI/CAN system can apply to the Ritter-type reaction of various alkylbenzenes and adamanatanes.

6.3 Functionalization of Alkanes Catalyzed by NHPI

Ph-CH2CH3 + CAN —[NHPI (10 mol%), EtCN, 80 °C, under Ar]→ Ph-CH(NHC(O)Et)-CH3

Conv. 61%   84%

(6.32)

HN-C(O)-Et on Ph-CR:
R = Et 69 (69)
R = iPr 93 (63)

Indanyl-NHC(O)Et: 74 (93)

Adamantyl-N(H)C(O)R:
R = Et   93 (57)
R = nPr 91 (82)
R = Ph   83 (72)

p-tolyl-C(Et)(NHC(O)Et): 56 (93)

Adamantyl (bridgehead)-NHC(O)Et: 80 (51)

tBu-NHC(O)Et: 28 (—)

Select. / % (Conv. / %)

It is reasonable to assume that the present reaction is initiated by the reaction of NHPI with CAN to form PINO, which is thought to be a key species for the generation of alkyl radicals (Scheme 6.17). Indeed, PINO is generated upon treatment of NHPI with CAN in MeCN at 70 °C. The resulting PINO abstracts a hydrogen atom from these hydrocarbons to generate the corresponding alkyl radicals (**A**), which undergo the one-electron oxidation by Ce(IV) to form carbocations (**B**). The carbocations **B** thus generated are trapped by nitriles, and this is followed by reaction with H$_2$O to afford amide derivatives [185].

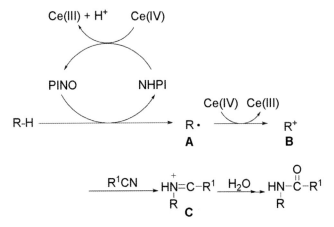

Scheme 6.17 A possible reaction path for the Ritter-type reaction by NHPI/CAN system.

6.4
Carbon-Carbon Bond-Forming Reaction via Catalytic Carbon Radicals Generation Assisted by NHPI

Additions of carbon radicals to alkenes, which can lead to the formation of new carbon-carbon bonds, are of major synthetic interest in organic chemistry because of the many advantages of the reactions over ionic processes [186]. Nowadays, numerous methods for the generation of carbon radicals and their inter- or intramolecular additions to alkenes for the synthesis of fine chemicals and natural products have been developed [186, 187]. For instance, reactions of alkyl halides with tributyltin hydride or tris(trimethylsilyl)silane [188] and the thermal decomposition of Barton esters [189] are the most common methodologies for the generation of alkyl radicals. Although the peroxide- and photo-initiated reactions are often used as practical synthetic means, major problems of these methods are the lack of selectivity, generality, and efficiency of the reaction [186]. Therefore, the carbon-carbon bond-forming reaction through the carbon radical generation from alkanes is a worthwhile target in free-radical chemistry.

6.4.1
Oxyalkylation of Alkenes with Alkanes and Dioxygen

The NHPI-catalyzed aerobic oxidation of alkanes proceeds through the formation of alkyl radicals, as mentioned previously. If alkyl radicals generated from alkanes could add to alkenes smoothly, this would provide a powerful strategy for the construction of a C–C bond in which alkanes can be directly used as alkyl sources. Furthermore, since the generation of PINO from NHPI is performed under a dioxygen atmosphere, the concomitant introduction of both an alkyl group and an oxygen function to alkenes is possible. This new reaction type may be regarded as a catalytic oxyalkylation of alkenes. An approach to oxyalkylation is illustrated in Figure 6.8. The reaction involves an alkyl radical generation by the NHPI/Co/O_2 system and the

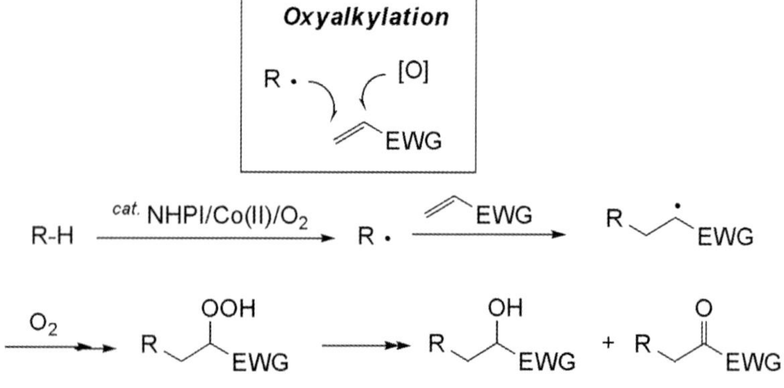

Figure 6.8 Oxyalkylation of alkenes with alkanes and O_2 catalyzed by NHPI/Co(II) system.

6.4 Carbon-Carbon Bond-Forming Reaction via Catalytic Carbon Radicals Generation Assisted by NHPI

addition of the resulting alkyl radical to an alkene to form an adduct radical which is readily trapped by O_2. The reaction of methyl acrylate with 1,3-dimethyladamantane under a mixed gas of O_2 (0.5 atm) and N_2 (0.5 atm) catalyzed by NHPI (20 mol%) in the presence of Co(acac)$_3$ (1 mol%) in acetonitrile at 75 °C for 16 h gave about a 7 : 3 mixture of oxyalkylated products, methyl 3-(3,3'-dimethyladamantyl)-2-hydroxypropionate, and methyl 3-(3,3'-dimethyladamantyl)-2-oxopropionate, in 91% yield (Eq. (6.33)) [190]. This is the first simultaneous introduction of alkyl and oxygen functions to alkenes through a catalytic process [191]. Fukunishi reported the peroxide-initiated simple radical addition of 1,3-dimethyladamantane to maleate and fumaronitrile affording the adduct [192]. Additionally, the reaction of cyclohexane with methyl acrylate under dioxygen (1 atm) gives rise to the corresponding three-component coupling product in 75% selectivity at 80% conversion.

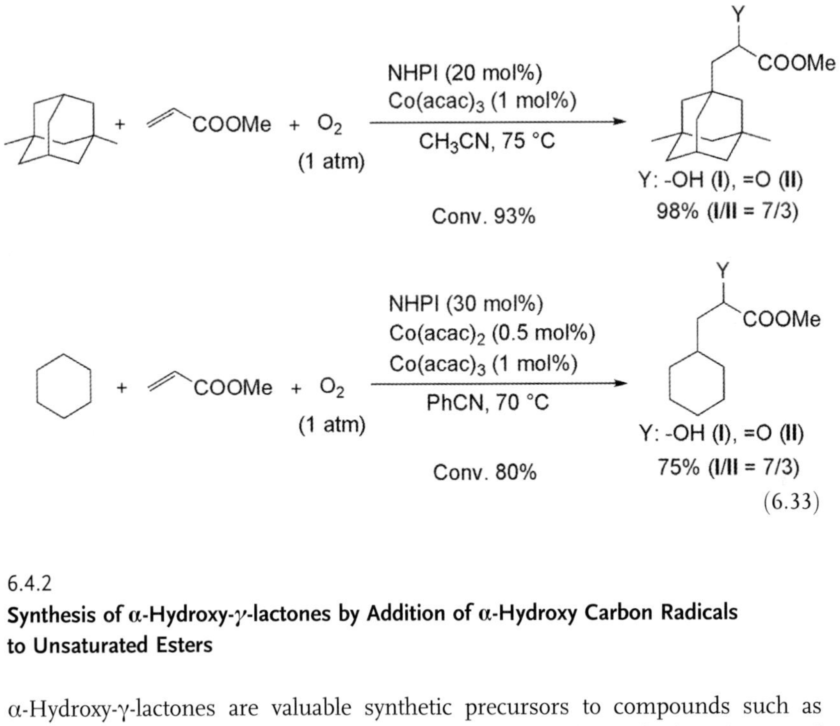

$$\text{1,3-dimethyladamantane} + \text{CH}_2\text{=CHCOOMe} + O_2 \quad (1 \text{ atm}) \xrightarrow[\text{CH}_3\text{CN, 75 °C}]{\text{NHPI (20 mol\%)} \atop \text{Co(acac)}_3 \text{ (1 mol\%)}} \text{product}$$

Conv. 93%
Y: -OH (I), =O (II)
98% (I/II = 7/3)

$$\text{cyclohexane} + \text{CH}_2\text{=CHCOOMe} + O_2 \quad (1 \text{ atm}) \xrightarrow[\text{PhCN, 70 °C}]{\text{NHPI (30 mol\%)} \atop {\text{Co(acac)}_2 \text{ (0.5 mol\%)} \atop \text{Co(acac)}_3 \text{ (1 mol\%)}}} \text{product}$$

Conv. 80%
Y: -OH (I), =O (II)
75% (I/II = 7/3)

(6.33)

6.4.2
Synthesis of α-Hydroxy-γ-lactones by Addition of α-Hydroxy Carbon Radicals to Unsaturated Esters

α-Hydroxy-γ-lactones are valuable synthetic precursors to compounds such as α,β-butenolides having potent biological activities [193], efficient food intake-control substances [194], and monomers of biodegradable polymers as well as fine chemicals [195]. However, there are few practical methods for the synthesis of these lactones [196]. The strategy for the oxyalkylation of alkenes with alkanes and O_2 was extended to the reaction of alkenes with alcohols and O_2 leading to α-hydroxy-γ-lactones [197], The reaction of 2-propanol with methyl acrylate under a dioxygen atmosphere was examined in the presence of NHPI (10 mol%) combined with Co(OAc)$_2$ (0.1 mol%) and Co(acac)$_3$ (1 mol%), and α-hydroxy-γ,γ-dimethyl-γ-butyrolactone was obtained in 78% yield (Figure 6.9). Mori et al. prepared the same lactone via three steps from isobutene and trichloroacetaldehyde in 14% yield [196a].

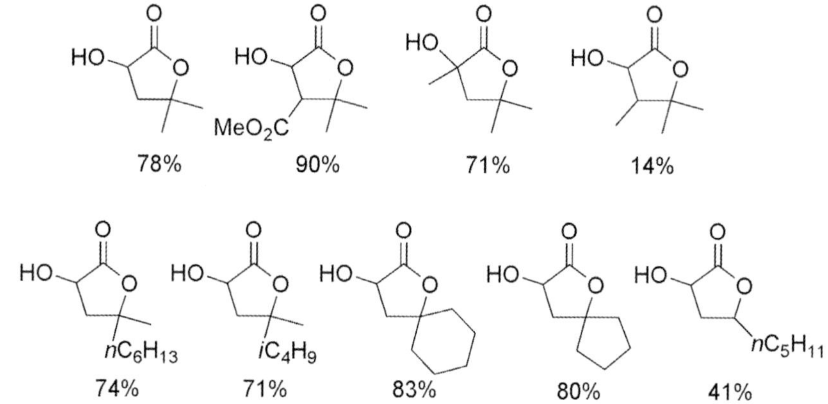

Figure 6.9 Synthesis of α-hydroxy-γ-lactones.

This new method for the construction of α-hydroxy-γ-lactones is quite general for a variety of alcohols and α,β-unsaturated esters (Figure 6.9). The preparation of α-hydroxy-γ-spirolactones from cyclic alcohols is especially notable, because these spirolactones have been very difficult to synthesize until now. The reaction can be explained by Scheme 6.18: (i) *in situ* generation of an α-hydroxy carbon radical from an alcohol assisted by NHPI/Co(II)/O_2, (ii) the addition of the radical to methyl acrylate, (iii) trapping of the adduct radical by O_2, and (iv) intramolecular cyclization to give α-hydroxy-γ-butyrolactone. Considering the low-cost material, reaction efficiency, and reaction simplicity, this method provides an innovative approach to α-hydroxy-γ-lactones which has considerable industrial potential.

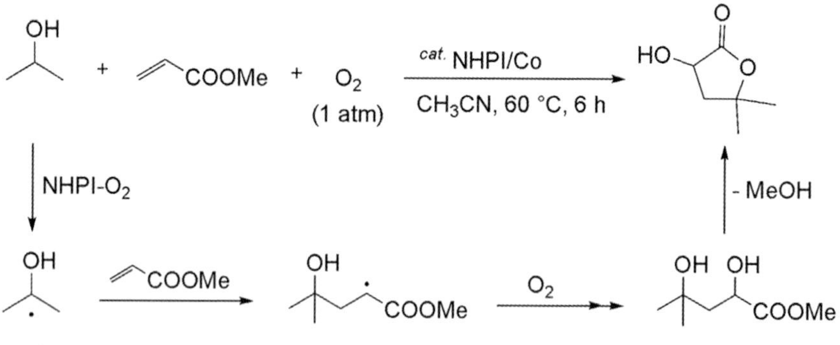

Scheme 6.18 Radical addition of 2-propanol to methyl acrylate under O_2 catalyzed by NHPI/Co(acac)$_2$ system.

6.4.3
Hydroxyacylation of Alkenes using 1,3-Dioxolanes and Dioxygen

Addition of aldehydes through cleavage of the aldehydic carbon-hydrogen bond to terminal alkenes is known as hydroacylation (Figure 6.10) [198]. If the concomitant

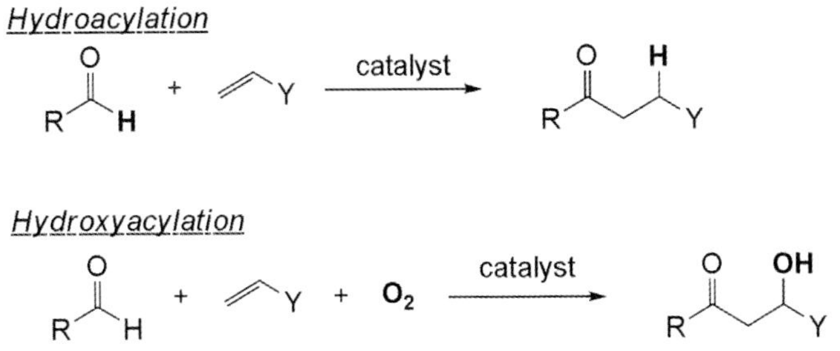

6.4 Carbon-Carbon Bond-Forming Reaction via Catalytic Carbon Radicals Generation Assisted by NHPI

Hydroacylation

$$R-C(O)H + CH_2=CHY \xrightarrow{catalyst} R-C(O)-CH_2-CH(H)-Y$$

Hydroxyacylation

$$R-C(O)H + CH_2=CHY + O_2 \xrightarrow{catalyst} R-C(O)-CH_2-CH(OH)-Y$$

Figure 6.10 Hydroxyacylation of alkenes.

introduction of acyl and hydroxy moieties to alkenes, which is referred to as hydroxyacylation, can be achieved by a cascade reaction, this would provide a novel route to β-hydroxy carbonyl compounds (Figure 6.10). β-Oxycarbonyl arrays constitute important structural subunits in a variety of natural and unnatural materials and in key intermediates leading to pharmaceuticals [199]. There is one report on the hydroxyacylation of acrylates with acyl radicals derived from aldehydes using dioxygen as a hydroxy source assisted by a cobalt(II) Schiff-base complex, but the attempt is not wholly successful due to the decarbonylation from the acyl radicals as well as the reaction of the acyl radicals with O_2 leading to carboxylic acids which cause undesired side reactions [200]. To overcome these drawbacks, 1,3-dioxolanes, masked aldehydes, were used as the acyl source [201].

Since α,α-dioxaalkyl radicals corresponding to acyl radical equivalents are expected to be generated from 1,3-dioxolanes by the use of the NHPI/O_2 system, the apparent hydroxyacylation of methyl acrylate using several 1,3-dioxolanes and O_2 was examined [202, 203]. Reaction of 2-methyl-1,3-dioxolane with methyl acrylate under O_2 (1 atm) in the presence of NHPI (5 mol%) and a small amount of Co(OAc)$_2$ (0.05 mol%) at room temperature for 3 h produced β-hydroxy ketal in 81% yield (Eq. (6.34)). The dioxolane moiety can be easily deprotected under acidic conditions in quantitative yield (Eq. (6.35)). This is a useful method for the introduction of formyl and hydroxy groups to alkenes. Although the direct use of formaldehyde in organic synthesis is restricted owing to its intractability, the reaction of 1,3-dioxolane, masked formaldehyde, with methyl acrylate followed by deprotection of the dioxolane moiety, gave an adduct in good yield. α-Hydroxy esters are reported to be valuable precursors for the synthesis of attractive compounds possessing a variety of pharmacological properties, for example, pyridazinones and the alkaloid epibatidine (Eq. (6.36)) [204].

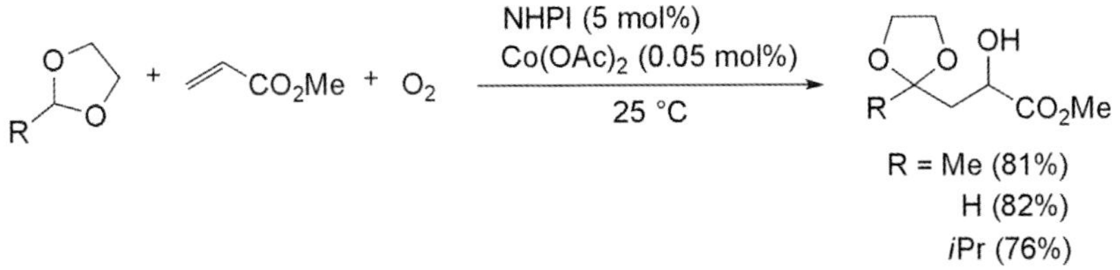

$$\text{R-dioxolane} + \text{CO}_2\text{Me} + O_2 \xrightarrow[25\,°C]{\text{NHPI (5 mol\%)}, \text{Co(OAc)}_2 \text{ (0.05 mol\%)}} \text{product}$$

R = Me (81%)
H (82%)
iPr (76%)

(6.34)

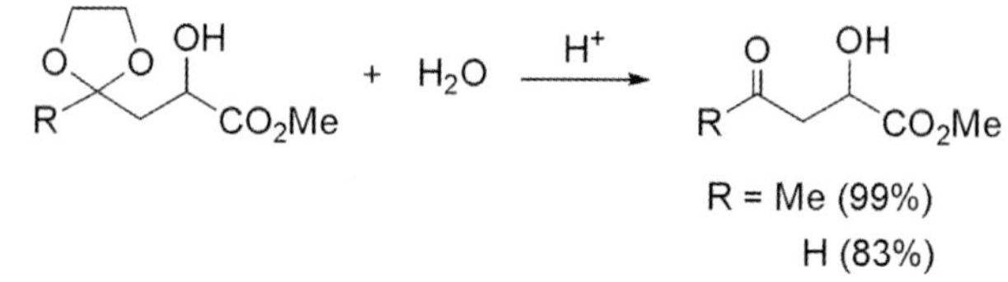

$$\underset{R}{\overset{O}{\bigcirc}}\underset{CO_2Me}{\overset{OH}{\bigcirc}} + H_2O \xrightarrow{H^+} \underset{R}{\overset{O}{\bigcirc}}\underset{CO_2Me}{\overset{OH}{\bigcirc}} \quad (6.35)$$

R = Me (99%)
H (83%)

$$\underset{R}{\overset{O}{\bigcirc}}\underset{CO_2Me}{\overset{OH}{\bigcirc}} + NH_2-NH_2 \xrightarrow{EtOH,\ r.t.} \text{pyridazinone} \quad (6.36)$$

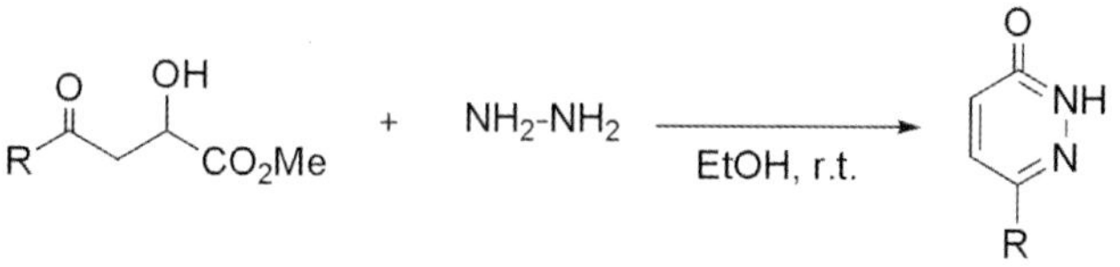

6.4.4
Hydroacylation of Alkenes Using NHPI as a Polarity Reversal Catalyst

Intermolecular radical-chain addition of aldehydes to alkenes is the simplest methodology for the synthesis of long-chain unsymmetrical ketones, but employment of this method is usually difficult for the synthesis of simple aliphatic ketones [205]. The hydroacylation between alkenes and aldehydes via a radical process involves the following reaction steps: (i) hydrogen abstraction from an aldehyde by a radical initiator to form an acyl radical (**A**), (ii) addition of the acyl radical to alkene leading to a β-oxocarbon radical (**B**), and (iii) abstraction of the aldehydic hydrogen atom from another aldehyde by **B**, generating ketone and acyl radical **A** (Scheme 6.19).

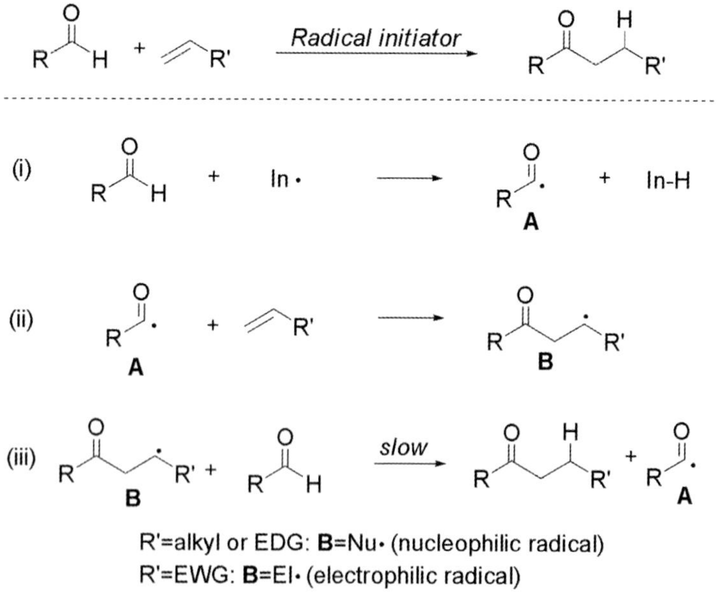

R'=alkyl or EDG: **B**=Nu• (nucleophilic radical)
R'=EWG: **B**=El• (electrophilic radical)

Scheme 6.19 Hydroacylation of alkenes via radical process.

6.4 Carbon-Carbon Bond-Forming Reaction via Catalytic Carbon Radicals Generation Assisted by NHPI

If the R' is an alkyl or an electron-donating group (EDG) in this radical-chain reaction, step (iii) becomes a sluggish process, since the abstraction of the aldehydic hydrogen atom by nucleophilic radical **B** (Nu•) proceeds with difficulty. In contrast, if the R' is an electron-withdrawing group (EWG), this step proceeds smoothly because of the ease of aldehydic hydrogen abstraction by an electrophilic radical **B** (El•) [206]. Acyl radicals, which are nucleophilic in nature, are known to add more easily to electron-deficient alkenes than normal alkenes [207]. As a result, the hydroacylation of simple alkenes with aldehydes via a radical process proceeds with difficulty. Recently, the hydroacylation of alkenes, particularly electron-rich alkenes, with aldehydes was reported to be achieved by the use of methyl thioglycolate ($HSCH_2CO_2Me$), which acts as a polarity-reversal catalyst. For instance, the addition of butanal to isopropenyl acetate using di-*tert*-butyl hyponitrite (TBHN) as an initiator and methyl thioglycolate ($HSCH_2CO_2Me$) as the polarity-reversal catalyst at 60 °C produces 1-acetoxyhexan-3-one in good yield [208, 209].

Figure 6.11 shows the hydroacylation of alkenes with aldehydes using NHPI as a polarity-reversal catalyst. The hydrogen atom of the hydroxylimide moiety adjacent to two carbonyl groups may be easily abstracted by a nucleophilic radical rather than an electrophilic one, and the resulting PINO would behave as an electrophilic radical that can efficiently abstract the aldehydic hydrogen atom. The radical-chain hydroacylation of simple alkenes with aldehydes assisted by NHPI is carried out as follows. Stirring a solution in dry toluene containing pentanal, oct-1-ene, NHPI, and dibenzoyl peroxide (BPO) at 80 °C under argon for 12 h, followed by addition of BPO in toluene, gives 5-tridecanone in 88% selectivity at 72% conversion (Figure 6.11). Several alkenes also react with pentanal under the influence of NHPI under an Ar atmosphere to form the corresponding ketones in moderate to good yields [210].

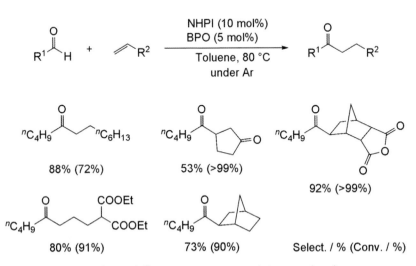

Figure 6.11 Hydroacylation of alkenes using NHPI as a polarity-reversal catalyst.

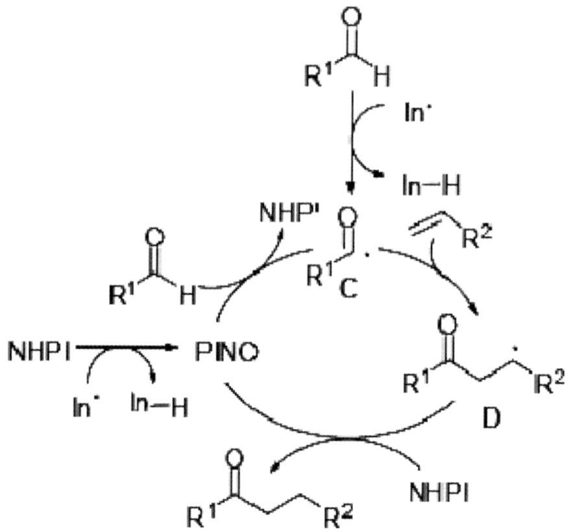

228 | 6 Aerobic Oxidations and Related Reactions Catalyzed by N-Hydroxyphthalimide

Scheme 6.20 A possible reaction path for hydroacylation using NHPI as polarity-reversal catalyst.

A possible reaction path for the NHPI-catalyzed hydroacylation of alkenes with aldehydes is shown in Scheme 6.20. The reaction may be initiated by a hydrogen atom abstraction from the aldehyde by the radical initiator (In•), giving an acyl radical (**C**) which then adds to an alkene to afford a β-oxocarbon radical (**D**). The resulting radical (**D**), having a nucleophilic character, abstracts the hydrogen atom from NHPI leading to ketone and PINO. The abstraction of the hydrogen atom from aldehyde by the PINO forms the acyl radical **C** and NHPI. An alternative formation of PINO from NHPI and radical initiator (In•) may also be possible.

6.4.5
Chiral NHPI Derivatives as Enantioselective Catalysts: Kinetic Resolution of Oxazolidines

A suitably designed chiral analog of NHPI can be used as an enantioselective oxidation catalyst. Einhorn reported the synthesis of axially chiral NHPI analogs (Scheme 6.21). These NHPI derivatives can be used in several catalytic asymmetric oxidation reactions, such as indane oxidation and oxidative deprotection of acetal,

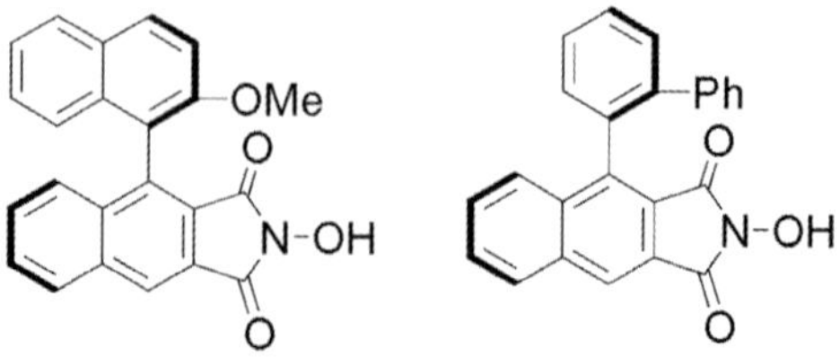

Scheme 6.21 Chiral NHPI derivatives.

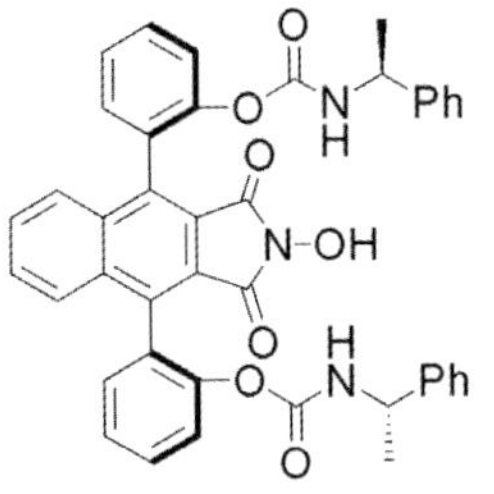

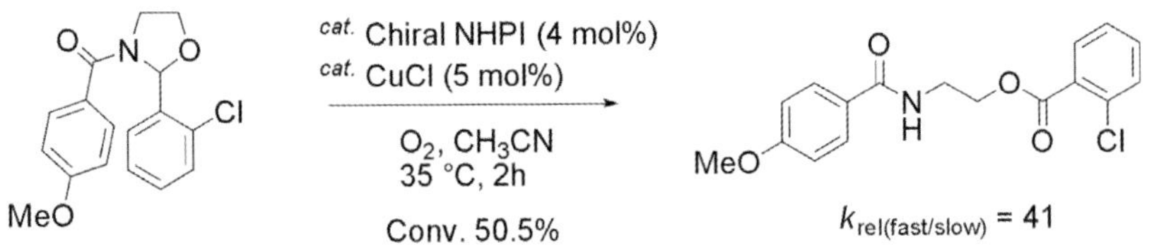

Scheme 6.22 Kinetic resolution of oxazolidines.

with moderate enantioselectivity (<24%ee) [211]. Highly enantioselective oxidative ring opening of oxazolidines by the chiral C_2-symmetrical NHPI was accomplished by efficient kinetic resolution of the oxazolizine (Scheme 6.22) [212]. Thus, 50.5% conversion was attained within 2 h, and the remaining oxazolidine had an ee value of 89% with $s = k_{rel(fast/slow)}$ of 41. This NHPI catalyst is useful for the synthesis of highly enantiomerically enriched oxazolidines.

6.5
Conclusions

The achievement of highly efficient and selective transformations of hydrocarbons to useful chemical substances is an ambitious goal in synthetic chemistry. It is interesting to open up a new vista in organic synthesis and to confirm the catalytic method for the carbon radical generation from a C-H bond of a wide variety of compounds by the use of N-hydroxyphthalimide (NHPI), which serves as a carbon radical-producing catalyst (CRPC). By employing NHPI as the catalyst, a novel aerobic oxidation of alkanes, which surpasses the conventional autoxidations in conversion and selectivity, has been achieved under mild conditions. This oxidation method provides entry to a diverse array of significant oxygen-containing compounds. In particular, a success in the direct conversion of cyclohexane to adipic acid with dioxygen in high conversion and selectivity has greatly benefited the chemical industry as an environmentally benign process, because the current production of adipic acid via nitric acid oxidation causes the evolution of nitrogen oxides that are serious air-polluting materials. In addition, the finding that the NHPI catalyzes the aerobic oxidation of alkylbenzenes even at room temperature is notable. The epoxidation of alkenes by in situ-generated hydroperoxides or hydrogen peroxide has been explored for the first time. This new methodology is applicable to the functionalization of alkanes to afford nitroalkanes, alkanesulfonic acids, and carboxylic acids by allowing them to react with NO_2, SO_2/O_2 and CO/O_2, respectively. Finally, a new type of reaction for the concomitant introduction of alkyl group and oxygen functions to alkenes, which is referred to as catalytic oxyalkylation of alkenes, has been established. α-Hydroxy-γ-lactones, which are very difficult to synthesize by conventional methods, are easily prepared by the reaction of alcohols, alkenes, and O_2 under the influence of the NHPI catalyst. The NHPI can be used as a polarity-reversal catalyst for hydroacylation of alkenes with aldehydes.

References

1 Satterfield, C.N. (1991) *Heterogeneous Catalysis in Industrial Practice*, McGraw-Hill, New York.
2 (a) Parshall, G.W. and Ittel, S.D. (1992) *Homogeneous Catalysis*, 2nd ed., Wiley, New York; (b) Parshall, G.W. (1978) *J. Mol. Catal.*, **4**, 243.
3 Emanuel, N.M., Denisov, E.T., and Marizus, Z.K. (1967) *Liquid-Phase Oxidation of Hydrocarbons*, Plenum Press, New York, pp. 309–346.
4 Wittcoff, H.A. and Reuben, B.G. (1996) *Industrial Organic Chemicals*, Wiley, New York.
5 (a) Sheldon, R.A. and Kochi, J.K. (1981) *Metal-Catalyzed Oxidation of Organic Compounds*, Academic Press, New York; (b) Hill, C.L. (1989) *Activation and Functionalization of Alkanes*, Academic Press, New York; (c) Sheldon, R.A. (1991) *CHEMTECH*, **21**, 566; (d) Simándi, L.I. (1992) *Catalytic Activation of Dioxygen by Metal Complexes*, Kluwer Academic, Dordrecht; (e) Barton, D.H.R., Martell, A.E., and Sawyer, D.T. (1993) *The Activation of Dioxygen and Homogeneous Catalytic Oxidation*, Plenum, New York.
6 (a) Murahashi, S.-I. (1995) *Angew. Chem. Int. Ed*, **34**, 2443; (b) Shilov, A.E. and Shul'pin, G.B. (1997) *Chem. Rev.*, **97**, 2912; (c) Shilov, A.E. and Shul'pin, G.B. (2000) *Activation and Catalytic Reactions of Saturated Hydrocarbons in the Presence of Metal Complexes*, Kluwer Academic, The Netherlands.
7 (a) Ishii, Y., Sakaguchi, S., and Iwahama, T. (2001) *Adv. Synth. Catal*, **343**, 379; (b) Ishii, Y. (1997) *J. Mol. Cat. A*, **117**, 123;
(c) Ishii, Y. and Sakaguchi, S. (1999) *Cat. Surv. Jpn.*, **3**, 27; (d) Recupero, F. and Punta, C. (2007) *Chem. Rev.*, **107**, 3800.
8 (a) Grochowski, E., Boleslawska, T., and Jurczak, J. (1977) *Synthesis*, 718; (b) Masui, M., Ueshima, T., and Ozaki, S. (1983) *Chem. Commun.*, 479; (c) Ozaki, S., Hamaguchi, T., Tsuchida, K., Kimata, Y., and Masui, M. (1989) *J. Chem. Soc., Perkin Trans. 2*, 951; (d) Foricher, J., Fuerbringer, C., and Pfoerther, K. (1986) EP198351.
9 Ishii, Y., Nakayama, K., Takeno, M., Sakaguchi, S., Iwahama, T., and Nishiyama, Y. (1995) *J. Org. Chem.*, **60**, 3934.
10 (a) Sheldon, R.A. (1994) *CHEMTECH* **38**, 24; (b) Arends, I.W.C.E., Sheldon, R.A., Wallau, M., and Schuchardt, U. (1997) *Angew. Chem. Int. Ed.*, **36**, 1144; (c) Mennier, B. (1992) *Chem. Rev.*, **92**, 1411; (d) Busch, D.H. and Alcock, N.W. (1994) *Chem. Rev.*, **94**, 585; (e) Simándi, L.I. (ed.) (1991) *Dioxygen Activation and Homogeneous Catalytic Oxidation*, Elsevier, Amsterdam, and references cited therein; (f) Centi, G. and Trifiro, F. (eds) (1990) *New Developments in Selective Oxidation*, Elsevier, Amsterdam.
11 (a) Tabushi, I. and Yazaki, A. (1979) *J. Am. Chem. Soc.*, **101**, 6456; (b) Tabushi, I. and Yazaki, A. (1981) *J. Am. Chem. Soc.*, **103**, 7371; (c) Tabushi, I. (1988) *Coord. Chem. Rev.*, **86**, 1.
12 (a) Barton, D.H.R., Boivin, J., Gastiger, M., Morzycki, K., Hay-Motherwell, R.S., Motherwell, W.B., Ozbalik, N., and Schwartzentruber, K.M. (1986) *J. Chem.*

Soc., Perkin Trans. 1, 947; (b) Barton, D.H.R. and Doller, D. (1992) *Acc. Chem. Res.*, **25**, 504; (c) Barton, D.H.R., Li, T., and Mackinnon, J. (1997) *Chem. Commun.*, 557, and references cited therein; (d) Groves, J.T. and Neumann, R. (1988) *J. Org. Chem.*, **53**, 3891; (e) Battioni, P., Bartoli, J.F., Leduc, P., Fontecave, M., and Mansuy, D. (1987) *Chem. Commun.*, 791; (f) Karasevich, E.I., Khenkin, A.M., and Shilov, A.E. (1987) *Chem. Commun.*, 731; (g) Kitajima, N., Ito, M., Fukui, H., and Moro-oka, Y. (1991) *Chem. Commun.*, 102; (h) Minisci, F. and Fontana, F. (1994) *Tetrahedron Lett.*, **35**, 1427. (i) Kurusu, Y. and Neckers, D.C. (1991) *J. Org. Chem.*, **56**, 1981.

13 (a) Murahashi, S.-I., Oda, Y., and Naota, T. (1992) *J. Am. Chem. Soc.*, **114**, 7913; (b) Komiya, N., Naota, T., and Murahashi, S.-I. (1996) *Tetrahedron Lett.*, **37**, 1633; (c) Komiya, N., Naota, T., Oda, Y., and Murahashi, S.-I. (1997) *J. Mol. Catal. A: Chemical*, **117**, 21; (d) Punniyamurthy, T., Karla, S.J.S., and Iqbal, J. (1995) *Tetrahedron Lett.*, **36**, 8497; (e) Battioni, P., Iwanejko, R., Mansuy, D., Mlodnicka, T., Poltowicz, J., and Sanches, F. (1996) *J. Mol. Catal. A: Chemical*, **109**, 91; (f) Bravo, A., Fontana, F., Minisci, F., and Serri, A. (1996) *Chem. Commun.*, 1843; (g) Dell'Anna, M.M., Mastrorilli, P., and Nobile, C.F. (1998) *J. Mol. Catal. A: Chemical*, **130**, 65.

14 (a) Lin, M. and Sen, A. (1992) *J. Am. Chem. Soc.*, **114**, 7307; (b) Fontecave, M. and Mansuy, D. (1984) *Tetrahedron*, **40**, 4297; (c) Lu, W.Y., Bartoli, J.F., Battioni, P., and Mansuy, D. (1992) *New J. Chem.*, **16**, 621; (d) Bartoli, J.F., Battioni, P., De Foor, W.R., and Mansuy, D. (1994) *Chem. Commun.*, 23; (e) Goldstein, A.S., Beer, R.H., and Drago, R.S. (1994) *J. Am. Chem. Soc.*, **116**, 2424.

15 (a) Chambers, R.C. and Hill, C.L. (1989) *Inorg. Chem.*, **28**, 2509; (b) Hill, C.L. and Prosser-McCartha, C.M. (1995) *Coord. Chem. Rev.*, **143**, 407.

16 Attanasio, D. and Suber, L. (1989) *Inorg. Chem.*, **28**, 3779.

17 (a) Shul'pin, G.B., Kats, M.M., and Lederer, P. (1989) *J. Gen. Chem. USSR*, **59**, 2450; (b) Maldotti, A., Bartocci, C., Amadelli, R., Polo, E., Battioni, P., and Mansuy, D. (1991) *Chem. Commun.*, 1487; (c) Shul'pin, G.B., Nizova, G.V., and Kozlov, Yu.N. (1996) *New. J. Chem.*, **20**, 1243, and references cited therein.

18 (a) Shul'pin, G.B., Guerreiro, M.C., and Schuchardt, U. (1996) *Tetrahedron*, **52**, 13051; (b) Nizova, G.V., Suss-Fink, G., and Shul'pin, G.B. (1997) *Chem. Commun.*, 397; (c) Nizova, G.V., Suss-Fink, G., and Shul'pin, G.B. (1997) *Tetrahedron*, **53**, 3603; (c) Nizova, G.V., Suss-Fink, G., Stanislas, S., and Shul'pin, G.B. (1998) *Chem. Commun.*, 1885.

19 (a) Lyons, J.E. and Ellis, P.E. Jr. (1994) *Metalloporphyrins in Catalytic Oxidations* (ed. R.A. Sheldon), Dekker, New York, p. 291, and references cited therein; (b) Ellis, P.E. Jr. and Lyons, J.E. (1989) *Chem. Commun.*, 1188; (c) Ellis, P.E. Jr. and Lyons, J.E. (1989) *Chem. Commun.*, 1190; (d) Ellis, P.E. Jr. and Lyons, J.E. (1989) *Chem. Commun.*, 1316; (e) Ellis, P.E. Jr. and Lyons, J.E. (1990) *Coord. Chem. Rev.*, **105**, 181; (f) Chen, H.L., Ellis, P.E. Jr., Wijesekera, T., Hagan, T.E., Groh, S.E., Lyons, J.E., and Ridge, D.P. (1994) *J. Am. Chem. Soc.*, **116**, 1086; (g) Grinstaff, M.W., Hill, M.G., Labinger, J.A., and Gray, H.B. (1994) *Science*, **264**, 1311.

20 (a) Mizuno, N., Tateishi, M., Hirose, T., and Iwamoto, M. (1993) *Chem. Lett.*, 2137; (b) Mizuno, N., Ishige, H., Seki, Y., Misono, M., Suh, D.-J., Han, W., and Kudo, T. (1997) *Chem. Commun.*, 1295; (c) Nozaki, C., Misono, M., and Mizuno, N. (1998) *Chem. Lett.*, 1263; (d) Mizuno, N. and Misono, M. (1998) *Chem. Rev.*, **98**, 199.

21 (a) Khan, M.M.T. and Shukla, R.S. (1988) *J. Mol. Catal.*, **44**, 85; (b) Khan, M.M.T., Bajaj, H.C., Shukla, R.S., Mirza, S.A., and Shukla, R.S. (1988) *J. Mol. Catal.*, **44**, 51.

22 Neumann, R., Khenkin, A.M., and Dahan, M. (1995) *Angew. Chem. Int. Ed.*, **34**, 1587.

23 Goldstein, A.S. and Drago, R.S. (1991) *Inorg. Chem.*, **30**, 4506.

24 (a) Davis, D.D. (1985) *Ullmann's Encyclopedia of Industrial Chemistry*, 5th edn, vol. A1 (ed. W. Gerhartz), Wiley, New York, pp. 270–272; (b) Davis, D.D. and Kemp, D.R. (1990) *Kirk-Othmer*

Encyclopedia of Chemical Technology, 4th ed., vol. 1 (eds J.I. Kroschwitz and M. Howe-Grant), Wiley, New York, 471–480, and references cited therein.

25 (a) Steeman, J.W.M., Kaarsemaker, S., and Hoftyzer, P. (1961) *J. Chem. Eng. Sci.*, **14**, 139; (b) Miller, S.A. (1969) *Chem. Proc. Eng. London*, **50**, 63.

26 Tanaka, K. (1974) *CHEMTECH*, 555, and references cited therein.

27 Sato, K., Aoki, M., and Noyori, R. (1998) *Science*, **281**, 1646.

28 Iwahama, T., Syojyo, K., Sakaguchi, S., and Ishii, Y. (1998) *Org. Proc. Res. Dev.*, **2**, 255.

29 Ishii, Y., Iwahama, T., Sakaguchi, S., Nakayama, K., and Nishiyama, Y. (1996) *J. Org. Chem.*, **61**, 4520.

30 Sawatari, N., Yokota, T., Sakaguchi, S., and Ishii, Y. (2001) *J. Org. Chem.*, **66**, 7889.

31 Guha, S.K., Obora, Y., Ishihara, D., Matsubara, H., Ryu, I., and Ishii, Y. (2008) *Adv. Synth. Catal.*, **350**, 1323.

32 (a) Lin, M., Hogan, T.E., and Sen, A. (1996) *J. Am. Chem. Soc.*, **118**, 4574; (b) Lin, M., Hogan, T.E., and Sen, A. (1992) *J. Am. Chem. Soc.*, **114**, 7307; (c) Sen, A. (1998) *Acc. Chem. Res.*, **31**, 550.

33 (a) Taniguchi, Y., Horie, S., Takai, K., and Fujiwara, Y. (1995) *J. Organomet. Chem.*, **504**, 137; (b) Taniguchi, Y., Kitamura, T., Fujiwara, Y., Horie, S., and Takai, K. (1997) *Catal. Today*, **36**, 85; (c) Jia, C., Kitamura, T., and Fujiwara, Y. (2001) *Acc. Chem. Res.*, **34**, 633.

34 Süss-Fink, G., Gonzalez, L., and Shul'pin, G.B. (2001) *App. Catal. A: General*, **217**, 111.

35 Shibamoto, A., Sakaguchi, S., and Ishii, Y. (2002) *Tetrahedron Lett.*, **43**, 8859.

36 Kroschwitz, J.I. and Howe-Grant, M. (eds) (1990) *Kirk-Othmer Encyclopedia of Chemical Technology*, 4th edn, vol. 4, Wiley, New York, pp. 691–700; pp. 463–474; pp. 358–405, and references cited therein.

37 Lyons, J.E., Ellis, P.E. Jr., and Myers, H.K. (1995) *J. Catal.*, **155**, 59.

38 Sakaguchi, S., Kato, S., Iwahama, T., and Ishii, Y. (1998) *Bull. Chem. Soc. Jpn.*, **71**, 1240.

39 (a) Bacha, J.D. and Kochi, J.K. (1965) *J. Org. Chem.*, **30**, 3272; (b) Greene, F.D., Savitz, M.L., Osterholts, F.D., Fau, H.H., Smith, W.H., and Zanet, P.M. (1963) *J. Org. Chem.*, **28**, 55; (c) Kochi, J.K. (1962) *J. Am. Chem. Soc.*, **84**, 1193.

40 Ishii, Y., Kato, S., Iwahama, T., and Sakaguchi, S. (1996) *Tetrahedron Lett.*, **37**, 4993.

41 We have found that quaternary ammonium bromides accelerate the NHPI-catalyzed aerobic oxidation even in the absence of any metal catalyst: Matsunaka, K., Iwahama, T., Sakaguchi, S., and Ishii, Y. (1999) *Tetrahedron Lett.*, **40**, 2165.

42 Ohtake, H., Higuchi, T., and Hirobe, M. (1992) *J. Am. Chem. Soc.*, **114**, 10660.

43 Sheldon, R.A. (1991) *Dioxygen Activation and Homogeneous Catalytic Oxidation* (ed. L.I. Simándi), Elsevier, Amsterdam, pp. 573–594.

44 Reichle, W.T., Konrad, F.M., and Brooks, J.R. (1975) *Benzene and its Industrial Derivatives* (ed. E.G. Hancock), Benn, London.

45 Yoshino, Y., Hayashi, Y., Iwahama, T., Sakaguchi, S., and Ishii, Y. (1997) *J. Org. Chem.*, **62**, 6810.

46 (a) Huyser, E.S. (1970) *Free-Radical Chain Reactions*, Wiley, New York; (b) Parsons, A.F. (2000) *An Introduction to Free Radical Chemistry*, Blackwell Science, Oxford.

47 Wentzel, B.B., Donners, M.P.J., Alsters, P.L., Feiters, M.C., and Nolte, R.J.M. (2000) *Tetrahedron*, **56**, 7797.

48 Mackor, A., Wajer, T.A.J.W., and de Boer, T. (1968) *Tetrahedron*, **24**, 1623.

49 Minisci, F., Punta, C., Recupero, F., Fontana, F., and Pedulli, G.F. (2002) *Chem. Commun.*, 688.

50 (a) Sheehan, J.R. (1999) *Ullmann's Encyclopedia of Industrial Organic Chemicals*, vol. 8, VCH, Weinheim, pp. 4573–4591; (b) Park, C. and Sheehan, J.R. (1995) *Kirk-Othmer Encyclopedia of Chemical Technology*, 4th ed., vol. 18 (eds J.I. Kroschwitz and M. Howe-Grant), Wiley, New York, pp. 991–1043.

51 (a) Brill, W.F. (1960) *Ind. Eng. Chem.*, **52**, 837; (b) Raghavendrchar, P. and Ramachandran, S. (1992) *Ind. Eng. Chem. Res.*, **31**, 453; (c) Roby, K.A. and Kingsley, P.J. (1996) *CHEMTECH*, 39; (d) Cincotti,

A., Orru, R., and Cao, G. (1999) *Catal. Today*, **52**, 331.

52 Partenheimer, W. (1995) *Catal. Today*, **23**, 69, and references cited therein.

53 Tashiro, Y., Iwahama, T., Sakaguchi, S., and Ishii, Y. (2001) *Adv. Synth. Catal.*, **343**, 220.

54 (a) Ravens, D.A.S. (1959) *Trans. Faraday. Soc.*, **55**, 1768; (b) Kamiya, Y., Nakajima, T., and Sakoda, K. (1966) *Bull. Chem. Soc. Jpn.*, **39**, 519.

55 (a) Chavan, S.A., Halligudi, S.B., and Ratnasamy, D.S.P. (2000) *J. Mol. Catal. A: Chemical*, **161**, 49, and references cited therein; (b) Partenheimer, W. and Grushin, V.V. (2001) *Adv. Synth. Catal.*, **343**, 102, and references cited therein.

56 (a) Hoelderich, W.F. (2000) *Appl. Catal. A: General*, **194–195**, 487; (b) Roy, A.N., Guha, D.K., and Bhattacharyya, D. (1982) *Indian Chem. Eng.*, **24**, 46, and references cited therein.

57 (a) Davis, D.D. (1985) *Ullmann's Encyclopedia of Industrial Chemistry*, 5th edn, vol. A27 (ed. W. Gerhartz), VCH, Weinheim, p. 584; (b) Paraskewas, P. (1974) *Synthesis*, 819; (c) Mukhopadhyay, S., and Chandalia, S.B. (1999) *Org. Proc. Res. Dev.*, **3**, 455.

58 Mukhopadhyay, S. and Chandalia, S.B. (1999) *Org. Proc. Res. Dev.*, **3**, 227.

59 Shibamoto, A., Sakaguchi, S., and Ishii, Y. (2000) *Org. Proc. Res. Dev.*, **4**, 505.

60 (a) Michael, J.P. (1997) *Nat. Prod. Rep.*, **14**, 605; (b) Jones, G. (1996) *Comprehensive Heterocyclic Chemistry II*, vol. 5 (eds A.R. Katritzky, C.W. Rees, and E.F.V. Scriven), Pergamon Press, Oxford, pp. 167–243.

61 (a) Kim, J.N., Lee, H.J., Lee, K.Y., and Kim, H.S. (2001) *Tetrahedron Lett.*, **42**, 3737; (b) Kobayashi, K., Nakahashi, R., Shimizu, A., Kitamura, T., Morikawa, O., and Konishi, H. (1999) *J. Chem. Soc., Perkin Trans. 1*, 1547, and references cited therein.

62 Cain, C.K., Plampin, J.N., and Sam, J. (1955) *J. Org. Chem.*, **20**, 466.

63 Ladner, D.W. (1986) *Synth. Commun.*, **16**, 157.

64 Paraskewas, S. (1974) *Synthesis*, 819.

65 Sakaguchi, S., Shibamoto, A., and Ishii, Y. (2002) *Chem. Commun.*, 180.

66 Ishii, Y., Iwahama, T., Sakaguchi, S., Nakayama, K., and Nishiyama, Y. (1996) *J. Org. Chem.*, **61**, 4520.

67 Einhorn, C., Einhorn, J., Marcadal, C., and Pierre, J.-L. (1997) *Chem. Commun.*, 447.

68 Einhorn, C., Einhorn, J., Marcadal, C., and Pierre, J.-L. (1999) *J. Org. Chem.*, **64**, 4542.

69 Landau, R., Sullivan, G.A., and Brown, D. (1979) *CHEMTECH*, 602.

70 (a) Hock, H. and Lang, S. (1944) *Ber. Dtsch. Chem. Ges.*, **B77**, 257; (b) Jordan, W., van Barneveld, H., Gerlich, O., Boymann, M.K., and Ullrich, J. (1985) *Ullmann's Encyclopedia of Industrial Organic Chemicals*, vol. A9, Wiley-VCH, Weinheim, pp. 299–312.

71 Fukuda, O., Sakaguchi, S., and Ishii, Y. (2001) *Adv. Synth. Catal.*, **343**, 809.

72 Arends, I.W.C.E., Sasidharan, M., Kühnle, A., Duda, M., Jost, C., and Sheldon, R.A. (2002) *Tetrahedron*, **58**, 9055.

73 Aoki, Y., Sakaguchi, S., and Isshii, Y. (2005) *Tetrahedron*, **61**, 5219.

74 Aoki, Y., Sakaguchi, S., and Isshii, Y. (2004) *Adv. Synth. Catal.*, **346**, 199.

75 Nakamura, R., Obora, Y., and Ishii, Y. (2008) *Chem. Commun.*, 3417.

76 Utimoto, K., Miwa, M., and Nozaki, H. (1981) *Tetrahedron. Lett.*, **22**, 4277.

77 Smith, A.B. III, Levenberg, P.A., and Suits, J.Z. (1986) *Synthesis*, 184.

78 Karpf, M., Huguet, J., and Dreiding, A.S. (1986) *Helv. Chim. Acta*, **65**, 13.

79 (a) Tam, S.T.-K., Klein, R.S., de las Heras, F.G., and Fox, J.J. (1979) *J. Org. Chem.*, **44**, 4854, and references cited therein; (b) Gupta, C.M., Jones, G.H., and Moffatt, J.G. (1976) *J. Org. Chem.*, **41**, 3000.

80 (a) Sayo, N., Azuma, K.-I., Mikawa, K., and Nakai, T. (1984) *Tetrahedron Lett.*, **25**, 565; (b) Midland, K. and Nguyen, N.H. (1981) *J. Org. Chem.*, **46**, 4107.

81 (a) Brown, H.C., Racherla, U.S., and Singh, S.M. (1984) *Tetrahedron Lett.*, **25**, 2411; (b) Normant, J.F. and Bourgan, M. (1970) *Tetrahedron Lett.*, **11**, 2659; (c) Davies, A.G. and Puddephatt, R.J. (1967) *Tetrahedron Lett.*, **8**, 2265.

82 Shaw, J.E. and Sherry, J.J. (1971) *Tetrahedron Lett.*, **12**, 4379.

83 Muzart, J. and Piva, O. (1988) *Tetrahedron Lett.*, **29**, 2321.

84 Umbreit, M.A. and Sharpless, K.B. (1977) *J. Am. Chem. Soc.*, **99**, 5527.

85 (a) Sheats, W.B., Olli, L.K., Stout, R., Lundzen, J.T., Justus, R., and Nigh, W.G. (1979) *J. Org. Chem.*, **44**, 4075; (b) McKillop, A., Oldenziel, O.H., Swann, B.P., Taylor, E.C., and Robey, R.L. (1973) *J. Am. Chem. Soc.*, **95**, 1296; (c) Schroder, M. and Griffith, W.P. (1978) *J. Chem. Soc., Dalton Trans.*, 1599; (d) Lee, D.G. and Chang, V.S. (1979) *J. Org. Chem.*, **44**, 2726; (e) Muller, P. and Godoy, A.J. (1981) *Helv. Chem. Acta*, **64**, 2531.

86 Golden, D.M. (1982) *Annu. Rev. Phys. Chem.*, **33**, 493.

87 Sakaguchi, S., Takase, T., Iwahama, T., and Ishii, Y. (1998) *Chem. Commun.*, 2037.

88 (a) Larock, R.C. (1989) *Comprehensive Organic Transformations*, VCH, New York, p. 604; (b) Ley, S. and Madin, A. (1991) *Comprehensive Organic Synthesis*, vol. 7 (eds B.M. Trost, I., Flemings and S.V. Ley), Pergamon, Oxford, p. 251.

89 (a) Simándi, L.L. (1992) *Catalytic Activation of Dioxygen by Metal Complexes*, Kluwer Academic, Dordrecht, pp. 297–317; (b) James, B.R. (1991) *Dioxygen Activation and Homogeneous Catalytic Oxidation* (ed. L.L. Simándi), Elsevier, Amsterdam, p. 195.

90 (a) Tang, R., Diamond, S.E., Neary, N., and Mares, F. (1978) *Chem. Commun.*, 562; (b) Matsumoto, M. and Watanabe, N. (1984) *J. Org. Chem.*, **49**, 3435; (c) Bilgrien, C., Davis, S., and Drago, R.S. (1987) *J. Am. Chem. Soc.*, **109**, 3786; (d) Murahashi, S.-I., Naota, T., and Hirai, J. (1993) *J. Org. Chem.*, **58**, 7318; (e) Kaneda, K., Yamashita, T., Matsushita, T., and Ebitani, K. (1998) *J. Org. Chem.*, **63**, 1750; (f) Blackburn, T.F. and Schwartz, J. (1977) *Chem. Commun.*, 158; (g) Kaneda, K., Fujie, Y., and Ebitani, K. (1997) *Tetrahedron Lett.*, **38**, 9023; (h) Mandal, A.K. and Iqbal, J. (1997) *Tetrahedron Lett.*, **53**, 7641; (i) Coleman, K.S., Coppe, M., Thomas, C., and Osborn, J.A. (1999) *Tetrahedron Lett.*, **40**, 3723; (j) Bäckvall, J.-E., Chowdhury, R.L., and Karlsson, U. (1991) *Chem. Commun.*, 473; (k) Wang, G.Z., Andreasson, U., and Bäckvall, J.-E. (1994) *Chem. Commun.*, 1037; (l) Inokuchi, T., Nakagawa, K., and Torii, S. (1995) *Tetrahedron Lett.*, **36**, 3223; (m) Dijksman, A., Marino-Gonzalez, A., Marita, A., Payeras, I., Arends, I.W.C.E., and Sheldon, R. (2001) *J. Am. Chem. Soc.*, **123**, 6826; (n) Csjernyik, G., Ell, A.H., Fadini, L., Pugin, B., and Bäckvall, J.-E. (2002) *J. Org. Chem.*, **67**, 1657.

91 (a) Ru: Markó, I.E., Giles, P.R., Tsukazaki, M., Chellé-Regnaut, M.I., Urch, C.J., and Brown, S.M. (1997) *J. Am. Chem. Soc.*, **119**, 12661; (b) Lenz, R. and Ley, S.V. (1997) *J. Chem. Soc., Perkin Trans. 1*, 3291; (c) Hanyu, A., Takezawa, E., Sakaguchi, S., and Ishii, Y. (1998) *Tetrahedron Lett.*, **39**, 5557; (d) Pd: Dijksman, A., Arends, I.W.C.E., and Sheldon F R.A. (1999) *Chem. Commun.*, 1591; (e) Peterson, K.P. and Larock, R.C. (1998) *J. Org. Chem.*, **63**, 3185; (f) Nishimura, T., Onoue, T., Ohe, K., and Uemura, S. (1998) *Tetrahedron Lett.*, **39**, 6011; (g) Co: Yamada, T. and Mukaiyama F T. (1989) *Chem. Lett.*, 519; (h) Other metals: Semmelhack, M.F., Schmid, C.R., Cortes, D.A., and Chou, C.S. (1984) *J. Am. Chem. Soc.*, **106**, 3374.

92 (a) Markó, I.E., Giles, P.R., Tsukazaki, M., and Urch, C.J. (1996) *Science*, **274**, 2044; (b) Markó, I.E., Gautier, A., Regnaut, I.C., Giles, P.R., Tsukazaki, M., Urch, C.J., and Brown, S.M. (1998) *J. Org. Chem.*, **63**, 7576; (c) Markó, I.E., Giles, P.R., Tsukazaki, M., Regnaut, I.C., Gautier, A., Brown, M., and Urch, C.J. (1999) *J. Org. Chem.*, **64**, 2433.

93 (a) Yamaguchi, K., Mori, K., Mizugaki, T., Ebitani, K., and Kaneda, K. (2000) *J. Am. Chem. Soc.*, **122**, 7144; (b) Yamaguchi, K. and Mizuno, N. (2002) *Angew. Chem. Int. Ed.*, **41**, 4538.

94 Nishimura, T., Kakiuchi, N., Inoue, M., and Uemura, S. (2000) *Chem. Commun.*, 1245.

95 (a) ten Brink, G.-J., Arends, I.W.C.E., and Sheldon, R.A. (2000) *Science*, **287**, 1636; (b) Uozumi, Y. and Nakao, R. (2003) *Angew. Chem. Int. Ed.*, **42**, 194.

96 Iwahama, T., Sakaguchi, S., Nishiyama, Y., and Ishii, Y. (1995) *Tetrahedron Lett.*, **36**, 6923.

97 Iwahama, T., Yoshino, Y., Keitoku, T., Sakaguchi, S., and Ishii, Y. (2000) *J. Org. Chem.*, **65**, 6502.

98 (a) Lee, T.V. (1991) *Comprehensive Organic Synthesis*, vol. 7 (eds B.M. Trost, I., Flemings and S.V. Ley), Pergamon, Oxford, p. 291; (b) Trahanovsky, W.S. (1978) *Oxidation in Organic Chemistry* (eds A.T. Blomquist and H. Wasserman), Academic Press, New York; (c) Hudlicky, M. (1986) Oxidation, in *Organic Chemistry*, ACS Monograph 186, American Chemical Society, Washington DC, and references cited therein.

99 (a) Venturello, C. and Ricci, M. (1986) *J. Org. Chem.*, **51**, 1599; (b) Sakata, Y. and Ishii, Y. (1991) *J. Org. Chem.*, **56**, 6233; (c) Sakata, Y., Katayama, Y., and Ishii, Y. (1992) *Chem. Lett.*, **671**, and references cited therein.

100 (a) Felthouse, T.R. (1987) *J. Am. Chem. Soc*, **109**, 7566; (b) Okamoto, T., Sakai, K., and Oka, S. (1988) *J. Am. Chem. Soc.*, **110**, 1187; (c) Vries, G.D. and Schors, S. (1968) *Tetrahedron Lett.*, **9**, 5689.

101 Nishimura, T., Onoue, T., Ohe, K., and Uemura, S. (1999) *J. Org. Chem.*, **64**, 6750.

102 Takezawa, E., Sakaguchi, S., and Ishii, Y. (1999) *Org. Lett.*, **1**, 713.

103 Oakley, M.A., Woodward, S., Coupland, K., Parker, D., and Temple-Heald, C. (1999) *J. Mol. Catal. A: Chemical*, **150**, 105.

104 Fetizon, M., Golfier, M., and Louis, J.M. (1969) *Chem. Commun.*, 1102.

105 Baskaran, S., Das, J., and Chandrasekaran, S. (1989) *J. Org. Chem.*, **54**, 5182.

106 Inokuchi, T., Matsumoto, S., Nishiyama, T., and Torii, S. (1990) *Synlett*, 57.

107 Iwahama, T., Skaguchi, S., Nishiyama, Y., and Ishii, Y. (1995) *Tetrahedron Lett.*, **36**, 1523.

108 (a) Ito, S., Inoue, K., and Matsumoto, M. (1982) *J. Am. Chem. Soc.*, **104**, 645; (b) Guengrich, F.P. and Macdonald, T.L. (1984) *Acc. Chem. Res.*, **17**, 9; (c) Tabushi, I. and Kodera, M. (1986) *J. Am. Chem. Soc.*, **108**, 1101. (d) Mansuy, D., Fontecve, M. and Bartoli, J.-F. (1981) *Chem. Commun.*, 874; (e) Marchon, J.-C. and Ramasseul, R. (1989) *Synthesis*, 389.

109 (a) Aldehydes as reducing agents: Mukaiyama, T., Takai, T., Yamada, T., and Rhode, O. (1990) *Chem. Lett.*, 1661; (b) Irie, R., Ito, Y., and Katsuki, T. (1991) *Tetrahedron Lett.*, **32**, 6891; (c) Kaneda, K., Haruna, S., Imanaka, T., Hamamoto, M., Nishiyama, Y., and Ishii, Y. (1991) *Tetrahedron Lett.*, **33**, 6827; (d) Murahashi, S.-I., Oda, Y., Naota, T., and Komiya, N. (1993) *Chem. Commun.*, 139; (e) Hamamoto, M., Nakayama, K., Nishiyama, Y., and Ishii, Y. (1993) *J. Org. Chem.*, **58**, 6421; (f) Mizuno, N., Hirose, T., Tateishi, M., and Iwamoto, M. (1993) *Chem. Lett.*, 1839; (g) Bouhlel, E., Laszlo, P., Levart, M., Montaufier, M.-T., and Singh, G.P. (1993) *Tetrahedron Lett.*, **34**, 1133; (h) Punniyamurthy, T., Bhatia, B., and Iqbal, J. (1994) *J. Org. Chem.*, **59**, 850; (i) Mastvorilli, P. and Nobile, C.F. (1994) *J. Mol. Catal.*, **94**, 19; (j) Mukaiyama, T. and Yamada, T. (1995) *Bull. Chem. Soc. Jpn.*, **68**, **17**, and references cited therein.

110 (a) Ketones as reducing agents: Takai, T., Hata, E., Yorozu, K., and Mukaiyama, T. (1992) *Chem. Lett.*, 2077; (b) Punniyamurthy, T., Bhatia, B., and Iqbal, J. (1993) *Tetrahedron Lett.*, **34**, 4657.

111 Acetals as reducing agents: Yorozu, K., Takai, T., Yamada, T., and Mukaiyama, T. (1994) *Bull. Chem. Soc. Jpn.*, **67**, 2195.

112 (a) Other reducing agents: Sakurai, H., Hataya, Y., Goromaru, T., and Matsuura, H. (1985) *J. Mol. Catal.*, **29**, 153; (b) Shimizu, M., Orita, H., Hayakawa, T., and Takehira, K. (1988) *J. Mol. Catal.*, **45**, 85.

113 Groves, J.T. and Quinn, R. (1985) *J. Am. Chem. Soc.*, **107**, 5790.

114 Neumann, R. and Dahan, M. (1997) *Nature*, **388**, 24.

115 Neumann, R. and Dahan, M. (1995) *Chem. Commun.*, 171.

116 (a) 2-Hydroperoxyhexafluoro-2-propanol is reported to be easily derived from HFA (or HFA-hydrate) and H_2O_2; Chambers, R.D. and Clark, M. (1970) *Tetrahedron Lett.*, **11**, 2741; (b) Ganem, B., Heggs, R.P., Biloski, A.J., and Schwartz, D.R. (1980) *Tetrahedron Lett.*, **21**, 685; (c) Ganem, B., Biloski, A.J., and Heggs, R.P. (1980) *Tetrahedron Lett.*, **21**, 689.

117 (a) Heggs, R.P. and Ganem, B. (1979) *J. Am. Chem. Soc.*, **101**, 2484; (b) Biloski, A.J., Heggs, R.P., and Ganem, B. (1980) *Synthesis*, 810; (c) van Vliet, M.C.A.,

Arends, I.W.C.E., and Sheldon, R.A. (1999) *Chem. Commun.*, 263.
118 For storage and transportation of H$_2$O$_2$; Goor, G. and Kunkel, W. (1989) *Ullmann's Encyclopedia of Industrial Chemistry*, 5th ed, vol. A13 (eds B. Elvers, S., Hawkins, M., Ravenscroft and G. Schulz), VCH, Weinheim, pp. 461–463.
119 (a) Iwahama, T., Sakaguchi, S., and Ishii, Y. (1999) *Chem. Commun.*, 727; (b) Iwahama, T., Sakaguchi, S., and Ishii, Y. (2000) *Heterocycles*, **52**, 693.
120 (a) Michael, T. and Musser, E.I. (1999) *Ullman's Encyclopedia Industrial Organic Chemicals*, vol. 3, Wiley-VCH, Weinheim, pp. 1807–1823; (b) Fisher, W.B. and Van Peppen, J.F. (1995) *Kirk-Othmer Encyclopedia of Chemical Technology*, 4th edn, vol. 7 (eds J.L. Kroshwite and M. Home-Grant), John Wiley and Sons, New York, 851–859.
121 (a) Krow, G.R. (1991) *Comprehensive Organic Synthesis*, vol. 7 (ed. B.M. Trost), Pergamon Press, New York, 671–688; (b) Swern, D. (ed.) (1971) *Organic Peroxides*, Wiley-Interscience, New York.
122 (a) Fischer, J. and Hölderich, W.F. (1999) *Appl. Catal. A: General*, **180**, 435–443; (b) Strukul, G., Varagnolo, A., and Pinna, F. (1997) *J. Mol. Catal. A: Chemical*, **117**, 413–423; (c) Wang, Z.B., Mizusaki, T., Sano, T., and Kawakami, Y. (1997) *Bull. Chem. Soc. Jpn.*, **70**, 25967–32570 (d) Frisone, M., Del, T., Pinna, F., and Strukul, G. (1993) *Organometallics*, **12**, 148–156.
123 Göttlich, R., Yamakoshi, K., Sasai, H., and Shibasaki, M. (1997) *Synlett*, 971–973.
124 (a) Bolm, A., Schlingloff, G., and Weickhardt, K. (1993) *Tetrahedron Lett.*, **34**, 3405–3408; (b) Hamamoto, M., Nakayama, K., Nishiyama, Y., and Ishii, Y. (1993) *J. Org. Chem.*, **58**, 6421–6425; (c) Murahashi, S., Oda, Y., and Naota, T. (1992) *Tetrahedron Lett.*, **33**, 7557–7560; (d) Yamada, T., Takahashi, K., Kato, K., Takai, T., Inoki, S., and Mukaiyama, T. (1991) *Chem. Lett.*, 641–644.
125 Fukuda, O., Sakaguchi, S., and Ishii, Y. (2001) *Tetrahedron Lett.*, **42**, 3479.
126 (a) Luedeke, V.D. (1978) *Encyclopedia of Chemical Processing and Design* (ed. J.J. McKetta), Marcel Dekker, New York, pp. 73–95; (b) Fisher, W.N. and Crescentini, L. (1992) *Kirk-Othmer Encyclopedia of Chemical Technology*, vol. 4 (ed. J.I. Kroschwitz), John Wiley & Sons, New York, pp. 827–839; (c) Ritz, J., Fuchs, H., Kiecza, H., and Moran, W.C. (1999) *Ullmann's Encyclopedia of Industrial Organic Chemicals*, vol. 2, Wiley-VCH, Weinheim, pp. 1013–1043.
127 About 2.8kg of ammonium sulfate is generated per kilogram of cyclohexanone oxime produced: Bellussi, G. and Perego, C. (2000) *Cattech*, **4**, 4.
128 Raja, R., Sankar, G., and Thomas, J.M. (2001) *J. Am. Chem. Soc.*, **123**, 8153.
129 (a) Cesana, A., Mantegazza, M.A., and Pastori, M. (1997) *J. Mol. Catal., A Chemical*, **117**, 367; (b) Mantegazza, M.A., Petrini, G., and Cesana, A. (1993) EP 00564040.
130 (a) Kakeya, N. and Takai, T. (2001) JP 2001122851. (b) Hawkins, E.G.E. (1976) US 3947406. (c) Capp, C.W., Harris, B.W., Hawkins, E.G.E., and Edwin, G.E. (1969) FR 1567559. (d) Reddy, J.S., Sivasanker, S., and Ratnasamy, P. (1991) *J. Mol. Catal*, **69**, 383.
131 (a) For recent reviews: Barton, D.H.R. and Doller, D. (1992) *Acc. Chem. Res.*, **25**, 504; (b) Sommer, J. and Bukala, J. (1993) *Acc. Chem. Res.*, **26**, 370; (c) Arndtsen, B.A., Bergman, R.G., Mobley, T.A., and Peterson, T.H. (1995) *Acc. Chem. Res.*, **28**, 154; (d) Hill, C.L. (1995) *Synlett*, 127; (e) Ryu, I. and Sonoda, N. (1996) *Angew. Chem. Int. Ed.*, **35**, 1051.
132 Sakakura, T., Sodeyama, T., Sasaki, K., Wada, K., and Tanaka, M. (1990) *J. Am. Chem. Soc.*, **112**, 7221.
133 Margl, P., Ziegler, T., and Blochl, P.E. (1996) *J. Am. Chem. Soc.*, **118**, 5412.
134 (a) Sato, K.-I., Watanabe, J., Takaki, K., and Fujiwara, Y. (1991) *Chem. Lett.*, 1433; (b) Fujiwara, Y., Takaki, K., and Taniguchi, Y. (1996) *Synlett*, 591, and references therein.
135 (a) Lin, M. and Sen, A. (1994) *Nature*, **368**, 613; (b) Lin, M., Hogan, T.E., and Sen, A. (1996) *J. Am. Chem. Soc.*, **118**, 4574.
136 Takeuchi, K.-I., Akiyama, F., Miyazaki, T., Kitagawa, I., and Okamoto, K. (1987) *Tetrahedron*, **43**, 701.

137 (a) Farooq, O., Marcelli, M., Prakash, G.K.S., and Olah, G.A. (1988) *J. Am. Chem. Soc.*, **110**, 864; (b) Akhrem, I.S., Bernadyuk, S.Z., and Vol'pin, M.E. (1993) *Mendeleev Commun.*, 188.

138 Brubaker, M.M., Coffman, D.D., and Hoehn, H.H. (1952) *J. Am. Chem. Soc.*, **74**, 1509.

139 (a) Coffman, D.D., Cramer, R., and Mochel, W.E. (1958) *J. Am. Chem. Soc.*, **80**, 2882; (b) Walling, C. and Savas, E.S. (1960) *J. Am. Chem. Soc.*, **82**, 1738; (c) Thaler, W.A. (1967) *J. Am. Chem. Soc.*, **89**, 1902; (d) Susuki, T. and Tsuji, J. (1970) *J. Org. Chem.*, **35**, 2982.

140 (a) Ryu, I., Kusano, K., Ogawa, A., Kambe, N., and Sonoda, N. (1990) *J. Am. Chem. Soc.*, **112**, 1295; (b) Ryu, I. and Sonoda, N. (1996) *Chem. Rev.*, **96**, 177; (c) Chatgilialoglu, C., Crich, D., Komatsu, M., and Ryu, I. (1999) *Chem. Rev.*, **99**, 1991, and references cited therein.

141 Sen, A. and Lin, M. (1992) *Chem. Commun.*, 892.

142 Boese, W.T. and Goldman, S. (1992) *J. Am. Chem. Soc.*, **114**, 350.

143 Hill, C.L. and Jaynes, B.S. (1995) *J. Am. Chem. Soc.*, **117**, 4704.

144 Ferguson, R.R. and Crabtree, R.H. (1991) *J. Org. Chem.*, **56**, 5503.

145 Kato, S., Iwahama, T., Sakaguchi, S., and Ishii, Y. (1998) *J. Org. Chem.*, **63**, 222.

146 (a) Albright, L.F. (1966) *Chem. Eng.*, **73**, 149, and references cited therein; (b) Spaeth, C.P. (1959) US Patent 2,883,432.

147 (a) Markofsky, S.B. (1999) *Ullmann's Encyclopedia of Industrial Organic Chemicals*, vol. 6 (eds B. Elvers and S. Hawkins), Wiley-VCH, Weinheim, pp. 3487–3501. (b) Albright, L.F. (1995) *Kirk-Othmer Encyclopedia of Chemical Technology*, vol. 17 (eds J.I. Kroschwite and M. Howe-Grant), Wiley, New York, pp. 68–107.

148 (a) Bachman, G.G., Hass, B.H., and Addison, M.L. (1952) *J. Org. Chem.*, **17**, 914; (b) Bachman, G.G., Hass, B.H., and Addison, M.L. (1952) *J. Org. Chem.*, **17**, 928; (c) Bachman, G.G., Hewett, V.J., and Millikan, A. (1952) *J. Org. Chem.*, **17**, 935. (d) Bachman, G.G. and Kohn, L. (1952) *J. Org. Chem.*, **17**, 942.

149 Sakaguchi, S., Nishiwaki, Y., Kitamura, T., and Ishii, Y. (2001) *Angew. Chem. Int. Ed.*, **40**, 222.

150 (a) Rita, J., Fuchs, H., Kieczka, H., and Moran, W.C. (1999) *Ullmann's Encyclopedia of Industrial Organic Chemicals*, vol. 2 (eds B. Elvers and S. Hawkins), Wiley-VCH, Weinheim, pp. 1013–1043. (b) Fisher, W.B. and Crescen'tini, L. (1995) *Kirk-Othmer Encyclopedia of Chemical Technology*, vol. 4 (eds J.I. Kroschwite and M. Howe-Grant), Wiley, New York, pp. 827–839.

151 The vapor-phase nitration of cyclohexane with NO_2 at $240\,°C$ forms nitrocyclohexane in only 16% yield together with large amounts of undesired nitrated products: Lee, R. and Albright, F.L. (1965) *Ind. Eng. Chem. Process Res. Dev.*, **4**, 411.

152 Jones, K. (1973) *Comprehensive Inorganic Chemistry* (ed. A.F. Trotman-Dickenson), Pergamon, Oxford, pp. 147–388.

153 Beringer, F.M. and Falk, R.A. (1959) *J. Am. Chem. Soc.*, **81**, 2997.

154 (a) Young, H.A. (1937) *J. Am. Chem. Soc.*, **59**, 811; (b) Murray, R.C. (1933) *J. Chem. Soc.*, 739.

155 (a) Bjellqvist, B. (1973) *Acta. Chem. Scand.*, **27**, 3180; (b) Ferguson, R.R. and Crabtree, R.H. (1991) *J. Org. Chem.*, **56**, 5503; (c) Crabtree, R.H. and Habib, A. (1991) *Comprehensive Organic Synthesis*, vol. 7 (ed. B.M. Trost), Pergamon, New York, and references cited therein.

156 Ishii, Y., Matsunaka, K., and Sakaguchi, S. (2000) *J. Am. Chem. Soc.*, **122**, 7390.

157 Smith, G.W. and Williams, H.D. (1961) *J. Org. Chem.*, **26**, 2207.

158 A review of the modern synthesis using vanadium species as a catalyst has been reported: Hirao, T. (1997) *Chem. Rev.*, **97**, 2707.

159 Yamamoto, K., Tsuchida, E., Nishide, H., Jikei, M., and Oyaizu, K. (1993) *Macromolecules*, **26**, 3432.

160 Kirihara, M., Ochiai, Y., Takizawa, S., Takahata, H., and Nemoto, H. (1999) *Chem. Commun.*, 1387.

161 Graf, R. (1952) *Justus Liebigs Ann. Chem.*, **578**, 50.

162 (a) Walling, C. (1957) *Free Radicals in Solution*, Wiley, New York; (b) Kochi, J.K.

(1973) *Free Radicals*, vol. II, Wiley, New York.
163 (a) Culotta, E. and Koshland, D.E. Jr. (1992) *Science*, **258**, 1862; (b) Wrightham, M.N., Cann, A.J., and Sewel, H.F. (1992) *Med. Hypotheses*, **38**, 236.
164 Kato, K. and Mukaiyama, T. (1990) *Chem. Lett.*, 1395.
165 Hata, E., Yamada, T., and Mukaiyama, T. (1995) *Bull. Chem. Soc. Jpn.*, **68**, 3629, and references cited therein.
166 (a) Nagano, T., Takizawa, H., and Hirobe, M. (1995) *Tetrahedron Lett*, **36**, 8239; (b) Itoh, T., Nagata, K., Matsuya, Y., Miyazaki, M., and Ohsawa, A. (1997) *J. Org. Chem.*, **62**, 3582, and references cited therein.
167 Bosch, E. and Kochi, J.K. (1995) *J. Chem. Soc., Perkin Trans.*, **1**, 1057.
168 Park, J.S.B. and Walton, J.C. (1997) *J. Chem. Soc., Perkin Trans.*, **2**, 2579.
169 Sakaguchi, S., Eikawa, M., and Ishii, Y. (1997) *Tetrahedron Lett.*, **38**, 7075.
170 Eikawa, M., Sakaguchi, S., and Ishii, Y. (1999) *J. Org. Chem.*, **64**, 4676.
171 Koch, V.R. and Miller, L.L. (1973) *J. Am. Chem. Soc.*, **95**, 8631.
172 Olah, G.A., Ramaiah, P., Rao, C.B., Stanford, G., Golam, R., Trivedi, N.J., and Olah, J.A. (1993) *J. Am. Chem. Soc.*, **115**, 7246.
173 (a) There have been few reports on practical methods for the synthesis of these dialdehydes so far: Cha, J.S., Lee, K.D., Kwon, O.O., Kim, J.M., and Lee, H.S. (1995) *Bull. Korean. Chem. Soc.*, **16**, 561; (b) Corriu, R.J.P., Lanneau, G.F., and Perrot, M. (1988) *Tetrahedron Lett.*, **29**, 1271; (c) Firouzabadi, H., Iranpoor, N., Kiaeezadeh, F., and Toofan, J. (1986) *Tetrahedron Lett.*, **42**, 719; (d) Green, G., Griffith, A.P., Hollinshead, D.M., Ley, S.V., and Schroder, M. (1984) *J. Chem. Soc., Perkin Trans.*, **1**, 681.
174 Bill, J.C. and Tarbell, D.S. (1963) *Organic Syntheses, Collect*, vol. IV, Wiley, New York, p. 807.
175 Suzuki, H. (1997) *Pharmacia*, **33**, 487.
176 Hirabayashi, T., Sakaguchi, S., and Ishii F Y. (2004) *Angew. Chem. Int. Ed.*, **43**, 1120.
177 (a) O'Neil, M.J., Smith, A. and Heckelman, P.E. (eds) (2001) *The Merck Index*, 13th edn, MERCK & Co., Whitehouse Station, p. 266; (b)Tracy, J.C. (1956) (Notre Dame Ind.) US 2739166. (1956) *Chem. Abstr.*, **50**, 82174. (c) Olan, F.S. (1990) US 4980496. (1991) *Chem. Abstr.*, **114**, 121474.
178 Olah, G.A., Ramaiah, P., Rao, C.B., Sandford, G., Golam, R., Trivedi, N.J., and Olah, G.A. (1993) *J. Am. Chem. Soc.*, **115**, 7246.
179 (a) Bach, R.D., Holubka, J.W., Badger, R.C., and Rajan, S.J. (1979) *J. Am. Chem. Soc.*, **101**, 1979; (b) Olah, G.A., Ramaiah, P., Rao, C.B., Sandford, G., Golam, R., Trivedi, N.J., and Olah, G.A. (1993) *J. Am. Chem. Soc.*, **115**, 7246.
180 Olah, G.A. and Wang, Q. (1992) *Synthesis*, 1090.
181 Koch, V.R. and Miller, L.L. (1973) *J. Am. Chem. Soc.*, **95**, 8631.
182 Jones, S.R. and Mellor, J.M. (1976) *J. Chem. Soc., Perkin Trans. 1*, 2576.
183 Bakke, J.M. and Storm, C.B. (1989) *Acta. Chem. Scand.*, **43**, 399.
184 Hill, C.L. (1995) *Synlett*, **2**, 127.
185 Sakaguchi, S., Hirabayashi, T., and Ishii, Y. (2002) *Chem. Commun.*, 516.
186 (a) Giese, B. (1986) *Radicals, Organic Synthesis: Formation of Carbon-Carbon Bonds*, Pergamon, Oxford; (b) Curran, D.P. (1991) *Comprehensive Organic Synthesis*, vol. 4 (eds B.M. Trost, I., Fleming and M.F. Semmelhock,), Pergamon, Oxford, p. 715; (c) Motherwell, W.B. and Crich, D. (1992) *Free-Radical Reactions in Organic Synthesis*, Academic Press, London; (d) Fossey, J., Lefort, D., and Sorba, J. (1995) *Free Radicals in Organic Synthesis*, Wiley, Chichester; (e) Chatgilialoglu, C. and Renaud, P. (1999) *General Aspects of the Chemistry of Radicals* (ed. Z.B. Alfassi), Wiley, Chichester, pp. 501–538.
187 (a) Hart, D.J. (1984) *Science*, **223**, 883; (b) Curran, D.P. (1988) *Synthesis*, 489. (c) Melikyan, G.G. (1993) *Synthesis*, **833**, and references therein; (d) Ollivier, C. and Renaud, P. (2000) *Angew. Chem. Int. Ed.*, **39**, 925; (e) Studer, A. and Amrein, S. (2000) *Angew. Chem. Int. Ed.*, **22**, 3080, and references cited therein.
188 (a) Reviews: Neumann, W.P. (1987) *Synthesis*, 665; (b) Curran, D.P. (1988) *Synthesis*, 417; (c) Davies, A.G. (1997) *Organotin Chemistry*, Wiley-VCH,

Weinheim; (d) Giese, B. (1983) *Angew. Chem. Int. Ed.*, **22**, 753; (e) Chatgilialoglu, C. (1992) *Acc. Chem. Res.*, **40**, 7019; (f) Dowd, P. and Zhang, W. (1993) *Chem. Rev.*, **93**, 2091.

189 (a) Barton, D.H.R., Crich, D., and Motherwell, W.B. (1985) *Tetrahedron*, **41**, 3901; (b) Crich, D. (1986) *Aldrichimica Acta*, **20**, 35; (c) Ramaiah, M. (1987) *Tetrahedron*, **43**, 3541.

190 Hara, T., Iwahama, T., Sakaguchi, S., and Ishii, Y. (2001) *J. Org. Chem.*, **69**, 6425.

191 (a) Acrylates are known to be easily polymerized by radical initiation. Fischer *et al.* have determined the accurate rate constant for the addition of $CH_2CO_2Bu^t$ radical to methyl acrylate ($k = 6 \times 10^5$ $M^{-1}\ s^{-1}$) [180a]. Thus, such alkenes may be difficult to use as an acceptor in the conventional radical additions of alkyl radicals [180b]. In contrast, the present oxyalkylation seems to provide the successful addition of 1,3-dimethyladamantane to acrylates, since O_2 existing *in situ* quickly quenches the radical intermediate to prevent the polymerization. (a) Beranek, I. and Fischer, H. (1989) *Free Radicals in Synthesis and Biology* (ed. Minisci, F.), Kluwer, Dordrecht, 303; (b) Itoh, M., Taguchi, T. Chung, V.V., Tokuda, M., and Suzuki, A. (1972) *J. Org. Chem.*, **37**, 2357.

192 Fukunishi, K. and Tabushi, I. (1988) *Synthesis*, 826.

193 (a) Hanson, R.L., Lardy, H.A., and Kupchan, S.M. (1970) *Science*, **168**, 378; (b) Rao, Y.S. (1976) *Chem. Rev.*, **76**, 625; (c) Hoffman, H.M.R. and Rabe, J. (1985) *Angew. Chem. Int. Ed.*, **24**, 94.

194 Puthuraya, K., Oomura, Y., and Shimizu, N. (1985) *Brain Res.*, **332**, 165.

195 Li, S. and Vert, M. (1995) *Degradable Polymers* (eds G. Scott and D. Gilead), Chapman & Hall, London, 43–52.

196 (a) Mori, K., Takigawa, T., and Matsuo, T. (1979) *Tetrahedron*, **35**, 933; (b) Garcia, G.A., Muñoz, H., and Tamariz, J. (1983) *Synth. Commun.*, **13**, 569; (c) Jarvis, B.B., Wells, K.M., and Kaufmann, T. (1990) *Synthesis*, 1079; (d) Munoz, A.H., Tamariz, J., Jiminez, R., and Mora, G.G. (1993) *J. Chem. Res. (S)*, 68. (e) Laurent-Robert, H., Le Roux, C., and Dubac, J. (1998) *Synlett*, 1138.

197 Iwahama, T., Sakaguchi, S., and Ishii, Y. (2000) *Chem. Commun.*, 613.

198 (a) Jun, C.-H., Hong, J.-B., and Lee, D.-Y. (1999) *Synlett*, 1, and references cited therein; (b) Schwartz, J. and Cannon, J.B. (1974) *J. Am. Chem. Soc.*, **96**, 4721; (c) Lochow, C.F. and Miller, R.G. (1976) *J. Am. Chem. Soc.*, **98**, 1281; (d) Tracy, M. and Patrick, J. (1952) *J. Org. Chem.*, **17**, 1009.

199 (a) Braun, M. (1987) *Angew. Chem. Int. Ed.*, **26**, 24; (b) Heathcock, C.H. (1984) Chapter 4, in *Comprehensive Carbanion Chemistry*, Part B (eds E. Buncel and T. Durst), Elsevier, Amsterdam; (c) Evans, D.A., Takacs, J.M., McGee, L.R., Ennis, M.D., Mathre, D.J., and Bartroli, J. (1981) *Pure Appl. Chem.*, **53**, 1109.

200 Punniyamurthy, T., Bhatia, B., and Iqbal, J. (1994) *J. Org. Chem.*, **59**, 850.

201 (a) The use of dioxolanes as masked aldehydes has been extensively studied: Dang, H.-S. and Roberts, B.P. (1999) *Tetrahedron Lett.*, **40**, 8929; (b) Gross, A., Fensterbank, L., Bogen, S., Thouvenot, R., and Malacria, M. (1997) *Tetrahedron*, **53**, 13797.

202 Hirano, K., Iwahama, T., Sakaguchi, S., and Ishii, Y. (2000) *Chem. Commun.*, 2457.

203 As another approach for the introduction of a carbonyl function to alkenes, we have reported the catalytic radical addition of ketones to terminal alkenes using a Mn/Co/O_2 redox system: Iwahama, T., Sakaguchi, S., and Ishii, Y. (2000) *Chem. Commun.*, 2317.

204 (a) Coates, W.J. and McKillop, A. (1993) *Synthesis*, 334; (b) Baraldi, P.G., Bigoni, A., Cacciari, B., Caldari, C., Manfredini, S., and Spalluto, G. (1994) *Synthesis*, 1158; (c) Hernandez, A., Marcos, M., and Rapoport, H. (1995) *J. Org. Chem.*, **60**, 2683.

205 Kharasch, M.S., Urry, W.H., and Kuderna, B.M. (1949) *J. Org. Chem.*, **14**, 248.

206 (a) Tracy, M. and Patrick, J.R. (1952) *J. Org. Chem.*, **17**, 1009; (b) Gottschalk, P. and Neckers, D.C. (1985) *J. Org. Chem.*, **50**, 3498.

207 (a) Dang, H.-S. and Roberts, B.P. (1996) *Chem. Commun.*, 2201; (b) Dang, H.-S. and Roberts, B.P. (1998) *J. Chem. Soc., Perkin Trans. 1*, 67.

208 (a) Paul, V., Roberts, B.P., and Willis, C.R. (1989) *J. Chem. Soc., Perkin Trans. 2*, 1953; (b) Allen, R.P., Roberts, B.P., and Willis, C.R. (1989) *Chem. Commun.*, 1387; (c) Roberts, B.P. (1999) *Chem. Soc. Rev.*, **28**, 25.

209 Another approach to understand controlling factors of the reactivity for the radical hydrogen abstraction has been made by A. A. Zavitsas; Zavitsas, A.A. (1998) *J. Chem. Soc., Perkin Trans. 2*, 499.

210 Tsujimoto, S., Iwahama, T., Sakaguchi, S., and Ishii, Y. (2001) *Chem. Commun.*, 2352.

211 Einhorn, C., Einhorn, J., Marcadal-Abbadi, C., and Pierre, J.-L. (1999) *J. Org. Chem.*, **64**, 4542.

212 Nechab, M., Kumar, D.N., Philouze, C., Einhorn, C., and Einhorn, J. (2007) *Angew. Chem. Int. Ed.*, **46**, 3080.

7
Ruthenium-Catalyzed Oxidation for Organic Synthesis
Shun-Ichi Murahashi and Naruyoshi Komiya

7.1
Introduction

Ruthenium complexes have great potential for catalytic oxidation reactions of various compounds [1–10]. The reactivity of ruthenium complexes can be controlled by its oxidation state and ligand. The highest-valent ruthenium complex is ruthenium (VIII) tetroxide (RuO_4), which is known as a strong oxidant and is useful for the cleavage of carbon-carbon double bonds. On the other hand, middle-valent oxo-ruthenium (Ru=O) species can be generated upon treatment of low-valent ruthenium complexes with a variety of oxidants. An important feature of these active species is their high capability to oxidize various substrates such as alkenes, alcohols, amines, amides, β-lactams, phenols, and unactivated hydrocarbons under mild conditions. Ruthenium-catalyzed oxidations and their applications to organic synthesis will be presented in this chapter.

7.2
RuO_4-Promoted Oxidation

RuO_4 has been widely used as a powerful oxidant for oxidative transformation of a variety of organic compounds [11]. RuO_4 can be generated upon treatment of $RuCl_3$ or RuO_2 with an oxidant. The oxidation reaction can be carried out conveniently in a biphasic system (Scheme 7.1) using a catalytic amount of $RuCl_3$ or RuO_2 with the combined use of an oxidant such as $NaIO_4$, HIO_4, $NaOCl$, or $NaBrO_3$, or under electrochemical conditions.

Problems such as very slow and incomplete reaction have been often encountered in the oxidations with RuO_4. These sluggish reactions are due to inactivation of ruthenium catalysts caused by the formation of low-valent ruthenium carboxylate complexes. The inactivation can be prevented by addition of CH_3CN. Thus, various oxidations with RuO_4 were improved considerably by employing a solvent system consisting of CCl_4-H_2O-CH_3CN [11c]. Typically, oxidative cleavage of (*E*)-5-decene (**1**)

Modern Oxidation Methods. Edited by Jan-Erling Bäckvall
Copyright © 2010 WILEY-VCH Verlag GmbH & Co. KGaA, Weinheim
ISBN: 978-3-527-32320-3

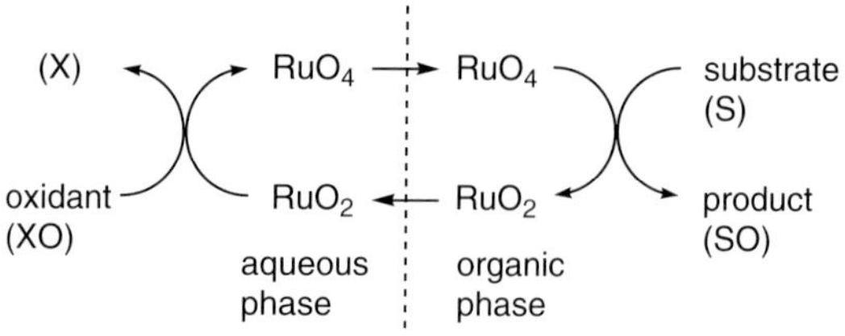

Scheme 7.1

with the RuCl$_3$/NaIO$_4$ system in a CCl$_4$-H$_2$O-CH$_3$CN system gave pentanoic acid in 88% yield, while the same reaction in a conventional CCl$_4$-H$_2$O system gave pentanal (17%) along with 80% of recovered **1** (Eq. (7.1)).

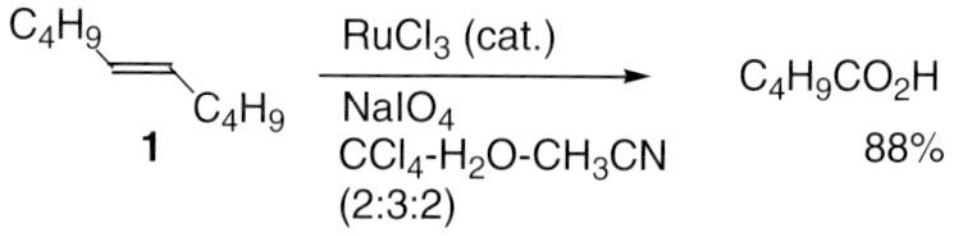

$$\text{C}_4\text{H}_9\text{-CH=CH-C}_4\text{H}_9 \xrightarrow[\substack{\text{NaIO}_4 \\ \text{CCl}_4\text{-H}_2\text{O-CH}_3\text{CN} \\ (2:3:2)}]{\text{RuCl}_3 \text{ (cat.)}} \text{C}_4\text{H}_9\text{CO}_2\text{H} \quad 88\% \tag{7.1}$$

Primary and secondary alcohols are oxidized to the corresponding carboxylic acids and ketones, respectively (Eqs. (7.2) and (7.3)) [12]. Alkenes undergo oxidative cleavage to afford the carbonyl compounds (Eqs. (7.4) and (7.5)) [13], while *cis*-dihydroxylation occurs selectively when the reaction is carried out over a short period of time (0.5 min) at 0 °C in EtOAc-CH$_3$CN-H$_2$O (Eq. (7.6)) [14, 15]. The latter reaction was applied to the oxidative cyclization of 1,5- and 1,6-dienes to give tetrahydrofuranyl and tetrahydropyranyl units (Eq. (7.7)) [15a]. The RuO$_4$ system can be applied to deprotection of allyl protective groups in amides and lactams [16].

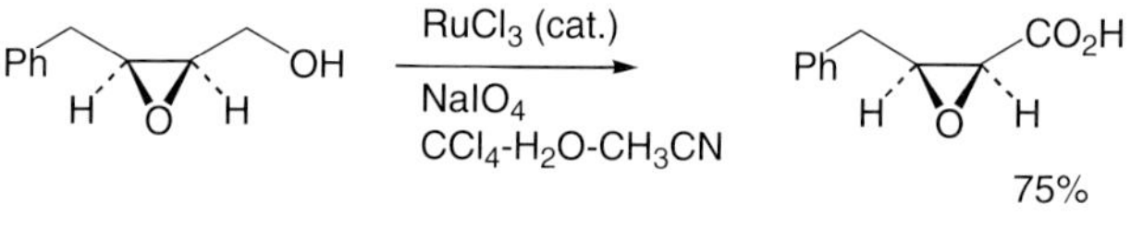

$$\text{(Eq. 7.2)} \quad 75\%$$

$$\text{(Eq. 7.3)} \quad 97\%$$

$$\text{(Eq. 7.4)} \quad 97\%$$

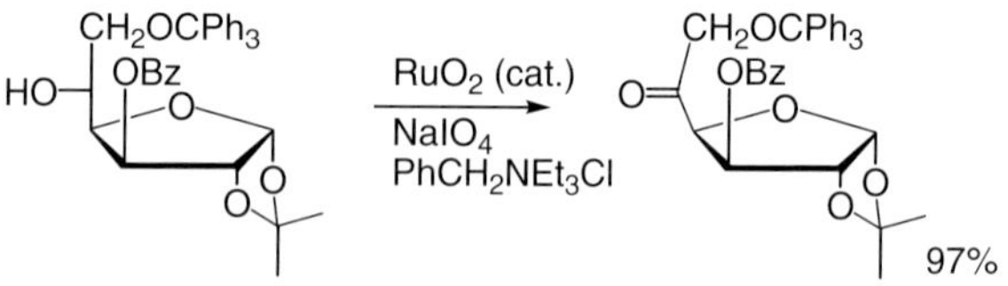

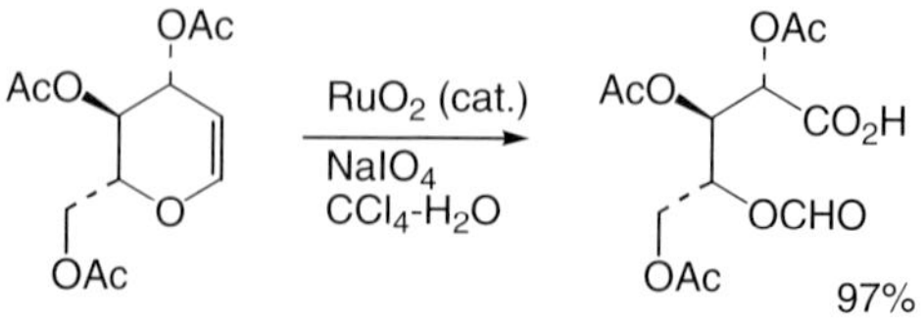

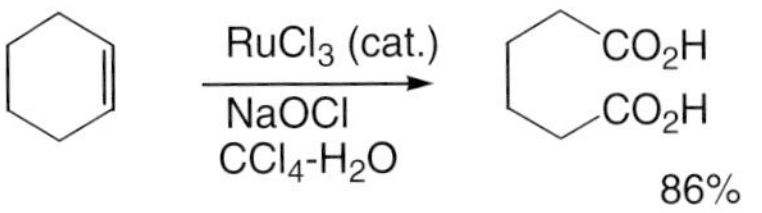

cyclohexene $\xrightarrow[\text{CCl}_4\text{-H}_2\text{O}]{\text{RuCl}_3 \text{ (cat.)} \atop \text{NaOCl}}$ adipic acid (CO$_2$H, CO$_2$H) 86% (7.5)

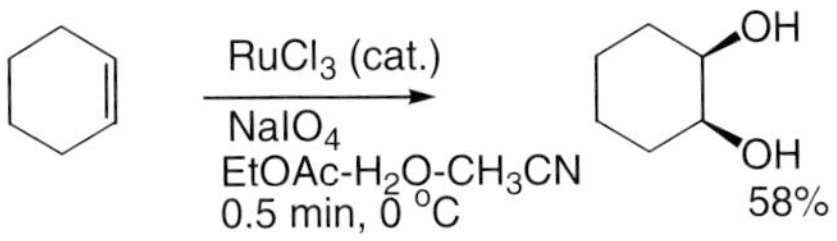

cyclohexene $\xrightarrow[\substack{\text{NaIO}_4 \\ \text{EtOAc-H}_2\text{O-CH}_3\text{CN} \\ 0.5 \text{ min, } 0\,°\text{C}}]{\text{RuCl}_3 \text{ (cat.)}}$ trans-1,2-cyclohexanediol 58% (7.6)

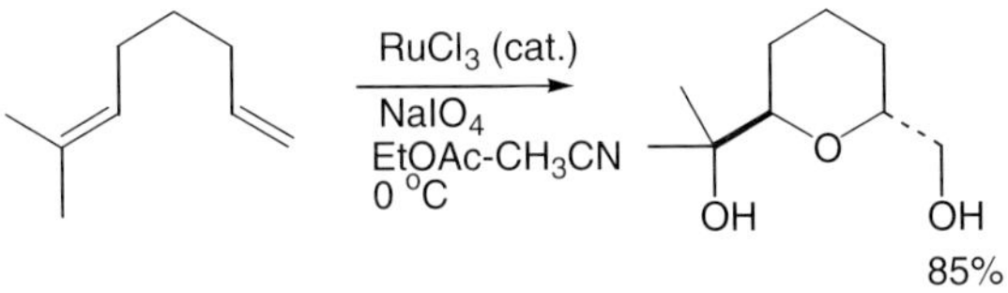

$\xrightarrow[\substack{\text{NaIO}_4 \\ \text{EtOAc-CH}_3\text{CN} \\ 0\,°\text{C}}]{\text{RuCl}_3 \text{ (cat.)}}$ tetrahydrofuran diol 85% (7.7)

Octavalent RuO$_4$ generated from an RuCl$_3$/hypochlorite or periodate system is usually too reactive, and the C=C bond cleavage is often a major reaction; however, the addition of a bipyridine ligand facilitates the epoxidation of alkenes, because it works as an electron-donating ligand to enhance the electron density on the metal and to modulate the reactivity of RuO$_4$ [17–20]. RuCl$_3$ associated with bipyridyl [17] and phenanthroline [18] catalyzes the epoxidation of alkenes with sodium periodate (Eq. (7.8)). The reactions are stereospecific for both *cis*- and *trans*-alkenes. The dioxoruthenium(IV) complex [RuO$_2$(bpy){IO$_3$(OH)$_3$}]·1.5H$_2$O (2) was isolated by the reaction of RuO$_4$ with bipyridyl in the presence of NaIO$_4$, and the complex acts as an efficient epoxidation catalyst under similar conditions (Eq. (7.8)) [19]. Ketohydroxylation of alkenes can be carried out by RuCl$_3$-catalyzed oxidation with oxone under buffered conditions [21].

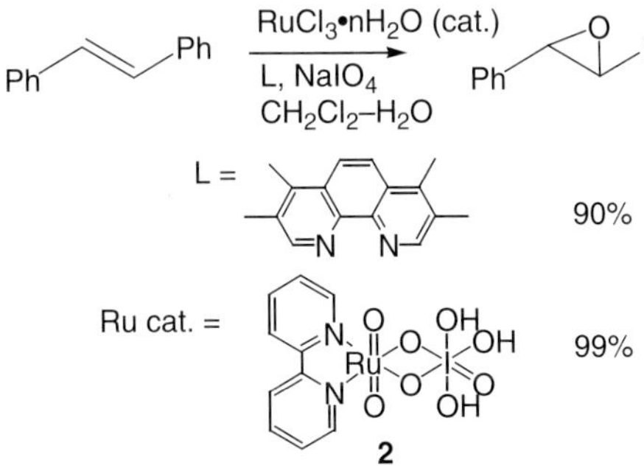

Ph–CH=CH–Ph $\xrightarrow[\substack{\text{L, NaIO}_4 \\ \text{CH}_2\text{Cl}_2\text{-H}_2\text{O}}]{\text{RuCl}_3\cdot n\text{H}_2\text{O (cat.)}}$ stilbene oxide

L = 2,9-dimethyl-1,10-phenanthroline, 90%

Ru cat. = 2, 99%   (7.8)

1,2-Dihaloalkenes are oxidized to α-diketones in a variety of norbornyl derivatives, which can serve as highly potent and inextricable templates for strained polycyclic unnatural compounds (Eq. (7.9)) [22].

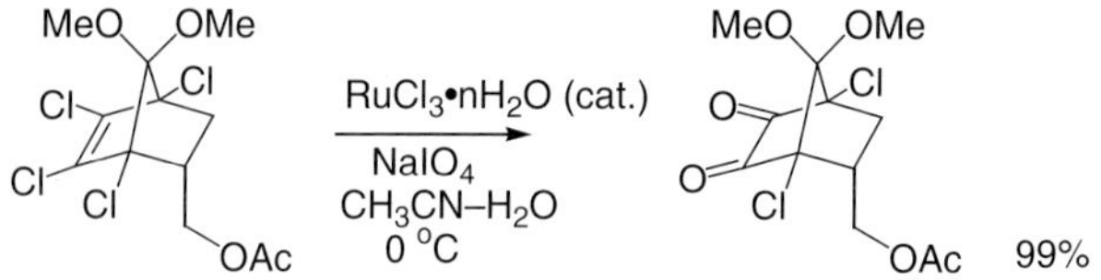

Aromatic and furan rings are smoothly converted into carboxylic acids (Eqs. (7.10) and (7.11)) [23]. Terminal alkynes undergo a similar oxidative cleavage to afford carboxylic acids, while internal alkynes are converted into diketones (Eq. (7.12)) [24, 25].

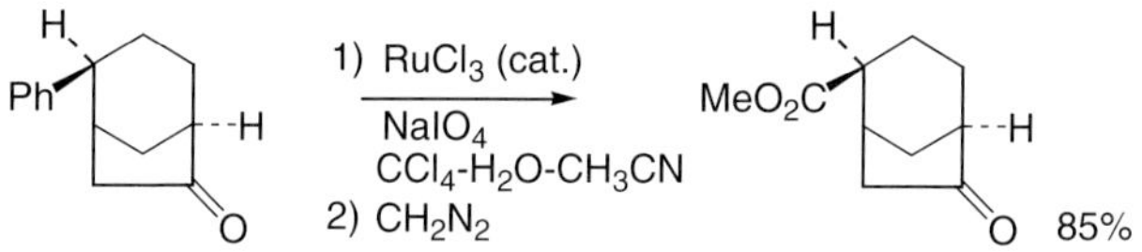

The oxidation of allenes gives α,α-dihydroxyketones (Eq. (7.13)) [26]. Various heteroatom-containing compounds undergo oxidation of methylene groups at the α-position. Ethers are converted into esters and lactones [27]. The efficiency of the α-oxidation of ethers can be improved by pH control using hypochlorite in biphasic media (Eq. (7.14)) [27a]. Tertiary amines [28] and amides [29] undergo similar oxygenation reactions at the α-position of nitrogen to afford the corresponding amides and imides, respectively (Eq. (7.15)). Carbon–carbon side-chain fragmentation occurs when N,C-protected serine and threonine are subjected to oxidation. The method has been successfully applied to the N–C bond scission of peptides in serine or threonine residue (Eq. (7.16)) [30].

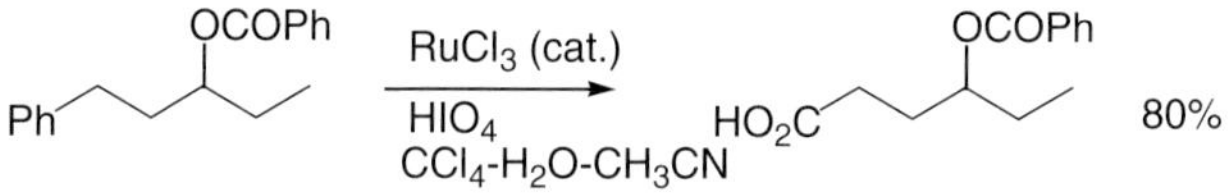

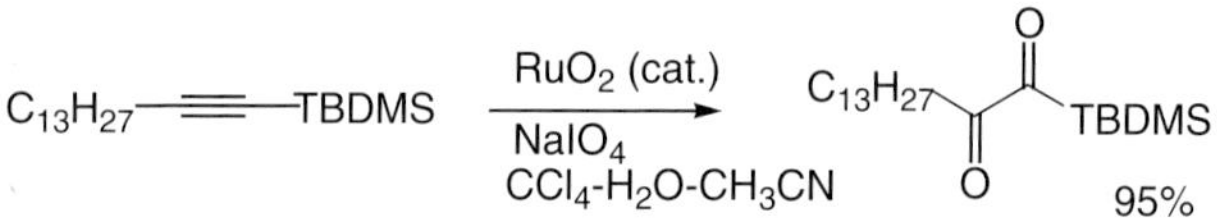

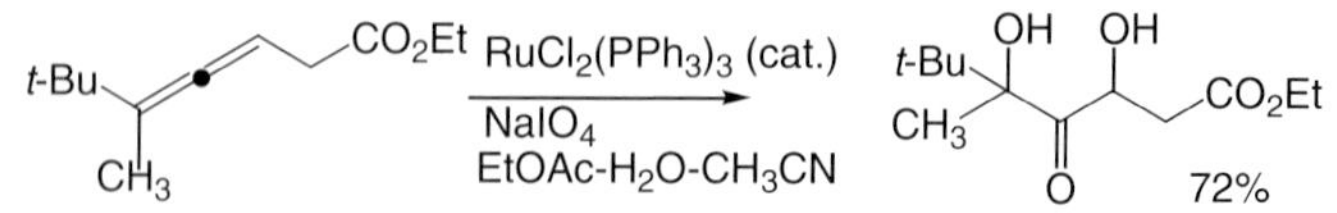

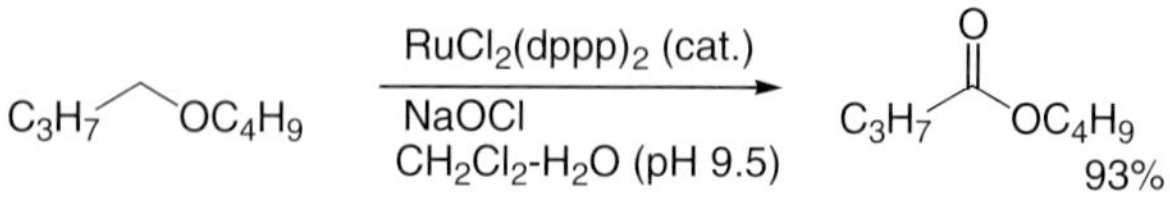

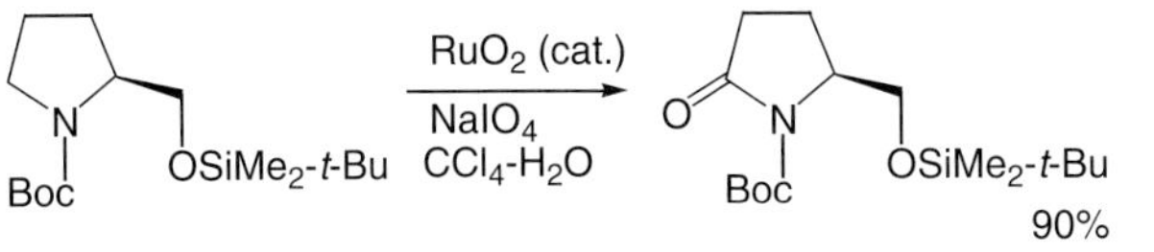

$$\text{(7.15)}$$

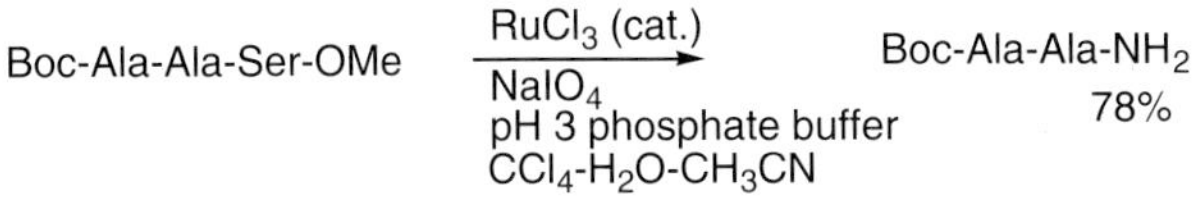

Pyrrolidine-Boc-OSiMe$_2$-t-Bu → lactam-Boc-OSiMe$_2$-t-Bu, RuO$_2$ (cat.), NaIO$_4$, CCl$_4$-H$_2$O, 90%

$$\text{(7.16)}$$

Boc-Ala-Ala-Ser-OMe $\xrightarrow[\text{pH 3 phosphate buffer}]{\text{RuCl}_3 \text{ (cat.), NaIO}_4}$ Boc-Ala-Ala-NH$_2$ 78%
CCl$_4$-H$_2$O-CH$_3$CN

Unactivated alkanes can also be oxidized with the RuCl$_3$/NaIO$_4$ system. Tertiary carbon–hydrogen bonds undergo chemoselective hydroxylation to afford the corresponding tertiary alcohols (Eq. (7.17)) [31].

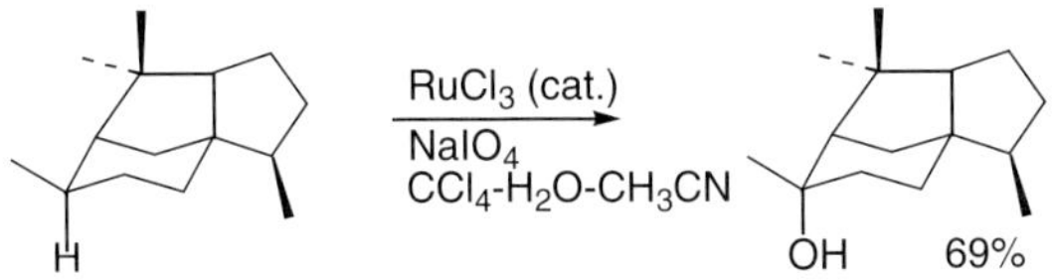

$$\text{(7.17)}$$

RuCl$_3$ (cat.), NaIO$_4$, CCl$_4$-H$_2$O-CH$_3$CN, 69% (OH)

Bridgehead carbons of adamantane [32], pinane [33], and fused norbornane [34] undergo selective hydroxylation under similar reaction conditions. Methylene groups of cycloalkanes undergo hydroxylation and then subsequent oxidation to afford the corresponding ketones [35]. In general, methyl groups of alkanes undergo no reaction with RuO$_4$, while the methyl group of toluene can be converted into the corresponding carboxylic acid (Eq. (7.18)) [36].

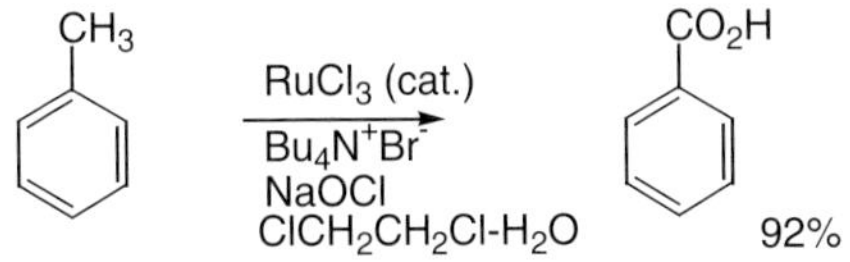

$$\text{(7.18)}$$

PhCH$_3$ $\xrightarrow[\text{Bu}_4\text{N}^+\text{Br}^-,\ \text{NaOCl},\ \text{ClCH}_2\text{CH}_2\text{Cl-H}_2\text{O}]{\text{RuCl}_3 \text{ (cat.)}}$ PhCO$_2$H, 92%

7.3
Oxidation with Low-Valent Ruthenium Catalysts and Oxidants

The treatment of a low-valent ruthenium catalysts with an oxidant generates middle-valent Ru=O species, which often show a different reactivity from that of the RuO$_4$ oxidation.

7.3.1
Oxidation of Alkenes

The epoxidation of alkenes with metalloporphyrins has been studied as model reactions of cytochrome P-450 [37]. Ruthenium porphyrins such as Ru(OEP)(PPh$_3$)Br (OEP = octaethylporphyrinato) have been examined for the catalytic oxidation of styrene with PhIO [38]. Hirobe and coworkers [39] and Groves and coworkers [40] reported that the ruthenium porphyrin-catalyzed oxidations of alkenes with 2,6-dichloropyridine N-oxide gave the corresponding epoxides in high yields (Eqs. (7.19) and (7.20)). The chlorine substituents at the 2- and 6- positions on pyridine N-oxide are necessary for high efficiency, because simple pyridine coordinates to the ruthenium more strongly and retards the catalytic activity. Ru(TMP)Cl$_2$-catalyzed oxidation of terminal alkenes with 2,6-dichloropyridine N-oxide gave a Wacker-type product of an aldehyde via epoxidation-isomerization mechanism (TMP = tetramethylporphyrinato) [41a]. Now, ruthenium porphyrin-catalyzed aerobic oxidation of terminal aryl alkenes to aldehyde can be carried out efficiently [41b]. Thus, the reaction of styrenes in the presence of Ru(TMP)Cl$_2$ (2 mol%), NaHCO$_3$, and H$_2$O in CDCl$_3$ gives the corresponding aldehydes by tandem epoxidation-isomerization. Nitrous oxide (N$_2$O) can also be used as an oxidant for the epoxidation of trisubstituted alkenes in the presence of ruthenium porphyrin catalyst [42]. Importantly, oxidative cleavage of alkenes to give dicarboxylic acid and cis-dihydroxylation can be carried out by either [Ru(Me$_3$tacn)(CF$_3$CO$_2$)$_2$(OH$_2$)]$^+$/H$_2$O$_2$ or [Ru(Me$_3$tacn)Cl$_3$]/H$_2$O$_2$ [43]. Terminal alkenes undergo selective C–C bond cleavage to give aldehydes with cis-Ru(dmp)(H$_2$O)$_2$/H$_2$O$_2$ [44].

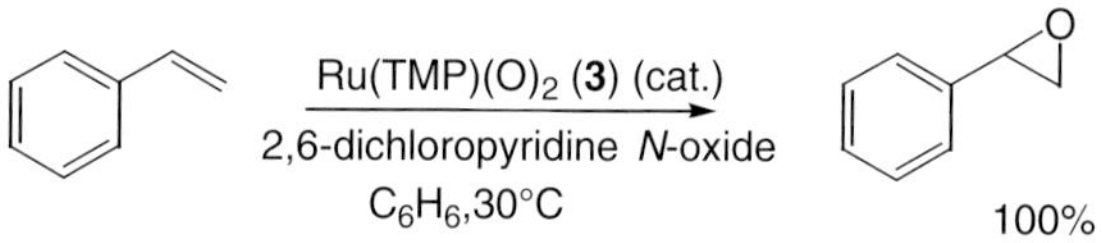

$$\text{PhCH=CH}_2 \xrightarrow[\substack{\text{2,6-dichloropyridine } N\text{-oxide} \\ \text{C}_6\text{H}_6, 30°\text{C}}]{\text{Ru(TMP)(O)}_2 \text{ (3) (cat.)}} \text{PhCH(O)CH}_2 \quad 100\% \tag{7.19}$$

$$\text{CH}_2\text{=CH(CH}_2\text{)}_5\text{CH}_3 \xrightarrow[\substack{\text{2,6-dichloropyridine } N\text{-oxide} \\ \text{CH}_2\text{Cl}_2, 65°\text{C}}]{\text{Ru(TPFPP)(CO) (4) (cat.)}} \text{epoxide} \quad 96\%$$

Ru(TMP)(O)$_2$ (3)
TMP: tetramesitylporphyrinato

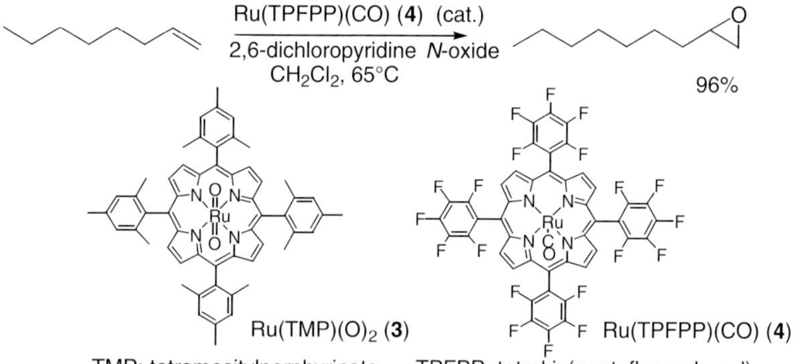

Ru(TPFPP)(CO) (4)
TPFPP: tetrakis(pentafluorophenyl)porphyrinato

(7.20)

Non-porphyrin ruthenium complexes such as [RuCl(DPPP)$_2$] (DPPP = 1,3-bis (diphenylphosphino)propane), [Ru(6,6-Cl$_2$bpy)$_2$(H$_2$O)$_2$], binaphthyl-ruthenium complex, and RuCl$_3$ catalyze oxidations of alkenes with PhIO [45], t-BuOOH [46], PhI(OAc)$_2$ [47], or H$_2$O$_2$ [48] to give the corresponding epoxides in moderate yields.

The ruthenium-catalyzed aerobic oxidation of alkenes has been explored by several groups. Groves and coworkers reported that Ru(TMP)(O)$_2$ (3)-catalyzed aerobic epoxidation of alkenes proceeds under 1 atm of molecular oxygen without any reducing agent [40b]. A ruthenium-containing polyoxometalate, $\{[WZnRu_2(OH)(H_2O)](ZnW_9O_{34})_2\}^{11-}$ [49], and a sterically hindered ruthenium complex, $[Ru(dmp)_2(CH_3CN)_2](PF_6)$ (dmp = 2,9-dimethyl-1,10-phenanthroline) [50], are also effective for epoxidation with molecular oxygen. Knochel and coworkers reported that the ruthenium catalyst bearing perfluorinated 1,3-diketone ligands catalyzes the aerobic epoxidation of alkenes in a perfluorinated solvent in the presence of i-C_3H_7CHO [51]. Asymmetric epoxidations have been reported using ruthenium complexes with oxidants such as PhIO [52], PhI(OAc)$_2$ [53], 2,6-dichloropyridine N-oxide [54–56], hydrogen peroxide [57], and molecular oxygen [58].

It was postulated that one of the possible intermediates for metalloporphyrin-promoted epoxidation is the intermediate **5** (Scheme 7.2) [59].

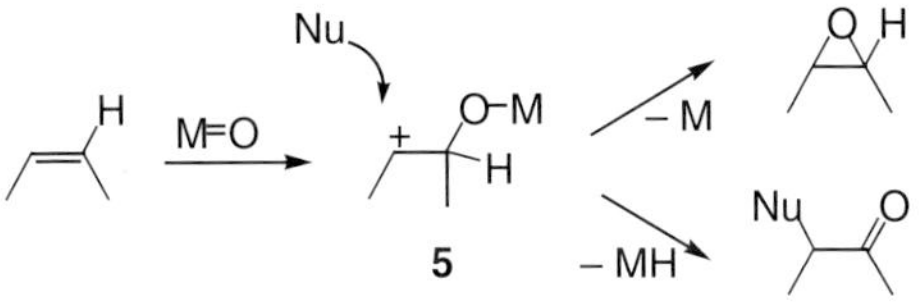

Scheme 7.2

If one could trap the intermediate **5** with external nucleophiles, such as water, a new type of catalytic oxidation of alkenes could be performed. Indeed, a transformation of alkenes into α-ketols was discovered to proceed highly efficiently. Thus, the low-valent ruthenium-catalyzed oxidation of alkenes with peracetic acid in an aqueous solution under mild conditions gives the corresponding α-ketols, which are important key structures of various biologically active compounds (Eq. (7.21)) [60].

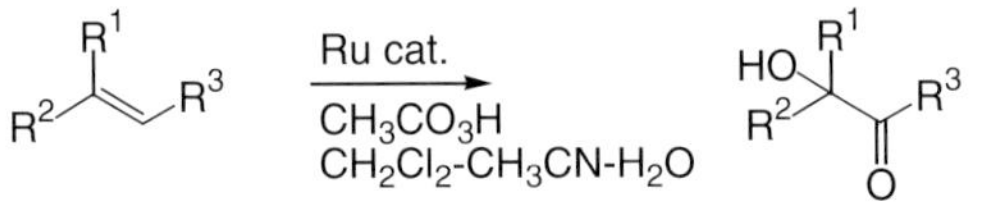

$$(7.21)$$

Typically, the RuCl$_3$-catalyzed oxidation of 3-acetoxy-1-cyclohexene (**6**) with peracetic acid in H_2O-CH_3CN-CH_2Cl_2 (1 : 1 : 1) gave (2R^*,3S^*)-3-acetoxy-2-hydroxycyclohexanone (**7**) chemo- and stereoselectively in 78% yield (Eq. (7.22)).

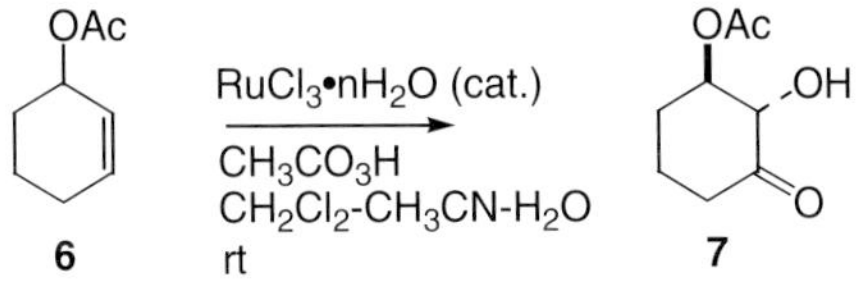

$$(7.22)$$

The oxidation is highly efficient and quite different from that promoted by RuO$_4$. Indeed, the oxidation of 1-methylcyclohexane (**8**) under the same conditions gives 2-hydroxy-2-methylcyclohexanone (**9**) (67%), while the oxidation of the same substrate **8** under conditions in which the RuO$_4$ is generated catalytically gives 6-oxoheptanoic acid (**10**) (91%) (Eq. (7.23)).

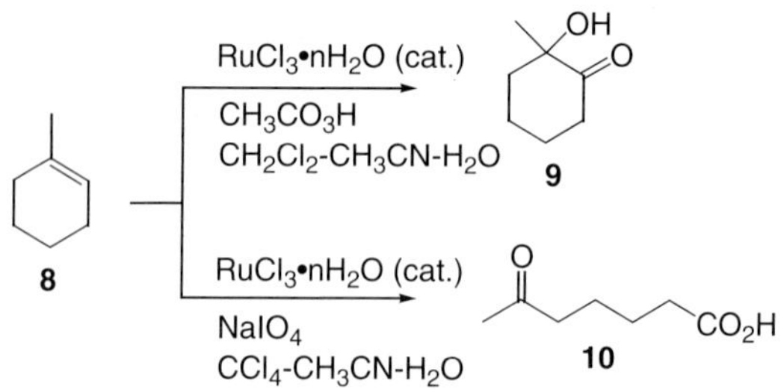

7 Ruthenium-Catalyzed Oxidation for Organic Synthesis

$$\text{(7.23)}$$

This particular oxidation can be applied to the oxidation of substituted alkenes having functional groups such as acetoxy, methoxycarbonyl, and azide groups to give the corresponding α-ketols in good to excellent yields. The oxidation of 3-azide-1-cyclohexene (**11**) gave ($2S^*$, $3R^*$)-3-azide-2-hydroxycyclohexanone (**12**) chemo- and stereoselectively (65%) (Eq. (7.24)).

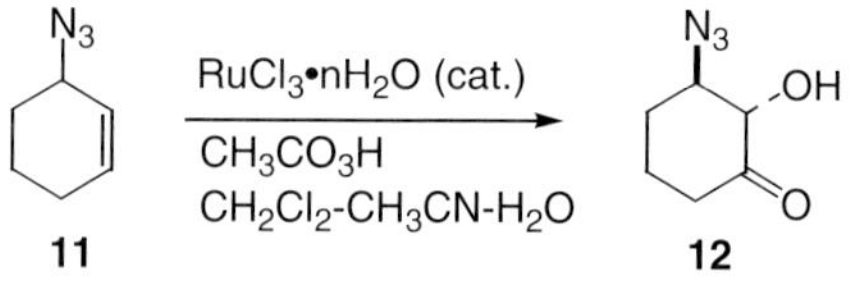

$$\text{(7.24)}$$

The efficiency of the present reaction has been demonstrated by the synthesis of cortisone acetate **15** [60], which is a valuable anti-inflammatory agent. The oxidation of 3β, 21-diacetoxy-5α-pregn-17-ene (**13**) proceeds stereoselectively to give 20-oxo-5α-pregnane-3β,17α,21-triol 3,21-diacetate (**14**) (57%) (Eq. (7.25)). Conventional treatment of **14** followed by microbial oxidation with *Rhizopus nigricaus* gave cortisone acetate (**15**).

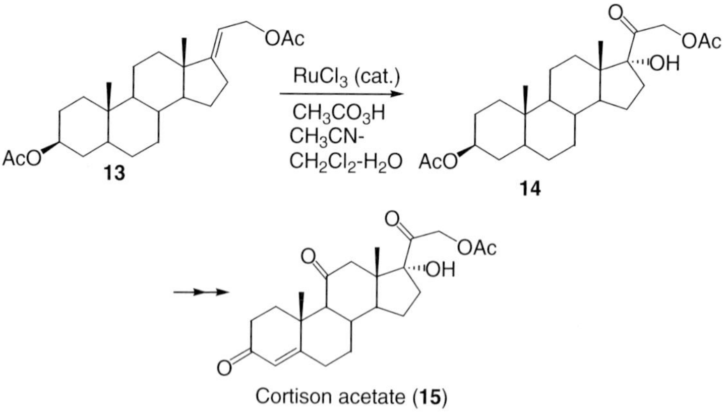

$$\text{(7.25)}$$

Cortison acetate (**15**)

Furthermore, the method can be applied to the synthesis of 4-demethoxyadriamycinone, which is the side-chain of cancer drug adriamycins such as idarubicin and annamycin (**16**). The ruthenium-catalyzed oxidation of allyl acetate **17** gives the corresponding α-hydroxyketone **18** in 60% yield (Eq. (7.26)) [61]. The reaction was also applied to the oxidation of α,β-unsaturated carbonyl compounds **19**, and this provides a new method for the synthesis of α-oxo-ene-diols **20** (Eq. (7.27)) [62].

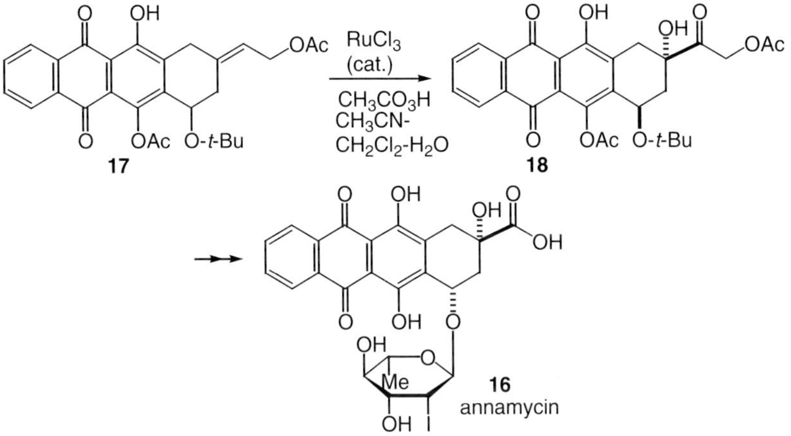

(7.26)

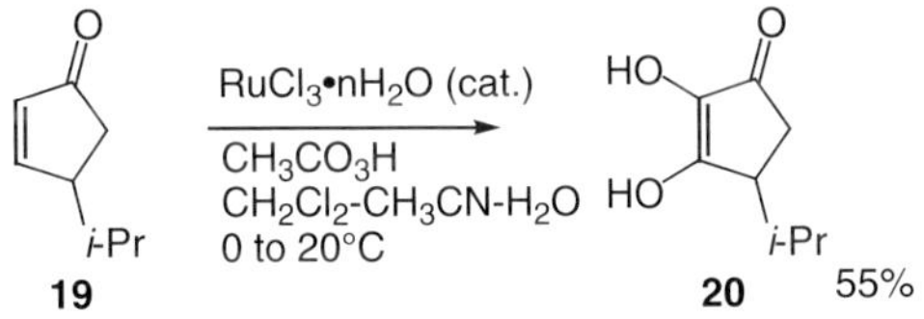

(7.27)

7.3.2
Oxidation of Alcohols

The ruthenium-catalyzed oxidation of alcohols has been reported using various catalytic systems, which include $RuCl_2(PPh_3)_3$ catalyst with oxidants such as N-methylmorpholine N-oxide (NMO) [63], iodosylbenzene [64], TMSOOTMS [65], $RuCl_3$ with hydrogen peroxide [66], K_2RuO_4 with peroxodisulfate [67], and Ru(pybox)(Pydic) complex with diacetoxyiodosylbenzene [68]. The salt of the perruthenate ion with a quaternary ammonium salt, $(n\text{-}Pr_4N)(RuO_4)$ (TPAP), which is soluble in a variety of organic solvents, shows far milder oxidizing properties than those of RuO_4 [69]. One of the key features of the TPAP system is its ability to tolerate other potentially reactive groups. For example, double bonds, polyenes, enones, halides, cyclopropanes, epoxides, and acetals all remain intact during TPAP oxidation [69]. The oxidation of primary and secondary alcohols with TPAP gives the corresponding aldehydes and ketones (Eqs. (7.28) and (7.29)), whereas oxidation with RuO_4 results in the formation of carboxylic acid. NaOCl can also be used as an oxidant for the TPAP-catalyzed oxidation of secondary alcohols [70].

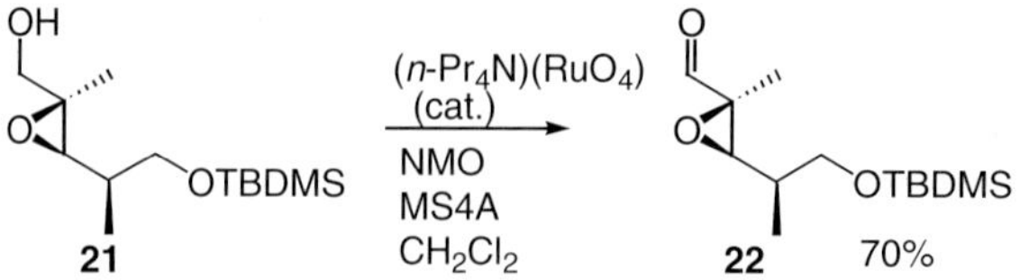

$$\text{(7.28)}$$

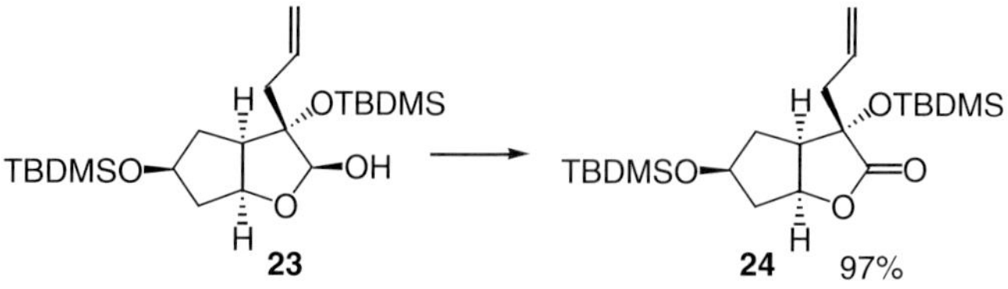

$$\text{(7.29)}$$

The RuCl$_2$(PPh$_3$)$_3$-catalyzed reaction of secondary alcohols with t-BuOOH gives ketones under mild conditions [71, 72]. This oxidation can be applied to the transformation of cyanohydrins into acyl cyanides [71], which are excellent acylating reagents. Typically, the oxidation of cyanohydrin **25** with two equiv. of t-BuOOH in dry benzene at room temperature gives benzoyl cyanide (**26**) in 92% yield (Eq. (7.30)). It is worth noting that the acyl cyanides thus obtained are excellent reagent for the chemoselective acylation reaction. The reaction of amino alcohols with acyl cyanides gives N-acylated amino alcohols selectively. Furthermore, primary amines are selectively acylated in the presence of secondary amines [73]. The utility of the reaction has been illustrated by the short-step synthesis of maytenine (**27**) (Eq. (7.30)). A ruthenium complex [Cn*Ru(CF$_3$CO$_2$)$_3$(H$_2$O)] (Cn* = N,N′,N″-trimethyl-1,4,7-triazacyclononane) catalyst can be used for the oxidation of alcohols with t-BuOOH [74].

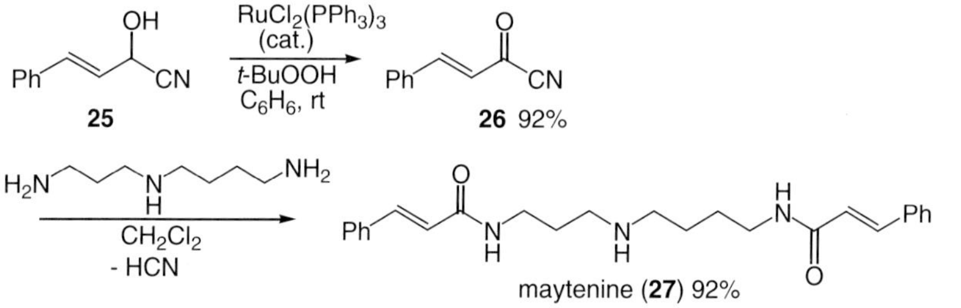

$$\text{(7.30)}$$

Various aliphatic and aromatic secondary alcohols can be oxidized with peracetic acid in the presence of RuCl$_3$ catalyst to give the corresponding ketones highly efficiently [75].

The generation of peracetic acid *in situ* provides an efficient method for the aerobic oxidation of alcohols. The oxidation of various aliphatic and aromatic alcohols can be carried out at room temperature with molecular oxygen (1 atm) in the presence of acetaldehyde and RuCl$_3$–Co(OAc)$_2$ bimetallic catalyst [76]. This method is highly convenient, because the products can be readily isolated simply by removal of both acetic acid and the catalyst by washing with a small amount of water.

An alternative method for the oxidation of alcohols is dehydrogenative oxidation via a hydrogen transfer reaction [2a]. The process of dehydrogenation of alcohols by metal catalysts is of importance from biological and industrial viewpoints. Alcohols can be activated efficiently with low-valent ruthenium complexes to give the carbonyl dihydridoruthenium intermediates (Eq. (7.31)). Capture of the intermediates with nucleophiles provides various catalytic oxidative condensations of alcohols. Thus, primary alcohols undergo oxidative condensation upon treatment with $RuH_2(PPh_3)_4$ [77] or $Ru(CO)_3(\eta^4$-tetracyclone) [78] catalyst to give esters along with evolution of molecular hydrogen (Eq. (7.32)). The $RuH_2(PPh_3)_4$-catalyzed reaction of 1,n-diol (n≠4,5) gives the corresponding polyesters and molecular hydrogen [2a]. Similar treatment of diols affords the corresponding lactones. In the presence of hydrogen acceptor the $RuH_2(PPh_3)_4$-catalyzed oxidative condensation of alcohols to esters proceeds at low temperature (Eq. (7.33)) [77]. As a hydrogen acceptor, α,β-unsaturated ketones and other substrates have been used; however, acetone was found to be the most effective and convenient hydrogen acceptor [77b–f]. The oxidative cross-esterification of alcohols with methanol is performed in the presence of $Ru(PPh_3)_3(CO)H_2$ catalyst (Eq. (7.34)) [79a]. Application of the present reaction provides novel catalytic reactions. Thus, the oxidative condensations of aldehydes with alcohols to esters and of aldehydes with water give acids or esters [2a,77b]. Hartwig and Milstein reported that dehydrogenative cyclization of diols to lactones can be carried out using $Ru(PMe_3)_2Cl_2$(eda) (eda: ethylenediamine) [79b] and RuHCl (PNN)(CO) (PNN: 2-(di-t-butylphosphinomethyl)-6-(diethylaminomethyl)pyridine) catalyst [79c]. Direct conversion of alcohols to acetals and H_2 catalyzed by an acridine-based ruthenium pincer complex is performed efficiently (Eq. (7.35)) [80].

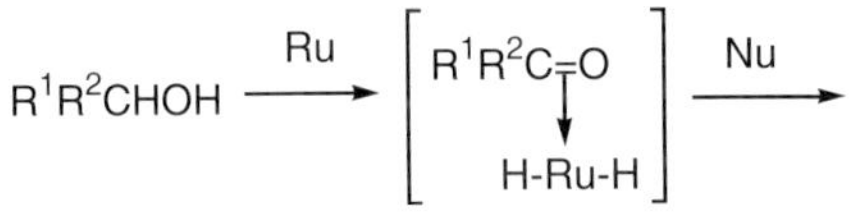

$$R^1R^2CHOH \xrightarrow{Ru} \begin{bmatrix} R^1R^2C=O \\ \downarrow \\ H-Ru-H \end{bmatrix} \xrightarrow{Nu} \quad (7.31)$$

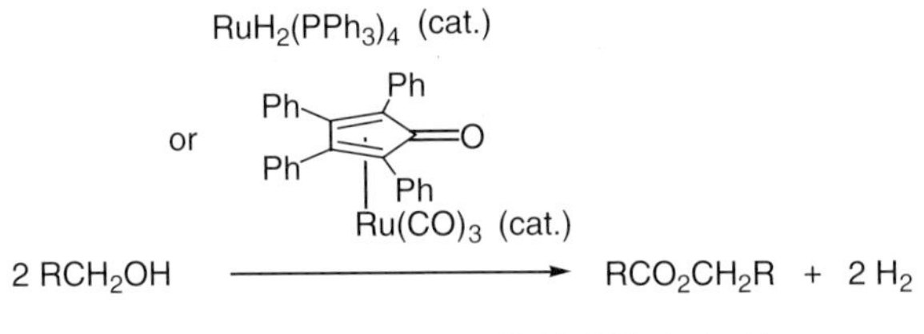

$$2\ RCH_2OH \xrightarrow[\text{or } Ru(CO)_3 \text{ (cat.)}]{RuH_2(PPh_3)_4 \text{ (cat.)}} RCO_2CH_2R + 2\ H_2 \quad (7.32)$$

(with tetraphenylcyclopentadienone ligand shown: Ph, Ph, Ph, Ph)

$$\text{HO-CH(CH}_2)\text{-CH}_2\text{OH} \xrightarrow[\substack{\text{acetone (hydrogen acceptor)} \\ \text{toluene} \\ 180\ °C,\ 3\ h}]{RuH_2(PPh_3)_3 \text{ (cat.)}} \text{lactone} \quad 98\% \quad (7.33)$$

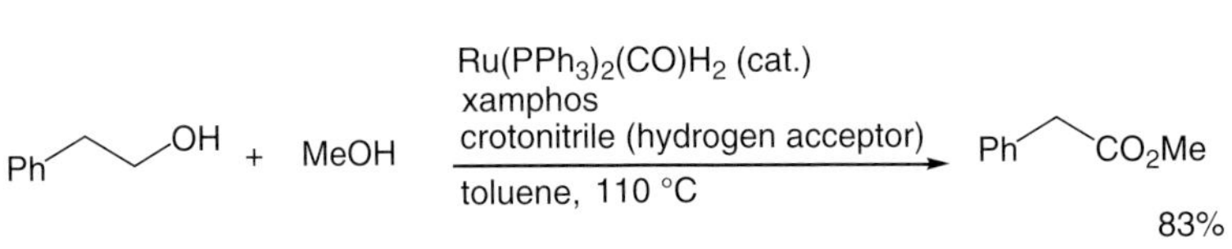

$$Ph\text{-CH}_2\text{OH} + MeOH \xrightarrow[\substack{\text{xamphos} \\ \text{crotonitrile (hydrogen acceptor)} \\ \text{toluene, } 110\ °C}]{Ru(PPh_3)_2(CO)H_2 \text{ (cat.)}} Ph\text{-CO}_2Me \quad 83\% \quad (7.34)$$

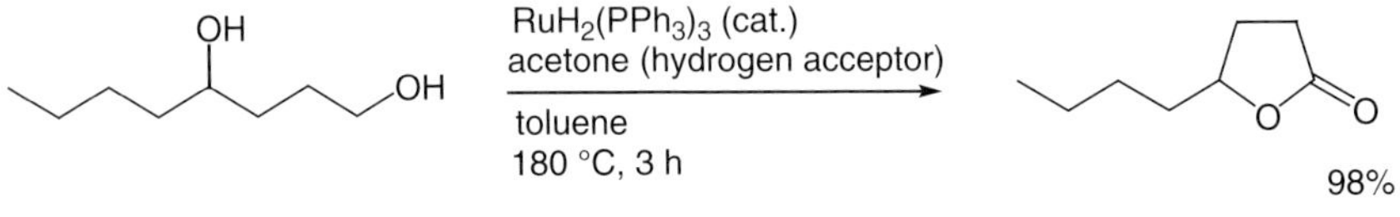

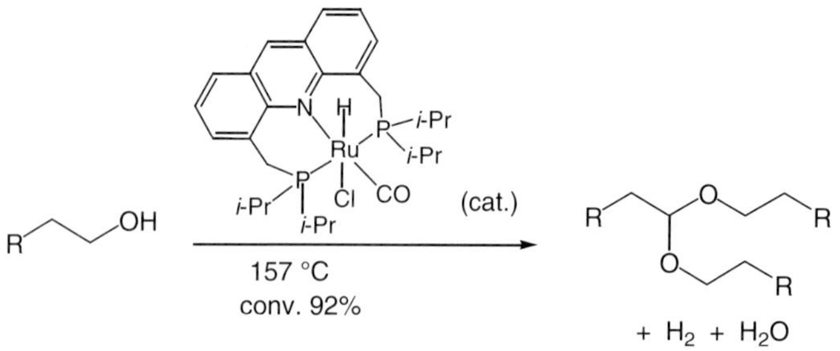

$$R\diagdown OH \xrightarrow[\text{conv. 92\%}]{\text{[Ru catalyst]} \text{ (cat.)} \atop 157\ °C} R\diagdown O\diagdown R + R\diagdown O\diagdown R + H_2 + H_2O \quad (7.35)$$

Direct methods for the synthesis of secondary amines from alcohols and amines using ruthenium catalysts were discovered in 1981–1982 [81a,b], although Pd catalyst [81c] and RhH(PPh$_3$)$_4$ [81d] are excellent catalysts for activation of alcohols in the presence of amines. The RuH$_2$(PPh$_3$)$_4$ catalyst is an excellent general catalyst for the activation of alcohols in the presence of amines, and can be used for the synthesis of aliphatic amines [81a]. On the other hand RuCl$_2$(PPh$_3$)$_3$ is a more reactive catalyst, but this catalyst can be used for synthesis of aryl amines [81b]. Recently, Beller reported that the catalytic system of Ru$_3$(CO)$_{12}$ and N-phenyl-2-(dicyclohexylphosphanyl)pyrrole can be used for amination of alcohols [81e]. These methods are now well documented as borrowing hydrogen [2a] (see excellent review in Ref.[81f]).

The first example of the ruthenium-catalyzed synthesis of amides from alcohols and amines was reported by Murahashi et al. in 1991 [82a]. The contrast results were obtained from the RuH$_2$(PPh$_3$)$_4$-catalyzed reaction of 5-aminopentanol. Thus, piperidine was obtained in 79% yield, while similar treatment in the presence of a hydrogen acceptor of 1-phenyl-1-buten-3-one gave piperidone in 65% yield (Eq. (7.36)). Recently, Williams reported the intermolecular amidation reaction of benzyl alcohols with amines in the presence of [Ru(p-cymene)Cl$_2$]$_2$ and 3-methyl-2-butanone [82b].

Milstein et al. discovered an extremely important result, that is, a ruthenium complex with a PNN ligand, where PNN is 2-(di-t-butylphosphinomethyl)-6-(diethylaminomethyl)pyridine, promotes the direct dehydrogenative acylation of amines with alcohols catalytically with liberation of H$_2$ (Eq. (7.37)) [83].

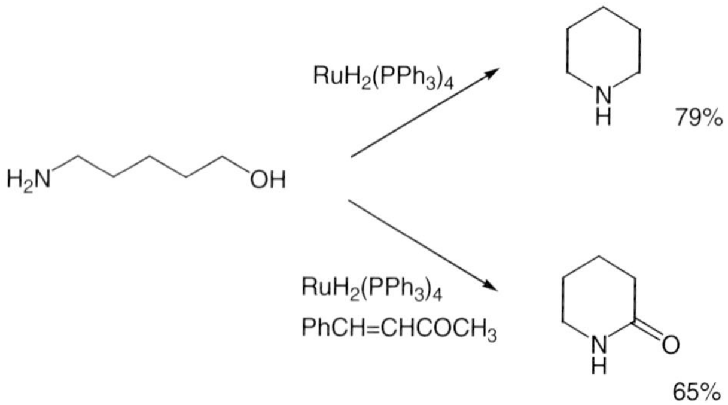

(7.36)

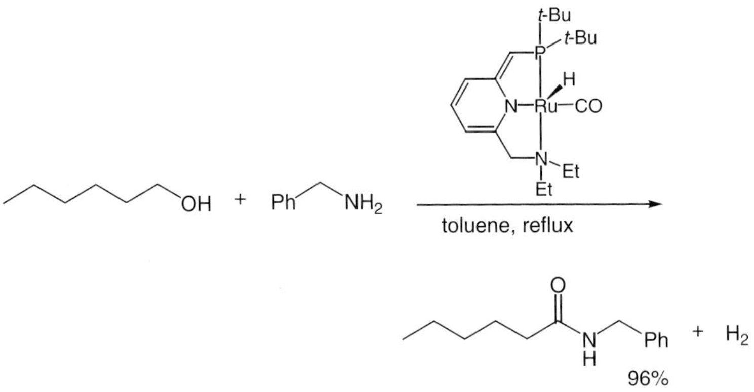

7.3 Oxidation with Low-Valent Ruthenium Catalysts and Oxidants | 253

$$\text{(7.37)}$$

By regenerating the hydrogen acceptor in the presence of co-catalyst and oxygen, the hydrogen transfer reaction can be extended to the catalytic aerobic oxidation. Bäckvall et al. found that the ruthenium hydride formed during the hydrogen transfer can be converted to ruthenium by a multistep electron transfer process including hydroquinone, the metal complex, and molecular oxygen (Scheme 7.3). On the basis of this process, aerobic oxidation of alcohols to aldehydes and ketones can be performed at ambient pressure of O_2 in the presence of a ruthenium-cobalt bimetallic catalyst and hydroquinone [84]. Typically, cycloheptanol is oxidized to cycloheptanone under O_2 atmosphere with a catalytic system consisting of ruthenium complex **29**, cobalt complex **30**, and 1,4-benzoquinone (Eq. (7.38)) [85].

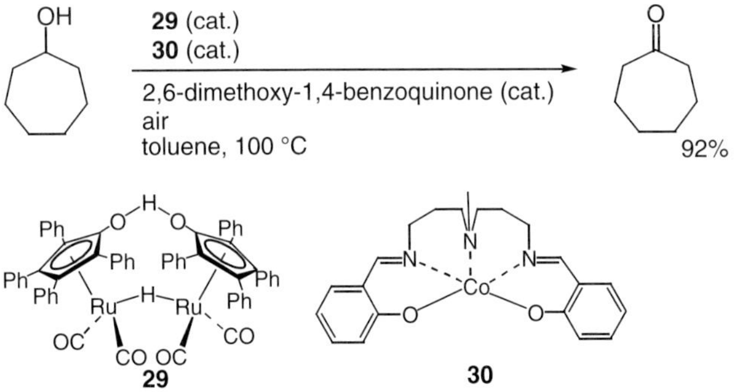

$$\text{(7.38)}$$

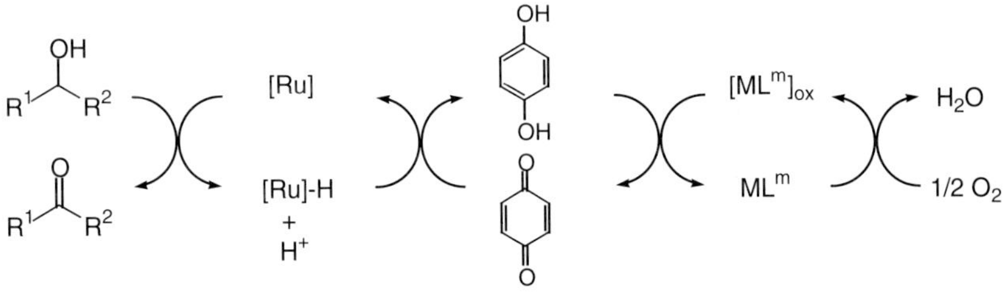

Scheme 7.3

Sheldon and Arends found that the combination $RuCl_2(PPh_3)_3$-TEMPO affords an efficient catalytic system for the aerobic oxidation of a broad range of primary and secondary alcohols at 100 °C, giving the corresponding aldehydes and ketones, respectively, in >99% selectivity in all cases [86]. The reoxidation of the ruthenium hydride species with TEMPO was proposed in the latter system [86c]. Allylic alcohols can be converted into α,β-unsaturated aldehydes with 1 atm of molecular oxygen in the presence of RuO_2 catalyst [87].

TPAP [69] can be used as an effective catalyst for the aerobic oxidation of alcohols to give the corresponding carbonyl compounds (Eq. (7.39)) [88]. A polymer-supported perruthenate (PSP) and a perruthenate immobilized within MCM-41 can be used for heterogeneous oxidation of alcohols [89]. Water-soluble diruthenium complex $[Ru_2(OAc)_3(CO_3)]$ is effective for aerobic oxidation of alcohols in water [90].

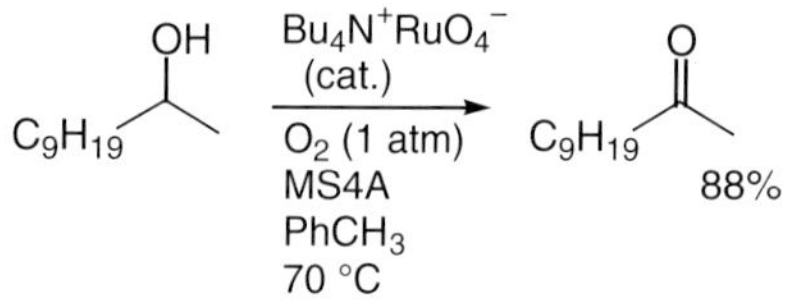

$$C_9H_{19}\text{-CH(OH)-CH}_3 \xrightarrow[\substack{O_2 \text{ (1 atm)} \\ MS4A \\ PhCH_3 \\ 70\ °C}]{Bu_4N^+RuO_4^- \text{ (cat.)}} C_9H_{19}\text{-C(O)-CH}_3 \quad 88\% \tag{7.39}$$

Heterogeneous catalysts such as Ru-Al-Mg-hydrotalcite, Ru-Co-Al-hydrotalcite, Ru-hydroxyapatite (RuHAP) (Eq. (7.40)) [91], Ru-Al$_2$O$_3$ [92a,b], and Ru/AlO(OH) [92c] are highly efficient catalysts for aerobic oxidation of alcohols. In these oxidation reactions, the key step is postulated as the reaction of Ru-H with O_2 to form Ru-OOH, in analogy to Pd-OOH, which has been shown to operate in the palladium-catalyzed Wacker-type asymmetric oxidation reaction [93]. It is noteworthy that ruthenium on carbon is simple and efficient for the oxidation of alcohols (Eq. (7.41)) [92d].

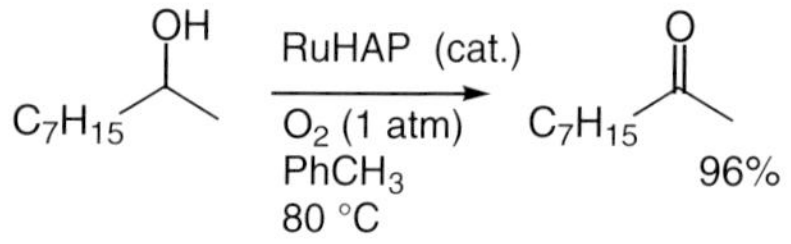

$$C_7H_{15}\text{-CH(OH)-CH}_3 \xrightarrow[\substack{O_2 \text{ (1 atm)} \\ PhCH_3 \\ 80\ °C}]{RuHAP \text{ (cat.)}} C_7H_{15}\text{-C(O)-CH}_3 \quad 96\% \tag{7.40}$$

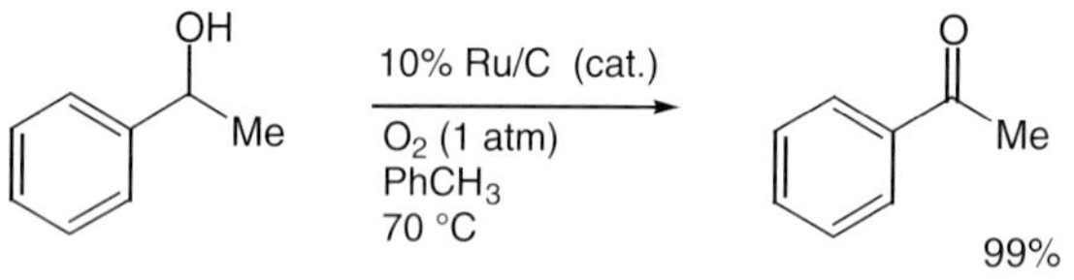

$$Ph\text{-CH(OH)-Me} \xrightarrow[\substack{O_2 \text{ (1 atm)} \\ PhCH_3 \\ 70\ °C}]{10\%\ Ru/C \text{ (cat.)}} Ph\text{-C(O)-Me} \quad 99\% \tag{7.41}$$

Polystyrene-based immobilized ruthenium catalysts can be used in oxidation of alcohols with NMO, TEMPO/O$_2$, or O$_2$ [94]. Catalytic oxidative cleavage of vicinal diols to aldehydes with dioxygen was reported with $RuCl_2(PPh_3)_3$ on active carbon [95]. Ionic liquids such as tetramethyl ammonium hydroxide and Aliquat® 336 can be used as the solvent for the $RuCl_2(PPh_3)_3$-catalyzed aerobic oxidation of alcohols [96].

7.3 Oxidation with Low-Valent Ruthenium Catalysts and Oxidants

Kinetic resolution of secondary alcohols and desymmetrization of diols were reported by asymmetric oxidation using chiral (nitrosyl)Ru(salen) chloride (31) (Eq. (7.42)) [97] in addition to the aerobic oxidation reaction [97d].

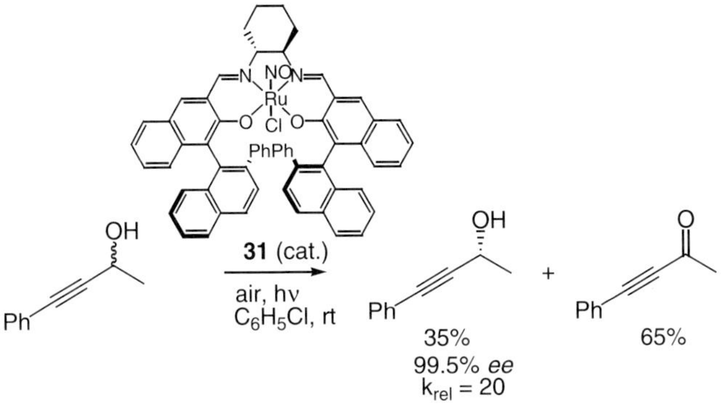

$$\text{Ph} \underset{}{\overset{OH}{\bigwedge}}\!\!\!\equiv \quad \xrightarrow[\text{air, } h\nu \;\; C_6H_5Cl, \text{ rt}]{31 \text{ (cat.)}} \quad \text{Ph} \underset{}{\overset{OH}{\bigwedge}}\!\!\!\equiv \; + \; \text{Ph} \underset{}{\overset{O}{\bigwedge}}\!\!\!\equiv \quad (7.42)$$

35%, 99.5% ee, $k_{rel} = 20$; 65%

7.3.3 Oxidation of Amines

Selective oxidative demethylation of tertiary methyl amines is one of the specific and important functions of cytochrome P-450. Novel cytochrome P-450-type oxidation behavior with tertiary amines has been found in the catalytic systems of low-valent ruthenium complexes with peroxides. These systems exhibit specific reactivity toward oxidations of nitrogen compounds such as amines and amides, differing from that with RuO_4. Low-valent ruthenium complex-catalyzed oxidation of tertiary methylamines with t-BuOOH gives the corresponding α-($tert$-butyldioxy)alkylamines efficiently (Eq. (7.43)) [98]. The hemiaminal-type product has a similar structure to that of the α-hydroxymethylamine intermediate derived from the oxidation with cytochrome P-450.

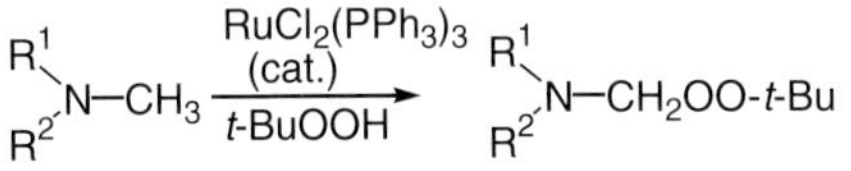

$$R^1R^2N\text{-}CH_3 \xrightarrow[t\text{-BuOOH}]{RuCl_2(PPh_3)_3 \text{ (cat.)}} R^1R^2N\text{-}CH_2OO\text{-}t\text{-Bu} \qquad (7.43)$$

As shown in Scheme 7.4, the catalytic oxidation reactions can be rationalized by assuming the formation of oxo-ruthenium species by the reaction of low-valent

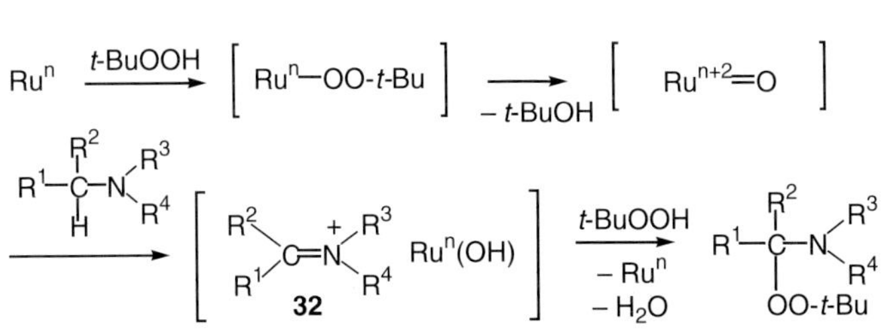

$$Ru^n \xrightarrow{t\text{-BuOOH}} [Ru^n\text{-OO-}t\text{-Bu}] \xrightarrow{-t\text{-BuOH}} [Ru^{n+2}=O]$$

$$R^1\underset{H}{\overset{R^2}{\underset{|}{C}}}\text{-}N\overset{R^3}{\underset{R^4}{\diagdown}} \longrightarrow \left[\underset{R^1}{\overset{R^2}{\diagdown}}C=N\overset{+}{\underset{R^4}{\diagup}}R^3 \;\; Ru^n(OH) \right] \xrightarrow[-\text{Ru}^n \; -H_2O]{t\text{-BuOOH}} R^1\underset{|}{\overset{R^2}{\underset{|}{C}}}\text{-}N\overset{R^3}{\underset{R^4}{\diagdown}}$$

32 ; OO-t-Bu

Scheme 7.4

ruthenium complexes with peroxides. α-Hydrogen abstraction from amines and the subsequent electron transfer gives the iminium ion ruthenium complex **32**. Trapping **32** with *t*-BuOOH would afford the corresponding α-*tert*-butyldioxyamine, water, and low-valent ruthenium complex to complete the catalytic cycle.

The oxidation of N-methylamines provides various useful methods for organic synthesis. Selective demethylation of tertiary methylamines can be performed by ruthenium-catalyzed oxidation and subsequent hydrolysis (Eq. (7.44)) [98]. This is the first practical synthetic method for the N-demethylation of tertiary amines. The methyl group is removed chemo-selectively in the presence of various alkyl groups.

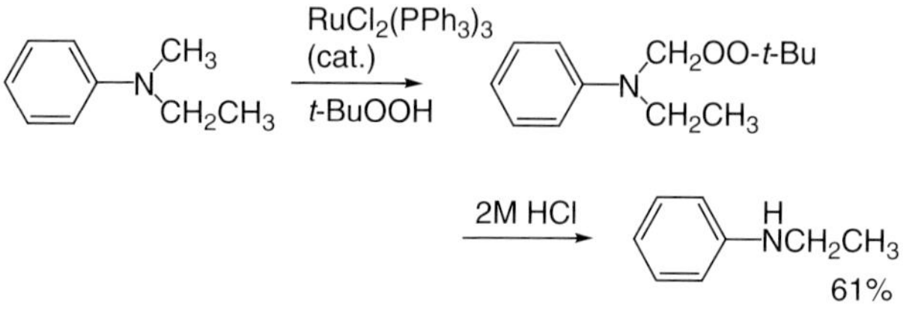

$$ \text{(7.44)} $$

Biomimetic construction of piperidine skeletons from N-methylhomoallylamines is performed by means of the ruthenium-catalyzed oxidation and subsequent alkene-iminium ion cyclization reaction. *trans*-1-Phenyl-3-propyl-4-chloropiperidine (**34**) was obtained from N-methyl-N-(3-heptenyl)aniline stereoselectively via **33** (55%) upon treatment with a 2 M HCl solution (Eq. (7.45)).

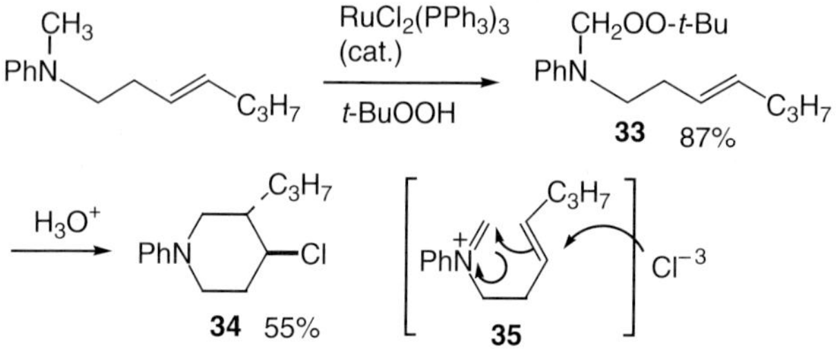

$$ \text{(7.45)} $$

This cyclization can be rationalized by assuming the formation of iminium ion **35** by protonation of the oxidation product **33**, subsequent elimination of *t*-BuOOH, and nucleophilic attack of an alkene, giving a carbonium ion, which is trapped with the Cl⁻ nucleophile from the less hindered side. Similar treatment using CF_3CO_2H in place of HCl gave the corresponding hydroxy derivative (Eq. (7.46)).

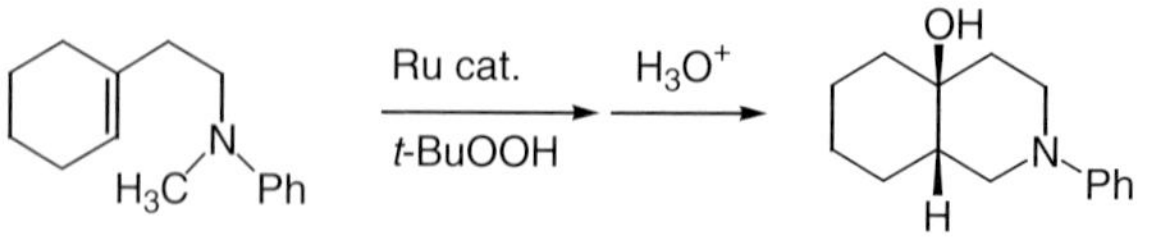

$$ \text{(7.46)} $$

α-Methoxylation of tertiary amines can be carried out upon treatment with hydrogen peroxide in the presence of $RuCl_3$ catalyst in MeOH [99]. Thus, the oxidation of tertiary amine **36** gave the corresponding α-methoxyamine **37** in 60% yield (Eq. (7.47)).

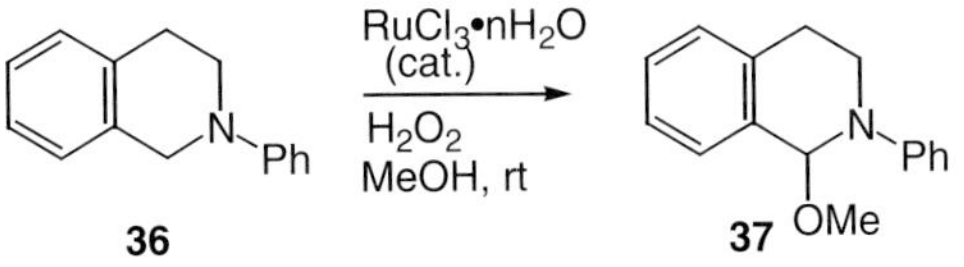

7.3 Oxidation with Low-Valent Ruthenium Catalysts and Oxidants

$$\text{36} \xrightarrow[\text{MeOH, rt}]{\substack{\text{RuCl}_3\cdot n\text{H}_2\text{O}\\ \text{(cat.)} \\ \text{H}_2\text{O}_2}} \text{37} \quad (7.47)$$

Oxidation with molecular oxygen and direct carbon-carbon bond formation by trapping the iminium ion intermediate with a carbon nucleophile can be performed under oxidative conditions. Thus, RuCl$_3$-catalyzed oxidative cyanation of tertiary amines with NaCN gives the corresponding α-aminonitriles, which are useful intermediates for the synthesis of various compounds such as α-amino acids and 1,2-diamines (Eq. (7.48)) [100]. Oxidative cyanation of pyrrolidine derivatives with H$_2$O$_2$ gives the corresponding α-cyanated products in good yields (Eq. (7.49)) [101]. Aminonitriles thus obtained are versatile synthetic intermediates for the construction of quinoline skeletons. For example, TiCl$_4$-promoted reaction of N-cyanomethyl-N-methylaniline (38) with allyltrimethylsilane gave 1-methyl-4-[(trimethylsilyl)methyl]-1,2,3,4-tetrahydroisoquinoline (39), which is an important precursor of isoquinoline alkaloids (Eq. (7.50)).

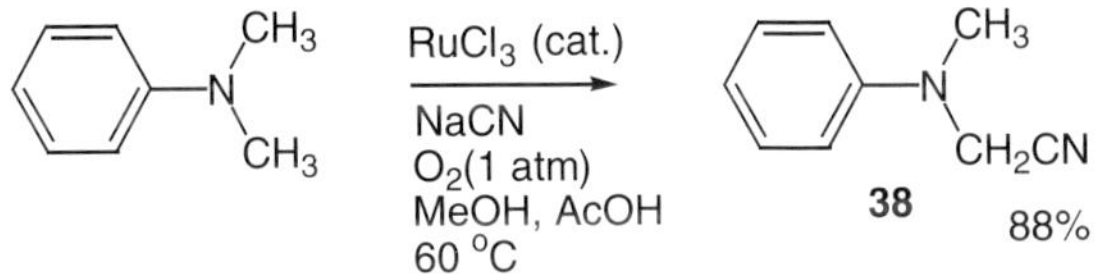

$$(7.48)$$

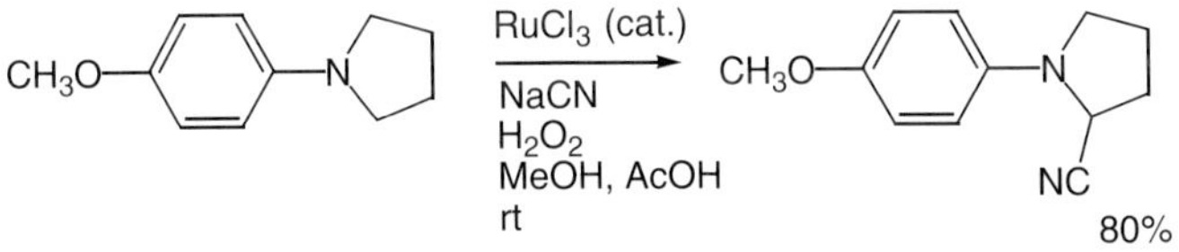

$$(7.49)$$

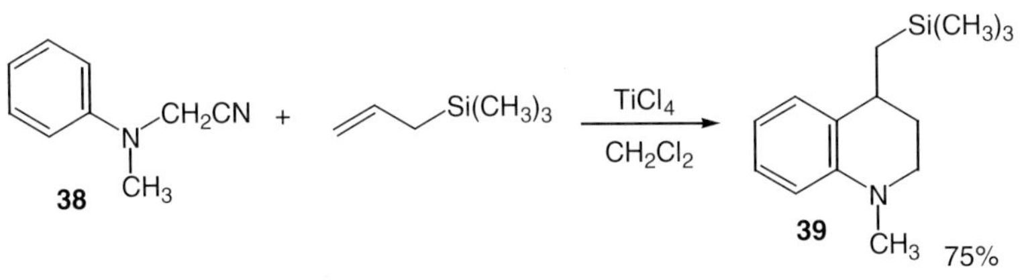

$$(7.50)$$

Tertiary amine N-oxides can be prepared from the corresponding tertiary amines by RuCl$_3$-catalyzed oxidation with molecular oxygen [102].

Secondary amines can be converted into the corresponding imines in a single step highly efficiently through treatment with 2 equiv. of t-BuOOH in benzene in the presence of RuCl$_2$(PPh$_3$)$_3$ catalyst at room temperature [103]. This is the first catalytic oxidative transformation of secondary amines to imines. The oxidations of tetrahydroisoquinoline 40 and allylamine 42 gave the corresponding cyclic imine 41 and azadiene 43 in 93% and 69% yields, respectively (Eqs. (7.51) and (7.52)).

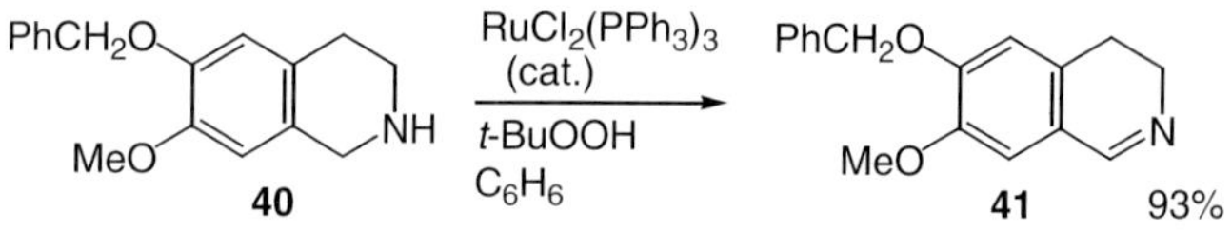

$$\text{PhCH}_2\text{O-}\underset{\underset{40}{\text{MeO}}}{\text{[tetrahydroisoquinoline-NH]}} \xrightarrow[\text{C}_6\text{H}_6]{\underset{\text{(cat.)}}{\text{RuCl}_2(\text{PPh}_3)_3} \; t\text{-BuOOH}} \text{PhCH}_2\text{O-}\underset{\underset{41 \quad 93\%}{\text{MeO}}}{\text{[dihydroisoquinoline=N]}} \quad (7.51)$$

$$\underset{42}{\text{Ph}\diagup\diagdown\diagup\text{N}(\text{H})\text{-C}_6\text{H}_4\text{-Cl}} \longrightarrow \underset{43}{\text{Ph}\diagup\diagdown\diagup\text{N}=\text{C}_6\text{H}_4\text{-Cl}} \quad 69\% \quad (7.52)$$

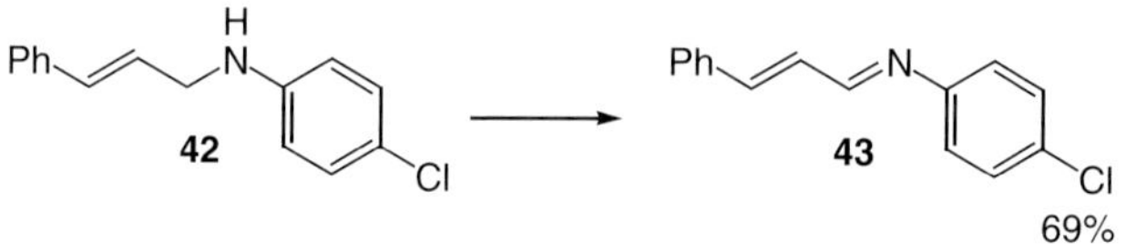

Aromatization takes place when an excess amount of t-BuOOH is used. For example, tetrahydroquinoline (**44**) can be converted into quinoline (**45**) (73%) (Eq. (7.53)).

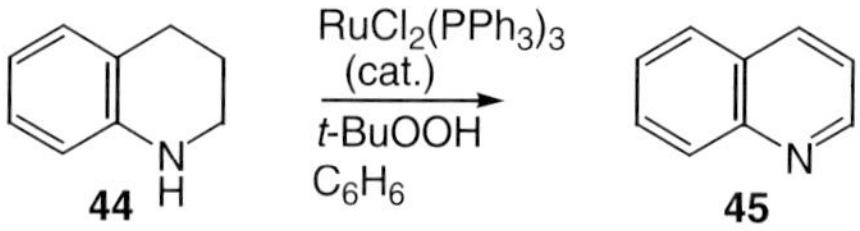

$$\underset{44}{\text{tetrahydroquinoline}} \xrightarrow[\text{C}_6\text{H}_6]{\underset{\text{(cat.)}}{\text{RuCl}_2(\text{PPh}_3)_3} \; t\text{-BuOOH}} \underset{45}{\text{quinoline}} \quad (7.53)$$

It is worth noting that tungstate-catalyzed oxidation of the secondary amine with hydrogen peroxide gives nitrone (**46**) (Eq. (7.54)) [104]. These two catalytic transformations of secondary amines (Eqs. (7.51) and (7.54)) are particularly useful for the introduction of a substituent at the α-position of the amines, because either imines or nitrones undergo diastereo and enantioselective reactions with nucleophiles to give chiral α-substituted amines highly efficiently [105].

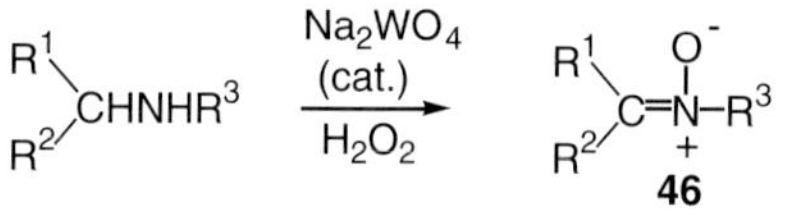

$$\underset{R^2}{\overset{R^1}{\diagdown}}\text{CHNHR}^3 \xrightarrow[\text{H}_2\text{O}_2]{\underset{\text{(cat.)}}{\text{Na}_2\text{WO}_4}} \underset{\underset{46}{R^2}}{\overset{R^1}{\diagdown}}\text{C}=\overset{\text{O}^-}{\underset{+}{\text{N}}}\text{-R}^3 \quad (7.54)$$

The catalytic system consisting of $(n\text{-Pr}_4\text{N})(\text{RuO}_4)$ and N-methylmorpholine N-oxide (NMO) can be also used for oxidative transformation of secondary amines to imines (Eq. (7.55)) [106].

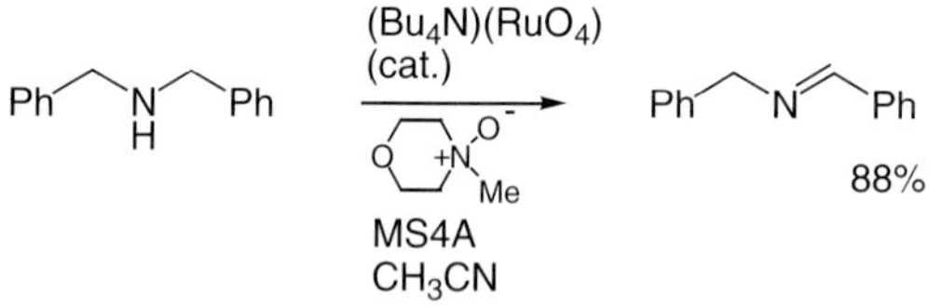

$$\text{Ph-N(H)-Ph} \xrightarrow[\substack{\text{MS4A} \\ \text{CH}_3\text{CN}}]{\substack{(\text{Bu}_4\text{N})(\text{RuO}_4) \\ \text{(cat.)} \\ \text{NMO}}} \text{Ph-N=CH-Ph} \quad 88\% \quad (7.55)$$

The ruthenium-catalyzed oxidation of diphenylmethylamine with t-BuOOH gave benzophenone (88%), which was formed by hydrolysis of the imine intermediate (Eq. (7.56)) [103].

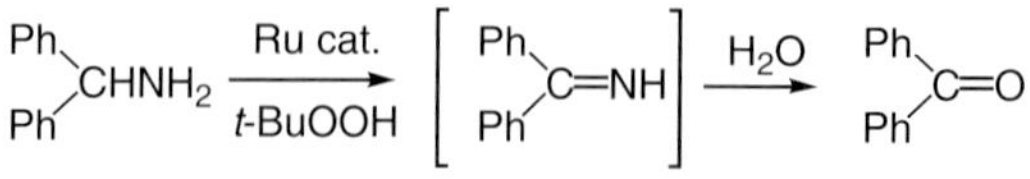

$$\underset{\text{Ph}}{\overset{\text{Ph}}{\diagdown}}\text{CHNH}_2 \xrightarrow[t\text{-BuOOH}]{\text{Ru cat.}} \left[\underset{\text{Ph}}{\overset{\text{Ph}}{\diagdown}}\text{C}=\text{NH} \right] \xrightarrow{\text{H}_2\text{O}} \underset{\text{Ph}}{\overset{\text{Ph}}{\diagdown}}\text{C}=\text{O} \quad (7.56)$$

7.3 Oxidation with Low-Valent Ruthenium Catalysts and Oxidants

Murahashi et al. found that aerobic oxidative transformation of secondary amines to the corresponding imines can be carried out highly efficiently using the bimetallic ruthenium catalyst $Ru_2(OAc)_4Cl$ (Eq. (7.57)) [107].

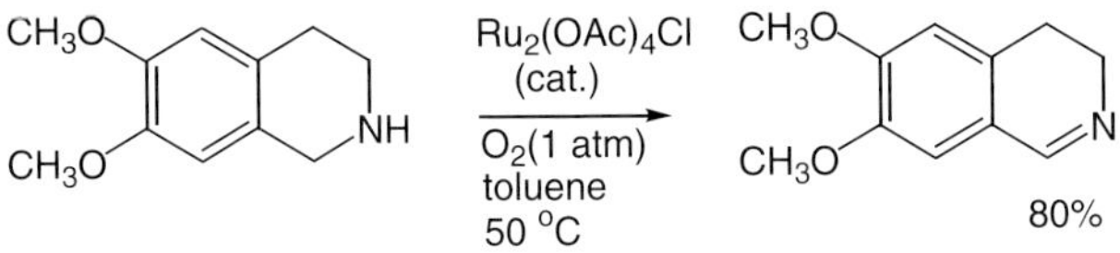

$$\text{6,7-dimethoxy-1,2,3,4-tetrahydroisoquinoline} \xrightarrow[\substack{O_2(1 \text{ atm}) \\ \text{toluene} \\ 50\,°C}]{Ru_2(OAc)_4Cl\ (\text{cat.})} \text{6,7-dimethoxy-3,4-dihydroisoquinoline} \quad 80\%$$

(7.57)

Bäckvall et al. found that the oxidation of secondary amines to imines can be performed by a hydrogen transfer reaction under mild conditions in the presence of 2,6-dimethoxybenzoquinone and the ruthenium catalyst **29** (Eq. (7.58)) [108].

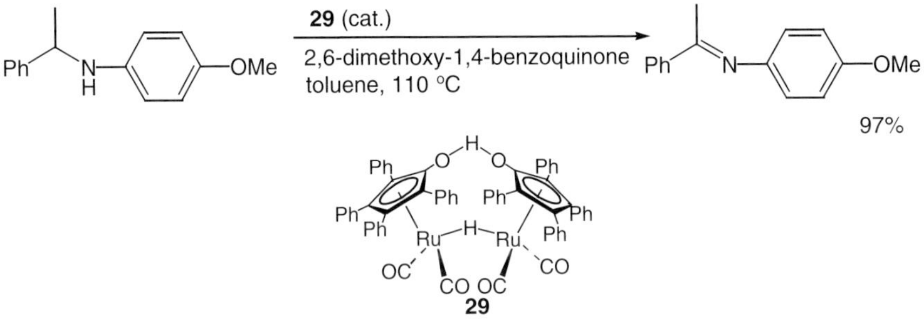

$$\text{Ph-NH-C}_6\text{H}_4\text{-OMe} \xrightarrow[\substack{\text{2,6-dimethoxy-1,4-benzoquinone} \\ \text{toluene, 110\,°C}}]{29\ (\text{cat.})} \text{Ph-N=C}_6\text{H}_4\text{-OMe} \quad 97\%$$

Catalyst **29**: $[Ru_2(\eta^5\text{-}C_5Ph_4)(\mu\text{-}H)(\mu\text{-}OH)(CO)_4]$ type complex

(7.58)

Potassium ruthenate (K_2RuO_4) was used as a catalyst for the oxidation of benzylamine with $K_2S_2O_8$ [109]. Primary amines can be converted to nitriles readily by aerobic oxidation reactions. James and coworkers reported that aerobic oxidative transformation of primary amines to nitriles can be carried out in the presence of a ruthenium porphyrin complex $Ru(TMP)(O)_2$ (**1**) (100%) (Eq. (7.59)) [110].

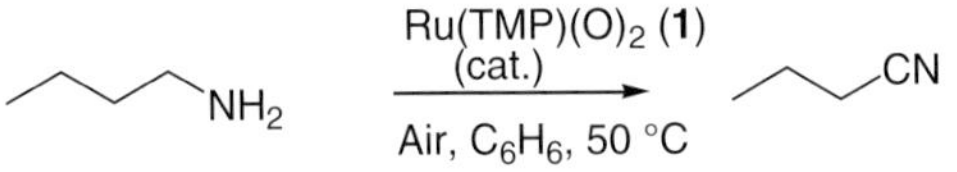

$$R\text{-}NH_2 \xrightarrow[\text{Air, }C_6H_6,\ 50\,°C]{Ru(TMP)(O)_2\ (\mathbf{1})\ (\text{cat.})} R\text{-}CN$$

(7.59)

Heterogeneous catalysts such as hydroxyapatite-bound Ru [111] (Eq. (7.60)) and Ru/Al_2O_3 can be also used for the aerobic oxidation of primary amines to nitriles (Eq. (7.61)) [112, 113].

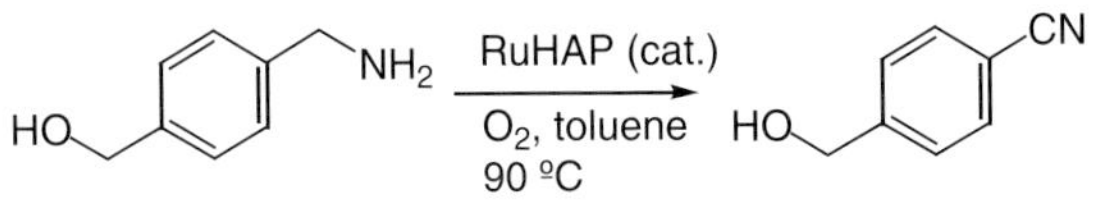

$$HO\text{-}C_6H_4\text{-}CH_2NH_2 \xrightarrow[\substack{O_2,\ \text{toluene} \\ 90\,°C}]{\text{RuHAP (cat.)}} HO\text{-}C_6H_4\text{-}CN$$

(RuHAP = hydroxyapatite-bound Ru)

(7.60)

$$o\text{-MeO-}C_6H_4\text{-}CH_2NH_2 \xrightarrow[\substack{O_2(1\text{ atm}) \\ PhCF_3,\ 100\,°C}]{Ru/Al_2O_3\ (\text{cat.})} o\text{-MeO-}C_6H_4\text{-}CN \quad 97\%$$

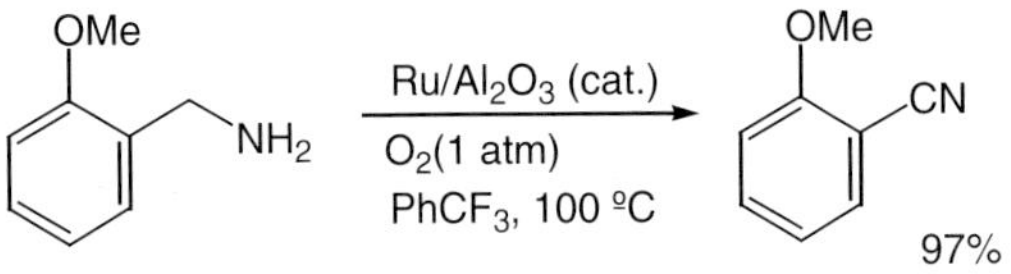

(7.61)

7.3.4
Oxidation of Amides and β-Lactams

The oxidation of the α-C—H bond of amides is an attractive strategy for the synthesis of biologically active nitrogen compounds. Selective oxidation of amides is difficult because of low reactivity. The $RuCl_2(PPh_3)_3$-catalyzed oxidation of amides with t-BuOOH proceeds under mild conditions to give the corresponding α-(tert-butyldioxy)amides **47** highly efficiently (Eq. (7.62)) [114].

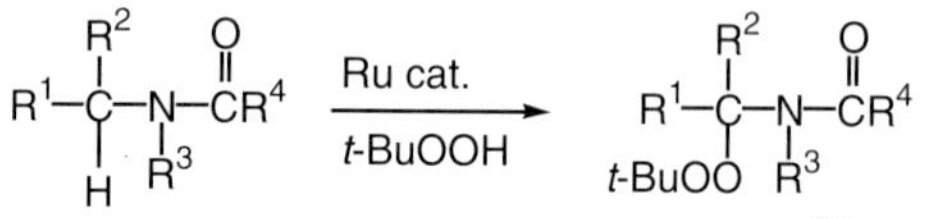

$$R^1-\underset{\underset{H}{|}}{\overset{\overset{R^2}{|}}{C}}-\underset{\underset{R^3}{|}}{N}-\overset{\overset{O}{\|}}{C}R^4 \xrightarrow[\text{t-BuOOH}]{\text{Ru cat.}} R^1-\underset{\underset{\text{t-BuOO}}{|}}{\overset{\overset{R^2}{|}}{C}}-\underset{\underset{R^3}{|}}{N}-\overset{\overset{O}{\|}}{C}R^4 \quad (7.62)$$

47

The ruthenium-catalyzed oxidation of 1-(methoxycarbonyl)pyrrolidine with t-BuOOH gives 2-(t-butyldioxy)-1-(methoxycarbonyl)pyrrolidine (**48**) in 60% yield (Eq. (7.63)).

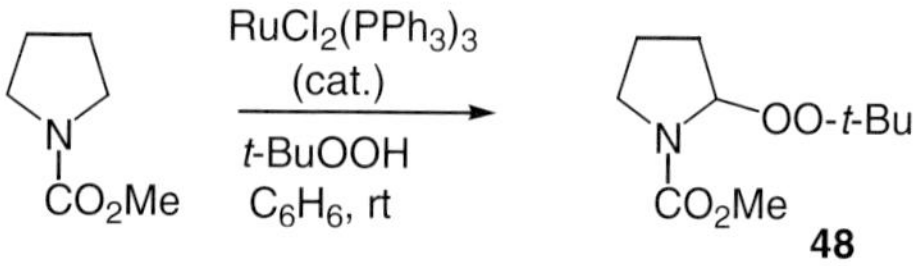

$$(7.63)$$

The *tert*-butyldioxy amides of the isoquinoline **49** and the indole **50**, which are important synthetic intermediates for the synthesis of natural products, were obtained in excellent yields (Eqs. (7.64) and (7.65)). Since the Lewis acid-promoted reactions of these oxidized products with nucleophiles give the corresponding N-acyl-α-substituted amines efficiently, the present reactions provide versatile methods for selective carbon-carbon bond formation at the α-position of amides [115].

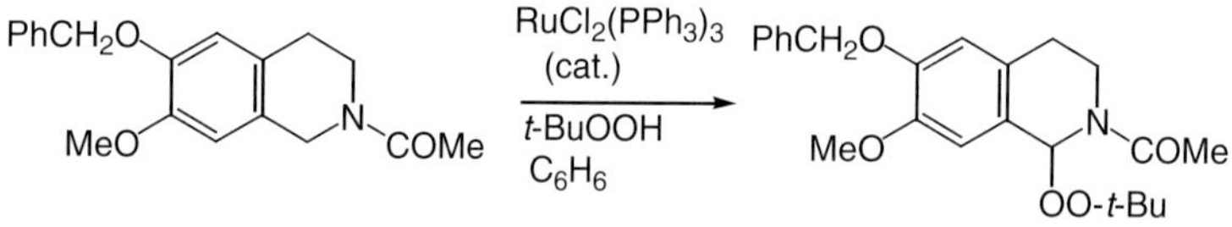

$$(7.64)$$

49 98%

$$(7.65)$$

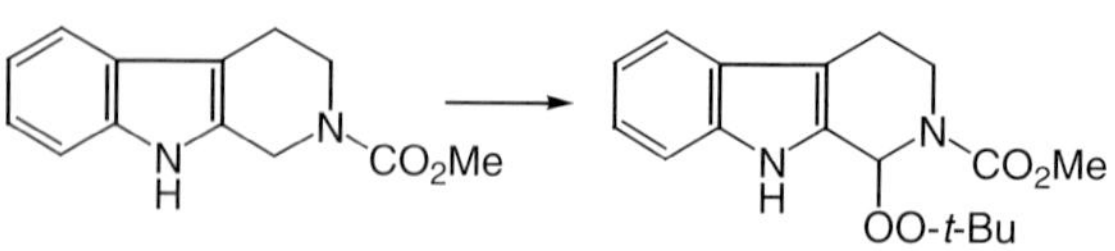

50 92%

Typically, the $TiCl_4$-promoted reaction of α-t-butyldioxypyrrolidine (**48**) with a silyl enol ether gave the keto amide **51** (81%), while a similar reaction with a less reactive 1,3-diene gave the α-substituted amide **52** (Eq. (7.66)).

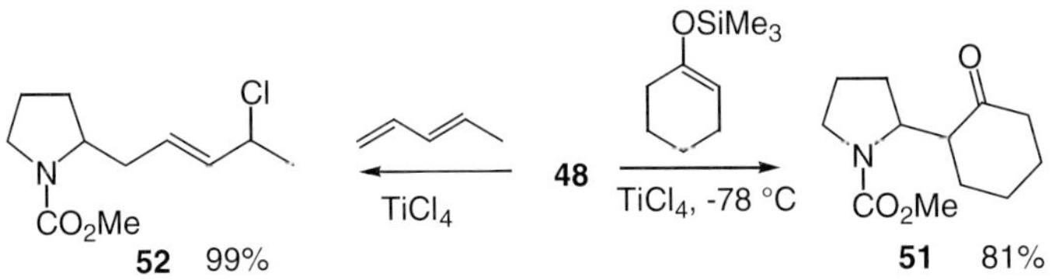

7.3 Oxidation with Low-Valent Ruthenium Catalysts and Oxidants | 261

$$\text{52} \xrightarrow{\text{TiCl}_4} \text{48} \xrightarrow[\text{TiCl}_4, -78\,°C]{\text{OSiMe}_3} \text{51} \tag{7.66}$$

(52, 99%; 51, 81%)

Oxidative modification of peptides has been performed by ruthenium-catalyzed oxidation with peracetic acid. For example, the reaction of N,C-protected peptides containing glycine residues with peracetic acid in the presence of RuCl$_3$ catalyst gives α-ketoamides derived from the selective oxidation at the C^α position of the glycine residue (81%, conv. 70%) (Eq. (7.67)) [116]. Direct conversion of N-acylproline to N-acylglutamate was achieved by Ru(TMP)Cl$_2$ and 2,6-dichloropyridine N-oxide (Eq. (7.68)) [117].

$$\text{Ac-Gly-Ala-OEt} \xrightarrow[\substack{\text{CH}_3\text{CO}_3\text{H} \\ \text{AcOH}}]{\text{RuCl}_3 \text{ (cat.)}} \text{AcNHCOCO-Ala-OEt} \tag{7.67}$$

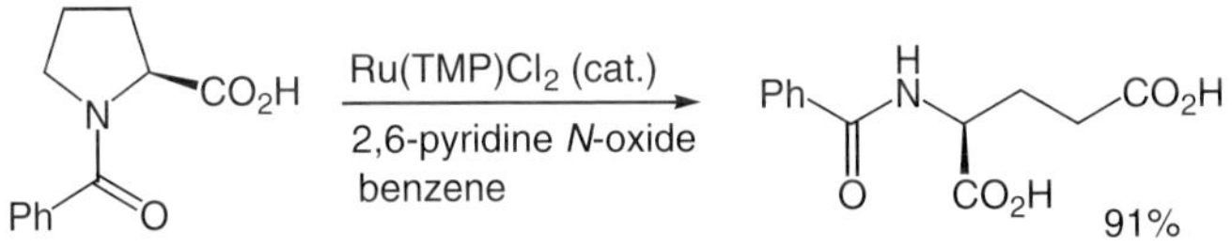

$$\tag{7.68}$$

(91%)

One of the most challenging topics among the oxidations of amides is the catalytic oxidation of β-lactams. Oxidation of β-lactams requires specific reaction conditions because of the high strain of the four-membered rings. Direct oxidative acyloxylation of β-lactams was successfully carried out by ruthenium-catalyzed oxidation with peracetic acid in acetic acid under mild conditions. The products obtained are highly versatile and are key intermediates in the synthesis of antibiotics. Thus, the ruthenium-catalyzed oxidation of 2-azetidinones with peracetic acid in acetic acid in the presence of sodium acetate at room temperature gives the corresponding 4-acetoxy-2-azetidinones **53** in 94% yield (Eq. (7.69)) [114]. One can use RuCl$_2$(PPh$_3$)$_3$ or RuCl$_3$, but for practical synthesis ruthenium on charcoal can be used conveniently. Although peracetic acid is the best oxidant, other oxidants such as m-chloroperbenzoic acid, methyl ethyl ketone peroxide, and iodosylbenzene can be used for the acyloxylation of β-lactams.

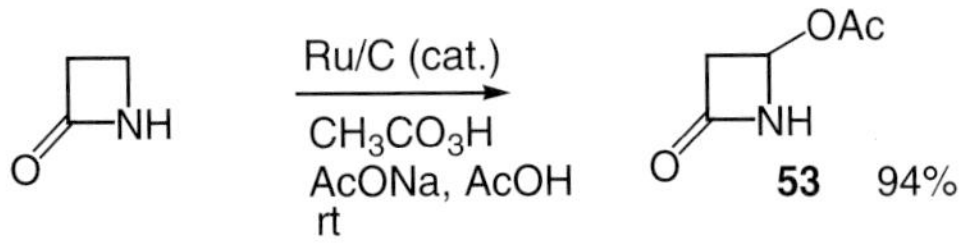

$$\tag{7.69}$$

(**53**, 94%)

Importantly, (1′R,3S)-3-[1′-(tert-butyldimethylsilyloxy)ethyl]azetidin-2-one (**54**) can be converted into the corresponding 4-acetoxyazetidinone **55** with extremely high diastereoselectivity (94%, >99%*de*) (Eq. (7.70)). The product (**55**) is a versatile and key intermediate for the synthesis of carbapenems of antibiotics. Now, 100 000 kg of the compound **55** is produced per year in the industry.

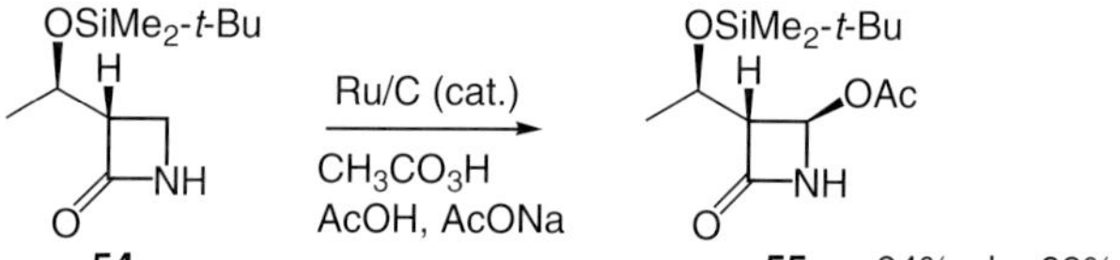

$$\text{54} \xrightarrow[\text{AcOH, AcONa}]{\underset{\text{CH}_3\text{CO}_3\text{H}}{\text{Ru/C (cat.)}}} \text{55} \quad 94\%, \text{ de}>99\% \tag{7.70}$$

This method was applied to the stereoselective synthesis of 3-amino-4-acetoxyazetidinone **56** in 85% yield (Eq. (7.71)) [118].

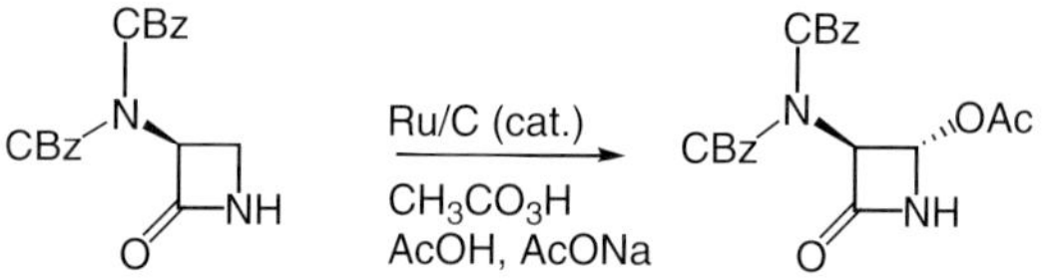

$$\xrightarrow[\text{AcOH, AcONa}]{\underset{\text{CH}_3\text{CO}_3\text{H}}{\text{Ru/C (cat.)}}} \text{56} \tag{7.71}$$

Aerobic oxidation of β-lactams can be performed highly efficiently in the presence of acetaldehyde, an acid, and sodium acetate [119]. Typically, the RuCl₃-catalyzed oxidation of β-lactam **54** with molecular oxygen (1 atm) in the presence of acetaldehyde and sodium carboxylate gave the corresponding 4-acyloxy β-lactam **55** in 91% yields (*de* >99%) (Eq. (7.72)). This aerobic oxidation shows similar reactivity to the ruthenium-catalyzed oxidation with peracetic acid.

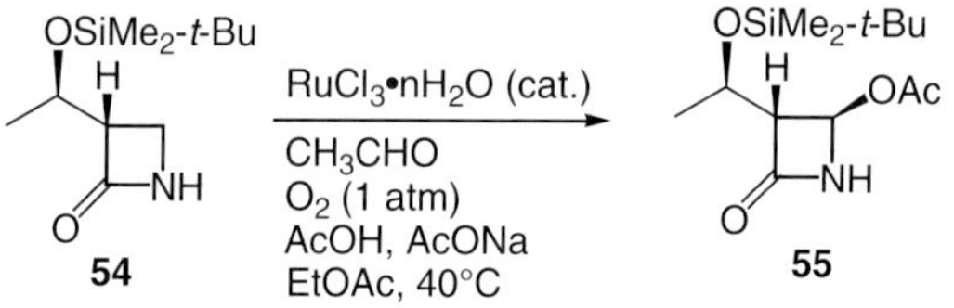

$$\text{54} \xrightarrow[\substack{\text{CH}_3\text{CHO}\\ \text{O}_2\text{ (1 atm)}\\ \text{AcOH, AcONa}\\ \text{EtOAc, 40°C}}]{\text{RuCl}_3 \cdot n\text{H}_2\text{O (cat.)}} \text{55} \quad 91\%, \text{ de}>99\% \tag{7.72}$$

7.3.5
Oxidation of Phenols

The oxidative transformation of phenols is of importance with respect to the biological and synthetic aspects. However, the oxidation of phenols generally lacks selectivity because of coupling reactions caused by phenoxyl radicals [120], and selective oxidation of phenols is limited to phenols bearing bulky substituents at the 2- and 6-positions [121]. Using ruthenium catalysts, a biomimetic and selective oxidation of phenols can be performed. Thus, the oxidation of *p*-substituted phenols bearing no substituent at the 2- and 6-positions with *t*-BuOOH in the presence of RuCl₂(PPh₃)₃ catalyst gives 4-(*tert*-butyldioxy)-4-alkylcyclohexadienones selectively (Eq. (7.73)) [122].

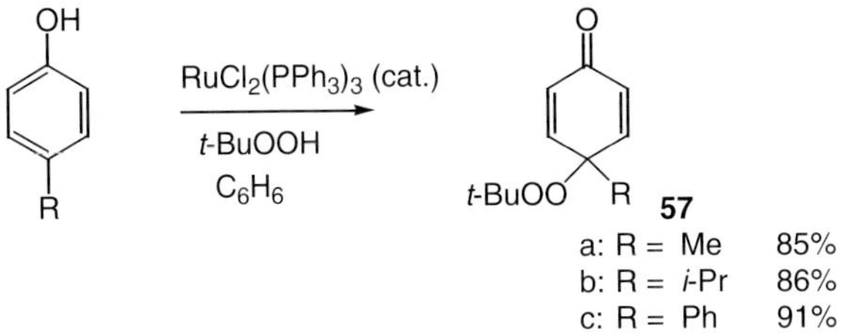

7.3 Oxidation with Low-Valent Ruthenium Catalysts and Oxidants

$$\text{(7.73)}$$

57
a: R = Me 85%
b: R = i-Pr 86%
c: R = Ph 91%

The reaction can be rationalized by assuming the mechanism which involves an oxo-ruthenium complex (Eq. (7.74)). Hydrogen abstraction with oxo-ruthenium species gives the phenoxyl radical **58**, which undergoes fast electron transfer to the ruthenium to give a cationic intermediate, **59**. Nucleophilic reaction with the second molecule of t-BuOOH gives the product, **57**.

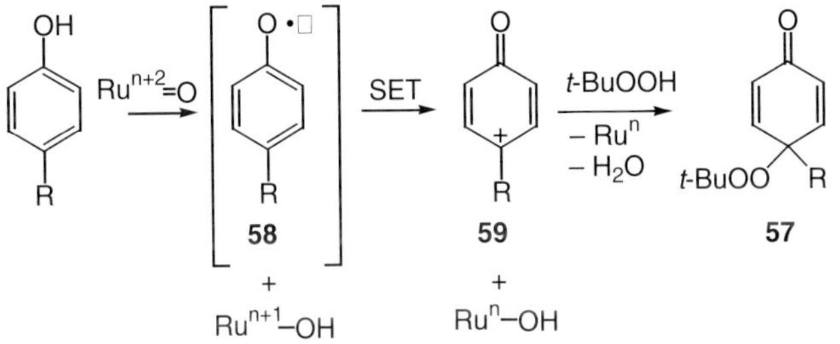

$$\text{(7.74)}$$

The 4-(*tert*-butyldioxy)-4-alkylcyclohexadienones **57** thus obtained are versatile synthetic intermediates. The $TiCl_4$-promoted transformation of **60**, obtained from the oxidation of 3-methyl-4-isopropylphenol, gives 2,6-disubstituted quinones **61**, which is derived from rearrangement of the i-Pr group (93%) (Eq. (7.75)).

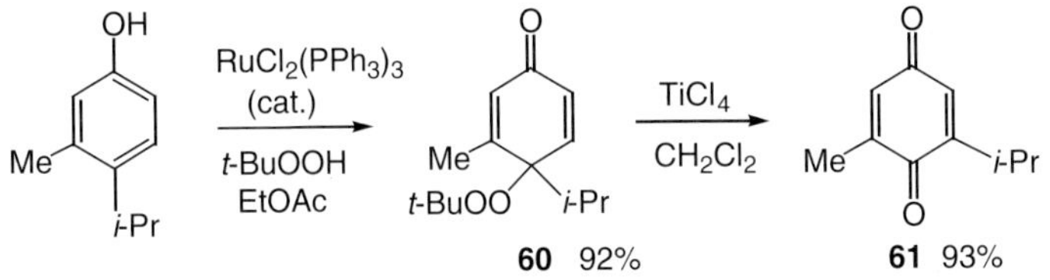

$$\text{(7.75)}$$

Interestingly, sequential migration–Diels–Alder reactions of *tert*-butyldioxy dienone **63** in the presence of *cis*-1,3-pentadiene gave cis-fused octahydroanthraquinone **64** stereoselectively (73%) (Eq. (7.76)).

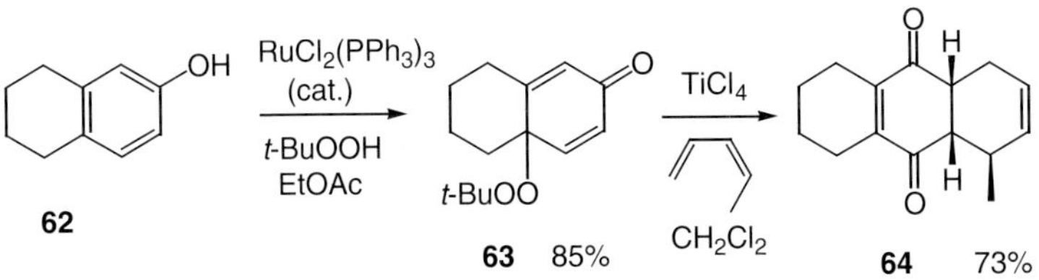

$$\text{(7.76)}$$

Ruthenium-catalyzed oxygenation of catechols gives muconic acid anhydride and 2H-pyran-2-one [123].

Oxidation of an aromatic ring bearing methoxy groups was performed using a ruthenium porphyrin catalyst. The Ru(TPP)(O)$_2$-catalyzed (TPP = 5,10,15,20-tetraphenylporphyrinato) oxidation of polymethoxybenzene with 2,6-dichloropyridine N-oxide gives the corresponding p-quinone derivatives (Eq. (7.77)) [124]. ^{18}O labeling experiments showed that the reaction proceeds via selective hydroxylation of the aromatic ring by oxo-ruthenium porphyrins to give phenol derivatives, which undergo subsequent oxidation to afford the corresponding quinones. Oxidation of anisole to p-benzoquinone can be performed by ruthenium complex of 1,4,7-trimethyl-1,4,7-triazacyclononane with tert-butyl hydroperoxide [125].

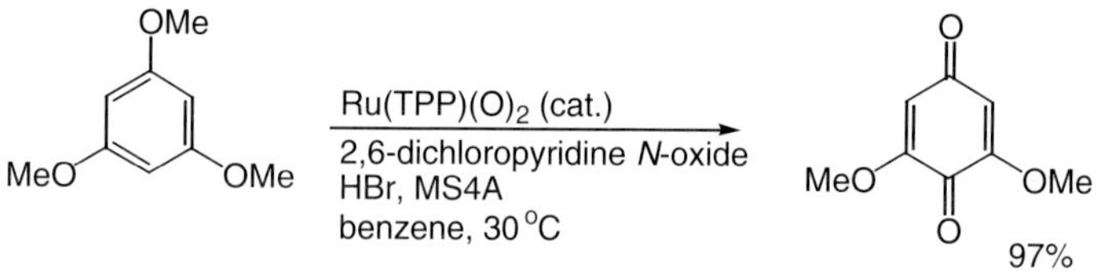

(7.77)

Oxidation of naphthalene derivatives gave naphthoquinones (Eq. (7.78)) [126]. Oxidative coupling can be carried out with supported ruthenium catalyst, Ru(OH)$_2$/Al$_2$O$_3$ (Eq. (7.79)) [127].

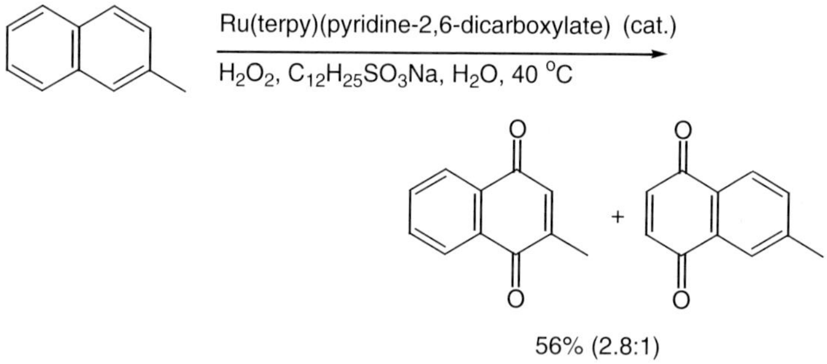

(7.78)

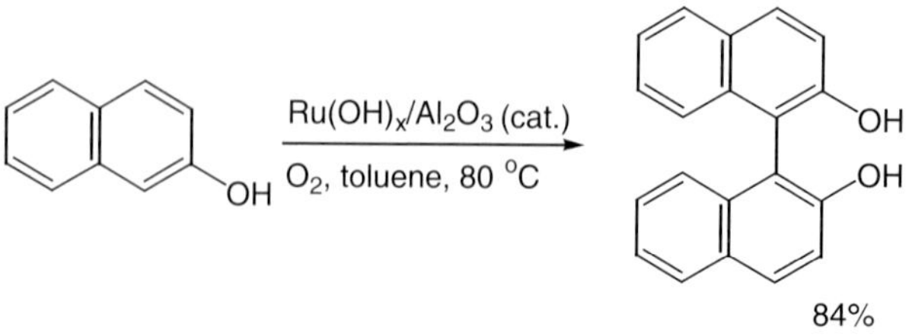

(7.79)

7.3.6
Oxidation of Hydrocarbons

The catalytic oxidation of unactivated hydrocarbons remains a challenging topic. Ruthenium porphyrins, such as Ru(OEP)(PPh$_3$)$_3$, show catalytic activity for the oxidation of alkanes with PhIO [38]. The oxidation of alkanes with 2,6-dichloropyridine N-oxide in the presence of Ru(TMP)(O)$_2$ (3) and HBr [128] and Ru(TPFPP)(CO) (TPFPP = tetrakis(pentafluorophenyl)porphyrinato) (4) [40a] gives the corresponding oxidized compounds. Hydroxylation of adamantane was achieved with high selectivity and high efficiency (12 300 turnovers) (Eq. (7.80)). Ru-porphyrin-catalyzed amidation of hydrocarbons can be performed with PhI=NTs [129]. Zeolite-encapsulated perfluorinated ruthenium phthalocyanines catalyze the oxidation of cyclohexane with t-BuOOH [130]. The addition of Lewis acids such as ZnCl$_2$ greatly accelerated the reaction rates in the stoichiometiric oxidation of alkanes by BaRu(O)$_2$(OH)$_3$ [131]. A dioxoruthenium complex with a D_4-chiral porphyrin ligand has been used for the enantioselective hydroxylation of ethylbenzene to give α-phenylethyl alcohol with 72% ee [132].

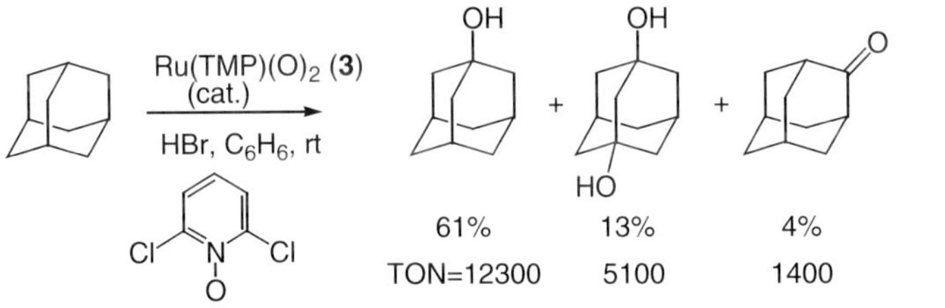

$$\text{adamantane} \xrightarrow[\text{HBr, C}_6\text{H}_6\text{, rt}]{\text{Ru(TMP)(O)}_2 \text{ (3) (cat.)}} \text{1-adamantanol} + \text{2-adamantanol} + \text{2-adamantanone} \quad (7.80)$$

61% (TON=12300), 13%, 4%
5100, 1400

Non-porphyrin ruthenium complexes can be used for the catalytic oxidation of alkanes with peroxides. BaRuO$_3$(OH)$_2$-catalyzed oxidation of cyclohexane with PhIO gives oxidation products [131a]. [RuCl(dpp)$_2$]$^+$ can be used for the oxidation of alkanes with PhIO or LiClO [45]. The combinations cis-[Ru(dmp)$_2$(MeCN)$_2$]$^{2+}$/H$_2$O$_2$ [50a] (Eq. (7.81)), cis-[Ru(Me$_3$tacn)(O)$_2$(CF$_3$CO$_2$)]$^+$/t-BuOOH (Me$_3$tacn = N,N′,N″-1,4,7-trimethyl-1,4,7-triazacyclononane) [46c], and cis-[Ru(6,6-Cl$_2$bpy)$_2$(OH$_2$)$_2$]$^{2+}$/t-BuOOH [46b] are efficient for the oxidation of cyclohexane.

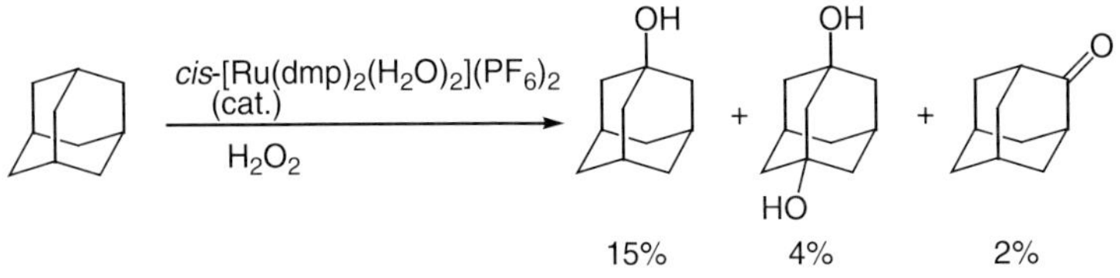

$$\text{adamantane} \xrightarrow[\text{H}_2\text{O}_2]{\textit{cis}\text{-[Ru(dmp)}_2\text{(H}_2\text{O)}_2\text{](PF}_6\text{)}_2 \text{ (cat.)}} \text{1-adamantanol} + \text{2-adamantanol} + \text{2-adamantanone} \quad (7.81)$$

15%, 4%, 2%

Ruthenium(III) complexes such as [RuCl$_2$(TPA)]$^+$ and [RuCl(Me$_2$SO)(TPA)]$^+$ bearing tripodal ligand TPA (TPA = tris(2-pyridylmethyl)amine) were synthesized, and catalytic oxidation of adamantane with m-chloroperbenzoic acid [133, 134a] or 2,6-dichloropyridine N-oxide [134b] was reported. Polyoxometalate [SiRu(H$_2$O)W$_{11}$O$_{39}$]$^{5-}$ also works as an oxidation catalyst using KHSO$_5$ [135a] and H$_2$O$_2$ [135b].

The oxidation of hydrocarbons with ruthenium catalysts bearing a simple ligand is highly effective. The oxidations of hydrocarbons with peroxides such as t-BuOOH and peracetic acid in the presence of ruthenium catalysts such as $RuCl_2(PPh_3)_3$ [136a,b] or Ru/C [136a,c] gave the corresponding ketones and alcohols. For example, $RuCl_2(PPh_3)_3$-catalyzed oxidation of fluorene with t-BuOOH gives fluorenone in 87% yield (Eq. (7.82)). The Ru/C-catalyzed oxidation of cyclohexane with peracetic acid in ethyl acetate gives cyclohexanone and cyclohexanol with 67% conversion [136c].

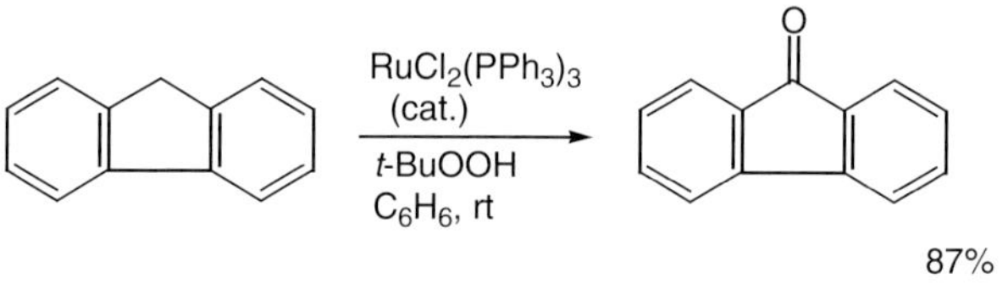

$$\text{fluorene} \xrightarrow[\substack{t\text{-BuOOH} \\ C_6H_6,\ rt}]{RuCl_2(PPh_3)_3\ (\text{cat.})} \text{fluorenone} \quad 87\% \tag{7.82}$$

It is expected that more reactive species will be generated in the presence of a strong acid. Indeed, the $RuCl_3 \cdot nH_2O$-catalyzed oxidation of cyclohexane in trifluoroacetic acid and dichloromethane (5:1) with peracetic acid gives cyclohexyl trifluoroacetate in 77% (Eq. (7.83)) [136a].

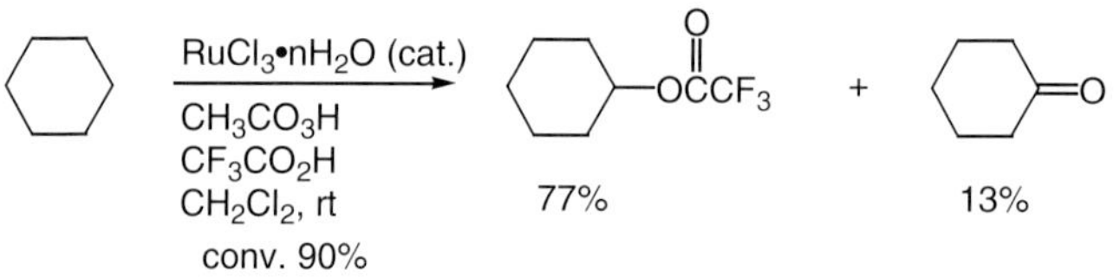

$$\text{cyclohexane} \xrightarrow[\substack{CH_3CO_3H \\ CF_3CO_2H \\ CH_2Cl_2,\ rt \\ \text{conv. } 90\%}]{RuCl_3 \cdot nH_2O\ (\text{cat.})} \text{C}_6\text{H}_{11}\text{-OCCF}_3\ (77\%) + \text{cyclohexanone}\ (13\%) \tag{7.83}$$

The ruthenium-catalyzed oxidation of nitriles takes place at the α-position to nitriles. For example, the $RuCl_3 \cdot nH_2O$-catalyzed oxidation of benzylcyanide with t-BuOOH gives benzoylcyanide in 94% yield [137].

The allylic position of steroidal alkene can be oxidized with t-BuOOH in the presence of $RuCl_3$ catalyst (Eq. (7.84)) [138].

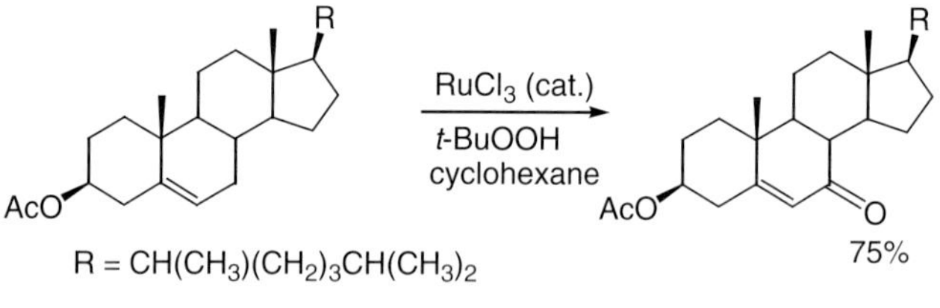

$$\text{steroidal alkene} \xrightarrow[\substack{t\text{-BuOOH} \\ \text{cyclohexane}}]{RuCl_3\ (\text{cat.})} \text{enone}\ 75\% \tag{7.84}$$

$R = CH(CH_3)(CH_2)_3CH(CH_3)_2$

The catalytic oxidation of alkanes with molecular oxygen under mild conditions is an especially rewarding goal, since direct functionalization of the unactivated C–H bonds of saturated hydrocarbons usually requires drastic conditions such as high temperature.

Oxo-metal species can be generated by the reaction of a low valent ruthenium complex with molecular oxygen in the presence of an aldehyde [119]. Thus, the ruthenium-catalyzed oxidation of alkanes with molecular oxygen in the presence of acetaldehyde gives alcohols and ketones efficiently [139a]. Typically,

7.3 Oxidation with Low-Valent Ruthenium Catalysts and Oxidants

$RuCl_3 \cdot nH_2O$-catalyzed oxidation of cyclooctane with molecular oxygen in the presence of an aldehyde give the corresponding alcohols and ketones selectively (Eq. (7.85)).

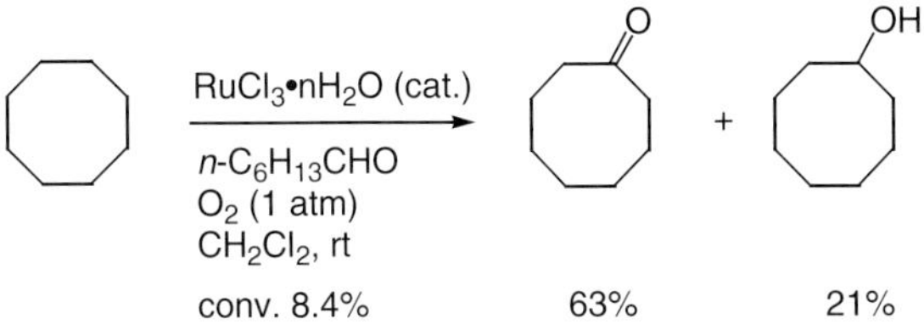

$$\text{cyclooctane} \xrightarrow[\substack{n\text{-}C_6H_{13}CHO \\ O_2 \text{ (1 atm)} \\ CH_2Cl_2, \text{ rt}}]{RuCl_3 \cdot nH_2O \text{ (cat.)}} \text{cyclooctanone} + \text{cyclooctanol} \quad (7.85)$$

conv. 8.4%   63%   21%

These aerobic oxidations can be rationalized by assuming the sequence shown in Scheme 7.5. The metal-catalyzed radical chain reaction of an aldehyde with molecular oxygen affords the corresponding peracid. The reaction of metal catalyst with the peracid thus formed would give an oxo-metal intermediate, followed by oxygen atom transfer to afford the corresponding alcohols. The alcohol is further oxidized to the corresponding ketone under these conditions.

$$RCHO + O_2 \xrightarrow{M \text{ cat.}} RCO_3H$$

$$M^n + RCO_3H \longrightarrow M^{n+2}=O + RCO_2H$$

$$M^{n+2}=O + RH \longrightarrow M^n + ROH$$

Scheme 7.5

A Ru(TPFPP)(CO) (4) complex has been prepared, and it was found that 4 is an efficient catalyst for the aerobic oxidation of alkanes using acetaldehyde [140]. Thus, the 4-catalyzed oxidation of cyclohexane with molecular oxygen in the presence of acetaldehyde gave cyclohexanone and cyclohexanol in 62% yields based on acetaldehyde with high turnover numbers of 14 100 (Eq. (7.86)). It is worth to note that iron [139] and copper [141] catalysts are also efficient for the oxidation of nonactivated hydrocarbons at room temperature under 1 atm of molecular oxygen.

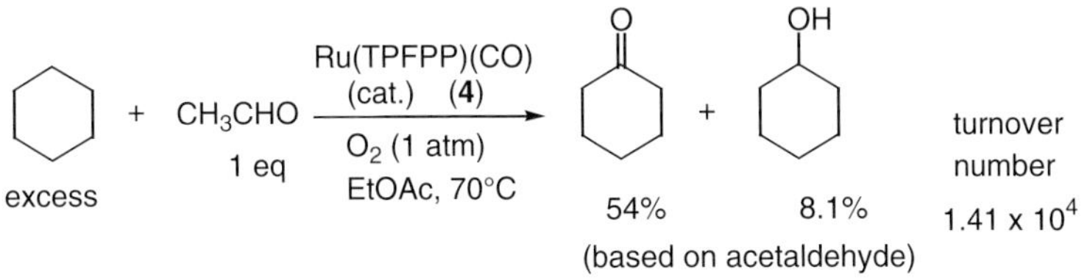

$$\text{cyclohexane (excess)} + CH_3CHO \text{ (1 eq)} \xrightarrow[\substack{O_2 \text{ (1 atm)} \\ \text{EtOAc}, 70°C}]{Ru(TPFPP)(CO) \text{ (cat.) (4)}} \text{cyclohexanone} + \text{cyclohexanol} \quad (7.86)$$

54%   8.1%   turnover number 1.41 × 10^4
(based on acetaldehyde)

These oxidation reactions provide a powerful strategy for the synthesis of carbonyl compounds such as cyclohexanone by combination of the Wacker oxidation of ethylene with the present metal-catalyzed oxidation of cyclohexane (Scheme 7.6).

Very few methods for direct aerobic oxidation of alkanes have been reported using a perfluorinated ruthenium catalyst $[Ru_3O(OCOCF_2CF_2CF_3)_6(Et_2O)_3]^+$ [50c], a ruthenium substituted polyoxometalate $[WZnRu_2(OH)(H_2O)(ZnW_9O_{34})_2]^{11-}$ (Eq. (7.87)) [[142, 143]], and $Ru(OH)_x/Al_2O_3$ catalyst [144].

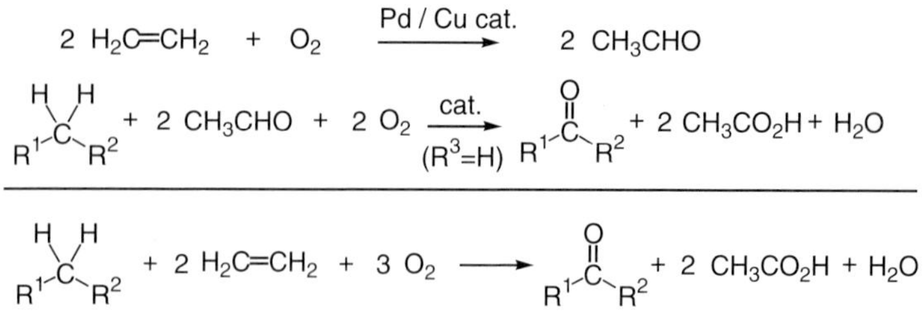

Scheme 7.6

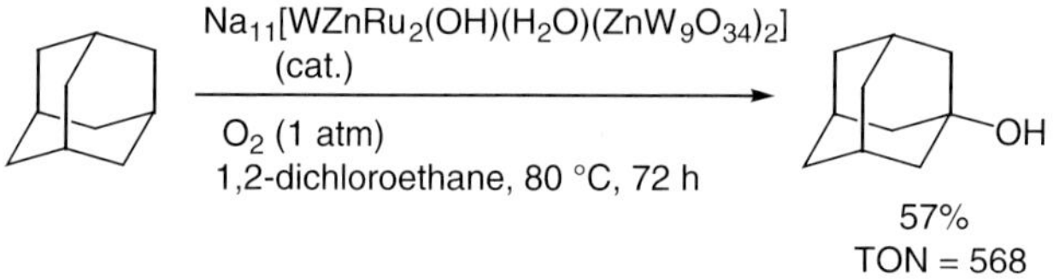

$$2\ H_2C{=}CH_2 + O_2 \xrightarrow{\text{Pd/Cu cat.}} 2\ CH_3CHO$$

$$R^1R^2CH_2 + 2\ CH_3CHO + 2\ O_2 \xrightarrow[\text{(R}^3\text{=H)}]{\text{cat.}} R^1C(O)R^2 + 2\ CH_3CO_2H + H_2O$$

$$R^1R^2CH_2 + 2\ H_2C{=}CH_2 + 3\ O_2 \longrightarrow R^1C(O)R^2 + 2\ CH_3CO_2H + H_2O$$

adamantane $\xrightarrow[\text{1,2-dichloroethane, 80 °C, 72 h}]{\text{Na}_{11}[\text{WZnRu}_2(\text{OH})(\text{H}_2\text{O})(\text{ZnW}_9\text{O}_{34})_2]\ (\text{cat.})}$ 1-adamantanol

57%
TON = 568

(7.87)

Recently enantioselective oxidative functionalization of C–H bonds was reported. Intramolecular C–H amination proceeds in the presence of ruthenium pybox catalyst with bis(acyloxy)iodobenzene (Eq. (7.88)) [145].

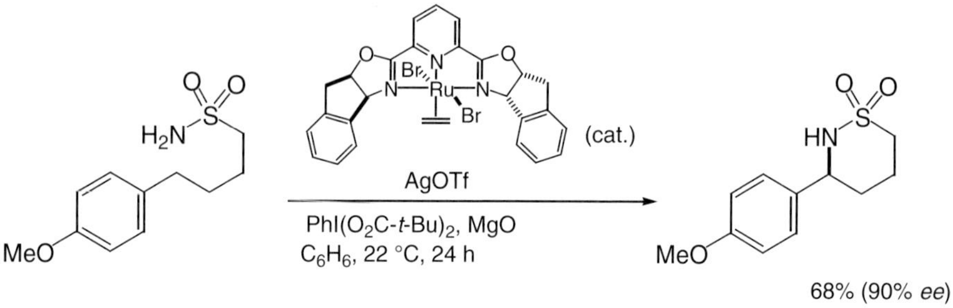

68% (90% ee)

(7.88)

References

1 Murahashi, S.-I. (1995) *Angew. Chem. Int. Ed. Engl.*, **34**, 2443–2465.
2 (a) Murahashi, S.-I. and Naota, T. (1994) *Advances in Metal-Organic Chemistry*, vol. 3 (ed. L.S. Liebeskind), JAI Press, London, pp. 225–254; (b) Murahashi, S.-I. and Naota, T. (1995) *Comprehensive Organometallic Chemistry II*, vol. 12 (eds. E.W. Abel, F.G.A. Stone, and G. Wilkinson), Pergamon, Oxford, pp. 1177–1192; (c) Naota, T., Takaya, H., and Murahashi, S.-I. (1998) *Chem. Rev.*, **98**, 2599–2660.
3 Murahashi, S.-I. and Komiya, N. (2000) *Biomimetic Oxidations Catalyzed by Transition Metal Complexes* (ed. B. Meunier), Imperial College Press, London, pp. 563–611.
4 Murahashi, S.-I. and Komiya, N. (2004) *Ruthenium in Organic Synthesis* (ed. S.-I. Murahashi), Wiley-VCH, Weinheim, pp. 53–93.

5 Murahashi, S.-I. and Zhang, D. (2008) *Chem. Soc. Rev.*, **37**, 1490–1501.
6 Seddon, E.A. and Seddon, K.R. (1984) *The Chemistry of Ruthenium*, Elsevier, Amsterdam.
7 Holm, R.H. (1987) *Chem. Rev.*, **87**, 1401–1449.
8 Jørgensen, K.A. (1989) *Chem. Rev.*, **89**, 431–458.
9 Griffith, W.P. (1992) *Chem. Soc. Rev.*, **21**, 179–185.
10 Pagliaro, M., Campestrini, S., and Ciriminna, R. (2005) *Chem. Soc. Rev.*, **34**, 837–845.
11 (a) Lee, D.G. and van den Engh, M. (1973) Part B, Chapter 4, in *Oxidation in Organic Chemistry* (ed. W.S. Trahanovski), Academic Press, New York (b) Courtney, J.T. (1984) Chapter 8, in *Organic Synthesis by Oxidation with Metal Compounds* (eds W.J. Mijs and C.R.H.I. de Jonge), Plenum Press, New York; (c) Carlsen, P.H.J., Katsuki, T., Martin, V.S., and Sharpless, K.B. (1981) *J. Org. Chem.*, **46**, 3936–3938; (d) Plietker, B. (2005) *Synthesis*, 2453–2472; (e) Plietker, B. and Niggemann, M. (2004) *Org. Biomol. Chem.*, **2**, 2403–2407.
12 (a) Morris, P.E. Jr. and Kiely, D.E. (1987) *J. Org. Chem.*, **52**, 1149–1152; (b) Yamamoto, Y., Suzuki, H., and Moro-oka, Y. (1985) *Tetrahedron Lett.*, **26**, 2107–2108; (c) Kanemoto, S., Tomioka, H., Oshima, K., and Nozaki, H. (1986) *Bull. Chem. Soc. Jpn.*, **59**, 105–108; (d) Giddings, S. and Mills, A. (1988) *J. Org. Chem.*, **53**, 1103–1107; (e) Torii, S., Inokuchi, T., and Sugiura, T. (1986) *J. Org. Chem.*, **51**, 155–161.
13 (a) Lee, D.G. and Chen, T. (1991) Chapter 3.8, in *Comprehensive Organic Synthesis*, vol. 7 (eds B.M. Trost and I. Fleming), Pergamon Press, Oxford, pp. 541–591; (b) Torii, S., Inokuchi, T., and Kondo, K. (1985) *J. Org. Chem.*, **50**, 4980–4982; (c) Webster, F.X., Rivas-Enterrios, J., and Silverstein, R.M. (1987) *J. Org. Chem.*, **52**, 689–691; (d) Albarella, L., Giordano, F., Lasalvia, M., Piccialli, V., and Sica, D. (1995) *Tetrahedron Lett.*, **36**, 5267–5270; (e) Denisenko, M.V., Pokhilo, N.D., Odinokova, L.E., Denisenko, V.A., and Uvarova, N.I. (1996) *Tetrahedron Lett.*, **37**, 5187–5190; (f) Orita, H., Hayakawa, T., and Takehira, K. (1986) *Bull. Chem. Soc. Jpn.*, **59**, 2637–2638.
14 (a) Shing, T.K.M., Tai, V.W.-F., and Tam, E.K.W. (1994) *Angew. Chem. Int. Ed. Engl.*, **33**, 2312–2313. (b) Ho, C.-M., Yu, W.-Y., and Che, C.-M. (2004) *Angew. Chem. Int. Ed.*, **43**, 3303–3307.
15 (a) Roth, S. and Stark, C.B.W. (2006) *Angew. Chem. Int. Ed.*, **45**, 6218–6221; (b) Roth, S., Göhler, S., Cheng, H., and Stark, C.B.W. (2005) *Eur. J. Org. Chem.*, 4109–4118; (c) Caserta, T., Piccialli, V., Gomez-Paloma, L., and Bifulco, G. (2005) *Tetrahedron*, **61**, 927–939.
16 Alcaide, B., Almendros, P., and Alonso, J.M. (2006) *Chem. Eur. J.*, **12**, 2874–2879.
17 Balavoine, G., Eskénazi, C., Meunier, F., and Rivière, H. (1984) *Tetrahedron Lett.*, **25**, 3187–3190.
18 Eskénazi, C., Balavoine, G., Meunier, F., and Rivière, H. (1985) *J. Chem. Soc., Chem. Commun.*, 1111–1113.
19 (a) Bailey, A.J., Griffith, W.P., White, A.J.P., and Williams, D.J. (1994) *J. Chem. Soc., Chem. Commun.*, 1833–1834; (b) Bailey, A.J., Griffith, W.P., and Savage, P.D. (1995) *J. Chem. Soc., Dalton Trans.*, 3537–3542.
20 Augier, C., Malara, L., Lazzeri, V., and Waegell, B. (1995) *Tetrahedron Lett.*, **36**, 8775–8778.
21 Plietker, B. (2004) *J. Org. Chem.*, **69**, 8287–8296.
22 (a) Khan, F.A., Dash, J., Sahu, N., and Sudheer, C. (2002) *J. Org. Chem.*, **67**, 3783–3787; (b) Khan, F.A. and Dash, J. (2002) *J. Am. Chem. Soc.*, **124**, 2424–2425; (c) Khan, F.A., Prabhudas, B., Sahu, N., and Dash, J. (2000) *J. Am. Chem. Soc.*, **122**, 9558–9559.
23 (a) Caputo, J.A. and Fuchs, R. (1967) *Tetrahedron Lett.*, 4729–4731; (b) Piatak, D.M., Herbst, G., Wicha, J., and Caspi, E. (1969) *J. Org. Chem.*, **34**, 116–120; (c) Chakraborti, A.K., and Ghatak, U.R. (1983) *Synthesis*, 746–748; (d) Kasai, M. and Ziffer, H. (1983) *J. Org. Chem.*, **48**, 2346–2349; (e) Spitzer, U.A. and Lee, D.G. (1974) *J. Org. Chem.*, **39**, 2468–2469; (f) Nuñez, M.T. and Martín, V.S. (1990) *J. Org. Chem.*, **55**, 1928–1932; (g) Wolfe, S., Hasan, S.K., and Campbell, J.R. (1970) *J.*

Chem. Soc., Chem. Commun., 1420–1421; (h) Noji, M., Sunahara, H., Tsuchiya, K., Mukai, T., Komakusa, A., and Ishii, K. (2008) *Synthesis*, 3835–3845.

24 (a) Torii, S., Inokuchi, T., and Hirata, Y. (1987) *Synthesis*, 377–379; (b) Zibuck, R. and Seebach, D. (1988) *Helv. Chim. Acta*, **71**, 237–240.

25 Yang, D., Chen, F., Dong, Z.-M., and Zhang, D.-W. (2004) *J. Org. Chem.*, **69**, 2221–2223.

26 Laux, M. and Krause, N. (1997) *Synlett*, 765–766.

27 (a) Gonsalve, L., Arends, I.W.C.E., and Sheldon, R.A. (2002) *Chem. Commun.*, 202–203; (b) Schuda, P.F., Cichowicz, M.B., and Heimann, M.R. (1983) *Tetrahedron Lett.*, **24**, 3829–3830; (c) Lee, J.S., Cao, H., and Fuchs, P.L. (2007) *J. Org. Chem.*, **72**, 5820–5823.

28 (a) Perrone, R., Bettoni, G., and Tortorella, V. (1976) *Synthesis*, 598–600; (b) Bettoni, G., Carbonara, G., Franchini, C., and Tortorella, V. (1981) *Tetrahedron*, **37**, 4159–4164.

29 (a) Sheehan, J.C. and Tulis, R.W. (1974) *J. Org. Chem.*, **39**, 2264–2267; (b) Yoshifuji, S., Tanaka, K., and Nitta, Y. (1985) *Chem. Pharm. Bull.*, **33**, 1749–1751.

30 Ranganathan, D., Vaish, N.K., and Shah, K. (1994) *J. Am. Chem. Soc.*, **116**, 6545–6557.

31 (a) Tenaglia, A., Terranova, E., and Waegell, B. (1989) *Tetrahedron Lett.*, **30**, 5271–5274; (b) Tenaglia, A., Terranova, E., and Waegell, B. (1992) *J. Org. Chem.*, **57**, 5523–5528.

32 Bakke, J.M. and Lundquist, M. (1986) *Acta. Chem. Scand. B*, **40**, 430–433.

33 Coudret, J.-L., Zöllner, S., Ravoo, B.J., Malara, L., Hanisch, C., Dörve, K., de Meijere, A., and Waegell, B. (1996) *Tetrahedron Lett.*, **37**, 2425–2428.

34 Tenaglia, A., Terranova, E., and Waegell, B. (1990) *J. Chem. Soc., Chem. Commun.*, 1344–1345.

35 Spitzer, U.A. and Lee, D.G. (1975) *J. Org. Chem.*, **40**, 2539–2540.

36 Sasson, Y., Zappi, G.D., and Neumann, R. (1986) *J. Org. Chem.*, **51**, 2880–2883.

37 (a) Montanari, F. and Casella, L. (eds) (1994) *Metalloporphyrins Catalyzed Oxidations*, Kluwer Academic Publishers, Dordrecht; (b) Sheldon, R.A. (ed.) (1994) *Metalloporphyrins in Catalytic Oxidations*, Marcel Dekker, New York; (c) Che, C.-M. and Huang, J.-S. (2009) *Chem. Commun.*, 3996–4015.

38 (a) Leung, T., James, B.R., and Dolphin, D. (1983) *Inorg. Chim. Acta*, **79**, 180–181; (b) Dolphin, D., James, B.R., and Leung, T. (1983) *Inorg. Chim. Acta*, **79**, 25–27.

39 (a) Higuchi, T., Ohtake, H., and Hirobe, M. (1989) *Tetrahedron Lett.*, **30**, 6545–6548; (b) Ohtake, H., Higuchi, T., and Hirobe, M. (1992) *Tetrahedron Lett.*, **33**, 2521–2524.

40 (a) Groves, J.T., Bonchio, M., Carofiglio, T., and Shalyaev, K. (1996) *J. Am. Chem. Soc.*, **118**, 8961–8962; (b) Groves, J.T., and Quinn, R. (1985) *J. Am. Chem. Soc.*, **107**, 5790–5792; (c) Wang, C., Shalyaev, K.V., Bonchio, M., Carofiglio, T., and Groves, J.T. (2006) *Inorg. Chem.*, **45**, 4769–4782.

41 (a) Chen, J. and Che, C.-M. (2004) *Angew. Chem. Int. Ed.*, **43**, 4950–4954; (b) Jiang, G., Chen, J., Thu, H.-Y., Huang, J.-S., Zhu, N., and Che, C.-M. (2008) *Angew. Chem. Int. Ed.*, **47**, 6638–6642; (c) Che, C.-M. and Huang, J.-S. (2009) *Chem. Commun.*, 3996–4015.

42 (a) Yamada, T., Hashimoto, K., Kitaichi, Y., Suzuki, K., and Ikeno, T. (2001) *Chem. Lett.*, 268–269; (b) Tanaka, H., Hashimoto, K., Suzuki, K., Kitaichi, Y., Sato, M., Ikeno, T., and Yamada, T. (2004) *Bull. Chem. Soc. Jpn.*, **77**, 1905–1914.

43 (a) Che, C.-M., Yip, W.-P., and Yu, W.-Y. (2006) *Chem. Asian J.*, **1**, 453–458; (b) Yip, W.-P., Ho, C.-M., Zhu, N., Lau, T.-C., and Che, C.-M. (2008) *Chem. Asian J.*, **3**, 70–77.

44 Kogan, V., Quintal, M.M., and Neumann, R. (2005) *Org. Lett.*, **7**, 5039–5042.

45 (a) Bressan, M. and Morvillo, A. (1988) *J. Chem. Soc., Chem. Commun.*, 650–651; (b) Bressan, M. and Morvillo, A. (1989) *J. Chem. Soc., Chem. Commun.*, 421–423; (c) Jitsukawa, K., Shiozaki, H., and Masuda, H. (2002) *Tetrahedron Lett.*, **43**, 1491–1494; (d) Okumura, T., Morishima, Y., Shiozaki, H., Yagyu, T., Funahashi, Y., Ozawa, T., Jitsukawa, K., and Masuda, H. (2007) *Bull. Chem. Soc. Jpn.*, **80**, 507–517; (e) Hamelin, O., Ménage, S., Charnay, F., Chavarot, M., Pierre, J.-L., Pécaut, J., and

Fontecave, M. (2008) *Inorg. Chem.*, **47**, 6413–6420.

46 (a) Cheng, W.-C., Yu, W.-Y., Cheung, K.-K., and Che, C.M. (1994) *J. Chem. Soc., Chem. Commun.*, 1063–1064; (b) Lau, T.C., Che, C.-M., Lee, W.O., and Poon, C.K. (1988) *J. Chem. Soc., Chem. Commun.*, 1406–1407; (c) Ho, P.K.K., Cheung, K.-K., and Che, C.-M. (1996) *Chem. Commun.*, 1197–1198.

47 Provins, L. and Murahashi, S.-I. (2007) *ARKIVOC*, 107–120.

48 Klawonn, M., Tse, M.K., Bhor, S., Döbler, C., and Beller, M. (2004) *J. Mol. Catal., A.*, **218**, 13–19.

49 (a) Neumann, R. and Dahan, M. (1997) *Nature*, **388**, 353–355; (b) Maayan, G. and Neumann, R. (2005) *Chem. Commun.*, 4595–4597.

50 (a) Goldstein, A.S., Beer, R.H., and Drago, R.S. (1994) *J. Am. Chem. Soc.*, **116**, 2424–2429; (b) Goldstein, A.S. and Drago, R.S. (1991) *J. Chem. Soc., Chem. Commun.*, 21–22; (c) Davis, S. and Drago, R.S. (1990) *J. Chem. Soc., Chem. Commun.*, 250–251; (d) Robbins, M.H. and Drago, R.S. (1996) *J. Chem. Soc., Dalton Trans.*, 105–110.

51 Klement, I., Lütjens, H., and Knochel, P. (1997) *Angew. Chem. Int. Ed. Engl.*, **36**, 1454–1456.

52 (a) Kureshy, R.I., Khan, N.H., Abdi, S.H.R., and Bhatt, K.N. (1993) *Tetrahedron: Asymmetry*, **4**, 1693–1701; (b) Kureshy, R.I., Khan, N.H., Abdi, S.H.R., and Bhatt, A.K. (1996) *J. Mol. Catal. A: Chemical*, **110**, 33–40.

53 Nishiyama, H., Shimada, T., Itoh, H., Sugiyama, H., and Motoyama, Y. (1997) *Chem. Commun.*, 1863–1864.

54 Gross, Z., Ini, S., Kapon, M., and Cohen, S. (1996) *Tetrahedron Lett.*, **37**, 7325–7328.

55 Berkessel, A. and Frauenkron, M. (1997) *J. Chem. Soc., Perkin Trans. 1*, 2265–2266.

56 Takeda, T., Irie, R., Shinoda, Y., and Katsuki, T. (1999) *Synlett*, 1157–1159.

57 (a) Tse, M.K., Döbler, C., Bhor, S., Klawonn, M., Mägerlein, W., Hugl, H., and Beller, M. (2004) *Angew. Chem. Int. Ed.*, **43**, 5255–5260; (b) Bhor, S., Tse, M.K., Klawonn, M., Döbler, C., Mägerlein, W., and Beller, M. (2004) *Adv. Synth. Catal.*, **346**, 263–267.

58 Lai, T.-S., Zhang, R., Cheung, K.-K., Kwong, H.-L., and Che, C.-M. (1998) *Chem. Commun.*, 1583–1584.

59 (a) Ostovic, D. and Bruice, T.C. (1992) *Acc. Chem. Res.*, **25**, 314–320; (b) Castellino, A.J. and Bruice, T.C. (1988) *J. Am. Chem. Soc.*, **110**, 158–162.

60 Murahashi, S.-I., Saito, T., Hanaoka, H., Murakami, Y., Naota, T., Kumobayashi, H., and Akutagawa, S. (1993) *J. Org. Chem.*, **58**, 2929–2930.

61 Hotopp, T., Gutke, H.-J., and Murahashi, S.-I. (2001) *Tetrahedron Lett.*, **42**, 3343–3346.

62 Beifuss, U. and Herde, A. (1998) *Tetrahedron Lett.*, **39**, 7691–7692.

63 Sharpless, K.B., Akashi, K., and Oshima, K. (1976) *Tetrahedron Lett.*, 2503–2506.

64 (a) Müller, P. and Godoy, J. (1981) *Tetrahedron Lett.*, **22**, 2361–2364; (b) Yusubov, M.S., Gilmkhanova, M.P., Zhdankin, V.V., and Kirschning, A. (2007) *Synlett*, 563–566; (c) Yang, Z.W., Kang, Q.X., Quan, F., and Lei, Z.Q. (2007) *J. Mol. Catal. A: Chemical*, **261**, 190–195.

65 Kanemoto, S., Oshima, K., Matsubara, S., Takai, K., and Nozaki, H. (1983) *Tetrahedron Lett.*, **24**, 2185–2088.

66 (a) Barak, G., Dakka, J., and Sasson, Y. (1988) *J. Org. Chem.*, **53**, 3553–3555; (b) Rothenberg, G., Barak, G., and Sasson, Y. (1999) *Tetrahedron*, **55**, 6301–6310; (c) Shi, F., Tse, M.K., and Beller, M. (2007) *Chem. Asian J.*, **2**, 411–415.

67 (a) Kim, K.S., Kim, S.J., Song, Y.H., and Hahn, C.S. (1987) *Synthesis*, 1017–1018; (b) Yusubov, M.S., Zagulyaeva, A.A., and Zhdankin, V.V. (2009) *Chem. Eur. J.*, **15**, 11091–11094.

68 Iwasa, S., Morita, K., Tajima, K., Fakhruddin, A., and Nishiyama, H. (2002) *Chem. Lett.*, 284–285.

69 (a) Griffith, W.P., Ley, S.V., Whitcombe, G.P., and White, A.D. (1987) *J. Chem. Soc., Chem. Commun.*, 1625–1627; (b) An excellent review, Ley, S.V., Norman, J., Griffith, W.P., and Marsden, S.P. (1994) *Synthesis*, 639–666, and references cited therein.

70 (a) Gonsalvi, L., Arends, I.W.C.E., and Sheldon, R.A. (2002) *Org. Lett.*, **4**, 1659–1661; (b) Gonsalvi, L., Arends,

I.W.C.E., Moilanen, P., and Sheldon, R.A. (2003) *Adv. Synth. Catal.*, **345**, 1321–1328.

71. (a) Murahashi, S.-I., Naota, T., and Nakajima, T. (1985) *Tetrahedron Lett.*, **26**, 925–928; (b) Murahashi, S.-I. and Naota, T. (1993) *Synthesis*, 433–440; (c) Murahashi, S.-I. and Naota, T. (1996) *Zhurnal Organicheskoi Khimii*, **32**, 223–232; (d) Murahashi, S.-I. and Naota, T. (1988) *Reviews on Heteroatom Chemistry*, vol. **1**, MYU Research, Tokyo, pp. 257–276.

72. (a) Tsuji, Y., Ohta, T., Ido, T., Minbu, H., and Watanabe, Y. (1984) *J. Organomet. Chem.*, **270**, 333–341; (b) Tanaka, M., Kobayashi, T., and Sakakura, T. (1984) *Angew. Chem. Int. Ed. Engl.*, **23**, 518–518.

73. Murahashi, S.-I., Naota, T., and Nakajima, N. (1987) *Chem. Lett.*, 879–882.

74. Fung, W.-H., Yu, W.-Y., and Che, C.-M. (1998) *J. Org. Chem.*, **63**, 2873–2877.

75. Murahashi, S.-I., Naota, T., Oda, Y., and Hirai, N. (1995) *Synlett*, 733–734.

76. Murahashi, S.-I., Naota, T., and Hirai, N. (1993) *J. Org. Chem.*, **58**, 7318–7319.

77. (a) Murahashi, S.-I., Ito, K., Naota, T., and Maeda, Y. (1981) *Tetrahedron Lett.*, **22**, 5327–5330; (b) Murahashi, S.-I., Naota, T., Ito, K., Maeda, Y., and Taki, H. (1987) *J. Org. Chem.*, **52**, 4319–4327; (c) Aranyos, A., Csjernyik, G., Szabó, K.J., and Bäckvall, J.-E. (1999) *Chem. Commun.*, 351–352; (d) Almeida, M.L.S., Beller, M., Wang, S.-Z., and Bäckvall, J.-E. (1996) *Chem. Eur. J.*, **2**, 1533–1536; (e) Yi, C.S., Zeczycki, T.N., and Guzei, I.A. (2006) *Organometallics*, **25**, 1047–1051; (f) Ito, M., Osaku, A., Shiibashi, A., and Ikariya, T. (2007) *Org. Lett.*, **9**, 1821–1824.

78. (a) Blum, Y., Reshef, D., and Shvo, Y. (1981) *Tetrahedron Lett.*, **22**, 1541–1544; (b) Blum, Y. and Shvo, Y. (1985) *J. Organomet. Chem.*, **282**, C7–C10.

79. (a) Owston, N.A., Parker, A.J., and Williams, J.M.J. (2008) *Chem. Commun.*, 624–625; (b) Zhao, J. and Hartwig, J.F. (2005) *Organometallics*, **24**, 2441–2446; (c) Zhang, J., Leitus, G., Ben-David, Y., and Milstein, D. (2005) *J. Am. Chem. Soc.*, **127**, 10840–10841.

80. Gunanathan, C., Shimon, L.J.W., and Milstein, D. (2009) *J. Am. Chem. Soc.*, **131**, 3146–3147.

81. (a) $RuH_2(PPh_3)4$: Murahashi, S.-I., Kondo, K., and Hakata, T. (1982) *Tetrahedron Lett.*, **23**, 229–232; (b) $RuCl2(PPh3)3$ for arylamines: Watanabe, Y., Tsuji, Y., and Ohsugi, Y. (1981) *Tetrahedron Lett.*, **22**, 2667–2670; (c) Murahashi, S.-I., Shimamura, T., and Moritani, I. (1974) *J. Chem. Soc., Chem. Commun.*, 931–932; (d) Grigg, R., Mitchell, T.R.B., Sutthivaiyakit, S., and Tongpenyai, N. (1981) *J. Chem. Soc., Chem. Commun.*, 611–612; (e) Hollmann, D., Tillack, A., Michalik, D., Jackstell, R., and Beller, M. (2007) *Chem. Asian J.*, **2**, 403–410, and references cited therein; (f) An excellent review: Haniti, M., Hamid, S.A., Slatford, P.A., and Williams, J.M.J. (2007) *Adv. Synth. Catal.*, **349**, 1555–1575.

82. (a) Murahashi, S.-I. and Naota, T. (1991) *Synlett*, 693–694; (b) Watson, A.J.A., Maxwell, A.C., and Williams, J.M.J. (2009) *Org. Lett.*, **11**, 2667–2670.

83. (a) Gunanathan, C., Ben-David, Y., and Milstein, D. (2007) *Science*, **317**, 790–792; (b) Nordstrøm, L.U., Vogt, H., and Madsen, R. (2008) *J. Am. Chem. Soc.*, **130**, 17672–17673.

84. (a) Bäckvall, J.-E., Chowdhury, R.L., and Karlsson, U. (1991) *J. Chem. Soc., Chem. Commun.*, 473–475; (b) Wang, G.-Z., Andreasson, U., and Bäckvall, J.-E. (1994) *J. Chem. Soc., Chem. Commun.*, 1037–1038; (c) A review for multistep electron transfer reaction, see: Piera, J. and Bäckvall, J.-E. (2008) *Angew. Chem. Int. Ed.*, **47**, 3506–3523.

85. Csjernyik, G., Éll, A.H., Fadini, L., Pugin, B., and Bäckvall, J.-E. (2002) *J. Org. Chem.*, **67**, 1657–1662.

86. (a) Dijksman, A., Arends, I.W.C.E., and Sheldon, R.A. (1999) *Chem. Commun.*, 1591–1592; (b) Dijksman, A., Marino-González, A., i Payeras, A.M., Arends, I.W.C.E., and Sheldon, R.A. (2001) *J. Am. Chem. Soc.*, **123**, 6826–6833; (c) Sheldon, R.A., Arends, I.W.C.E., ten Brink, G.J., and Dijksman, A. (2002) *Acc. Chem. Res.*, **35**, 774–781.

87. Matsumoto, M. and Watanabe, N. (1984) *J. Org. Chem.*, **49**, 3435–3436.

88 Markó, I.E., Giles, P.R., Tsukazaki, M., Chellé-Regnaut, I., Urch, C.J., and Brown, S.M. (1997) *J. Am. Chem. Soc.*, **119**, 12661–12662.

89 (a) Hinzen, B., Lenz, R., and Ley, S.V. (1998) *Synthesis*, 977–979; (b) Bleloch, A., Johnson, B.F.G., Ley, S.V., Price, A.J., Shephard, D.S., and Thomas, A.W. (1999) *Chem. Commun.*, 1907–1908.

90 Komiya, N., Nakae, T., Sato, H., and Naota, T. (2006) *Chem. Commun.*, 4829–4831.

91 (a) Kaneda, K., Yamashita, T., Matsushita, T., and Ebitani, K. (1998) *J. Org. Chem.*, **63**, 1750–1751; (b) Matsushita, T., Ebitani, K., and Kaneda, K. (1999) *Chem. Commun.*, 265–266; (c) Yamaguchi, K., Mori, K., Mizugaki, T., Ebitani, K., and Kaneda, K. (2000) *J. Am. Chem. Soc.*, **122**, 7144–7145; (d) Ebitani, K., Motokura, K., Mizugaki, T., and Kaneda, K. (2005) *Angew. Chem. Int. Ed.*, **44**, 3423–3426.

92 (a) Yamaguchi, K. and Mizuno, N. (2002) *Angew. Chem. Int. Ed. Engl.*, **41**, 4538–4541; (b) Oishi, T., Yamaguchi, K., and Mizuno, N. (2009) *Angew. Chem. Int. Ed.*, **48**, 6286–6288; (c) Kim, W.-H., Park, I.S., and Park, J. (2006) *Org. Lett.*, **8**, 2543–2545; (d) Mori, S., Takubo, M., Makida, K., Yanase, T., Aoyagi, S., Maegawa, T., Monguchi, Y., and Sajiki, H. (2009) *Chem. Commun.*, 5159–5161.

93 (a) Hosokawa, T. and Murahashi, S.-I. (1990) *Acc. Chem. Res.*, **23**, 49–54; (b) Hosokawa, T., Uno, T., Inui, S., and Murahashi, S.-I. (1981) *J. Am. Chem. Soc.*, **103**, 2318–2323.

94 (a) Kobayashi, S., Miyamura, H., Akiyama, R., and Ishida, T. (2005) *J. Am. Chem. Soc.*, **127**, 9251–9254; (b) Matsumoto, T., Ueno, M., Kobayashi, J., Miyamura, H., Mori, Y., and Kobayashi, S. (2007) *Adv. Synth. Catal.*, **349**, 531–534; (c) Matsumoto, T., Ueno, M., Wang, N., and Kobayashi, S. (2008) *Chem. Asian J.*, **3**, 196–214.

95 Takezawa, E., Sakaguchi, S., and Ishii, Y. (1999) *Org. Lett.*, **1**, 713–715.

96 Wolfson, A., Wuyts, S., De Vos, D.E., Vankelecom, I.F.J., and Jacobs, P.A. (2002) *Tetrahedron Lett.*, **43**, 8107–8110.

97 (a) Shimizu, H., Onitsuka, S., Egami, H., and Katsuki, T. (2005) *J. Am. Chem. Soc.*, **127**, 5396–5413; (b) Nakamura, Y., Egami, H., Matsumoto, K., Uchida, T., and Katsuki, T. (2007) *Tetrahedron*, **63**, 6383–6387; (c) Chatterjee, D. (2008) *Coord. Chem. Rev.*, **252**, 176–198; (d) Mizoguchi, H., Uchida, T., Ishida, K., and Katsuki, T. (2009) *Tetrahedron Lett.*, **50**, 3432–3435.

98 Murahashi, S.-I., Naota, T., and Yonemura, K. (1988) *J. Am. Chem. Soc.*, **110**, 8256–8258.

99 Murahashi, S.-I., Naota, T., Miyaguchi, N., and Nakato, T. (1992) *Tetrahedron Lett.*, **33**, 6991–6994.

100 (a) Murahashi, S.-I., Komiya, N., Terai, H., and Nakae, T. (2003) *J. Am. Chem. Soc.*, **125**, 15312–15313; (b) Murahashi, S.-I., Komiya, N., Terai, H., and Nakae, T. (2008) *J. Am. Chem. Soc.*, **130**, 11005–11012; (c) North, M. (2004) *Angew. Chem. Int. Ed.*, **43**, 4126–4128.

101 Murahashi, S.-I., Komiya, N., and Terai, H. (2005) *Angew. Chem. Int. Ed.*, **44**, 6931–6933.

102 (a) Riley, D.P. (1983) *J. Chem. Soc., Chem. Commun.*, 1530–1532; (b) Jain, S.L. and Sain, B. (2002) *Chem. Commun.*, 1040–1041.

103 Murahashi, S.-I., Naota, T., and Taki, H. (1985) *J. Chem. Soc., Chem. Commun.*, 613–614.

104 (a) Mitsui, H., Zenki, S., Shiota, T., and Murahashi, S.-I. (1984) *J. Chem. Soc., Chem. Commun.*, 874–875; (b) Murahashi, S.-I., Mitsui, H., Shiota, T., Tsuda, T., and Watanabe, S. (1990) *J. Org. Chem.*, **55**, 1736–1744; (c) Murahashi, S.-I., Shiota, T., and Imada, Y. (1991) *Org. Synth.*, **70**, 265–271; (d) Murahashi, S.-I., Imada, Y., and Ohtake, H. (1994) *J. Org. Chem.*, **59**, 6170–6172; (e) Murahashi, S.-I. and Shiota, T. (1987) *Tetrahedron Lett.*, **28**, 2383–2386.

105 (a) Kawakami, T., Ohtake, H., Arakawa, H., Okachi, T., Imada, Y., and Murahashi, S.-I. (2000) *Bull. Chem. Soc. Jpn.*, **73**, 2423–2444; (b) Murahashi, S.-I., Imada, Y., Kawakami, T., Harada, K., Yonemushi, Y., and Tomita, T. (2002) *J. Am. Chem. Soc.*, **124**, 2888–2889; (c) Murahashi, S.-I., Tsuji, T., and Ito, S. (2000) *Chem. Commun.*, 409–410.

106 (a) Goti, A. and Romani, M. (1994) *Tetrahedron Lett.*, **35**, 6567–6570; (b) Goti, A., De Sarlo, F., and Romani, M. (1994) *Tetrahedron Lett.*, **35**, 6751–6574.

107 Murahashi, S.-I., Okano, Y., Sato, H., Nakae, T., and Komiya, N. (2007) *Synlett*, 1675–1678.

108 (a) Éll, A.H., Samec, J.S.M., Brasse, C., and Bäckvall, J.E. (2002) *Chem. Comm.*, 1144–1145; (b) Samec, J.S.M., Éll, A.H., and Bäckvall, J.-E. (2005) *Chem. Eur. J.*, **11**, 2327–2334; (c) Samec, J.S.M., Bäckvall, J.-E., Anderson, P.G., and Braudt, P. (2006) *Chem. Soc. Rev.*, **35**, 237–248.

109 Schröder, M. and Griffith, W.P. (1979) *J. Chem. Soc., Chem. Commun.*, 58–59.

110 (a) Bailey, A.J. and James, B.R. (1996) *Chem. Commun.*, 2343–2344; (b) Cheng, S.Y.S., Rajapakse, N., Rettig, S.J., and James, B.R. (1994) *J. Chem. Soc., Chem. Commun.*, 2669–2670.

111 Mori, K., Yamaguchi, K., Mizugaki, T., Ebitani, K., and Kaneda, K. (2001) *Chem. Commun.*, 461–462.

112 (a) Yamaguch, K. and Mizuno, N. (2003) *Angew. Chem. Int. Ed.*, **42**, 1480–1483; (b) Yamaguch, K. and Mizuno, N. (2003) *Chem. Eur. J.*, **9**, 4353–4361; (c) Kotani, M., Koike, T., Yamaguchi, K., and Mizuno, N. (2006) *Green Chem.*, **8**, 735–741; (d) Kim, J.W., Yamaguchi, K., and Mizuno, N. (2008) *Angew. Chem. Int. Ed.*, **47**, 9249–9251.

113 Li, F., Chen, J., Zhang, Q., and Wang, Y. (2008) *Green Chem.*, **10**, 533–562.

114 Murahashi, S.-I., Naota, T., Kuwabara, T., Saito, T., Kumobayashi, H., and Akutagawa, S. (1990) *J. Am. Chem. Soc.*, **112**, 7820–7822.

115 Naota, T., Nakato, T., and Murahashi, S.-I. (1990) *Tetrahedron Lett.*, **31**, 7475–7478.

116 Murahashi, S.-I., Mitani, A., and Kitao, K. (2000) *Tetrahedron Lett.*, **41**, 10245–10249.

117 Ito, R., Umezawa, N., and Higuchi, T. (2005) *J. Am. Chem. Soc.*, **127**, 834–835.

118 Cainelli, G., DaCol, M., Galletti, P., and Giacomini, D. (1997) *Synlett*, 923–924.

119 Murahashi, S.-I., Saito, T., Naota, T., Kumobayashi, H., and Akutagawa, S. (1991) *Tetrahedron Lett.*, **32**, 5991–5994.

120 (a) Ingold, K.U. (1973) *Free Radicals*, vol. **1** (ed. J.K. Kochi), John Wiley & Sons, New York, pp. 37–112; (b) Brovo, A., Fontana, F., and Minisci, F. (1996) *Chem. Lett.*, 401–402.

121 Shimizu, M., Orita, H., Hayakawa, T., Watanabe, Y., and Takehira, K. (1991) *Bull. Chem. Soc. Jpn.*, **64**, 2583–2584.

122 Murahashi, S.-I., Naota, T., Miyaguchi, N., and Noda, S. (1996) *J. Am. Chem. Soc.*, **118**, 2509–2510.

123 Matsumoto, M. and Kuroda, K. (1982) *J. Am. Chem. Soc.*, **104**, 1433–1434.

124 Higuchi, T., Satake, C., and Hirobe, M. (1995) *J. Am. Chem. Soc.*, **117**, 8879–8880.

125 Cheung, W.-H., Yip, W.-P., Yu, W.-Y., and Che, C.-M. (2005) *Can. J. Chem.*, **83**, 521–526.

126 Shi, F., Tse, M.K., and Beller, M. (2007) *Adv. Synth. Catal.*, **349**, 303–308.

127 Matsushita, M., Kamata, K., Yamaguchi, K., and Mizuno, N. (2005) *J. Am. Chem. Soc.*, **127**, 6632–6640.

128 (a) Ohtake, H., Higuchi, T., and Hirobe, M. (1992) *J. Am. Chem. Soc.*, **114**, 10660–10662; (b) Shingaki, T., Miura, K., Higuchi, T., Hirobe, M., and Nagano, T. (1997) *Chem. Commun.*, 861–862.

129 Zhang, J.-L., Huang, J.-S., and Che, C.-M. (2006) *Chem. Eur. J.*, **12**, 3020–3031.

130 Balkus, K.J. Jr., Eissa, M., and Levado, R. (1995) *J. Am. Chem. Soc.*, **117**, 10753–10754.

131 (a) Lau, T.-C. and Mak, C.-K. (1995) *J. Chem. Soc., Chem. Commun.*, 943–944; (b) Lau, T.-C. and Mak, C.-K. (1993) *J. Chem. Soc., Chem. Commun.*, 766–767.

132 Zhang, R., Yu, W.-Y., Lai, T.-S., and Che, C.-M. (1999) *Chem. Commun.*, 1791–1792.

133 Kojima, T., Hayashi, K., Iizuka, S., Tani, F., Naruta, Y., Kawano, M., Ohashi, Y., Hirai, Y., Ohkubo, K., Matsuda, Y., and Fukuzumi, S. (2007) *Chem. Eur. J.*, **13**, 8212–8222.

134 (a) Yamaguchi, M., Kousaka, H., Izawa, S., Ichii, Y., Kumano, T., Masui, D., and Yamagishi, T. (2006) *Inorg. Chem.*, **45**, 8342–8354; (b) Yamaguchi, M., Kumano, T., Masui, D., and Yamagishi, T. (2004) *Chem. Commun.*, 798–799.

135 (a) Neumann, R. and Abu-Gnim, C. (1989) *J. Chem. Soc., Chem. Commun.*, 1324–1325; (b) Matsumoto, Y., Asami, M., Hashimoto, M., and Misono, M. (1996) *J. Mol. Catal. A: Chemical*, **114**, 161–168.

136 (a) Murahashi, S.-I., Komiya, N., Oda, Y., Kuwabara, T., and Naota, T. (2000) *J. Org. Chem.*, **65**, 9186–9193; (b) Murahashi, S.-I., Oda, Y., Naota, T., and Kuwabara, T. (1993) *Tetrahedron Lett.*, **34**, 1299–1302; (c) Murahashi, S.-I., Oda, Y., Komiya, N., and Naota, T. (1994) *Tetrahedron Lett.*, **35**, 7953–7956.

137 Murahashi, S.-I., Naota, T., and Kuwabara, T. (1989) *Synlett*, 62–63.

138 Miller, R.A., Li, W., and Humphrey, G.R. (1996) *Tetrahedron Lett.*, **37**, 3429–3432.

139 (a) Murahashi, S.-I., Oda, Y., and Naota, T. (1992) *J. Am. Chem. Soc.*, **114**, 7913–7914; (b) Murahashi, S.-I., Zhou, X.-G., and Komiya, N. (2003) *Synlett*, 321–324.

140 Murahashi, S.-I., Naota, T., and Komiya, N. (1995) *Tetrahedron Lett.*, **36**, 8059–8062.

141 (a) Murahashi, S.-I., Oda, Y., Naota, T., and Komiya, N. (1993) *J. Chem. Soc., Chem. Commun.*, 139–140; (b) Komiya, N., Naota, T., and Murahashi, S.-I. (1996) *Tetrahedron Lett.*, **37**, 1633–1636; (c) Komiya, N., Naota, T., Oda, Y., and Murahashi, S.-I. (1997) *J. Mol. Catal. A: Chemical*, **117**, 21–37.

142 Neumann, R., Khenkin, A.M., and Dahan, M. (1995) *Angew. Chem. Int. Ed. Engl.*, **34**, 1587–1589.

143 Neumann, R. and Dahan, M. (1998) *J. Am. Chem. Soc.*, **120**, 11969–11976.

144 Kamata, K., Kasai, J., Yamaguchi, K., and Mizuno, N. (2004) *Org. Lett.*, **6**, 3577–3580.

145 Milczek, E., Boudet, N., and Blakey, S. (2008) *Angew. Chem. Int. Ed.*, **47**, 6825–6828.

8
Selective Oxidation of Amines and Sulfides
Jan-E. Bäckvall

8.1
Introduction

Heteroatom oxidation is of great importance in organic synthesis, amines and sulfides being the compounds most commonly involved in such reactions. These compounds can be oxidized to a number of different products, and various reagents have been developed for these transformations. This chapter deals with selective oxidations of sulfides (thioethers) to sulfoxides and of tertiary amines to N-oxides.

8.2
Oxidation of Sulfides to Sulfoxides

The oxidation of sulfides has been reviewed previously [1–5], one of these reviews being focused on asymmetric sulfoxidation [5]. Organosulfur compounds such as sulfoxides and sulfones are useful synthetic reagents in organic chemistry. In particular, sulfoxides are important intermediates in the synthesis of natural products and biologically significant molecules [6]. Optically active sulfoxides are also of importance in medicinal and pharmaceutical chemistry [5], and chiral sulfoxides have been extensively used in asymmetric synthesis [7]. They have also been employed as ligands in asymmetric catalysis [8] and as oxotransfer reagents [9]. The synthesis and utilization of chiral sulfoxides has been reviewed [8a]. A large number of methods are available for the oxidation of sulfides to sulfoxides, and an important issue is to obtain high selectivity for sulfoxide over sulfone. Usually there is a reasonably good selectivity for oxidation to sulfoxide, since the sulfide is much more nucleophilic than the sulfoxide and hence reacts faster with the electrophilic reagent/catalyst. However, there are large variations between the different oxidation systems.

In this review, oxidations of sulfides to sulfoxides have been divided into three subsections: (i) stoichiometric reactions, (ii) chemocatalytic reactions, and (iii) biocatalytic reactions.

Modern Oxidation Methods. Edited by Jan-Erling Bäckvall
Copyright © 2010 WILEY-VCH Verlag GmbH & Co. KGaA, Weinheim
ISBN: 978-3-527-32320-3

8.2.1
Stoichiometric Reactions

A large number of methods are known in the literature for the oxidation of sulfides (thioethers) to sulfoxides by electrophilic reagents. These include the use of peracids [10], NaIO$_4$, hypochlorites, MnO$_2$, CrO$_3$, SeO$_2$, and iodosobenzene. The use of these reagents has been reviewed up to 2005 [3, 4]. Hydrogen peroxide can also be used as a direct oxidant, although this reaction is slow in the absence of catalyst. Its use in catalytic reactions is discussed below in Section 8.2.2.

8.2.1.1 Peracids
Peracids are commonly used oxidants in the oxidation of sulfides to sulfoxides [10]. The reaction proceeds at a good rate at room temperature, giving the sulfoxide together with some sulfone. Usually the selectivity for sulfoxide over sulfone is good enough for synthetic purposes, since the oxidation of sulfoxide is considerably slower than that of sulfide.

8.2.1.2 Dioxiranes
Dioxiranes have been successfully used for the selective oxidation of sulfides to sulfoxides (Eq. (8.1)) [11].

$$R^1\text{-S-}R^2 + \underset{\times}{\text{O-O}} \longrightarrow R^1\text{-S(=O)-}R^2 + \text{acetone} \tag{8.1}$$

These reactions are rapid, and the sulfide is efficiently oxidized to sulfoxide as the only product with no overoxidation to sulfone. The reaction is thought to proceed via direct oxygen transfer from the oxirane to the sulfide. In a mechanistic study [12], it was found that a hypervalent sulfurane is an intermediate in the reaction, being in equilibrium with the electrophilic zwitterionic intermediate formed from electrophilic attack by the peroxide on the sulfide.

The oxirane oxidation of sulfides has found applications in organic synthesis [13, 14]. For example, sulfoxidation of disulfide **1** with dioxirane afforded disulfoxide **2** in 98% yield (Eq. (8.2)) [13].

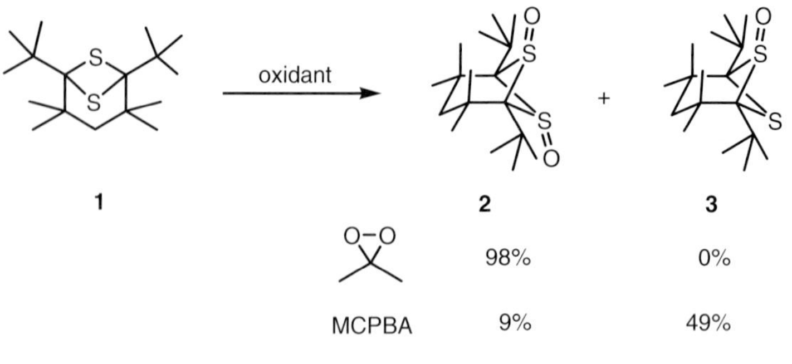

	1	2	3
O-O (dioxirane)		98%	0%
MCPBA		9%	49%

(8.2)

The corresponding sulfoxidation of **1** with MCPBA was unsuccessful and gave only 9% of disulfoxide **2** together with substantial amounts of degraded dicarbonyl compound (29%) and monosulfoxide **3** (49%). Other applications of sulfoxidation with the use of dioxirane are given in Ref. [14].

8.2.1.3 Oxone and Derivatives

Oxone, which is commercially available as a 2 : 1 : 1 mixture of $KHSO_5$, $KHSO_4$, and K_2SO_4, has been used for the oxidation of sulfides to sulfoxides [15]. It shows a tendency to overoxidation, and was originally used for the oxidation of sulfides to sulfones [16]. More recently, some improvements were obtained with surface-mediated oxone oxidations, the oxone being bound to a silica gel surface (Eq. (8.3)). However, some sulfone was still formed.

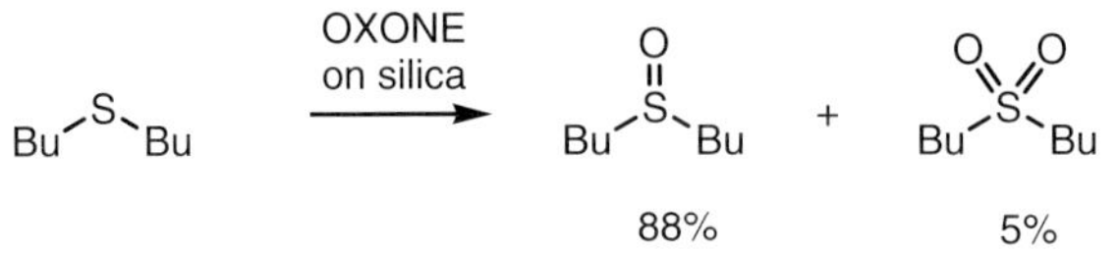

$$Bu-S-Bu \xrightarrow{\text{OXONE on silica}} Bu-S(=O)-Bu \;(88\%) + Bu-S(=O)_2-Bu \;(5\%) \quad (8.3)$$

In a recent modification of the oxone-type oxidation, the quaternary salt benzyltriphenyl phosphonium peroxymonosulfate $PhCH_2Ph_3P^+HSO_5^-$ was employed (Eq. (8.4)). No overoxidation to sulfone was detected, according to the authors.

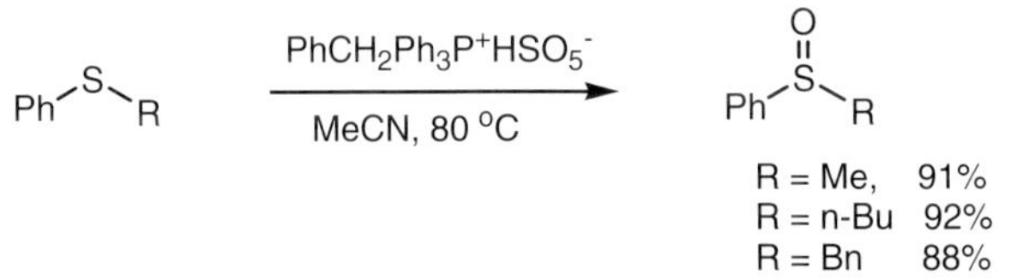

$$Ph-S-R \xrightarrow[\text{MeCN, 80 °C}]{PhCH_2Ph_3P^+HSO_5^-} Ph-S(=O)-R \quad (8.4)$$

R = Me, 91%
R = n-Bu 92%
R = Bn 88%

8.2.1.4 H_2O_2 in "Fluorous Phase" and Related Reactions

Oxidation of sulfides to sulfoxides by H_2O_2 in hexafluoro-2-propanol has been reported to occur with an exceptionally high rate and selectivity [17]. Reaction of ethyl phenyl sulfide with 2 equiv. of 30% aqueous H_2O_2 in hexafluoro-2-propanol at 25 °C was complete within 5 min and gave the corresponding sulfoxide in 97% yield. No sulfone was formed; in a control experiment the sulfoxide was stirred with 2 equiv. of 30% H_2O_2 for 8 h without any formation of sulfone. The normally slow-reacting sulfides **4**, **5**, and **6**, reacted fast, and the procedure tolerates double bonds (Table 8.1). Sulfides having double bonds also underwent a selective sulfoxidation (e.g., **7** and **8**, entries 4 and 5). This seems to be an excellent method for efficient and highly selective oxidation of sulfides (thioethers) to the corresponding sulfoxides, the only limitation being the toxicity of the hexafluoro-2-propanol.

A highly selective oxidation of sulfides to sulfoxides takes place by 30% aqueous H_2O_2 in phenol at room temperature [18]. Methyl phenyl sulfide was oxidized to the sulfoxide within 0.5 minutes in 99% yield. No sulfone could be detected. In a control experiment methyl phenyl sulfoxide was allowed to react for 14 h with 30% aqueous H_2O_2 in phenol at room temperature; the sulfoxide was recovered unchanged.

Table 8.1 Hydrogen peroxide oxidation of sulfides to sulfoxides in 'fluorous phase'.

Entry	Substrate	Reaction time (min)	% Yield of sulfoxide
1	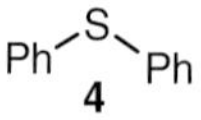 Ph-S-Ph (4)	5	99
2	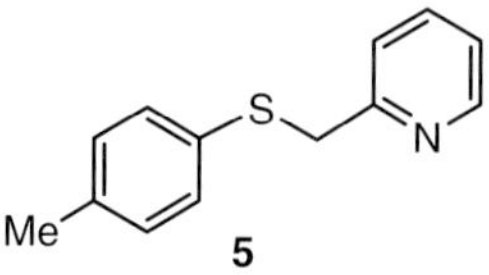 Me-C6H4-S-CH2-(2-pyridyl) (5)	10	93
3	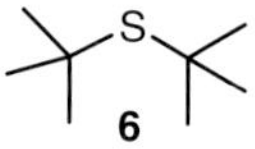 tBu-S-tBu (6)	20	97
4	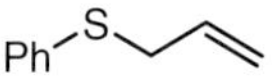 Ph-S-allyl (7)	5	99
5	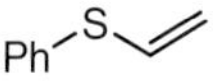 Ph-S-vinyl (8)	15	94

A range of sulfides including diphenyl sulfide were oxidized to sulfoxides in high yield (93–99%) in less than 5 min.

The high rates of the H_2O_2-based sulfoxidation in hexafluoro-2-propanol and phenol are ascribed to hydrogen-bonded H_2O_2 (see Chapter 4).

8.2.2
Chemocatalytic Reactions

In almost all catalytic reactions reported for oxidation of sulfides to sulfoxides a peroxide compound or molecular oxygen is employed. The most common oxidant is H_2O_2, usually as a 30% aqueous solution.

8.2.2.1 H_2O_2 as Terminal Oxidant

A large number of catalysts have been reported in the literature for the H_2O_2 oxidation of sulfides to sulfoxides. These include various metal complexes (of transition metals and lanthanides), flavins, and benzthiazoles.

Transition Metals as Catalysts Oxidation of various sulfides to sulfones by H_2O_2 mediated by $TiCl_3$ was reported by Watanabe [19]. Quantitative yields of sulfoxide were obtained within 5–15 min with no overoxidation to sulfone. However, a sevenfold excess of hydrogen peroxide and 2 equivalents of $TiCl_3$ per mole of substrate were employed.

Tungsten-based catalytic systems for H_2O_2 oxidations of sulfides have attracted considerable interest, and some early reports include the use of H_2WO_4 [20]. More recently, various tungsten-catalyzed methods have been used [21–25].

8.2 Oxidation of Sulfides to Sulfoxides

The Venturello-type peroxo complex $Q_3\{(PO_4)[W(O)(O_2)_2]_4\}$, where $Q = N\text{-}(n\text{-}C_{16}H_{33})$ pyridinium, was employed as catalyst for the oxidation of sulfides to sulfoxides and sulfones by hydrogen peroxide [21]. The selectivity for sulfoxides was low with this catalyst, which gave only sulfone. The corresponding molybdenum complex $Q_3\{(PO_4)[M(O)(O_2)_2]_4\}$ and $Q_3PMo_{12}O_{40}$ as catalyst in the H_2O_2 oxidations gave mixtures of sulfoxide and sulfone that ranged from 3:1 to 1:3 depending on the substrate. Finally, $Q_3PW_{12}O_{40}$ as catalyst in the H_2O_2 oxidations gave a good selectivity [21]. Related peroxo-tungstates and -molybdates $Ph_2PO_2)[MO(O_2)_2]_2^-$ (M = W or Mo) were studied as catalysts for H_2O_2-oxidation of sulfoxides. The selectivity for sulfoxide was low for these catalysts [22].

Polyoxometalates $WZnMn_2(ZnW_9O_{34})_2^{2-}$ were employed to oxidize sulfides to sulfones in moderate selectivity (85–90% selectivity with 10–15% of sulfone) [23]. The catalytic effect was strong for the oxidation of aromatic sulfides but weak for the oxidation of aliphatic sulfides. Recently, Mizuno has reported that by using the selenium-contaning peroxotungstate $[SeO_4\{WO(O_2)_2\}_2]^{2-}$, various sulfides were efficiently oxidized to sulfoxides by H_2O_2 [26]. The selectivity and yield of these reactions were very high. A larger-scale oxidation (100 mmol) of thioanisole was demonstrated, which afforded 12.73 g (91%) of the corresponding sulfoxide (≥99% pure). The reaction gives very high turnover numbers.

PhSMe + H_2O_2 → PhS(O)Me

(TBA)$_2$[SeO$_4${WO(O$_2$)$_2$}] (0.005 mol%), CH$_3$CN (40 ml), 20 °C, 4h

100 mmol + 100 mmol → 12.73 g (91%)

Noyori has reported on a tungsten-based halogen-free system for the oxidation to sulfoxides with 30% aqueous H_2O_2 that gives sulfoxides with good to moderate selectivity [24]. The reactions were run without organic solvent, and the catalyst employed was Na_2WO_4 together with $PhPO_3H_2$ and $CH_3(n\text{-}C_8H_{17})NHSO_4$. For example, thioanisole gave sulfoxide and sulfone in a ratio of 94:6 after 9 h at 0 °C with the use of a substrate-to-catalyst ratio of 1000:1.

Choudary and coworkers [25] subsequently extended the tungsten system to the use of layered double hydroxides (LDHs) with water as solvent and 30% H_2O_2 as the oxidant. Under these conditions the new catalysts LDH-WO_4^{2-} gave good turnover rates; however, the selectivity of the oxidation of thioanisole was moderate, with a sulfoxide:sulfone ratio of 88:12 (Eq. (8.5)). Other sulfide oxidations also gave a moderate selectivity (Eq. (8.5))

Ph-S-R →[LDH-WO_4^{2-}] Ph-S(O)-R + Ph-S(O)$_2$-R   (8.5)

R = Me	94	:	6	ref [24]
R = Me	88	:	12	ref [25]
R = vinyl	71	:	29	ref [25]

The advantage with the latter system, in spite of the moderate selectivity for sulfoxide over sulfone, is that the immobilized catalyst can be recovered and reused. The catalyst showed consistent activity and selectivity for six recyclings.

The catalytic cycle with the WO_4^{2-} catalysts is thought to involve the generation of tungsten peroxy complexes (Scheme 8.1). These react fast with the sulfide with transfer of an oxygen to give the sulfoxide.

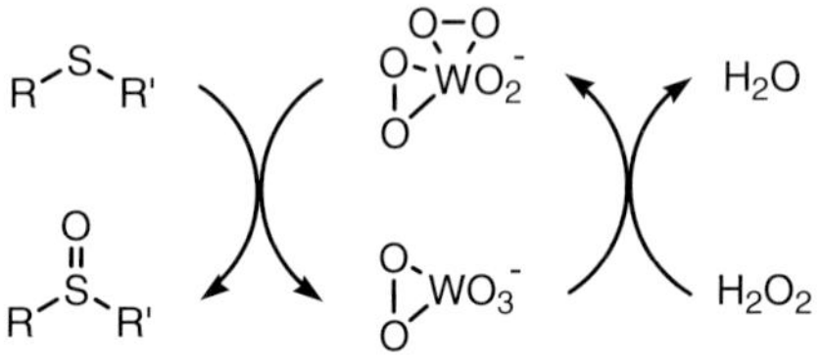

Scheme 8.1

Feringa investigated various nitrogen ligands for the selective Mn-catalyzed oxidation of sulfides to sulfoxides with 30% aqueous H_2O_2 [27]. The use of Mn $(OAc)_3 \cdot 2H_2O$ with bipyridine ligand **9** in the oxidation of thioanisole gave sulfoxide free from sulfone in 55% yield. With ligand **10** the same yield was obtained, but now with the sulfoxide in 18% ee.

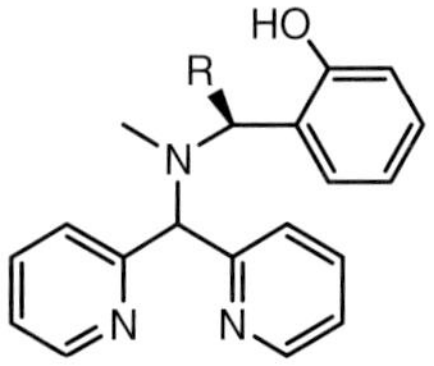

9 R = H
10 R = Me

Titanium-catalyzed oxidations with 35% aqueous H_2O_2 using Schiff-base (salen) titanium oxo complexes as catalysts showed very high activity [28]. The oxidation of methyl phenyl sulfide required only 0.1 mol% of catalyst. The use of chiral salen complexes gave low enantioselectivity (<20% ee).

Vanadium-catalyzed H_2O_2 oxidations of sulfides to sulfoxides have been reported by several groups. These reactions have been shown to work well for asymmetric sulfoxidation. In 1995 Bolm and Bienewald reported on the use of vanadium-chiral Schiff base complexes as catalysts for asymmetric sulfoxidation by 30% H_2O_2 [29]. Oxidation of thioanisole using ligand **11** afforded the corresponding sulfoxide in 70% ee in high yield without any significant overoxidation to sulfone (Eq. (8.6)).

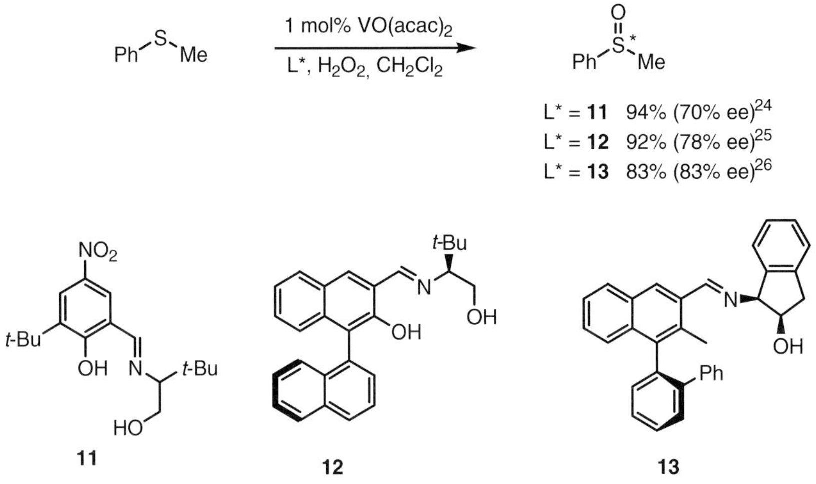

[Reaction scheme 8.6: Ph-S-Me + 1 mol% VO(acac)₂, L*, H₂O₂, CH₂Cl₂ → Ph-S(=O)*-Me]

L* = **11**   94% (70% ee)[24]
L* = **12**   92% (78% ee)[25]
L* = **13**   83% (83% ee)[26]

[Structures of ligands **11**, **12**, and **13**]

(8.6)

Vetter and Berkessel later improved this reaction by changing the ligand to **12**, which afforded 78% ee (Eq. (8.6)) [30]. Further improvement of this protocol was reported by Katsuki and coworkers [31], who used ligand **13** to obtain 83% ee in the oxidation of thioanisole to sulfoxide (Eq. (8.6)). A further increase in the enantioselectivity with ligand **13** was obtained with methanol as an additive (2% methanol in methylene chloride) [31]. With this protocol ees up to 93% were obtained for aryl methyl sulfides (Eq. (8.7)). In all of the reactions, except for Ar = p-NO$_2$C$_6$H$_4$, only traces of sulfone was formed. The latter substrate gave 10% sulfone.

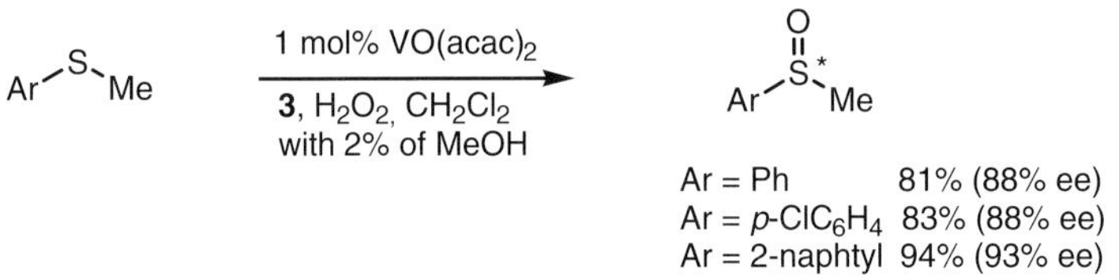

[Reaction scheme 8.7: Ar-S-Me + 1 mol% VO(acac)₂, **3**, H₂O₂, CH₂Cl₂ with 2% of MeOH → Ar-S(=O)*-Me]

Ar = Ph   81% (88% ee)
Ar = p-ClC$_6$H$_4$ 83% (88% ee)
Ar = 2-naphtyl   94% (93% ee)

(8.7)

More recently, Bolm and coworkers extended their original work on the Schiff-base ligands to iron-catalyzed reactions [5, 32]. With the use of ligand (S)-**14** they obtained high ee in the oxidation of various sulfides. In the original report the yields of sulfoxide were moderate (up to 44%), and sulfoxides could be obtained in up to 90% ee. The reaction was later improved by the addition of additives, and with benzoic acid as additive methyl tolyl sulfide was oxidized to the (S)-sulfoxide in 78% yield and 92% ee [32b]. The highest ee obtained with the improved procedure was 96%, but then the yield was more moderate.

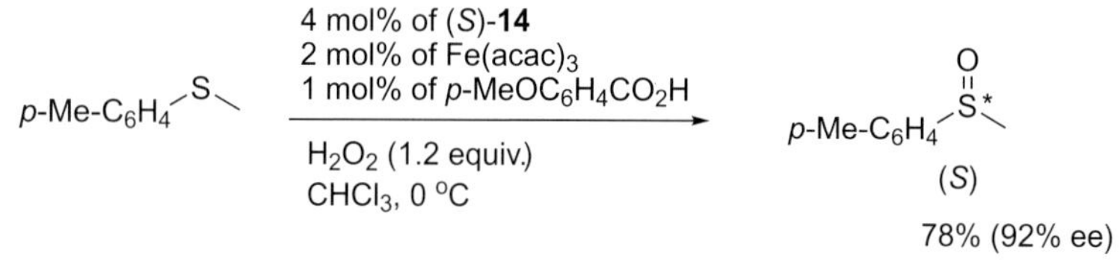

p-Me-C$_6$H$_4$-S- $\xrightarrow{\begin{array}{c}\text{4 mol\% of (S)-14}\\\text{2 mol\% of Fe(acac)}_3\\\text{1 mol\% of } p\text{-MeOC}_6\text{H}_4\text{CO}_2\text{H}\\\hline \text{H}_2\text{O}_2 \text{ (1.2 equiv.)}\\\text{CHCl}_3, 0\,°\text{C}\end{array}}$ p-Me-C$_6$H$_4$-S(=O)-* (S)

78% (92% ee)

(8.8)

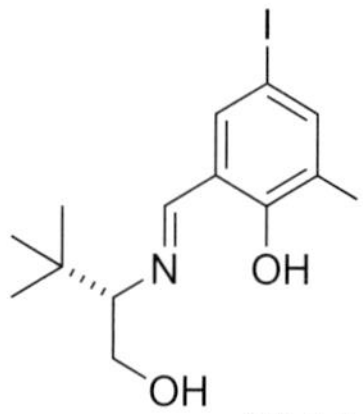

(S)-14

Jackson and coworkers [33] developed immobilized Schiff-base ligands inspired by those used by Bolm. A peptide Schiff-base library with ligands **15** bound to solid support was investigated, where two amino acids (AA1 and AA2) and the salicylaldehyde were varied. A library of 72 ligands was prepared using six different salicylaldehydes, six different amino acids as aminoacid 1 (AA1) and two different aminoacids as aminoacid 2 (AA2). Screening of these ligands in the VO(acac)$_2$-catalyzed H$_2$O$_2$ oxidation of sulfides in CH$_2$Cl$_2$ gave only a moderate enantioselectivity of 11% for thioanisole with the best ligand (R^1 = Ph, R^2 = H). Screening the ligands with Ti(O^{i}Pr)$_4$ as catalyst gave a better result, and thioanisole afforded the sulfoxide in quantitative yield in 64% ee with ligand **16**. The best result with this ligand was obtained with 2-naphthyl methyl sulfide, which gave 72% ee in a high yield (Eq. (8.9)).

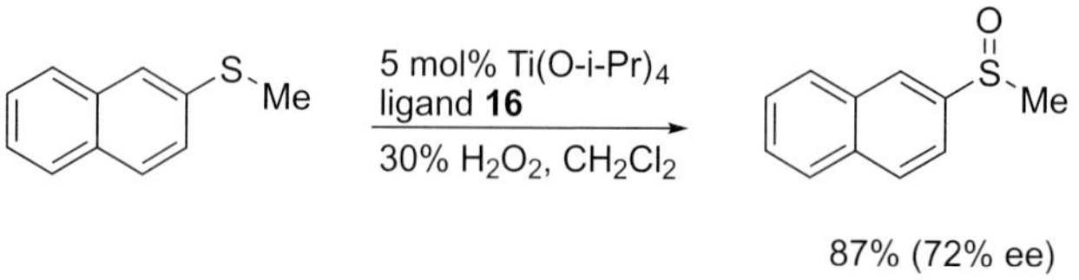

Naphthyl-S-Me $\xrightarrow{\begin{array}{c}\text{5 mol\% Ti(O-i-Pr)}_4\\\text{ligand } \mathbf{16}\\\hline \text{30\% H}_2\text{O}_2, \text{CH}_2\text{Cl}_2\end{array}}$ Naphthyl-S(=O)-Me

87% (72% ee)

(8.9)

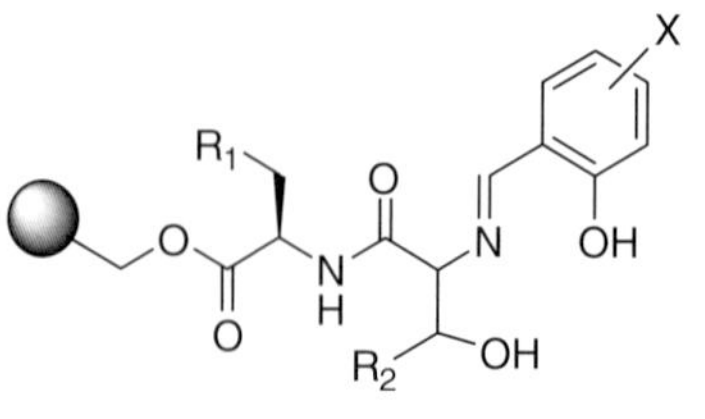

15

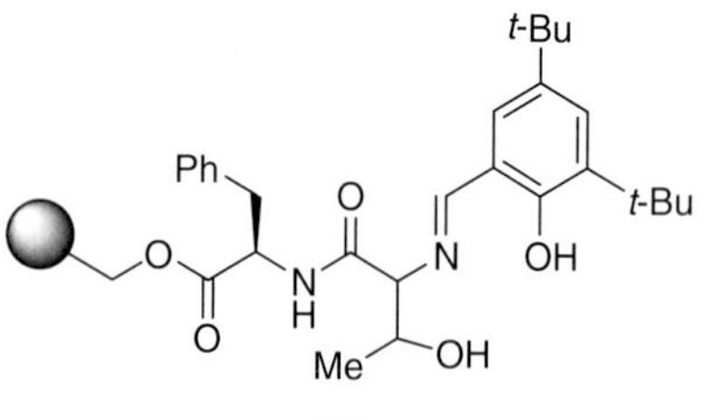

16

A vanadium-catalyzed enantioselective sulfoxidation using related Schiff-base ligands was later developed by the Jackson group. In this reaction, selective oxidation of the minor enantiomer of the sulfoxide to sulfone occurs, which improves the *ee* of the product. For example, methyl tolyl sulfide was oxidized to its (*R*)-sulfoxide in 99.5% *ee* with a yield of 82% using ligand (*R*)-**17** [34].

$$p\text{-Me-C}_6H_4\text{-S-} \xrightarrow[\text{CHCl}_3,\ 0\,°\text{C}]{\substack{1.5\ \text{mol\%\ of\ }(R)\text{-}\mathbf{17}\\1.0\ \text{mol\%\ of\ VO(acac)}_2\\ \text{H}_2\text{O}_2\ (1.2\ \text{equv})}} p\text{-Me-C}_6H_4\text{-S(=O)*-}$$

(*R*)
82% (99.5% *ee*)

(8.10)

(*R*)-**17**

Chiral Mn(III)(salen) complexes have been used as catalysts in the oxidation of sulfides to sulfoxides by 30% aqueous H_2O_2 in acetonitrile [35]. The use of 2–3 mol% of catalyst led to an efficient reaction with enantioselectivities up to 68% *ee*. The use of Ti(salen) complexes as catalysts and aqueous H_2O_2 in methanol afforded sulfoxides in 76% *ee* from the oxidation of sulfides [36]. Katsuki has recently improved these types of catalysts with other metals for H_2O_2-based sulfoxidation and obtained highly efficient reactions with excellent enantioselectivity [37, 38]. With a chiral Al(salalene) complex as catalyst, methyl phenyl sulfide was oxidized to the corresponding sulfoxide in 90% yield and 98% *ee* [37a]. The sulfoxidation was accompanied by formation of 9% of sulfone. This overoxidation was shown to account for the high *ee* by removing some of the minor enantiomer of the sulfoxide. The reaction was subsequently improved to give 99% *ee* under solvent-free or highly concentrated conditions [37b]. An iron-based catalyst, Fe(salan) was used for the enantioselective oxidation of various sulfides in high yields and enantioselectivities up to 96% *ee*.

Platinum-catalyzed asymmetric sulfoxidation of thioethers with hydrogen peroxide in water was reported to give up to 88% *ee*. (*R*)-BINAP was used as chiral ligand on the metal [39].

Oxidation of sulfides in the presence of electron-rich double bonds is problematic with many of the traditional oxidants such as MCPBA, $NaIO_4$, and oxone because of interference with double bond oxidation (e.g., epoxidation). Koo and coworkers [40] addressed this problem and studied the selective oxidation of allylic sulfides having multiple alkyl substituents. They tested various stoichiometric oxidants and a number of catalytic reactions with 30% aqueous H_2O_2 as the oxidant. Of all the oxidation systems tested for the sulfoxidation, they found that the use of $LiNbMoO_6$ as catalyst with H_2O_2 as the oxidant gave the best result. With this system no epoxidation took place and a reasonably good selectivity for sulfoxide over sulfone was obtained (Table 8.2).

Lanthanides as Catalysts Catalytic amounts of scandium triflate (Sc(OTf)$_3$) was found to greatly increase the rate of oxidation of sulfides by 60% H_2O_2 (Table 8.3) [41].

Table 8.2 Selective oxidation of allylic sulfides with electron-rich double bonds.

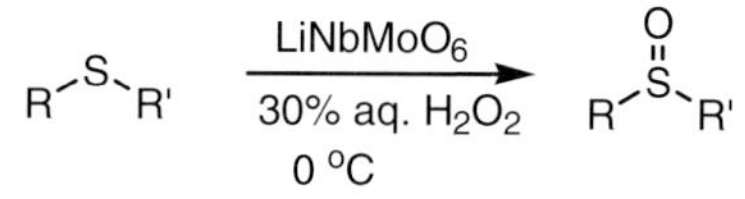

$$R-S-R' \xrightarrow[\substack{30\% \text{ aq. } H_2O_2 \\ 0\,°C}]{LiNbMoO_6} R-S(=O)-R'$$

Entry	Sulfide	% Yield Sulfoxide	% Yield Sulfone
1	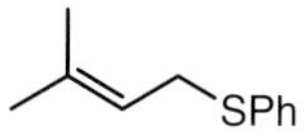 ⸺⸺SPh (prenyl SPh)	77	14
2	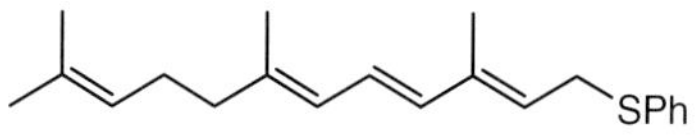 geranylgeranyl-SPh	80	4
3	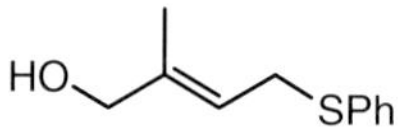 HO-⸺-SPh	75	12
4	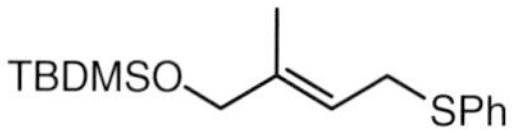 TBDMSO-⸺-SPh	82	12
5	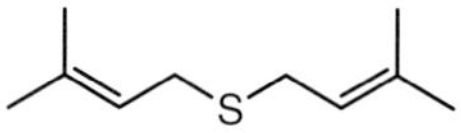 bis-allyl sulfide	54	8

The reaction is run at room temperature in methylene chloride containing 10% of ethanol. The reaction shows quite a high selectivity for sulfoxide, with only 2–4% sulfones being formed.

Table 8.3 Oxidation of sulfides in the presence of catalytic amounts of scandium triflate.

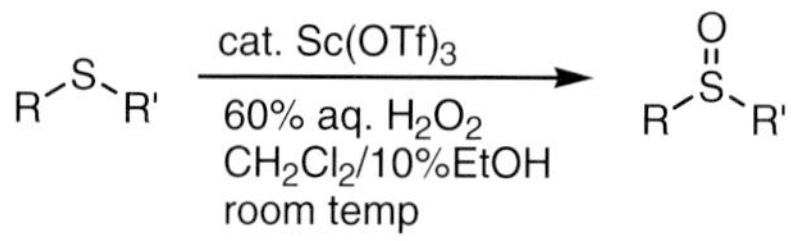

$$R-S-R' \xrightarrow[\substack{60\% \text{ aq. } H_2O_2 \\ CH_2Cl_2/10\%EtOH \\ \text{room temp}}]{cat.\ Sc(OTf)_3} R-S(=O)-R'$$

Entry	Substrate	Reaction time (h)	% Yield Sulfoxide	% Yield Sulfone
1	Ph-S-Me	3	94	4
2	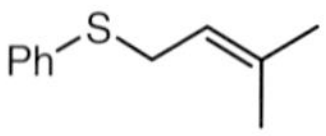 Ph-S-CH₂CH=C(Me)₂	2	98	2
3	p-Br-C₆H₄-S-Me	3	98	2
4	p-MeO-C₆H₄-S-Me	1.3	98	2

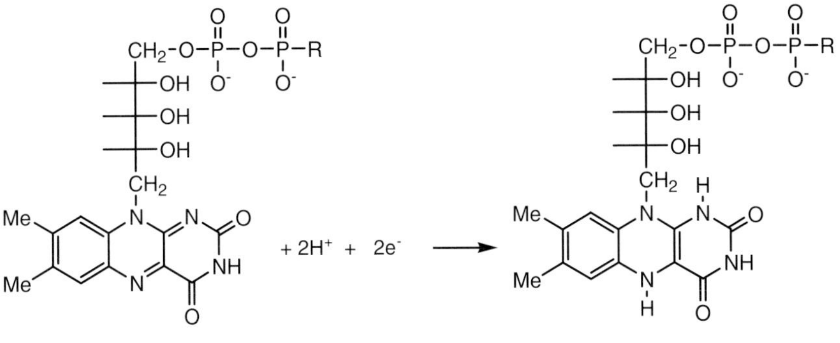

Figure 8.1 The FAD/FADH$_2$ redox system is a cofactor in flovoenzymes (R = adenosine).

The reaction was applied to the oxidation of various cysteine derivatives to their corresponding sulfoxides (Eq. (8.11)).

$$\text{Boc-Phe-Cys(Me)Ala-OAll} \xrightarrow[\substack{60\% \text{ aq. } H_2O_2 \\ 15 \text{ min}}]{20 \text{ mol\% Sc(OTf)}_3} \text{Boc-Phe-Cys(O)(Me)Ala-OAll} \quad 100\%$$

(8.11)

Flavins as Catalysts Flavins are organic molecules that are part of the FADH$_2$ cofactor of flavoenzymes (Figure 8.1). In FAD-containing monooxygenases (FADMOs), molecular oxygen is activated to generate a flavin hydroperoxide.

Model compounds of the natural flavins were studied by Bruice in the late 1970s [42]. In these studies N,N,N,-3,5,10-trialkylated flavins **16** and **17** were used. It was demonstrated that reactive flavin hydroperoxides can be generated from the reduced form **16** and molecular oxygen or from the oxidized form **17** and H$_2$O$_2$ (Scheme 8.2).

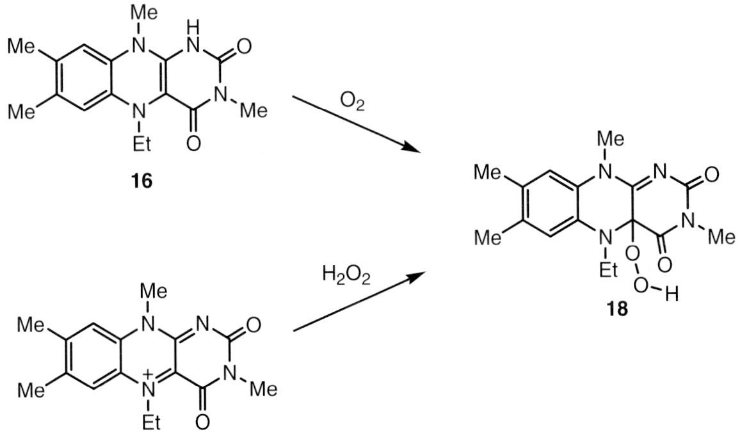

Scheme 8.2

8 Selective Oxidation of Amines and Sulfides

Stoichiometric oxidation reactions with hydroperoxide **18** were studied, and it was found that **18** oxidizes sulfides to sulfoxides in a highly selective manner [42, 43]. It was later demonstrated that these flavins can participate as catalysts for the H_2O_2 oxidation of sulfides to sulfoxides [44, 45].

More recently, a modification of the structure of these flavins gave more efficient and robust organocatalysts for the H_2O_2-based sulfoxidations [46]. These new flavin catalysts (**19**) are superior to the previous 'natural-based' flavin catalysts and have the advantage that they also give excellent results for the oxidation of tertiary amines to amine oxides [47] (see Section 8.3.2).

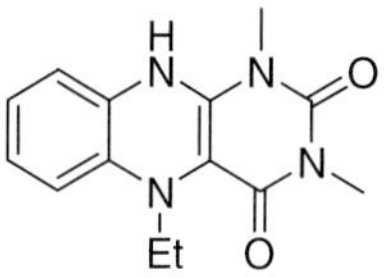

19

Various thioethers were oxidized with the use of flavin **19** as the catalyst (Table 8.4) [46]. Only sulfoxide was formed and no overoxidation to sulfone could be detected.

Table 8.4 Oxidation of thioethers using flavin **19** as catalyst.

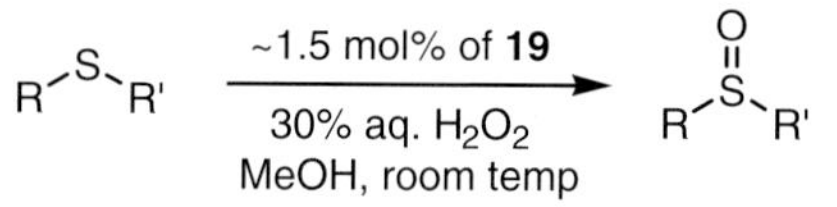

$$R-S-R' \xrightarrow[\text{MeOH, room temp}]{\sim 1.5 \text{ mol\% of } \mathbf{19} \\ 30\% \text{ aq. } H_2O_2} R-S(=O)-R'$$

Entry	Substrate	Mol%	Reaction time	% Yield of sulfoxide
1	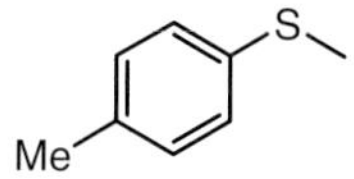 Me–C₆H₄–S–Me	1.8	1 h	100
2	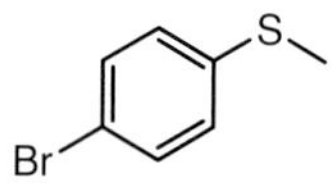 Br–C₆H₄–S–Me	1.6	2 h 40 min	96
3	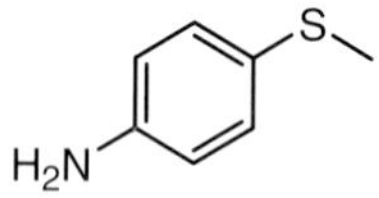 H₂N–C₆H₄–S–Me	1.3	23 min	99
4	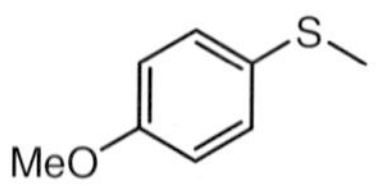 MeO–C₆H₄–S–Me	1.6	45 min	92
5	 1,3-dithiane	1.7	20 min	99
6	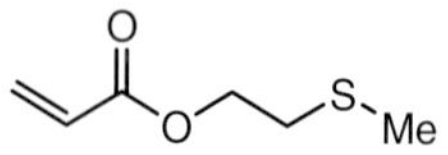 CH₂=CH–C(O)–O–CH₂–S–Me	1.1	30 min	99

8.2 Oxidation of Sulfides to Sulfoxides

The structure of the flavin was studied, and the new flavin structures (N,N,N-1,3,5-trialkyl) were compared with those previously used (N,N,N-3,5,10-trialkyl). It was found that the new flavins were between one and two orders of magnitude faster than the previously used flavins.

The flavin-catalyzed sulfoxidation was later extended to allylic and vinylic sulfides. It was found that the oxidation of allylic sulfides having electron-rich double bonds proceed with an exceptional selectivity for sulfoxidation (Table 8.5) [48]. Sulfone formation was depressed below detection (<0.5%) and only in one single case could sulfone be observed (1.5% relative yield, for entry 6). No epoxide could be detected.

The mechanism of the flavin-catalyzed oxidation of sulfides by hydrogen peroxide is shown in Scheme 8.3.

The reaction is initiated by reaction of catalyst **19** with molecular oxygen to give flavin hydroperoxid **20**. Once in the cycle, this hydroperoxide can be regenerated by H_2O_2. The hydroperoxide **20** transfers an oxygen to the sulfide via the hydrogen-bonded transition state **21** to give sulfoxide and hydroxyflavin intermediate **22**.

Table 8.5 Flavin-catalyzed sulfoxidation of sulfides.

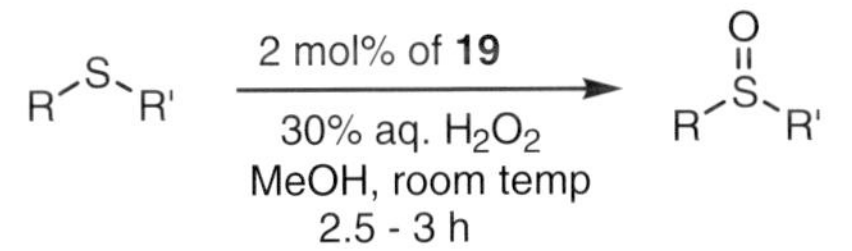

$$R\text{-}S\text{-}R' \xrightarrow[\text{MeOH, room temp} \\ 2.5\text{-}3\text{ h}]{\text{2 mol\% of }\textbf{19} \\ 30\%\text{ aq. }H_2O_2} R\text{-}S(=O)\text{-}R'$$

Entry	Sulfide	% Yield Sulfoxide	Sulfone[a]
1	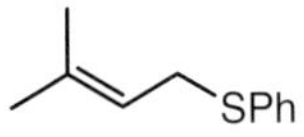 (prenyl-SPh)	92	n.d.
2	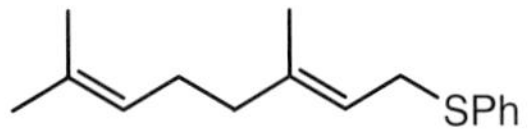 (geranyl-SPh)	76	n.d.
3	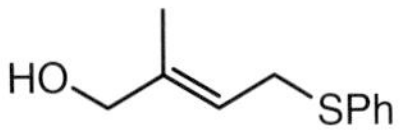 HO-substituted allyl SPh	77	n.d.
4	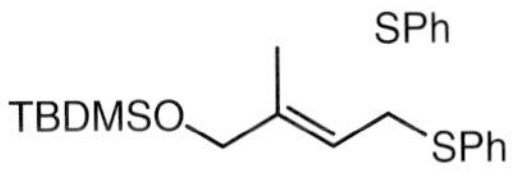 TBDMSO-substituted bis-SPh	87	n.d.
5	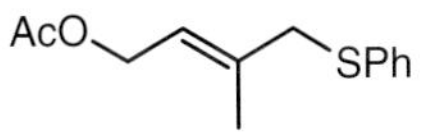 AcO-substituted allyl SPh	96	n.d.
6	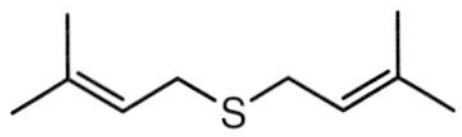 diallyl sulfide derivative	85	1.3

a) n.d. = not detected.

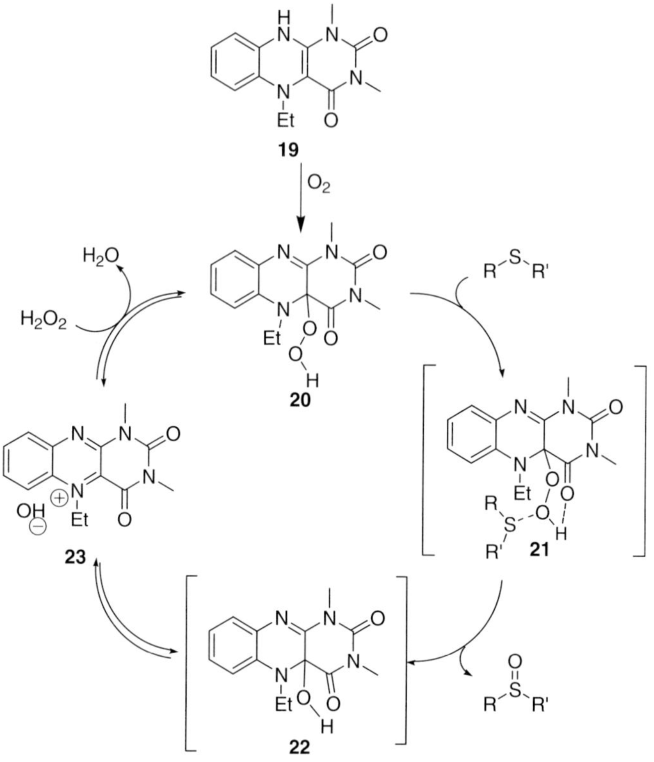

Scheme 8.3

Elimination of OH⁻ from **22** produces the aromatic 1,4-diazine **23**. The latter species, which becomes the catalytic intermediate, reacts with hydrogen peroxide to regenerate the catalytic flavin hydroperoxide. The advantage of the N,N,N-1,3,5-trialkyl flavin system over the N,N,N-3,5,10-trialkylated analogs is that the elimination of OH⁻ (**22** → **23**) is fast because of aromatization. In the N,N,N-3,5,10-trialkylated system this step is slow and has been found to be the rate-determining step [44].

The redox properties of eight different N,N,N-1,3,5-trialkylated flavins and their activity as redox catalysts for the oxidation of sulfides were studied [49]. The redox potentials (E_0) were determined by cyclic voltammetry, and for the eight flavins studied E_0 varied from −0.414 to −0.162 V. There was a linear free-energy relationship between the oxidation potentials and the Hammett σ value. The redox potentials of the flavins also correlated well with their efficiency as catalysts in the H₂O₂ oxidation of methyl p-tolyl sulfide. Interestingly, it was found that the electron rich 7,8-dimethyl derivative **24** ($E_0 = -0.414$) was oxidation sensitive

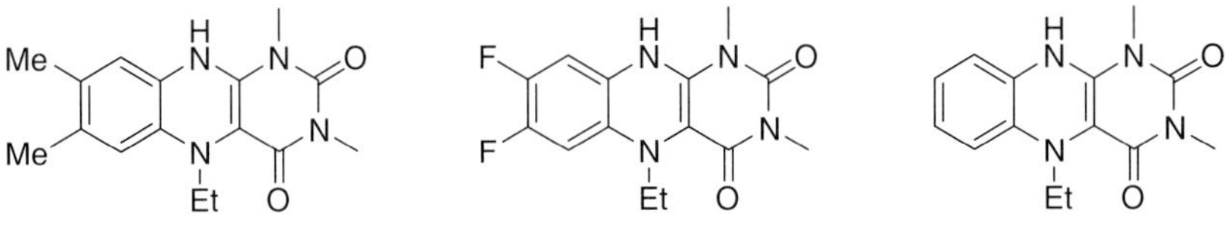

8.2 Oxidation of Sulfides to Sulfoxides

24, **25**, **19**

and underwent degradation, whereas the 7,8-difluoro derivative **25** ($E_0 = -0.207$) was very stable. The latter derivative showed an induction period in the oxidation of p-tolyl sulfide with H_2O_2 due to the slow generation of hydroperoxide (cf. **19** → **20** in Scheme 8.3). Once formed, the difluoro flavin **25** is a highly efficient and robust catalyst. Catalyst **19** is also quite stable, and furthermore it does not show any detectable induction period, which is consistent with the lower redox potential (−0.305) compared to that of **25** (−0.207).

A derivative of the N,N,N-1,3,5-trialkylated flavins with a carboxylate group in either the 7 or 8 position (**26**) was studied as a recyclable catalyst for H_2O_2-based sulfoxidation in ionic liquid [BMIm]PF_6 [50]. It was found that flavin **26** is a highly efficient catalyst for the

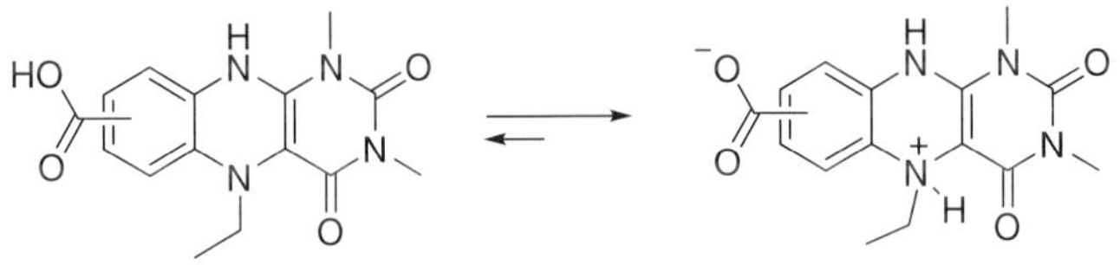

26

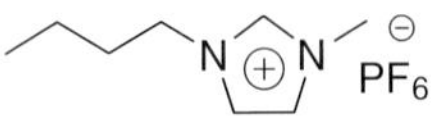

[BMIm]PF_6

H_2O_2 oxidation of various sulfides to sulfoxides. The sulfoxides were obtained in good to high yields and high selectivity without any detectable overoxidation to sulfone. More importantly, the flavin catalyst **26** in the ionic liquid was recycled up to seven times in the sulfoxidation of some representative sulfides without loss of activity or selectivity (Table 8.6) [50]. According to ^{1}H NMR, the zwitterionic form of flavin **26** predominates, which explains why **26** stays in the ionic liquid after the extraction of the product.

Heterogeneous catalytic oxidation of sulfides in ionic liquids by anhydrous H_2O_2 or urea hydroperoxide with MCM-41 and related mesoporous catalysts containing Ti or Ti and Ge was studied by Hardacre and coworkers [51]. The Ti-based catalyst gave a quite selective sulfoxidation. Addition of Ge to Ti increased the rate of the oxidation but reduced the selectivity toward sulfoxide [51a].

Chiral flavins have been used to obtain an asymmetric sulfoxidation with H_2O_2 as the oxidant [45a, 52]. Flavin **27**, with planar chirality, was used to oxidize different aryl methyl sulfides with 35% H_2O_2. The hydrogen peroxide was added slowly over 5 days at −20 °C to the substrate and the catalyst. The best result was obtained with the p-tolyl derivative, which gave 65% ee (Eq. (8.12)).

8 Selective Oxidation of Amines and Sulfides

Table 8.6 Recycling of the flavin-catalyst in ionic liquid.[a),b)]

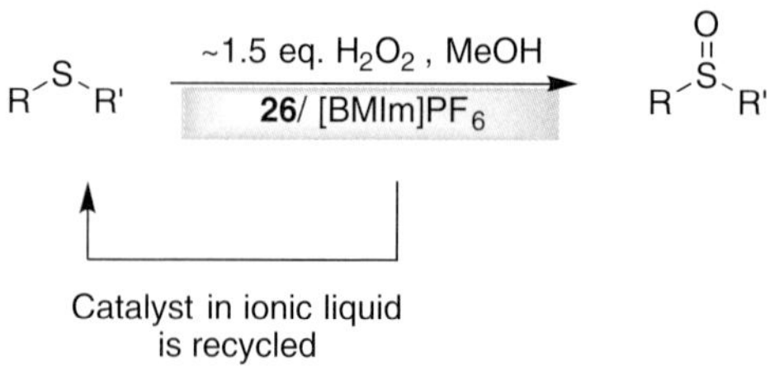

$$R\text{-}S\text{-}R' \xrightarrow[26/\,[BMIm]PF_6]{\sim 1.5\ \text{eq. } H_2O_2,\ MeOH} R\text{-}S(=O)\text{-}R'$$

Catalyst in ionic liquid is recycled

Entry	Sulfide	Yield of sulfoxide[c)] (%)					
		Run 1	Run 2	Run 3	Run 4	Run 5	Run 6/7
1	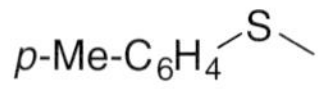 p-Me-C$_6$H$_4$-S-	78	84	80	88	91	89/-
2[d)]	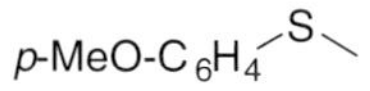 p-MeO-C$_6$H$_4$-S-	87[e)]	87[e)]	87[e)]	87[e)]	87[e)]	87[e)]/88
3	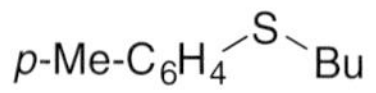 p-Me-C$_6$H$_4$-S-Bu	95	98	89	86	92	94/-
4	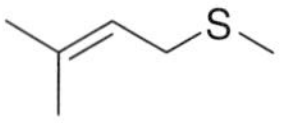 (prenyl)-S-	91	94	88	—	—	-/-

a) Unless otherwise noted: The reactions were run using 1 mmol of sulfide in 0.5 mL [BMIm]PF$_6$ and 3.2 mL MeOH. Flavin **26** (6.3 mg, 2 mol%) and H$_2$O$_2$ (150–170 μL, ~1.5 equiv.) were added and reaction stirred for 1.5–2.5 h.
b) Conversion determined by ^{1}H-NMR and was 96–99%.
c) Isolated yield.
d) A seventh run gave 88% yield.
e) The crude products of runs 1–6 were combined to give 522% isolated yield converted on one run, which gives in average 87% yield per run.

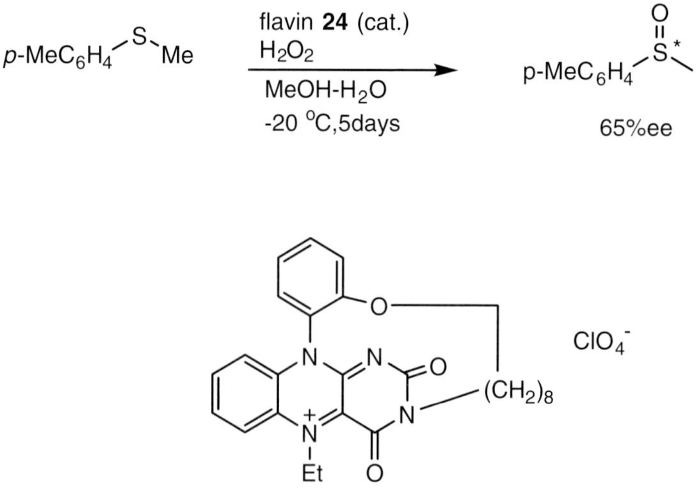

$$p\text{-MeC}_6H_4\text{-S-Me} \xrightarrow[\text{MeOH-H}_2O,\ -20\,°C,\ 5\,\text{days}]{\text{flavin } \mathbf{24}\ (\text{cat.}),\ H_2O_2} p\text{-MeC}_6H_4\text{-S}^*(=O)\text{-Me} \quad 65\%\ ee$$

(8.12)

27 (flavin structure with ClO$_4^-$ counterion, (CH$_2$)$_8$ tether, Et on N)

More recently, Murahashi has used flavin **28** to oxidize naphthyl methyl sulfide to its sulfoxide in 72% ee (Eq. (8.13)) [52].

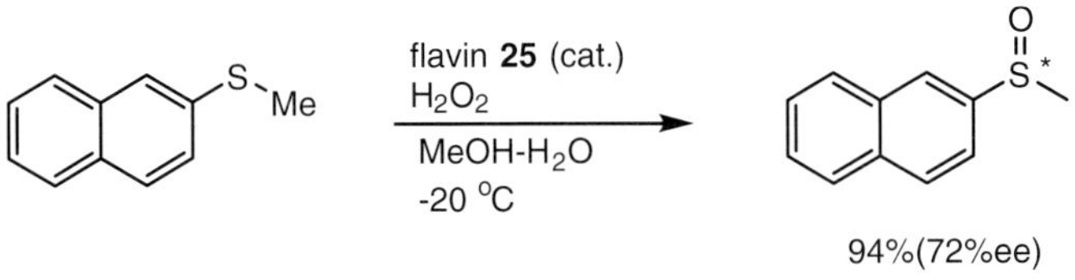

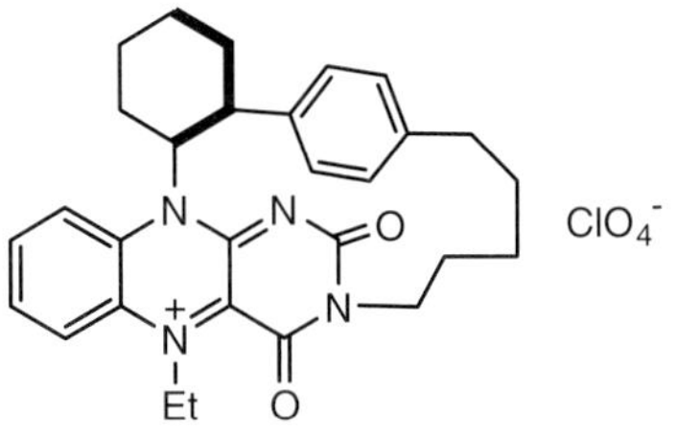

$$\text{Naphthyl-S-Me} \xrightarrow[\text{MeOH-H}_2\text{O}, -20\,°\text{C}]{\text{flavin 25 (cat.)}, \text{H}_2\text{O}_2} \text{Naphthyl-S(=O)-Me} \quad 94\%\,(72\%\,ee)$$

(8.13)

28 (with ClO_4^- counterion)

The use of flavins as organocatalysts for environmentally benign molecular transformations has been reviewed [53].

8.2.2.2 Molecular Oxygen as Terminal Oxidant

Aerobic oxidation of sulfides to sulfoxides by molecular oxygen is of importance from an environmental point of view. This transformation can be achieved by noncatalyzed direct reaction with molecular oxygen, but only at high oxygen pressure and elevated temperatures [54]. More recently, catalytic procedures that work at atmospheric pressure of molecular oxygen have been reported [55–59]. Various alkyl and aryl thioethers were selectively oxidized to sulfoxides by molecular oxygen in the presence of catalytic amounts of nitrogen dioxide (NO_2) [55]. The catalytic amount of NO_2 employed was between 4 and 36 mol%. Some examples are given in Eq. (8.14). The reaction, which was run at room temperature, is highly selective for sulfoxide over sulfone, and no sulfone could be detected.

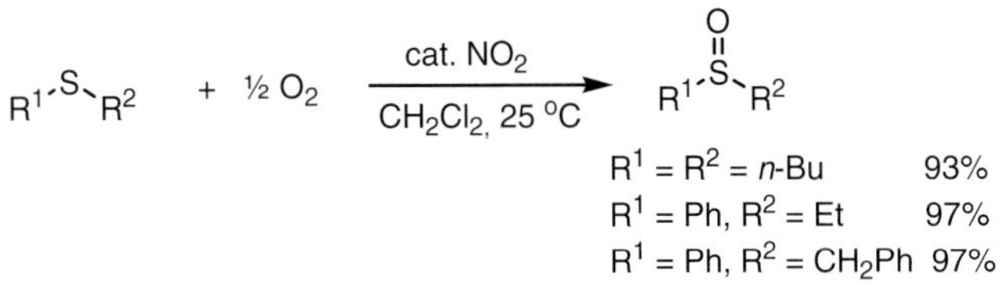

$$R^1\text{-S-}R^2 + \tfrac{1}{2}O_2 \xrightarrow[\text{CH}_2\text{Cl}_2,\,25\,°\text{C}]{\text{cat. NO}_2} R^1\text{-S(=O)-}R^2$$

$R^1 = R^2 = n\text{-Bu}$ 93%
$R^1 = \text{Ph},\, R^2 = \text{Et}$ 97%
$R^1 = \text{Ph},\, R^2 = \text{CH}_2\text{Ph}$ 97%

(8.14)

Ishi reported on aerobic oxidation of sulfides in the presence of N-hydroxyphthalimide (NHPI) and alcohols [56]. The reaction works at atmospheric pressure of oxygen; however, it requires 80–90 °C, and the selectivity for sulfoxide over sulfone is moderate (~85–90%).

The binary system $Fe(NO_3)_3$-$FeBr_3$ was used as an efficient catalytic system for the selective aerobic oxidation of sulfides to sulfoxides [57]. The reaction works with air at room temperature at ambient pressure and employs 10 mol% of Fe(NO$_3$)$_3$·9H$_2$O and 5 mol% of FeBr$_3$ (Eq. (8.15)).

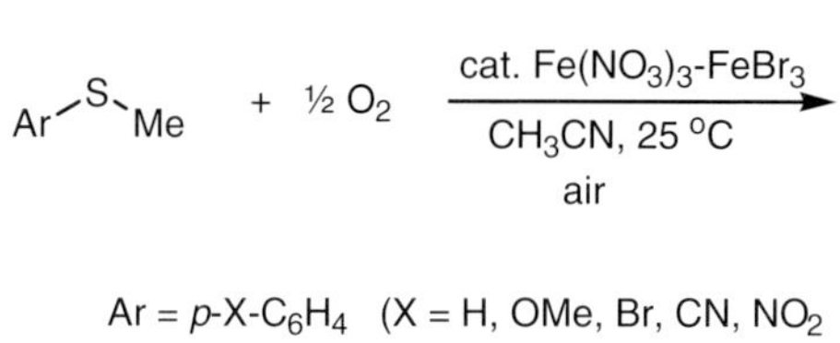

$$\text{Ar}-\text{S}-\text{Me} + \tfrac{1}{2}\text{O}_2 \xrightarrow[\text{CH}_3\text{CN, 25 °C, air}]{\text{cat. Fe(NO}_3\text{)}_3\text{-FeBr}_3} \text{Ar}-\text{S(=O)}-\text{Me} \quad 91\text{-}92\%$$

(8.15)

Ar = p-X-C$_6$H$_4$ (X = H, OMe, Br, CN, NO$_2$)

The mechanism of this aerobic oxidation involves the oxidation of bromide to bromine. The procedure may therefore be limited to sulfides that lack olefinic functionality.

A method for mild and efficient aerobic oxidation of sulfides catalyzed by HAuCl$_4$/AgNO$_3$ was reported by Hill [58]. The active catalyst is thought to be Au(III)Cl$_2$NO$_3$(thioether). A very high selectivity for sulfoxide was observed in these oxidations and no sulfone was detected. Isotope labeling studies with H$_2$^{18}O shows that water and not O$_2$ is the source of oxygen in the sulfoxide product.

Murahashi has reported on an interesting flavin-catalyzed aerobic oxidation of sulfides to sulfoxides (Scheme 8.4) [59]. Flavin hydroperoxides can be generated from reaction of the lowest reduced form of the flavin (**16**) and molecular oxygen. These hydroperoxides (**18**) have been studied in stoichiometric oxidation of sulfides to sulfoxides by Bruice [42].

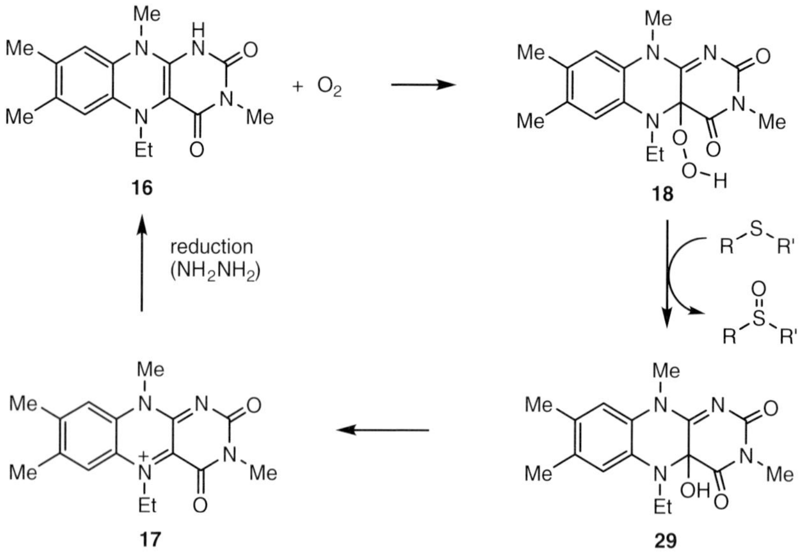

Scheme 8.4

They react rapidly with sulfides with transfer of an oxygen to give sulfoxides. The 4a-hydroxyflavin **29** generated in this process can lose a hydroxide to give **17**. The latter flavin, which is the 2-electron-oxidized flavin (compared to **16**), is unreactive toward molecular oxygen. On the other hand, it can react with a hydrogen peroxide to give **18**, but in an aerobic process it is inert. In nature, the corresponding molecule in the flavoenzyme (FAD-containing monooxygenase) is reduced by NADPH. In the process developed by Murahashi, hydrazine (NH_2NH_2) is employed for the reduction of the flavin **17** back to the reduced form (**16**). In this way a flavin-catalyzed aerobic oxidation of sulfides to sulfoxides was obtained (Eq. (8.16))

$$R\text{-}S\text{-}R' + O_2 + \tfrac{1}{2}NH_2NH_2 \xrightarrow[\substack{35\,°C \\ 1\,atm\,O_2}]{1\,mol\%\,\mathbf{17}} R\text{-}S(=O)\text{-}R' + \tfrac{1}{2}N_2 + H_2O \quad (96\text{–}99\%)$$

(8.16)

This catalytic system constitutes a mimic of the flavoenzymatic aerobic oxidation and the reaction is highly selective for sulfoxide without overoxidation to sulfone.

8.2.2.3 Alkyl Hydroperoxides as Terminal Oxidant

Alkyl hydroperoxides are known to oxidize sulfides slowly in a noncatalyzed reaction [3, 15b, 60]. If silica gel is present there is a significant acceleration of the rate of reaction, showing that there is a catalytic effect by the silica [15b].

Most applications of sulfide oxidations by alkyl hydroperoxides have involved titanium catalysis together with chiral ligands for enantioselective transformations. The groups of Kagan in Orsay [61] and Modena in Padova [62] reported independently on the use of chiral titanium complexes for the asymmetric sulfoxidation by the use of ^{t}BuOOH as the oxidant. A modification of the Sharpless reagent with the use of $Ti(O^iPr)_4$ and (*R*,*R*)-diethyl tartrate (*R*,*R*)-DET afforded chiral sulfoxides with up to 90% ee (Eq. (8.17)).

$$p\text{-tolyl}\text{-}S\text{-}Me \xrightarrow[\substack{^t BuOOH}]{Ti(O^iPr)_4,\,(R,R)\text{-DET}} p\text{-tolyl}\text{-}S^*(=O)\text{-}Me \quad (89\%\,ee) \quad (8.17)$$

The outcome of the reaction was later improved by replacing ^{t}BuOOH by cumene hydroperoxide [63].

An improved catalytic reaction with the use of 10 mol% of titanium using a ratio $Ti(O^iPr)_4/(R,R)\text{-DET}/^i PrOH = 1:4:4$ in the presence of molecular sieves gave an efficient sulfoxidation with *ee*s up to 95% with various aryl methyl sulfoxides [64]. The asymmetric Ti-catalyzed sulfoxidations with alkyl hydroperoxides have been reviewed by Kagan [65].

The asymmetric titanium-catalyzed sulfoxidation with ^{t}BuOOH also works with chiral diols as ligands [66–68]. Various 1,2-diaryl-1,2-ethanediols were employed as ligands, and the use of 15 mol% of $Ti(O^iPr)_4$ with 1,2-diphenyl-1,2-ethanediol gave ees up to 90% [66b]. Also, the use of BINOL (1,1′-binaphtalene-2,2′-diol) and derivatives gave asymmetric sulfoxidations [67].

The use of (S,S)-1,2-bis-^{t}butyl-1,2-ethanediol (S,S)-30 in the titanium-catalyzed oxidation of various aryl methyl sulfides by cumene hydroperoxide afforded sulfoxides in ees up to 95% (Scheme 8.5) [68]. Interestingly, the authors observed that the ee of the sulfoxide increased with the reaction time, indicating a kinetic resolution of the sulfoxide product. A control experiment with racemic p-tolyl sulfoxide showed that the (R)-enantiomer is oxidized to sulfone three times faster than the (S)-enantiomer by the catalytic system employed. For this reason, the yields of the chiral sulfoxides are moderate and in the range of 40–50%.

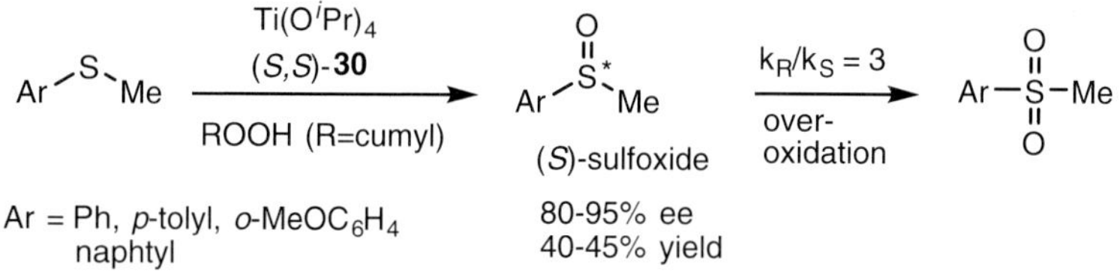

$$Ar-S-Me \xrightarrow[ROOH (R=cumyl)]{Ti(O^iPr)_4, (S,S)-30} Ar-S(=O)-Me \xrightarrow[\text{over-oxidation}]{k_R/k_S = 3} Ar-S(=O)_2-Me$$

Ar = Ph, p-tolyl, o-MeOC$_6$H$_4$, naphtyl

(S)-sulfoxide
80–95% ee
40–45% yield

Scheme 8.5

A similar observation of kinetic resolution of the sulfoxide by overoxidation to sulfone leading to an amplification of the ee has been previously reported by Uemura [67].

An application of the Kagan-Modena procedure for the synthesis of the enantiomerically pure S enantiomer of omeprazol was reported by Cotton et al. [69] This enantiomer is called esomeprazol and is the active component of Nexium®. It is a highly potent gastric acid secretion inhibitor.

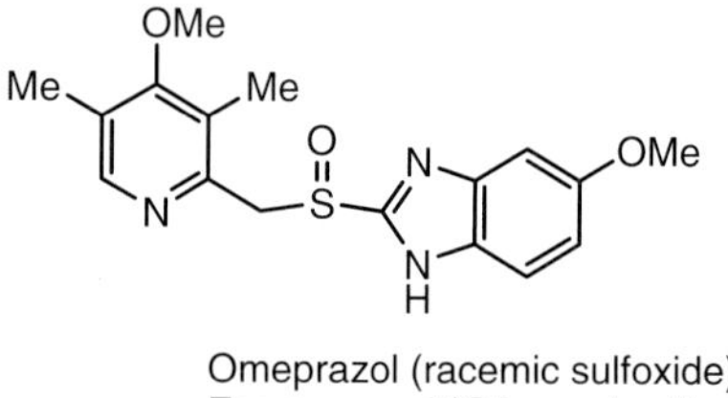

Omeprazol (racemic sulfoxide)
Esomeprazol (S-form of sulfoxide)

It was found that a modification of the original procedure by addition of N,N-diisopropyl ethylamine had a dramatic effect on the enantioselectivity of the reaction. The role of the added amine is unclear. A large-scale oxidation of sulfide 31 (6.2 kg) using 30 mol% of the titanium catalyst in the presence of (S,S)-DET and N,N-diisopropyl ethylamine gave 92% of a crude product, which was >94% ee (Scheme 8.6). The ratio of sulfoxide to sulfone was 76:1. Recrystallization gave 3.83 kg of a product that was >99.5% ee. It is possible to run with less catalyst, but this gives a slightly lower ee. Thus, the use of 4 mol% Ti(O^{i}Pr)$_4$ with Ti/(S,S)-DET/EtN(^{i}Pr)$_2$ = 1:2:1 gave esomeprazol in 96% yield that was 91% ee. The ratio of sulfoxide to sulfone was 35:1.

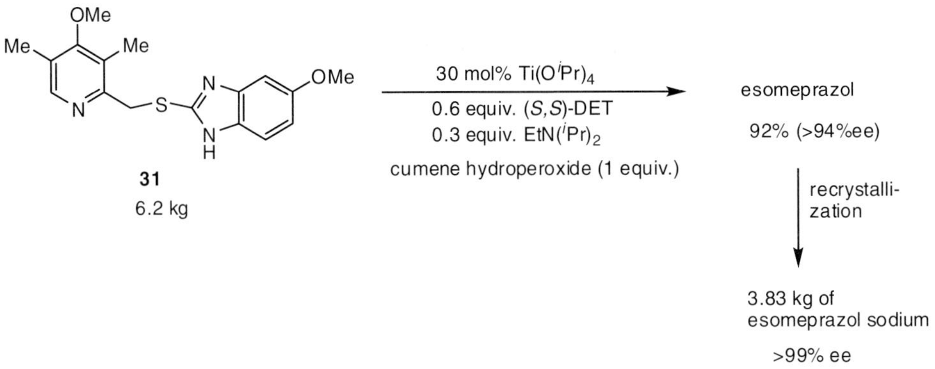

Scheme 8.6

Compound **31** (6.2 kg) → esomeprazol

Conditions: 30 mol% Ti(O^{i}Pr)$_4$, 0.6 equiv. (S,S)-DET, 0.3 equiv. EtN(^{i}Pr)$_2$, cumene hydroperoxide (1 equiv.)

92% (>94% ee)

recrystallization → 3.83 kg of esomeprazol sodium, >99% ee

Artificial metalloenzymes based on vanadyl-loaded streptavidin was used for catalytic enantioselective sulfoxidation with t-butyl hydroperoxide [70]. Incorporation of a vanadyl ion into the biotin-binding pocket of streptavidin resulted in a catalyst that transformed both dialkyl and alkyl-aryl sulfides to their sulfoxides in high *ee*.

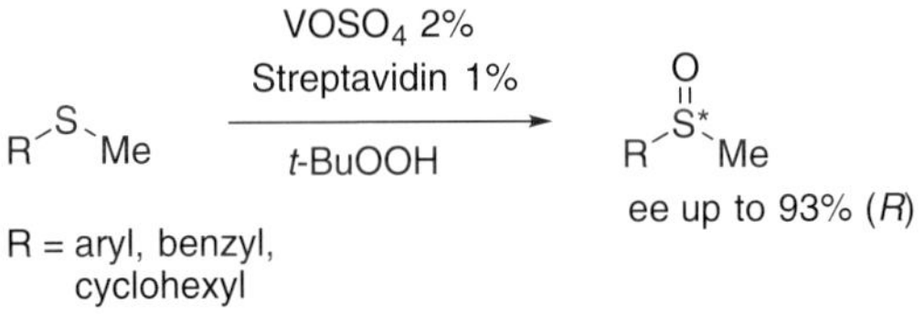

R–S–Me → R–S*(=O)–Me

Conditions: VOSO$_4$ 2%, Streptavidin 1%, t-BuOOH

ee up to 93% (R)

R = aryl, benzyl, cyclohexyl

8.2.2.4 Other Oxidants in Catalytic Reactions

Several chemocatalytic systems for sulfoxidation that employ other oxidants than hydrogen peroxide, molecular oxygen, or alkyl hydroperoxide have been reported. Manganese-catalyzed oxidation of sulfides with iodosobenzene (PhIO) using chiral Mn(salen) complexes was used to obtain chiral sulfoxides in up to 94% *ee* [71]. PhIO was also employed as the oxidant in sulfoxidations catalyzed by quaternary ammonium salts [72]. The use of cetyltrimethylammonium bromide (n-C$_{16}$H$_{33}$Me$_3$N$^+$ Br$^-$) gave the best result, and with 5–10 mol% of this catalyst, high yields (90–100%) of sulfoxide were obtained from various sulfides.

A mild and chemoselective oxidation of sulfides to sulfoxides by o-iodooxybenzoic acid (IBX) catalyzed by tetraethylammonium bromide (TEAB) has been reported [73]. The reaction is highly selective, and no overoxidation to sulfone was observed. Simple aryl alkyl sulfides are oxidized in 93–98% yield in 0.3–2 h at room temperature with the use of 5 mol% of TEAB. Diphenyl sulfide and phenyl benzyl sulfide took 30 and 36 h, respectively, to go to completion under these conditions.

8.2.3
Biocatalytic Reactions

Various peroxidases and monooxygenases have been used as biocatalysts for the oxidation of sulfides to sulfoxides [74, 75]. Haloperoxidases have been studied in oxidations of sulfides, and these reactions work with hydrogen peroxide as the oxidant. Baeyer-Villiger monooxygenases, whose natural role is to oxidize ketones to esters, are NAD(P)H-dependent flavoproteins that have been used for sulfoxidations. Until recently only cyclohexanone monooxygenase (CHMO) had been cloned and overexpressed, but new developments have made a number of other Bayer-Villiger monooxygenases available.

8.2.3.1 Peroxidases

Oxidation of sulfides catalyzed by haloperoxidases has been reviewed [74]. The natural biological role of haloperoxidases is to catalyze oxidation of chloride, bromide, or iodide by hydrogen peroxide. Three classes of haloperoxidases have been identified: (i) those without a prosthetic group, found in bacteria, (ii) heme-containing peroxidases such as chloroperoxidase (CPO), and (iii) vanadium-containing peroxidases.

Asymmetric H_2O_2 oxidations of aryl methyl sulfides catalyzed by CPO occur in excellent enantioselectivity (Eq. (8.18)) [76, 77]. Electronic and, in particular, steric factors dramatically affect the yield of the reaction. Thus, small aromatic groups gave high yields in 99% ee, whereas a slight increase in size led to a dramatic drop in yield, though still in high ee (99%).

$$Ar-S-Me + H_2O_2 \xrightarrow{CPO} Ar-S(=O)^*-Me + H_2O$$

99% ee (R-form)

Ar = Ph, (thiophene), (thiazole)   100% yield

Ar = o-Me-C$_6$H$_4$ 3% yield

Ar = p-Br-C$_6$H$_4$ 15% yield

(8.18)

The analogous oxidations of cyclic sulfides with the same biocatalyst (CPO) were studied by Allenmark and coworkers [78]. Only 1-thiaindane gave a synthetically useful outcome with high yield (99.5%) and high enantioselectivity (99% ee).

Allenmark and coworkers also studied the asymmetric sulfoxidation catalyzed by vanadium-containing bromoperoxidase (VBrPO) from *Corallina officinalis* [79, 80]. The practical use of this reaction is limited since the enzyme accepts very few substrates, such as 2,3-dihydrobenzo[b]thiophene, 2-(carboxy)phenyl methyl sulfide and 2-(carboxy)vinyl methyl sulfides [74, 79].

Some peroxidases are sensitive to excess hydrogen peroxide, which may complicate synthetic procedures with these enzymes. For example *Coprinus cinerem* peroxidase (Cip) has been used for the enantioselective oxidation of sulfides to sulfoxides, either by continuous slow addition of hydrogen peroxide [81] or by the use of an alkyl hydroperoxide [82]. Sulfoxidation with Cip as the catalyst has been developed into an aerobic procedure by combining the peroxidase (Cip) with a glucose oxidase [83]. The glucose oxidase and molecular oxygen gives a slow production of hydrogen peroxide which is slow enough to avoid degradation of the enzyme. This is a convenient procedure, and aryl methyl sulfides, where the aryl group is phenyl, p-MeC$_6$H$_4$ or naphthyl, gave good yields (85–91%) in 79, 88, and 90% *ee*, respectively.

The *Coprinus cinerem* peroxidase (Cip) was later used in ionic liquid [BMIm]PF$_6$ as a reaction medium in peroxidase-catalyzed oxidation of sulfides to sulfoxides in high enantioselectivity (92% *ee*) [84]. However the reaction was very slow (14% conversion after 16 h). Glucose oxidase from *Aspergillus niger* was employed to generate H$_2$O$_2$ from O$_2$.

8.2.3.2 Ketone Monooxygenases

A number of ketone monooxygenases are available for synthetic transformations today [85]. Cyclohexanone monooxygenase (CHMO) was cloned and overexpressed as early as 1988 and has until recently been the only ketone monooxygenase extensively studied. In new developments, a number of other monooxygenases such as cyclopentanone monooxygenase (CPMO), cyclododecanone monooxygenase (CDMO), steroid monooxygenase (SMO), and 4-hydroxyacetophenone monooxygenase (HAPMO) have been cloned [85]. Of the ketone monooxygenases known today CHMO, CPMO, and HAPMO have been used for sulfoxidation.

Oxidation of sulfides by molecular oxygen catalyzed by cyclohexanone monooxygenase (CHMO) has been studied by an Italian team [75, 86, 87]. CHMO is a flavin-dependent enzyme of about 60 KDa and is active as a monomer. It has found application in Baeyer-Villiger oxidation [85, 88] and in the oxidation of sulfides to sulfoxides [89]. The aerobic oxidation with these monooxygenases requires NADPH to reduce the oxidized flavin back to the reduced form so that it can react again with molecular oxygen. CHMO-catalyzed oxidation of various sulfides by molecular oxygen in the presence of NADPH afforded sulfoxides in high yields and in most cases good to high enantioselectivity. The NADPH was employed in catalytic amounts by recycling of NADP by glucose-6-phosphate or L-malate. Results from aerobic oxidations of some methyl-substituted sulfides are given in Eq. (8.19).

$$R-S-Me + \tfrac{1}{2} O_2 \xrightarrow[\text{NADPH}]{\text{CHMO}} R-S(=O)^*-Me \qquad (8.19)$$

R = Ph   88% (99% ee)
R = *o*-MeC$_6$H$_4$ 90% (87% ee)
R = 2-pyridyl   86% (87% ee)
R = *p*-FC$_6$H$_4$ 96% (92% ee)
R = ^{t}Bu   98% (99% ee)
R = *p*-MeC$_6$H$_4$ 94% (37% ee)

Also, the corresponding ethyl derivatives p-FC$_6$H$_4$SEt and p-MeC$_6$H$_4$SEt gave high yields in 93 and 89% *ee*, respectively. On the other hand the p-MeC$_6$H$_4$SMe behaved differently and gave the corresponding sulfoxide in only 37% *ee*. The CHMO system with NADPH and glucose-6-phosphate was subsequently applied to the oxidation of various dialkyl sulfides. Thus, methyl sulfides RSMe with R = cyclopentyl, cyclohexyl, and allyl gave the corresponding sulfoxides in 82–86% yield and in >98% *ee* [86].

More recently, Jansen and coworkers [90] used 4-hydroxyacetophenone monooxygenase (HAPMO) for aerobic oxidation of sulfides. Interestingly, both PhSMe and p-MeC$_6$H$_4$SMe gave the corresponding sulfoxides in >99% ee, which should be compared with the 99 and 37% ee, respectively, obtained with CHMO [86]. The flavoenzyme HAPMO, which has been cloned [90, 91], is a promising biocatalyst for enantioselective oxidation of sulfides to sulfoxides.

Recombinant of baker's yeast expressing CHMO from *Actinobactersp.* NCIP 9871 have been used as whole-cell biocatalysts for oxidation of sulfides to their corresponding sulfoxides (Eq. (8.20)) [92].

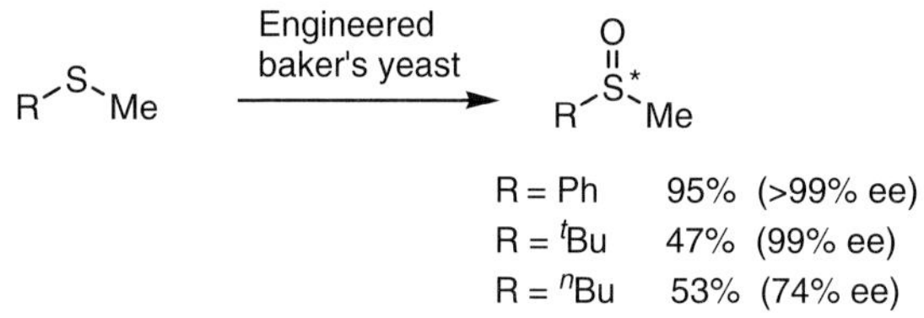

$$R\text{-S-Me} \xrightarrow{\text{Engineered baker's yeast}} R\text{-S(=O)*-Me} \quad (8.20)$$

R = Ph   95% (>99% ee)
R = ^{t}Bu   47% (99% ee)
R = ^{n}Bu   53% (74% ee)

8.3
Oxidation of Tertiary Amines to N-Oxides

Previous reviews have dealt with metal-catalyzed [93] and stoichiometric [94] oxidation of amines in a broad sense. This section will be limited to the selective oxidation of tertiary amines to N-oxides. Amine N-oxides are synthetically useful compounds [95, 96] and are frequently used as stoichiometric oxidants in osmium- [97–99] manganese- [100] and ruthenium-catalyzed [101, 102] oxidations, as well as in other organic transformations [103–105]. Aliphatic *tert*-amine N-oxides are useful surfactants [96] and are essential components in hair conditioners, shampoos, toothpaste, cosmetics, and so on [106]. Chiral N-oxides have been used in asymmetric catalysis involving metal-free catalytic transformations [107] as well as metal-catalyzed reactions where the N-oxide serves as a ligand [107, 108]. Chiral tertiary amine N-oxides were recently used as reagents in asymmetric epoxidation of α,β-unsaturated ketones [109].

Because of their importance, various methods have been reported for the oxidation of tertiary amines to N-oxides. The oxidations of amines are discussed below under the following headings: (i) stoichiometric reactions, (ii) chemocatalytic reactions, and (iii) biocatalytic reactions. Finally we provide some examples where N-oxides are generated *in situ* as catalytic oxotransfer species in catalytic transformations.

8.3.1
Stoichiometric Reactions

Amine N-oxides can be prepared from amines with 30% aqueous hydrogen peroxide in a noncatalytic slow reaction [97, 110]. At elevated temperatures this oxidation proceeds at a reasonable rate and has been used in industrial applications.

8.3 Oxidation of Tertiary Amines to N-Oxides

Various other oxidants have also been employed for N-oxidation of tertiary amines such as peracids [111], 2-sulfonyloxaziridines [112], and α-azohydroperoxides [113].

Messeguer and coworkers reported the use of dioxiranes for the oxidation of amines to N-oxides [114]. Oxidation of various tertiary aromatic amines with dimethyldioxirane (DMD) afforded amine N-oxides in quantitative yields. A few examples are given in Eq. (8.21).

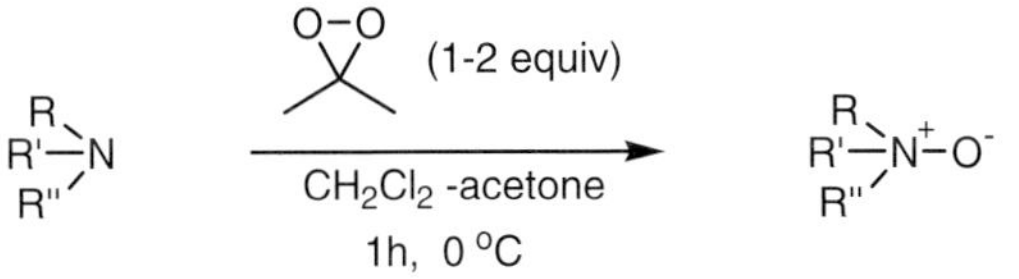

$$\underset{R''}{\overset{R}{\underset{|}{R'-N}}} \xrightarrow[\text{1h, 0 °C}]{\text{DMD (1-2 equiv)} \atop \text{CH}_2\text{Cl}_2\text{-acetone}} \underset{R''}{\overset{R}{\underset{|}{R'-\overset{+}{N}-O^-}}} \tag{8.21}$$

R = R' = R" = Bu
R = R' = Me; R" = PhCH$_2$
R = PhCH$_2$; R', R" = -(CH$_2$)$_5$-
R = Me; R', R" = -(CH$_2$)$_2$O(CH$_2$)$_2$-

Interestingly, the reaction is chemoselective, and oxidation of aminoalkenes gave selectivily the N-oxides without any epoxide formation. One example is given in Eq. (8.22). A number of substituted pyridines were also oxidized to the pyridine N-oxides by DMD in quantitative yields [114].

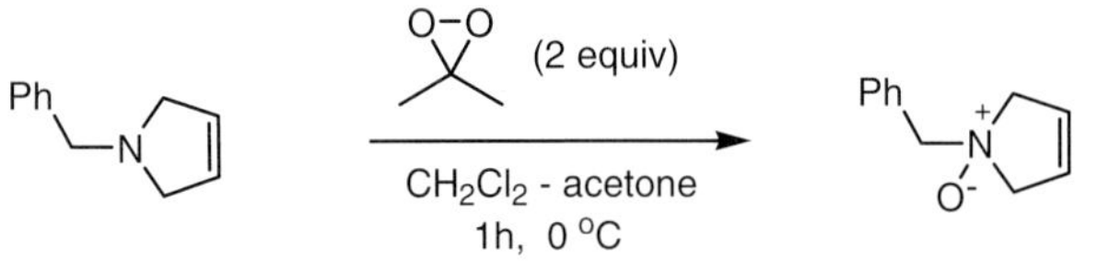

$$\text{Ph-N-pyrrole} \xrightarrow[\text{1h, 0 °C}]{\text{DMD (2 equiv)} \atop \text{CH}_2\text{Cl}_2 \text{ - acetone}} \text{Ph-N}^+(\text{O}^-)\text{-pyrrole} \tag{8.22}$$

Selective oxidation of tertiary amines to N-oxides by HOF·CH$_3$CN was reported by Rozen and coworkers [115]. The reaction is rapid, and amine N-oxides were isolated in high yields (Eq. (8.23)). The HOF·CH$_3$CN complex was also employed to oxidize a number of substituted pyridines to their corresponding N-oxides (Eq. (8.24)).

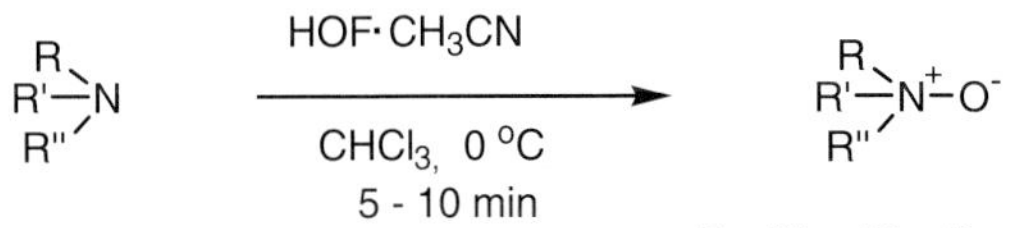

$$\underset{R''}{\overset{R}{\underset{|}{R'-N}}} \xrightarrow[\text{CHCl}_3, 0°C \atop \text{5 - 10 min}]{\text{HOF·CH}_3\text{CN}} \underset{R''}{\overset{R}{\underset{|}{R'-\overset{+}{N}-O^-}}} \tag{8.23}$$

R = R' = R" = Bu   82%
R = R' = C$_8$H$_{17}$; R" = Me   95%
R = R' = cyclohexyl; R" = Me 95%
R = Ph; R' = Bn, R" = Et   85%
R = Ph R' = R" = -(CH$_2$)$_5$- 85%

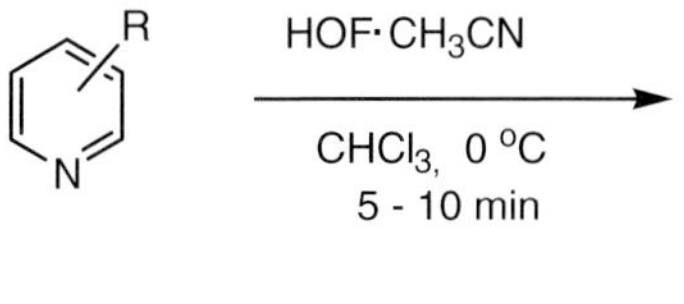

8 Selective Oxidation of Amines and Sulfides

$$\text{pyridine-R} \xrightarrow[\text{5 - 10 min}]{\substack{\text{HOF·CH}_3\text{CN} \\ \text{CHCl}_3, \ 0\,°\text{C}}} \text{pyridine-R N-oxide} \quad (8.24)$$

8.3.2
Chemocatalytic Oxidations

Some early work describes the vanadium-catalyzed oxidations of tertiary amines to N-oxides by t-butylhydroperoxide, but these reactions require elevated temperature [116]. In 1998 a mild flavin-catalyzed oxidation of tertiary amines to N-oxides by 30% aqueous H_2O_2 was reported [47]. The flavin **19** was employed as the catalyst, and the reaction occurs at room temperature. With the use of 2.5 mol% of the flavin the reaction takes 25–60 min to go to high conversion (Eq. (8.25), Table 8.7). The flavin hydroperoxide generated *in situ* (cf. Scheme 8.2) shows a very high reactivity toward amines, and it was estimated that the flavin hydroperoxide reacts >8000 times faster than H_2O_2 with the amine in entry 4, Table 8.7. The success with N,N,N-1,3,5-trialkylated flavin **19** is that the flavin-OH (**22**, Scheme 8.3) formed after oxo transfer to the amine from flavin-OOH can easily lose the OH group and give the aromatized flavin **23**, which is the active catalyst.

Table 8.7 Oxidation of tertiary amines according to Eq. (8.25).

Entry	Amine	Reaction time for >85% conversion	Product	Rate enhancement Catalyzed: noncatalyzed[a]
1	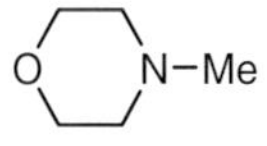 morpholine N–Me	1 h	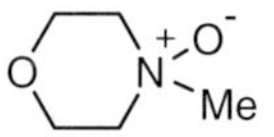 morpholine N(+)(O–)–Me	61:1 (6344:1)[b]
2	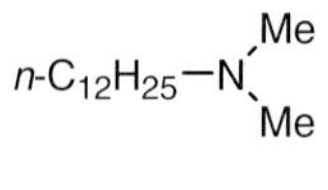 $n\text{-}C_{12}H_{25}\text{-}N(\text{Me})_2$	27 min	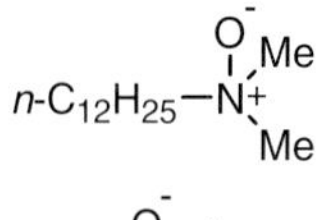 $n\text{-}C_{12}H_{25}\text{-}N^+(\text{O}^-)(\text{Me})_2$	49:1 (5096:1)[b]
3	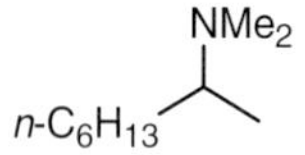 $n\text{-}C_6H_{13}\text{-}C(\text{NMe}_2)$	25 min	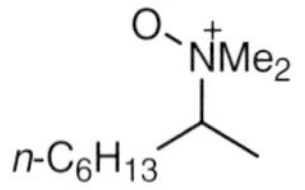 $n\text{-}C_6H_{13}\text{-}C(\text{N}^+(\text{O}^-)\text{Me}_2)$	51:1 (5304:1)[b]
4	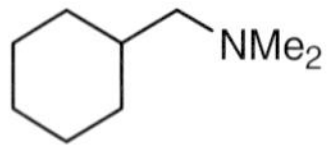 cyclohexyl-NMe$_2$	50 min	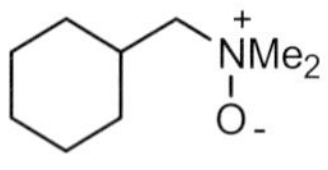 cyclohexyl-N$^+$(O$^-$)Me$_2$	83:1 (8632:1)[b]
5	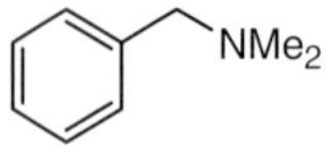 phenyl-NMe$_2$	31 min	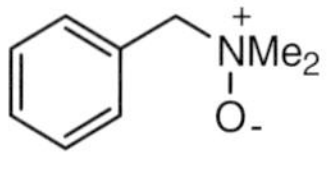 phenyl-N$^+$(O$^-$)Me$_2$	67:1 (6968:1)[b]
6	Et$_3$N	54 min	Et$_3$N$^+$–O$^-$	61:1 (2507:1)[b]

a) Calculated by division of the initial rates of catalyzed and non-catalyzed reactions.
b) Estimated ratio of the reactivities of catalytic flavin hydroperoxide and H_2O_2.

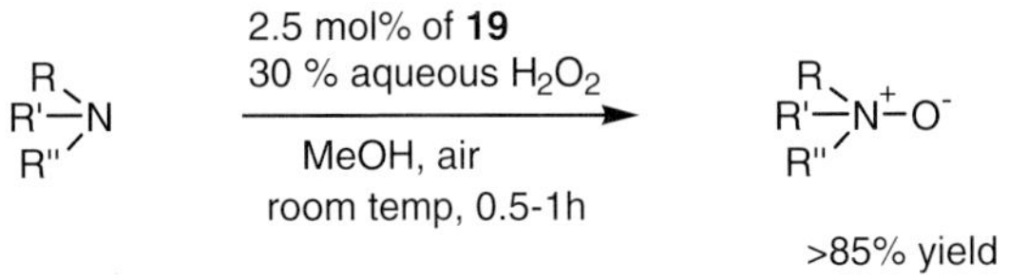

8.3 Oxidation of Tertiary Amines to N-Oxides

$$\underset{R''}{\overset{R}{\underset{|}{R'-N}}} \xrightarrow[\text{MeOH, air} \atop \text{room temp, 0.5-1h}]{2.5 \text{ mol\% of } \mathbf{19} \atop 30\% \text{ aqueous } H_2O_2} \underset{R''}{\overset{R}{\underset{|}{R'-\overset{+}{N}-O^-}}} \quad (8.25)$$

>85% yield

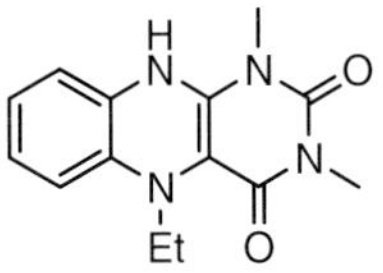

19

In preparative experiments, the N,N-dimethylamino derivatives **32a** and **32b** (Eq. (8.26)) were oxidized by H_2O_2 at room temperature for 2 h to the corresponding N-oxides **33a** and **33b** in 85 and 82% yield, respectively, using 1 mol% of flavin **19** [47].

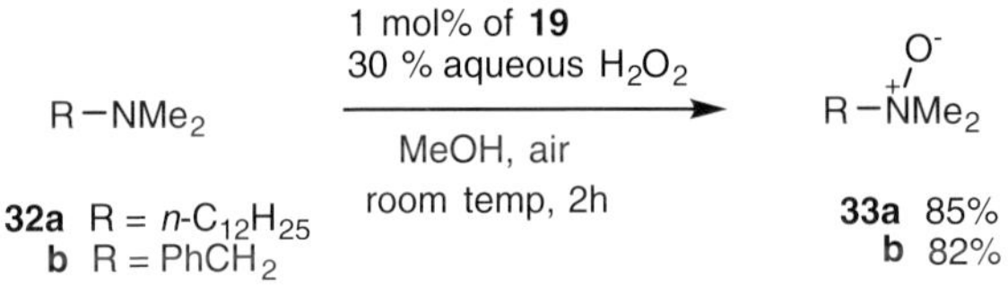

$$R-NMe_2 \xrightarrow[\text{MeOH, air} \atop \text{room temp, 2h}]{1 \text{ mol\% of } \mathbf{19} \atop 30\% \text{ aqueous } H_2O_2} R-\overset{+}{N}Me_2 \atop \overset{|}{O^-} \quad (8.26)$$

32a R = n-C$_{12}$H$_{25}$
b R = PhCH$_2$

33a 85%
b 82%

A tungstate-exchanged Mg-Al layered double hydroxide (LDH) was employed to catalyze oxidation of aliphatic tertiary amines to N-oxides by 30% aqueous H_2O_2 (Table 8.8) [117]. The reaction takes 1–3 h at room temperature. The use of dodecylbenzenesulfonic acid sodium salt as an additive increased the rate of the oxidation by a factor of 2–3 except for N-methyl morpholine. An advantage with the LDH-WO_4^{2-} catalyst is that it can be reused.

A mild Pt-catalyzed oxidation of tertiary amines to their N-oxides by aqueous H_2O_2 was recently reported by Strukul [118]. Platinum(II) complexes of the general formula **34** were used, and the reaction is proposed to involve tertiary amine-Pt-peroxide complexes in which an oxygen transfer to the amine nitrogen occurs. Various tertiary amines were oxidized in good yields and high selectivity with **34c** as the catalyst.

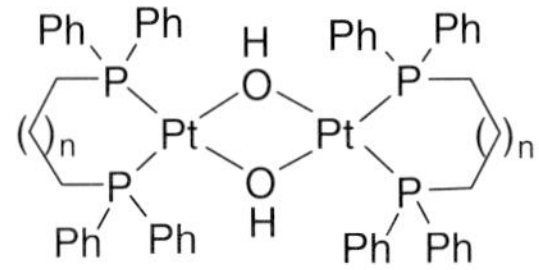

34a (n=0)
b (n=1)
c (n=2)

N-Oxidation of tertiary amines with H_2O_2/bicarbonate and H_2O_2/CO_2 was reported by Richardson, and the mechanism for the CO_2-catalysis was investigated [119].

Table 8.8 Oxidation of tert-amines catalyzed by LDH-WO$_4^{2-}$.

Entry	Amine	Procedure[a]	Product	Reaction time (h)	% Yield
1	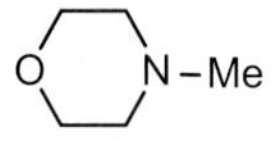 morpholine-N-Me	I	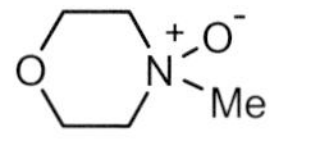 morpholine-N$^+$(Me)-O$^-$	1.0	96
		II		1.0	96
2	n-C$_{10}$H$_{21}$-N(Me)$_2$	I	n-C$_{10}$H$_{21}$-N$^+$(Me)$_2$-O$^-$	1.0	97
3	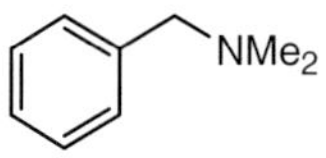 PhCH$_2$-NMe$_2$	I	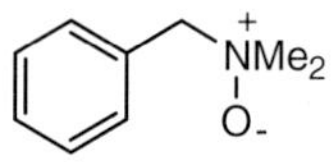 PhCH$_2$-N$^+$Me$_2$-O$^-$	1.5	96
		II		1.0	96
4	Et$_3$N	I	Et$_3$N$^+$-O$^-$	3.0	96
		II		1.5	96
5	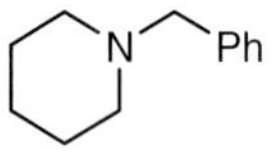 N-Ph piperidine	I	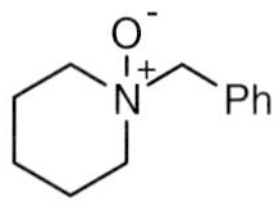 N$^+$(O$^-$)-Ph piperidine	3.0	97
		II		1.0	97

a) I: 2 mmol of substrate was oxidized by H$_2$O$_2$ in water with LDH-WO$_4^{2-}$ as catalyst; II: as in I but 6 mg of dodecylbenzenesulfonic acid sodium salt was added.

The aerobic flavin system with NH$_2$NH$_2$ as a reducing agent that was employed for the sulfoxidation (Section 8.2.2.2) can also be used for the N-oxidation of tertiary amines [59]. However the reaction requires elevated temperature (60 °C) (Eq. (8.27)), most likely because elimination of OH from the flavin-OH (cf. **26** → **17**, Scheme 8.4) is difficult for the N,N,N-3,5,10-trialkylated flavin at the higher pH caused by the amine.

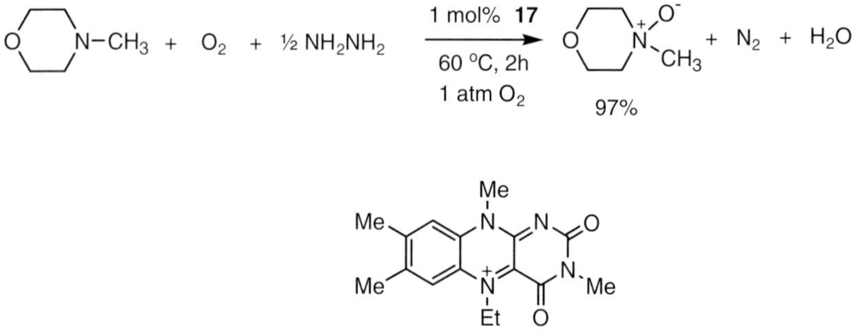

$$\text{morpholine-N-CH}_3 + \text{O}_2 + \tfrac{1}{2}\,\text{NH}_2\text{NH}_2 \xrightarrow[\substack{60\,°C,\ 2\text{h} \\ 1\ \text{atm}\ O_2}]{1\ \text{mol\%}\ \mathbf{17}} \text{morpholine-N}^+(\text{CH}_3)\text{-O}^- + \text{N}_2 + \text{H}_2\text{O}$$

97%

17 (N,N,N-trialkyl flavinium)

(8.27)

8.3 Oxidation of Tertiary Amines to N-Oxides

An aerobic oxidation of tertiary amines catalyzed by Cobalt Schiff-base complexes afforded N-oxides in high yields [120]. The reaction was run at room temperature with 0.5 mol% of the cobalt catalyst **35** (Eq. (8.28)). The presence of molecular sieves (5 Å) enhanced the rate of the reaction. With this procedure, various pyridines were oxidized to their corresponding N-oxides in yields ranging from 50 to 85%. Electron-deficient pyridines such as 4-cyanopyridine gave a slow reaction with only 50% yield.

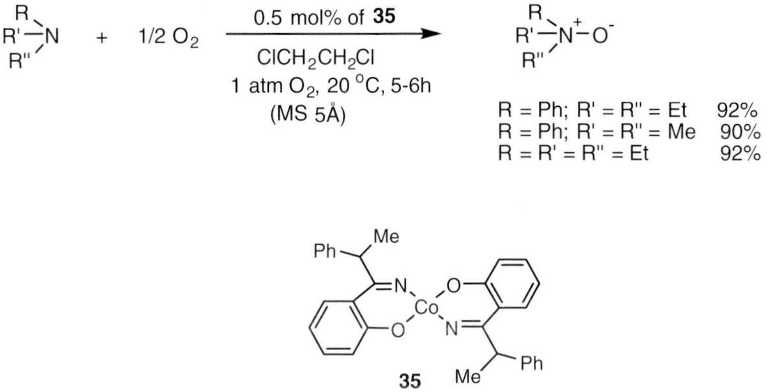

$$RR'R''N + 1/2\,O_2 \xrightarrow[\text{ClCH}_2\text{CH}_2\text{Cl},\ 1\ \text{atm}\ O_2,\ 20\ ^\circ C,\ 5\text{-}6h\ (\text{MS 5Å})]{0.5\ \text{mol\%}\ \text{of}\ \mathbf{35}} R R'R''\overset{+}{N}-O^-$$

R = Ph; R' = R'' = Et   92%
R = Ph; R' = R'' = Me   90%
R = R' = R'' = Et   92%

35 (Co Schiff-base complex)

(8.28)

The same author also reported on an aerobic oxidation of tertiary amines and pyridines to their corresponding N-oxides catalyzed by ruthenium trichloride [121].

Aerobic oxidation of tertiary amines to N-oxide on a heterogeneous gold catalyst (Au/C) was reported to give high yields and selectivity in aqueous solution [122].

The oxidation of pyridines to their corresponding N-oxides catalyzed by methyltrioxorhenium (MTO) was reported to occur with various substituted pyridines [123, 124]. The oxidant employed is either H_2O_2 [123] or $Me_3SiOOSiMe_3$ [124]. The reaction gives high yields with both electron-rich and electron-deficient pyridines (Eq. (8.29)).

$$\text{R-pyridine} \xrightarrow[\text{CH}_2\text{Cl}_2,\ \text{room temp}]{\substack{0.5\ \text{mol\%}\ \text{MeReO}_3 \\ 30\%\ H_2O_2\ \text{or} \\ Me_3SiOOSiMe_3}} \text{R-pyridine N-oxide}$$

80–99%

(8.29)

More recently, an oxidation procedure with sodium percarbonate as the oxidant and MTO as the catalyst for the efficient oxidation of pyridines to the N-oxides was reported [125]. The latter system also worked well with N,N-dialkylated anilines to give the corresponding N-oxides. The same research group had previously reported a related Re-catalyzed N-oxidation with bromamine-T as the terminal oxidant [126].

8.3.3
Biocatalytic Oxidation

Only a limited number of examples of the biocatalytic oxidation of tertiary amines have been reported. Colonna et al. used bovine serum albumin (BSA) as a biocatalyst for the asymmetric oxidation of tertiary amines to N-oxides [127]. Oxone, $NaIO_4$, H_2O_2, and MCPBA were tested as oxidants, and the best results were obtained with $NaIO_4$ and H_2O_2 (Eq. (8.30)). Thus, BSA-catalyzed N-oxidation of **36** with these oxidants afforded N-oxide **37** in high yield in 64–67% ee. The reaction is formally a dynamic kinetic resolution, since the enantiomers of the starting material are in rapid equilibrium.

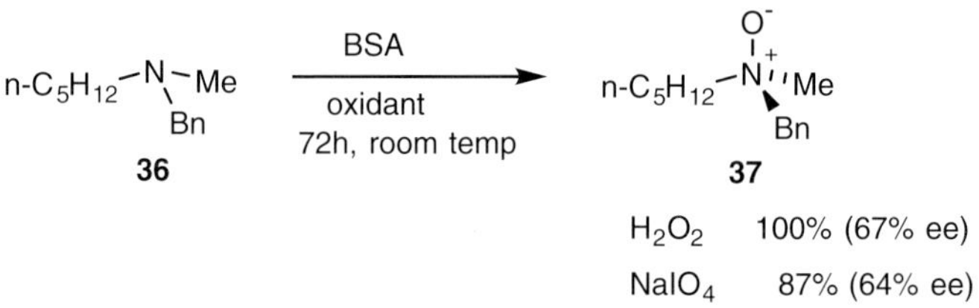

$$n\text{-}C_5H_{12}\text{-}\underset{Bn}{\underset{|}{N}}\text{-}Me \quad \xrightarrow[\text{72h, room temp}]{\text{BSA, oxidant}} \quad n\text{-}C_5H_{12}\text{-}\underset{Bn}{\overset{O^-}{\underset{|}{\overset{+}{N}}}}\text{-}Me \quad\quad (8.30)$$

36 → **37**

H_2O_2 100% (67% ee)
$NaIO_4$ 87% (64% ee)

More recently, Colonna and coworkers reported that cyclohexanone monooxygenase (CHMO) from *Actinobacter calcoaceticus* NCIMB9871 catalyzed the aerobic oxidation of amines [128].

8.3.4
Applications of Amine N-Oxidation in Coupled Catalytic Processes

In 1976 N-methyl morpholine N-oxide (NMO) was introduced as a stoichiometric oxidant in osmium-catalyzed dihydroxylation of alkenes by VanRheenen (The Upjohn procedure) [129]. This made it possible to use osmium tetroxide in only catalytic amounts. However, more environmentally friendly oxidants were searched for, and one idea was to recycle the amine to N-oxide *in situ* either by hydrogen peroxide or molecular oxygen. In this way it would be possible to use the NMO or even better the N-methyl morpholine (NMM) in only catalytic amounts. In 1999 a biomimetic dihydroxylation of alkenes based on this principle was reported in which the tertiary amine (NMM) in catalytic amounts is continuously reoxidized to N-oxide (NMO) by a catalytic flavin/H_2O_2 system (Eq. (8.31), Scheme 8.7) [130]. The coupled electron transfer resembles oxidation processes occurring in biological systems. OsO_4, NMM, and flavin **19** are used in catalytic amounts, and this leads to an efficient H_2O_2-based dihydroxylation of alkenes in high yields at room temperature.

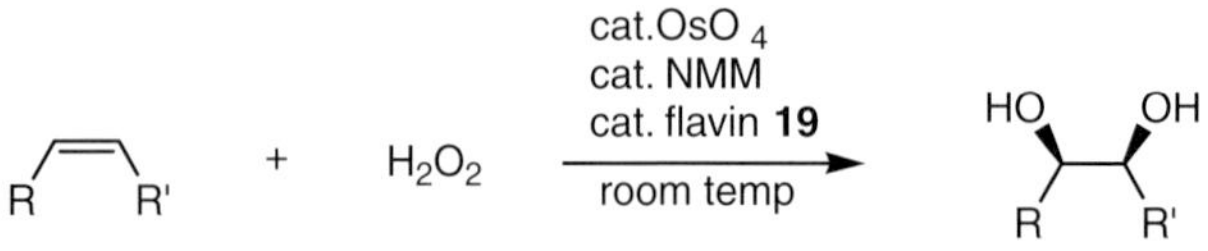

$$R\text{-}CH=CH\text{-}R' + H_2O_2 \quad \xrightarrow[\text{room temp}]{\text{cat. }OsO_4, \text{ cat. NMM, cat. flavin }\mathbf{19}} \quad R\text{-}CH(OH)\text{-}CH(OH)\text{-}R' \quad\quad (8.31)$$

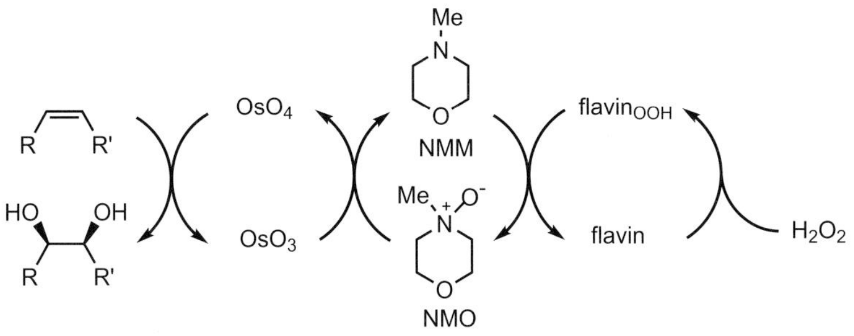

8.3 Oincluded Oxidation of Tertiary Amines to N-Oxides | 307

Scheme 8.7 Dihydroxylation of alkenes via coupled electron transfer.

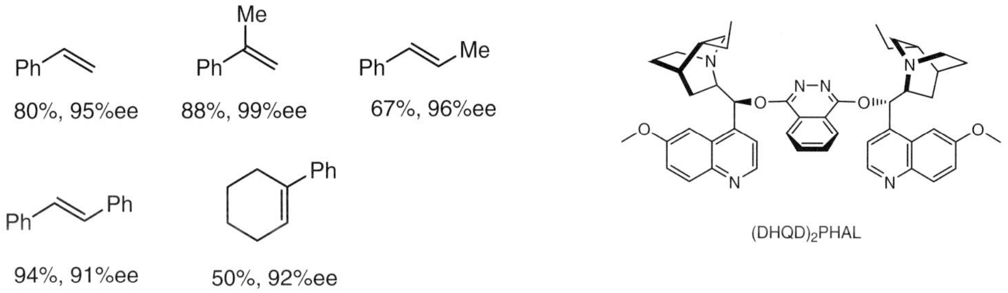

Figure 8.2 Biomimetic dihydroxylation of alkenes to diols using the biomimetic system of Scheme 8.7 with chiral ligand (DHQD)$_2$PHAL.

In the presence of chiral ligand (DHQD)$_2$PHAL, high enantioselectivity was obtained in the dihydroxylation with hydrogen peroxide (Figure 8.2) [131].

In a model study it was demonstrated that the catalytic system also works with MCPBA as the oxidant in place of catalytic flavin/H$_2$O$_2$ [132].

It was later found that the chiral amine can be used as the tertiary amine generating the N-oxide needed for reoxidation of Os(VI). Thus, a simplified procedure for osmium-catalyzed asymmetric dihydroxylation of alkenes by H$_2$O$_2$ was developed in which the tertiary amine NMM is omitted [133]. A robust version of this reaction, where the flavin **19** has been replaced by MeReO$_3$ (MTO), was reported (Scheme 8.8) [134]. The chiral ligand has a dual role in these reactions: it acts as

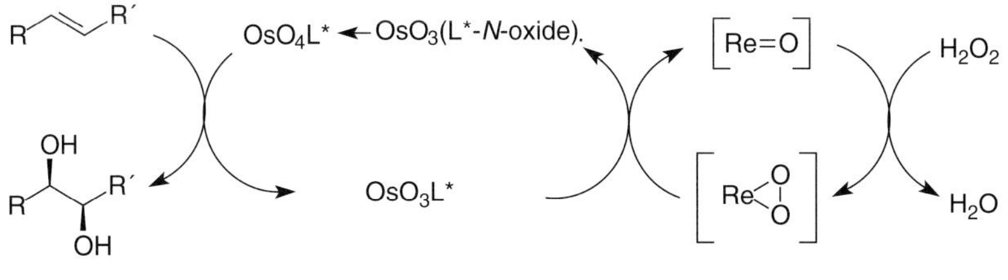

Scheme 8.8 MTO-based dihydroxylation involving chiral ligand N-oxide formation.

a chiral inductor as well as an oxotransfer mediator. The amine oxide of the chiral amine ligand was isolated and characterized by high-resolution mass spectrometry [134].

An efficient osmium-catalyzed dihydroxylation of alkenes by H_2O_2 with NMM and MTO as electron transfer mediators under acidic conditions was subsequently reported [135]. Under these conditions, alkenes that are normally difficult to dihydroxylate gave high yields of diol (Eq. (8.32)). The N-oxide NMO is generated from NMM by catalytic MTO/H_2O_2 in analogy with Schemes 8.7 and 8.8.

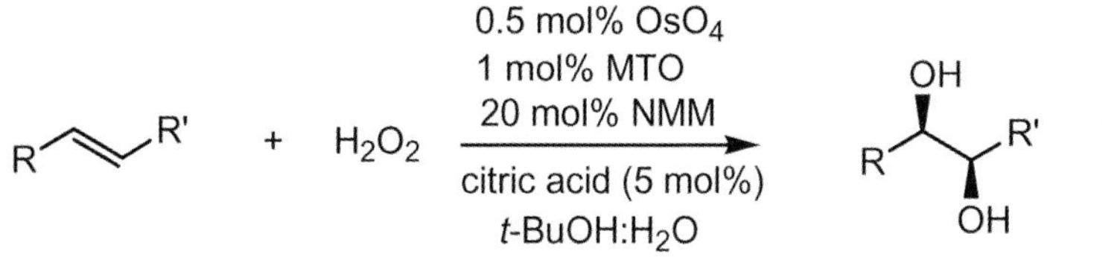

$$R\diagup\!\!\!\diagdown R' + H_2O_2 \xrightarrow[\substack{\text{citric acid (5 mol\%)} \\ t\text{-BuOH:}H_2O}]{\substack{0.5 \text{ mol\% OsO}_4 \\ 1 \text{ mol\% MTO} \\ 20 \text{ mol\% NMM}}} R\diagup\!\!\!\diagup\!\!\!\diagdown R' \text{(OH)(OH)} \quad (8.32)$$

R = H, R' = CH$_2$SO$_2$Ph	93%
R = Me, R' = CO$_2$Et	90%
R = CO$_2$Et, R' = CO$_2$Et	85%
R = Ph, R' = CO$_2$Et	84%

The coupled catalytic system of Scheme 8.7 was more recently immobilized in an ionic liquid, [bmim]PF$_6$ [136]. After completion of the reaction, the product diol is extracted from the ionic liquid, and the osmium, NMO, and flavin stay in the ionic liquid. The immobilized catalytic system was reused 6 times without any loss of activity. In a subsequent study, ionic liquid [bmim]PF$_6$ was employed to immobilize a robust system where the flavin of the previous system had been replaced by VO(acac)$_2$ or MeReO$_3$ (MTO) [137]. A range of alkenes were dihydroxylated with this system, and it was demonstrated that for some of the alkenes the catalytic system can be recycled up to five times.

Choudary and coworkers [138] have also used the principle of *in situ* generation of N-oxide in catalytic amounts in osmium-catalyzed dihydroxylation of alkenes. They used LDH-WO$_4^{2-}$ as the catalyst for H_2O_2 reoxidation of the amine to N-oxide. Recently, an osmium-catalyzed asymmetric dihydroxylation of alkenes was reported in which the N-oxide of N-methyl morpholine (NMM) was generated from H_2O_2 with CO$_2$ as a catalyst [139]. In the latter system, NMM is not used in catalytic amounts; instead, 2 equiv. relative to the alkene was required.

8.4
Concluding Remarks

The selective oxidation of sulfides and amines to sulfoxide and amine oxides, respectively, can be obtained with a variety of oxidants. In particular, catalytic oxidations employing environmentally benign oxidants such as hydrogen peroxide and molecular oxygen have recently attracted considerable interest. A number of catalytic methods that give highly selective and mild oxidations with the latter oxidants are known today. Organocatalysts (e.g., flavins), biocatalysts, and metal catalysts have been used for these transformations.

References

1. Patai, S. and Rappoport, Z. (1994) *Synthesis of Sulfones, Sulfoxides and Cyclic Sulfides*, Wiley, Chichester.
2. Madesclair, M. (1986) *Tetrahedron*, **42**, 5459–5495.
3. Hudlický, M. (1990) *Oxidation in Organic Chemistry*, ACS, Washington D.C.
4. (a) Kowalski, P., Mitka, K., Ossowska, K., and Kolarska, Z. (2005) *Tetrahedron*, **61**, 1933–1953; (b) Kaszorowska, K., Kolarska, Z., Mitka, K., and Kowalski, P. (2005) *Tetrahedron*, **61**, 8315–8327.
5. Legros, J., Delhi, J.R., and Bolm, C. (2005) *Adv. Synth. Catal.*, **347**, 19–31.
6. (a) Prilezhaeva, E.N. (2001) *Russ. Chem. Rev.*, **70**, 897–920; (b) Carreño, M.C. (1995) *Chem. Rev.*, **95**, 1717–1760; (c) Burrage, S., Raynham, T., Williams, G., Essez, J.W., Allen, C., Cardno, M., Swali, V., and Bradley, M. (2000) *Chem. Eur. J.*, **6**, 1455–1466; (d) Padwa, A. and Danca, M.D. (2002) *Org. Lett.*, **4**, 715–717.
7. (a) Pellissier, H. (2006) *Tetrahedron*, **62**, 5559–5601; (b) Colobert, F., Tito, A., Khiar, N., Denni, D., Medina, M.A., Martin-Lomas, M., Garcia Ruano, J.-L., and Solladié, G. (1998) *J. Org. Chem.*, **63**, 8918–8921.
8. (a) Fernandéz, I. and Khiar, N. (2003) *Chem. Rev.*, **103**, 3651; (b) Thorarensen, A., Palmgren, A., Itami, K., and Bäckvall, J.E. (1997) *Tetrahedron Lett.*, **38**, 8541; (c) Tokunoh, R., Sodeoka, M., Aoe, K., and Shhibasaki, M. (1995) *Tetrahedron Lett.*, **36**, 8035–8038; (d) Petra, D.G.I., Kamer, P.C.J., Spek, A.L., Schoemaker, H.E., and Van Leeuwen, P.W.N.M. (2000) *J. Org. Chem.*, **65**, 3010–3017; (e) Rowlands, J.G. (2003) *Synlett*, 236–240; (f) Dorta, R., Shimon, L.J.W., and Milstein, D. (2003) *Chem. Eur. J.*, **9**, 5237–5249; (g) Covell, D.J. and White, M.C. (2008) *Angew. Chem. Int. Ed.*, **47**, 6448–6451.
9. Khenkin, A.M. and Neumann, R. (2002) *J. Am. Chem. Soc.*, **124**, 4198–4199.
10. (a) Nicolau, K.C., Magolda, R.L., Sipio, W.J., Barnette, W.E., Lyzenko, Z., and Joullie, m.M. (1980) *J. Am. Chem. Soc.*, **102**, 3784–3793; (b) Paquette, L.A. and Carr, R.V.C. (1985) *Org. Synth.*, **64**, 157.
11. (a) Murray, R.W. and Jeyaraman, R. (1985) *J. Org. Chem.*, **50**, 2487–2853; (b) Murray, R.W. (1989) *Chem. Rev.*, **89**, 1187; (c) Colonna, S. and Gaggero, N. (1989) *Tetrahedron Lett.*, **30**, 6233.
12. Gonzàlez-Núñez, M.E., Mello, R., Royo, J., Rios, J.V., and Asensio, G. (2002) *J. Am. Chem. Soc.*, **124**, 9154–9163.
13. Ishii, A., Tsuchiya, C., Shimada, T., Furusawa, K., Omata, T., and Nakayama, J. (2000) *J. Org. Chem.*, **65**, 1799–1806.
14. (a) Gildersleeve, J., Smith, A., Sakurai, K., Raghavan, S., and Kahne, D. (1999) *J. Am. Chem. Soc.*, **121**, 6176–6182; (b) Jin, Y.-N.J., Ishii, A., Sugihara, Y., and Nakayama, J. (1998) *Tetrahedron Lett.*, **39**, 3525–3528; (c) Beddoes, R.L., Painter, J.E., Quayle, P., and Patel, P. (1997) *Tetrahedron*, **53**, 17297–17306; (d) Schenk, W.A., Frisch, J., Durr, M., Burzlaff, N., Stalke, D., Fleischer, R., Adam, W., Prechtl, F., and Smerz, A.K. (1997) *Inorg. Chem.*, **36**, 2372–2378; (e) Schenk, W.A. and Durr, M. (1997) *Chem. Eur. J.*, **3**, 713.
15. Foti, C.J. and Fields, J.D. (1999) *Org. Lett.*, **1**, 903–904. Kropp, P.J., Breton, G.W., Fields, J.D., Tung, J.C., and Loomis, B.R. (2000) *J. Am. Chem. Soc.*, **122**, 4280–4285.
16. (a) Trost, B.M. and Curran, D.P. (1981) *Tetrahedron Lett.*, **22**, 1287–1290; (b) Trost, B.M., and Braslau, R. (1988) *J. Org. Chem.*, **53**, 532.

17 (a) Ravikumar, K.S., Bégué, J.-P., and Bonnet-Delpon, D. (1998) *Tetrahedron Lett.*, **39**, 3141–3144; (b) Ravikumar, K.S., Zhang, Y.M., Bégué, J.-P., and Bonnet-Delpon, D. (1998) *Eur. J. Org. Chem.*, 2937–2940.

18 Xu, W.L., Li, Y.Z., Zhang, Q.S., and Zhu, H.S. (2004) *Synthesis*, 227–232.

19 Watanabe, Y., Numata, T., and Oae, S. (1981) *Synthesis*, 204.

20 Schultz, H.S., Freyermuth, H.B., and Bue, S.R. (1963) *J. Org. Chem.*, **28**, 1140.

21 Ishi, Y., Tanaka, H., and Nisiyama, Y. (1994) *Chem. Lett.*, 1.

22 Gresley, N.M., Griffith, W.P., Laemmel, A.C., Nogueira, H.I.S., and Parkin, B.C. (1997) *J. Mol. Catal.*, **117**, 185–198.

23 Neumann, R. and Juviler, D. (1996) *Tetrahedron*, **52**, 8781–8788.

24 Sato, K., Hyodo, M., Aoki, M., Zheng, X., and Noyori, R. (2001) *Tetrahedron*, **57**, 2469–2476.

25 Choudary, B.M., Bharathi, B., Venkat Reddy, Ch., and Lakshmi Kantam, M. (2002) *J. Chem. Soc. Perkin Trans. 1*, 2069–2074.

26 Kamamta, K., Hirano, T., and Mizuno, N. (2009) *Chem. Commun*, 3958–3960.

27 Brinksma, J., La Crois, R., Feringa, B.L., Donnoli, M.I., and Rosini, C. (2001) *Tetrahedron Lett.*, **42**, 4049–4052.

28 Colombo, A., Marturano, G., and Pasini, A. (1986) *Gazz. Chim. Ital.*, **116**, 35–40.

29 (a) Bolm, C. and Bienewald, F. (1995) *Angew. Chem. Int. Ed.*, **34**, 2640–2642; (b) Bolm, C. and Bienewald, F. (1998) *Synlett*, 1327–1328.

30 Vetter, A.H. and Berkessel, A. (1998) *Tetrahedron Lett.*, **39**, 1741–1744.

31 Ohta, C., Shimizu, H., Kondo, A., and Katsuki, T. (2002) *Synlett*, 161–163.

32 (a) Legros, J. and Bolm, C. (2003) *Angew. Chem. Int. Ed.*, **42**, 5487; (b) Legros, J. and Bolm, C. (2005) *Chem. Eur. J.*, **11**, 1086–1092.

33 Green, S.D., Monti, C., Jackson, R.F.W., Anson, M.S., and Macdonald, S.J.F. (2001) *Chem. Commun.*, 2594–2595.

34 Drago, C., Caggiano, L., and Jackson, R.F.W. (2005) *Angew. Chem. Int. Ed.*, **44**, 7721–7723.

35 Palucki, M., Hanson, P., and Jacobsen, E.N. (1992) *Tetrahedron Lett.*, **33**, 7177–7114.

36 Saito, B. and Katsuki, T. (2001) *Tetrahedron Lett.*, **42**, 3873–3876.

37 (a) Yamaguchi, T., Matsumoto, K., Saito, B., and Katsuki, T. (2007) *Angew. Chem. Int. Ed.*, **46**, 4729–4731; (b) Matsumoto, K., Yamaguchi, T., and Katsuki, T. (2008) *Chem. Commun.*, 1704–1706.

38 (a) Egami, H. and Katsuki, T. (2007) *J. Am. Chem. Soc.*, **129**, 8940–8941; (b) Egami, H. and Katsuki, T. (2008) *Synlett*, 1543–1546.

39 Scarso, A. and Strukul, G. (2005) *Adv. Synth. Catal.*, **347**, 1227–1234.

40 Choi, S., Yang, J.-D., Ji, M., Choi, H., Kee, M., Ahn, K.-H., Byeon, S.-H., Baik, W., and Koo, S. (2001) *J. Org. Chem.*, **66**, 8192–8198.

41 Matteucci, M., Bhalay, G., and Bradley, M. (2003) *Org. Lett.*, **5**, 235–237.

42 (a) Ball, S. and Bruice, T.C. (1979) *J. Am. Chem. Soc.*, **101**, 4017–4019; (b) Bruice, T.C. (1980) *Acc. Chem. Res.*, **13**, 256–262.

43 (a) Ball, S. and Bruice, T.C. (1980) *J. Am. Chem. Soc.*, **102**, 6498–6503; (b) Miller, A.E., Bischoff, J.J., Bizup, C., Luninoso, P., and Smiley, S. (1986) *J. Am. Chem. Soc.*, **108**, 7773–7778; (c) Oae, S., Asada, K., and Yoshimura, T. (1983) *Tetraedron Lett.*, **24**, 1265–1268.

44 Murahashi, S.-I., Oda, T., and Masui, Y. (1989) *J. Am. Chem. Soc.*, **111**, 5002–5003.

45 (a) Shinkai, S., Yamaguchi, T., Manabe, O., and Toda, F. (1988) *J. Chem. Soc., Chem. Commun.*, 1399–1401; (b) Shinkai, S., Yamaguchi, T., Kawase, A., Kitamura, A., and Manabe, O. (1987) *J. Chem. Soc., Chem. Commun.*, 1506–1508.

46 Minidis, A.B.E. and Bäckvall, J.E. (2001) *Chem. Eur. J.*, **7**, 297–302.

47 Bergstad, K. and Bäckvall, J.E. (1998) *J. Org. Chem.*, **63**, 6650–6655.

48 Lindén, A.A., Krüger, L., and Bäckvall, J.E. (2003) *J. Org. Chem.*, **68**, 5890–5896.

49 Lindén, A.A., Hermanns, N., Ott, S., and Bäckvall, J.E. (2005) *Chem. Eur. J.*, **11**, 112–119.

50 Lindén, A.A., Johansson, M., Hermanns, N., and Bäckvall, J.E. (2006) *J. Org. Chem.*, **71**, 3849–3853.

51 (a) Cimpeanu, V., Parvulescu, V.I., Amorós, P., Beltrán, D., Thompson, J.M., and Hardacre, C. (2004) *Chem. Eur. J.*, **10**, 4640–4646; (b) Cimpeanu, V., Parvulescu, A.N., Parvulescu, V.I., On, D.T., Kaliaguine, S., Thompson, J.M., and Hardacre, C. (2005) *J. Catal.*, **232**, 60–67.
52 Murahashi, S.-I. (1995) *Angew. Chem. Int. Ed.*, **34**, 2443–2465.
53 Imada, Y. and Naota, T. (2007) *Chem. Rec.*, **7**, 354–361.
54 Correa, P.E. and Riley, D.P. (1985) *J. Org. Chem.*, **50**, 1787–1788.
55 Bosch, E. and Kochi, J.K. (1995) *J. Org. Chem.*, **60**, 3172–3183.
56 Iwahama, T., Sakaguchi, S., and Ishi, Y.P. (1998) *Tetrahedron Lett.*, **39**, 9059–9062.
57 Martín, S.E. and Rossi, L.I. (2001) *Tetrahedron Lett.*, **42**, 7147–7151.
58 Boring, E., Geletii, Y.V., and Hill, G.L. (2001) *J. Am. Chem. Soc.*, **123**, 1625–1635.
59 (a) Imada, Y., Iida, H., Ono, S., and Murahashi, S.-I. (2003) *J. Am. Chem. Soc.*, **125**, 2868–2869; (b) Imada, Y., Iida, H., Ono, S., Masui, Y., and Murahashi, S.-I. (2006) *Chem. Asian J.*, **1**, 136–147.
60 Jonsons, C.R. and McCants, D. Jr. (1965) *J. Am. Chem. Soc.*, **87**, 1109–1114.
61 Pitchen, P., Dunach, E., Deshmukh, M.N., and Kagan, H.B. (1984) *J. Am. Chem. Soc.*, **106**, 8188–8193.
62 Di Furia, F., Modena, G., and Seraglia, R. (1984) *Synthesis*, 325–326.
63 Zhao, S., Samuel, O., and Kagan, H.B. (1989) *Org. Synth.*, **68**, 49–56.
64 (a) Brunel, J.M. and Kagan, H.B. (1996) *Synlett*, 404–406; (b) Brunel, J.M. and Kagan, H.B. (1996) *Bull. Soc. Chim. Fr.*, **133**, 1109–1115.
65 (a) Kagan, H. and Lukas, T. (1998) *Transition Metals for Organic Synthesis* (eds M. Beller and C. Bolm), VCH-Wiley, pp. 361–373. (b) Kagan, H.B. (2000) *Catalytic Asymmetric Synthesis* (ed. I. Ojima), Wiley-VCH, pp. 327–356.
66 (a) Yamamoto, K., Ands, H., Shuetas, T., and Chikamatsu, H. (1989) *Chem. Comm.*, 754–755; (b) Superchi, S. and Rosini, C. (1997) *Tetrahedron: Asymmetry*, **8**, 349–352.
67 (a) Komatsu, N., Hashizume, M., Sugita, T., and Uemura S. (1993) *J. Org. Chem.*, **58**, 4529–4533; (b) Reetz, M.T., Merk, C., Naberfeld, G., Rudolph, J., Griebenow, N., and Goddard, R. (1997) *Tetrahedron Lett.*, **38**, 5273–5276.
68 Yamanoi, Y. and Imamoto, T. (1997) *J. Org. Chem.*, **62**, 8560–8564.
69 Cotton, H., Elebring, T., Larsson, M., Li, L., Sörensen, H., and von Unge, S. (2000) *Tetrahedron: Asymmetry*, **11**, 3819–3825.
70 Pordea, A., Creus, M., Panek, J., Duboc, C., Mathis, D., Novic, M., and Ward, T.R. (2008) *J. Am. Chem. Soc.*, **130**, 8085–8088.
71 (a) Noda, K., Hosoya, N., Yanai, R., Irie, T., and Katsuki, T. (1994) *Tetrahedron Lett.*, **35**, 1887–1890; (b) Noda, K., Hosoya, N., Yanai, R., Irie, Y., Yamashita, T., and Katsuki, T. (1994) *Tetrahedron*, **50**, 9609–9618; (c) Kokubo, C. and Katsuki, T. (1996) *Tetrahedron*, **52**, 13895–13900.
72 Tohma, H., Takizawa, S., Watanabe, H., and Kita, Y. (1998) *Tetrahedron Lett.*, **39**, 4547–4550.
73 Shukla, V.G., Slagaonkar, D., and Akamanchi, K.G. (2003) *J. Org. Chem.*, **68**, 5422–5425.
74 Dembitsky, V.M. (2003) *Tetrahedron*, **59**, 4701–4720.
75 Collonna, S., Gaggero, N., Pasta, P., and Ottolina, G. (1996) *Chem. Commun.*, 2303–2307.
76 Collonna, S., Gaggero, N., Richelmi, C., and Pasta, P. (1999) *Trends Biotechnol*, **17**, 163–168.
77 Van Deurzen, M.P.J., Remkes, I.J., Van Rantwijk, F., and Sheldon, R.A. (1997) *J. Mol. Catal. A: Chem.*, **117**, 329–337.
78 Allenmark, S.G. and Andersson, M.A. (1996) *Tetrahedron: Asymmetry*, **7**, 1089–1094.
79 (a) Andersson, M.A., Willets, A., and Allenmark, S.G. (1997) *J. Org. Chem.*, **62**, 8455–8458; (b) Andersson, M.A. and Allenmark, S.G. (1998) *Tetrahedron*, **54**, 15293–15304.
80 Andersson, M.A. and Allenmark, S.G. (2000) *Biocatalysis and Biotransformation*, **18**, 79–86.
81 Tuynman, A., Vink, M.K.S., Dekker, H.L., Schoemaker, H.E., and Wever, R. (1998) *Eur. J. Biochem.*, **258**, 906–913.

82 Adam, W., Mock-Knoblauch, C., and Saha-Möller, C.R. (1999) *J. Org. Chem.*, **64**, 4834–4839.
83 Okrasa, K., Guibé-Jampel, E., and Therisod F Michel (2000) *J. Chem. Soc. Perkin Trans. 1*, 1077–1079.
84 Okrasa, K., Guibe-Jampel, E., and Therisod, M. (2003) *Tetrahedron: Asymmetry*, **14**, 2487–2490.
85 Kamerbeek, N.M., Janssen, D.B., van Berkel, W.J.H., and Fraaije, M.W. (2003) *Adv. Synth. Catal.*, **345**, 667–678.
86 Colonna, S., Gaggero, N., Carrea, G., and Pasta, P. (1997) *Chem. Commun.*, 439–440.
87 Secundo, F., Carrea, G., Dallavalle, S., and Franzosi, G. (1993) *Tetrahedron: Asymmetry*, **4**, 1981–1982.
88 Walsh, C.T. and Chen, Y.C.J. (1988) *Angew. Chem. Int. Ed. Engl.*, **27**, 333.
89 Carrea, G., Redigolo, B., Riva, S., Collonna, S., Gaggero, N., Battistel, E., and Bianchini, D. (1992) *Tetrahedron: Asymmetry*, **3**, 1063–1068.
90 Kamerbeek, N.M., Olsthoorn, A.J.J., Fraaije, M.W., and Janssen, D.B. (2003) *Appl. Environ. Microbiol.*, **69**, 419–426.
91 Kamerbeek, N.M., Moonen, M.J., Van Der Ven, J.G., Van Berkel, W.J.H., Fraaije, M.W., and Janssen, D.B. (2001) *Eur. J. Biochem.*, **268**, 2547–2557.
92 Chen, G., Kayser, M.M., Mihovilovic, M.D., Mrstik, M.E., Martinez, C.A., and Stewart, J.D. (1999) *New J. Chem.*, **23**, 827–832.
93 (a) Murahashi, S.-I. and Imada, Y. (1998) *Transition Metals for Organic Synthesis*, vol. 2 (eds M. Beller and C. Bolm), VCH-Wiley, pp. 373–383. (b) Chapter 5 of this book, Section 8.3.3.
94 Gilchrist, T.L. (1991) *Comprehensive Organic Synthesis*, vol. 7 (eds B.M. Trost and I. Fleming), Pergamon Press, London, pp. 735–756.
95 Albini, A. (1993) *Synthesis*, 263–267.
96 Sauer, J.D. (1990) *Surfactant Sci. Ser.*, **34**, 275.
97 VanRheenen, V., Cha, D.Y., and Hartley, W.M. (1988) *Org. Syn. Coll*, **VI**, 342.
98 Schröder, M. (1980) *Chem. Rev.*, **80**, 187.
99 Ahrgren, L. and Sutin, L. (1997) *Org. Proc. Res. Dev.*, **1**, 425.
100 Cicchi, S., Cardona, F., Brandi, A., Corsi, M., and Goti, A. (1999) *Tetrahedron Lett.*, **40**, 1989.
101 (a) Griffith, W.P., Ley, S.V., Whitcombe, G.P., and White, A.D. (1987) *J. Chem. Soc., Chem. Commun.*, 1625.
102 Ley, S.V., Norman, J., Griffith, W.P., and Marsden, S.P. (1994) *Synthesis*, 639.
103 Franzen, V. and Otto, S. (1961) *Chem. Ber.*, **94**, 1360.
104 Suzuki, S., Onishi, T., Fujita, Y., Misawa, H., and Otera, J. (1986) *Bull. Chem. Soc. Jpn.*, **59**, 3287.
105 Godfrey, A.G. and Ganem, B. (1990) *Tetrahedron Lett.*, **31**, 4825.
106 (a) Kirk-Othmer (1997) *Encyclopedia of Chemical Technoogy*, 4th edn, vol. **23**, John Wiley and Sons, Wiley-Interscience, New York, p. 524. (b) Isbell, T.A., Abbott, T.P., and Dvorak, J.A. (2000) US Pat. 6,051,214.
107 Malkov, A.V. and Kočovský, P. (2007) *Eur. J. Org. Chem.*, 29–36.
108 Chelucci, G., Murineddu, G., and Pinna, G.A. (2004) *Tetrahedron: Asymmetry*, **15**, 1373–1389.
109 Oh, K. and Ryu, J. (2008) *Tetrahedron Lett.*, **49**, 1935–1938.
110 Cope, A.C. and Ciganek, E. (1963) *Org. Syn. Coll. Vol.*, **IV**, 612.
111 (a) Mosher, H.S., Turner, L., and Carlsmith, A. (1963) *Org. Syn. Coll.*, **IV**, 828; (b) Brougham, P., Cooper, M.S., Cummerson, D.A., Heaney, H., and Thompson, N. (1987) *Synthesis*, 1015; (c) Craig, J.C. and Purushothaman, K.K. (1970) *J. Org. Chem.*, **35**, 1721.
112 Zajac, W.W., Walters, T.R., and Darcy, M.G. (1988) *J. Org. Chem.*, **53**, 5856.
113 Baumstark, A.L., Dotrong, M., and Vasquez, P.C. (1987) *Tetrahedron Lett.*, **28**, 1963.
114 Ferrer, M., Sanchez-Baeza, F., and Messeguer, A. (1997) *Tetrahedron*, **53**, 15877–15888.
115 Dayan, S., Kol, M., and Rozen, S. (1999) *Synthesis* (Spec Issue), 1427–1430.
116 (a) Kuhnen, L. (1966) *Chem. Ber.*, **99**, 3384; (b) Sheng, M.N. and Zajacek, J.G. (1988) *Org. Synth., Collective*, **6**, 501.
117 Choudary, B.M., Bharathi, B., Reddy, V., Kantam, M.L., and Raghavan, K.V. (2001) *Chem. Commun.*, 1736–1738.

118 Colladon, M., Scarso, A., and Strukul, G. (2008) *Green. Chem.*, **10**, 793–798.
119 Balagam, B. and Richardson, D.E. (2008) *Inorg. Chem.*, **447**, 1173–1178.
120 Jain, S.L. and Sain, B. (2003) *Angew. Chem. Int. Ed.*, **42**, 1265–1267.
121 Jain, S.L. and Sain, B. (2002) *Chem. Commun.*, 1040–1041.
122 Della Pina, C., Falletta, E., and Rossi, M. (2007) *Top. Catal.*, **44**, 325–329.
123 Coperet, C., Adolfsson, H., Khuong, T.-A.V., Yudin, K.A., and Sharpless, K.B. (1998) *J. Org. Chem.*, **63**, 1740–1741.
124 Coperet, C., Adolfsson, H., Chiang, J.P., Yudin, A.K., and Sharpless, K.B. (1998) *Tetrahedron Lett.*, **39**, 761–764.
125 Jain, S.L., Joseph, J.J., and Sain, B. (2006) *Synlett*, 2661–2663.
126 Sharma, V.B., Jain, S.L., and Sain, B. (2004) *Tetrahedron Lett.*, **45**, 4281–4283.
127 Colonna, S., Gaggero, N., Drabowicz, J., Lyzwa, P., and Mikolajczyk, M. (1999) *Chem. Commun.*, 1787–1788.
128 (a) Colonna, S., Pironti, V., Pasta, P., and Zambianchi, F. (2003) *Tetrahedron Lett.*, **44**, 869–871; (b) Colonna, S., Pironti, V., Carrea, G., Pasta, P., and Zambianchi, F. (2004) *Tetrahedron*, **60**, 569–575.
129 VanRheenen, V., Kelly, R.C., and Cha, D.Y. (1976) *Tetrahedron Lett.*, **17**, 1973–1976.
130 Bergstad, K., Jonsson, S.Y., and Bäckvall, J.E. (1999) *J. Am. Chem. Soc.*, **121**, 10424–10425.
131 Jonsson, S.Y., Färnegårdh, K., and Bäckvall, J.E. (2001) *J. Am. Chem. Soc.*, **123**, 1365–1371.
132 Bergstad, K., Piet, J.J.N., and Bäckvall, J.E. (1999) *J. Org. Chem.*, **64**, 2545–2548.
133 Jonsson, S.Y., Adolfsson, H., and Bäckvall, J.-E. (2001) *Org. Lett.*, **3**, 3463–3466.
134 Jonsson, S.Y., Adolfsson, H., and Bäckvall, J.E. (2003) *Chem. Eur. J.*, **9**, 2783–2788.
135 Éll, A.H., Closson, A., Adolfsson, H., and Bäckvall, J.E. (2003) *Adv. Synth. Catal.*, **345**, 1012–1016.
136 Closson, A., Johansson, M., and Bäckvall, J.E. (2004) *Chem. Commun.*, 1494–1495.
137 Johansson, M., Lindén, A.A., and Bäckvall, J.E. (2005) *J. Organometal. Chem.*, **690**, 3614–3619.
138 Choudary, B.M., Showdari, N.S., Madhi, S., and Kantam, M.L. (2001) *Angew. Chem. Int. Ed.*, **40**, 4619–4623.
139 Balagam, B., Mira, R., and Richardson, D.E. (2008) *Tetrahedron Lett.*, **49**, 1071–1075.

9
Liquid Phase Oxidation Reactions Catalyzed by Polyoxometalates
Ronny Neumann

9.1
Introduction

Environmentally benign and sustainable transformations in organic chemistry are now considered to be basic requirements in the development of modern organic syntheses. In order to meet these requirements, reactions should be free of dangerous waste, have high atom economy, and use solvents that are as environmentally friendly as possible. These societally dictated requirements for the preparation of organic compounds have led to the introduction of mainly homogeneous liquid phase catalysis into the arena of organic chemistry. Therefore, it is not surprising that in the area of oxidative transformations one major goal is the replacement of stoichiometric procedures using classical toxic waste-producing oxidants with catalytic procedures using environmentally benign oxidants. Oxidants whose use is being contemplated include molecular oxygen, hydrogen peroxide, nitrous oxide, and several others where there is either no by-product, the by-product is environmentally benign, for example, water or nitrogen, or the by-product can be easily recovered and recycled. For synthetic utility, where high conversion and selectivity are desirable, these oxidants will require activation by appropriate, usually metal-based catalysts. Furthermore, it would be preferable if possible to carry out these reactions in aqueous media or in organic media with low environmental load. As always in research involving catalysis, attention must also be paid to the vital issue of catalyst integrity, recovery, and recycling. In order to succeed in carrying out the necessary catalytic reactions, new catalysts must obviously be developed. Beyond the description of the practical utility of the new catalysts, significant insights into the mechanism of oxidant and substrate activation are key to the understanding of catalytic activity and selectivity. Such an understanding will improve the likelihood of our finding yet better catalysts. Therefore, in this chapter we will survey in detail, with some emphasis on our own research, the study and use of an interesting class of oxidation catalysts, polyoxometalates, and describe their utility for oxidative transformations. Emphasis will be put on both mechanistic and synthetic aspects of their use, together with a discussion of catalytic systems designed to facilitate the recovery

Modern Oxidation Methods. Edited by Jan-Erling Bäckvall
Copyright © 2010 WILEY-VCH Verlag GmbH & Co. KGaA, Weinheim
ISBN: 978-3-527-32320-3

and recycling of the catalyst. Following a short introduction to polyoxometalates, their structure and general properties, the major parts of this review deal with three different types of oxidants: (i) mono-oxygen donors, (ii) peroxides and (iii) molecular oxygen. Finally, various systems for catalyst 'engineering' designed to aid in catalyst manipulation, recovery and recycling are discussed.

9.2
Polyoxometalates (POMs)

Research applications of polyoxometalates have over the past two decades become very apparent and important, as reflected by the recent publication of a special volume of Chemical Reviews [1] devoted to these compounds. The diversity of research in the polyoxometalate area is significant and includes their application in many fields, including structural chemistry, analytical chemistry, surface chemistry, medicine, electrochemistry, and photochemistry. However, the most extensive research on the application of polyoxometalates appears to have been in the area of catalysis, where their use as Brønsted acid catalysts and oxidation catalysts has been going on since the late 1970s. Research published over the past decade or two has firmly established the significant potential of polyoxometalates as homogeneous oxidation catalysts. Through the development of novel ideas and concepts, polyoxometalates have been shown to have significant diversity of activity and mechanism that in the future may lead to important practical applications. In recent years, a number of excellent general reviews on the subject of catalysis by polyoxometalates have already been published [2], and while some repetition is inevitable we will attempt to keep the redundancy in the present review to a minimum.

A basic premise behind the use of polyoxometalates in oxidation chemistry is the fact that these compounds are oxidatively stable. In fact, as a class of compounds they are thermally stable generally to at least 350–450 °C in the presence of molecular oxygen. This, *a priori*, leads to the conclusion that for practical purposes polyoxometalates would have distinct advantages over widely investigated organometallic compounds that are vulnerable to decomposition due to oxidation of the ligand bound to the metal center. Polyoxometalates, previously also called heteropolyanions, are oligooxide clusters of discrete structure with a general formula $[X_xM_mO_y]^{q-}$ ($x \leq m$) where X is defined as the heteroatom and M are the addenda atoms. The addenda atoms are usually either molybdenum or tungsten in their highest oxidation state, 6^+, while the heteroatom can be any number of elements, transition metals and main group elements phosphorus and silicon being the most common heteroatoms. A most basic polyoxometalate, $[XM_{12}O_{40}]^{q-}$ where (X = P, Si, etc.; M = W, Mo), is that possessing the Keggin structure, Figure 9.1. Such Keggin type polyoxometalates, commonly available in their protic form, are significant for catalysis only as Brønsted acid catalysts.

However, since polyoxometalate synthesis is normally carried out in water by mixing the stoichiometrically required amounts of monomeric metal salts and adjusting the pH to a specific acidic value, many other structure types are accessible

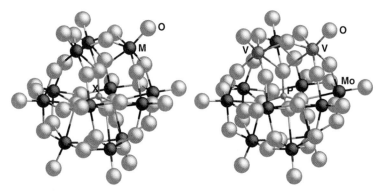

Figure 9.1 Polyoxometalates with the Keggin structure (a) $[XM_{12}O_{40}]^{q-}$ and (b) $[PM_{12}V_2O_{40}]^{5-}$.

by variation of the reaction stoichiometry, replacement of one or more addenda atoms with other transition or main group metals, and pH control. Oxygen atoms may also be partially substituted by fluorine, nitrogen, or sulfur. In the discussion below, additional polyoxometalate structures used in catalytic applications will be described, but already at this juncture it is important to note that there are important relationships beween polyoxometalate structure and oxidation reactivity and selectivity that represent significant advantages in the use of polyoxometalates as oxidation catalysts. As may be noted from the general formula of polyoxometalates, they are also polyanionic species. This property enables fine control of polyoxometalate solubility by simple variation of the cation. A summary of this aspect of the solution chemistry of polyoxometalates is given in Figure 9.2.

9.3
Oxidation with Mono-Oxygen Donors

One class of polyoxometalate compounds is that in which transition or main group metals, often in a lower oxidation state, substitute for a tungsten- or molybdenum-oxo group at the polyoxometalate surface. The substituting metal center is thus penta-coordinated by the 'parent' polyoxometalate. The octahedral coordination sphere is

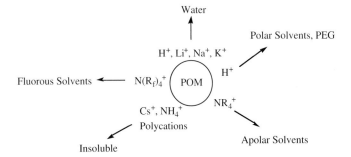

Figure 9.2 Solubility of polyoxometalates as a function of the counter cation.

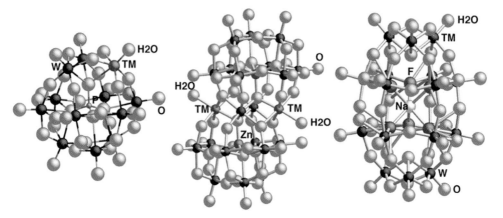

Figure 9.3 Various transition metal-substituting polyoxometalates.

completed by an additional sixth labile ligand, L (usually L = H_2O). This lability of the sixth ligand allows the interaction of the substituting transition metal atom with a reaction substrate and/or oxidant leading to reaction at the transition metal center. In analogy to organometallic chemistry the 'pentadentate' polyoxometalate acts as an inorganic ligand. This analogy, in earlier years, led to such transition metal-substituted polyoxometalates being referred to as 'inorganic metalloporphyrins'. Many structural variants of such transition metal-substituted polyoxometalates are known. For example, some of those used in catalytic applications are: (a) the transition metal-substituted 'Keggin'-type compounds, $[XTM(L)M_{11}O_{39}]^{q-}$ (X = P, Si; TM = Co(II,III), Zn(II), Mn(II,III), Fe(II,III), Cu(I,II), Ni(II), Ru(III), etc.; M = Mo, W) (Figure 9.3a), (b) the so-called 'sandwich'-type polyoxometalates, $\{[(WZnTM_2(H_2O)_2][(ZnW_9O_{34})_2]\}^{q-}$ (Figure 9.3b) having a ring of transition metals between two truncated Keggin 'inorganic ligands', and (c) the polyfluorooxometalates, Figure 9.3c, of a quasi Wells-Dawson structure. In particular, these latter two compounds classes often have superior catalytic activity and stability.

One of the first uses of transition metal-substituted polyoxometalates (TMSPs) was in the context of a comparison of the catalytic activity of such TMSPs with their metalloporphyrin counterparts. Thus, it seemed natural at the time to evaluate the activity of such TMSPs with iodosobenzene and pentafluoroiodosobenzene as oxidant [3]. Especially the manganese(II/III)-substituted Keggin $[PMn(H_2O)W_{11}O_{39}]^{5-}$ (see Figure 9.3a) and Wells-Dawson $[\alpha\text{-}P_2Mn(H_2O)W_{17}O_{61}]^{7-}$ polyoxometalates showed good activity and very high selectivity for the epoxidation of alkenes and some activity for the hydroxylation of alkanes that usually compared favorably to the activity of the manganese(III) tetra-2,6-dichlorophenylporphyrin. The stereoselectivities and regioselectivities observed in epoxidation of probe substrates such as limonene, cis-stilbene, naphthalene, and others, lead to the hypothesis that a reactive manganese(V) oxo intermediate was the catalytically active species. A comparison was also made between Wells-Dawson-type polyoxometalates incorporating a metal center, for example, $[\alpha\text{-}P_2Mn(H_2O)W_{17}O_{61}]^{7-}$, and a polyoxometalate supporting a manganese center, $\{Mn(II)(CH_3CN)_x/[P_2W_{15}Nb_3O_{62}]^{9-}\}$, and it was found that the incorporating analog was about twice as active [4]. Other manganese nonpolyoxometalate supporting compounds showed very significantly reduced activity. Recently, with use of both $[PMn(H_2O)W_{11}O_{39}]^{5-}$ for DFT computational studies and $[P_2Mn(H_2O)W_{17}O_{61}]^{7-}$ for experimental research it was shown that both a triplet and a singlet Mn(V)=O species are formed as active oxidants [5]. Although it was originally computationally predicted for both manganese- and iron-substituted polyoxometalates that they would be as reactive as or more reactive than their analogous metalloporphyrins [6]; this is in fact not the case because a highly negative entropy of activation due to strong solvation of the anionic polyoxometalate leads to a higher free energy of activation for the polyoxometalate analogs compared to the metalloporphyrins. In addition, it was observed that Fe(V)=O species were not accessible via the use of iodosobenzene [6b]. High-valent chromium(V) oxo species, $[XCr^V=OW_{11}O_{39}]^{(9-n)-}$ (X = P(V), Si(IV)), thought to serve as analogs to the proposed manganese intermediate, were also prepared and evaluated as stoichiometric oxidants for alkene epoxidation [7]. Unfortunately, the selectivity in favor of the epoxide product was relatively low, usually at best only ~10%, indicating at the least that the chromium-oxo species reacted via a pathway different from that proposed for the manganese compound. By the use of a different mono-oxygen donor, p-cyano-N,N-dimethylaniline-N-oxide and other N-oxides, it was later observed that the family of tetrametal substituted 'sandwich' polyoxometalates, $[M_4(H_2O)_2P_2W_{18}O_{68}]^{10-}$, catalyzed epoxidation of alkenes [8]. Surprisingly, the cobalt-substituted polyoxometalate was much more reactive than the corresponding manganese, iron, or nickel analogs, but the $[PCo(H_2O)W_{11}O_{39}]^{5-}$ polyoxometalate was as reactive as $[Co_4(H_2O)_2P_2W_{18}O_{68}]^{10-}$. From mostly kinetic studies it was concluded that the active catalytic intermediate was a relatively rarely observed Co(IV)=O species. This conclusion, if correct, raises the question of how the different active species in the manganese and chromium systems are formed since the manganese-substituted compounds are inferior catalysts with N-oxides by two orders of magnitude.

The observations, discussed above, that transition metal-substituted polyoxometalates could be used to catalyze oxygen transfer from iodosobenzene to organic substrates raised the possibility of activation of other mono-oxygen donors. This conceptual advance was then coupled with the ability to incorporate noble metals into the polyoxometalate framework to utilize periodate as an oxidant. The $Ru^{III}(H_2O)SiW_{11}O_{39}]^{5-}$ (see Figure 9.3a) compound catalyzed the oxidation of alkenes and alkanes using various oxygen donors [9]. In particular, sodium periodate proved to be a mild and selective oxidant enabling the oxidative cleavage of carbon-carbon double bonds in styrene derivatives to yield the corresponding benzaldehyde products in practically quantitative yields. A further kinetic and spectroscopic investigation led to the proposition of a reaction mechanism that included the formation of a metallocyclooxetane intermediate, which is transformed into a Ru(V) cyclic diester in the rate-determining step requiring water. The diester is then rearranged via carbon-carbon bond cleavage to form the benzaldehyde product [10]. The periodate-$[Ru^{III}(H_2O)SiW_{11}O_{39}]^{5-}$ system was also nicely adapted in an electrochemical two-phase system where iodate, formed from spent periodate, is reoxidized at a lead oxide electrode [11]. Highly selective and efficient synthesis of aldehydes by carbon-carbon bond cleavage was possible. These ruthenium-substituted Keggin compounds were also used for the oxidation of alkanes to alcohols and ketones using sodium hypochlorite [12], and the oxidation of alcohols and aldehydes to carboxylic acids with potassium chlorate as oxidant [13]. The stability of Keggin type compounds to basic sodium hypochlorite conditions is, however, suspect, and research should be carried out to validate the catalyst integrity under such conditions.

Ozone is a highly electrophilic oxidant attractive as an environmentally benign oxidant (33% active oxygen, O_2 by-product). Although its reactivity with nucleophilic substrates such as alkenes is well known to be very high, noncatalytic reactions at saturated carbon-hydrogen bonds are of very limited synthetic value. The catalytic activation of ozone in aqueous solution was shown to be possible with the manganese-substituted polyoxometalate $\{[(WZnMn(II)_2(H_2O)_2][(ZnW_9O_{34})_2]\}^{12-}$ [14]. Good conversions of alkanes to ketones, for example, cyclohexane to cyclohexanone, were observed at 2 °C within 45 min. Low-temperature, *in situ* observation of an emerald green intermediate by UV-vis and EPR led to the suggestion that a high-valent manganese ozonide was the catalytically active species. The manganese ozonide complex was formulated as POM−Mn^{IV}−O−O−O$^•$ although other canonical or tautomeric forms may be pictured, for example, POM−MnIII−O−O−O$^+$ or POM−Mn^V−O−O−O$^-$. In the absence of a substrate, the ozonide intermediate very quickly decays by reduction or disproportionation to a brown manganese(IV) oxo or hydroxy species. In the presence of alkanes, reactivity was observed typical of radical intermediates, while with alkenes stereoselective epoxidation was noted. The possible reaction pathways are compiled in Scheme 9.1.

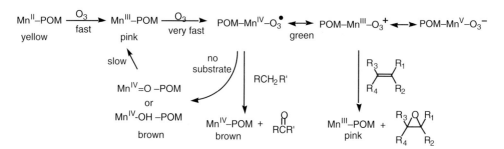

Scheme 9.1 Activation of ozone and proposed catalytic pathways in the presence of $\{[(WZnMn(II)_2(H_2O)_2][(ZnW_9O_{34})_2]\}^{12-}$.

Nitrous oxide is also potentially attractive as an environmentally benign oxidant (36% active oxygen, N_2 by-product) but it is usually considered to be an inert [15] and poor ligand toward transition metals [16]. In practice, there are only a few catalytic systems that have been shown to be efficient for the activation of N_2O for selective hydrocarbon oxidation. Notable among them are the iron-containing acidic zeolites [17], which at elevated temperatures are thought to yield surface-activated iron-oxo species (α-oxygen) [18] capable of transferring an oxygen atom to inert hydrocarbons such as benzene [19]. Iron oxide on basic silica has been shown to catalyze, albeit nonselectively, propene epoxidation [20]. Also, stoichiometric oxygen transfer from nitrous oxide to a ruthenium porphyrin [21] yielded a high-valent ruthenium-dioxo species capable of oxygen transfer to nucleophiles such as alkenes and sulfides. Under more extreme conditions (140 °C, 10 atm N_2O), the ruthenium porphyrin has just been shown to catalyze oxygen transfer to trisubstituted alkenes [22] and the oxidation of alcohols to ketones [23]. We have shown that a manganese-substituted polyoxometalate of the 'sandwich' structure, $^8Q_{10}[Mn^{III}_2ZnW(ZnW_9O_{34})_2]$ (see Figure 9.3b, $^8Q = (C_8H_{17})_3CH_3N^+$), is capable of selectively (>99.9%) catalyzing the epoxidation of alkenes (Scheme 9.2) [24]. It appears that there is little correlation between the catalytic activity and the nucleophilicity of the alkene. In addition, reactions were stereoselective.

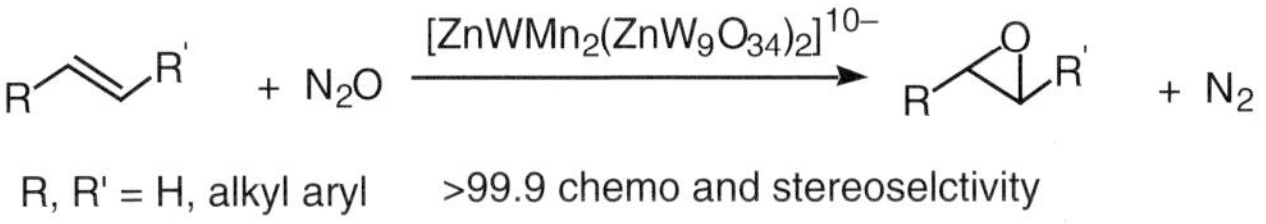

Substrate	TON
1-octene	10
E-2-octene	14
cyclooctene	19
1-decene	8
cyclohexene	9
Z-2-hexen-1-ol	21
E-2-hexen-1-ol	19
Z-stilbene	15
E-stilbene	23

Scheme 9.2 Results for epoxidation of alkenes with N_2O catalyzed by $Q_{10}[Mn^{III}_2ZnW(ZnW_9O_{34})_2]$.

Also, the vanadium-substituted polyoxomolybdate, $^4Q_5[PV_2Mo_{10}O_{40}]$ (see Figure 9.1b, $^4Q = (C_4H_9)_4N^+$) was catalytically active and highly selective in the oxidation of primary and secondary alcohols with nitrous oxide to yield aldehydes and ketones, respectively, Scheme 9.3 [25]. In addition, the same catalyst under similar conditions catalyzed the oxidation of alkylarenes, Scheme 9.3. In the oxidation of alkylarenes with hydrogen atoms at β positions, the substrate is dehydrogenated, for example, cumene to α-methylstyrene, whereas in the absence of β hydrogen atoms ketones are formed, for example, diphenylmethane to benzophenone.

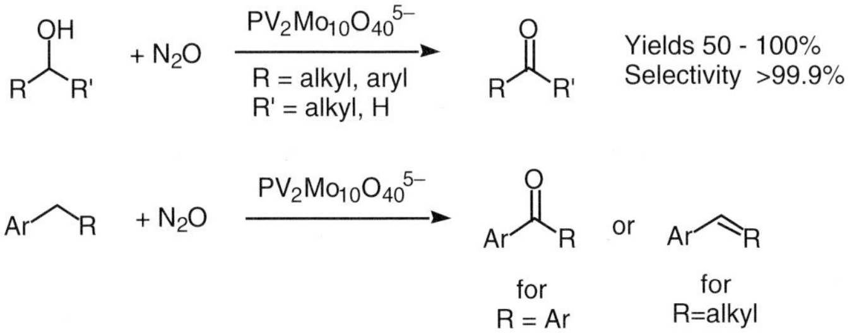

Scheme 9.3 Oxidation of alkylarenes and alcohols with N_2O catalyzed by $Q_5[PV_2Mo_{10}O_{40}]$.

It is important to note that the catalysts are orthogonal in their catalytic activity, that is, $^8Q_{10}[Mn^{III}_2ZnW(ZnW_9O_{34})_2]$ is active only for alkene epoxidation but not for alcohol or alkylarene oxidation, while for $^4Q_5[PV_2Mo_{10}O_{40}]$ the opposite is true. The mechanistic understanding of the catalytic activation of nitrous oxide is still quite rudimentary; however, one very notable observation is that nitrous oxide appears to be oxidized in the presence of both polyoxometalates as was measured by the reduction (UV-vis and EPR) of the polyoxometalates in its presence under reaction conditions without substrate. Considering the high oxidation potential of nitrous oxide as measured by its ionization potential in the gas phase, this result is surprising. A further important observation in reactions catalyzed by $^4Q_5[PV_2Mo_{10}O_{40}]$ and from correlation of rates with homolytic benzylic carbon-hydrogen bond dissociation energies and Hammett plots indicate that perhaps an intermediate polyoxometalate–nitrous oxide complex leads to carbon-hydrogen bond cleavage. Further experiments are required to further understand the mechanistic picture.

Sulfoxides are potentially interesting oxidants and/or oxygen donors, notably used in numerous variants of Swern type oxidations of alcohols in the presence of a stoichiometric amount of an electrophilic activating agent [26]. The deoxygenation of sulfoxides to sulfides catalyzed by metal complexes with oxygen transfer to the metal complex or to reduced species such as hydrohalic acids, phosphines, carbenes, and carbon monoxide is also well established [27]. In this context it has been demonstrated in our group that sulfoxides can be used as oxygen donors/oxidants in polyoxometalate-catalyzed reactions. For the first time an oxygen transfer from a sulfoxide to an alkylarene hydrocarbon to yield sulfide and a carbonyl product was demonstrated; in certain cases oxidative dehydrogenation was observed, Scheme 9.4 [28].

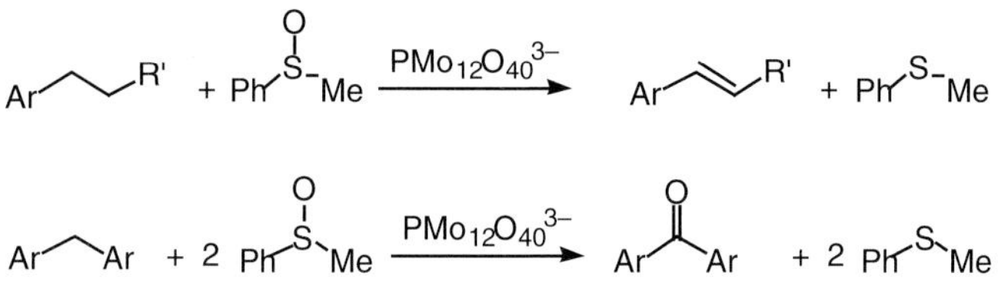

Substrate	TON
xanthene	145
diphenylmethane	142
fluorene	67
triphenylmethane	70
Isochroman	61
dibenzyl	26
dihydroanthracene	300
dihydrophenathrene	55

Scheme 9.4 Oxidation of alkylarenes with phenylmethylsulfoxide catalyzed by $Q_3[PMo_{12}O_{40}]$.

Further research into the reaction mechanism revealed that the reaction rate was correlated with the electron structure of the sulfoxide; the more electropositive sulfoxides were the better oxygen donors. Excellent correlation of the reaction rates with the heterolytic benzylic carbon-hydrogen bond dissociation energies indicated a hydride abstraction mechanism in the rate-determining step to yield a carbocation intermediate. The formation of 9-phenylfluorene as by-product in the oxidation of triphenylmethane supports this suggestion. Further kinetic experiments and ^{17}O NMR showed the formation of a polyoxometalate–sulfoxide complex before the oxidation reaction, this complex being the active oxidant in these systems. Subsequently, in a similar reaction system, sulfoxides were used to facilitate the aerobic oxidation of alcohols [29]. In this manner, benzylic, allylic, and aliphatic alcohols were all oxidized to aldehydes and ketones in a reaction catalyzed by Keggin-type polyoxomolybdates, $[PV_xMo_{12-x}O_{40}]^{-(3+x)}$ ($x = 0$, 2), with DMSO as solvent. The oxidation of benzylic alcohols was quantitative within hours and selective to the corresponding benzaldehydes, but the oxidation of allylic alcohols was less selective. The oxidation of aliphatic alcohols was slower but selective. In mechanistic studies considering oxidation of benzylic alcohols, similar to the oxidation of alkylarenes, a polyoxometalate–sulfoxide complex appears to be the active oxidant. Further isotope-labeling experiments, kinetic isotope effects, and especially Hammett plots showed that oxidation occurs by oxygen transfer from the activated sulfoxide and elimination of water from the alcohol. However, the exact nature of the reaction pathway is dependent on the identity of substituents on the phenyl ring.

Summarizing the information disclosed in the section above, one notes that polyoxometalates appear to be versatile oxidation catalysts capable of activating various mono-oxygen donors such as iodosobenzene, periodate, ozone, nitrous oxide, and sulfoxides. Some of these reactions are completely new from both a synthetic and mechanistic point of view. The various reaction pathways expressed are also rather unusual and point to the many options and reaction pathways available for oxidation catalyzed by polyoxometalates.

9.4
Oxidation with Peroxygen Compounds

Before specifically discussing oxidation by peroxygen compounds using polyoxometalates as catalysts, a few general comments concerning peroxygen compounds as oxidants or oxygen donors are worth making. First, from a practical point of view, hydrogen peroxide is certainly the most sustainable oxidant of this class, since it has a high percentage (47%) of active oxygen, it is inexpensive, and the by-product of oxidation is water. On the down side, its use as an aqueous solution presents problems of compatibility and reactivity with hydrophobic organic substrates or solvents, and some precautions must be taken such as working under reasonably low concentrations (usually <20 wt% in polar organic solvent) to prevent safety hazards. Various 'solid' forms of hydrogen peroxide such as urea hydroperoxide, sodium perborate, and sodium percarbonate, are also available. Alkyl hydroperoxides, notably *tert*-butylhydroperoxide, have the advantage that they are freely soluble in organic media and can thus be used in strictly nonaqueous solvents. The alcohol by-product resulting from the use of alkyl hydroperoxides as oxygen donors can often be easily recovered, for example, by distillation, and at least in principle the alkyl hydroperoxide can be re-synthesized from the alcohol. There are also other peroxygen oxidants readily available; one notable inorganic compound is monoperoxosulfate, HSO_5^-, normally available as a triple salt, Oxone™.

From a mechanistic point of view it is important to realize that polyoxometalates may interact with peroxygen oxidants in several different ways depending on the composition, structure, and redox potential of the polyoxometalate compounds. On the one hand, one may expect reaction pathways typical for any oxotungstate or oxomolybdate compounds with formation of peroxo or hydroperoxy (alkylperoxy) intermediates capable of oxygen transfer reactions with nucleophilic substrates such as alkenes to yield epoxides. On the other hand, depending on the redox potential of the polyoxometalate, a varying degree of homolytic cleavage of oxygen-oxygen and hydrogen-oxygen bonds will lead to hydroxy (alkoxy) and peroxy (peralkoxy) intermediate radical species. The trend of increased formation of radical species as a function of increasing oxidation potential is clearly evident in the series of Keggin type heteropoly acids: $H_5PV_2Mo_{10}O_{40} > H_3PMo_{12}O_{40} > H_3PW_{12}O_{40}$. In particular, hydroxy or alkoxy radicals will lead to further hydrogen abstraction from the substrate molecules and formation of additional radical species. The rate of formation and fate of these latter radical species will determine the conversion, selectivity, and identity of the products formed in the reaction. This tendency for homolytic cleavage in the peroxygen compounds can also be expected to be strongly influenced by the presence of substituting transition metals in the polyoxometalate structure. A high oxidation potential of the polyoxometalate and/or presence of redox-active transition metals will also lead to dismutation reactions and thus nonproductive decomposition of the peroxygen oxidant and low yields based on the oxidant. There is also a more remote possibility that intermediate hydroperoxide species of a transition metal substituted in polyoxometalate structure, for example a Fe(III)-OOH intermediate, will lead to an Fe(V)=O species or equivalent. To date, there has been no observation of such a biomimetic transformation in polyoxometalate catalytic chemistry, although we have recently isolated such an Fe(III)-OOH intermediate that, however, acts as a reducing agent (benzoquinone to hydroquinone) and does yield higher-valent oxo species [30].

Originally, *tert*-butylhydroperoxide was used together with transition metal-substituted Keggin type compounds and then later on more effectively with transition metal substituted 'sandwich' compounds for the oxidation of alkanes to alcohols and ketones [31]. The oxidation of alkenes went with low selectivity. Although the mechanism was not rigorously studied, it would seem quite certain that these reactions proceed by a radical mechanism via hydrogen abstraction by alkoxy radicals from the substrate. Oxone has been similarly used for the oxidation of benzylic and aliphatic alcohols [32]. Interestingly, it has been observed that oxidation of alkanes with *tert*-butylhydroperoxide catalyzed by a polyoxomolybdate, $H_3PMo_{12}O_{40}$, may be redirected from oxygenation to oxydehydrogenation yielding alkenes as the major products [33]. Thus, both acyclic and cyclic alkanes were oxidized to alkenes by *tert*-butylhydroperoxide in acetic acid with $H_3PMo_{12}O_{40}$ as catalyst with reaction selectivity generally $\geq 90\%$. Some minor amounts of alcohols, ketones, and hydroperoxide products formed via oxygenation with molecular oxygen were also obtained, as were some acetate esters. The alkene product selectively tended toward the kinetically favored product rather than the thermodynamically more stable alkenes. Therefore, oxidation of 1-methylcyclohexane yielded mostly 3- and 4-methylcyclohexene rather than 1-methylcyclohexene. Similarly, in the oxidation of 2,2,4-trimethylpentane, the terminal alkene, 2,2,4-trimethyl-4-pentene, was formed in fourfold excess relative to the internal alkene, 2,2,4-trimethyl-3-pentene. A reaction scheme to explain the reaction selectivity is presented in Scheme 9.5.

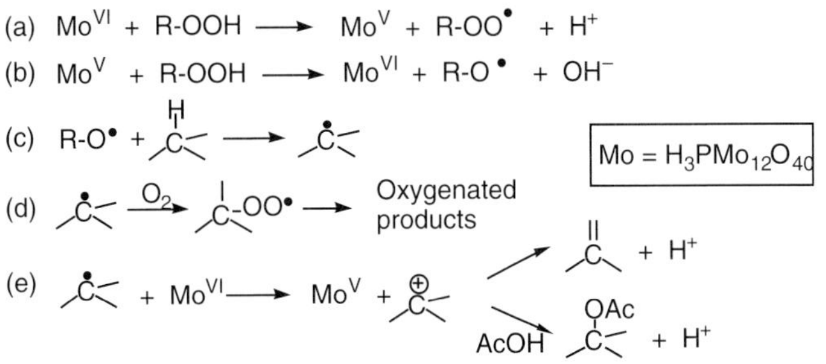

(a) $Mo^{VI} + R\text{-}OOH \longrightarrow Mo^{V} + R\text{-}OO^{\bullet} + H^{+}$

(b) $Mo^{V} + R\text{-}OOH \longrightarrow Mo^{VI} + R\text{-}O^{\bullet} + OH^{-}$

(c) R-O• + >C-H → >C•

(d) >C• —O₂→ >C-OO• → Oxygenated products

(e) >C• + Mo^{VI} → Mo^{V} + >C⁺ → ...

$Mo = H_3PMo_{12}O_{40}$

Scheme 9.5 Oxidation of alkanes with *tert*-butylhydroperoxide catalyzed by $H_3PMo_{12}O_{40}$.

Tert-butylhydroperoxide reacts with the $H_3PMo_{12}O_{40}$ catalyst to yield alkoxy and alkylperoxy radicals (reactions a and b). The alkoxy radical, which can be trapped by spin traps and observed by EPR, homolytically abstracts hydrogen from a reactive carbon–hydrogen moiety (reaction c). Instead of the usual diffusion rate-controlled oxygenation with molecular oxygen (reaction d), oxidative electron transfer occurs yielding a carbocation that in turn is dehydrogenated to yield an alkene or is attacked by acetic acid to give the acetate ester as by-product (reaction e). *Tert*-butylhydroperoxide has also been used for the highly selective oxidation of thioethers, for example, tetrahydrothiophene, to the corresponding sulfoxides without further oxidation to sulfones using $H_5PV_2Mo_{10}O_{40}$ as the catalyst [34]. Although one may automatically assume that such an oxidation would take place by an oxygen transfer reaction from a polyoxometalate-alkylperoxy intermediate to the sulfide, the evidence presented indicates that in fact oxidation occurs via electron transfer from the thioether to the polyoxometalate, where the role of the *tert*-butylhydroperoxide is to re-oxidize the reduced polyoxometalate. This type of mechanism is in line with what is known about oxidation catalyzed by $H_5PV_2Mo_{10}O_{40}$ with oxygen as terminal oxidant, as discussed in Section 9.5 below (see Scheme 9.9).

Enantioselective oxidation catalysis to yield chiral products from prochiral substrates had not until recently been observed using polyoxometalate catalysts. However, in a combined effort of several research groups it has been shown that the racemic vanadium-substituted 'sandwich' type polyoxometalate, $[(V^{IV}O)_2ZnW(ZnW_9O_{34})_2]^{12-}$, is an extremely effective catalyst (up to 40 000 turnovers) at near ambient temperatures, for the enantioselective epoxidation of allylic alcohols to the 2R,3R-epoxyalcohol with the sterically crowded chiral hydroperoxide, TADOOH, as oxygen donor [35], Scheme 9.6. The enantiomeric excesses, *ee*, attained using aryl-substituted allylic alcohols was quite high, generally 70–90%, and at high conversions >95%. However, less sterically hindered allylic alcohols such as geraniol gave a low enantiomeric excess (20%) of chiral 2R,3R-epoxygeraniol. The chiral induction observed in the reaction is thought to be due to the presence of a vanadium template for the chiral hydroperoxide and the allylic alcohol. Thus, nonfunctionalized alkenes, for example, 1-phenylcyclohexene, showed essentially negligible enantioselectivity. Also, substitution of vanadium with other transition metals yielded significantly lower enantioselectivity and low conversion of allylic alcohols.

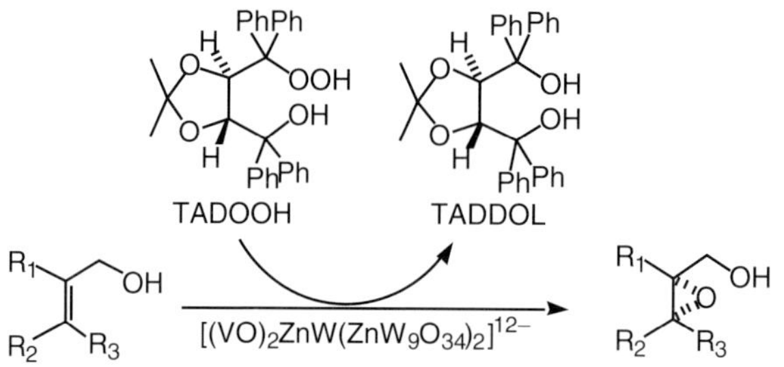

R_1	R_2	R_3	%ee
Ph	Ph	H	82
Me	Ph	H	84
Me	4-MeOPh	H	70
H	Ph	H	50
Ph	H	H	44

Scheme 9.6 Enantioselective epoxidation of allylic alcohols with a chiral hydroperoxide, TADOOH, catalyzed by $[(V^{IV}O)_2ZnW(ZnW_9O_{34})_2]^{12-}$.

As noted above, the oxotungstate or oxomolybdate nature of polyoxometalate compounds boded well for their activation of hydrogen peroxide. Thus, Ishii and his coworkers described the first use of polyoxometalates with 30–35% aqueous hydrogen peroxide as the oxidant. They used the commercially available phosphotungstic acid, $H_3PW_{12}O_{40}$, as catalyst. In order to utilize a biphasic reaction medium (organic substrate/aqueous oxidant) they added a quaternary ammonium salt, hexadecylpyridinium bromide, to dissolve the $[PW_{12}O_{40}]^{3-}$ in the organic apolar solvent reaction phase. Ishii's group and also others gave numerous examples of oxidation reactions typical for the use of reactions with hydrogen peroxide in the presence of tungsten-based catalysts. The first examples dealt with the epoxidation of allylic alcohols [36] and alkenes [37]. Generally, high epoxide yields, >90%, were obtained with only a relatively small excess of hydrogen peroxide. An evaluation of the catalytic activity for epoxidation reveals turnover frequencies of 5–15 h^{-1} per tungsten atom. Under more acidic conditions and at higher temperatures the epoxides are sensitive to hydrolysis leading to formation of vicinal diols, which are subsequently oxidized to keto-alcohols, α,β-diketones [38] or, at longer reaction times, undergo oxidative carbon-carbon bond cleavage to yield carboxylic acids and ketones. The phosphotungstate polyoxometalate was also effective for oxidation of secondary alcohols to ketones, while primary alcohols were not reactive allowing for the high-yield regioselective oxidation of nonvicinal diols to the corresponding keto-alcohols; α,ω-diols did, however, react to give lactones (e.g., γ-butyrolactone from 1,4-butanediol) in high yields [39]. Other research showed that alkynes [40], amines [41], and sulfides [37], could be oxidized efficiently to ketones, N-oxides, and sulfoxides and sulfones, respectively. Various quinones were also synthesized from active arene precursors [42].

While the synthetic applications involving oxidation of the various substrate types was being investigated mostly by Ishii's group [36–42] using $[PW_{12}O_{40}]^{3-}$ as catalyst, other researchers have actively pursued studies aiming at an understanding of the identity of the true catalyst in these reactions [44–47]. In this context it should be noted that the isostructural compound $[SiW_{12}O_{40}]^{4-}$ showed almost no catalytic activity compared to $[PW_{12}O_{40}]^{3-}$ under identical conditions [36]. At practically the same time, Csanyi and Jaky [43], and the groups of Brégault [44], Griffith [45], and Hill [46] suggested, and convincingly proved, using various spectroscopic and kinetic probes, that the $[PW_{12}O_{40}]^{3-}$ and even more so the lacunary $[PW_{11}O_{39}]^{7-}$ polyoxometalate formed at pH $\geq \sim 3$–4 was unstable in the presence of aqueous hydrogen peroxide, leading mainly to the formation of the peroxophosphotungstate, $\{PO_4[WO(O_2)_2]\}^{3-}$. This compound had been previously synthesized and characterized by Venturello et al. and had been shown to have very similar catalytic activity in various oxidation reactions with hydrogen peroxide [47]. A general conclusion resulting from these studies of the groups of Ishii, Venturello, Csanyi and Jaky, Brégeault, Griffith, and Hill is that the $\{PO_4[WO(O_2)_2]\}^{3-}$ peroxophosphotungstate compound is one of the best catalysts, especially from the point of view of synthetic versatility of all of the many peroxotungstates that have been studied. A more extensive review of the complex phosphotungstate solution chemistry in the presence of hydrogen peroxide is beyond the scope of this present chapter. In more recent years, it has been shown by Xi and coworkers that by the careful choice of the quaternary ammonium counter cation and reaction solvent a possibly technologically practical process for the epoxidation of propene to propene oxide could be envisioned using $\{PO_4[WO(O_2)_2]\}^{3-}$ as catalyst [48]. For example, in the presence of hydrogen peroxide using a combination of toluene and tributylphosphate as solvent, a soluble $\{PO_4[WO(O_2)_2]\}^{3-}$ compound was obtained. Once the hydrogen peroxide is used up, a $\{PO_4[WO_3]\}^{3-}$ compound is formed that is insoluble in the reaction medium, allowing simple recovery for recycling of the phosphotungstate species. Importantly, it was claimed that the system could be coupled with the synthesis of hydrogen peroxide from hydrogen and oxygen using the classic ethylanthraquinone process for hydrogen peroxide preparation.

The hydrolytic instability of the simple and lacunary Keggin-type polyoxometalates, $[PW_{12}O_{40}]^{3-}$ and $[PW_{11}O_{39}]^{7-}$, in the presence of aqueous hydrogen peroxide, leading to formation in solution of various peroxotungstate species of varying catalytic activity, led to two intertwined issues. The first issue that came up was the necessity to carefully analyze the stability of polyoxometalates under hydrogen peroxide/hydrolytic conditions. For example, it had been claimed that lanthanide-containing polyoxometalates, $[LnW_{10}O_{36}]^{9-}$, were active catalysts for alcohol oxidation [49]; however, subsequent research showed that they in fact decomposed to smaller and known peroxotungstate species that were the catalytically active species [50]. On the other hand, other Keggin compounds appeared to be stable in the presence of aqueous hydrogen peroxide. For example, a stable peroxo species based on the Keggin structure, $[SiW_9(NbO_2)_3O_{37}]^{7-}$, was synthesized, characterized, and used for epoxidation of reactive allylic alcohols but not alkenes [51]. The $[PZnMo_2W_9O_{39}]^{5-}$ polyoxometalate was used to oxidize sulfides to sulfoxides [52]. The $Q_5[PV_2Mo_{10}O_{40}]$ (Q = quaternary ammonium cation) in aqueous hydrogen peroxide/acetic acid was stable and catalyzed the oxidation of alkylaromatic substrates in the benzylic position [53], while $Q_5[PV_2W_{10}O_{40}]$ used for the oxidation of benzene to phenol also remained intact during the reaction [54]. Likewise, titanium-substituted Keggin-type phosphotungstates are apparently stable in the presence of hydrogen peroxide [55]. Kholdeeva and her coworkers have viewed such titanium-substituted compounds as models for hydrogen peroxide-based oxidation on titanium centers [56]. The research kinetically followed the formation of the titanium peroxo species. 2,3,6-Trimethyl phenol is oxidized to yield the oxygenated product, 2,3,5-trimethylbenzoquinone and the oxidatively coupled product, 2,2′,3,3′,5,5′-hexmethyl-4,4′-biphenol, presumably obtained by electron transfer oxidation. The presence of a second proton on the peroxo moiety is crucial for determining the ability of the titanium Keggin polyoxometalate to oxidize an alkene such as cyclohexene by a heterolytic oxygen transfer, although, because of the acidic condition, the 1,2-*trans*-cyclohexanediol is the major product obtained. More recently, another group has carried out similar research [57]. It would also appear that various iron-substituted Keggin compounds reported by Mizuno and coworkers for alkene and alkane oxidation are also stable in the presence of hydrogen peroxide, although the study was not completely definitive [58]. From these examples and others not noted, it is clear that certain Keggin-type polyoxometalates can be stable under certain reaction conditions. Parameters to consider in this context are pH and the relative stability of the specific polyoxometalate at such a pH, also the solvent and temperature.

In recent years, a γ-$[SiW_{10}O_{34}(H_2O)_2]^{4-}$ polyoxometalate with a 'defect' site, originally prepared by Mizuno and coworkers, has been shown to have similar activity (i.e. highly effective epoxidation of primary alkenes) to $\{PO_4[WO(O_2)_2]\}^{3-}$ (normalized per tungsten atom) [59]. The catalyst was formulated to have two aqua and two oxo ligands at the 'active site'. The fact that there was a significant induction period prior to epoxidation using γ-$[SiW_{10}O_{34}(H_2O)_2]^{4-}$ with H_2O_2 and the disappearance of this induction period by pretreatment with the oxidant indicated that γ-$[SiW_{10}O_{34}(H_2O)_2]^{4-}$ was in fact the catalyst precursor. Hammett correlations indicate the formation of a very electrophilic oxidant, and the trans selectivity observed in the epoxidation of 3-methyl-1-cyclohexene was explained by strong steric control at the active site. NMR and MS spectroscopic measurements showed that addition of H_2O_2 to γ-$[SiW_{10}O_{34}(H_2O)_2]^{4-}$ yielded an inactive compound that only later formed the active species.

Others have been also been intrigued by this research. In a computational study, it was concluded that the catalyst precursor is better formulated as having four hydroxy terminal ligands rather than two aqua and two oxo ligands [60]. Further computational studies by the same team suggested that the reactivity of

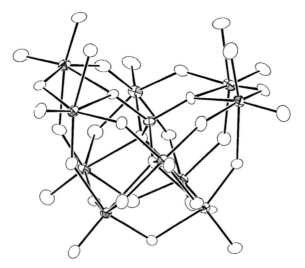

Figure 9.4 Ortep representation of the γ-$[SiW_{10}O_{34}(H_2O)_2]^{4-}$ polyoxometalate.

γ-$[SiW_{10}O_{34}(H_2O)_2]^{4-}$, formulated by them as γ-$[SiW_{10}O_{32}H_4]^{4-}$, could be explained by an active WOOH end-on hydroperoxo species rather than a side-on s = active common usually invoked in tungsten-based catalysts for the activation of H_2O_2 [61]. Others based on titration/^{183}W NMR experiments together with DFT calculations also demonstrate the importance of acidity for reactivity but conclude that the active precursor should be defined as originally presented by Mizuno and coworkers [62]. It should finally be noted that it has been reported that substitution of phenyl phosphonate moieties at the defect site of γ-$[SiW_{10}O_{34}(H_2O)_2]^{4-}$ leads to a catalyst of similar reactivity but greater stability [63].

There have been some other notable developments related to the oxidative catalytic behavior of γ-$[SiW_{10}O_{34}(H_2O)_2]^{4-}$-based catalysts. One observation has been that partial protonation of γ-$[SiW_{10}O_{34}(H_2O)_2]^{4-}$ in an organic solvent leads to dehydration of γ-$[SiW_{10}O_{34}(H_2O)_2]^{4-}$ and formation of an S-shaped disilicoicostungstate that showed good activity for the Baeyer-Villiger oxidation of cycloalkanes to lactones, although it appears quite unclear what the actual catalyst is although it appears to be quite acidic in nature [64]. Substitution of vanadium into the 'defect' sites of γ-$[SiW_{10}O_{34}(H_2O)_2]^{4-}$ to yield a γ-$[H_2SiV_2W_{10}O_{40}]^{4-}$ compound has also led to interesting results [65]. This compound also led to a highly electrophilic oxidant and in addition catalyzed chemoselective epoxidation with H_2O_2. Thus, cis-2-octene was 32 times more reactive than its geometric isomer, trans-2-octene. Terminal alkenes were much more reactive than the substituted ones, for example, 1,4-hexadiene yielded mostly 1,2-epoxy-4-hexene. Furthermore, high diastereoselectivity was also observed; epoxidation of 3-methyl-1-cyclohexene and 2-cyclohexen-1-ol led to the preferred formation of the trans oxirane, showing selective anti addition to the double bond. The chemoselectivity and diasteroselectivity were attributed to steric constraints at the active site, and reactivity is explained via formation of a hydroperoxo and perhaps a μ-$η_2$-$η_2$ peroxo moiety. Similar substitution of titanium into γ-$[SiW_{10}O_{34}(H_2O)_2]^{4-}$ yielded a μ-oxo-bridged dimeric [{γ-$H_2SiTi_2W_{10}O_{38}$}$_2$(μ-O)$_2$]$^{8-}$ compound that showed similar but much diminished reactivity to that observed with γ-$[H_2SiV_2W_{10}O_{40}]^{4-}$ [66].

Another issue is whether there are polyoxometalate structures that are intrinsically stable toward the hydrolytic conditions of aqueous hydrogen peroxide. We observed that, in general, larger polyoxometalates, specifically polyoxotungstates of various 'sandwich'-type structures were solvolytically stable toward hydrogen peroxide. Unfortunately, often the substituting transition metal catalyzes the fast decomposition of hydrogen peroxide leading to low reaction yields and nonselective reactions of little synthetic value. However, there is now a considerable body of research into several types of transition metal-substituted polyoxometalates that are synthetically useful. Various iron-containing polyoxometalates of 'sandwich'-type structures have been investigated by the Hill group and found to have good activity for alkene oxidation with only moderate nonproductive decomposition of hydrogen peroxide [67]. A relatively new class of transition metal-substituted compounds, polyfluorooxometalates, $[TM(H_2O)H_2W_{17}O_{55}F_6]^{q-}$, which have a quasi Wells-Dawson structure (see Figure 9.3c) and where there is partial replacement of oxygen by fluorine proved to be very active and stable oxidation catalysts, which can be monitored by ^{19}F NMR, for epoxidation of alkenes and allylic alcohols with hydrogen peroxide [68]. The nickel-substituted compound was the most active of the series studied. Previously, our group observed that a far more catalytically active class of compounds that were also stable in oxidation reactions using aqueous hydrogen peroxides were the $\{[(WZnTM_2(H_2O)_2][(ZnW_9O_{34})_2]\}^{q-}$ 'sandwich'-type polyoxometalates. Originally we observed that, among this class of compounds, the manganese and analogous rhodium derivatives dissolved in the organic phase were uniquely active when reactions were carried out in biphasic systems, preferably 1,2-dichloroethane–water [69]. At lower temperatures, highly selective epoxidation could be carried out even with cyclohexene, which is normally highly susceptible to allylic oxidation. Nonproductive decomposition of hydrogen peroxide at low temperatures was minimal but increased with temperature and was also dependent on the reactivity of the substrate. The rhodium compound was preferable in terms of minimization of hydrogen peroxide decomposition, but of course it is more expensive. Up to tens of thousands of turnovers could be attained for reactive hydrocarbon substrates [70]. The synthetic utility of the $\{[(WZnMn(II)_2(H_2O)_2][(ZnW_9O_{34})_2]\}^{12-}$ polyoxometalate as catalyst for hydrogen peroxide activation was then extended to additional substrate classes having various functional units [71]. Thus, allylic primary alcohols were oxidized selectively to the corresponding epoxides in high yields and >90% selectivity. Allylic secondary alcohols were oxidized to a mixture of α- and β-unsaturated ketones (the major product) and epoxides (the minor product). Secondary alcohols were oxidized to ketones, and sulfides to a mixture of sulfoxides and sulfones. The reactivity of simple alkenes is inordinately affected by the steric bulk of the substrate. For example, the general reactivity scale for the epoxidation of alkenes indicates a strong correlation between the rate of the epoxidation and the nucleophilicity of the alkene, which is in turn correlated with the degree of substitution at the double bond. Thus, it was expected and observed that 2,3-dimethyl-2-butene would be more reactive than 2-methyl-2-heptene; however, other more bulky substrates such as 1-methylcyclohexene were found to be less reactive than cyclohexene, in contrast to what would normally be expected. Furthermore, α-pinene did not react at all. This led, for example, to unusual reaction selectivity in limonene epoxidation, where both epoxides were formed in equal amounts in contrast to the usual situation where epoxidation at the endo double bond is highly preferred [72]. In these catalytic systems high turnover conditions can be easily achieved and high conversions are attained for reactive substrates, but sometimes for less reactive substrates such as terminal alkenes conversions and yields are low. The conversion can be increased by continuous or semi-continuous addition of hydrogen peroxide and removal of spent aqueous phases.

After the original studies on the activity of the $\{[(WZnMn(II)_2(H_2O)_2][(ZnW_9O_{34})_2]\}^{12-}$ polyoxometalates in the mid 1990s, recent industrial interest revived research in this area. Originally, the large size of the 'sandwich'-type structure was thought to be disadvantageous for the large scale and practical applications because, even at low molar percent loads of catalyst, relatively large amounts of polyoxometalate would be required. However, the large molecular size (high molecular weight) have an under-appreciated advantage in that they significantly simplify catalyst recovery from homogeneous solutions via easily applied nano-filtration techniques [73]. This reverses some of the previous thinking in this area. The newly initiated reinvestigation of the use of 'sandwich'-type polyoxometalates, $\{[(WZnTM_2(H_2O)_2][(ZnW_9O_{34})_2]\}^{q-}$, showed that for a significant series of transition metals, most notably the zinc analog, these were exceptionally active catalysts for epoxidation of allylic alcohols using toluene or ethyl acetate as solvent [74]. The identity of the transition metal did not affect the reactivity, chemoselectivity, or stereoselectivity of the allylic alcohol epoxidation by hydrogen peroxide. These selectivity features support a conclusion that a tungsten peroxo complex rather than a high-valent transition-metal-oxo species operates as the key intermediate in the sandwich-type POM-catalyzed epoxidations. The marked enhancement of reactivity and selectivity of allylic alcohols versus simple alkenes was explained by a template formation in which the allylic alcohol is coordinated through metal-alcoholate bonding, and the hydrogen-peroxide oxygen source is activated in the form of a peroxo tungsten complex. 1,3-Allylic strain expresses a high preference for the formation of the *threo* epoxy alcohol, whereas in substrates with 1,2-allylic strain the *erythro* diastereomer was favored. In contrast to acyclic allylic alcohols the $\{[(WZnTM_2(H_2O)_2][(ZnW_9O_{34})_2]\}^{q-}$-catalyzed oxidation of the cyclic allylic alcohols by hydrogen peroxide yielded significant amounts of enone rather than epoxides. A comprehensive comparison of $\{[(WZn_3(H_2O)_2][(ZnW_9O_{34})_2]\}^{12-}$ with other tungsten-based catalysts taking into account activity per tungsten atom, reaction selectivity, time for formation of the active species, and recycle showed appreciable advantages of $\{[(WZn_3(H_2O)_2][(ZnW_9O_{34})_2]\}^{12-}$ [75].

In the present section we have highlighted research that has been carried out using polyoxometalates as catalysts for oxidation with peroxygen compounds. Not all of the synthetic applications have been noted, but those missing have been previously reviewed [2]. It is important to stress that from a synthetic point of view various substrates with varying functional groups can be effectively transformed to desired products. In addition, interesting reaction selectivity can be obtained in certain cases. In this sense polyoxometalates are one class of compounds among others that may be considered for such transformations. In general, the often-simple preparations of catalytically significant polyoxometalates along with conceivable recovery from solution by nano-filtration present a conceptual advantage in the use of polyoxometalates. From a mechanistic point of view, the wide range of properties available in the various classes of polyoxometalate compounds allows one to express reactivity in a number of ways ranging from nucleophilic–electrophilic reactions between peroxo or hydroperoxy intermediates and organic substrates to radical and radical chain reactions via alkoxy or hydroxy radicals formed by homolytic cleavage of peroxygen compounds by polyoxometalates.

9.5
Oxidation with Molecular Oxygen

The basic ecological and economic advantage and impetus for the use of oxygen from air as primary oxidant for catalytic oxidative transformations are eminently clear. Yet, the chemical properties of ground state molecular oxygen limit its usefulness as an oxidant for wide-ranging synthetic applications. The limiting properties are the radical nature of molecular oxygen, the strong oxygen-oxygen bond, and the fact that one-electron reduction of oxygen is generally not thermodynamically favored ($\Delta G > 0$). The ground state properties of molecular oxygen lead to the situation that under typical liquid phase conditions, reactions proceed by the well-known autooxidation pathways, Scheme 9.7.

(a) $RH + M^{n+} \longrightarrow RH^{+\bullet} + M^{(n-1)+} \longrightarrow R^\bullet + M^{(n-1)+} + H^+$ electron and proton transfer
or
(b) $RH + M^{n+} \longrightarrow R^\bullet + M^{(n-1)+} + H^+$ hydrogen abstraction

(c) $R^\bullet + O_2 \longrightarrow ROO^\bullet$
propagation
(d) $ROO^\bullet + RH \longrightarrow ROOH + R^\bullet$

(e) $ROOH + M^{(n-1)+} \longrightarrow RO^\bullet + M^{n+} + OH^-$
hydroperoxide decomposition
(f) $ROOH + M^{n+} \longrightarrow ROO^\bullet + M^{(n-1)+} + H^+$

Scheme 9.7 Metal-catalyzed autooxidation pathways.

Metal-based catalysts may affect such pathways in various, but most notably have an influence in initiating the radical chain propagation and decomposing intermediate alkylhydroperoxide species to alkoxy and peraalkoxy radicals as discussed in Section 9.4 above. It is also very instructive to note that in nature common monooxygenase enzymes such as cytochrome P-450 and methane monooxygenase use reducing agents to activate molecular oxygen (Scheme 9.8).

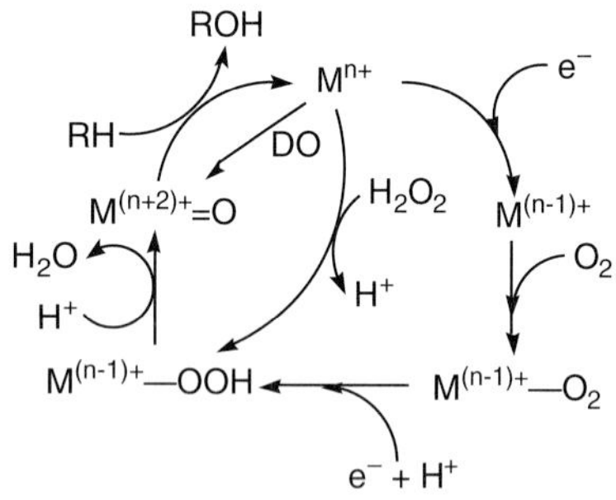

Scheme 9.8 Oxidation under reducing conditions – monooxygenase-type reactions.

The scheme depicted is not presented as an exact mechanistic representation, but rather to illustrate several basic points. First, one may observe that oxygen is a unique oxidant compared to other oxygen donors – the oxygen donors being in principle reduced relative to molecular oxygen. In fact, even the active oxidizing intermediate in metal-catalyzed autooxidation pathways is the reduced peroxo intermediate (Scheme 9.7, reaction d). In addition, only one oxygen atom of molecular oxygen in both schemes is incorporated in the product; the other atom is reduced coupled with formation of water. Second, the requirement of a reducing agent for activation negates the basic ecological and economic impetus for the use of molecular oxygen since the reducing agent becomes in fact a limiting or sacrificial reagent. These observations lead to the conclusion that newer and preferred methods of molecular oxygen activation should employ a superbiotic or abiotic approach. Polyoxometalates have played a part in such approaches to oxidative transformations.

Polyoxometalates have been investigated as catalysts for aerobic oxidation reactions that are based on various mechanistic motifs. As indicated above, one way to utilize molecular oxygen is to oxidize a hydrocarbon in the presence of a reducing agent in a reaction that proceeds by an autoxidation-type mechanism with the appropriate radical species as intermediates. In the most synthetically interesting case, a polyoxometalate may initiate a radical chain reaction between oxygen and an aldehyde as the reducing and sacrificial reagent. Aldehydes are practical sacrificial reagents because the relatively low carbon-hydrogen homolytic bond energy allows easy formation of the initial intermediates, the acylperoxo radical or an acylhydroperoxide (peracid). Also some aldehydes such as isobutyraldehyde are readily available and inexpensive. As for all peroxygen species, these active intermediates may then be used for the epoxidation of alkenes, the oxidation of alkanes to ketones and alcohols, and for the Baeyer-Villiger oxidation of ketones to esters. This has been demonstrated using both vanadium- ($H_5PV_2Mo_{10}O_{40}$) and cobalt- ($[Co(II)PW_{11}O_{39}]^{5-}$) containing Keggin-type polyoxometalates as catalysts, with isobutyraldehyde as the preferred acylperoxo/peracid precursor, with substrates such as alkenes and sulfides being most investigated [76]. Significant yields at very high selectivities were obtained in most examples. In this context, various polyoxometalates and O_2 have also been used to purposely oxidize aldehydes to the corresponding carboxylic acids for both synthetic applications and to eliminate pollutants in the air such as formaldehyde [77].

Polyoxometalates with the required redox properties can also be used in a straightforward manner as autoxidation catalysts. In this way the trisubstituted Keggin compound, $[M_3(H_2O)_3PW_9O_{37}]^{6-}$ (M = Fe(III) and Cr(III)) and $[Fe_2M(H_2O)_3PW_9O_{37}]^{7-}$ (M = Ni(II), Co(II), Mn(II) and Zn(II)) were used in the autoxidation of alkanes such as propane and isobutane to acetone and t-butyl alcohol [78]. Later $[Fe_2Ni(OAc)_3PW_9O_{37}]^{10-}$ and others were prepared and used to oxidize alkanes such as adamantane, cyclohexane, ethylbenzene and n-decane, where the reaction products (alcohol and ketone) and regioselectivities were typical for metal-catalyzed autoxidations [79]. Sulfides [78d, 80], arenes [81], and alkenes [82] have also been oxidized in this manner. An interesting recent application of such an autoxidation is the oxidation of 3,5-di-*tert*-catechol by iron- and/or vanadium-substituted polyoxometalates [83]. In this reaction there is a very high turnover number, >100 000. In this case the polyoxometalates are excellent mimics of catechol dioxygenase. Further research by the same group appeared to indicate that, contrary to the original hypothesis that the vanadium-substituted polyoxometalate was the active catalyst, the results could be better explained via the *in situ* formation of a previously characterized vanadyl semiquinone catechol dimer complex [84]. Another use of a polyoxometalate, mainly $[PCo(II)Mo_{11}O_{39}]^{5-}$, was to catalyze autooxidation of cumene to the hydroperoxo/peroxo intermediate by the Co(II) component of the polyoxometalate followed by oxygen transfer to an alkene such as 1-octene to yield epoxide, catalyzed by the molybdate component [85]. With the analogous tungsten polyoxometalate there was negligible oxygen transfer. In these reactions the cumene acts as a sacrificial reducing agent.

As indicated above, other mechanistic motifs have been utilized in aerobic oxidation catalyzed by polyoxometalates. Perhaps the oldest and possibly most developed of all the mechanistic motifs considered is an abiotic approach whereby the polyoxometalate activates the reaction substrate, both organic and inorganic, rather than the oxygen that serves as the ultimate oxidant. In such catalytic reactions the polyoxometalate undergoes a redox-type interaction involving electron transfer with the reaction substrate leading to its oxidation and concomitant reduction of the polyoxometalate. Generally, the initial electron transfer is rate determining, but exceptions are known. Molecular oxygen is used to reoxidize the reduced polyoxometalate. The mechanistic approach is summarized in Scheme 9.9.

(a) $RH_2 + POM^{n-} \xrightarrow{slow} RH_2^{+\bullet} + POM^{(n+1)-}$

(b) $RH_2^{+\bullet} + POM^{(n+1)-} \xrightarrow{fast} R + POM^{(n+2)-} + 2H^+$

RH_2 = substrate; R = product

(c) $POM^{(n+2)-} + 2H^+ + 1/2\, O_2 \longrightarrow POM^{n-} + H_2O$

Scheme 9.9 Redox-type mechanism for oxidation with polyoxometalates.

The basic requirement for a catalyst for such a reaction is that the oxidation potential be sufficient for oxidation of organic substrates. Yet a too high oxidation potential is also not desirable, because then it will not be possible to reoxidize the polyoxometalate with molecular oxygen. For example, $[Co(III)W_{12}O_{40}]^{5-}$ has a high oxidation potential, enabling oxidation of substrates such as xylene, but the resulting $[Co(II)W_{12}O_{40}]^{6-}$ is not oxidized by molecular oxygen and thus can be used only as a stoichiometric oxidant [86]. It turns out that most commonly used catalysts for the reaction sequence described in Scheme 9.9 are the phosphovanadomolybdates, $[PV_xMo_{12-x}O_{40}]^{(3+x)-}$, especially but not exclusively when $x=2$. This compound in its acid form has an oxidation potential of ~ 0.7 V as measured by cyclic voltammetry. The use of $H_5PV_2Mo_{10}O_{40}$ was first described as a co-catalyst in the Wacker reaction [87]. The Wacker reaction oxidation of terminal alkenes is a reaction that epitomizes the mechanistic motif as expressed in Scheme 9.9. The $H_5PV_2Mo_{10}O_{40}$ polyoxometalate acts to reoxidize the palladium species, which in fact in the absence of a cocatalyst is a stoichiometric oxidant of alkenes. The use of $H_5PV_2Mo_{10}O_{40}$ replaces the classic $CuCl_2$ system, which, because of the high chloride concentration, both is corrosive and forms chlorinated side-products. In the 1990s, Grate and coworkers at Catalytica significantly improved the Wacker-type oxidation of ethylene to acetaldehyde [88]. Afterwards, longer-chain alkenes were also oxidized in this way [89]. An interesting extension of the use of $H_5PV_2Mo_{10}O_{40}$ in palladium-catalyzed Wacker reactions has been to add benzoquinone as an additional co-catalyst to reoxidize the primary palladium catalyst; the resulting hydroquinone is in-turn re-oxidized by the polyoxometalate. This catalytic sequence has been used for the palladium-catalyzed oxidation of alkenes [90] and conjugated dienes [91]. In recent years there have been additional interesting reports with co-catalytic systems involving Pd/Pd^{2+} and various phosphovanadomolybdates, $[PV_xMo_{12-x}O_{40}]^{(3+x)-}$. One example is the oxidative coupling reaction of arenes (benzene and derivatives and furan) with acrylate esters to yield cinnamate esters as the major product in moderate yields. Similar reactions with ethylene were less successful but did give styrene and stilbene in low yields [92]. In the absence of an alkene, benzene was oxidatively coupled to biphenyl in the presence of $Pd(OAc)_2[PV_xMo_{12-x}O_{40}]^{(3+x)-}$, although conversions were low [93]. A system with three co-catalysts, namely Pd $(OAc)_2[PV_1Mo_{11}O_{40}]^{4-}/CeCl_3$, catalyzed the aerobic addition of aldehydes to acrylate esters or acrylic acid in the presence of methanol and acetic acid to yield furoates in moderate to high yields [94]. In a typical example, methyl acrylate was reacted with propanal to yield 2-ethyl-4-carboxymethylfuran. Several reactions catalyzed by $Pd/Pd^{2+}/[PV_xMo_{12-x}O_{40}]^{(3+x)-}$ utilizing a gaseous mixture of CO/O_2 have also been reported. The first example was the dicarboxylation of cyclopentene to give a mixture of dimethyl cis-1,2-cyclopentanedicarboxylate and dimethyl cis-1,3-cyclopentanedicarboxylate in moderate yields [95]. The $Pd^{2+}/[PV_xMo_{12-x}O_{40}]^{(3+x)-}/CO/O_2$ catalyst-oxidant combination was then further examined in the hydroxylation of benzene to phenol as major product (~25% yield) and benzoquinone as minor product (~5% yield) [96]. Isotope labeling experiments with $^{18}O_2$ appear to indicate that H_2O_2 that possibly could be formed in the reaction does not appear to be the active species. Further work on substituted benzene derivatives gave similar results; for example, toluene gave a mixture of o-, m-, and p-cresols although the combined yield was low. Interestingly, with biphenyl both hydroxylation and carboxylation reactions were observed, with prominent formation of hydroxybiphenylcarboxylic acids, biphenylcarboxylic acids being the first products formed. Strangely, however, the reaction of benzoic acid yielded mostly phthalic acid in preference to terephthalic acid [97].

$H_5PV_2Mo_{10}O_{40}$ was also used to oxidize gaseous hydrogen bromide to molecular bromine that was utilized *in situ* for the selective bromination of phenol to 4-bromophenol [98]. More recently, $H_5PV_2Mo_{10}O_{40}$ has been used in a similar way with molecular iodine to carry out catalytic quantitative iodination of a wide range of aromatic substrates without formation of any hydrogen iodide as by-product [99]. The aerobic oxidative iodination reaction was also used for the iodoacetoxylation of alkenes. The iodoacetate could be further reacted *in situ* to yield predominantly the cis-acetate, which then hydrolyzed to yield the cis-diol [100]. Another early interest in the catalytic chemistry of $H_{3+x}PV_xMo_{12-x}O_{40}$ was in the oxidation of sulfur-containing compounds of interest in purification of industrial waste and natural gas. Oxidation included that of oxidation of H_2S to elemental sulfur, sulfur dioxide to sulfur trioxide (sulfuric acid), mercaptans to disulfides, and sulfides to sulfoxides and sulfones [101]. Hill and his group have continued the investigation of the oxidation chemistry of sulfur compounds and have shown that, for H_2S oxidation, catalysts of low oxidation potential are sufficient, because the oxidation of H_2S to elemental sulfur is thermodynamically favored ($\Delta G < 0$) [102].

In our opinion, a significant challenge for the use of the mechanistic motif indicated in Scheme 9.9 is the use of $[PV_2Mo_{10}O_{40}]^{5-}$ for direct oxidation of hydrocarbon substrates coupled with the suppression of autooxidation pathways. Perhaps an early use of $[PV_2Mo_{10}O_{40}]^{5-}$ in this context was the reaction described by Brégeault and coworkers where $H_5PV_2Mo_{10}O_{40}$ was used in combination with dioxygen to oxidatively cleave vicinal diols [103] and ketones [104]. For example, 1-phenyl-2-propanone can be cleaved to benzaldehyde (benzoic acid) and acetic acid, ostensibly through the α,β-diketone intermediate, 1-phenyl-1,2-propane dione. Similarly, cycloalkanones can be cleaved to keto-acids and di-acids. In general, the conversions and selectivities are very high. Both vanadium centers and acidic sites appeared to be a requisite for the reaction. It would be interesting to carry out the oxidative cleavage of diols also under nonacidic conditions as a possible pathway to the formation of a chiral pool from natural carbohydrate sources. In this context, nearly neutral forms of iodomolybdates, $[IMo_6O_{24}]^{5-}$, have been found to show some activity for aerobic carbon-carbon bond cleavage reactions of diols with phenyl substituents, but unfortunately aliphatic diols are less reactive [105]. Just recently, we have extended the use of $[PV_2Mo_{10}O_{40}]^{5-}$ for the oxidation cleavage of primary aliphatic alcohols [106]. Thus, instead of typical oxidation via C–H bond activation, $[PV_2Mo_{10}O_{40}]^{5-}$ reacted with primary alcohols to yield the C–C bond cleavage products. In this way, 1-butanol reacted to give propanal and formaldehyde through a reaction mechanism involving an electron transfer (from the alcohol to $[PV_2Mo_{10}O_{40}]^{5-}$) and oxygen transfer (from $[PV_2Mo_{10}O_{40}]^{5-}$ to the alcohol). The aldehydes formed apparently reacted immediately with excess primary alcohol to yield the hemiacetals; these were oxidized to the corresponding carboxylic acid esters (butylformate and butylpropionate), which were the isolated products from the reaction. In the late 1980s to early 1990s the $[PV_2Mo_{10}O_{40}]^{5-}$ polyoxometalate was shown to be active in a series of oxidative dehydrogenation reactions such as the oxydehydrogenation of cyclic dienes to the corresponding aromatic derivatives [107] and the selective oxydehydrogenation of alcohol compounds to aldehydes with no over-oxidation to the carboxylic acids [108]. Significantly, autoxidation of the aldehyde to the carboxylic acid was strongly inhibited, in fact especially at higher concentrations (0.1–1 mol%), $[PV_2Mo_{10}O_{40}]^{5-}$ can be considered an excellent autoxidation inhibitor. Similarly to alcohol dehydrogenation to aldehydes, amines may be dehydrogenated to intermediate and unstable imines [78]. In the presence of water, aldehyde is formed, which may immediately undergo further reaction with the initial amine to yield a Schiff base. Since the Schiff base is formed under equilibrium conditions, aldehydes are eventually the sole products. Under the careful exclusion of water, the intermediate imine was efficiently dehydrogenated to the corresponding nitrile. It should be noted that several ruthenium- and osmium-substituted polyoxometalates also catalyzed the oxidation of benzylic alcohols to their benzaldehyde derivatives; however, there is no certainty that these reactions proceed by the same mechanism [109]. During this period, the oxydehydrogenation of activated phenols to quinones was also demonstrated. In this way, oxidation of activated phenols in alcohol solvents yielded only oxidative dimerization products, diphenoquinones. Unfortunately, under these mild conditions, the less reactive phenols did not react. It was observed that there was a clear correlation of the reaction rate with the oxidation potential of the phenol, which indicated that an electron transfer step was rate determining. This type of electron transfer oxidation was further investigated in a careful mechanistic investigation with similar catalysts [110]. An interesting extension of this work is the oxidation of 2-methyl-1-naphthol to 2-methyl-1,4-naphthaquinone (Vitamin K$_3$, menadione) in fairly high selectivity (about 83% at atmospheric O$_2$) [111]. This work could lead to a new environmentally favorable process to replace the stoichiometric CrO$_3$ oxidation of 2-methylnaphthalene used today. Also, the finding that $[PV_2Mo_{10}O_{40}]^{5-}$ could catalyze the oxydehydrogenation of hydroxylamine to nitrosium cations led to an effective and general method for aerobic selective oxidation of alcohols to aldehydes or ketones by the use of nitroxide radicals and $[PV_2Mo_{10}O_{40}]^{5-}$ as cocatalysts. Typically, quantitative yields were obtained for the oxidation of aliphatic, allylic, and benzylic alcohols to the corresponding ketones or aldehydes with very high selectivity [112] Based mostly on kinetic evidence and some spectroscopic support, a reaction scheme was formulated (Scheme 9.10). The results indicated that the polyoxometalate oxidizes the nitroxyl radical to the nitrosium cation. The latter oxidizes the alcohol to the ketone/aldehyde and is reduced to the hydroxylamine, which is then reoxidized by $[PV_2Mo_{10}O_{40}]^{5-}$.

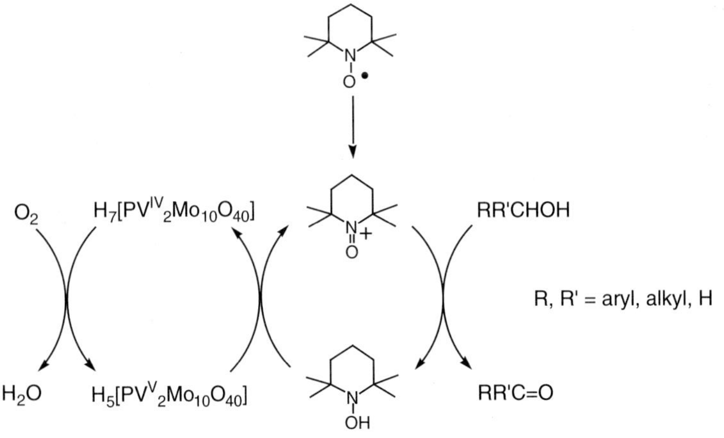

Scheme 9.10 Aerobic oxidation of alcohols with TEMPO and H$_5$PV$_2$Mo$_{10}$O$_{40}$.

Another very important example of the use of polyoxometalates for oxydehydrogenation is the technology proposed by Hill and Weinstock for the delignification of wood pulp [113]. In the first step, lignin is oxidized selectively in the presence of cellulose, and the polyoxometalate is reduced. The now oxidized and water-soluble lignin component is separated from the whitened pulp and mineralized at high temperature with oxygen to CO$_2$ and H$_2$O. During the mineralization process, the polyoxometalate is re-oxidized by molecular oxygen (air) and can be used for an additional process cycle.

A mechanistic exploration of $[PV_2Mo_{10}O_{40}]^{5-}$-catalyzed oxydehydrogenations utilizing kinetic and spectroscopic tools was also carried out [114]. The room temperature oxydehydrogenation of α-terpinene to p-cymene was chosen as a model reaction. Dehydrogenation was explained by a series of fast electron and proton transfers leading to the oxidized or dehydrogenated product and the reduced polyoxometalate. Interestingly, there were clear indications that the re-oxidation of the reduced polyoxometalate by molecular oxygen went through an inner-sphere mechanism, presumably via formation of a μ-peroxo intermediate. Subsequent research has given conflicting but still inconclusive evidence that the re-oxidation might occur via an outer-sphere mechanism [115].

In the reactions reviewed in the paragraphs immediately above, the oxidation of the hydrocarbon substrate by the polyoxometalate catalyst is purely a dehydrogenation reaction and no oxygenation of the substrate was observed, as is implicit in Scheme 9.9. An important extension of this mechanistic theme would be to couple electron transfer from the hydrocarbon to the polyoxometalate with oxygen transfer from the polyoxometalate to the reduced hydrocarbon substrate. This type of reactivity is known in an important area of gas phase heterogeneous oxidation reactions, whereby a metal oxide compound at high temperature (about 450 °C) transfers oxygen from the lattice of the oxide to a hydrocarbon substrate. Mars and Van Krevelen originally proposed this type of mechanism, and the reaction is important in several industrial applications such as the oxidation of propene to acrolein and of butane to maleic anhydride. Recently, it was shown by us that with the $PV_2Mo_{10}O_{40}^{5-}$ catalyst, electron transfer–oxygenation reactions were possible for oxidation of hydrocarbons at moderate temperatures (<80 °C) [116]. Substrates oxygenated in this manner include polycyclic aromatic compounds and alkyl aromatic compounds. Thus, anthracene was oxidized to anthraquinone, and active secondary alkyl arenes were oxidized to ketones. Use of $^{18}O_2$ and isotopically labeled polyoxometalates, as well as carrying out stoichiometric reactions under anaerobic conditions, provided strong evidence for a homogeneous Mars–van Krevelen-type mechanism and also provided evidence against autooxidation and oxidative nucleophilic substitution as alternative possibilities (Scheme 9.11).

(a) $RH_2 + H_5[PV_2Mo_{10}O_{40}]^{5-} \xrightarrow{slow} RH_2^{+\bullet} + H_5[PV_2Mo_{10}O_{40}]^{6-}$

for RH = anthracene

(b) $RH_2^{+\bullet} + H_5[PV_2Mo_{10}O_{40}]^{6-} \longrightarrow R=O + H_7[PV_2Mo_{10}O_{39}]^{5-}$

for RH = xanthene

(b') $RH_2^{+\bullet} + H_5[PV_2Mo_{10}O_{40}]^{6-} \longrightarrow R^+ + H_7[PV_2Mo_{10}O_{40}]^{7-}$

$R^+ + H_7[PV_2Mo_{10}O_{40}]^{7-} \longrightarrow R=O + H_7[PV_2Mo_{10}O_{39}]^{5-}$

(c) $H_7[PV_2Mo_{10}O_{39}]^{5-} + O_2 \longrightarrow H_5[PV_2Mo_{10}O_{40}]^{5-} + H_2O$

Scheme 9.11 Mars–van Krevelen-type oxygenation of anthracene and xanthene.

Evidence for the activation of the hydrocarbon by electron transfer was inferred from the excellent correlation of the reaction rate with the oxidation potential of the substrate. For anthracene the intermediate cation radical was observed by ESR spectroscopy, whereas for xanthene the cation radical quickly underwent additional electron and proton transfer yielding a benzylic cation species observed by ^{1}H NMR. Comparison of the oxidation potentials of the organic substrates (1.35–1.50 V) with that of the catalyst (about 0.7 V) and analysis of the reaction rates led to the conclusion that the electron transfer step from the hydrocarbon to the polyoxometalate occurs through an outer-sphere mechanism. The reactions are thermodynamically feasible because of the high negative charge of the polyoxometalate catalyst. As shown by Marcus theory, this introduces a large electrostatic work function and lowers the free energy of the reaction. In another recent application of $H_5PV_2Mo_{10}O_{40}$ it was shown that in a reaction with neat nitrobenzene there was selective formation of 2-nitrophenol in the presence of O_2 [117]. Evidence was provided from ESR experiments, use of labeled $^{18}O_2$ and $H_2^{18}O$, and competitive kinetic isotope experiments, that there was formation of an $H_5PV_2Mo_{10}O_{40}$-nitrobenzene complex leading to C–H bond activation at the *ortho* position followed by reaction with O_2.

The propensity of polyoxometalates in general and that of $H_5PV_2Mo_{10}O_{40}$ in particular to act as redox catalysts can be further extended to facilitate catalytic cycles with oxidants such as O_2 that were previously not possible. This is possible through the preparation of metallorganic-polyoxometalate hybrid catalysts wherein the polyoxometalate can modulate the oxidation state or reactivity of the metallorganic 'partner' in the catalytic compound. Thus, it was first shown that covalent attachment of an $SiW_{11}O_{39}^{8-}$ moiety to a metallosalen compound changed the oxidation state of the metallosalen moiety, for example, manganese(III) to manganese(IV) [118]. Similar effects were observed on attaching hexamolybdate groups via phenyl spacers to a phenanthroline ligand [119]. This hybrid concept was then realized in a catalytic cycle. Thus, it had been known from work by Shilov and coworkers and then others that methane can be oxidized to methanol by platinum(II) catalysts using oxidants such as platinum(IV) and then later on SO_3. In our hands, preparation of a platinum(II)methylpyrimidinium-$H_5PV_2Mo_{10}O_{40}$ catalyst allowed the *aerobic* oxidation of methane with significant turnover. Equal amounts of methanol and acetaldehyde were formed, Scheme 9.12 [120].

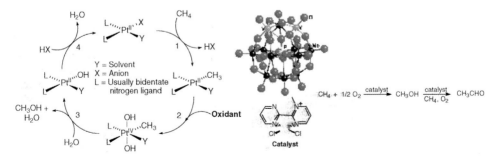

Scheme 9.12 Consensus catalytic scheme for the oxidation of methane and the catalyst and reactions observed in the aerobic oxidation of methane in water.

Since polyoxometalates are polyanions, they are able to stabilize colloidal or nanoparticles via electrostatic repulsion of the particles that prevents their aggregation. In this context, there is interest to see if the catalytic properties of such nanoparticles will be affected by the presence of such polyoxometalates. In one example it was shown that there is a synergistic effect in the oxidation of carbon monoxide by carbon-supported gold nanoparticles in the presence of polyoxometalates in solution [121]. Silver and ruthenium nanoparticles stabilized by $H_5PV_2Mo_{10}O_{40}$ supported on alumina catalyzed the aerobic epoxidation of alkenes at low conversions, presumably via prevention of autooxidation pathways, although at higher conversions autooxidation reduced the epoxide yield [122]. Although similar platinum particles stabilized by $H_5PV_2Mo_{10}O_{40}$ were inactive for alkene epoxidation, they catalyzed the aerobic oxidation of secondary alcohols to ketones under acidic conditions [123]. Finally, polyoxometalates that were covalently modified with alkylthiol groups were used to stabilize palladium nanoparticles. Interestingly, such particles lead to oxydehydrogenation of vinylcylohexene and vinylcyclohexane to styrene via activation of the tertiary carbon-hydrogen bond. Palladium nanoparticles not stabilized in this way led to predominant formation of ethylbenzene via an isomerization-dehydrogenation pathway [124].

Beyond the two mechanistic themes presented above, that is, autooxidation and redox-type reactions involving electron transfer, a ruthenium-substituted 'sandwich'-type polyoxometalate was shown to be a catalyst for oxidation by a 'dioxygenase'-type mechanism as outlined in Scheme 9.13 [125].

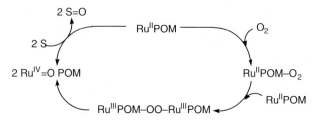

Scheme 9.13 Oxygen activation by an Ru-polyoxometalate: A dioxygenase mechanism.

A range of supporting evidence for such a mechanism in the hydroxylation of adamantane and for alkene epoxidation was obtained by providing evidence against autooxidation reactions (radical traps, isotope effects, and other reaction probes), and by substantiating the 'dioxygenase' mechanism by confirming the reaction stoichiometry and isolating and characterizing a ruthenium-oxo intermediate. The intermediate was also shown to be viable for oxygen transfer in a quantitative and stereoselective manner. The catalytic cycle was also supported by kinetic data. Others have recently observed different results, thus leading to different mechanistic conclusions; this may be a result of research being carried out with different catalysts [126].

As can be concluded from the details presented in this section of the review, the variety of properties available in polyoxometalate compounds enables them to be used for aerobic oxidation that may proceed by a number of mechanistic schemes. In some cases, practical synthetic techniques are already available, especially for aerobic alcohol oxidation and other oxidative dehydrogenation reactions. In other cases, it is hoped that mechanistic possibilities that have been put forward will also lead to new synthetic capabilities in the future.

9.6
Heterogenization of Homogeneous Reactions – Solid-Liquid, Liquid-Liquid, and Alternative Reaction Systems

Beyond questions of catalytic activation of oxidants and/or substrates by catalysts in general and by polyoxometalates in particular to achieve oxidation, an important part of catalysis research is connected with questions of catalyst recovery and recycling. In general one can distinguish between two broad approaches. The first basic approach is to immobilize a catalyst with proven catalytic properties onto a solid support, leading to catalytic system that may be filtered and reused. Such approaches include concepts such as simple use of catalysts as insoluble bulk material, impregnation of a catalyst onto a solid and usually inert matrix, attachment through covalent or ionic bonds of a catalyst to a support, inclusion of a catalyst in a membrane or other porous material, and several others. The second basic approach is to use biphasic liquid-liquid systems, such that at separation temperatures, which are usually ambient, the catalyst and product phases may be separated by a suitable phase separation process, the catalyst phase is reused, and the product is worked up in the usual manner. Numerous biphasic media have been discussed in the literature, including using catalysts in aqueous, fluorous, and ionic liquids, supercritical fluids, and other liquid phases. Some research in this general area of catalyst recovery has also been carried out using polyoxometalates as catalysts, with emphasis naturally being placed on reactions with oxygen or hydrogen peroxide as the most attractive oxidants for large-scale applications.

9.6.1
Solid-Liquid Reactions

The first application of liquid phase oxidation involving heterogenization of the homogeneous catalyst was impregnation onto a solid support. Impregnation onto various supports has been reported. Thus, impregnating phosphovanadomolybdate catalysts, $[PV_xMo_{12-x}O_{40}]^{(3+x)-}$ ($x = 2, 3, 4$), onto active carbon proved to be uniquely active for aerobic oxidation. In this way, first $[PV_2Mo_{10}O_{40}]^{5-}$ and then $[PV_6Mo_6O_{40}]^{9-}$ on carbon were used to catalyze oxidation of alcohols, amines, and phenols [78, 127]. Recently, a ruthenium-containing polyoxometalate has also been used for alcohol oxidation [128]. Toluene is a good solvent for many reactions and does not lead to measurable leaching. On the other hand, polar solvents tend to dissolve the catalysts into solution. Similarly, $[PV_2Mo_{10}O_{40}]^{5-}$ on similar supports, such as carbon or textile fibers, was found to be active for oxidation of various odorous volatile organics such as acetaldehdye, 1-propanethiol, and thiolane [129]. The impetus of such research was not preparative (synthetic) but rather to deodorize air. The unique and high activity of active carbon versus other supports, such as silica or alumina, led to the suggestion that the support may be actively involved in the catalysis. A subsequent study led to the idea that quinones, likely formed on the active carbon surface through the presence of the polyoxometalate and oxygen, might play a role as an intermediate oxidant [130]. Thus, a catalytic cycle may be considered whereby a surface quinone oxidizes the alcohol to the aldehyde and is reduced to a hydroquinone, which is reoxidized in the presence of the catalyst and molecular oxygen.

Silica in various forms has also been reported to be a useful support. For example, an iron-substituted polyoxometalate supported on cationic silica was also found to be active for oxidation of sulfides and aldehydes [131]. The γ-$[H_2SiV_2W_{10}O_{40}]^{4-}$ polyoxometalate with an N-octyldihydroimidazolium cation supported on silica showed retention of the catalytic properties of γ-$[H_2SiV_2W_{10}O_{40}]^{4-}$, that is, chemoselective and diastereoselective epoxidation of alkenes with H_2O_2 in solution, with effective catalyst recycle and minimal leaching [132]. Polyoxometalates were supported on mesoporous silicates or modified by silanization with amino groups [133]. Thus, $H_5PV_2Mo_{10}O_{40}$, supported on a mesoporous molecular sieve both by adsorption to MCM-41 and by electrostatic binding to MCM-41 modified with amino groups, was active in the aerobic oxidation of alkenes and alkanes in the presence of isobutyraldehyde as sacrificial reagent. Cyclohexane was oxidized by O_2, and alkenes were epoxidized by H_2O_2 on a supported γ-$[SiW_{10}O_{34}(H_2O)_2]^{4-}$ catalyst. In similar fashion, known catalysts were supported on fluoroapatite [134], hydrotalcites [135], and cross-linked polystyrene beads [136] to catalyze typical transformations described above.

Since heteropoly acids can form complexes with crown ether type complexes [137], an interesting twist, especially useful in oxidation with the acidic $H_5PV_2Mo_{10}O_{40}$, was to use the inexpensive polyethylene glycol as solvent [138]. Upon cooling the reaction mixture, the $H_5PV_2Mo_{10}O_{40}$–polyethylene glycol phase separates from the product. In this way, previously known reactions with $H_5PV_2Mo_{10}O_{40}$, such as aerobic oxidation of alcohols, dienes and sulfides, and Wacker-type oxidations were demonstrated. Beyond the simple use of polyethylene glycol as a solvent, the attachment of both hydrophilic polyethylene glycol and hydrophobic polypropylene glycol to silica by the sol-gel synthesis leads to solid particles that upon dispersion in organic solvents lead to liquid-like phases (Scheme 9.14). Addition of $H_5PV_2Mo_{10}O_{40}$ leads to what we have termed solvent-anchored supported liquid phase catalysis and reactivity typical for this catalyst [139]. The balance of hydrophilicity-hydrophobicity of the surface is important for tweaking the catalytic activity.

Catalysts useful for reactions with hydrogen peroxide have also been heterogenized on a solid support. Since polyoxometalates are anionic, preparation of silica particles with quaternary ammonium moieties on the surface led to a useful catalytic assembly with $\{[(WZnMn_2(H_2O)_2][(ZnW_9O_{34})_2]\}^{12-}$ as the active species. Importantly, using the sol-gel synthesis for the preparation of silica, the surface hydrophobicity could be controlled by choice of the organosilicate precursors [140]. This control of surface hydrophobicity led to the tuning of the catalytic activity and gave

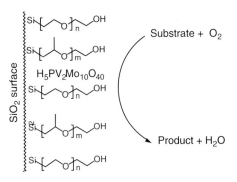

Scheme 9.14 Solvent-anchored supported liquid phase catalysis: Silica–PEG/PPG–$H_5PV_2Mo_{10}O_{40}$.

essentially the same reactivity as that in the previously reported biphasic liquid-liquid reaction medium. No organic solvent was needed. Reactions can be carried out by mixing aqueous hydrogen peroxide and the organic substrate with a solid catalyst particle, which is easily recoverable. Instead of supporting polyoxometalates on materials, notably mesoporous materials, it is possible by judicious choice of counter cations to the polyanionic polyoxometalates to prepare catalytic materials where the polyoxometalate is part of the material backbone. In this, tripodal polycations mixed with $\{[(WZnMn_2(H_2O)_2][(ZnW_9O_{34})_2]\}^{12-}$ yielded insoluble mesoporous materials of moderate surface area that were active heterogeneous catalysts for epoxidation of allylic alcohols and oxidation of secondary alcohols to ketones [141]. The reactivity was similar to that of the homogeneous catalyst. Heterogeneous catalysts based on polyoxometalates could also by prepared in another way. Thus, deposition of Keggin-type acidic polyoxometalates onto micelles of cesium dodecylsulfate in water lead to the formation of 10-nm spherical polyoxometalate particles. Interestingly, these clustered polyoxometalate assemblies showed higher activity than nonclustered polyoxometalates in the aerobic oxidation of sulfides, showing the importance of cooperative effects [142].

9.6.2
Liquid-Liquid Reactions and Reactions in 'Alternative' Media

As noted above, various liquid-liquid and alternative biphasic media can be considered for oxidation using polyoxometalates as catalysts. For example, an ionic liquid such as 1-butyl-3-methyl imidazolium hexafluorophosphate was used as the solvent for the epoxidation of alkenes with hydrogen peroxide and $H_3PW_{12}O_{40}$ as catalyst [143]. A significant enhancement in the turnover frequency ($\times 290$) was reported in the epoxidation of cyclooctene compared to traditional chlorinated hydrocarbon solvents. The active species was the Venturello compound discussed above. The use of supercritical carbon dioxide, sCO_2, as solvent was also demonstrated for the $H_5PV_2Mo_{10}O_{40}$-catalyzed aerobic oxidation of benzylic alcohols and the dehydrogenation of activated arenes [144]. Catalyst recycle was very effective. By preparing fluorinated quaternary ammonium cations, anionic polyoxometalates such as $\{[(WZn_3(H_2O)_2][(ZnW_9O_{34})_2]\}^{12-}$ were dissolved in fluorous reaction media and tested for the oxidation of alkenes and alcohols with aqueous hydrogen peroxide [145].

In oxidation reactions with aqueous hydrogen peroxide, the common methodology is to dissolve the polyoxometalate catalyst in a nonaqueous miscible organic phase by use of a hydrophobic quaternary ammonium cation. Some alternative strategies have also been reported. For example, water in oil microemulsions can be prepared with polyoxometalates using mixtures of cationic and non-ionic surfactants, aqueous hydrogen peroxide, and n-octane [146]. Although these media were effective for alkene epoxidation with the Venturello compound, the reaction activity and selectivity was apparently inferior to that observed in the typical biphasic reaction media. Dendrimers with catalytically active polyoxometalates both at the focal point (core) and periphery were prepared and shown to be very active for the epoxidation of alkenes and the oxidation of sulfides [147]. These catalytic systems represent an interesting interface between homogeneous and heterogeneous catalysis and provide new vistas for catalyst separation, recovery, and recycle.

Although polyoxometalates are commonly synthesized as water-soluble salts of alkali metals, the polyoxometalate-water/organic substrate system was realized only recently after the idea of carrying out reactions in aqueous biphasic media was proposed. Thus, $\{[(WZn_3(H_2O)_2][(ZnW_9O_{34})_2]\}^{12-}$ in water catalyzes the oxidation of alcohols with hydrogen peroxide (Scheme 9.15) [148]. The catalytic system is quite effective for oxidation of secondary and primary alcohols to ketones and carboxylic acids, respectively. An important and key characteristic of this catalytic system is that the catalyst, $Na_{12}[(WZn_3(H_2O)_2][(ZnW_9O_{34})_2]$ does not have to be prepared beforehand. Assembly, in situ, of the polyoxometalate by mixing sodium tungstate, zinc nitrate, and nitric acid in water is sufficient to attain a fully catalytically active system. After completion of the reaction and phase separation, the catalyst-water solution may be reused without loss of activity. In a continuation of this study, other functional groups such as amines and diols were also reacted in such biphasic media [149]. Further, a combination of ammonia and hydrogen peroxide could be reacted in such systems to produce hydroxylamine in situ. The dangerous hydroxylamine, kept at low concentrations in this way, reacted further with ketones and aldehydes to produce the

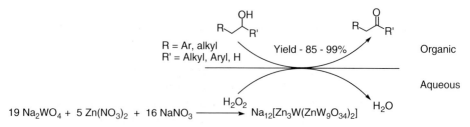

Scheme 9.15 Aqueous biphasic reactions via self-assembly for oxidation of alcohols with hydrogen peroxide.

corresponding oximes, which are useful for the synthesis of amides by the Beckman rearrangement in high yields [150].

Despite the fact that the recovery and recycling of the catalyst-containing aqueous phase is by simple phase separation, which also allows separate purification of the organic product, catalyst compatibility with water and the slow mass transfer of the substrate to the catalytic site have limited its applicability. These above-mentioned reactions were successful because the more hydrophilic functional groups that were oxidized were available to the catalytic site at the interface of the organic and aqueous phases. Thus, although the polyoxometalate solubilized into an organic medium was very active for epoxidation of hydrophobic alkenes, this reaction failed completely in the analogous aqueous biphasic reaction. In order to overcome the obstacle of lack of availability of the hydrophobic substrate to the catalytic site and to allow aqueous biphasic oxidation of hydrophobic alkenes with H_2O_2 catalyzed by polyoxometalates, the idea that alkylated polyethyleneimine could be considered a very primitive enzyme or synzyme having a hydrophobic core and a hydrophilic surface was introduced. Indeed, alkylation of polyethyleneimine led to an amorphous structure, soluble in water, that retained hydrophobic cores that intercalated polyoxometalate catalysts and permitted solubilization of the hydrophobic substrates in the alkylated polyethyleneimine-catalyst construct. This allowed for the efficient aqueous biphasic oxidation of even very hydrophobic substrates such as methyl oleate with hydrogen peroxide [151]. The recovery and recycle of the aqueous catalyst phase was simple and efficient. The concept is pictured in Scheme 9.16.

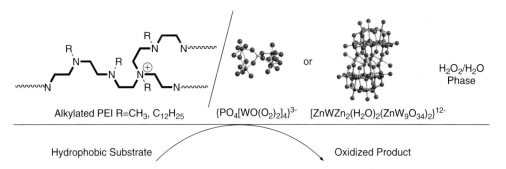

Scheme 9.16 Aqueous biphasic reactions mediated by alkylated polyethyleneimine.

As a consequence of showing that alkylated polyethyleneimines can facilitate aqueous biphasic catalysis, we extended the concept by the preparation of a cross-linked polyethyleneimine assembly that encapsulated a polyoxometalate catalyst, which resulted in the *lipophiloselective* oxidation of secondary alcohols (Scheme 9.17) [152]. Thus, even though reactions were carried out in water, competitive oxidation of a more hydrophobic alcohol in the presence of a hydrophilic alcohol significantly favored the former. The lipophiloselectivity was proportional to the relative partition coefficient of the substrates.

Concepts and techniques to utilize polyoxometalates in an efficient way are only in their infancy. As the number of synthetic uses of polyoxometalates increases and as

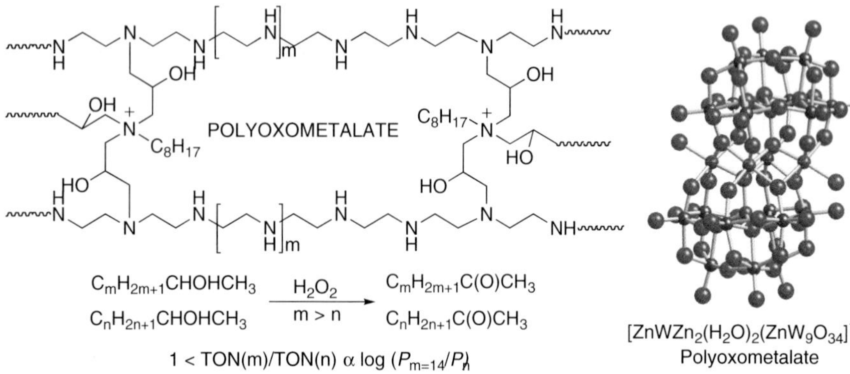

Scheme 9.17 Lipophiloselective reactions with polyoxometalates encapsulated in cross-linked polyethyleneimine.

the practical potential of polyoxometalate oxidation catalysis becomes a reality, one may expect a number of new methods for catalyst 'engineering' to aid in recovery and recycling.

9.7
Conclusion

Liquid phase oxidation catalysis by polyoxometalates became a research topic only about thirty years ago. Since then various applications of polyoxometalates as practical oxidation catalysts useful for synthesis have been demonstrated. Additional synthetic procedures are not far away owing to the wide variety of polyoxometalates that can be prepared and the important structure-activity relationships that have been shown. In fact, it is the mechanistic research that has been carried out that points to many new and possibly exciting synthetic applications. Although polyoxometalates are of high molecular weight, efficient methods of catalyst recycle such as nanofiltration and some use of supports and biphasic media are already available. This bodes well for the eventual use of polyoxometalate catalysts along with benign oxidants as an attractive platform for replacing the still common use of environmentally damaging stoichiometric oxidants.

References

1 (1998) *Chem. Rev.*, **1**, 98.
2 (a) Pope, M.T. (1983) *Isopoly and Heteropoly Anions*, Springer, Berlin, Germany; (b) Müller, A. (2001) *Polyoxometalate Chemistry*, Kluwer Academic, Dordrecht. The Netherlands; (c) Kozhevnikov, I.V. (2002) *Catalysis by Polyoxometalates*, Wiley, Chichester, England; (d) Hill, C.L. and Prosser-McCartha, C.M. (1995) *Coord. Chem. Rev*, **143**, 407; (e) Mizuno, N. and Misono, M. (1998) *Chem. Rev.*, **98**, 171; (f) Neumann,

R. (1998) *Prog. Inorg. Chem.*, **47**, 317; (g) Neumann, R. and Khenkin, A.M. (2006) *Chem. Commun.*, 2529.

3 (a) Hill, C.L. and Brown, R.B. (1986) *J. Am. Chem. Soc.*, **108**, 536; (b) Mansuy, D., Bartoli, J.F., Battioni, P., Lyon, D.K., and Finke, R.G. (1991) *J. Am. Chem. Soc.*, **113**, 7222.

4 Weiner, H., Hayashi, Y., and Finke, R.G. (1999) *Inorg. Chem.*, **38**, 2579.

5 Khenkin, A.M., Kumar, D., Shaik, S., and Neumann, R. (2006) *J. Am. Chem. Soc.*, **128**, 15451.

6 (a) de Visser, S.P., Kumar, D., Neumann, R., and Shaik, S. (2004) *Angew. Chem. Int. Ed.*, **43**, 5661; (b) Kumar, D., Derat, E., Khenkin, A.M., Neumann, R., and Shaik, S. (2005) *J. Am. Chem. Soc.*, **127**, 17712; (c) Derat, E., Kumar, D., Neumann, R., and Shaik, S. (2006) *Inorg. Chem.*, **45**, 8655.

7 (a) Katsoulis, D.E. and Pope, M.T. (1986) *J. Chem. Soc., Chem. Commun.*, 1186; (b) Khenkin, A.M. and Hill, C.L. (1993) *J. Am. Chem. Soc.*, **115**, 8178–8186.

8 Zhang, X., Sasaki, K., and Hill, C.L. (1996) *J. Am. Chem. Soc.*, **118**, 4809.

9 Neumann, R. and Abu-Gnim, C. (1989) *J. Chem. Soc., Chem. Commun.*, 1324.

10 Neumann, R. and Abu-Gnim, C. (1990) *J. Am. Chem. Soc.*, **112**, 6025.

11 Steckhan, E. and Kandzia, C. (1992) *Synlett*, 139.

12 Bressan, M., Morvillo, A., and Romanello, G. (1992) *J. Mol. Catal.*, **77**, 283.

13 Kuznetsova, L.I., Likholobov, V.A., and Detusheva, L.G. (1993) *Kinet. Catal.*, **34**, 914.

14 Neumann, R. and Khenkin, A.M. (1998) *Chem. Commun.*, 1967.

15 Banks, R.G.S., Henderson, R.J., and Pratt, J.M. (1968) *J. Chem. Soc. A*, 2886–2889.

16 Bottomly, F., Lin, I.J.B., and Mukaida, M. (1980) *J. Am. Chem. Soc.*, **102**, 5238.

17 (a) Panov, G.I. (2000) *CATTECH*, **4**, 18; (b) Panov, G.I., Uriate, A., Rodkin, M.A., and Sobolev, V.I. (1998) *Catal. Today*, **41**, 365.

18 Notte, N.P. (2000) *Top. Catal.*, **13**, 387.

19 Rhodkin, M.A., Sobolev, V.I., Dubkov, K.A., Watkins, N.H., and Panov, G.I. (2000) *Stud. Surf. Sci. Catal.*, **130**, 881.

20 Duma, V. and Hönicke, D. (2000) *J. Catal.*, **191**, 93.

21 Groves, J.T. and Roman, J.S. (1995) *J. Am. Chem. Soc.*, **117**, 5594.

22 Yamada, T., Hashimoto, K., Kitaichi, Y., Suzuki, K., and Ikeno, T. (2001) *Chem. Lett.*, 268.

23 Hashimoto, K., Kitaichi, Y., Tanaka, H., Ikeno, T., and Yamada, T. (2001) *Chem. Lett.*, 922.

24 Ben-Daniel, R., Weiner, L., and Neumann, R. (2002) *J. Am. Chem. Soc.*, **124**, 8788.

25 Ben-Daniel, R. and Neumann, R. (2003) *Angew. Chem. Int. Ed.*, **42**, 92.

26 (a) Mancuso, A.J. and Swern, D. (1981) *Synthesis*, 165; (b) Tidwell, T.T. (1990) *Org. React.*, **39**, 297; (c) Tidwell, T.T. (1990) *Synthesis*, 857.

27 Kukushkin, V.Y. (1995) *Coord. Chem. Rev.*, **139**, 375–407.

28 Khenkin, A.M. and Neumann, R. (2002) *J. Am. Chem. Soc.*, **124**, 4198.

29 Khenkin, A.M. and Neumann, R. (2002) *J. Org. Chem.*, **67**, 7075.

30 Barats, D., Popovitz-Biro, R., Lietus, G., and Neumann, R. (2008) *Angew. Chem. Int. Ed.*, in press.

31 (a) Faraj, M. and Hill, C.L. (1987) *J. Chem. Soc., Chem. Commun.*, 1487; (b) Neumann, R. and Khenkin, A.M. (1995) *Inorg. Chem.*, **34**, 5753; (c) Cramarossa, M.R., Forti, L., Fedotov, M.A., Detusheva, L.G., Likholobov, V.A., Kuznetsova, L.I., Semin, G.L., Cavani, F., and Trifiró, F. (1997) *J. Mol. Catal.*, **127**, 85; (d) Matsumoto, Y., Asami, M., Hashimoto, M., and Misono, M. (1996) *J. Mol. Catal.*, **114**, 161.

32 Maradur, S.P., Halligudi, S.B., and Gokavi, G.S. (2004) *Catal. Lett.*, **68**, 165.

33 Khenkin, A.M. and Neumann, R. (2001) *J. Am. Chem. Soc.*, **123**, 6437.

34 Gall, R.D., Faraj, M., and Hill, C.L. (1994) *Inorg. Chem.*, **33**, 5015.

35 (a) Adam, W., Alsters, P.L., Neumann, R., Saha-Möller, C.R., Seebach, D., and Zhang, R. (2003) *Org. Lett.*, **5**, 725; (b) Adam, W., Alsters, P.L., Neumann, R., Saha-Möller, C.R., Seebach, D., and Zhang, R. (2003) *J. Org. Chem.*, **68**, 8222.

36 (a) Matoba, Y., Ishii, Y., and Ogawa, M. (1984) *Synth. Commun.*, **14**, 865;

(b) Wang, J., Yan, L., Li, G., Wang, X., Ding, Y., and Suo, J. (2005) *Tetrahedron Lett.*, **46**, 7123; (c) Wang, J., Yan, L., Qian, G., Li, S., Yang, K., Liu, H., and Wang, X. (2007) *Tetrahedron*, **63**, 1826.

37 (a) Ishii, Y., Yamawaki, K., Ura, T., Yamada, H., Yoshida, T., and Ogawa, M. (1988) *J. Org. Chem.*, **53**, 3587; (b) Oguchi, T., Sakata, Y., Takeuchi, N., Kaneda, K., Ishii, Y., and Ogawa, M. (1989). *Chem. Lett.*, 2053; (c) Schwegler, M., Floor, M., and van Bekkum, H. (1988) *Tetrahedron Lett.*, **29**, 823.

38 (a) Sakata, Y., Katayama, Y., and Ishii, Y. (1992) *Chem. Lett.*, 671; (b) Sakata, Y. and Ishii, Y. (1991) *J. Org. Chem.*, **56**, 6233; (c) Iwahama, T., Sakaguchi, S., Nishiyama, Y., and Ishii, Y. (1995) *Tetrahedron Lett.*, **36**, 1523.

39 Ishii, Y., Yamawaki, K., Yoshida, T., and Ogawa, M. (1988) *J. Org. Chem.*, **53**, 5549.

40 (a) Ballistreri, F.P., Failla, S., Spina, E., and Tamaselli, G.A. (1989) *J. Org. Chem.*, **54**, 947; (b) Bordoloi, A., Vinu, A., and Halligudi, S.B. (2007) *Appl. Catal. A Gen.*, **333**, 143.

41 Sakaue, S., Sakata, Y., Nishiyama, Y., and Ishii, Y. (1992) *Chem. Lett.*, 289.

42 (a) Orita, H., Shimizu, M., Haykawa, T., and Takehira, K. (1991) *React. Kinet. Catal. Lett.*, **44**, 209; (b) Petrov, L.A., Lobanova, N.P., Volkov, V.L., Zakharova, G.S., Kolenko, I.P., and Buldakova, L.Yu. (1989) *Izv. Akad. Nauk SSSR, Ser. Khim.*, 1967; (c) Shimizu, M., Orita, H., Hayakawa, T., and Takehira, K. (1989) *Tetrahedron Lett.*, **30**, 471.

43 (a) Csanyi, L.J. and Jaky, K. (1990) *J. Mol. Catal.*, **61**, 75; (b) Csanyi, L.J. and Jaky, K. (1991) *J. Catal.*, **127**, 42.

44 (a) Salles, L., Aubry, C., Robert, F., Chottard, G., Thouvenot, R., Ledon, H., and Brégault, J.-M. (1993) *New J. Chem.*, **17**, 367; (b) Aubry, C., Chottard, G., Platzer, N., Brégault, J.-M., Thouvenot, R., Chauveau, F., Huet, C., and Ledon, H. (1991) *Inorg. Chem.*, **30**, 4409; (c) Salles, L., Aubry, C., Thouvenot, R., Robert, F., Dorémieux-Morin, C., Chottard, G., Ledon, H., Jeannin, Y., and Brégault, J.-M. (1994) *Inorg. Chem.*, **33**, 871.

45 (a) Dengel, A.C., Griffith, W.P., and Parkin, B.C. (1993) *J. Chem. Soc., Dalton Trans.*, 2683; (b) Bailey, A.J., Griffith, W.P., and Parkin, B.C. (1995) *J. Chem. Soc., Dalton Trans.*, 1833.

46 Duncan, D.C., Chambers, R.C., Hecht, E., and Hill, C.L. (1995) *J. Am. Chem. Soc.*, **117**, 681.

47 Venturello, C., D'Aloiso, R., Bart, J.C., and Ricci, M. (1985) *J. Mol. Catal.*, **32**, 107.

48 (a) Xi, Z., Zhou, N., Sun, Y., and Li, K. (2001) *Science*, **292**, 1139; (b) Sun, Y., Xi, Z., and Cao, G. (2001) *J. Mol. Catal. A*, **166**, 219; (c) Xi, Z., Wang, H., Sun, Y., Zhou, N., Cao, G., and Li, M. (2001) *J. Mol. Catal. A*, **168**, 299.

49 (a) Shiozaki, R., Goto, H., and Kera, Y. (1993) *Bull. Chem. Soc. Jpn.*, **66**, 2790; (b) Shiozaki, R., Kominami, H., and Kera, Y. (1996) *Synth. Commun.*, **26**, 1663.

50 Griffith, W.P., Morley-Smith, N., Nogueira, H.I.S., Shoair, A.G.F., Suriaatmaja, M., White, A.J.P., and Williams, D.J. (2000) *J. Organomet. Chem.*, **607**, 146.

51 Droege, M.W. and Finke, R.G. (1991) *J. Mol. Catal.*, **69**, 323.

52 Yadollahi, B. (2003) *Chem. Lett.*, **32**, 1066.

53 Neumann, R. and de la Vega, M. (1993) *J. Mol. Catal.*, **84**, 93.

54 Nomiya, K., Yanagibayashi, H., Nozaki, C., Kondoh, K., Hiramatsu, E., and Shimizu, Y. (1996) *J. Mol. Catal. A*, **114**, 181.

55 (a) Yamase, T., Ozeki, T., and Motomura, S. (1992) *Bull. Chem. Soc. Jpn.*, **65**, 1453; (b) Kholdeeva, O.A., Maksimov, G.M., Maksimovskyaya, R.I., Kovaleva, L.A., Fedetov, M.A., Grogoriev, V.A., and Hill, C.L. (2000) *Inorg. Chem.*, **39**, 3828.

56 (a) Kholdeeva, O.A., Trubitsina, T.A., Maksimov, G.M., Golovin, A.V., and Maksimovskaya, R.I. (2005) *Inorg. Chem.*, **44**, 1635; (b) Kholdeeva, O.A., Trubitsina, T.A., Timfeeva, M.A., Maksimov, G.M., Maksimovskaya, R.I., and Rogov, V.A. (2005) *J. Mol. Catal. A Chem.*, **232**, 173; (c) Kholdeeva, O.A. (2006) *Topics Catal.*, **40**, 229.

57 Hayashi, K., Kato, C.N., Shinohara, A., Sakai, Y., and Nomiya, K. (2007) *J. Mol. Catal. A Chem.*, **262**, 30.

58 (a) Seki, Y., Min, J.S., Misono, M., and Mizuno, N. (2000) *J. Phys. Chem. B*, **104**,

5940; (b) Mizuno, N., Seki, Y., Nishiyama, Y., Kiyoto, I., and Misono, M. (1999) *J. Catal.*, **184**, 550; (c) Mizuno, N., Kiyoto, I., Nozaki, C., and Misono, M. (1999) *J. Catal.*, **181**, 171; (d) Mizuno, N., Nozaki, C., Kiyoto, L., and Misono, M. (1998) *J. Am. Chem. Soc.*, **120**, 9267.

59 (a) Kamata, K., Yonehara, K., Sumida, Y., Yamaguchi, K., Hikichi, S., and Mizuno, N. (2003) *Science*, **300**, 964; (b) Kamata, K., Nakagawa, Y., Yamaguchi, K., and Mizuno, N. (2004) *J. Catal.*, **224**, 224; (c) Kamata, K., Kotani, M., Yamaguchi, K., Hikichi, H., and Mizuno, N. (2007) *Chem. Eur. J.*, **13**, 639.

60 Musaev, D.G., Morokuma, K., Geletii, Y.V., and Hill, C.L. (2004) *Inorg. Chem.*, **43**, 7702.

61 Prabhakar, R., Morokuma, K., Hil, C.L., and Musaev, D.G. (2006) *Inorg. Chem.*, **45**, 5703.

62 (a) Sartorel, A., Carraro, M., Bagno, A., Scorrano, G., and Bonchio, M. (2007) *Angew. Che, Int. Ed.*, **46**, 3255; (b) Sartorel, A., Carraro, M., Bagno, A., Scorrano, G., and Bonchio, M. (2008) *J. Phys. Org. Chm.*, **21**, 596.

63 (a) Carraro, M., Sandei, L., Sartorel, A., Scorrano, G., and Bonchio, M. (2006) *Org. Lett.*, **8**, 3671; (b) Berardi, S., Bonchio, M., Carraro, M., Conte, V., Sartorel, A., and Scorrano, G. (2007) *J. Org. Chem.*, **72**, 8954.

64 (a) Yoshida, A., Yoshimura, M., Masayuki, U., Kazuhiro, H., Hikichi, S., and Mizuno, N. (2006) *Angew. Chem. Int. Ed.*, **45**, 1956; (b) Yoshida, A., Uehara, K., Hikichi, S., and Mizuno, N. (2007) *Stud. Surf. Sci. Catal.*, **172**, 205.

65 (a) Nakagawa, Y., Kamata, K., Kotani, M., Yamaguchi, K., and Mizuno, N. (2005) *Angew. Chem. Int. Ed.*, **44**, 5136; (b) Nakagawa, Y. and Mizuno, N. (2007) *Inorg. Chem.*, **46**, 1727.

66 Goto, Y., Kamata, K., Yamaguchi, K., Uehara, K., Hikichi, S., and Mizuno, N. (2006) *Inorg. Chem.*, **45**, 2347.

67 (a) Khenkin, A.M. and Hill, C.L. (1993) *Mendeleev Commun.*, 140; (b) Zhang, X., Chen, Q., Duncan, D.C., Lachicotte, R.J., and Hill, C.L. (1997) *Inorg. Chem.*, **36**, 4381; (c) Zhang, X., Chen, Q., Duncan, D.C., Campana, C.F., and Hill, C.L. (1997) *Inorg. Chem.*, **36**, 4208; (d) Zhang, X., Anderson, T.M., Chen, Q., and Hill, C.L. (2001) *Inorg. Chem.*, **40**, 418.

68 Ben-Daniel, R., Khenkin, A.M., and Neumann, R. (2000) *Chem. Eur. J.*, **6**, 3722.

69 (a) Neumann, R. and Gara, M. (1994) *J. Am. Chem. Soc.*, **116**, 5509; (b) Neumann, R. and Khenkin, A.M. (1996) *J. Mol. Catal.*, **114**, 169.

70 Neumann, R. and Gara, M. (1995) *J. Am. Chem. Soc.*, **117**, 5066.

71 (a) Neumann, R. and Juwiler, D. (1996) *Tetrahedron*, **47**, 8781; (b) Neumann, R., Khenkin, A.M., Juwiler, D., Miller, H., and Gara, M. (1997) *J. Mol. Catal.*, **117**, 169.

72 Bösing, M., Nöh, A., Loose, I., and Krebs, B. (1998) *J. Am. Chem. Soc.*, **120**, 7252.

73 (a) Witte, P.T., Chowdury, S.R., ten Elshof, J.E., Sloboda-Rozner, D., Neumann, R., and Alsters, P.L. (2005) *Chem. Commun.*, 1206; (b) Chowdury, S.R., Witte, P.T., Blank, D.H.A., Alsters, P.L., and ten Elshof, J.E. (2006) *Chem. Eur. J.*, **12**, 3061.

74 (a) Adam, W., Alsters, P.L., Neumann, R., Saha-Möller, C.R., Sloboda-Rozner, D., and Zhang, R. (2002) *Synlett*, 2011; (b) Adam, W., Alsters, P.L., Neumann, R., Saha-Möller, C.R., Sloboda-Rozner, D., and Zhang, R. (2003) *J. Org. Chem.*, **68**, 1721.

75 (a) Witte, P.T., Alsters, P.L., Jary, W., Muellner, R., Poechlauer, P., Sloboda-Rozner, D., and Neumann, R. (2004) *Org. Proc. Res. Develop.*, **8**, 524; (b) Nardello, V., Aubry, J.-M., de Vos, D.E., Neumann, R., Adam, W., Zhang, R., ten Elshof, J.E., Witte, P.T., and Alsters, P.L. (2006) *J. Mol. Catal. A Chem.*, **251**, 185.

76 (a) Hamamoto, M., Nakayama, K., Nishiyama, Y., and Ishii, Y. (1993) *J. Org. Chem.*, **58**, 6421; (b) Mizuno, N., Hirose, T., Tateishi, M., and Iwamoto, M. (1993) *Chem. Lett.*, 1839; (c) Mizuno, N., Tateishi, M., Hirose, T., and Iwamoto, M. (1993) *Chem. Lett.*, 1985; (d) Mizuno, N., Hirose, T., Tateishi, M., and Iwamoto, M. (1994) *Stud. Surf. Sci. Catal.*, **82**, 593; (e) Khenkin, A.M., Rosenberger, A., and Neumann, R. (1999) *J. Catal.*, **182**, 82; (f) Maksimchuk, N.V., Melguniv, M.S.,

Chesalov, Y.A., Mrowiec-Bialon, J., Jarzebski, A.B., and Kholdeeva, O.A. (2007) *J. Catal.*, **246**, 241; (g) Lu, H., Gao, J., Jiang, X., Yang, Y., Song, B., and Li, C. (2007) *Chem. Commun.*, 150.

77 (a) Kholdeeva, O.A., Vanina, M.P., Timofeeva, M.N., Maksimovskaya, R.I., Trubitsina, T.A., Melgunov, M.S., Burgina, E.B., Mrowiec-Bialon, J., Jarzebski, A.B., and Hill, C.L. (2004) *J. Catal.*, **226**, 363; (b) Kholdeeva, O.A., Timofeeva, M.N., Maksimov, G.M., Maksimovskaya, R.I., Neiwart, W.A., and Hill, C.L. (2005) *Inorg. Chem.*, **44**, 666; (c) Sloboda-Rozner, D., Neiman, K., and Neumann, R. (2007) *J. Mol. Catal. A Chem.*, **262**, 109; (d) Okun, N.A., Anderson, T.M., and Hill, C.L. (2003) *J. Am. Chem. Soc.*, **125**, 3194.

78 Lyons, J.E., Ellis, P.E., and Durante, V.A. (1991) *Stud. Surf. Sci. Catal.*, **67**, 99.

79 (a) Mizuno, N., Hirose, T., Tateishi, M., and Iwamoto, M. (1994) *J. Mol. Catal.*, **88**, L125; (b) Mizuno, N., Tateishi, M., Hirose, T., and Iwamoto, M. (1993) *Chem. Lett.*, 2137; (c) Shinachi, S., Mitsunori, M., Yamaguchi, K., and Mizuno, N. (2005) *J. Catal.*, **233**, 81.

80 (a) Okun, N.A., Anderson, T.M., and Hill, C.L. (2003) *J. Mol. Catal. A Chem.*, **197**, 283; (b) Okun, N.A., Tarr, J.C., Hillesheim, D.A., Zhang, L., Hardcastle, K.I., and Hill, C.L. (2006) *J. Mol. Catal. A Chem.*, **246**, 11; (c) Okun, N.A., Anderson, T.M., Hardcastle, K.I., and Hill, C.L. (2003) *Inorg. Chem.*, **42**, 6610.

81 (a) Razi, R., Abedini, M., Kharat, A.N., and Amini, M.M. (2007) *Catal. Commun.*, **9**, 245; (b) Liu, Y., Muratam, K., and Inaba, M. (2005) *Catal. Commun.*, **6**, 679.

82 (a) Weiner, H., Trovareli, A., and Finke, R.G. (2003) *J. Mol. Catal. A Chem.*, **191**, 253; (b) Bonchio, M., Carrao, M., Farinazzo, A., Sartorel, A., Scorrano, G., and Kortz, U. (2007) *J. Mol. Catal. A Chem.*, **262**, 36; (c) Bonchio, M., Carrao, M., Sartorel, A., Scorrano, G., and Kortz, U. (2006) *J. Mol. Catal. A Chem.*, **251**, 93.

83 Weiner, H. and Finke, R.G. (1999) *J. Am. Chem. Soc.*, **121**, 9831.

84 (a) Yin, C.-X. and Finke, R.G. (2005) *J. Am. Chem. Soc.*, **127**, 9003; (b) Yin, C.-X., Sasaki, Y., and Finke, R.G. (2005) *Inorg. Chem.*, **44**, 8521.

85 Neumann, R. and Dahan, M. (1995) *J. Chem. Soc., Chem. Commun.*, 2277.

86 (a) Chester, A.W. (1970) *J. Org. Chem.*, **35**, 1797; (b) Eberson, L. and Wistrand, L.-G. (1980) *Act. Chem. Scand. B*, **34**, 349; (c) Eberson, L. (1983) *J. Am. Chem. Soc.*, **105**, 3192.

87 (a) Matveev, K.I. (1977) *Kinet. Catal.*, **18**, 716; (b) Matveev, K.I. and Kozhevnikov, I.V. (1980) *Kinet. Catal.*, **21**, 855.

88 Grate, J.R., Mamm, D.R., and Mohajan, S. (1993) *Mol. Eng.*, **3**, 205. Grate, J.R., Mamm, D.R., and Mohajan, S. (1993) *Polyoxometalates: From Platonic Solids to Anti-Retroviral Activity* (eds M.T. Popeand A. Müller), Kluwer, The Netherlands, p. 27.

89 Yokota, T., Sakakura, A., Tani, M., Sakaguchi, S., and Ishii, Y. (2002) *Tetrahedron Lett.*, **43**, 8887.

90 (a) Greenberg, H., Bergstad, K., and Bäckvall, J.-E. (1996) *J. Mol. Catal.*, **113**, 355; (b) Yokota, T., Fujibayashi, S., Nishiyama, Y., Sakaguchi, S., and Ishii, Y. (1996) *J. Mol. Catal.*, **114**, 113.

91 Bergstad, K., Greenberg, H., and Bäckvall, J.-E. (1998) *Organometallics*, **17**, 45.

92 (a) Yokota, T., Tani, M., Sakaguchi, S., and Ishii, Y. (2003) *J. Am. Chem. Soc.*, **125**, 1476; (b) Yamada, T., Sakakura, A., Sakaguchi, S., Obora, Y., and Ishii, Y. (2008) *New J. Chem.*, **32**, 738.

93 Yokota, T., Sakaguchi, S., and Ishii, Y. (2002) *Adv. Catal. Synth.*, **344**, 849.

94 Tamaso, K.I., Hatamoto, Y., Obora, Y., Sakaguchi, S., and Ishii, Y. (2007) *J. Org. Chem.*, **72**, 8820.

95 Yokota, Y., Sakaguchi, S., and Ishii, Y. (2002) *J. Org. Chem.*, **67**, 5005.

96 Tani, M., Sakamoto, T., Mita, S., Sakaguchi, S., and Ishii, Y. (2005) *Angew. Chem. Int. Ed.*, **44**, 2586.

97 (a) Mita, S., Sakamoto, T., Yamada, S., Sakaguchi, S., and Ishii, Y. (2005) *Tetrahedron Lett.*, **46**, 7729; (b) Yamada, S., Sakaguchi, S., and Ishii, Y. (2007) *J. Mol. Catal. A Chem.*, **262**, 48; (c) Yamada, S., Ohashi, S., Obora, Y., Sakaguchi, S., and Ishii, Y. (2008) *J. Mol. Catal. A Chem.*, **282**, 22.

98 Neumann, R. and Assael, I. (1998) *J. Chem. Soc., Chem. Commun.*, 1285.
99 Branytska, O.V. and Neumann, R. (2003) *J. Org. Chem.*, **69**, 9510.
100 Branytska, O. and Neumann, R. (2005) *Synlett*, 2525.
101 (a) Kozhevnikov, I.V., Simagina, V.I., Varnakova, G.V., and Matveev, K.I. (1979) *Kinet. Catal.*, **20**, 506; (b) Dzhumakaeva, B.S. and Golodov, V.A. (1986) *J. Mol. Catal.*, **35**, 303; (c) Karbanko, V.E., Sidelnikov, V.N., Kozhevnikov, I.V., and Matveev, K.I. (1982) *React. Kinet. Catal. Lett.*, **21**, 209.
102 (a) Harrup, M.K. and Hill, C.L. (1994) *Inorg., Chem.*, **33**, 5448; (b) Harrup, M.K. and Hill, C.L. (1996) *J. Mol. Catal. A*, **106**, 57; (c) Hill, C.L. and Gall, R.D. (1996) *J. Mol. Catal. A*, **114**, 103.
103 Brégeault, J.-M., El Ali, B., Mercier, J., Martin, J., and Martin, C. (1989) *C. R. Acad. Sci. II*, **309**, 459.
104 (a) El Ali, B., Brégeault, J.-M., Martin, J., and Martin, C. (1989) *New J. Chem.*, **13**, 173; (b) El Ali, B., Brégeault, J.-M., Mercier, J., Martin, J., Martin, C., and Convert, O. (1989) *J. Chem. Soc., Chem. Commun.*, 825; (c) Atlamsani, A., Ziyad, M., and Brégeault, J.-M. (1995) *J. Chim. Phys., Phys.-Chim. Biol.*, **92**, 1344; (d) Vennat, M., Herson, P., Brégeault, J.-M., and Shul'pin, G.B. (2003) *Eur. J. Inorg. Chem.*, 908.
105 Khenkin, A.M. and Neumann, R. (2002) *Adv. Syn. Catal.*, **344**, 1017.
106 Khenkin, A.M. and Neumann, R. (2008) *J. Am. Chem. Soc.*, **130**, 14474.
107 Neumann, R. and Lissel, M. (1989) *J. Org. Chem.*, **54**, 4607–4610.
108 Neumann, R. and Levin, M. (1991) *J. Org. Chem.*, **56**, 5707–5710.
109 (a) Khenkin, A.M. and Neumann, R. (2003) *Inorg. Chem.*, **42**, 3331; (b) Oonaka, T., Hashimoto, K., Kominami, H., Matsubara, Y., and Kera, Y. (2006) *J. Jpn. Pet. Inst.*, **49**, 43.
110 Gallo, C., Gentili, P., Nuna Pontes, A.S., Gamelas, J.A.F., and Evtuguin, D.V. (2007) *New J. Chem.*, **31**, 1461.
111 Matveev, K.I., Zhizhina, E.G., and Odyakov, V.F. (1995) *React. Kinet. Catal. Lett.*, **55**, 47.
112 Ben-Daniel, R., Alsters, P.L., and Neumann, R. (2001) *J. Org. Chem.*, **66**, 8650.
113 (a) Weinstock, I.A., Atalla, R.H., Reiner, R.S., Moen, M.A., Hammel, K.E., Houtman, C.J., and Hill, C.L. (1996) *New J. Chem.*, **20**, 269; (b) Weinstock, I.A., Atalla, R.H., Reiner, R.S., Moen, M.A., Hammel, K.E., Houtman, C.J., Hill, C.L., and Harrup, M.K. (1997) *J. Mol. Catal. A-Chem.*, **116**, 59; (c) Weinstock, I.A., Atalla, R.H., Reiner, R.S., Houtman, C.J., and Hill, C.L. (1998) *Holzforschung*, **52**, 304.
114 Neumann, R. and Levin, M. (1992) *J. Am. Chem. Soc.*, **114**, 7278.
115 Duncan, D.C. and Hill, C.L. (1997) *J. Am. Chem. Soc.*, **119**, 243.
116 (a) Khenkin, A.M. and Neumann, R. (2000) *Angew. Chem. Int. Ed.*, **39**, 4088; (b) Khenkin, A.M., Weiner, L., Wang, Y., and Neumann, R. (2001) *J. Am. Chem. Soc.*, **123**, 8531.
117 Khenkin, A.M., Weiner, L., and Neumann, R. (2005) *J. Am. Chem. Soc.*, **127**, 9988.
118 Bar-Nahum, I., Cohen, H., and Neumann, R. (2003) *Inorg. Chem.*, **42**, 3677.
119 Bar-Nahum, I., Narasimhulu, K.V., Weiner, L., and Neumann, R. (2005) *Inorg. Chem.*, **44**, 4900.
120 Bar-Nahum, I., Khenkin, A.M., and Neumann, R. (2004) *J. Am. Chem. Soc.*, **126**, 10236.
121 Kim, B.W., Rodriguez-Rivera, G.J., Evans, S.T., Voitl, T., Einspahr, J.J., Voyles, P.M., and Dumesic, J. (2005) *J. Catal.*, **235**, 327.
122 Maayan, G. and Neumann, R. (2005) *Chem. Commun.*, 4595.
123 Maayan, G. and Neumann, R. (2008) *Catal. Lett.*, **123**, 41.
124 De bruyn, M. and Neumann, R. (2007) *Adv. Synth. Catal.*, **349**, 1624.
125 (a) Neumann, R. and Dahan, M. (1997) *Nature*, **388**, 353; (b) Neumann, R. and Dahan, M. (1998) *J. Am. Chem. Soc.*, **120**, 11969.
126 Ying, C.-X. and Finke, R.G. (2005) *Inorg. Chem.*, **44**, 4175.
127 (a) Fujibayashi, S., Nakayama, K., Hamamoto, M., Sakaguchi, S., Nishiyama, Y., and Ishii, Y. (1996) *J. Mol. Catal. A*, **110**, 105; (b) Nakayama, K.,

Hamamoto, M., Nishiyama, Y., and Ishii, Y. (1993) *Chem. Lett.*, 1699.
128. Yamaguchi, K. and Mizuno, N. (2002) *New J. Chem.*, **26**, 972.
129. Xu, L., Boring, E., and Hill, C.L. (2000) *J. Catal.*, **195**, 394.
130. Khenkin, A.M., Vigdergauz, I., and Neumann, R. (2000) *Chem. Eur. J.*, **6**, 875–882.
131. (a) Okun, N.M., Anderson, T.M., and Hill, C.L. (2003) *J. Am. Chem. Soc.*, **125**, 3194; (b) Okun, N.M., Anderson, T.M., and Hill, C.L. (2003) *J. Mol. Catal. A*, **197**, 283.
132. Kasai, J., Nakagawa, Y., Uchida, S., Yamaguchi, K., and Mizuno, N. (2006) *Chem. Eur. J.*, **12**, 4176.
133. (a) Khenkin, A.M., Neumann, R., Sorokin, A.B., and Tuel, A. (1999) *Catal. Lett.*, **63**, 189; (b) Li, H., Perkas, N., Li, Q., Gofer, Y., Koltypin, Y., and Gedanken, A. (2003) *Langmuir*, **19**, 10409; (c) Yu, X., Xu, L., Yang, X., Guo, Y., Li, K., Hu, J., Li, W., Ma, F., and Guo, Y. (2008) *Appl. Surf. Sci.*, **254**, 4444.
134. Ichihara, J., Iteya, K., Kambara, A., Akihiro, S., and Sasaki, Y. (2003) *Catal. Today*, **87**, 163.
135. (a) Liu, Y., Murata, K., and Inaba, M. (2006) *Chem. Lett.*, **35**, 436; (b) Carriazo, D., Lima, S., Martin, C., Pillinger, M., Vante, A.A., and Rives, V. (2007) *J. Chem. Phys. Solids*, **68**, 1872; (c) Jana, S.K., Kubota, Y., and Tatsumi, T. (2008) *J. Catal.*, **255**, 40; (d) Liu, P., Wang, H., Feng, Z., Ying, P., and Li, C. (2008) *J. Catal.*, **256**, 345.
136. Lang, X., Li, Z., and Xia, C. (2008) *Synth. Commun.*, **38**, 1610.
137. Neumann, R. and Assael, I. (1989) *J. Chem. Soc., Chem. Commun.*, 547.
138. Haimov, A. and Neumann, R. (2002) *Chem. Commun.*, 876.
139. (a) Neumann, R. and Cohen, M. (1997) *Angew. Chem. Int. Ed.*, **36**, 1738; (b) Cohen, M. and Neumann, R. (1999) *J. Mol. Catal. A*, **146**, 293.
140. Neumann, R. and Miller, H. (1995) *J. Chem. Soc., Chem. Commun.*, 2277.
141. Vasylyev, M.V. and Neumann, R. (2004) *J. Am. Chem. Soc.*, **126**, 884.
142. Maayan, G., Popovitz-Biro, R., and Neumann, R. (2006) *J. Am. Chem. Soc.*, **128**, 4968.
143. Liu, L., Chen, C., Hu, X., Mohamood, T., Ma, W., Lin, J., and Zhao, J. (2008) *New J. Chem.*, **32**, 283.
144. Maayan, G., Ganchegui, B., Leitner, W., and Neumann, R. (2006) *Chem. Commun.*, 2230.
145. Maayan, G., Fish, R.H., and Neumann, R. (2003) *Org. Lett.*, **5**, 3547.
146. (a) Lambert, A., Plucinski, P., and Kozhevnilkov, I.V. (2003) *Chem. Commun.*, 714; (b) Kaur, J. and Kozhevnikov, I.V. (2004) *Catal. Commun.*, **5**, 709.
147. (a) Plaut, L., Hauseler, A., Nlate, S., Astruc, D., Ruiz, J., Gatard, S., and Neumann, R. (2004) *Angew. Chem. Int. Ed.*, **43**, 2924; (b) Nlater, S., Astruc, D., and Neumann, R. (2004) *Adv. Synth. Catal.*, **346**, 1445; (c) Nlate, S., Plaut, L., and Astruc, D. (2006) *Chem. Eur. J.*, **12**, 903.
148. Sloboda-Rozner, D., Alsters, P.L., and Neumann, R. (2003) *J. Am. Chem. Soc.*, **125**, 5280.
149. Sloboda-Rozner, D., Witte, P., Alsters, P.L., and Neumann, R. (2004) *Adv. Synth. Catal.*, **346**, 339.
150. Sloboda-Rozner, D. and Neumann, R. (2006) *Green Chem.*, **8**, 679.
151. Haimov, A., Cohen, H., and Neumann, R. (2004) *J. Am Chem. Soc.*, **126**, 11762.
152. Haimov, A. and Neumann, R. (2006) *J. Am Chem. Soc.*, **128**, 15697.

10
Oxidation of Carbonyl Compounds

Eric V. Johnston and Jan-E. Bäckvall

10.1
Introduction

Oxidation of carbonyl compounds is an important area of oxidation and involves a variety of different reaction types. This area has been reviewed by Bolm in Chapter 9 of the book 'Modern Oxidation Methods' (2004), and the present chapter is an update of this review. We have focused on two areas of carbonyl oxidation where important developments have occurred: (i) oxidation of aldehydes to carboxylic acids and (ii) Baeyer-Villiger oxidation.

10.2
Oxidation of Aldehydes to Carboxylic Acids

The transformation of aldehydes to carboxylic acids is a fundamental reaction in organic synthesis. Many successful methods have been developed for these types of oxidations [1], but most of them have limitations as they require stoichiometric amounts of oxidants such as chlorite [2], chromium(VI) reagents [3], potassium permanganate [4], or peroxides [5]. The use of organic solvents e.g., acetonitrile, dichloromethane, cyclohexane, formic acid, or benzene is also usually necessary. Despite the fact that these methods have several disadvantages, such as low selectivity and production of waste, some of them have been widely used in industry and are still in use today. The growing awareness of the environment has created a demand for efficient oxidation processes with environmentally friendly oxidants under mild conditions ('green' chemistry) [6].

Molecular oxygen and hydrogen peroxide are desirable oxidants for these transformations, since they are inexpensive and environmentally friendly, with water as the only by-product. Therefore, various improved methodologies using molecular oxygen or hydrogen peroxide directly as the oxidants have been explored in recent years and reported in the literature.

Modern Oxidation Methods. Edited by Jan-Erling Bäckvall

ISBN: 978-3-527-32320-3

10.2.1
Metal-Free Oxidation of Aldehydes to Carboxylic Acids

A 'metal-free' general procedure for aerobic oxidation of a variety of aldehydes that takes place 'on water' was reported by Shapiro and Vigalok [7]. Its use in the consecutive three-component Passerini reaction both 'on water' and 'in water' was described. It was found that hydrophobic aliphatic aldehydes (branched and linear), as well as aromatic aldehydes, undergo facile oxidation upon simply stirring their aqueous emulsions in air, affording the corresponding carboxylic acids in high yields (Eq. (10.2)) [7, 8]. The starting materials and products are insoluble in water and can easily be isolated. The reactions were found to proceed smoothly both on a small scale (1 mmol) and on a relatively large laboratory scale (50 mmol).

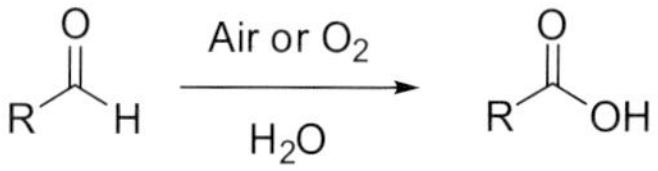

$$R-CHO \xrightarrow[H_2O]{Air\ or\ O_2} R-COOH \qquad (10.1)$$

Aromatic and aliphatic aldehydes have been successfully oxidized to their corresponding carboxylic acids by the use of a H_2O_2/HCl system in the presence of hydroxylamine hydrochloride (Eq. (10.2)) [9]. The method is selective and tolerates the presence of other functional groups such as, carbon-carbon double bonds, hydroxyl groups, and other heteroatoms.

$$R-CHO \xrightarrow[CH_3CN,\ reflux]{H_2O_2/HCl/NH_2OH \cdot HCl} R-COOH \qquad (10.2)$$

Recently, N-heterocyclic carbene-catalyzed (NHC-catalyzed) oxidative transformations have gained increasing attention, and this area was reviewed in 2007 [10]. Miyashita [11] has demonstrated that NHCs derived from triazolium and benzimidazolium salts are capable of catalyzing the oxidation of aryl aldehydes, employing oxidants such as Oxone [12], pyridinium hydrobromide perbromide [13], or peroxides [14]. However, these early investigations have mainly focused on ionic processes [15]. Recently, biomimetic transition metal-free organocatalytic oxidation methods were developed [16]. One of these methods employs the 2,2,6,6-tetramethylpiperidine N-oxyl radical (TEMPO) as the oxidant, generating a TEMPO ester (Eq. (10.3)), which can be hydrolyzed to give the acid. The synthetic potential of these reactions is far from being fully realized [17], although this type of aldehyde activation opens up a new field of organocatalysis [18].

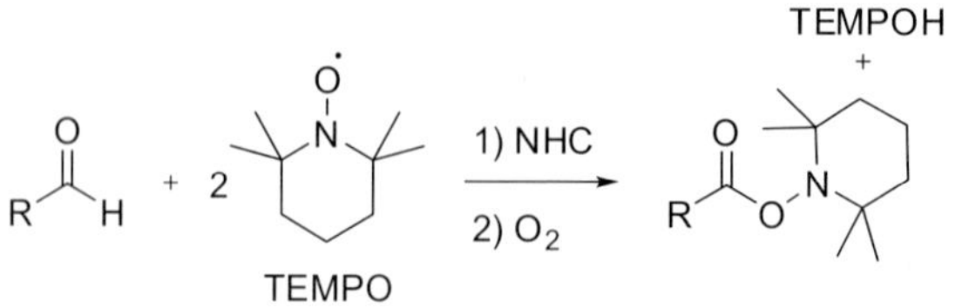

$$R-CHO + 2\ TEMPO \xrightarrow[2)\ O_2]{1)\ NHC} R-C(O)-O-TEMP + TEMPOH \qquad (10.3)$$

10.2.2
Metal-Catalyzed Oxidation of Aldehydes to Carboxylic Acids

Copper complexes as catalysts have previously been applied in the oxidation of carbonyl compounds to carboxylic acids [19]. Recently, a CuCl-catalyzed oxidation of aldehydes to the corresponding carboxylic acids by aqueous *tert*-butyl hydroperoxide in acetonitrile at room temperature was reported by Mannam and Sekar (Eq. (10.4)) [20]. This new procedure proved to be simple and mild, and works exceptionally well without any additives. Aromatic, vinylic, and aliphatic aldehydes were oxidized to the corresponding acids with short reaction times in excellent yields. Aromatic dialdehydes such as phthaldehyde and terephthaldehyde were also oxidized to their corresponding diacids at room temperature in high yields. Interestingly, aliphatic aldehydes such as cyclohexanecarboxaldehyde, palmitaldehyde and the aliphatic dialdehyde glutaraldehyde could also be transformed into the corresponding carboxylic acids and diacids, respectively.

$$\underset{R}{\overset{O}{\|}}\underset{H}{\overset{}{\diagup}} \xrightarrow[\text{MeCN, rt}]{\text{CuCl (5 mol \%)} \atop \text{aq. 70\% tBuOOH (1 equiv.)}} \underset{R}{\overset{O}{\|}}\underset{OH}{\overset{}{\diagup}} \quad (10.4)$$

An established procedure for the conversion of an aldehyde to a carboxylic acid involves the use of Ag_2O in the presence of NaOH [21, 22]. This has inspired researchers to employ different forms of Ag(I) compounds as catalysts in combination with an oxidant. Recently, a mild and selective method for the transformation of aldehydes into carboxylic acids with a silver catalyst was reported [23]. Thus, in the presence of 10 mol% $AgNO_3$ and 5 equiv. of 30% aq. H_2O_2 in acetonitrile at 50 °C, various aromatic, conjugated, and aliphatic aldehydes were readily oxidized to the corresponding carboxylic acids in good yields.

The use of 150 nm-size Ag_2O/CuO and CuO catalysts was found to catalyze the oxidation of aromatic and aliphatic aldehydes to their corresponding acids by molecular oxygen in good yields [24]. This method has been used in the industrial production of furoic acid, although the obvious choice in this case would be CuO because of easy collection and regeneration of the catalyst.

Over the past decade, the utilization of gold catalysis has emerged as an important field in oxidation [25]. Gold supported on CeO_2 and TiO_2 nanoparticles has given promising results with high activity and selectivity in the oxidation of a variety of aldehydes to carboxylic acids [26, 27].

Oxidation of organic substrates by nickel oxide has been known for more than a century. The substrate scope of the nickel oxide hydroxide oxidation method is quite broad, and includes alcohols, aldehydes, phenols, amines, and oximes. However, an extensive review covering the major part of these reactions demonstrates the requirement of stoichiometric amounts of nickel oxide hydroxide [28]. Recent findings have led to a new oxidation method utilizing catalytic amounts of nickel(II) salts and excess of bleach (5% aqueous sodium hypochlorite) under ambient conditions [29]. Notably, the oxidation of aldehydes to the corresponding carboxylic acids also proceeds without the use of an organic solvent, giving estimated products in high yields (70–95%) and with high purities (90–100%).

10.3
Oxidation of Ketones

Although there are a number of different reaction types for the oxidation of ketones (see Chapter 9 of 'Modern Oxidation Methods' by Bolm), the present update chapter will only deal with the Baeyer-Villiger reaction.

10.3.1
Baeyer-Villiger Reactions

Ever since the discovery by Adolf Baeyer and Victor Villiger [30] that ketones can be oxidatively converted into the corresponding esters or lactones, substantial progress has been made in understanding the mechanism and predicting the migratory preference [31, 32]. However, the classical approach to the Baeyer-Villiger reaction has a major drawback: a potent oxidant, which in most cases is hazardous, is needed in stoichiometric amounts. In addition to the fact that these reactive oxidants have to be handled with care, they are relatively expensive. This has led to an increased interest in developing more gentle catalytic systems that are enantioselective and environmentally benign 'green' (chemistry).

10.3.2
Catalytic Asymmetric Baeyer-Villiger Reactions

In the late 20th century, chemists witnessed remarkable advancements in catalytic asymmetric synthesis. Significant efforts were devoted to the development of efficient oxidation catalysts, and a range of enantioselective metal-catalyzed reactions were developed [33]. Despite the long history of the Baeyer-Villiger (BV) oxidation, the enantioselective version of the process was not studied until recently.

After the pioneering work by the groups of Bolm [34] and Strukul [35] in 1994, a number of chiral metal complexes and organic molecules were developed as catalysts for the BV reaction of various ketones [36]. Although the use of aqueous hydrogen peroxide as an environmentally benign oxidant has been a long-sought goal for the BV oxidation, there are only a few cases in which catalytic reactions can be carried out with this oxidant [34, 37]. Impressive results have been achieved for the catalytic enantioselective BV reaction in the work reported by the groups of Bolm [34, 36], Katsuki [38], and Murahashi [37b]. However, the development of asymmetric BV reactions has been rather slow when compared to the rapid development of other catalytic asymmetric transformations [39], and significant efforts are required in this area to meet the strong demand for the development of more efficient catalytic methods where environmentally friendly oxidants are employed.

10.3.2.1 Chemocatalytic Versions

At present, high conversions together with high enantioselectivity can be achieved either with chiral enantiopure organocatalysts [37b,40] or with chiral metal complexes.

The BV transformations with chiral metal complexes are based on either transition metals such as Cu [37b, 40, 41], Pt [42, 43], Co [44], Pd [45], Zr [46], Hf [47] or nontransition metals such as Mg [48] and Al [49]. For example, Kočovský and coworkers have developed a series of new terpene-derived pyridinephosphine ligands, whose complexes with Pd(II) have been proven to catalyze the BV oxidation. Prochiral cyclobutanones (1) were oxidized at low temperature ($-40\,°C$), with 5 mol% catalyst loading and the urea-H_2O_2 complex as the stoichiometric oxidant to give lactones 2 in good yields and up to 81% ee (Eq. (10.5)) [50].

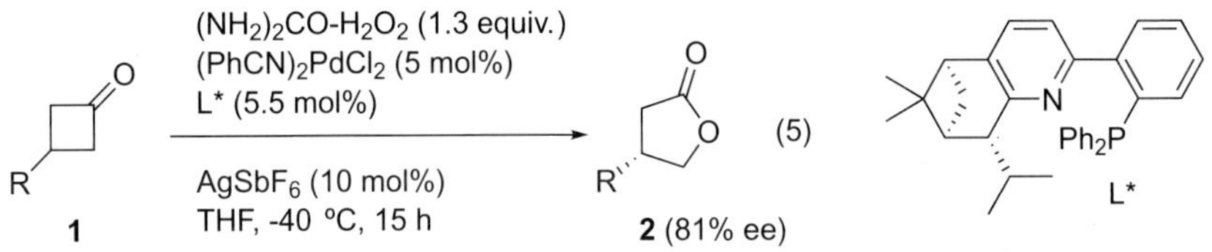

$$(10.5)$$

All these complexes are active towards meso or chiral cyclobutanone substrates 1, which are significantly more reactive than the larger cyclic ketones. Oxidation of the latter were only successful with Pt(II) [42, 46, 51, 52] or Cu(II) [53] catalysts, resulting in moderate to good enantiomeric excess.

Recently, the first example of an environmentally benign enantioselective BV oxidation, utilizing a strong chiral Brønsted acid as catalyst was reported [42]. With a chiral phosphoric acid and 30% aqueous H_2O_2 as the oxidant, various of 3-substituted cyclobutanones were converted into their corresponding γ-lactones in excellent yields and enantioselectivities up to 93% ee.

Among the synthetic catalysts used for cyclohexanones, the organocatalyst developed by Peris and Miller is noteworthy [54]. Recently, an efficient system for the oxidation of 4-substituted cyclohexanones (3) to lactones (4) was reported by Strukul and coworkers [55], in which they use a chiral enantiopure Pt(II) catalyst in water with added surfactant and hydrogen peroxide as the terminal oxidant (up to 92% ee), (Eq. (10.6)). The surfactant enables solubilization of the otherwise insoluble chiral catalyst and enhances the enantioselectivity of the reaction because of tighter catalyst–substrate interactions favored by the micellar supramolecular aggregate.

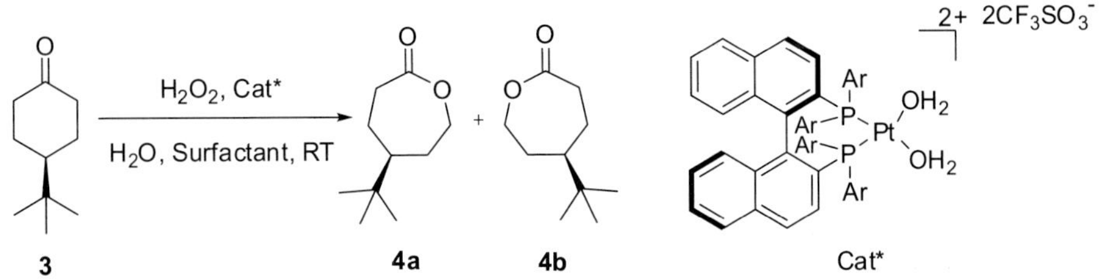

$$(10.6)$$

10.3.2.2 Biocatalytic Versions

Many of today's target molecules in oxidation are asymmetric and require enantioselective reactions for their preparation. An obvious approach is to use a biocatalyst together with molecular oxygen as the oxidant. The rapid development of molecular engineering during the past decade has enabled biocatalysis to become a well-established field of research for chemists, and has provided access to a wide range of interesting bioreagents.

The first indication of the existence of so-called Baeyer-Villiger monooxygenases (BVMOs) was reported in the late 1940s [56]. It was observed that certain fungi were able to oxidize steroids via a BV reaction [56], but two decades elapsed before the first BVMOs were isolated and characterized [57, 58]. All characterized BVMOs contain a flavin cofactor that is vital for the catalytic activity of the enzyme, Furthermore, NADH or NADPH cofactors are needed as electron donors. Careful inspection of all available biochemical data on BVMOs has revealed that at least two discrete classes of BVMOs exist, types I and II [59].

Type I and Type II BVMOs Type I BVMOs consist of only one polypeptide of about 500 amino acids with the cofactor flavin adenine dinucleotide (FAD) tightly bound into the structure, and is dependent on NADPH for activity [60]. Most of the presently known BVMOs are of this type, and are typically soluble proteins located in the cytosol of bacteria or fungi, which is in contrast to many other monooxygenase systems that often are found to be membrane bound or membrane associated. They contain two Rossmann sequence motifs, GxGxxG, which indicate that these enzymes bind the two cofactors (FAD or NADPH) using separate dinucleotide binding domains [61].

Type II BVMOs on the other hand are typically composed of two different subunits, and use flavin mononucleotide (FMN) as flavin cofactor and NADH as electron donor. At the time of the preliminary classification, the respective N-terminus sequences did not provide any clue concerning the structure of these two-component monooxygenases. However, sequence data suggest a relationship with the flavin-dependent luciferases [62]. In contrast to the Type I BVMOs, for which the genes have been frequently reported, Type II BVMOs have only been explored to a limited extent, and so far no cloning of Type II genes has been described in the literature. In fact only one Type II BVMO sequence (limonene monooxygenase, gi47116765) has been deposited in the database.

Other BVMOs Recent findings of several enzymes that display Baeyer-Villiger activity, although they do not resemble the above-mentioned BVMOs, suggest that at least two other BVMO classes exist [63].

A flavoprotein monooxygenase (MtmOIV) involved in the biosynthesis of mithramycin, and a heme-containing BVMO belonging to the cytochrome P450 superfamily have recently been reported [64]. Both enzymes were shown to catalyze BV oxidations [65]. In addition, a plant enzyme turned out to be capable of modifying steroids via BV oxidation [66]. Further studies will reveal more mechanistic details concerning these newly identified BVMOs that may be useful for the development of new biocatalytic applications in the future.

10.3 Oxidation of Ketones

Cyclohexanone Monooxygenase (CHMO$_{Acineto}$) Most biochemical and biocatalytic studies have been performed on Type I BVMOs [67], since they represent relatively uncomplicated monooxygenase systems, and several expression systems have been developed for a number of Type I BVMOs.

The isolation and characterization of cyclohexanone monooxygenase (CHMO) from *Acinetobacter* sp. NCIB 9871 was reported in 1976 [68]. In addition to cyclohexanone and cyclopentanone, CHMO was shown to be able to catalyze the oxidation of a variety of cyclic ketones to the corresponding lactones [69]. This attracted the attention of several other groups, which led to investigations of its mechanism [70, 71] as well as sequencing and cloning [72]. It is by far the most extensively studied BVMO, and it has been used as a model system for up-scaling BVMO-mediated biocatalysis.

A comprehensive list of ketonic substrates was first published in 1998 and has grown significantly in recent years [73]. Substrate profiling studies have shown that BVMOs have wide substrate specificity that often overlap. In 2002 it was reported that over 100 different substrates could be converted by CHMO [74], including cyclic, bicyclic, tricyclic, and heterocyclic ketones with a variety of substitution patterns [73, 75]. The oxidizing capacity of CHMO is not only limited to Baeyer-Villiger reactions, but has also been successfully employed in the enantioselective oxidation of a wide range of sulfides [76, 77], dithienes, dithiolanes [78], and in the conversion of tertiary amines, secondary amines, and hydroxylamines to their *N*-oxides, hydroxylamines, and nitrones, respectively [79].

Isolated Enzymes versus Designer Organisms In the early 1990s, biotransformations were performed with either isolated enzymes or whole-cell native organisms. All type I BVMOs are strictly dependent on cofactors, which complicates their utilization in organic synthesis. In particular, the consumption of stoichiometric quantities of NAD(P)H requires appropriate recycling strategies in order to allow for a cost-effictive process on an industrial scale [80].

To overcome this obstacle, two different approaches have been exploited, both possessing benefits and disadvantages. For the isolated enzyme, a closed-loop system has been developed, where an auxiliary substrate has been added to regenerate NAD(P)H. The second approach is based on whole-cell-mediated biotransformations.

A comparative study of the isolated CHMO enzyme with that overexpressed in *E. coli* has shown that they are practically identical. The slight difference in pH and thermal stability was due to the differences in trace impurities with proteolytic enzymes present in the *E. coli* host [81]. At present, the main focus has been on using intact cells expressing CHMO as a biocatalyst. Intact cells of 'designer' bioreagents are more accessible and more chemical friendly than the isolated enzymes used for biotransformations.

Biocatalytic Properties of Available Recombinant BVMOs The construction of recombinant overexpression systems for CHMO in *E. coli* and their success in regio- and enantioselective oxidation has inspired the search for other BVMOs. Several new

BVMOs have been overexpressed in *E. coli.*, and the list of reported Type I BVMOs has grown significantly during recent years (Table 10.1).

They include several cyclohexanone monooxygenases [83, 84], a steroid monooxygenase (SMO) [85], and cyclodecanone monooxygenase (CDMO). The latter was the first characterized enzyme to catalyze BV oxidation of large cyclic compounds [86]. In 2001, overexpression systems in *E. coli* were engineered for 4-hydroxyacetophenone monooxygenase (HAPMO) [87] and for cyclopentanone monooxygenase (CPMO) [88, 89]. Quite recently, BVMOs that readily accept phenylacetone derivatives (PAMO) have been described [90]. Also, variants specific for acetone (ACMO) [91], linear aliphatic ketones (AKMO) [92], and for the large cyclic ketone cyclopentadecanone (CPDMO) [93] were reported.

Substrate-profiling studies suggest that BVMOs have a rather broad specificity and often display overlapping substrate specificities. However, catalytic efficiencies and regio- and/or enantioselectivities can differ significantly when comparing BVMOs.

Comparative biocatalytic studies using highly similar enzymes have revealed that all studied CHMOs and CPMOs cover a similar substrate range [94–96], although it was observed that CPMOs and CHMOs often display opposite enantioselectivities [96].

Although BVMOs display broad substrate specificity, each type of BVMO has a certain preference for a specific type of substrate. CHMO and CPMO are highly active with a range of smaller cyclic aliphatic ketones, whereas HAPMO and PAMO prefer aromatic substrates [87, 90, 97–99].

Directed Evolution of Enantioselective Enzymes for Catalysis in Baeyer-Villiger Reactions
Reetz and co-workers have demonstrated that the methods of directed evolution can be applied successfully to the creation of enantioselective cyclohexanone monooxygenases (CHMOs) as catalysts in Baeyer-Villiger reactions of several different substrates, for which the enantioselectivity ranges between 90–99% [100]. Ketone **5** gives a very poor enantioselectivety (9% *ee*, *R*-selective) with the wild-type CHMO. The enantioselectivety for **5** was significantly improved by directed evolution, and an *S*-selective variant gave 79% *ee* (Scheme 10.1).

Recently, Reetz has devised a new strategy in directed evolution in order to construct a robust experimental platform for asymmetric Baeyer-Villiger reactions based on the thermostable phenylacetone monooxygenase (PAMO) [101]. Unfortunately, the substrate scope of the wild-type (WT) PAMO is very limited, only accepting phenylacetone and structurally similar linear phenyl-substituted ketones. By exploiting bioinformatics data derived from sequence alignment of eight different BVMOs, in conjunction with the known X-ray structure of PAMO, this problem could be circumvented. Their goal was to expand the substrate scope, to increase the reaction rate, and to reach high enantioselectivity without compromising thermostability. Mutants were evolved which showed unusually high activity and enantioselectivity in the oxidative kinetic resolution of a variety of structurally different 2-substituted aryl- and alkylcyclohexanone derivatives and of a structurally unrelated bicyclic ketones. It is interesting to note that WT PAMO favors the formation of **8** as the (1*S*,5*R*)-enantiomer and also produces some **9** as the (1*S*,5*R*)-enantiomer, whereas the

Table 10.1 List of BVMOs that have been overexpressed in *E. coli*. [82].

BVMO	Abbreviation	Primary substrate	Origin	Year of cloning
Cyclohexanone monooxygenase	CHMO	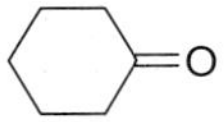 cyclohexanone	*Acinetobacter* sp. NCIMB 9871	1988
Steroid monooxygenase	STMO	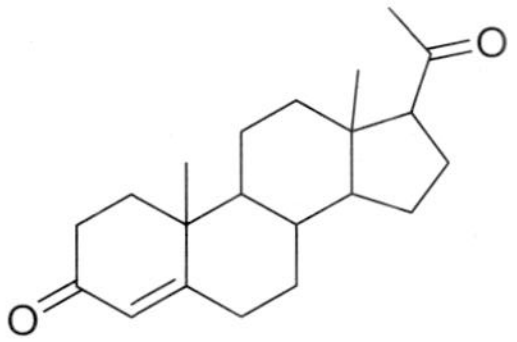 progesterone	*Rhodococcus rhodochrous*	1999
4-Hydroxyacetophenone monooxygenase	HAPMO	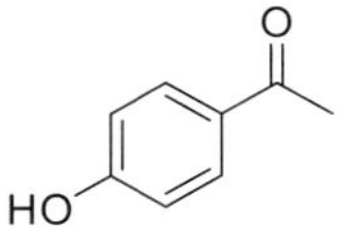 4-hydroxyacetophenone	*Pseudomonas fluorescens* ACB	2001
Cyclododecanone monooxygenase	CDMO	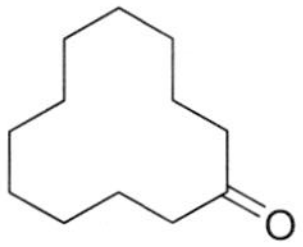 cyclododecanone	*Rhodococcus ruber*	2001
Cyclopentanone monooxygenase	CPMO	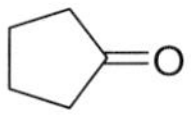 cyclopentanone	*Comamonas* sp. NCIMB 9872	2002

(*Continued*)

Table 10.1 (Continued)

BVMO	Abbreviation	Primary substrate	Origin	Year of cloning
Phenylacetone monooxygenase	PAMO	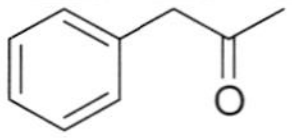 phenylacetone structure	*Thermobifida fusca*	2005
Acetone monooxygenase	ACMO	 acetone structure	*Gordonia* sp. TY-5	2006
Alkyl ketone monooxygenase	AKMO	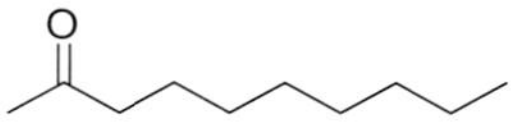 alkyl ketone structure	*Pseudomonas fluorescens*	2006
Cyclopentadecanone monooxygenase	CPDMO	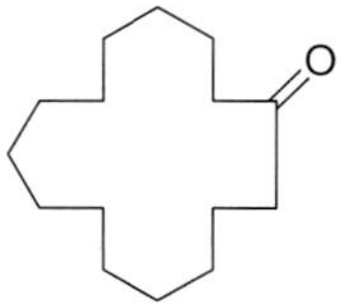 cyclopentadecanone structure	*Pseudomonas* sp. strain HI-70	2006

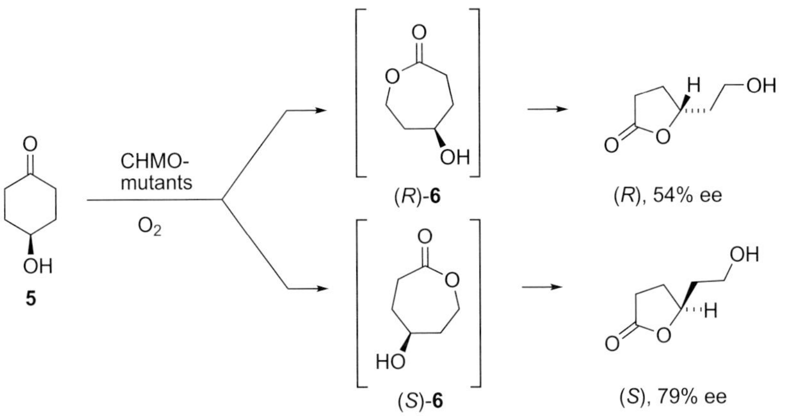

Scheme 10.1 Directed evolution of cyclohexanone monooxygenases (CHMOs) improves enantioselectivety in Baeyer-Villiger reactions.

mutants lead to a reversal of enantioselectivity in the case of regioisomer **9** (Scheme 10.2).

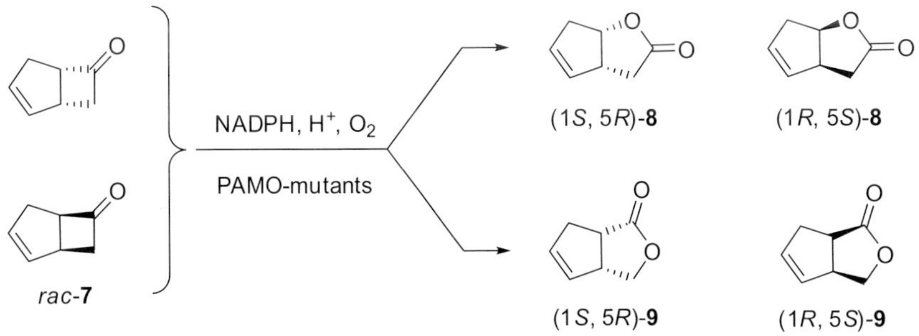

Scheme 10.2 PAMO-mutants with high activity and enantioselectivity in the oxidative kinetic resolution of bicyclic ketone **7**.

Advances in Regeneration Systems of BVMOs Additional contributions to the proliferation of the Baeyer-Villiger biotransformation platform is the combination of the catalytic activity of a redox biocatalyst with concomitant coenzyme recycling in a single fusion protein, reported by Mihovilovic and Fraaije (Scheme 10.3) [102].

This self-sufficient two-in-one redox biocatalyst enables the use of phosphite (HPO_3^{2-}) as an inexpensive and sacrificial electron donor, and does not require an additional catalytic entity for coenzyme recycling. As model BVMOs, PAMO, CHMO, and CPMO were selected. No inhibition of the activity of either the BVMO or phosphite dehydrogenase (PTDH) by the substrate or product of the other subunit could be observed.

The use of expensive stoichiometric amounts of NADPH or complicated regeneration systems (Scheme 10.4 A) can be avoided by employing light as a source of

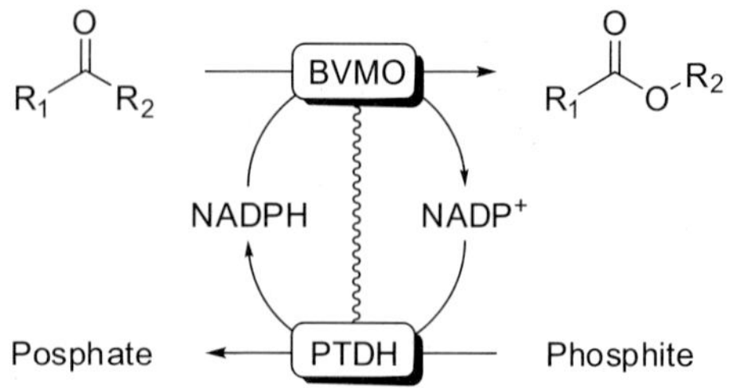

Scheme 10.3 Coenzyme regeneration by CRE/BVMO fusion enzymes.

energy (Scheme 10.4 B). In 2007, Reetz and coworkers devised and implemented experimentally for the first time a catalytic scheme for a light-driven flavin-dependent enzymatic reaction, using EDTA as a reservoir of electrons [103].

As the flavin-dependent enzyme, a mutant of phenylacetone monooxygenase, PAMO-P3, was chosen [104], which had previously been engineered to catalyze the enantioselective BV reaction of substituted ketones. The unnatural regeneration conditions did not impair the inherent stereoselectivity of PAMO-P3, suggesting an unaltered mechanism of the actual oxidation reaction. The present light-driven BV reaction is expected to work just as well using other flavin-dependent BVMOs such as CHMO, which accepts a broader range of substrates.

Despite their efficiency and many successes, the bioreagents discussed so far do not fulfill all the needs of synthetic chemistry. There are still a lot of substrates that

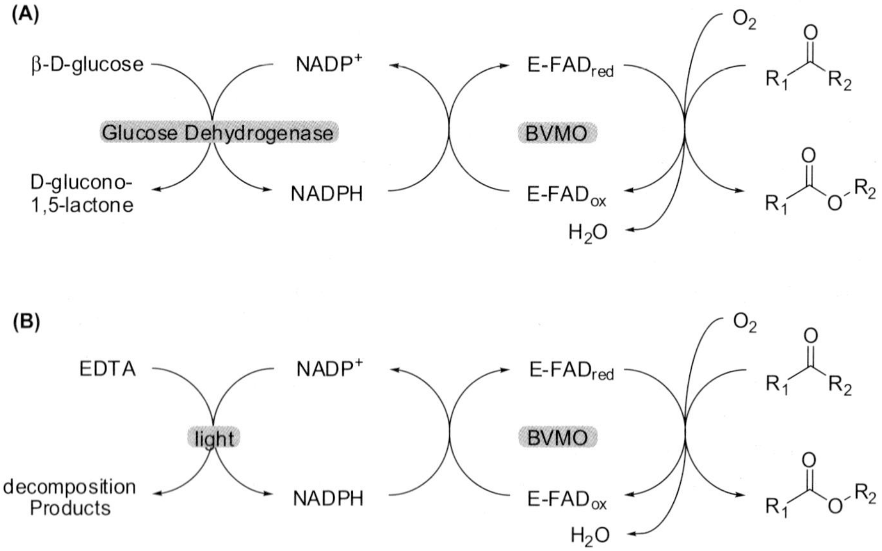

Scheme 10.4 Comparison of the conventional regeneration of a Baeyer-Villiger monooxygenase (BVMO, e.g., PAMO) for catalysis through enzymatic regeneration of NADPH (A) and the novel simplified light-driven pathway using a flavin (B).

are not accepted by any of the currently available BVMOs. In other cases, the regio- and enantio-selectivities of the reactions are unacceptably low. While the number of recombinant BVMOs has increased vastly over the past few years, there is still a demand for other BVMOs to expand the biocatalytic diversity. To meet this demand new BVMOs have to be discovered or engineered by enzyme redesign. This truly illustrates the need for a large library of BVMOs to meet the demand for enantio- and/or regioselective reactions.

References

1. (a) Larock, R.C. (1989) *Comprehensive Organic Transformations*, VCH Publishers, New York; (b) Sagar, S., Nilotpal, B., and Jubaraj, B.B. (2005) *J. Mol. Catal. A: Chem.*, **229**, 171; (c) Joshua, H. (2000) *Tetrahedron Lett.*, **41**, 6627.
2. (a) Dalcanale, E. and Montanari, F. (1986) Selective oxidation of aldehydes to carboxylic acids with sodium chlorite-hydrogen peroxide. *J. Org. Chem.*, **51**, 567–569; (b) Bayle, J.P., Perez, F., and Courtieu, J. (1990) Oxidation of aldehydes by sodium chlorite. *Bull. Soc. Chim. Fr.*, **4**, 565–567; (c) Babu, B.R. and Balasubramaniam, K.K. (1994) Simple and facile oxidation of aldehydes to carboxylic acids. *Org. Prep. Proced. Int.*, **26**, 123–125; (d) Hase, T. and Wähälä, K. (1995) *Reagents for Organic Synthesis*, vol. 7 (ed. L.A. Paquette), Wiley, Chichester, p. 4533.
3. (a) Hurd, C.D., Garrett, J.W., and Osborne, E.N. (1933) Furan reactions. IV. Furoic acid from furfural. *J. Am. Chem. Soc.*, **55**, 1082–1084; (b) Gupta, K.K.S., Dey, S., Gupta, S.S., Adhikari, M., and Banerjee, A. (1990) Evidence of esterification during the oxidation of some aromatic aldehydes by chromium (VI) in acid medium and the mechanism of the oxidation process. *Tetrahedron*, **46**, 2431–2444; (c) Cainelli, G. and Cardillo, G. (eds) (1984) *Chromium Oxidations in Organic Chemistry*, Springer, New York, (d) Bowden, K., Heilbron, I.M., Joes, E.R.H., and Weedon, B.C.L. (1946) *J. Chem. Soc.*, 39.
4. (a) Jursic, B. (1989) Surfactant-assisted permanganate oxidation of aromatic compounds. *Can. J. Chem.*, **67**, 1381–1383; (b) Takemoto, T., Yasuda, K., and Ley, S.V. (2001) Solid-supported reagents for the oxidation of aldehydes to carboxylic acids. *Synlett*, **10**, 1555–1556; (c) Arndt, D. (ed.) (1980) *Manganese Compounds as Oxidizing agents in Organic Chemistry*, Open Court, La Salle, (d) Fatiadi, A.J. (1987) *Synthesis*, 85.
5. (a) Sato, K., Hyodo, M., Takagi, J., Aoki, M., and Noyori, R. (2000) Hydrogen peroxide oxidation of aldehydes to carboxylic acids: an organic solvent-, halide- and metal-free procedure. *Tetrahedron Lett.*, **41**, 1439–1442; (b) Heaney, H. and Newbold, A.J. (2001) The oxidation of aromatic aldehydes by magnesium monoperoxyphthalate and urea-hydrogen peroxide. *Tetrahedron Lett.*, **42**, 6607–6610; (c) Ando, W. (ed.) (1992) *Organic Peroxides*, Wiley, Chichester.
6. Ji, H.B. and She, Y.B. (2005) *Green Oxidation and Reduction*, China Petrochemical Press, Beijing.
7. Shapiro, N. and Vigalok, A. (2008) *Angew. Chem. Int. Ed.*, **47**, 2849–2852.
8. As the aldehyde oxidation reactions can be influenced by even trace amounts of metal (see for example: Loeker, F. and Leitner, W. (2000) *Chem. Eur. J.*, **6**, 2011.), the reactions in Ref. 7 were performed under strict metal-free conditions.
9. Kiumars, B., Mehdi, K.M., and Shahab, K. (2008) *Chin. J. Chem.*, **26**, 1119–2112.
10. Enders, D., Niemeier, O., and Henseler, A. (2007) *Chem. Rev.*, **107**, 5606.
11. Miyashita, A., Suzuki, Y., Nagasaki, I., Ishiguro, C., Iwamoto, K., and Higashino, T. (1997) *Chem. Pharm. Bull.*, **45**, 1254–1258.

12 Travis, B.R., Sivakumar, M., Hollist, G.O., and Borhan, B. (2003) *Org. Lett*, **5**, 1031–1034.

13 Sayama, S. and Onami, T. (2004) *Synlett*, 2739–2745.

14 (a) Gopinath, R., Barkakaty, B., Talukdar, B., and Patel, B.K. (2003) *J. Org. Chem.*, **68**, 2944–2947; (b) Yoo, W.J. and Li, C.J. (2007) *Tetrahedron Lett.*, **48**, 1033–1035.

15 (a) NHC-catalyzed reactions of aldehydes leading to acid derivatives: Chiang, P.-C., Kaeobamrung, J., and Bode, J.W. (2007) *J. Am. Chem. Soc.*, **129**, 3520; (b) Bode, J.W. and Sohn, S.S. (2007) *J. Am. Chem. Soc.*, **129**, 13798.

16 Guin, J., Sarkar, S.D., Grimme, S., and Studer, A. (2008) *Angew. Chem. Int. Ed.*, **47**, 8727–8730.

17 For relevant examples, see: (a) Castells, J., Llitjos, H., and Morenomanas, M. (1977) *Tetrahedron Lett.*, **18**, 205–206; (b) Inoue, H. and Higashiura, K. (1980) *Chem. Commun.*, 549–550; (c) Miyashita, A., Suzuki, Y., Nagasaki, I., Ishiguro, C., Iwamoto, K., and Higashino, T. (1997) *Chem. Pharm. Bull.*, **45**, 1254–1258; (d) Tam, S.W., Jimenez, L., and Diederich, F. (1992) *J. Am. Chem. Soc.*, **114**, 1503–1505; (e) Chan, A., and Scheidt, K.A. (2006) *J. Am. Chem. Soc.*, **128**, 4558–4559.

18 Radical chemistry in organocatalysis: (a) Dressel, M., Aechtner, T., and Bach, T. (2006) *Synthesis*, 2206; (b) Bauer, A., Westkaemper, F., Grimme, S., and Bach, T. (2005) *Nature*, **436**, 1139; (c) Aechtner, T., Dressel, M., and Bach, T. (2004) *Angew. Chem.*, **116**, 5974; (2004) *Angew. Chem. Int. Ed.*, **43**, 5849; (d) Cho, D.H. and Jang, D.O. (2006) *Chem. Commun.*, 5045; (e) Sibi, M.P. and Hasegawa, M. (2007) *J. Am. Chem. Soc.*, **129**, 4124; (f) Jang, H.-Y., Hong, J.-B., and MacMillan, D.W.C. (2007) *J. Am. Chem. Soc.*, **129**, 7004; (g) Beeson, T.D., Mastracchio, A., Hong, J.-B., Ashton, K., and MacMillan, D.W.C. (2007) *Science*, **316**, 582; (h) Kim, H. and MacMillan, D.W.C. (2008) *J. Am. Chem. Soc.*, **130**, 398.

19 Jallabert, C., Lapinte, C., and Riviere, H. (1982) *J. Mol. Catal., B Enzym.*, **14**, 75–86.

20 Mannam, S. and Sekar, G. (2008) *Tetrahedron Lett.*, **49**, 1083–1086.

21 Pearl, I.A. (1963) *Org. Synth.*, **4**, 972.

22 Campaigne, E. and LeSuer, W.M. (1963) *Org. Synth.*, **4**, 919.

23 Chakraborty, D., Gowda, R.R., and Malik, P. (2009) *Tetrahedron Lett.*, **50**, 6553–6556.

24 Tian, Q., Shi, D., and Sha, Y. (2008) *Molecules*, **13**, 948–957.

25 Stephen, A., Hashmi, K., and Hutchings, G.J. (2006) *Angew. Chem., Int. Ed.*, **45**, 7896.

26 Corma, A. and Domine, M.E. (2005) *Chem. Commun.*, 4042–4044.

27 Marsden, C., Taarning, E., Hansen, D., Johansen, L., Klitgaard, S.K., Egeblad, K., and Christensen, C.H. (2008) *Green Chem.*, **10**, 168–170.

28 George, M.V. and Balachandran, K.S. (1975) *Chem. Rev.*, **75**, 491–519.

29 Grill, M.J., Ogle, J.W., and Miller, S.A. (2006) *J. Org. Chem.*, **71**, 9291–9296.

30 Baeyer, A. and Villiger, V. (1899) *Chem. Ber.*, **32**, 3625–3633.

31 Krow, G.R. (1993) *Org. React.*, **43**, 251–798.

32 Renz, M. and Meunier, B. (1999) *Eur. J. Org. Chem.*, 737–750.

33 (a) Jacobsen, E.N., Pfaltz, A., and Yamamoto, H. (1999) *Comprehensive Asymmetric Catalysis*, vol. II, Springer, Berlin, (b) Ojima, I. (2000) *Catalytic Asymmetric Synthesis*, 2nd edn, Wiley-VCH, New York.

34 (a) Bolm, C., Schlingloff, G., and Weickhardt, K. (1994) *Angew. Chem.*, **106**, 1944–1946; (1994) *Angew. Chem. Int. Ed. Engl.*, **33**, 1848–1849.

35 Gusso, A., Baccin, C., Pinna, F., and Strukul, G. (1994) *Organometallics*, **13**, 3442–3451.

36 For reviews, see: (a) Bolm, C. and Beckmann, O. (1999) *Comprehensive Asymmetric Catalysis*, vol. 2 (eds E.N. Jacobsen, A., Pfaltz, and H. Yamamoto), Springer, Berlin, pp. 803–810; (b) Bolm, C., Palazzi, C., and Bechmann, O. (2004) *Transition Metals for Organic Chemistry: Building Blocks and Fine Chemicals*, 2nd edn, vol. 2 (eds M. Beller and C. Bolm), Wiley-VCH, Weinheim, pp. 267–274; (c) ten Brink, G.-J., Arends, I.W.C.E., and Sheldon, R.A. (2004) *Chem. Rev.*, **104**, 410–4123; (d) Bolm, C. (2006) *Asymmetric Synthesis-The Essentials* (eds M.

Christmann and S. Bräse), Wiley-VCH, Weinheim, pp. 57–61.

37 For selected examples, see: (a) Paneghetti, C., Cavagnin, R., Pinna, F., and Strukul, G. (1999) *Organometallics*, **18**, 5057–5065; (b) Murahashi, S., Ono, S., and Imada, Y. (2002) *Angew. Chem.*, **114**, 2472–2474; (2002) *Angew. Chem. Int. Ed.*, **41**, 2366–2368. (c) Colladon, M., Scarso, A., and Strukul, G. (2006) *Synlett*, 3515–3520.

38 (a) Uchida, T. and Katsuki, T. (2001) *Tetrahedron Lett.*, **42**, 6911–6914; (b) Watanabe, A., Uchida, T., Ito, K., and Katsuki, T. (2002) *Tetrahedron Lett.*, **43**, 4481–4485; (c) Ito, K., Ishii, A., Kuroda, T., and Katsuki, T. (2003) *Synlett*, 643–646. (d) Watanabe, A., Uchida, T., Irie, Y., and Katsuki, T. (2004) *Proc. Natl. Acad. Sci. USA*, **101**, 5737–5742.

39 For examples, see: (a) Noyori, R. (1994) *Asymmetric Catalysis in Organic Synthesis*, Wiley-Interscience, New York; (b) Ojima, I. (ed.) (2000) *Catalytic Asymmetric Synthesis*, 2nd edn, Wiley-VCH, Weinheim; (c) Jacobsen, E.N., Pfaltz, A., and Yamamoto, H. (eds) (1999) *Comprehensive Asymmetric Catalysis*, vol. **I–III**, Springer, Berlin; (d) Yamamoto, H. (ed.) (2001) *Lewis Acids in Organic Synthesis*, Wiley-VCH, Weinheim; (e) Blaser, H.U. and Schmidt, E. (eds) (2004) *Asymmetric Catalysis on Industrial Scale: Challenges, Approaches and Solutions*, Wiley-VCH, Weinheim; (f) Mikami, K. and Lautens, M. (eds) (2007) *New Frontiers in Asymmetric Catalysis*, Wiley-VCH, Weinheim.

40 Wang, B., Shen, Y.-M., and Shi, Y. (2006) *J. Org. Chem.*, **71**, 9519–9521.

41 Bolm, C., Khanh Luong, T.K., and Schlingloff, G. (1997) *Synlett*, 1151–1152.

42 Xu, S., Wang, Z., Zhang, X., Zhang, X., and Ding, K. (2008) *Angew. Chem.*, **120**, 2882; (2008) *Angew. Chem. Int. Ed.*, **47**, 2840–2843.

43 Paneghetti, C., Gavagnin, R., Pinna, F., and Strukul, G. (1999) *Organometallics*, **18**, 5057–5065.

44 Uchida, T. and Katsuki, T. (2001) *Tetrahedron Lett.*, **42**, 6911–6914.

45 Ito, K., Ishii, A., Kuroda, T., and Katsuki, T. (2003) *Synlett*, 0643–0646.

46 (a) Watanabe, A., Uchida, T., Irie, R., and Katsuki, T. (2004) *Proc. Natl. Acad. Sci. USA*, **101**, 5737–5742; (b) Watanabe, A., Uchida, T., Ito, K., and Katsuki, T. (2002) *Tetrahedron Lett.*, **43**, 4481–4485; (c) Bolm, C. and Beckmann, O. (2000) *Chirality*, **12**, 523–525.

47 Matsumoto, K., Watanabe, A., Uchida, T., Ogi, K., and Katsuki, T. (2004) *Tetrahedron Lett.*, **45**, 2385–2388.

48 Bolm, C., Beckmann, O., Cosp, A., and Palazzi, C. (2001) *Synlett*, 1461–1463.

49 (a) Bolm, C., Frison, J.-C., Zhang, Y., and Wulff, W.D. (2004) *Synlett*, 1619–1621; (b) Frison, J.-C., Palazzi, C., and Bolm, C. (2006) *Tetrahedron*, **62**, 6700–6706; (c) Bolm, C., Beckmann, O., K_hn, T., Palazzi, C., Adam, W., Bheema Rao, P., and Saha-Mçller, C.R. (2001) *Tetrahedron: Asymmetry*, **12**, 2441–2446; (d) Bolm, C., Beckmann, O., and Palazzi, C. (2001) *Can. J. Chem.*, **79**, 1593–1597.

50 Malkov, A.V., Friscourt, F., Bell, M., Swarbrick, M.E., and Kocovsky, P. (2008) *J. Org. Chem.*, **73**, 3996–4003.

51 Del Todesco Frisone, M., Pinna, F., and Strukul, G. (1993) *Organometallics*, **12**, 148–156.

52 (a) Strukul, G., Varagnolo, A., and Pinna, F. (1997) *J. Mol. Catal. A*, **117**, 413–423; (b) Gavagnin, R., Cataldo, M., Pinna, F., and Strukul, G. (1998) *Organometallics*, **17**, 661–667; (c) Brunetta, A. and Strukul, G. (2004) *Eur. J. Inorg. Chem.*, 1030–1038; (d) Michelin, R.A., Pizzo, E., Sgarbossa, P., Scarso, A., Strukul, G., and Tassan, A. (2005) *Organometallics*, **24**, 1012–1017; (e) Conte, V., Floris, B., Galloni, P., Mirruzzo, V., Sordi, D., Scarso, A., and Strukul, G. (2005) *Green Chem.*, **7**, 262–266; (f) Sgarbossa, P., Scarso, A., Pizzo, E., Mazzega Sbovata, S., Michelin, R.A., Strukul, G., Mozzon, M., and Benetollo, F. (2006) *J. Mol. Catal. A*, **243**, 202–206; (g) Sgarbossa, P., Scarso, A., Michelin, R.A., and Strukul, G. (2007) *Organometallics*, **26**, 2714–2719; (h) Greggio, G., Sgarbossa, P., Scarso, A., Michelin, R.A., and Strukul, G. (2008) *Inorg. Chim. Acta*, **361**, 3230–3236.

53 Crudden, C.M., Chen, A.C., and Calhoun, L.A. (2000) *Angew. Chem.*, **112**, 2973–2977; (2000) *Angew. Chem. Int. Ed.*, **39**, 2851–2855.

54 Peris, G. and Miller, S. (2008) *Org. Lett.*, **10**, 3049–3052.
55 Cavarzan, A., Bianchini, G., Sgarbossa, P., Lefort, L., Gladiali, S., Scarso, A., and Strukul, G. (2009) *Chem. Eur. J.*, **15**, 7930–7939.
56 Turfitt, G.E. (1948) *Biochem. J.*, **42**, 376–383.
57 Conrad, H.E., DuBus, R., Namtvedt, M.J., and Gunsalus, I.C. (1965) *J. Biol. Chem.*, **240**, 495–503.
58 Forney, F.W. and Markovetz, A.J. (1969) *Biochem. Biophys. Res. Commun.*, **37**, 31–38.
59 Willetts, A. (1997) *Trends Biotechnol.*, **15**, 55–62.
60 Kamerbeek, N.M., Janssen, D.B., Berkel, W.J.H., and Fraaije, M.W. (2003) *Adv. Synth. Catal.*, **345**, 667–678.
61 Wierenga, R.K., Terpstra, P., and Hol, W.G.J. (1986) *J. Mol. Biol.*, **187**, 101–107.
62 van Berkel, W.J.H., Kamerbeek, N.M., and Fraaije, M.W. (2006) *J. Biotechnol.*, **124**, 670–689.
63 Schmid, R.D. and Urlacher, V.B. (eds) (2007) *Modern Biooxidation: Enzymes, Reactions and Applications*, Wiley-VCH, Weinheim, p. 79.
64 Nomura, T., Kushiro, T., Yokota, T., Kamiya, Y., Bishop, G.J., and Yamaguchi, S. (2005) *J. Biol. Chem.*, **280**, 17873–17879.
65 Rodríguez, D., Quirós, L.M., Braña, A.F., and Salas, J.A. (2003) *J. Bacteriol.*, **185**, 3962–3965.
66 Swinney, D.C. and Mak, A.Y. (1994) *Biochemistry*, **33**, 2185–2190.
67 Kamerbeek, N.M., Janssen, D.B., van Berkel, W.J.H., and Fraaije, M.W. (2003) *Adv. Synth. Catal.*, **345**, 667–678.
68 Trudgill, P.W. and Gibson, D.T. (eds) (1984) *Microbial Degradation of Organic Compounds.*, Marcel Dekker, New York, NY, pp. 131–180.
69 Donoghue, N., Norris, D., and Trudgill, P.W. (1976) *Eur. J. Biochem.*, **63**, 175–192.
70 (a) Ryerson, C.C., Ballou, D.P., and Walsh, C.T. (1982) *Biochemistry*, **21**, 2644–2655; (b) Branchaud, B.P., and Walsh, C.T. (1985) *J. Am. Chem. Soc.*, **107**, 2153–2161; (c) Latham, J.A., and Walsh, C.T. (1987) *J. Am. Chem. Soc.*, **109**, 3421–3427.
71 (a) Schwab, J.M. (1981) *J. Am. Chem. Soc.*, **103**, 1876–1878; (b) Schwab, J.M., Li, W.-B., and Thomas, L.P. (1983) *J. Am. Chem. Soc.*, **105**, 4800–4808.
72 Chen, Y.-C.J., Peoples, O.P., and Walsh, C.T. (1988) *J. Bacteriol.*, **170**, 781–789.
73 Stewart, J.D. (1998) *Curr. Org. Chem.*, **2**, 195–216.
74 Mihovilovic, M.D., Müller, B., and Stanetty, P. (2002) *Eur. J. Org. Chem.*, **22**, 3711–3730.
75 (a) Gonzalez, D., Schapiro, V., Seoane, G., and Hudlicky, T. (1997) *Tetrahedron: Asymmetry*, **8**, 975–977; (b) Hudlicky, T., and Reed, J.W. (1995) *Advances in Asymmetric Synthesis*, vol. **1** (ed. A. Hassner), JAI, pp. 271–312.
76 Colonna, S., Gaggero, N., Carrea, G., and Pasta, P. (1997) *Chem. Commun.*, 439–440.
77 Ottolina, G., Pasta, P., Varley, D., and Holland, H.L. (1996) *Tetrahedron: Asymmetry*, **7**, 3427–3430.
78 (a) Colonna, S., Gaggero, N., Pasta, P., and Ottolina, G. (1996) *Chem. Commun.*, 2303–2307; (b) Colonna, S., Gaggero, N., Bertinotti, A., Carrea, G., Pasta, P., and Bernardi, A. (1995) *J. Chem. Soc. Chem. Commun.*, 1123–1124; (c) Colonna, S., Gaggero, N., Carrea, G., and Pasta, P. (1998) *Chem. Commun.*, 415–416.
79 Colonna, S., Pironti, V., Pasta, P., and Zambianchi, F. (2003) *Tetrahedron Lett.*, **44**, 869–871.
80 Alphand, V., Carrea, G., Wohlgemuth, R., Furstoss, R., and Woodley, J.M. (2003) *Trends Biotechnol.*, **21**, 318–323.
81 Secundo, F., Zambianchi, F., Crippa, G., Carrea, G., and Tedeschi, G.J. (2005) *Mol. Catal. B:Enzym.*, **34**, 1–6.
82 Pazmiño, D.E.T. (2008) Molecular Redesign of Baeyer-Villiger Monooxygenases, Dissertation, Groningen.
83 Chen, Y.C., Peoples, O.P., and Walsh, C.T. (1988) *J. Bacteriol.*, **170**, 781–789.
84 Donoghue, N.A., Norris, D.B., and Trudgill, P.W. (1976) *Eur. J. Biochem.*, **63**, 175–192.
85 Morii, S., Sawamoto, S., Yamauchi, Y., Miyamoto, M., Iwami, M., and

Itagaki, E. (1999) *J. Biochem.*, **126**, 624–631.

86 Kostichka, K., Thomas, S.M., Gibson, K.J., Nagarajan, V., and Cheng, Q. (2001) *J. Bacteriol.*, **183**, 6478–6486.

87 Kamerbeek, N.M., Moonen, M.J.H., van Der Ven, J.G.M., van Berkel, W.J.H., Fraaije, M.W., and Janssen, D.B. (2001) *Eur. J. Biochem.*, **268**, 2547–2557.

88 Griffin, M. and Trudgill, P.W. (1976) *Eur. J. Biochem.*, **63**, 199–209.

89 Iwaki, H., Hasegawa, Y., Wang, S., Kayser, M.M., and Lau, P.C.K. (2002) *Appl. Environ. Microbiol*, **68**, 5671–5684.

90 Fraaije, M.W., Wu, J., Heuts, D.P.H.M., van Hellemond, E.W., Lutje Spelberg, J.H., and Janssen, D.B. (2005) *Appl. Microbiol. Biotechnol.*, **66**, 393–400.

91 Kotani, T., Yurimoto, H., Kato, N., and Sakai, Y. (2007) *J. Bacteriol.*, **189**, 886–893.

92 Kirschner, A., Altenbuchner, J., and Bornscheuer, U.T. (2006) *Appl. Microbiol. Biotechnol.*, **75**, 1065–1072.

93 Iwaki, H., Wang, S., Grosse, S., Bergeron, H., Nagahashi, A., Lertvorachon, J., Yang, J., Konishi, Y., Hasegawa, Y., and Lau, P.C.K. (2006) *Appl. Environ. Microbiol.*, **72**, 2707–2720.

94 Mihovilovic, M.D., Snajdrova, R., and Grötzl, B. (2006) *J. Mol. Catal. B*, **39**, 135–140.

95 Kyte, B.G., Rouvière, P.E., Cheng, Q., and Stewart, J.D. (2004) *J. Org. Chem.*, **69**, 12–17.

96 Mihovilovic, M.D., Rudroff, F., Grötzl, B., Kapitan, P., Snajdrova, R., Rydz, J., and Mach, R. (2005) *Angew. Chem. Int. Ed.*, **44**, 3609–3613.

97 de Gonzalo, G., Torres Pazmiño, D.E., Ottolina, G., Fraaije, M.W., and Carrea, G. (2005) *Tetrahedron: Asymmetry*, **16**, 3077–3083.

98 de Gonzalo, G., Torres Pazmiño, D.E., Ottolina, G., Fraaije, M.W., and Carrea, G. (2006) *Tetrahedron: Asymmetry*, **17**, 130–135.

99 Mihovilovic, M.D., Kapitan, P., Rydz, J., Rudroff, F., Ogink, F.H., and Fraaije, M.W. (2005) *J. Mol. Catal. B: Enzym.*, **32**, 135–140.

100 Reetz, M.T., Brunner, B., Schneider, T., Schulz, F., Clouthier, C.M., and Kayser, M.M. (2004) *Angew. Chem. Int. Ed.*, **43**, 4075–4078.

101 Reetz, M.T. and Wu, S. (2009) *J. Am. Chem. Soc.*, **131**, 15424–15432.

102 Pazmiño, D.E.T., Snajdrova, R., Baas, B.-J., Ghobrial, M., Mihovilovic, M.D., and Fraaije, M.W. (2008) *Angew. Chem. Int. Ed.*, **47**, 2275–2278.

103 Hollmann, F., Taglieber, A., Schulz, F., and Reetz, M.T. (2007) *Angew. Chem. Int. Ed.*, **46**, 2903–2906.

104 (a) Fraaije, M.W., Wu, J., Heuts, D.P.H.M., van Hellemond, E.W., Spelberg, J.H.L., and Janssen, D.B. (2005) *Appl. Microbiol. Biotechnol.*, **66**, 393–400; (b) crystal structure of PAMO: Malito, E., Alfieri, A., Fraaije, M.W., and Mattevi, A. (2004) *Proc. Natl. Acad. Sci. USA*, **101**, 13157–13162.

11
Manganese-Catalyzed Oxidation with Hydrogen Peroxide
Wesley R. Browne, Johannes W. de Boer, Dirk Pijper, Jelle Brinksma, Ronald Hage, and Ben L. Feringa

11.1
Introduction

Oxidation reactions are among the most important synthetic transformations in organic chemistry [1] and offer core methods for the introduction and modification of functional groups [2]. Despite its central role, the continued use of stoichiometric oxidants including nitric acid, chromic acid and its derivatives, alkyl hydroperoxides, permanganate, osmium tetroxide, hypochlorite bleach, hydrogen peroxide, and peracids is remarkable considering the extensive use of catalytic methods in most other areas of chemical synthesis and the high costs, formation of toxic waste, and low atom efficiency that the use of stoichiometric methods involves. The increasing role that catalysis plays in organic chemistry in general and the ever-increasing consideration given to environmental aspects provide a strong incentive to develop sustainable catalytic alternatives to stoichiometric reagents [3–6].

Oxidations with molecular dioxygen [7], the primary oxidant in biologic systems [8], are desirable and indeed already employed in large-scale industrial processes [9], albeit being frequently hampered by low conversion, modest selectivity, and safety issues. Although both oxygen atoms of O_2 can be transferred in dioxygenation or reactions with singlet dioxygen, frequently stoichiometric amounts of a reducing agent are required to convert one of the oxygen atoms of O_2 to water. Hydrogen peroxide is particularly attractive, as it has a high active-oxygen content, yields H_2O as its only waste product, and is a relatively low-cost oxidant. One of the problems frequently encountered in metal-catalyzed oxidations with hydrogen peroxide is the concomitant decomposition of H_2O_2 (catalase activity), which often requires that a large excess of H_2O_2 is employed to reach full conversion [10]. Application of H_2O_2 [11], in particular to metal-catalyzed epoxidations in general [12] and green oxidations [4, 5], have been reviewed elsewhere. In this chapter we will focus on Mn-catalyzed oxidations with H_2O_2.

$KMnO_4$, the archetypal stoichiometric manganese based oxidant, still retains its place as a classic oxidation reagent in nearly every introductory course on organic

Modern Oxidation Methods. Edited by Jan-Erling Bäckvall
Copyright © 2010 WILEY-VCH Verlag GmbH & Co. KGaA, Weinheim
ISBN: 978-3-527-32320-3

chemistry [13, 14]. Ironically, this oxidizing agent is of limited use in practical synthetic chemistry and is often seen as a tool of last resort. It is only in recent years that several discoveries have revealed the potential of manganese catalysts for selective oxidations. This has evolved in part from studies on the structure and function of Mn-based redox enzymes [15, 16]. In this chapter we will review biomimetic manganese systems briefly and then proceed to discuss oxidative transformations with H_2O_2 catalyzed by Mn complexes in more depth, with special consideration of mechanistic insights gained over recent years.

11.2
Bio-inspired Manganese Oxidation Catalysts

Manganese can frequently be found in the catalytic redox center of several enzymes [10, 15] including superoxide dismutases [17], catalases [18], and the dioxygen-evolving complex photosystem II (PSII) [19, 20]. Superoxide ($O_2^{\bullet-}$), a radical harmful to living organisms, is the product of the one-electron reduction of dioxygen [21]. Its high reactivity requires that it is converted to less reactive species rapidly. Superoxide dismutase metalloenzymes catalyze the dismutation of the superoxide ($O_2^{\bullet-}$) to dioxygen (O_2) and hydrogen peroxide (H_2O_2) [22]. The latter product can be disproportionated by catalase enzymes to water and dioxygen (see above). Superoxide dismutase (SOD) enzymes can be classified into two major structural families; copper-zinc SOD and manganese or iron SOD [21, 23]. The active site of manganese SOD contains a mononuclear five-coordinate Mn^{III}-ion bound to three histidines, one aspartate residue, and one water or hydroxide ligand. The mechanism of the catalytic conversion of superoxide to dioxygen starts by binding of the superoxide radical anion to the Mn^{III}-monomer leading to the reduction to Mn^{II} and oxidation of superoxide to dioxygen [17, 24]. Subsequently, the catalytic cycle is closed by binding of a second superoxide to the Mn^{II}-ion resulting in the oxidation to Mn^{III} and reduction of the superoxide anion to H_2O_2.

In photosystem II (PSII), located in the thylakoid membrane of chloroplasts in green plants, algae, and a number of cyanobacteria, two water molecules are oxidized to dioxygen, that is, water splitting [19]. PSII consists of light-harvesting pigments, a water oxidation center (WOC), and electron transfer components [19]. Based on detailed spectroscopic analyses, it has been recognized that a tetranuclear Mn cluster is the active catalyst for the dioxygen evolution, and this assignment is supported by X-ray analysis [25]. Although the detailed mechanism of the water oxidation component remains to be elucidated fully, current opinion highlights the cooperativity of the manganese ions in the cluster with calcium in achieving the transfer of four electrons and liberation of dioxygen [26].

Manganese-containing catalases have been isolated from three species of bacteria; *Lactobacillus plantarum* [27], *Thermus thermophilus* [28], and *Thermoleophilum album* [18]. X-ray crystallographic structure analysis [29] has shown that these catalases contain a dinuclear manganese core. During catalysis, the dinuclear manganese active site cycles between the Mn_2^{II}- and Mn_2^{III} oxidation states [30].

EPR [31], NMR [32], and UV-Vis [32a] spectroscopic studies have identified that in the disproportionation of H_2O_2 both Mn_2^{II} and Mn_2^{III} oxidation states are involved [33]. The proposed catalase mechanism is depicted in Scheme 11.1. Hydrogen peroxide decomposition is initiated by (a) the binding of H_2O_2 to the Mn^{III}-Mn^{III} dinuclear center, and this is followed by (b) reduction to the Mn^{II}-Mn^{II} intermediate and concomitant oxidation of the peroxide to O_2 [33, 34]. Subsequent binding of a second molecule of H_2O_2 to the Mn^{II}-Mn^{II} species (c) is followed by the reduction of H_2O_2 to H_2O and the oxidation of the Mn^{II}-Mn^{II} species (d), which closes the catalytic cycle [17].

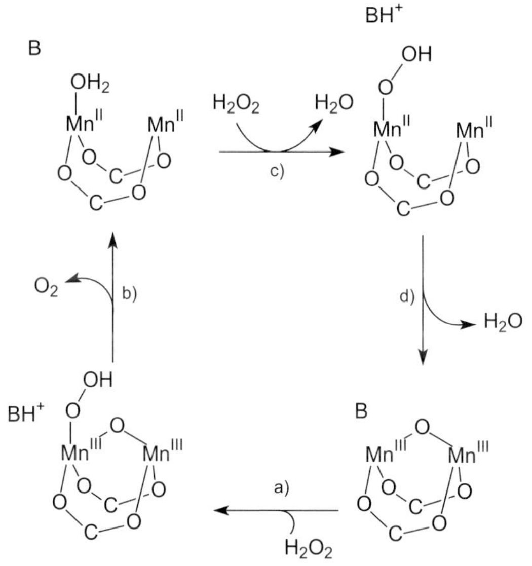

Scheme 11.1 Proposed mechanism for H_2O_2 decomposition by manganese catalase [17].

A series of Mn complexes that mimic the active site have been developed to gain insight into the mechanisms by which these enzymes operate [34]. Dismukes and coworkers reported the first functional catalase model that exhibited high activity toward H_2O_2 decomposition; even after turnover numbers of 1000, no loss of activity toward H_2O_2 decomposition was observed [35]. The dinuclear Mn^{II} complex is based on ligand **1** (Figure 11.1). EPR and UV-Vis spectroscopic investigations indicate that under conditions of H_2O_2 decomposition both Mn^{III}–Mn^{III} and Mn^{II}–Mn^{II} oxidation states are present, as was observed for the related manganese catalase enzymes [34].

Sakiyama and coworkers have explored several dinuclear manganese complexes based on the ligand 2,6-bis[N-(2-dimethylamino)ethyl]iminomethyl-4-methylphenolate) (**2**, Figure 11.1) and related ligands as catalase mimics. Employing UV-Vis and MS techniques both mono- and di-nuclear Mn^{IV}-oxo intermediates could be detected [36]. Notably, the proposed mechanism (Scheme 11.2) is different from that reported for the manganese catalases and model compounds containing ligand **1** involving high-valent Mn=O species [36].

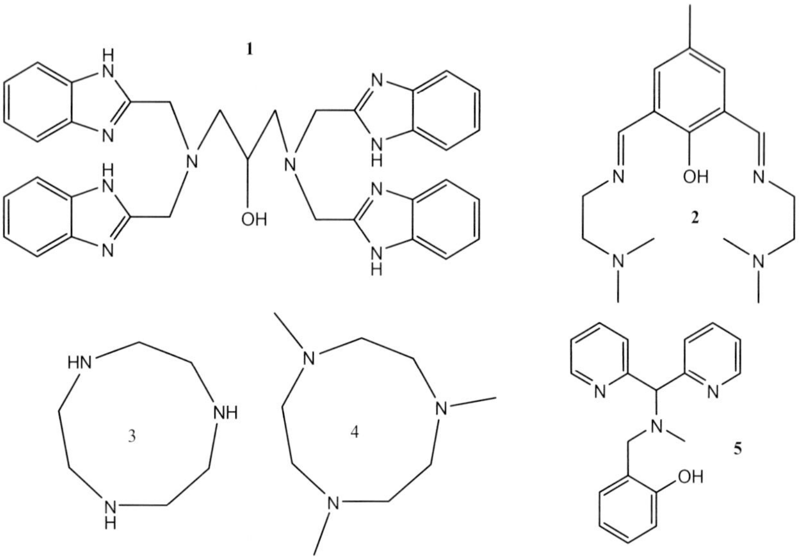

Figure 11.1 Ligand sets developed for manganese complexes as catalase mimics.

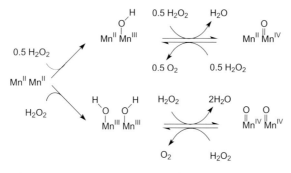

Scheme 11.2 Proposed mechanism of H_2O_2 decomposition catalyzed by Mn complexes based on ligand **2** [36].

Manganese complexes of 1,4,7-triazacyclononane (**3**, tacn) or 1,4,7-trimethyl-1,4,7-triazacyclononane (**4**, tmtacn, Figure 11.1) were studied by Wieghardt and coworkers as models for the dioxygen-evolving center of photosystem II as well as manganese catalase [37]. Turnover numbers for the decomposition of H_2O_2 as high as 1300 are reached readily [37d]. More recently Krebs and Pecoraro used the tripodal bpia ligand (bpia = bis-(picolyl)(N-methylimidazol-2-yl)amine) as a Mn catalase model system. Several Mn complexes based on this ligand were found to be structural mimics for certain catalase enzymes. Remarkably, the catalytic activity was found to be within 2 to 3 orders of magnitude relative to these enzymes [38]. More recently, dinuclear manganese-based complexes based on ligand **5**, that were developed as an oxidation catalyst, were found to be highly efficient in H_2O_2 decomposition both in solution and when immobilized on solid supports [39]. Several of the manganese oxidation catalysts which were developed from these systems will be discussed in the following sections in their application to oxidation catalysis with H_2O_2.

11.3
Manganese-Catalyzed Bleaching

Bleaching processes in the paper industry and the bleaching of stains on textiles have been studied intensively over the last century. Conventional bleaching procedures for laundry cleaning employ H_2O_2 and high temperatures [6, 34]. Several catalysts have been investigated to achieve bleaching at lower temperatures (i.e., 40–60 °C) and even under ambient conditions [34, 40]. Manganese complexes based on 1,4,7-trimethyl-1,4,7-triazacyclononane, that is, $[Mn_2O_3(tmtacn)_2](PF_6)_2$ (**6**) (Figure 11.2), were studied extensively by Unilever Research in the 1990s as bleach catalysts for stain removal at reduced temperatures [41]. Unfortunately, following press releases on textile damage as a result of using this catalyst for laundry cleaning, this catalyst is no longer used in laundry detergent products [41]. Nevertheless, this and related catalysts have seen new life in bulk processes including wood pulp and raw cotton bleaching and in dishwasher formulations, areas where multi-wash textile damage is less relevant [6, 42].

11.4
Epoxidation and *cis*-Dihydroxylation of Alkenes

Epoxides are an important and extremely versatile class of organic compounds, and the development of new methods for the selective epoxidation of alkenes continues to be a major challenge [2, 12, 43]. The epoxidation of alkenes can be achieved by

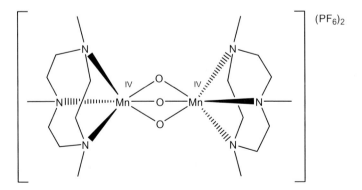

Figure 11.2 The complex $[Mn_2O_3(tmtacn)_2](PF_6)_2$ (**6**).

applying one of a number of oxidants including peroxycarboxylic acids [44], dioxiranes [45], alkylhydroperoxides [46], hypochlorite [47], iodosylbenzene [47], dioxygen [48], and hydrogen peroxide [12,43c,46]. With a few exceptions, most of the oxidants have the disadvantage that, in addition to the oxidized products, stoichiometric amounts of waste products are formed which have to be separated from the often sensitive epoxides. The use of H_2O_2 in combination with Mn complexes offers several advantages including the high reactivity of the catalytic systems, although the oxidant is often partially destroyed by the catalase-type activity typically associated with Mn catalysts [34]. It should be noted also that nonselective side reactions can occur because of the formation of hydroxyl radicals formed by homolytic cleavage of H_2O_2 [49].

11.4.1
Manganese Salts

Ligand-free epoxidation systems are attractive, particularly in the context of the development of green oxidation procedures and in terms of cost [3, 50]. A remarkably simple and effective 'ligand-free' Mn-based epoxidation system, using 0.1–1.0 mol% of $MnSO_4$ and 30% aqueous H_2O_2 as the oxidant in the presence of bicarbonate, was introduced by Burgess and coworkers [12, 51]. Bicarbonate and H_2O_2 form the actual oxidant peroxy monocarbonate (Scheme 11.3), which is proposed to react with the Mn ion to generate the active epoxidation catalyst, as was indicated by EPR studies [51, 52].

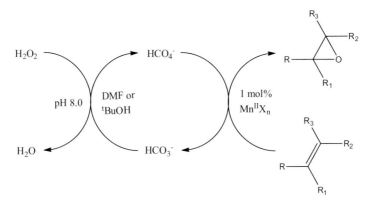

Scheme 11.3 $MnSO_4$-catalyzed epoxidation with bicarbonate/hydrogen peroxide, where X_n is an undefined ligand.

A series of cyclic alkenes and aryl- and trialkyl-substituted alkenes are converted into their corresponding epoxides in high yields using 10 equiv. of H_2O_2. Notably, monoalkyl alkenes were unreactive with this system. A range of additives were tested in an effort to increase the H_2O_2 efficiency by enhancing the activity for epoxidation and suppressing H_2O_2 decomposition. The use of 6 mol% of sodium acetate in ^tBuOH or 4 mol% of salicylic acid in DMF as solvent resulted in an improved epoxidation system with higher epoxide yields, decreased reaction times, and amount of H_2O_2 used (5 equiv., Table 11.1) [51].

The Mn-salt/bicarbonate system is also catalytically active in ionic liquids. Epoxidation of a range of alkenes with 30% aqueous H_2O_2 can be accomplished with

Table 11.1 Epoxidation of alkenes using $MnSO_4$/salicylic acid catalyst [51].

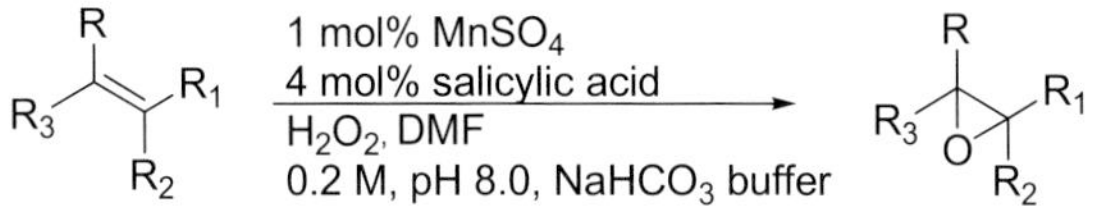

Alkene	Epoxide	Equiv. H_2O_2	Yield (%)
 cyclohexene	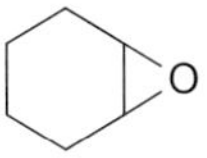 cyclohexene oxide	2.8	96
 bicyclic alkene	 bicyclic epoxide	5	89
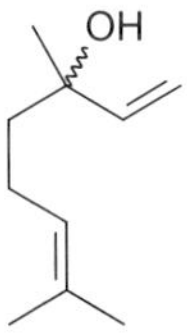 allylic alcohol	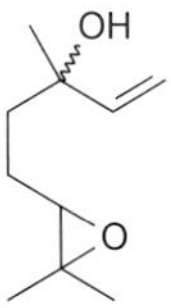 allylic alcohol epoxide	5	91
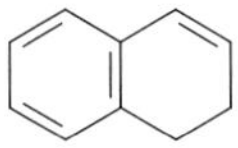 dihydronaphthalene	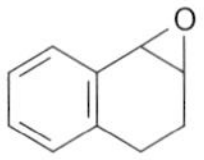 dihydronaphthalene epoxide	5	97
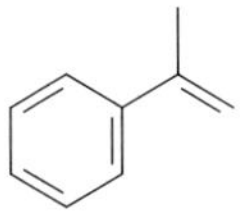 α-methylstyrene derivative	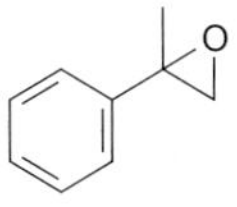 styrene oxide derivative	5	95
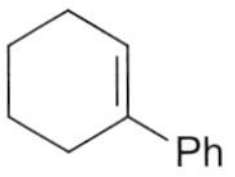 1-phenylcyclohexene	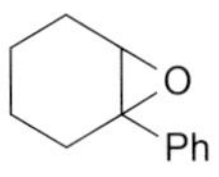 1-phenylcyclohexene oxide	5	95
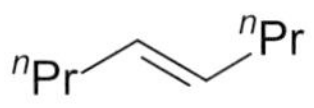 (E)-4-octene	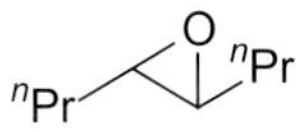 trans-4,5-epoxyoctane	25	75
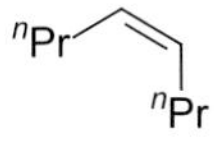 (Z)-4-octene	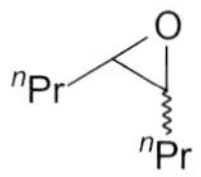 cis-4,5-epoxyoctane	25	75[a]

a) Approximately 1:1 cis/trans mixture.

catalytic amounts of MnSO$_4$ in combination with TMAHC (tetramethylammonium hydrogen carbonate) in the ionic liquid [bmim][BF$_4$] (1-butyl-3-methylimidazolium tetrafluoroborate). Moderate to excellent yields are obtained for internal alkenes, and the ionic liquid can be reused at least 10 times when fresh amounts of the Mn salt and bicarbonate are added [53].

11.4.2
Porphyrin-Based Catalysis

Metallo-porphyrins and several other metal porphyrin complexes, in particular those of Mn, Fe, and Cr, have been studied extensively as catalysts in the epoxidation of alkenes [47, 48]. Although the terminal oxidants iodosylarenes, alkylhydroperoxides, peracids, and hypochlorite have received the most attention, reactions using H$_2$O$_2$ have been reported also [47, 48]. The earliest porphyrin-based catalysts were limited by rapid deactivation due to oxidative degradation of the ligand. Substantial improvements were achieved in catalyst robustness and activity in both alkene epoxidation and alkane hydroxylation through the introduction of halogen substituents on the porphyrin ligands [54]. Nevertheless, porphyrin-based epoxidation catalysts suffer general disadvantages compared with other systems in regard to their synthesis, and purification, which is often tedious.

Initial attempts to use H$_2$O$_2$ as an oxidant for alkene epoxidation with porphyrin-based catalysts were unsuccessful due to dismutation of H$_2$O$_2$ into H$_2$O and O$_2$, leading to rapid depletion of the oxidant. Introduction of bulky groups on the porphyrin ligand enabled the use of aqueous H$_2$O$_2$, albeit with only low conversions being achieved. It was demonstrated, however, that this catalytic system could be improved by performing the oxidation reaction in the presence of excess imidazole [55, 56]. The role of the imidazole is proposed to be twofold: (a) in acting as a stabilizing axial ligand, and (b) in promoting the formation of the Mn^V=O intermediate (the oxygen transfer agent) through heterolysis of an Mn^{III}-OOH intermediate. This catalytic system provides epoxides in yields of up to 99%. The excess in axial ligand could be reduced significantly through addition of a catalytic amount of carboxylic acid [57, 58]. Under two-phase reaction conditions with addition of benzoic acid the oxidation reaction was accelerated significantly, and high conversions could be obtained in less than 10 min at 0 °C (Scheme 11.4, Table 11.2) [57].

The carboxylic acids and nitrogen-containing additives are generally considered to facilitate the heterolytic cleavage of the O–O bond in the manganese hydroperoxy intermediate to provide a catalytically active manganese(V)-oxo species [59]. DFT (Density Functional Theory) calculations reported by Balcells *et al.* have highlighted the potential importance of the axial ligand in determining the activity of the Mn^V-oxo species formed in engaging in C–H abstraction [60]. However, competing homolytic cleavage of the O–O bond leads to the formation of hydroxyl radicals and nonselective oxidation reactions – a serious challenge encountered in general in using H$_2$O$_2$ in metal-catalyzed oxidations [48]. The proposed catalytic cycle for epoxidation of alkenes using manganese porphyrin **7** begins with conversion to the well-established

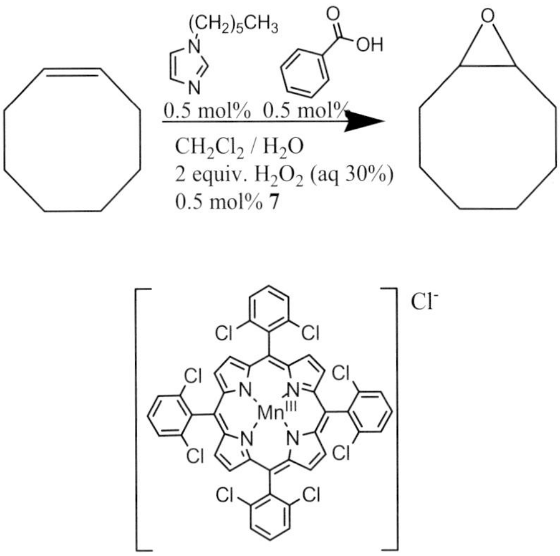

Scheme 11.4 Manganese porphyrin complex **7** – an effective catalyst in the epoxidation of alkenes with H_2O_2 [57].

Mn^V-oxo species (Scheme 11.5) [49a,61]. Subsequently, the oxygen atom is transferred to the alkene via a concerted- (path **a**) or stepwise- (path **b**) pathway followed by release of the Mn^{III}-species and formation of the epoxide. In the stepwise route **b**, which involves a neutral carbon radical intermediate, rotation around the former double bond can result in cis/trans isomerization leading to *trans*-epoxides from *cis*-alkenes, as is observed experimentally [61].

Improvement in the stereoselectivity of the oxidation of *cis*-stilbene was observed by increasing the number of substituents on the aryl groups of the porphyrin ligand, pointing to an enhanced preference for a concerted pathway. In general, it should be noted, however, that *trans*-alkenes are poor substrates for these catalysts [57, 62].

Enhanced epoxidation rates were observed using a Mn-porphyrin complex **8** in which the carboxylic acid and imidazole groups are both linked to the ligand

Table 11.2 Oxidation of alkenes with Mn^{III}-porphyrin complex **7** [57].

Substrate	Conversion (%)	Epoxide (%)	Reaction time (min)
Cyclooctene	100	100	10
Dodec-1-ene	96	92	15
α-Methylstyrene	100	100	7
cis-Stilbene	90	85 (*cis*)	20
trans-Stilbene	0	0	300
trans-4-Octene	75	54 (*trans*)	15

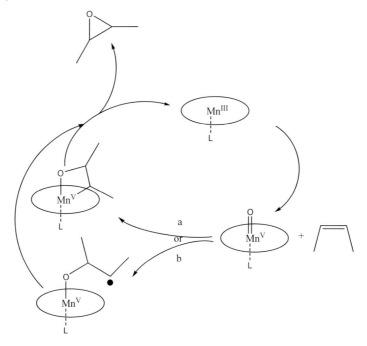

Scheme 11.5 The proposed (a) concerted pathway, (b) stepwise pathway in the epoxidation of alkenes by manganese-porphyrin complexes.

covalently (Scheme 11.6) [63]. When 0.1 mol% of the Mn complex and 2 equiv. of H_2O_2 were employed, cyclooctene was converted in only 3 min to the corresponding epoxide with 100% conversion and selectivity. Analogous results were obtained for alkenes such as α-methylstyrene, p-chlorostyrene, α-pinene, and camphene, with turnover numbers of up to 1000. The proposed mechanism is similar to oxidation reactions with porphyrin-based catalysts in the presence of the imidazole and carboxylic acid co-catalysts, with a prominent role played by the pendent carboxylic acid group in heterolysis of the hydroperoxide (Scheme 11.6) [63].

Following the first report by Groves and Myers [72] on asymmetric oxidation using a chiral metalloporphyrin, a wide range of porphyrin ligands linked to chiral functional groups have been reported [64]. Although high enantioselectivities were observed with iodosylbenzene as terminal oxidant, with H_2O_2 only moderate enantioselectivity has been achieved to date [65]. For a discussion of the design of chiral ligands and the stereochemical issues involved, the reader is referred to detailed reviews on the topic [64, 66].

The immobilization of homogeneous Mn-porphyrin epoxidation catalysts on silica to achieve facile catalyst recovery has been realized through anchoring of the porphyrin ligand A [67] or the axial imidazole ligand B (Figure 11.3) [68]. The advantages of the supported catalysts are to some extent lost, however, because of reduced epoxidation activity compared to the analogous homogeneous system.

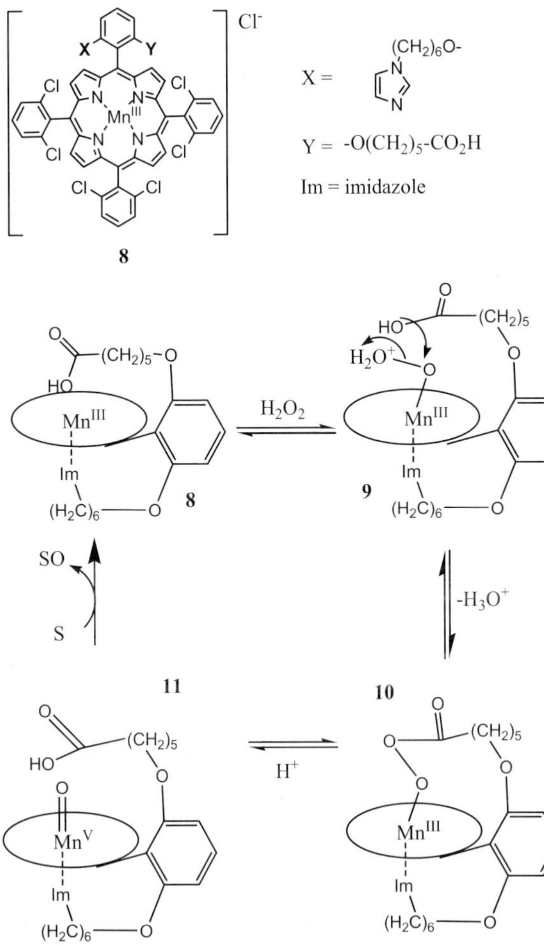

Scheme 11.6 Mn-porphyrin complex **6** with tethered carboxylate and imidazole groups and their proposed role in catalyzed oxidation [63].

Recently, Hulsken et al. have employed STM to probe the mechanism by which manganese porphyrin complexes achieve epoxidation of alkenes with dioxygen [69]. Importantly, this study demonstrates the positive role of solid-support immobilization in precluding the formation of inactive oxido-bridged catalyst dimers (a common deactivation pathway).

11.4.3
Salen-Based Systems

Following the seminal report by Kochi and coworkers on the use of Mn-salen complexes as epoxidation catalysts [70], considerable attention has been directed

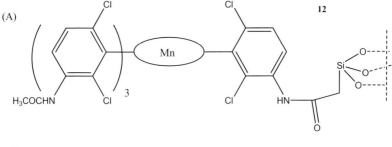

Figure 11.3 Immobilized Mn-porphyrin epoxidation catalysts.

toward this family of catalysts, as they offer synthetic advantages over the porphyrin systems described in the previous section. The breakthroughs made by the groups of Jacobsen [71a], Katsuki [71b], and coworkers in Mn-catalyzed alkene epoxidation by the introduction of a chiral diamine functionality in the salen ligand (Figure 11.4) have paved the way for effective enantioselective epoxidation catalysts.

Compared to chiral manganese porphyrin complexes [72], in general the use of the Mn-salen catalysts can provide enantioselectivities greater than 90% and yields exceeding 80% [47, 73]. A wide range of oxidants, including hypochlorite [73], iodosylbenzene [73], or m-chloroperbenzoic acid (m-CPBA), can be applied [74]. Excellent enantioselectivities are observed in the epoxidation of cis-disubstituted alkenes and trisubstituted alkenes catalyzed by the Mn-salen complexes **13** and **14**,

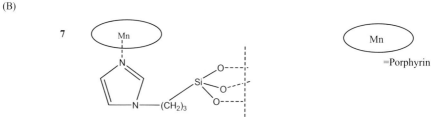

Figure 11.4 Chiral manganese complexes introduced by Jacobsen (**13**) and Katsuki (**14, 15**) for asymmetric epoxidation of nonfunctionalized alkenes [71].

employing iodosylbenzene as the terminal oxidant. In sharp contrast, the epoxidation of *trans*-alkenes showed moderate selectivities (*ee* <60%); however, the enantioselectivities could be improved by the introduction of additional chiral groups at the 3,3′-positions of the phenolate part of the ligand. For the conversion of *trans*-stilbene, enantioselectivities of up to 80% have been reported using these modified salen ligands [73].

The oxidizing species in this catalytic epoxidation is proposed to be a Mn^V-oxo intermediate [74d,e] similar to that observed in the Mn-porphyrin-catalyzed epoxidation, as based on electrospray ionization mass spectrometry [75]. An extensive discussion of the stereoselectivity, mechanism, and scope of this asymmetric epoxidation [73] using the preferred oxidant iodosylbenzene is beyond the scope of this chapter, although several mechanistic features apply to epoxidation with H_2O_2. Despite the fact that there is consensus on the nature of the active species (i.e., a Mn^V-oxo intermediate), some controversy remains on the exact mechanism by which enantioselection is achieved. Three key issues can be distinguished: (i) the catalyst structure (i.e., whether the salen ligand is planar, bent, or twisted), (ii) the trajectory of approach of the reacting alkene, and (iii) the mode of oxygen transfer from the salen $Mn^V=O$ to the alkene (involving a concerted pathway, a stepwise radical pathway, or a metallaoxetane intermediate) [47, 73, 76]. Cumulative experimental evidence indicates that, in addition to the catalyst structure being either planar or twisted, the substituents at the C_2-symmetric diimine bridge and bulky substituents at the 3,3′-positions play an important role in governing the trajectory of the side-on approach of the alkene and, as a consequence, the asymmetric induction. With the five-membered chelate ring, consisting of the ethylenediamine and the Mn^V-ion, being non-planar, the approach of the alkene over the downwardly bent benzene ring of the salen ligand along one of the Mn—N bonds can be envisioned (Scheme 11.7). The largest substituent of the alkene is then pointing away from the 3,3′-substituents, and this would then govern the stereochemical outcome of the reaction between the Mn^V-oxo intermediate and the alkene [73].

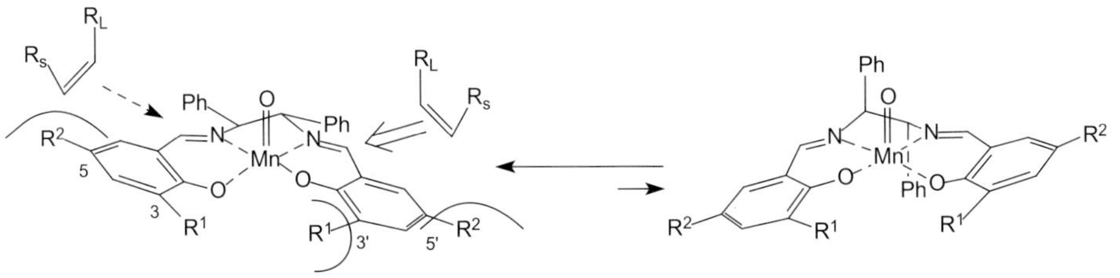

Scheme 11.7 Model rationalizing the stereocontrol achieved in Mn-salen-catalyzed epoxidation [73c].

Although high enantioselectivities are obtained for a wide range of substrates, the stability of the Mn-salen complexes is often a severe problem, and turnover numbers are usually in the range 40–200. More recently, a robust salen catalyst, was introduced by Katsuki and coworkers [77] based on ligand **15** with a carboxylic acid functionality attached to the diamine bridge (Figure 11.4). With this catalyst,

2,2-dimethylchromene was converted to the corresponding epoxide in 99% *ee* with iodosylbenzene as oxidant. Turnover numbers as high as 9200 after 6 h were reached, but results with H_2O_2 as the terminal oxidant have not been reported to date [77].

While iodosylbenzene and hypochlorite are the most common oxidants, considerable effort has been devoted to the use of H_2O_2 in epoxidations with Mn-salen catalysts. Promising results have been reported for certain substrates (see below), although low turnover numbers (generally up to 20–50) were obtained with H_2O_2 as the terminal oxidant for a limited scope of substrates. Employing H_2O_2 as the oxidant, the manganese-salen systems were found to be only catalytically active in the presence of additives such as imidazole and derivatives thereof or carboxylates [49b, 78, 79, 80]. The role of these additives is considered to be in inhibiting O–O bond homolysis, which leads to radical pathways and destruction of the catalyst, as has been discussed for the Mn-porphyrin based catalysts (see above).

Berkessel and coworkers have designed a chiral dihydrosalen ligand with a covalently attached imidazole group. This salen complex (**16**) was used to convert 1,2-dihydronaphthalene to the corresponding epoxide in 72% yield and with promising enantioselectivities (up to 64%; Table 11.3) using a dilute (1%) aqueous solution of H_2O_2 as oxidant. An important feature of this system is that epoxidation reactions can be performed in the absence of the additives usually employed [49b].

Using Mn-salen **17** together with *N*-methylimidazole as an axial ligand, Katsuki and coworkers obtained up to 96% *ee* in the epoxidation of a substituted chromene with 30% aqueous H_2O_2 as oxidant, although the yield of the epoxide was only 17%. With excess of H_2O_2 (10 equiv.) and increased concentration of the reactants, the yield was increased to 98%, with only a slight decrease in the enantioselectivity to 95% (Table 11.3) [79]. Although only a limited number of substrates were tested, these excellent results (*ee* 88–98%) demonstrate that full enantioselectivity with H_2O_2 can be achieved.

Pietikäinen reported that, in the presence of carboxylate salts, 30% aqueous H_2O_2 could be used as an oxidant for asymmetric epoxidation with chiral Mn-salen catalysts (64–96% *ee*, Table 11.3) [80b]. Furthermore it was shown that the use of *in situ* prepared peroxycarboxylic acids, from the corresponding anhydrides and anhydrous H_2O_2, gives improved enantioselectivity in the epoxidation of alkenes compared with the use of aqueous H_2O_2 in the presence of a carboxylate salt [81]. In particular, good results are obtained with maleic anhydride and UHP (urea-H_2O_2) in combination with Mn(III)-salen complex **18a** together with the additive *N*-methylmorpholine *N*-oxide (NMO). Although the substrate scope tested is again limited, in general 3–5% higher enantioselectivities were obtained, and the reaction time was reduced under these conditions. The use of urea-H_2O_2 for the Mn(III)-salen catalyzed epoxidation of alkenes has also been described by Kureshy and coworkers [82]. Although for styrene only 39% *ee* was obtained, moderate to excellent enantioselectivities were reported for chromene derivatives (55–99%) in the presence of ammonium acetate.

Recently, immobilization of salen-based catalysts has been demonstrated both on solid supports [83] and on dendritic molecular frameworks, which allow for enantioselective catalysis with good to excellent *ee* over several cycles [84].

Table 11.3 Epoxidation of alkenes by Mn(III)-salen complexes employing H$_2$O$_2$ as oxidant.

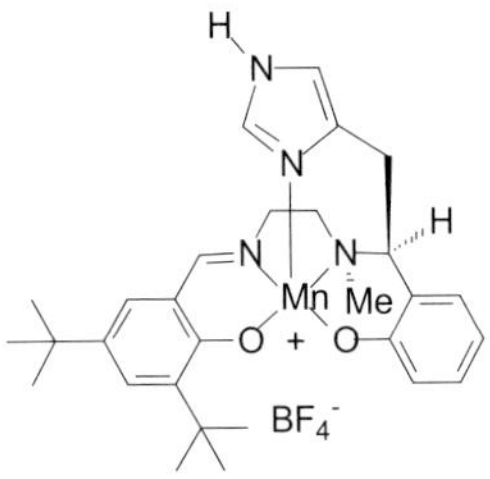

16

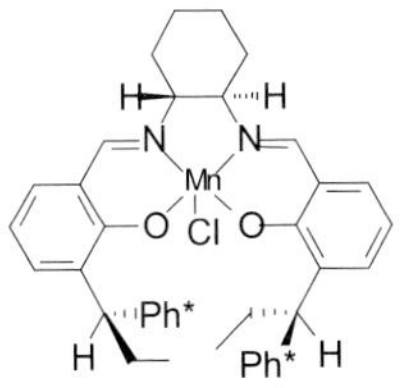

17 Ph* = 4-t-butylphenyl

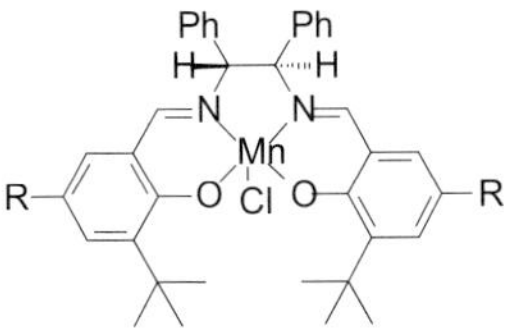

18 a R = Me
b R = CH$_2$N(C$_8$H$_{17}$)$_2$

Substrate	Mn-salen	Oxidant (equiv.)	Epoxide yield (%)	ee (%)	Ref.
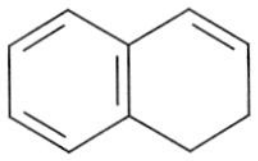 1,2-dihydronaphthalene	16	1% H$_2$O$_2$ (10)	72	64	[49b]
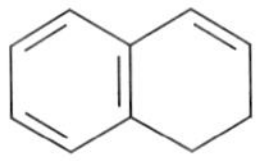 1,2-dihydronaphthalene	18a	30% H$_2$O$_2$ (1.5)	74	69	[81]
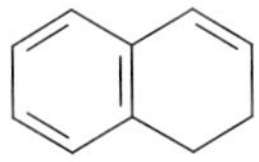 1,2-dihydronaphthalene	18a	Urea.H$_2$O$_2$/maleic anhydride (1.5)	70	73	[81]
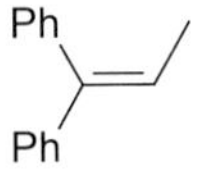 Ph–C(=CHMe)–Ph	13	30% H$_2$O$_2$ (4)	84	96	[80b]
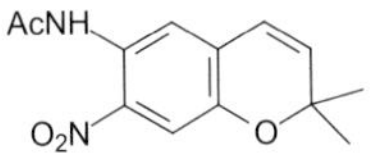 AcNH/O$_2$N-chromene	17	30% H$_2$O$_2$ (1)	17	96	[79]
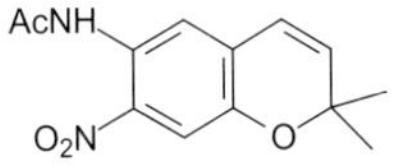 AcNH/O$_2$N-chromene	17	30% H$_2$O$_2$ (10)	98	95	[79]
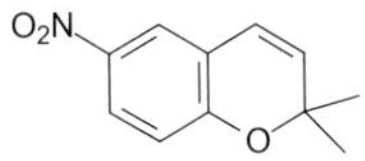 O$_2$N-chromene	18b	Urea.H$_2$O$_2$ (2)	>99	>99	[82b]

11.4.4
Tri- and Tetra-azacycloalkane Derivatives

Over the last two decades several groups have focused attention on tri- and tetra-azacycloalkane-based manganese complexes in the catalyzed oxidation of organic

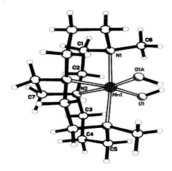

Figure 11.5 X-Ray crystallographic structure of $[(Me_2EBC)Mn^{III}(OH)_2]^+$. Reproduced from Ref. [86] with permission. Copyright ACS 2006.

substrates including alkanes, alkenes, alcohols, aldehydes, and sulfides, albeit with the majority of studies focusing on alkene epoxidation [6, 10].

11.4.4.1 Tetra-azacyloalkane Derivatives

Busch and coworkers have reported a number of catalytic and mechanistic studies on the manganese complexes formed with the ligand 4,11-dimethyl-1,4,8,11-tetra-azabicyclo[6.6.2]hexadecane (Me$_2$EBC) [85–89] and related ligands [90]. The ligand Me$_2$EBC is designed specifically to preclude formation of manganese di- and multi-nuclear complexes and allowed for the isolation of an Mn^{III}-(OH)$_2$ complex such as that shown in Figure 11.5.

The complex could achieve up to 45 turnovers in the oxidation of alkenes (Table 11.4); however, despite its low activity relative to other systems, it serves as an excellent platform for mechanistic studies given the relative stabilities of the higher-oxidation-state manganese complexes formed and isolated using this ligand. Haras and Ziegler have recently proposed, based on DFT calculations, that the mode of action of these complexes is via the Mn^{IV}-OOH adduct, with oxygen transfer of the

Table 11.4 Epoxidation of alkenes with H_2O_2 catalyzed by Mn(Me$_2$EBC)Cl$_2$[a].

Substrate	Product	Yield (%)
Cyclohexene	Cyclohexene oxide	18.0
	Cyclohexen-1-one	13.3
Styrene	Styrene oxide	45.5
	Benzaldehyde	2.8
Norbornylene	Norbornylene oxide	32.0
cis-Stilbene	cis-Stilbene oxide	17.5
	trans-Stilbene oxide	2.0
	Benzaldehyde	2.6

a) Reaction conditions: acetone/water (4:1), catalyst (1 mM), alkene (0.1 M), 50% H$_2$O$_2$ (1 mL), added stepwise by 0.2 mL/0.5 h, rt, yield determined by GC with internal standard. From Ref. [86].

distal oxygen to the alkene substrate, that is, via a classic inorganic peracid oxidation-type mechanism [91]. This mechanism places this class of complex in sharp contrast with the salen- and porphyrin-based systems discussed above, in which $Mn^{IV}=O$ and $Mn^{V}=O$ species are widely accepted to be the oxygen transfer agents.

11.4.4.2 Triazacyclononane Derivatives

Manganese complexes such as **6** ([Mn_2O_3(tmtacn)$_2$](PF$_6$)$_2$, Figure 11.2), **19** ([Mn(OMe)$_3$(tmtacn)]) and **20** [Mn_2O(OAc)$_2$(tmtacn)$_2$](PF$_6$)$_2$ based on the N,N',N''-trimethyl-1,4,7-triazacyclononane ligand (**4**, tmtacn, Figure 11.1) were developed by Wieghardt and coworkers in the late 1980s and 1990s as functional models for biotic manganese systems [10, 37], in particular, dinuclear manganese-based catalase enzymes [92, 93] and PSII [19]. The activity of these complexes as oxidation catalysts with H_2O_2 in both aqueous [40] and nonaqueous [94] media have, however, made this family of complexes the focus of considerable interest for a broad range of oxidative transformations including textile stain bleaching [6], benzyl alcohol oxidation [94c], C—H bond activation [94f], sulfoxidation [95] and the cis-dihydroxylation, and epoxidation of alkenes [94, 117].

Epoxidation with H_2O_2 In combination with H_2O_2, the dinuclear manganese complex **6** was found to be a highly active and selective epoxidation catalyst [40,94g,96]. High turnover numbers (>400) were obtained in the oxidation of styrene and vinylbenzoic acid. In methanol carbonate-buffered solutions, conversions of 99% were reached without notable catalyst degradation [94g,96]. The scope for epoxidation reactions was considerably expanded by De Vos and Bein through the use of acetone as the solvent [94b,97]. Although the procedure is not suitable for the epoxidation of electron-deficient alkenes, high turnover numbers of up to 1000 have been reported for the conversion of a range of alkenes to their corresponding epoxides using Mn-tmtacn complexes prepared *in situ* with MnSO$_4$ and tmtacn (Table 11.5). For example, with styrene, complete conversion with 98% epoxide selectivity is achieved at 0 °C in acetone with 2 equiv. of H_2O_2 (Scheme 11.8) [97]. It should be noted that cleavage of the double bond was not observed, although some cis/trans isomerization was observed in the oxidation of cis-alkenes. Furthermore, with alkenes such as cyclohexene only minor amounts of allylic oxidation products are obtained.

Table 11.5 Oxidation of selected alkenes with Mn-tmtacn [94b].

Substrate	Turnover number[a]	Selectivity (%)[b]
Cyclohexene	290	87
Styrene	1000	>98
cis-2-Hexene	540	>98
1-Hexene	270	>98
trans-β-Methylstyrene	850	90

a) In acetone (0.6 g), alkene (1 mmol), tmtacn (1.5 mmol), Mn^{2+} (1 mmol). H_2O_2 (2 mmol 30% diluted in acetone 3 times) was added over 1h at room temperature. Turnover number in moles product per mole catalyst (after 3 h).
b) Selectivity: moles of epoxide per mole of converted substrate.

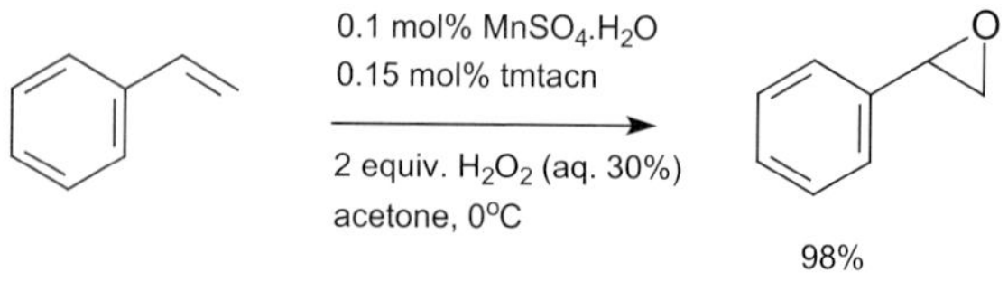

Scheme 11.8 Oxidation of styrene catalyzed by an *in-situ*-formed manganese complex in acetone [97].

The turnover numbers for this reaction are high; however, there remains a need to develop catalytic systems that employ H_2O_2 efficiently. With many Mn or Fe catalysts, decomposition of H_2O_2 is a significant secondary, if not primary, process, and hence often a large excess of H_2O_2 is required to reach full conversion [34]. Supression of catalase-type activity is indeed possible by performing the oxidation reactions in acetone at subambient temperatures. In contrast, the use of other solvents results in rapid competitive oxidant decomposition. The oxidation characteristics of the Mn-tmtacn complexes in acetone compared with other solvents were rationalized by a mechanism involving the nucleophilic addition of H_2O_2 to acetone, resulting in the formation of 2-hydroperoxy-2-hydroxypropane (hhpp, **21**) as depicted in Scheme 11.9 [98]. Most probably, due to the reduction of the H_2O_2 concentration in acetone, the epoxidation reaction is favored over oxidant decomposition. It was proposed that at low temperature hhpp serves as an oxidant reservoir, which gradually releases H_2O_2 and maintains a low steady-state oxidant concentration [97b], although direct involvement of peroxide **21** in the epoxidation pathway was not excluded. However, the combination of acetone and H_2O_2 can also result in the formation of potentially explosive cyclic peroxides, and therefore this solvent is not acceptable for industrial applications involving H_2O_2.

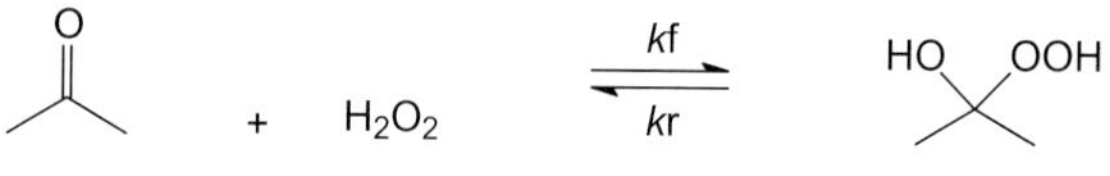

Scheme 11.9 Reaction of acetone with H_2O_2.

Hydrogen peroxide decomposition by Mn-tmtacn complexes in CH_3CN was shown to be suppressed effectively by addition of oxalate [94d] or ascorbic acid [94a] as co-catalysts. Indeed the addition of these additives even in co-catalytic amounts resulted in a dramatic enhancement in the epoxidation activity of the *in situ* prepared Mn-tmtacn complex [94d]. In general, full conversion was reached with less than 1 mol% of catalyst within 1 h. In addition to oxalic acid, several other bi- or polydentate additives, for example, diketones or diacids, in combination with Mn-tmtacn and H_2O_2 were found to favor alkene epoxidation over oxidant decomposition [94d]. Employing this mixed catalytic system, allylic alkenes (e.g., allyl

11.4 Epoxidation and cis-Dihydroxylation of Alkenes

acetate) and especially terminal alkenes (e.g., 1-hexene, see Scheme 11.10 and Table 11.6) could be converted to their corresponding epoxides in high yields with only 1.5 equiv. of H_2O_2 [94d].

Scheme 11.10 Selective epoxidation of 1-hexene by Mn-tmtacn using H_2O_2 in the presence of oxalate [94d].

Reaction conditions: 0.1 mol% $MnSO_4 \cdot H_2O$, 0.15 mol% tmtacn, 0.3 mol% oxalate buffer, 1.5 eq. H_2O_2 (35% aq.), MeCN, 0 °C, >99%.

Importantly, the isomerization of cis- and trans-alkenes is suppressed in the presence of oxalate. The epoxidation of 2-hexene was found to be completely stereospecific (>98%) using only 1.5 equiv. of the oxidant, in contrast to the reaction in acetone [97], in which the Mn-tmtacn catalyst produced as much as 34% trans-epoxide from cis-2-hexene. Furthermore, several functional groups, including $-CH_2OH$, $-CH_2OR$, $-COR$, and $-CO_2R$, are tolerated. Despite the high reactivity

Table 11.6 Representative examples of epoxidation of terminal and deactivated alkenes with the Mn-tmtacn/oxalate system [94d].

Substrate	Epoxide yield (%)	Epoxide selectivity (%)[a]
 1-hexene	>99	>99
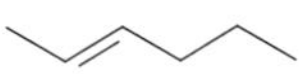 trans-2-hexene	35	>98 (trans)
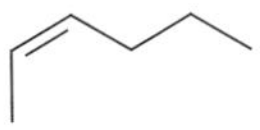 cis-2-hexene	72	>98 (cis)
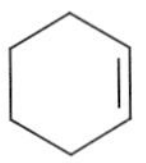 cyclohexene	83	92
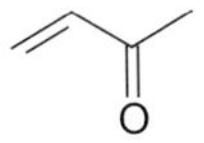 vinyl methyl ketone (methyl vinyl ketone)	66	96
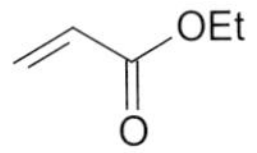 ethyl acrylate (OEt ester)	55	94
 allyl alcohol (OH)	88	91
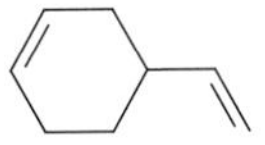 4-vinylcyclohexene	89	91 (diepoxide) / 8 (monoepoxide)

a) Selectivity: moles of epoxide/moles of converted substrate.

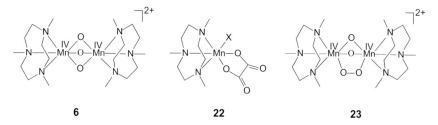

Figure 11.6 Dinuclear Mn-tmtacn complexes **6** and **23**, and the proposed structure for the Mn-tmtacn/oxalate oxidation catalyst **22** (X = activated "O" to be transferred [99].

of the catalytic system, epoxidation is preferred over alcohol oxidation in the case of alkenes bearing alcohol moieties. This procedure is also suitable for the oxidation of dienes resulting in bis-epoxidation. For example, 4-vinylcyclohexene yields the corresponding bis-epoxide.

Although the precise role of the oxalate co-catalyst remains unclear (see below), the formation of a Mn-tmtacn/oxalate species (**22**, Figure 11.6) [99], related to known Cu^{2+}- and Cr^{3+}-structures [100], has been proposed. It has been suggested that the addition of a catalytic amount of the bidentate oxalate impedes the formation of the μ-peroxo-bridged dimer **23**, and as a result the catalase-type decomposition of H_2O_2, typical of dinuclear complexes, is suppressed [34].

Nevertheless, the system of De Vos and coworkers based on the use of oxalate buffer represented a major advance in the application of this catalyst and stimulated the search for other additives that could further improve the reactivity of this system.

Subsequently, additives such as ascorbic acid (Scheme 11.11) and squaric acid facilitated extension in the epoxidation activity of the Mn-tmtacn complex with H_2O_2 [94a]. Although only a limited number of substrates were examined, nearly

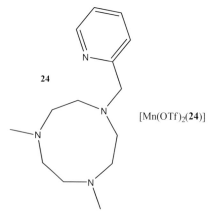

Figure 11.7 Ligand **24** used in the catalyzed oxidation of alkenes with H_2O_2 and acetic acid [101, 102].

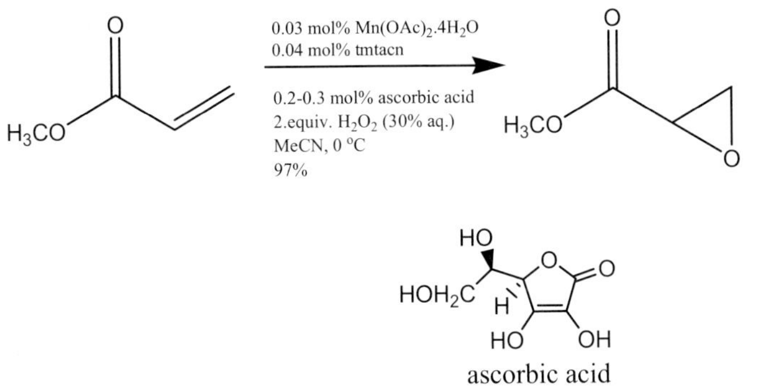

11.4 Epoxidation and cis-Dihydroxylation of Alkenes

Scheme 11.11 Epoxidation promoted by ascorbic acid [94a].

quantitative yields of epoxides with retention of alkene configuration were found employing catalyst loadings as low as 0.03 mol%. The electron-deficient methyl acrylate and the terminal alkene 1-octene were converted to the corresponding epoxides with yields of 97% and 83%, respectively. Typical conditions are shown in Scheme 11.11. Importantly, the H_2O_2 efficiency with this Mn catalytic system was one of the highest reported thus far.

Recently, Costas and coworkers reported a modified manganese-tmtacn system for the epoxidation of alkenes that has shown to be particularly effective where the terminal oxidant is a peracetic acid (0.1 mol% catalyst, 500–1000 t.o.n., Table 11.7) [101]. This system is based on a tmtacn-type ligand in which one of the methyl groups is replaced by a pyridyl group (24, Figure 11.7) and shows a broad substrate scope including cyclic, aromatic, and allylic cis- and trans-alkenes and terminal alkenes. Isotope labeling studies indicated that the incorporated oxygen was derived solely from the peracid. This system was subsequently found to be highly effective when H_2O_2 was employed as terminal oxidant also, and, importantly, the key to its success was found to be the addition of excess acetic acid. As in the case of the related Mn-tmtacn based systems, the acetic acid suppresses the catalase activity of the complex, allowing epoxidation to compete effectively [102]. Under these conditions, normally acid-sensitive substrates such as stilbenes and 1-phenylcyclohexene could be epoxidised effectively (Table 11.7), which was not possible with peracetic acid. This difference has been ascribed to the presence of traces of strong acid impurities such as H_2SO_4 in commercial peracetic acid. From a mechanistic perspective, the absence of cis-dihydroxylation (see below) indicates that this system, although structurally similar to Mn-tmtacn, represents a new class of manganese oxidation catalysts. Importantly, the addition of the pyridyl groups results in distinct chemoselectivity, which allows, for example, for selective monoepoxidation of dienes. As noted by Costas and coworkers, a key point in understanding the increase in efficiency in the case of acid-sensitive substrates, seen upon switching from peracetic acid to acetic acid/H_2O_2, highlights the point that commercial peracetic acid contains substantial levels of H_2O_2 and is considerably more acidic.

Table 11.7 Epoxidation of alkenes with H_2O_2/CH_3CO_2H in CH_3CN catalyzed by $[Mn(OTf)_2(L)]$[a].

Alkene	L = 24 Conv. (yield)	L = 37 Conv. (yield)
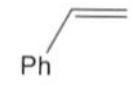 Ph-CH=CH$_2$	100 (94)	89 (77)
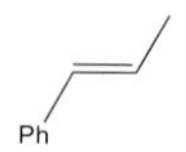 Ph-CH=CH-CH$_3$ (cis)	91 (91)	—
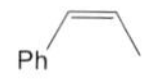 Ph-CH=CH-CH$_3$ (trans)	100 (94)	31 (31)
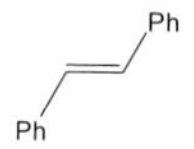 Ph-CH=C(Ph)-CH$_3$	100 (92)	83 (78)
 Ph-CH=CH-CH$_2$-Ph	100 (86)	26 (26)
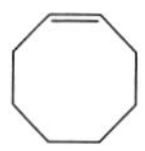 cyclooctene	100 (95)	36 (28)
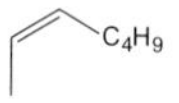 CH$_2$=CH-C$_4$H$_9$	100 (83)	100 (97)
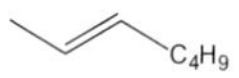 CH$_3$-CH=CH-C$_4$H$_9$	96 (81)	95 (75)
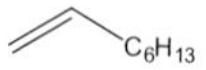 CH$_2$=CH-C$_6$H$_{13}$	90 (90)	95 (70)
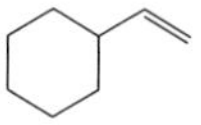 vinylcyclohexane	84 (84)	89 (89)
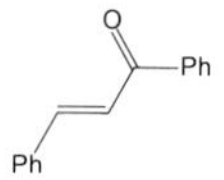 Ph-CH=CH-C(O)-Ph	87 (87)	61 (53)
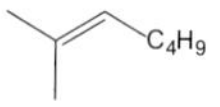 (CH$_3$)$_2$C=CH-C$_4$H$_9$	100 (89)	100 (87)
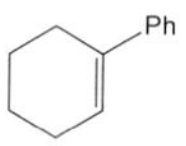 1-phenyl-cyclohexene	100 (90)	59 (51)
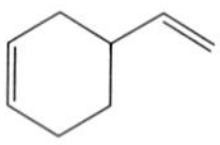 3-vinylcyclohexene	97 (83)	—
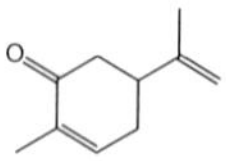 carvone derivative	100 (92)	—
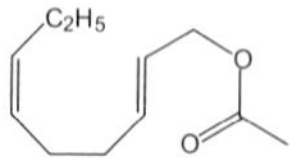 ethyl-dienyl acetate	95 (86)	—

a) Typical reaction conditions: alkene (1.66 mmol), catalyst (1.66 μmol), acetic acid 14 equiv. (23.3 mmol), CH_3CN (15 mL), 1.1–1.4 equiv. H_2O_2 added over 30 min at 0 °C.

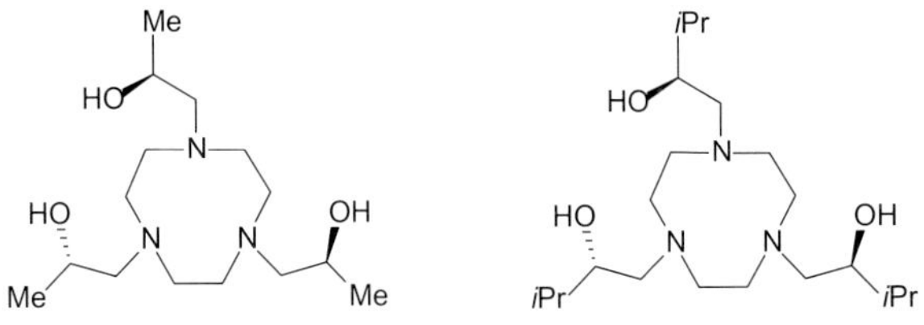

Figure 11.8 C$_3$-symmetric chiral ligands [103a].

Enantioselective Alkene Epoxidation Although ascorbic acid is chiral and plays a key role in boosting the activity of **6**, enantioselective epoxidation was not observed [94a]. By contrast, Beller, Bolm, and coworkers reported that enantiomerically enriched epoxides were obtained with manganese complexes based on chiral analogs of the tmtacn ligand (Figure 11.8) [94f,103,104]. The first asymmetric epoxidation reactions were catalyzed by *in situ* prepared Mn catalysts from N-substituted chiral tacn ligands [103a]. The chirality was introduced via alkylation of the secondary amine moieties to generate the C$_3$-symmetric ligands depicted in Figure 11.8.

Styrene was converted to the corresponding epoxide with an *ee* of 43%, albeit in only 15% yield after 5 h using H$_2$O$_2$ and the Mn complex based on ligand **25**. Using longer reaction times, higher temperatures and higher catalyst loadings, the yield was increased but at the price of a decrease in enantioselectivity [103a]. With the sterically more demanding ligand **26**, enantioselectivities in the range 13–38% were observed for styrene and chromene. Higher enantioselectivity was achieved with *cis*-β-methylstyrene as substrate. The *trans*-epoxide was found as the major product in 55% *ee* whereas the *cis*-epoxide was formed as the minor product with an *ee* of 13% (Scheme 11.12).

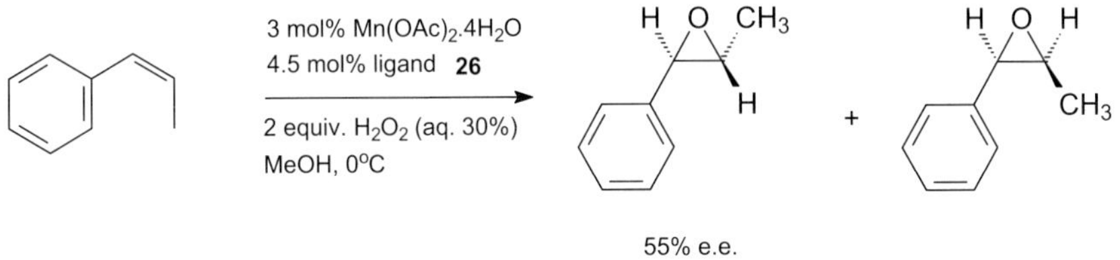

Scheme 11.12 Asymmetric epoxidation with Mn complex of ligand **26** [103a].

The enantiopure C$_3$-symmetric tris-pyrrolidine-1,4,7-triazacyclononane ligand **27** has been reported by Bolm and coworkers (Figure 11.9) [94f]. The tacn derivative was obtained by reduction of an L-proline-derived cyclotripeptide, and the corresponding dinuclear manganese complex was applied in the catalytic enantioselective epoxidation of vinylarenes with H$_2$O$_2$ as oxidant. For the epoxidation of styrene, 3-nitrostyrene, and 4-chlorostyrene excellent conversions (up to 88%) and *ee*'s of up to 30% were obtained [94f]. In addition to the chiral C$_3$- and C$_1$-symmetric ligands,

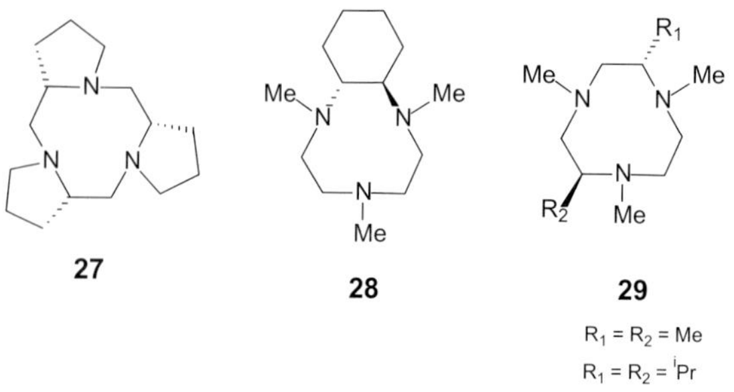

Figure 11.9 Chiral tmtacn based ligands.

C_2-symmetric tacn analogs have been tested (Figure 11.9) [103]. Thus far, modest enantioselectivity has been obtained, but the potential of these tmtacn analogs by further fine-tuning of the chiral ligand structure is evident.

Recently Süss-Fink, Schul'pin and coworkers have reported two tmtacn detivatives where one of the methyl groups has been replaced by a 2-methyl-butyl or 2-hydroxybutyl group [105]. The catalytic activity was examined in the oxidation of indene and phenylethanol (racemic) in the presence of ascorbic acid [94] or oxalate [97]. Although moderate activity was observed (up to 300 t.o.n.), unfortunately only very low levels of enantioselectivity were achieved (<17%).

Kilic et al. have reported an alternative approach to controlling the stereochemical outcome of the epoxidation catalyzed by Mn-tmtacn complexes using substrate control via the hydroxyl group of allylic alcohols and by varying the steric nature of the substituents on the alkene [106].

Manganese-tacn Derivatives on Solid Supports Several successful attempts to improve catalyst selectivity have been made by encapsulation of the Mn-tmtacn complex in zeolites [107]. Immobilization of the triazacyclononane ligand on an inorganic support provided a new class of heterogeneous manganese catalysts with increased epoxidation selectivity [108]. A novel approach to immobilizing Mn-tmacn-based catalysts has been taken by Veghini et al. who employed tungstosilicic acid (as a large anion) to render the catalyst insoluble but still allowing it to be dispersed in solution [109]. However, as is often the case with immobilized homogenous catalysts, the conversions were lower than those obtained with the analogous homogenous catalysts [108].

A notable exception to this is found in the seminal work of De Vos and Jacobs on immobilization of the tmtacn catalyst onto silica. The heterogenization procedure of the tacn ligand started with the conversion of dimethyl tacn (dmtacn, **30**, Figure 11.10) to the silylated compound **31** with 3-(glycidyloxy)propyltrimethoxysilane followed by immobilization on an SiO_2 surface and subsequently metalation of the heterogenized ligand with $MnSO_4 \cdot H_2O$ [99]. Not only was this system highly effective, but serendipitously it led to the discovery of a previously unexplored field of oxidation catalysis for the tmtacn family of catalysts, that of cis-dihydroxylation of alkenes. This will be discussed in more depth in the following section. A similar approach to

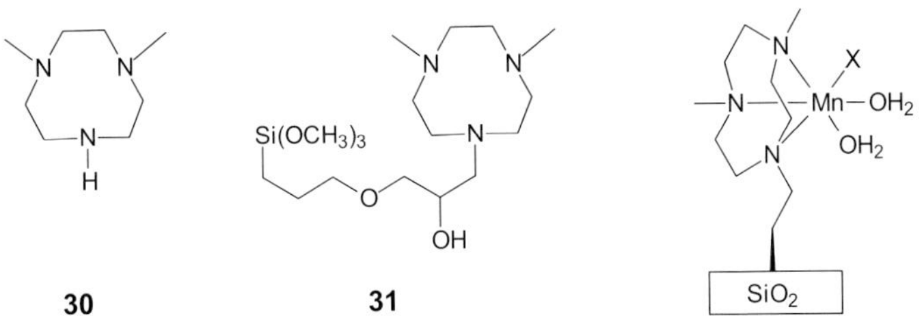

Figure 11.10 Structures of dmtacn (**30**), heterogenizable ligand **31**, and the proposed active complex **32** (X = activated "O" to be transferred) [99].

immobilizing manganese polypyridyl catalysts has been taken by Stack and coworkers recently, which may be of relevance to the tmtacn-based systems [110]. By using a copper templating approach, the immobilization of the ligands in sufficient proximity to allow for two ligands to bind each manganese ion was found to be essential to achieve full catalytic activity. Although the system employed peracetic acid as the oxidant, it nevertheless is of direct relevance to the future design of immobilized manganese-based oxidation catalysts. For example, where a binuclear catalyst is the active species under homogenous conditions, low surface coverage of the ligand will limit the ability to form such catalysts in a heterogenized form. The templating approach allows for two ligands to be immobilized in proximity to each other and thereby overcomes this problem.

cis-Dihydroxylation of Alkenes cis-Dihydroxylation is an important synthetic transformation, and several reagents can be used for the addition of two hydroxyl groups to an alkene. Both OsO_4 and alkaline $KMnO_4$ are suitable for cis-dihydroxylation [1, 14], but the catalytic versions of the OsO_4 method using O_2 [111], H_2O_2 [112], or other oxidants [113] is the method of choice for this transformation [114]. The introduction of the highly enantioselective OsO_4-catalyzed dihydroxylation by Sharpless and its extensive use in synthetic chemistry in recent years have provided ample demonstration of the key role of this oxidation reaction [115]. The toxicity of OsO_4 and the stoichiometric nature of the $KMnO_4$ cis-dihydroxylation, which usually provides only modest yields of diols, are strong incentives to develop catalytic cis-dihydroxylation reactions with H_2O_2 such as those based on Fe [116] and Mn (see below).

When using the heterogenized Mn-tmtacn system for the oxidation of alkenes (see above), De Vos, Jacobs and coworker noted that substantial amounts of cis-diol were formed in addition to the expected epoxide [99]. In oxidation reactions with **32**, with H_2O_2 as oxidant and CH_3CN as solvent, alkenes were converted to the corresponding cis-diols albeit with the catalyst activity with respect to cis-diol formation being modest (10–60 mol cis-diol/mol Mn) and epoxides being the major products. For internal alkenes, for example, 2-hexene, retention of configuration was found for both epoxide and cis-diol. Control experiments with dmtacn **30** showed severe peroxide decomposition, and no oxidation products were obtained. A sufficiently long-lived mononuclear complex (**32**) was postulated as the active species for both epoxidation and *cis*-dihydroxylation. This complex contains cis coordination sites for labile ligands (e.g., H_2O and X), and both oxygen atoms from H_2O and X (the activated oxygen) are proposed to be transferred to the alkene to produce the *cis*-diol [99]. Although the proposed structure was speculative, it did correspond with the results obtained experimentally.

Prompted by the use of oxalate and other additives and by the observation of *cis*-diol formation in the hetereogenized system of De Vos and coworkers, a wider search for effective additives both to suppress the catalase type activity of the Mn-tmtacn complexes and to promote *cis*-dihydroxylation was initiated. An early success was the strongly enhanced *cis*-dihydroxylation activity observed using $[Mn_2O_3(tmtacn)_2]$ $(PF_6)_2$ (**6**) (Figure 11.2) in the combination with aldehydes such as glyoxylic acid methylester methyl hemiacetal (gmha (**33**), Scheme 11.13) [94k].

This mixed Mn-tmtacn/activated carbonyl system provided for a highly active and H_2O_2 efficient catalyst for the epoxidation of alkenes as well as the first homogeneous Mn-based catalytic *cis*-dihydroxylation system using H_2O_2. Catalytic oxidations were performed by slow addition of aqueous 50% H_2O_2 (1.3 equiv. with respect to the substrate) to a mixture of alkene, the Mn-tmtacn catalyst (0.1 mol%), and gmha (25 mol%) in CH_3CN at $0\,°C$. Under these reaction conditions high conversions are reached, whereas only 30% excess of oxidant with respect to substrate was required (Table 11.8). The H_2O_2 efficiency represented a dramatic improvement compared to previous Mn-based systems [132]. In most cases the conversions were also significantly higher than those obtained with oxalate as co-catalyst using 1.3 equiv. of H_2O_2. Substantial amounts of *cis*-diols were formed with the *cis*-diol/epoxide ratio depending heavily on the alkene structure. The highest selectivity for *cis*-dihydroxylation was obtained with cyclooctene (Scheme 11.13), which afforded the *cis*-diol as the main product (42%, 420 t.o.n.). Minor amounts of 2-hydroxycyclooctanone were observed due to overoxidation of the diol. The ring size of cycloalkenes had a profound influence on the *cis*-diol/epoxide product ratio. For cyclic alkenes, almost no *trans*-diol could be detected (ratio *cis*-diol/*trans*-diol >99.5/0.5). *cis*-Diol formation is also observed for aliphatic acyclic alkenes. Yields of diol were significantly lower for *trans*-2-hexene than for *cis*-2-hexene, but the *cis*-diol/epoxide ratio was similar for both substrates. The aryl-substituted alkenes yielded nearly exclusively epoxide under these conditions. Importantly, diols were not formed when the substrate was replaced by the corresponding epoxide in the reaction, excluding epoxide hydrolysis as a source of diol.

Since *cis*-diol formation through Mn-catalyzed epoxide hydrolysis can be excluded it was proposed that the *cis*-diols were formed by reaction of the alkene with a Mn oxo-hydroxo species. As in the case of oxalate, activated carbonyl compounds such as gmha[1] could in principle inhibit the catalase-active dinuclear Mn complex **6** (Figure 11.2) [34] through formation of mononuclear Mn species via complexation

---

1) Gmha is an equilibrium mixture, which also contains some hydrated methyl glyoxylate. NMR experiments indicated that the formation peroxyhydrate from gmha and aqueous H_2O_2 is established slowly.

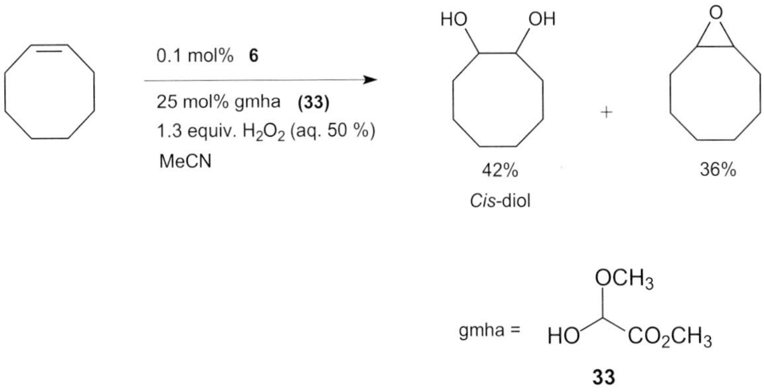

Scheme 11.13 Dihydroxylation of alkenes in the presence of glyoxylic acid methyl ester methyl hemiacetal (**33**) [94k].

Table 11.8 cis-Dihydroxylation and epoxidation of selected alkenes by H_2O_2 with Mn-tmtacn/gmha catalysts [132].

Substrate	Conversion %	Product	Turnover number (t.o.n.)[a]
Cyclohexene	88	Epoxide	590
		cis-Diol	90
		2-Cyclohexenone	80
Cyclooctene	90	Epoxide	360
		cis-Diol	420
		2-HO-Cyclooctanone	220
Norbornylene	95	exo Epoxide	540
		exo-cis-Diol	180
trans-2-Hexene	77	trans-Epoxide	210
		cis-Epoxide	50
		RR/SS-Diol	150
		RS/SR-Diol	0
cis-2-Hexene	93	cis-Epoxide	450
		trans-Epoxide	40
		SR/RS-Diol	280
		RR/SS-Diol	10
cis-Stilbene	82	cis-Epoxide	260
		trans-Epoxide	200
		meso-Hydrobenzoin	40
		Hydrobenzoin	40
Styrene	97	Epoxide	860
		Ph(CH)(OH)CH$_2$OH	60
		PhC(O)CH$_2$OH	10

a) 50% aqueous H_2O_2 (1.3 equiv. with respect to substrate) was added over 6 h to a mixture of alkene (40 mmol), Mn$_2$O$_3$ (tmtacn)$_2$ (PF$_6$)$_2$ (0.1 mol%), and GMHA (25 mol%) in MeCN (40 ml) with internal standard (1.2-dichlorobenzene, 20 mmol) at 0 °C. Analysis by GC 1 h after addition of oxidant was completed.

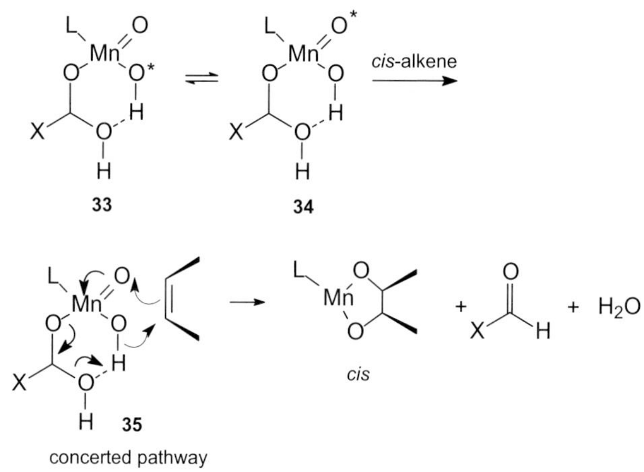

Scheme 11.14 Early proposal for the mechanism of cis-dihydroxylation by **6** (L=tmtacn, X=CO$_2$Me). Note that gmha is an equilibrium mixture which also contains some hydrated methyl glyoxylate.

to the Mn center. cis-Diol formation from an Mn oxo-hydroxo species with a coordinated hydrated carbonyl ligand could be induced through a hydrogen-bonded 6-membered ring transition state (concerted pathway, Scheme 11.14). Reoxidation of the Mn center by H$_2$O$_2$, release of the diol from Mn, and hydration of the carbonyl compound closes the catalytic cycle. It should be noted that this mechanism was tentative and based on general principles of manganese coordination chemistry rather than empirical data, which, once it became available (see below), as is so often the case, demonstrated that a thorough understanding of a system is essential to unraveling mechanistic aspects. Nevertheless, the use of activated carbonyl compounds in combination with Mn-tmtacn not only provides for a highly active (up to 860 t.o.n.) and H$_2$O$_2$-efficient epoxidation system (see above), but also was the most active Os-free homogeneous catalyst for cis-dihydroxylation (up to 420 t.o.n.), reported at that stage. Due to competing cis-dihydroxylation and epoxidation pathways it was not suitable for routine application in synthesis, however. Nevertheless it proved an important lead to develop highly selective Mn-catalyzed cis-dihydroxylation systems employing H$_2$O$_2$.

Subsequent to these reports on the activity of aldehydes in enhancing the reactivity of Mn-tmtacn-based catalysts, Feringa and coworkers demonstrated that carboxylic acids at co-catalytic levels were actually responsible for the enhancement, and that the aldehydes served solely as a source of these acids [117]. Indeed, in the presence of carboxylic acids, complex **6** [Mn$_2^{IV}$(μ-O)$_3$(tmtacn)$_2$]$^{2+}$ proved to be highly efficient in catalyzing the oxidation of alkenes to the corresponding cis-diol and epoxide products using H$_2$O$_2$. The selectivity of the catalytic system both in terms of cis-dihydroxylation and epoxidation of alkenes could be tuned readily by judicious choice of the carboxylic acid employed. High turnover numbers (t.o.n.

>2000) were achieved, especially for cis-dihydroxylation, that is, with 2,6-dichlorobenzoic acid, the highest t.o.n. reported thus far for cis-dihydroxylation of alkenes catalyzed by an osmium free system. Furthermore, the catalyst formed (see below) is almost completely efficient in regard to the use of H_2O_2 for oxidation of substrates [118].

Recent Mechanistic Insights   Until recently, most attention focused either on the coordination chemistry of the Mn-tmtacn family of complexes or on their application in functional group transformation, such as the oxidation of alkenes. A key challenge faced was probing the molecular nature of the catalysts under the reaction conditions. In part this was a consequence of their high activity and hence the low catalyst loadings employed. The complex coordination and redox chemistry, not least ligand exchange and disproportionation reactions, complicates unraveling the mode of action of this versatile oxidation catalyst. By analogy with manganese porphyrin and permanganate chemistry, the obvious candidates for the 'active catalytic species' are high-valent manganese-oxo species as well as radical intermediates (the latter being readily excluded by the retention of stereochemistry observed during catalysis). The first concerted efforts to explore the mechanistic aspects of the catalysis were reported by Hage and coworkers [119], who identified the formation of carboxylic acid-bridged dinuclear manganese complexes upon reduction of **6** in aqueous media and by Lindsay-Smith and coworkers in their studies on the oxidation of cinnamates in buffered aqueous media, primarily using mass spectrometry [94i,k, 120, 121, 122].

A key feature of catalysis with complex **6** was the frequent observation of an induction period prior to initiation of substrate conversion. This indicated that the original $[Mn_2O_3(tmtacn)_2](PF_6)_2$ complex (**6**) has first to be converted to a catalytically active species. Indeed, it was found that the catalytic activity of Mn-tmtacn was significantly increased when it was pre-treated with excess of H_2O_2 prior to the addition of the substrate (in the case of benzyl alcohol oxidation) [94c]. From the 16-line spectrum obtained by electron paramagnetic resonance spectroscopy (EPR) measurements it was inferred that the $Mn^{IV}-Mn^{IV}$ dimer was reduced by H_2O_2 to a dinuclear $Mn^{III}-Mn^{IV}$ mixed-valent species in acetone. The intensity of the mixed-valent species gradually diminished, with the subsequent appearance of an Mn^{II} species. EPR studies of the catalysts under comparable catalytic oxidation conditions using alkenes as substrates instead of alcohols showed again the mixed-valence $Mn^{III}-Mn^{IV}$ dimer [40a, 97]. Based on EPR data, similar manganese species were identified during related phenol oxidation experiments [120]. Barton et al. proposed the formation of an $Mn^V=O$ intermediate during the oxidation of 2,6-di-tert-butylphenol with Mn-tmtacn and H_2O_2 [121]. In electrospray mass spectrometry (ESI/MS) experiments the mononuclear $Mn^V=O$ species could indeed be assigned [122]. This species was also generated in oxidation reactions using a mononuclear Mn^{IV} complex and from an *in situ* prepared Mn^{II} complex using $MnSO_4$ and free tmtacn ligand [94g]. A key question arises, however, as to the relevance of these species. It was subsequently shown that the features observed by EPR spectroscopy were only in part related to the catalysis itself but were mostly related to disproportionation reactions at the high catalyst concentrations employed in mechanistic studies [118]. Furthermore, it was not demonstrated that the $Mn^V=O$ species observed by ESI-MS were active in the catalytic oxidation of alkenes.

For the combination of **6** and alkyl and aromatic carboxylic acids, the high activity and selectivity was found to be due to the *in situ* formation of bis-μ-carboxylato-bridged dinuclear manganese(III,III) complexes [117, 118]. These complexes were formed through a series of redox processes (Scheme 11.15) in which H_2O_2 acts as a terminal reductant of the dimeric Mn^{IV} complex **6**. The ability of different carboxylate ligands to tune the activity of the catalyst was found to be dependent on both the electron-withdrawing nature of the ligand and on steric effects; that is, bulky electron-deficient ligands gave the highest activity. By contrast, the *cis*-diol/epoxide selectivity is determined primarily by steric factors; that is, more bulky carboxylic acids lead to a higher ratio in favor of the *cis*-diol product. A mechanistic study of the roles played

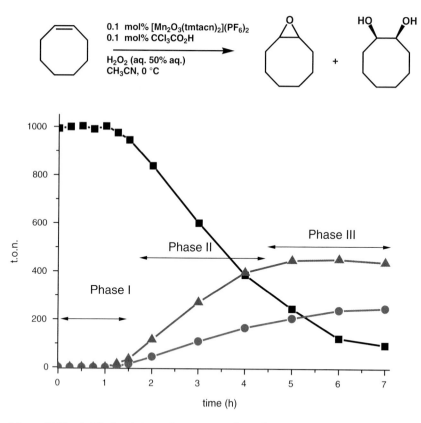

Scheme 11.15 *cis*-Dihydroxylation and epoxidation of alkenes by **6**: the multiple roles of carboxylic acids in forming the active catalysts for homogenous alkene oxidation from the catalyst **6** allow for the effect of variations in the composition of the reaction mixture on catalytic activity to be understood at a molecular level.

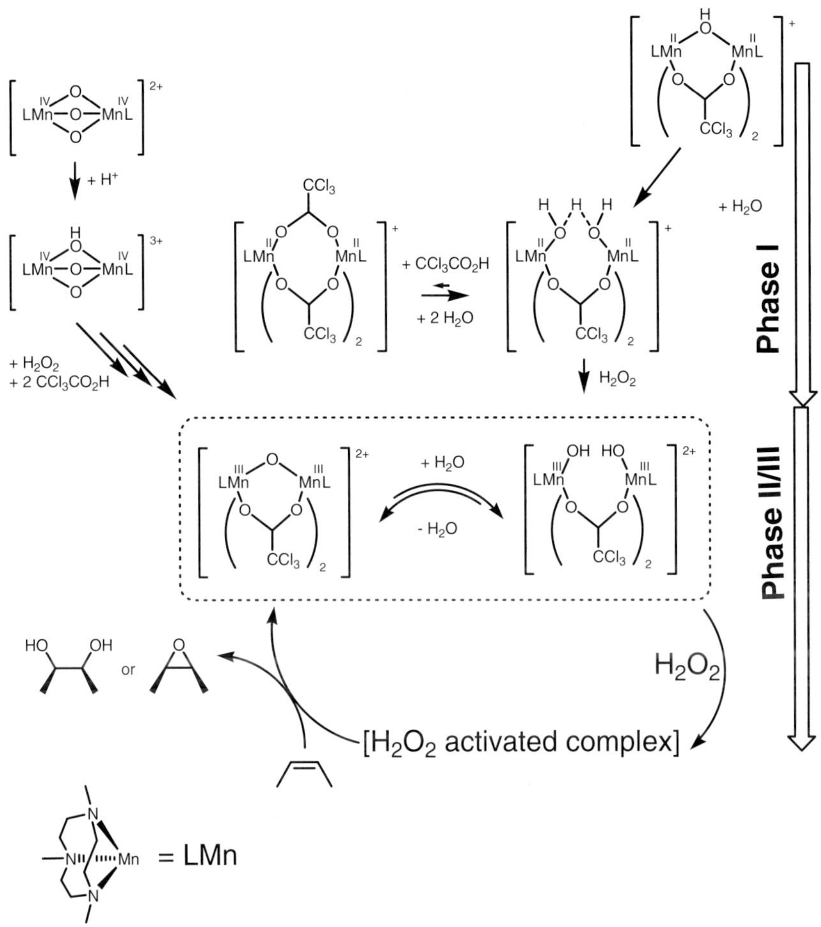

11.4 Epoxidation and cis-Dihydroxylation of Alkenes

Scheme 11.15 (Continued).

by solvent, initial catalyst oxidation state, water, carboxylic acid concentration, and the nature of the carboxylic acid employed in determining both the activity and the selectivity observed was reported [118]. The resting state of the active form of the catalyst, that is, $[Mn_2^{III}O(\mu\text{-}O)(\mu\text{-}RCO_2)_2(tmtacn)_2]^{2+}$, was identified through a combination of speciation analysis employing a range of spectroscopic techniques and isotope-labeling studies. These $[Mn_2^{III}(\mu\text{-}O)(\mu\text{-}RCO_2)_2(tmtacn)_2]^{2+}$ complexes show coordination chemistry dependent on redox state and solvent (Scheme 11.15), but the available evidence indicates that redox changes or change in the dinuclear structure do not occur during catalysis.

The report by de Boer et al. [118] does not exclude the possibility of the transient formation of high-valent mononuclear species in the catalytic cycle as proposed by several groups [94i,k, 120, 121, 122]. However, the data available up to now support an alternative view of the role of the tmtacn-based manganese catalysts as Lewis acid activators of H_2O_2, which holds considerable implications for the study of related homogeneous manganese oxidation catalysts and in building a conceptual bridge with related biological systems such as dinuclear catalase and arginase enzymes (see above) [10].

An important question, however, is whether all systems based on the manganese-tmtacn catalyst system operate via a similar mechanism, that is, the *in situ* formation of $[Mn^{III}_2O(\mu\text{-}O)(\mu\text{-}RCO_2)_2(tmtacn)_2]^{2+}$ complexes. The recent spectroscopic identification of carboxylate-bridged complexes as active catalysts does not necessarily mean that the first systems based on **6** in which oxalic acid or ascorbic acid were used by De Vos, Berkessel, and coworkers to promote oxidation catalysis involve the same general mechanism [94a,b,d]. Recently de Boer et al. compared the catalytic activity of $[Mn^{IV}_2(\mu\text{-}O)_3(tmtacn)_2]^{2+}$ using salicylic acid, ascorbic acid, and oxalic acid as additives by observing the spectroscopic changes during the course of the catalyzed reactions [123]. In the case of salicyclic acid, the electronic absorption spectra of the reaction mixture were quite different from what is observed using other carboxylic acids. Furthermore, a mononuclear complex was isolated in which the salicylato dianion is bound as a chelate. However, it was demonstrated through other spectroscopic and electrochemical techniques that these differences were not directly relevant to the catalysis itself. It was found that the role of the salicylic acid in the catalysis, that is, to act as a carboxylato ligand for binuclear complexes, was the same as that for other acids despite the presence of a potentially chelating hydroxyl group.

In the case of ascorbic acid and oxalic acid as additives, their redox activity adds an additional dimension to the catalysis they promote with **6** [123]. For both these acids the most notable observation was the absence of an induction period; that is, catalysis commenced immediately upon addition of H_2O_2. This is not surprising as both oxalic acid and ascorbic acid are reductants. In the case of ascorbic acid, although a $[Mn^{III}_2O(\mu\text{-}O)(\mu\text{-}RCO_2)_2(tmtacn)_2]^{2+}$-type complex could be identified by UV-Vis spectroscopy, EPR spectroscopy indicated that $Mn^{III,IV}_2$ dinuclear species are present in the reaction mixture during catalysis also. Hence, although in principle ascorbic acid may promote the system both by serving as a reductant of $[Mn^{IV}_2(\mu\text{-}O)_3(tmtacn)_2]^{2+}$ as well as generating $[Mn^{III}_2O(\mu\text{-}O)(\mu\text{-}RCO_2)_2(tmtacn)_2]^{2+}$-type complexes, it is also possible that a distinct mechanism is in operation. This is especially the case considering the systems ability to oxidize electron-deficient alkenes.

In the case of oxalic acid an even more complex picture emerged [123]. Initially only epoxidation is observed. However, after a certain period of time, *cis*-dihydroxylation begins. This is in stark contrast to the other carboxylic acid-based systems where epoxidation and *cis*-dihydroxylation proceed concurrently. Furthermore, the reaction mixture does not contain EPR-active species at 77 K, and the UV-vis absorption spectra are unstructured. Concurrently with the initiation of *cis*-dihydroxylation activity, however, a change in the UV-vis absorption spectrum is observed, and ultimately the characteristic absorption spectrum of a $[Mn^{III}_2O(\mu\text{-}O)(\mu\text{-}RCO_2)_2(tmtacn)_2]^{2+}$-type complex appears. Notably, further addition of oxalic acid at the point where the selectivity of the system changes suppresses cis-dihydroxylation completely, and only epoxidation is observed. These studies demonstrate that it is not necessarily the case that a single mechanism is in operation for all additives studied. It is clear that systems containing ascorbic or oxalic acid are more complex, and while insight gained with one additive/$[Mn^{IV}_2(\mu\text{-}O)_3(tmtacn)_2]^{2+}$ combination is relevant to other similar systems, it must be recognized that distinctly different catalytically active species can be generated with different additives. The occurrence of different catalytic mechanisms implies that it may be possible to tune the activity of **6** further through the use of other additive classes than those tested to date.

Asymmetric cis-dihydroxylation of Alkenes with Mn-tmtacn Although readily accessible through the osmium-based catalytic methods developed by Sharpless and coworkers [114, 124, 125], asymmetric cis-dihydroxylation (AD) of alkenes still poses a key challenge industrially because of the cost and toxicity of the osmium-based AD systems, which has precluded their widespread industrial application [126]. This has provided a strong driving force behind the identification and development of economically viable and environmentally benign methods based on first-row transition metals and H_2O_2. Despite considerable effort over the past two decades the search for first-row transition metal-catalyzed alkene cis-dihydroxylation methods themselves remains a key challenge. Recent success in using iron-based systems by Que and coworkers [127] demonstrated that the challenge may yet be met. Recently, Feringa and coworkers demonstrated [128] that the Mn-tmtacn based catalytic system employing carboxylic acids can be used for asymmetric cis-dihydroxylation of an electron-rich cis-alkene – ironically a notably challenging substrate in the case of the osmium-based systems. Importantly, the manganese-based catalytic system for AD of alkenes developed employs H_2O_2 as the oxidant. With 0.4 mol% catalyst loading, the enantioselective cis-dihydroxylation of 2,2′-dimethylchromene proceeds with full conversion and high selectivity toward the cis-diol product with an enantiomeric excess of up to 54% (Scheme 11.16). In contrast to previous efforts to achieve enantioselectivity with tmtacn-type catalysts by modifying the tmtacn ligand, in this approach chiral carboxylato ligands were employed allowing for rapid screening of a large library of potential chiral ligands.

Although the reactivity and selectivity is readily tunable by variation in the carboxylic acid employed, the preference of this system toward electron-rich cis-alkenes limits its scope. Nevertheless, the high turnover numbers and efficiency achieved, the tunablity of the system, and its use of H_2O_2 as the terminal oxidant demonstrate that a sustainable and synthetically useful method for first-row transition metal-catalyzed AD is realizable.

11.4.4.3 Manganese Complexes for Alkene Oxidation Based on Pyridyl Ligands

A drawback associated with the 1,4,7-trimethyl-1,4,7-triazacyclononane based catalyst is the often tedious procedure to achieve modifications of the ligand

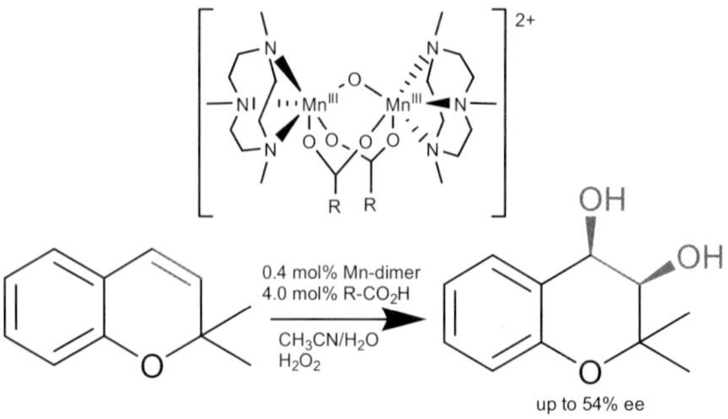

Scheme 11.16 Asymmetric cis-dihydroxylation (AD) of 2,2-dimethylchromene with H_2O_2 catalyzed by the system **6**/R-CO_2H (=acetyl-D-phenylglycine) [128].

structure [129]. In view of the excellent catalytic activity of the Mn-tmtacn systems, a major challenge is therefore the design of novel dinucleating ligands featuring the three-N-donor facial coordinating set (as for the tmtacn ligand) for each manganese site [131] while retaining the high oxidation activity. Stack, Chan, and coworkers have employed pyridyl-amine based manganese complexes as highly efficient catalysts in the epoxidation of alkenes using peracetic acid and isobutyraldehyde/O_2, respectively [130]. The development of equally effective catalysts employing H_2O_2, however, is a major challenge.

High epoxidation activity was found for manganese complexes based on dinucleating ligand N,N,N′,N″-tetrakis(2-pyridylmethyl)-1,3-propanediamine (tptn) [132]. The ligand contains a three-carbon spacer between the three-N-donor sets. This type of ligand is readily accessible, and facile modification of the ligand structure can be achieved. Complexes of this type have also been reported as mimics for PSII [131].

Complex **34** [(Mn$_2$O(OAc)$_2$tptn)]$^{2+}$ is able to catalyze the oxidation of a range of alkenes including styrene, cyclohexene, and trans-2-octene to the corresponding epoxides in good yields and turnovers up to 870 (Scheme 11.17). In sharp contrast,

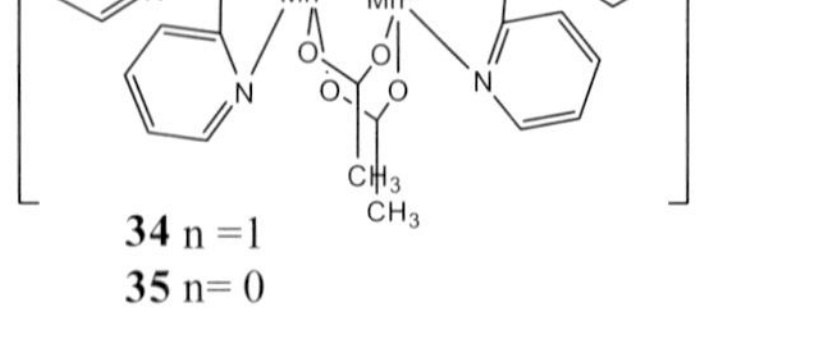

Scheme 11.17 Epoxidation with Mn-tptn catalyst and structures of manganese complexes [132].

complex **35** based on *N*,*N*,*N′*,*N″*-tetrakis(2-pyridylmethyl)-1,2-ethylenediamine (tpen), featuring a two-carbon spacer between the three-N-donor sets in the ligand, was unreactive in epoxidation reactions [132].

High selectivity was observed in the epoxidation of cyclic alkenes (especially cyclohexene) with the important feature that allylic oxidation products were not obtained. Excellent results are reported for internal alkenes, for example, *trans*-2-octene and *trans*-4-octene, whereas terminal linear alkenes give slightly lower yields. The oxidation of *cis*-β-methylstyrene with H_2O_2 in the presence of $[Mn_2O(OAc)_2tptn]^{2+}$ catalyst (**34**) provides in addition to the corresponding *cis*-epoxide also a considerable amount of *trans*-epoxide. Cis/trans isomerization has been frequently observed in mechanistic studies using porphyrin and manganese salen catalysts and is usually attributed to the formation of a radical intermediate (**a**, Scheme 11.18) with a lifetime sufficient for internal rotation before ring closure via reaction path B occurs to give the thermodynamically more stable *trans*-epoxide (**b**) [133]. In the event of a rapid collapse of the radical intermediate (via reaction path A) retention of configuration will be observed. The Mn-**39**-based catalyst [34] provides a viable alternative to the Mn-tmtacn and Mn-salen systems, with high activity for epoxidation and the distinct advantage that ligand variation for further catalyst fine-tuning is readily accomplished.*

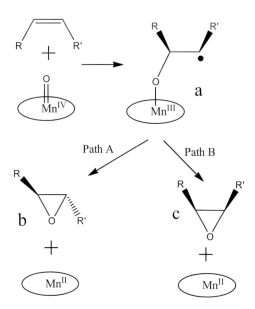

Scheme 11.18 Radical-type pathways to epoxidation of alkenes [133].

*A very recent study has identified that fort a wide range of ligands containing the (pyridin-2-yl)methyl unit, decomposition to pyridine-2-carboxylic acid occurs in the presence of hydrogen peroxide and acetone. It is the pyridine-2-carboxylic acid thus formed that is responsible for the activity observed. See Pijper et al. (2010) *Dalton*, DOI:10.1039/C0DT00452A and Saisaha et al. (2010) *Org. Biomol. Chem.*, in press.

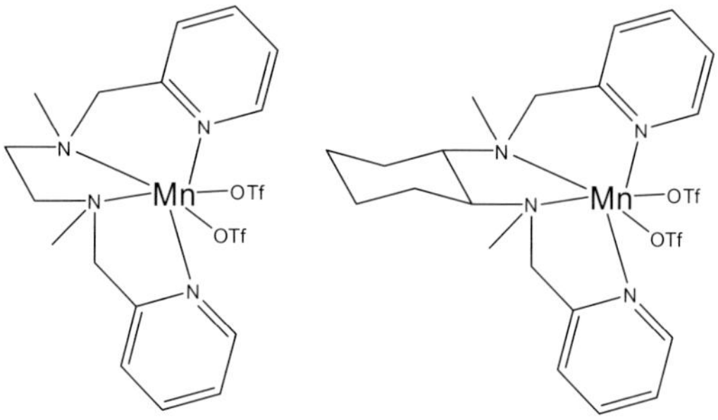

Figure 11.11 BPMEN type ligands **36** and **37** employed by Costas and coworkers for the epoxidation of alkenes with H_2O_2/acetic acid [102].

Costas and coworkers have shown recently that manganese complexes based on BPMEN-type ligands (where BPMEN is bis((pyridin-2-yl)methyl)ethylene-1,2-diamine, Figure 11.11) show enhanced activity in the epoxidation of alkenes with H_2O_2 in the presence of excess acetic acid, albeit with lower activity and with a narrower substrate scope than for the pyridyl-tmtacn based systems reported in the same study (see above, Table 11.7) [102].

Recently, Zhong et al. have reported a series of quinolin-8-ol-based manganese catalysts which show remarkably high efficiency in the epoxidation of a wide range of alkenes with H_2O_2 in acetone/water [134]. A key finding in this study is that the activity was highly pH dependent and the reaction could be controlled by regulating pH.

11.5
Manganese Catalysts for the Oxidation of Alkanes, Alcohols, and Aldehydes

11.5.1
Oxidation of Alkanes

From an industrial perspective, alkanes represent one of the most challenging and one of the most important classes of substrates with regard to oxidation catalysis. C—H activation with manganese-based catalysts is an area of continuing interest, with considerable attention being given to systems employing peracetic acid [135] and iodosylarenes [136]. Nevertheless, several groups have demonstrated that H_2O_2 can also be employed to effect alkane oxidation with the help of Mn catalysts [137] and Fe catalysts [138].

In general, the focus on alkane oxidation has been on C—H activation of standard alkane substrates; however, recently focus has shifted toward selective reactions on real synthetic targets. An impressive example of this can be seen in the report of Chen and White employing an iron-based catalyst and H_2O_2 as the terminal oxidant [139]. More recently, Macleod et al. have used the Jacobsen salen catalyst to achieve selective

C—H activation of pharmaceutical intermediates including phenylethylmalondiamide, primidone, and phenobarbital [140].

Although focused on the oxidation of alkenes, de Boer et al. [117] noted in their report from 2005 that $[Mn_2^{IV}(\mu\text{-O})_3(tmtacn)_2]^{2+}$ together with 1 mol% of CCl_3CO_2H was moderately effective in the oxidation of linear and cyclic alkanes with H_2O_2, for example, with cyclooctane: 50% conversion to cyclooctanone with 1.3 equiv. H_2O_2. Unfortunately, the high activity of this class of complex for alcohol oxidation precludes selective C—H oxidation to the alcohol (see below).

11.5.2
Oxidation of Alcohols and Aldehydes

In the common repertoire of synthetic methods, the selective oxidation of alcohols to aldehydes holds a prominent position. A number of catalytic procedures have been introduced in recent years, and the Ley-Griffith system employing $[NBu_4][RuO_4]$/N-methylmorpholine oxide has proven to be particularly valuable in synthetic applications [141]. Selective catalytic aldehyde formation using H_2O_2 as terminal oxidant is highly warranted, however. The Mn-tmtacn [94c] catalysts and several *in situ* prepared complexes [142] of $Mn(OAc)_3$ with tptn-type ligands (Scheme 11.19) have been shown to be selective catalysts for the oxidation of benzyl alcohols as well as secondary alcohols to the corresponding aldehydes and ketones. In the case of oxidation of alcohols with Mn-tmtacn-based catalysts, aldehydes are formed initially, but the reaction continues with formation of the corresponding carboxylic acids [117]. Mn complexes based on ligands **39–43** (Scheme 11.19), show high activity and selectivity (t.o.n. up to 850, Table 11.9), depending on the ligand structure. Ligands **40**

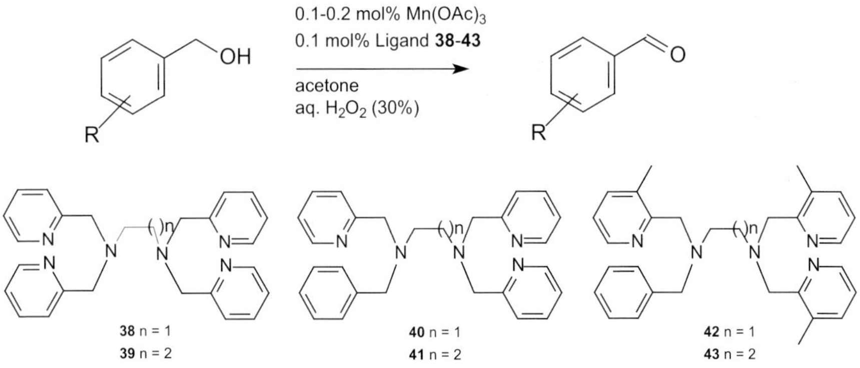

Scheme 11.19 Ligands used for the manganese-catalyzed alcohol oxidation at ambient temperature [142].

and **41**, containing a two-carbon or three-carbon spacer and lacking one pyridine compared to tptn, were found to form only moderately active catalysts, however, with as longer induction periods observed than the case of **6**.

Table 11.9 Oxidation of selected alcohols with in-situ-prepared Mn catalysts based on ligands **39** and **43** [142].

Substrate	T.o.n.[a]	Selectivity (%)[a]	T.o.n.[a]	Selectivity (%)[a]
	39		43	
Benzyl alcohol	326	95	303	99
4-Methoxybenzyl alcohol	201	80	291	75
4-Chlorobenzyl alcohol	449	99	414	99
4-Trifluoromethylbenzyl alcohol	329	70	258	70
4-Fluorobenzyl alcohol	233	90	248	70
2,5-Dimethoxybenzyl alcohol	90	99	63	99
Cyclohexanol	363	95	593	80
Cycloheptanol	849	85	688	99
1-Octanol	108	85	46	90
2-Octanol	680	95	480	95
sec-Phenylethyl alcohol	657	90	715	95

a) Turnover numbers (t.o.n.) after 4 h and selectivity (%) with ligands **39** and **43**.

Using in situ prepared complexes based on ortho-methyl-substituted ligands **42** and **43**, excellent results were found, and, remarkably, the induction period was strongly reduced. A strong 16-line EPR signal was observed immediately after mixing ligand **43** with Mn(OAc)$_3$, H$_2$O$_2$, and substrate, which points to the involvement of dinuclear species in the oxidation reaction. The catalyst based on the ligands with the three-carbon spacers show in all cases much higher reactivity (shorter induction period) than the two-carbon spacer analogs, which is likely to be a result of faster formation of the dinuclear species in the former case. The primary kinetic isotope effects (k_H/k_D) for the Mn-catalyzed oxidation of benzyl alcohol and benzyl-d_7 alcohol observed are in the range of 2.2–4.3. These values strongly indicate that cleavage of the benzylic C−H bond is involved in the rate-determining step [143]. It has been concluded on the basis of these data that hydroxyl radicals are not involved in these processes, as, owing to the high reactivity of these radicals, a much lower isotopic effect would be expected [144]. Accordingly, no indication of hydroxylation of aromatic rings was observed for any of the substrates. At this point it has not been established which active species (e.g., high-valent Mn$=$O or Mn−OOH) is involved in the selective aldehyde formation.

11.5.3
Sulfides, Sulfoxides, and Sulfones

Although sufides are among the most straightforward of substrates to oxidize, the selective oxidation of sulfides to sulfoxides and to a lesser extent sulfones is often challenging for substrates other than simple dialkyl-, diaryl-, and alkylaryl-sulfides. In transition metal-catalyzed oxidations the relatively low reactivity of electron-deficient sulfides and their insolubility in polar solvents is problematic when H$_2$O$_2$ is employed as the terminal oxidant.

11.5 Manganese Catalysts for the Oxidation of Alkanes, Alcohols, and Aldehydes

The selective catalytic oxidation of sulfides to sulfoxides has been a challenge for many years, not unexpectedly in view of the importance of sulfoxides as intermediates in organic synthesis [145]. The undesired sulfone is a common by-product in sulfide oxidation using H_2O_2 as oxidant, and its formation must be suppressed in any system intended for widespread application. Recently, Bagherzadeh et al. have reported the selective oxidation of sulfides to sulfoxides using a manganese(III) complex based on bidentate oxazoline ligands [146]. In this system, a water-free source of H_2O_2 was employed (UHP) in methanol/dichloromethane. Notably the selectivity toward oxidation to the sulfoxide or sulfone was determined both by the solvent employed and by the nature of the axial ligand added (e.g., imidazole).

The use of aqueous H_2O_2 is, however, advantageous because of the reduction in the waste products that need to be dealt with. In addition, considerable effort has been devoted to the development of catalytic methods for the preparation of optically active sulfoxides [145]. Jacobsen and coworkers have applied manganese(III) salen complexes to sulfide oxidation [71a]. The high reactivity encountered with sodium hypochlorite precluded the selective oxidation of sulfides. With iodosylbenzene, by contrast, complete selective oxidation to sulfoxide could be achieved [147]. As with hypochlorite and UHP, iodosylbenzene shows poor solubility and low atom efficiency, while it is costly in large-scale applications. Suprisingly, with H_2O_2 high yields and identical enantioselectivities (34–68% ee) compared with iodosylbenzene could be achieved [147]. In acetonitrile, 2–3 mol% of catalyst and 6 equiv. of oxidant resulted in minimum formation of the sulfone [147]. Ligands with bulky substituents at the 3,3′- and 5,5′-positions yielded the highest enantioselectivities (Scheme 11.20). The enantioselectivity for sulfoxidation was found to be generally lower than that observed for epoxidation using the same catalysts. Para-tert-butyl-substituted salen **44** (R = ^{t}Bu, Scheme 11.20) was the most selective for a range of substrates. Mn complexes derived from ligands with electron-withdrawing substituents showed lower

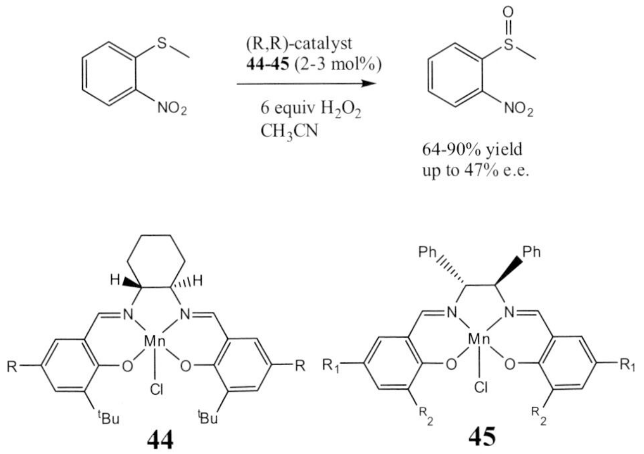

Scheme 11.20 Manganese(III)-salen complexes for sulfide oxidation introduced by Jacobsen and coworkers [147].

enantioselectivities, and, in the case of the *para*-nitro-substituted complex 44 (R = NO$_2$), asymmetric sulfide oxidation was not observed.

Katsuki and coworkers have employed the related chiral manganese salen complexes, in particular the so-called second-generation Mn(salen) **46**, in sulfide oxidation [148]. This complex was found to serve as an efficient catalyst in asymmetric sulfoxidation, albeit the less atom-efficient iodosylbenzene was required as oxidant (Scheme 11.21).

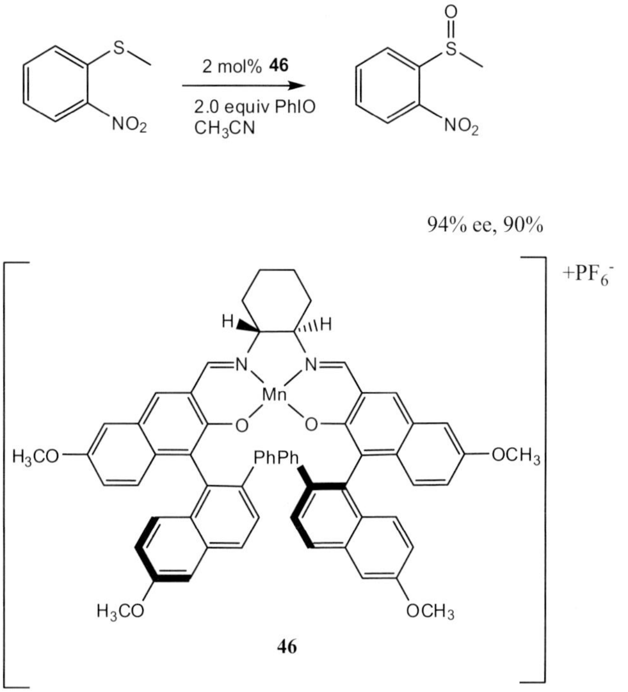

Scheme 11.21 Mn(III)-salen complexes for asymmetric sulfide oxidation reported by Katsuki and coworkers [148c].

Katsuki and Saito reported that di-μ-oxo titanium complexes of chiral salen ligands serve as efficient catalysts for asymmetric oxidation of a range of sulfides using H$_2$O$_2$ or UHP as terminal oxidants [149]. Enantioselectivities as high as 94% were achieved [149]. A monomeric peroxo titanium species was proposed as the active oxidant based on MS and NMR studies [150].

Mn catalysts that show activity in alkene or alcohol oxidation with H$_2$O$_2$ are potentially active in the oxidation of sulfides also. The Mn-tmtacn catalysts and a number of *in-situ*-formed complexes employing ligands such as **39** are examples of such catalysts (see above). These complexes were found to be highly active in the oxidation of sulfides to sulfoxides. For example, the dinuclear manganese complex based on tmtacn (**6**) performs efficiently in the oxidation of aryl alkyl sulfides and generally results in full conversion within 1 h. Unfortunately, as is often the case, in

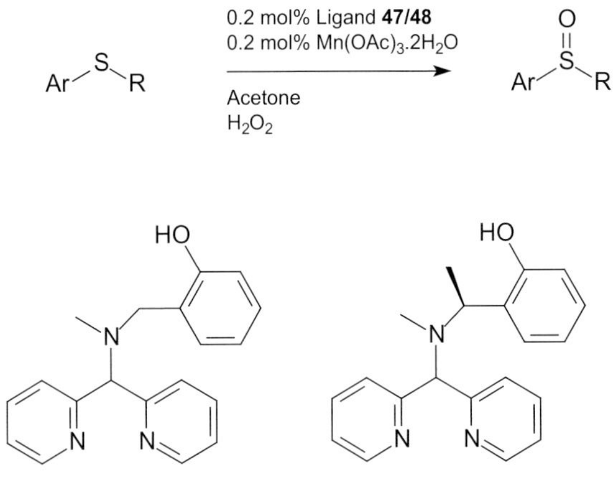

Scheme 11.22 Nitrogen- and oxygen-based donor ligands used in sulfide oxidation by Feringa and coworkers [152].

addition to the desired sulfoxide, the sulfone was also formed. Similar reactivity patterns were observed with manganese complexes based on tpen (**38**) and tptn (**39**). With ligand **47** (Scheme 11.22), however, only slight over-oxidation to sulfone was observed [151]. Ligand **47** combines structural features of both tptn and salen ligands. With ligand **48**, a chiral version of ligand **47**, the Mn-catalyzed conversion of sulfides to sulfoxides with H_2O_2 proceeded in yields ranging from 48 to 55%, albeit with low enantioselectivities (<18%) [152]. Schoumacker et al. have employed manganese complexes of chiral bis-pyridyl-cyclohexane-1,2-diamine-based ligands in the asymmetric oxidation of several aryl alkyl sulfides using H_2O_2 as the terminal oxidant [153]. Importantly, although the enantiomeric excess achieved was modest (ee <34%), the presence or absence of acetylacetonate in the reaction determined which sulfoxide enantiomer was formed. This is an indication that, as for epoxidation, the combination of metal and ligand is not the only factor that must be considered in catalyst design. Other components may either act as additional ligands or affect the coordination mode of the ligand to the metal, thereby influencing the reactivity of the catalyst as has been observed in the Mn-tmtacn based systems, vide supra.

11.6
Conclusions

Hydrogen peroxide is a particularly attractive oxidant and holds a prominent position in the development of benign catalytic oxidation procedures. In recent years a number of highly versatile catalytic oxidation methods based on, for example, polyoxometalates [154], methyltrioxorhenium [155] or tungstate [4, 156] complexes in the presence of phase transfer catalysts, all using hydrogen peroxide as the terminal oxidant, have been introduced. Mn-catalyzed epoxidations, aldehyde formation, and sulfoxidation with H_2O_2 have emerged as effective and practical alternatives. In particular, recently developed epoxidation catalysts based on a combination of Mn-tmtacn and additives show high activity and excellent selectivity in the epoxidation of a wide range of alkenes. Despite considerable progress in enantioselective epoxidation with Mn-salen systems using H_2O_2 as the oxidant, a general catalytic epoxidation method based on chiral Mn complexes remains a highly warranted goal. Particularly promising are the findings that significant *cis*-dihydroxylation can be achieved with Mn catalysts. These studies could provide guiding principles for the design of Mn catalysts as an alternative to current Os-based chiral *cis*-dihydroxylation systems. For industrial application, further improvement with respect to hydrogen peroxide efficiency and catalytic activity is needed for most of the Mn systems developed so far. The delicate balance between oxygen transfer to the substrate and hydrogen peroxide decomposition remains a critical issue in all these systems. Other challenges include determination of the nature of the Mn complexes in solution and identification of the actual active species involved in oxygen transfer and the mechanisms of the Mn-catalyzed oxidations with hydrogen peroxide, and elucidation of the key role of the additives in several cases. It is likely that detailed insight into these aspects of the catalytic systems developed recently will bring major breakthroughs in Mn-catalyzed oxidations with hydrogen peroxide in the near future.

References

1 (a) Sheldon, R.A. and Kochi, J.K. (1981) *Metal-Catalyzed Oxidations of Organic Compounds*, Academic Press, New York; (b) Piera, J. and Bäckvall, J.-E. (2008) *Angew. Chem. Int. Ed.*, **47**, 3506–3523.

2 (a) Mijs, W.J. and de Jonge, C.R.H.I. (eds) (1986) *Organic Synthesis by Oxidation with Metal Compounds*, Plenum Press, New York, (b) Trost, B.M. and Fleming, I. (eds) (1991) *Comprehensive Organic Synthesis*, vol. 7, Pergamon Press, Oxford, (c) Hudlicky, M. (1990) *Oxidations in Organic Chemistry*, ACS Monograph Ser. 186, American Chemical Society, Washington D.C.

3 Anastas, P.T. and Warner, J.C. (1998) *Green Chemistry, Theory and Practice*, Oxford University Press, Oxford.

4 Noyori, R., Aoki, M., and Sato, K. (2003) *Chem. Commun.*, 1977–1986.

5 Sibbons, K.F., Shastri, K., and Watkinson, M. (2006) *Dalton Trans.*, 645–661.

6 Hage, R. and Lienke, A. (2006) *Angew. Chem. Int. Ed.*, **45**, 206–222.

7 Simándi, L.I. (ed.) (1992) *Advances in Catalytic Activation of Dioxygen by Metal Complexes*, Kluwer Academic, Dordrecht.

8 Costas, M., Mehn, M.P., Jensen, M.P., and Que, L. Jr. (2004) *Chem. Rev.*, **104**, 939–986.

9 Weissermel, K. and Arpe, H.-J. (1993) *Industrial Organic Chemistry*, VCH, Weinheim.

10 de Boer, J.W., Browne, W.R., Feringa, B.L., and Hage, R. (2007) *C. R. Chimie*, **10**, 341–354.

11 (a) Jones, C.W. (1999) *Applications of Hydrogen Peroxide and Derivatives*, MPG Books Ltd., Cornwall, UK, (b) Hage, R. and Lienke, A. (2006) *J. Mol. Catal. A – Chem.*, **251**, 150–158.

12 Lane, B.S. and Burgess, K. (2003) *Chem. Rev.*, **103**, 2457–2474.

13 Vollhardt, K.P.C. and Schore, N.E. (2003) *Organic Chemistry*, 4th edn, W. H. Freeman, New York.

14 March, J. (1985) *Advanced Organic Chemistry*, 3rd edn, John Wiley & Sons, New York.

15 Lippard, S.J. and Berg, J.M. (1994) *Principles of Bioinorganic Chemistry*, University Science Books, Mill Valley, California, USA.

16 Que, L. Jr. and Tolman, W.B. (2008) *Nature*, **455**, 333–340.

17 Pecoraro, V.L. (ed.) (1992) *Manganese Redox Enzymes*, VCH Publisher, New York.

18 (a) Allgood, G.S. and Perry, J.J. (1986) *J. Bacteriol.*, **168**, 563–567; (b) Wu, A.J., Penner-Hahn, J.E., and Pecoraro, V.L. (2004) *Chem. Rev.*, **104**, 903–938.

19 Waldo, G.S. and Penner-Hahn, J.E. (1995) *Biochemistry*, **34**, 1507–1512.

20 Mullins, C.S. and Pecoraro, V.L. (2008) *Coord. Chem. Rev.*, **252**, 416–443.

21 Riley, D.P. (1999) *Chem. Rev.*, **99**, 2573–2588.

22 Fridovich, I. (1989) *J. Biol. Chem.*, **264**, 7761–7764.

23 Jakoby, W.B. and Ziegler, D.M. (1990) *J. Biol. Chem.*, **265**, 20715–20718.

24 Pick, M., Rabani, J., Yost, F., and Fridovich, I. (1974) *J. Am. Chem Soc.*, **96**, 7329–7333.

25 Zouni, A., Witt, H.-T., Kern, J., Fromme, P., Krauss, N., Saenger, W., and Orth, P. (2001) *Nature*, **409**, 739–743.

26 Pecoraro, V.L. and Hsieh, W.-Y. (2008) *Inorg. Chem.*, **47**, 1765–1778, and references therein.

27 Kono, Y. and Fridovich, I. (1983) *J. Biol. Chem.*, **258**, 6015–6019.

28 Barynin, V.V. and Grebenko, A.I. (1986) *Dokl. Akad. Nauk SSSR*, **286**, 461–464.

29 *Thermus thermophilus*: Antonyuk, S.V., Melik-Adamyan, V.R., Popov, A.N., Lamzin, V.S., Hempstead, P.D., Harrison, P.M., Artymyuk, P.J., and Barynin, V.V. (2000) *Crystallogr. Rep.*, **45**, 105–116; *Lactobacillus plantarum*: Barynin, V.V., Whittaker, M.M., Antonyuk, S.V., Lamzin, V.S., Harrison, P.M., Artymiuk, P.J., and Whittaker, J.W. (2001) *Structure*, **9**, 725–738.

30 Ghanotakis, D.F. and Yocum, C.F. (1990) *Annu. Rev. Plant Physiol. Plant Mol. Biol.*, **41**, 255–276.

31 Dexheimer, S.L., Gohdes, J.W., Chan, M.K., Hagen, K.S., Armstrong, W.H., and Klein, M.P. (1989) *J. Am. Chem. Soc.*, **111**, 8923–8925.

32 (a) Sheats, J.E., Czernuszewicz, R.S., Dismukes, G.C., Rheingold, A.L., Petrouleas, V., Stubbe, J., Armstrong, W.H., Beer, R.H., and Lippard, S.J. (1987) *J. Am. Chem. Soc.*, **109**, 1435–1444; (b) Wu, F.-J., Kurtz, D.M. Jr., Hagen, K.S., Nyman, P.D., Debrunner, P.G., and Vankai, V.A. (1990) *Inorg. Chem.*, **29**, 5174–5183.

33 (a) Waldo, G.S., Yu, S., and Penner-Hahn, J.E. (1992) *J. Am. Chem. Soc.*, **114**, 5869–5870; (b) Pessiki, P.J. and Dismukes, G.C. (1994) *J. Am. Chem. Soc.*, **116**, 898–903.

34 Hage, R. (1996) *Recl. Trav. Chim. Pays-Bas*, **115**, 385–395.

35 Mathur, P., Crowder, M., and Dismukes, G.C. (1987) *J. Am. Chem. Soc.*, **109**, 5227–5233.

36 (a) Sakiyama, H., Okawa, H., and Isobe, R. (1993) *J. Chem. Soc., Chem. Commun.*, 882–884; (b) Sakiyama, H., Okawa, H., and Suzuki, M. (1993) *J. Chem. Soc., Dalton Trans.*, 3823–3825; (c) Higuchi, C., Sakiyama, H., Okawa, H., and Fenton, D.E. (1995) *J. Chem. Soc., Dalton Trans.*, 4015–4020; (d) Yamami, M., Tanaka, M., Sakiyama, H., Koga, T., Kobayashi, K., Miyasaka, H., Ohba, M., and Okawa, H. (1997) *J. Chem. Soc., Dalton Trans.*, 4595–4601.

37 (a) Wieghardt, K., Bossek, U., Ventur, D., and Weiss, J. (1985) *J. Chem. Soc., Chem. Commun.*, 347–349; (b) Wieghardt, K., Bossek, U., Nuber, B., Weiss, J., Bonvoisin, J., Corbella, M., Vitols, S.E., and Girerd, J.J. (1988) *J. Am. Chem. Soc.*, **110**, 7398–7411; (c) Bossek, U., Weyhermüller, T., Wieghardt, K., Nuber, B., and Weiss, J. (1990) *J. Am. Chem. Soc.*, **112**, 6387–6388; (d) Bossek, U., Saher, M., Weyhermüller, T., and Wieghardt, K. (1992) *J. Chem. Soc., Chem. Commun.*, 1780–1782; (e) Stockheim, C., Hoster, L., Weyhermüller, T., Wieghardt, K., and Nuber, B. (1996) *J. Chem. Soc., Dalton Trans.*, 4409–4416; (f) Burdinski, D., Bothe, E., and Wieghardt, K. (2000) *Inorg. Chem.*, **39**, 105–116.

38 Triller, M.U., Hsieh, W.-Y., Pecoraro, V.L., Rompel, A., and Krebs, B. (2002) *Inorg. Chem.*, **41**, 5544–5554.

39 (a) Vicario, J., Eelkema, R., Browne, W.R., Meetsma, A., La Crois, R.M., and Feringa, B.L. (2005) *Chem. Commun.*, 3936–3938; (b) Heureux, N., Lusitani, F., Browne, W.R., Pshenichnikov, M.S., van Loosdrecht, P.H.M., and Feringa, B.L. (2008) *Small*, **4**, 476–480; (c) Stock, C., Heureux, N., Browne, W.R., and Feringa, B.L. (2008) *Chem. Eur. J.*, **14**, 3146–3153.

40 (a) Hage, R., Iburg, J.E., Kerschner, J., Koek, J.H., Lempers, E.L.M., Martens, R.J., Racherla, U.S., Russell, S.W., Swarthoff, T., Van Vliet, M.R.P., Warnaar, J.B., Van der Wolf, L., and Krijnen, B. (1994) *Nature*, **369**, 637–639; (b) Gilbert, B.C., Lindsay Smith, J.R., Newton, M.S., Oakes, J., and Pons i Prats, R. (2003) *Org. Biomol. Chem.*, **1**, 1568–1577; (c) Alves, V., Capanema, E., Chen, C.-L., and Gratzl, J. (2003) *J. Mol. Catal. A: Chem.*, **206**, 37–51.

41 (a) Verall, M. (1994) *Nature*, **369**, 511; (b) Comyns, A.E. (1994) *Nature*, **369**, 609–610.

42 Shastri, K., Cheng, E.W.C., Motevalli, M., Schofield, J., Wilkinson, J.S., and Watkinson, M. (2007) *Green Chem.*, **9**, 996–1007.

43 (a) Gorzynski Smith, J. (1984) *Synthesis*, 629–656; (b) Bonini, C. and Righi, G. (1994) *Synthesis*, 225–238; (c) Grigoropoulou, G., Clark, J.H., and Elings, J.A. (2003) *Green Chem.*, **5**, 1–7.

44 James, A.P., Johnstone, R.A.W., McCarron, M., Sankey, J.P., and Trenbirth, B. (1998) *Chem. Commun.*, 429–430, and references cited therein.

45 Denmark, S.E. and Wu, Z.C. (1999) *Synlett*, 847–859.

46 Hill, C.L. and Prosser-McCartha, C.M. (1995) *Coord. Chem. Rev.*, **143**, 407–455.

47 Katsuki, T. (1995) *Coord. Chem. Rev.*, **140**, 189–214.

48 Meunier, B. (1992) *Chem. Rev.*, **92**, 1411–1456.

49 (a) Finney, N.S., Pospisil, P.J., Chang, S., Palucki, M., Konsler, R.G., Hansen, K.B., and Jacobsen, E.N. (1997) *Angew. Chem. Int. Ed.*, **36**, 1720–1723; (b) Berkessel, A., Frauenkron, M., Schwenkreis, T., and Steinmetz, A. (1997) *J. Mol. Catal. A: Chem.*, **117**, 339–346; (c) Moiseev, I.I. (1997) *J. Mol. Catal. A: Chem.*, **127**, 1–23.

50 Ho, K.-P., Wong, W.-L., Lam, K.-M., Lai, C.-P., Chan, T.H., and Wong, K.-Y. (2008) *Chem. Eur. J.*, **14**, 7988–7996.

51 (a) Lane, B.S. and Burgess, K. (2001) *J. Am. Chem. Soc.*, **123**, 2933–2934; (b) Lane, B.S., Vogt, M., DeRose, V.J., and Burgess, K. (2002) *J. Am. Chem. Soc.*, **124**, 11946–11954.

52 (a) Richardson, D.E., Yao, H., Frank, K.M., and Bennett, D.A. (2000) *J. Am. Chem. Soc.*, **122**, 1729–1739; (b) Yao, H., and Richardson, D.E. (2000) *J. Am. Chem. Soc.*, **122**, 3220–3221.

53 Tong, K.-H., Wong, K.-Y., and Chan, T.H. (2003) *Org. Lett.*, **5**, 3423–3425.

54 Mansuy, D. (1993) *Coord. Chem. Rev.*, **125**, 129–141.

55 (a) Renaud, J.-P., Battioni, P., Bartoli, J.F., and Mansuy, D. (1985) *J. Chem. Soc., Chem. Commun.*, 888–889; (b) Battioni, P., Renaud, J.-P., Bartoli, J.F., Momenteau, M., and Mansuy, D. (1987) *Recl. Trav. Chim. Pays-Bas*, **106**, 332; (c) Battioni, P., Renaud, J.-P., Bartoli, J.F., Reina-Artiles, M., Fort, M., and Mansuy, D. (1988) *J. Am. Chem. Soc.*, **110**, 8462–8470; (d) Poriel, C., Ferrand, Y., Le Maux, P., Rault-Berthelot, J., and Simonneaux, G. (2003) *Tetrahedron Lett.*, **44**, 1759–1761.

56 For the oxidation of monoterpenes and alkylarenes with good conversions, albeit as a mixtures of products, see: (a) Martins, R.R.L., Neves, M.G.P.M.S., Silvestre, A.J.D., Simões, M.M.Q., Silva, A.M.S., Tomé, A.C., Cavaleiro, J.A.S., Tagliatesta, P., and Crestini, C. (2001) *J. Mol. Catal. A: Chem.*, **172**, 33–42; (b) Rebelo, S.L.H., Simões, M.M.Q., Neves, M.G.P.M.S., and Cavaleiro, J.A.S. (2003) *J. Mol. Catal. A: Chem.*, **201**, 9–22.

57 Anelli, P.L., Banfi, S., Montanari, F., and Quici, S. (1989) *J. Chem. Soc., Chem. Commun.*, 779–780.

58 De Paula, R., Simões, M.M.Q., Neves, M.G.P.M.S., and Cavaleiro, J.A.S. (2008) *Catal. Commun.*, **10**, 57–60.

59 Groves, J.T., Watanabe, Y., and McMurry, T.J. (1983) *J. Am. Chem. Soc.*, **105**, 4489–4490.

60 Balcells, D., Raynaud, C., Crabtree, R.H., and Eisenstein, O. (2008) *Inorg. Chem.*, **47**, 10090–10099.

61 (a) Ostovic, D. and Bruice, T.C. (1992) *Acc. Chem. Res.*, **25**, 314–320; (b) Arasasingham, R.D., He, G.-X., and Bruice, T.C. (1993) *J. Am. Chem. Soc.*, **115**, 7985–7991.

62 Baciocchi, E., Boschi, T., Cassioli, L., Galli, C., Jaquinod, L., Lapi, A., Paolesse, R., Smith, K.M., and Tagliatesta, P. (1999) *Eur. J. Org. Chem.*, 3281–3286.

63 (a) Banfi, S., Legramandi, F., Montanari, F., Pozzi, G., and Quici, S. (1991) *J. Chem. Soc., Chem. Commun.*, 1285–1287; (b) Anelli, P.L., Banfi, S., Legramandi, F., Montanari, F., Pozzi, G., and Quici, S. (1993) *J. Chem. Soc., Perkin Trans.*, **1**, 1345–1357.

64 (a) Campbell, L.A. and Kodadek, T. (1996) *J. Mol. Catal. A: Chem.*, **113**, 293–310; (b) Collman, J.P., Zhang, X., Lee, V.J., Uffelman, E.S., and Brauman, J.I. (1993) *Science*, **261**, 1404–1411.

65 Collman, J.P., Lee, V.J., Kellen-Yuen, C.J., Zhang, X., Ibers, J.A., and Brauman, J.I. (1995) *J. Am. Chem. Soc.*, **117**, 692–703.

66 Naruta, Y. (1994) Chapter 8, in *Metalloporphyrins in Catalytic Oxidations* (ed. R.A. Sheldon), Marcel Dekker, New York.

67 Martinez-Lorente, M.A., Battioni, P., Kleemiss, W., Bartoli, J.F., and Mansuy, D. (1996) *J. Mol. Catal. A: Chem.*, **113**, 343–353.

68 Doro, F.G., Lindsay Smith, J.R., Ferreira, A.G., and Assis, M.D. (2000) *J. Mol. Catal. A: Chem.*, **164**, 97–108.

69 Hulsken, B., van Hameren, R., Gerritsen, J.W., Khoury, T., Thordarson, P., Crossley, M.J., Rowan, A.E., Nolte, R.J.M., Elemans, J.A.A.W., and Speller, S. (2007) *Nature Nanotech.*, **2**, 285–289.

70 Srinivasan, K., Michaud, P., and Kochi, J.K. (1986) *J. Am. Chem. Soc.*, **108**, 2309–2320.

71 (a) Zhang, W., Loebach, J.L., Wilson, S.R., and Jacobsen, E.N. (1990) *J. Am. Chem. Soc*, **112** 2801–2803; (b) Irie, R., Noda, K., Ito, Y., Matsumoto, N., and Katsuki, T. (1990) *Tetrahedron Lett.*, **31**, 7345–7348.

72 Groves, J.T. and Myers, R.S. (1983) *J. Am. Chem. Soc.*, **105**, 5791–5796.

73 (a) Jacobsen, E.N. and Wu, M.H. (1999) in *Comprehensive Asymmetric Catalysis*, vol. II (eds E.N. Jacobsen, A. Pfaltz, and H. Yamamoto), Springer, Berlin, pp. 649–677; (b) Katsuki, T. (1996) *J. Mol. Catal. A: Chem.*, **113**, 87–107; (c) Katsuki, T. (2002) *Adv. Synth. Catal.*, **344**, 131–147; (d) Katsuki, T. (2003) *Synlett*, 281–297.

74 (a) Palucki, M., Pospisil, P.J., Zhang, W., and Jacobsen, E.N. (1994) *J. Am. Chem. Soc.*, **116**, 9333–9334; (b) Palucki, M., McCormick, G.J., and Jacobsen, E.N. (1995) *Tetrahedron Lett.*, **36**, 5457–5460; (c) Van der Velde, S.L. and Jacobsen, E.N. (1995) *J. Org. Chem.*, **60**, 5380–5381; (d) Jacobsen, E.N., Deng, L., Furukawa, Y., and Martinez, L.E. (1994) *Tetrahedron*, **50**, 4323–4334; (e) Hughes, D.L., Smith, G.B., Liu, J., Dezeny, G.C., Senanayake, C.H., Larsen, R.D., Verhoeven, T.R., and Reider, P.J. (1997) *J. Org. Chem.*, **62**, 2222–2229.

75 Feichtinger, D. and Plattner, D.A. (1997) *Angew. Chem. Int. Ed.*, **36**, 1718–1719.

76 (a) Dalton, C.T., Ryan, K.M., Wall, V.M., Bousquet, C., and Gilheany, D.G. (1998) *Top. Catal.*, **5**, 75–91; (b) Canali, L. and Sherrington, D.C. (1999) *Chem. Soc. Rev.*, **28**, 85–93.

77 Ito, Y.N. and Katsuki, T. (1998) *Tetrahedron Lett.*, **39**, 4325–4328.

78 (a) Schwenkreis, T. and Berkessel, A. (1993) *Tetrahedron Lett.*, **34**, 4785–4788; (b) Berkessel, A., Frauenkron, M., Schwenkreis, T., Steinmetz, A., Baum, G., and Fenske, D. (1996) *J. Mol. Catal. A: Chem.*, **113**, 321–342.

79 Irie, R., Hosoya, N., and Katsuki, T. (1994) *Synlett*, 255–256.

80 (a) Pietikäinen, P. (1994) *Tetrahedron Lett.*, **35**, 941–944; (b) Pietikäinen, P. (1998) *Tetrahedron*, **54**, 4319–4326.

81 Pietikäinen, P. (2001) *J. Mol. Catal. A: Chem.*, **165**, 73–79.

82 (a) Kureshy, R.I., Khan, N.H., Abdi, S.H.R., Patel, S.T., and Jasra, R.V. (2001) *Tetrahedron: Asymmetry*, **12**, 433–437; (b) Kureshy, R.I., Khan, N.H., Abdi, S.H.R., Singh, S., Ahmed, I., Shukla, R.S., and Jasra, R.V. (2003) *J. Catal.*, **219**, 1–7.

83 Vartzouma, C., Evaggellou, E., Sanakis, Y., Hadjiliadis, N., and Louloudi, M. (2007) *J. Mol. Catal. A, Chem.*, **263**, 77–85.

84. Beigi, M., Roller, S., Haag, R., and Liese, A. (2008) *Eur. J. Org. Chem.*, **12**, 2135–2141.
85. Yin, G., Buchalova, M., Danby, A.M., Perkins, C.M., Kitko, D., Carter, J.D., Scheper, W.M., and Busch, D.H. (2005) *J. Am. Chem. Soc.*, **127**, 17170–17171.
86. Yin, G., McCormick, J.M., Buchalova, M., Danby, A.M., Rodgers, K., Day, V.W., Smith, K., Perkins, C.M., Kitko, D., Carter, J.D., Scheper, W.M., and Busch, D.H. (2006) *Inorg. Chem.*, **45**, 8052–8061.
87. Yin, G., Danby, A.M., Kitko, D., Carter, J.D., Scheper, W.M., and Busch, D.H. (2007) *J. Am. Chem. Soc.*, **129**, 1512–1513.
88. Yin, G., Danby, A.M., Kitko, D., Carter, J.D., Scheper, W.M., and Busch, D.H. (2007) *Inorg. Chem.*, **46**, 2173–2180.
89. Yin, G., Danby, A.M., Kitko, D., Carter, J.D., Scheper, W.M., and Busch, D.H. (2008) *J. Am. Chem. Soc.*, **130**, 16245–16253.
90. He, H.T., Yin, G., Hiler, G., Kitko, D., Carter, J.D., Scheper, W.M., Day, V., and Busch, D.H. (2008) *J. Coord. Chem.*, **61**, 45–59.
91. Haras, A. and Ziegler, T. (2009) *Can. J. Chem.*, **87**, 33–38.
92. (a) Pecoraro, V.L., Baldwin, M.J., and Gelasco, A. (1994) *Chem. Rev.*, **94**, 807–826; (b) Jackson, T.A. and Brunold, T.C. (2004) *Acc. Chem. Res.*, **37**, 461–470; (c) Boelrijk, A.E.M., Khangulov, S.V., and Dismukes, G.C. (2000) *Inorg. Chem.*, **39**, 3009–3019; (d) Boelrijk, A.E.M. and Dismukes, G.C. (2000) *Inorg. Chem.*, **39**, 3020–3028.
93. Kanyo, Z.F., Scolnick, L.R., Ash, D.E., and Christianson, D.W. (1996) *Nature*, **383**, 554–557.
94. (a) Berkessel, A. and Sklorz, C.A. (1999) *Tetrahedron Lett.*, **40**, 7965–7968; (b) De Vos, D.E. and Bein, T. (1996) *Chem. Commun.*, 917–918; (c) Zondervan, C., Hage, R., and Feringa, B.L. (1997) *Chem. Commun.*, 419–420; (d) De Vos, D.E., Sels, B.F., Reynaers, M., Subba Rao, Y.V., and Jacobs, P.A. (1998) *Tetrahedron Lett.*, **39**, 3221–3224; (e) Woitiski, C.B., Kozlov, Y.N., Mandelli, D., Nizova, G.V., Schuchardt, U., and Shul'pin, G.B. (2004) *J. Mol. Catal. A: Chem.*, **222**, 103–119; (f) Bolm, C., Meyer, N., Raabe, G., Weyhermuller, T., and Bothe, E. (2000) *Chem. Commun.*, 2435–2436; (g) Quee-Smith, V.C., DelPizzo, L., Jureller, S.H., Kerschner, J.L., and Hage, R. (1996) *Inorg. Chem.*, **35**, 6461–6465; (h) Gilbert, B.C., Lindsay Smith, J.R., Mairata i Payeras, A., Oakes, J., and Pons i Prats, R. (2004) *J. Mol. Catal. A: Chem.*, **219**, 265–272; (i) Ryu, J.Y., Kim, S.O., Nam, W., Heo, S., and Kim, J. (2003) *Bull. Korean Chem. Soc.*, **24**, 1835–1837; (j) Lindsay Smith, J.R., Gilbert, B.C., Mairata i Payeras, A., Murray, J., Lowdon, T.R., Oakes, J., Pons i Prats, R., and Walton, P.H. (2006) *J. Mol. Catal. A: Chem.*, **251**, 114–122, and references cited herein; (k) Brinksma, J., Schmieder, L., Van Vliet, G., Boaron, R., Hage, R., De Vos, D.E., Alsters, P.L., and Feringa, B.L. (2002) *Tetrahedron Lett.*, **43**, 2619–2622.
95. Barker, J.E. and Ren, T. (2004) *Tetrahedron Lett.*, **45**, 4681–4683.
96. For epoxidation reactions with Ru-complexes based on the tmtacn ligand, see: Cheng, W.-C., Fung, W.-H., and Che, C.-M. (1996) *J. Mol. Catal. A: Chem.*, **113**, 311–319.
97. De Vos, D.E. and Bein, T. (1996) *J. Organomet. Chem.*, **520**, 195–200.
98. Sauer, M.C.V. and Edwards, J.O. (1971) *J. Phys. Chem.*, **75**, 3004–3011.
99. De Vos, D.E., De Wildeman, S., Sels, B.F., Grobet, P.J., and Jacobs, P.A. (1999) *Angew. Chem. Int. Ed.*, **38**, 980–983.
100. (a) Chaudhuri, P. and Oder, K. (1990) *J. Chem. Soc., Dalton Trans.*, 1597–1605; (b) Niemann, A., Bossek, U., Haselhorst, G., Wieghardt, K., and Nuber, B. (1996) *Inorg. Chem.*, **35**, 906–915.
101. Garcia-Bosch, I., Company, A., Fontrodona, X., Ribas, X., and Costas, M. (2008) *Org. Lett.*, **10**, 2095–2098.
102. Garcia-Bosch, I., Ribas, X., and Costas, M. (2009) *Adv. Synth. Catal.*, **351**, 348–352.
103. (a) Bolm, C., Kadereit, D., and Valacchi, M. (1997) *Synlett*, 687–688; (b) Koek, J.H., Kohlen, E.W.M.J., Russell, S.W., Van der Wolf, L., Ter Steeg, P.F., and Hellemons, J.C. (1999) *Inorg. Chim. Acta*, **295**, 189–199; (c) Golding, S.W., Hambley, T.W., Lawrance, G.A., Luther, S.M.,

Maeder, M., and Turner, P. (1999) *J. Chem. Soc., Dalton Trans.*, 75–80; (d) Kim, B., So, S.M., and Choi, H.J. (2002) *Org. Lett.*, **4**, 949–952; (e) Scheuermann, J.E.W., Ilyashenko, G., Griffiths, D.V., and Watkinson, M. (2002) *Tetrahedron: Asymmetry*, **13**, 269–272; (f) Scheuermann, J.E.W., Ronketti, F., Motevalli, M., Griffiths, D.V., and Watkinson, M. (2002) *New J. Chem.*, **26**, 1054–1059; (g) Argouarch, G., Gibson, C.L., Stones, G., and Sherrington, D.C. (2002) *Tetrahedron Lett.*, **43**, 3795–3798.

104 (a) Beller, M., Tafesch, A., Fischer, R.W., and Scharbert, B. (1996) DE19523891. (b) Bolm, C., Kadereit, D., and Valacchi, M. (1998) DE19720477.

105 Romakh, V.B., Therrien, B., Süss-Fink, G., and Shul'pin, G.B. (2007) *Inorg. Chem.*, **46**, 1315–1331.

106 Kilic, H., Adam, W., and Alsters, P.L. (2009) *J. Org. Chem.*, **74**, 1135–1140.

107 De Vos, D.E., Meinershagen, J.L., and Bein, T. (1996) *Angew. Chem. Int. Ed.*, **35**, 2211–2213.

108 Subba Rao, Y.V., De Vos, D.E., Bein, T., and Jacobs, P.A. (1997) *Chem. Commun.*, 355–356.

109 Veghini, D., Bosch, M., Fischer, F., and Falco, C. (2008) *Catal. Commun.*, **10**, 347–350.

110 Terry, T.J. and Stack, T.D.P. (2008) *J. Am. Chem. Soc.*, **130**, 4945–4953.

111 Döbler, C., Mehltretter, G.M., Sundermeier, U., and Beller, M. (2000) *J. Am. Chem. Soc.*, **122**, 10289–10297.

112 Jonsson, S.Y., Färnegårdh, K., and Bäckvall, J.E. (2001) *J. Am. Chem. Soc.*, **123**, 1365–1371.

113 Milas, N.A. and Sussman, S. (1936) *J. Am. Chem. Soc.*, **58**, 1302–1304.

114 Kolb, H.C., VanNieuwenhze, M.S., and Sharpless, K.B. (1994) *Chem. Rev.*, **94**, 2483–2547.

115 Markó, I.E. and Svendsen, J.S. (1999) in *Comprehensive Asymmetric Catalysis*, vol. II (eds E.N. Jacobsen, A. Pfaltz, and H. Yamamoto), Springer, Berlin, pp. 711–787.

116 (a) Chen, K. and Que, L. Jr. (1999) *Angew. Chem. Int. Ed.*, **38**, 2227–2229; (b) Chen, K., Costas, M., Kim, J., Tipton, A.K., and Que, L. Jr. (2002) *J. Am. Chem. Soc.*, **124**, 3026–3035; (c) Fujita, M., Costas, M., and Que, L. Jr. (2003) *J. Am. Chem. Soc.*, **125**, 9912–9913; (d) Feng, Y., Ke, C.-Y., Xue, G., and Que, L. Jr. (2009) *Chem. Commun.*, 50–52; (e) Ryu, J.Y., Kim, J., Costas, M., and Chen, K., Nam, W. Que, L. (2002) *Chem. Commun*, 1288–1289; (f) Hirao, H., Que, L., Nam, W., and Shaik, S. (2008) *Chem. Eur. J.*, **14**, 1740–1756.

117 De Boer, J.W., Brinksma, J., Browne, W.R., Meetsma, A., Alsters, P.L., Hage, R., and Feringa, B.L. (2005) *J. Am. Chem. Soc.*, **127**, 7990–7991.

118 de Boer, J.W., Browne, W.R., Brinksma, J., Alsters, P.L., Hage, R., and Feringa, B.L. (2007) *Inorg. Chem.*, **46**, 6353–6372.

119 Hage, R., Krijnen, B., Warnaar, J.B., Hartl, F., Stufkens, D.J., and Snoeck, T.L. (1995) *Inorg. Chem.*, **34**, 4973–4978.

120 Gilbert, B.C., Kamp, N.W.J., Lindsay Smith, J.R., and Oakes, I. (1997) *J. Chem. Soc., Perkin Trans.*, **2**, 2161–2166.

121 Barton, D.H.R., Choi, S.-Y., Hu, B., and Smith, J.A. (1998) *Tetrahedron*, **54**, 3367–3378.

122 Gilbert, B.C., Kamp, N.W.J., Lindsay Smith, J.R., and Oakes, J. (1998) *J. Chem. Soc., Perkin Trans.*, **2**, 1841–1844.

123 de Boer, J.W., Alsters, P.L., Meetsma, A., Hage, R., Browne, W.R., and Feringa, B.L. (2008) *Dalton Trans.*, 6283–6295.

124 Wang, Z.-M., Kakiuchi, K., and Sharpless, K.B. (1994) *J. Org. Chem.*, **59**, 6895–6897.

125 (a) R., Noyori (2005) *Chem. Commun.*, 1807–1811; (b) Chang, D.L., Zhang, J., Witholt, B., and Li, Z. (2004) *Biocatal. Biotransfor.*, **22**, 113–130.

126 Beller, M. (2004) *Adv. Synth. Catal.*, **346**, 107–108.

127 Suzuki, K., Oldenburg, P.D., and Que, L. Jr. (2008) *Angew. Chem. Int. Ed.*, **47**, 1887–1889.

128 de Boer, J.W., Browne, W.R., Harutyunyan, S., Bini, L., Tiemersma-Wegman, T.D., Alsters, P.L., Hage, R., and Feringa, B.L. (2008) *Chem. Commun.*, 3747–3749.

129 Zondervan, C. (1997) Homogeneous Catalytic Oxidation, A Ligand Approach, Ph.D. Thesis, University of Groningen, Chapter 4.

130 (a) Murphy, A., Dubois, G., and Stack, T.D.P. (2003) *J. Am. Chem. Soc.*, **125**,

5250–5251; (b) Murphy, A., Pace, A., and Stack, T.D.P. (2004) *Org. Lett.*, **6**, 3119–3122; (c) Qi, J.-Y., Li, Y.-M., Zhou, Z.-Y., Che, C.-M., Yeung, C.-H., and Chan, A.S.C. (2005) *Adv. Synth. Catal.*, **347**, 45–49.

131 (a) Toftlund, H. and Yde-Andersen, S. (1981) *Acta Chem. Scand., Ser. A*, **35**, 575–585; (b) Toftlund, H., Markiewicz, A., and Murray, K.S. (1990) *Acta Chem. Scand.*, **44**, 443–446; (c) Mandel, J.B., Maricondi, C., and Douglas, B.E. (1988) *Inorg. Chem.*, **27**, 2990–2996; (d) Pal, S., Gohdes, J.W., Wilisch, W.C.A., and Armstrong, W.H. (1992) *Inorg. Chem.*, **31**, 713–716.

132 Brinksma, J., Hage, R., Kerschner, J., and Feringa, B.L. (2000) *Chem. Commun.*, 537–538.

133 Zhang, W., Lee, N.H., and Jacobsen, E.N. (1994) *J. Am. Chem. Soc.*, **116**, 425–426.

134 Zhong, S., Fu, Z., Tan, Y., Xie, Q., Xie, F., Zhou, X., Ye, Z., Peng, G., and Yin, D. (2008) *Adv. Synth. Catal.*, **350**, 802–806.

135 Kozlov, Y.N., Nizova, G.V., and Shul'pin, G.B. (2008) *J. Phys. Org. Chem.*, **21**, 119–126.

136 Nakayama, N., Tsuchiya, S., and Ogawa, S. (2007) *J. Mol. Catal. A: Chem.*, **277**, 61–71.

137 (a) Wessel, J. and Crabtree, R.H. (1996) *J. Mol. Catal. A: Chem.*, **113**, 13–22; (b) Shul'pin, G.B., Matthes, M.G., Romakh, V.B., Barbosa, M.I.F., Aoyagi, J.L.T., and Mandelli, D. (2008) *Tetrahedron*, **64**, 2143–2152; (c) Nizova, G.V. and Shul'pin, G.B. (2007) *Tetrahedron*, **63**, 7997–8001; (d) Mandelli, D., Kozlov, Y.N., Golfeto, C.C., and Shul'pin, G.B. (2007) *Catal. Lett.*, **118**, 22–29.

138 Pleitker, B. (2008) *Iron Catalysis in Organic Chemistry*, Wiley VCH, Weinheim.

139 Chen, M.S. and White, M.C. (2007) *Science*, **318**, 783–787.

140 MacLeod, T.C.O., Faria, A.L., Barros, V.P., Queiroz, M.E.C., and Assis, M.D. (2008) *J. Mol. Catal. A: Chem.*, **296**, 54–60.

141 Griffith, W.P., Ley, S.V., Whitcombe, G.P., and White, A.D. (1987) *J. Chem. Soc., Chem. Commun.*, 1625–1627.

142 Brinksma, J., Rispens, M.T., Hage, R., and Feringa, B.L. (2002) *Inorg. Chim. Acta*, **337**, 75–82.

143 Wang, Y., DuBois, J.L., Hedman, B., Hodgson, K.O., and Stack, T.D.P. (1998) *Science*, **279**, 537–540.

144 Khenkin, A.M. and Shilov, A.E. (1989) *New J. Chem.*, **13**, 659–667.

145 (a) Solladié, G. (1981) *Synthesis*, 185–196; (b) Carreño, M.C. (1995) *Chem. Rev.*, **95**, 1717–1760; (c) Colobert, F., Tito, A., Khiar, N., Denni, D., Medina, M.A., Martin-Lomas, M., Garcia Ruano, J.-L., and Solladié, G. (1998) *J. Org. Chem.*, **63**, 8918–8921; (d) Bravo, P., Crucianelli, M., Farina, A., Meille, S.V., Volonterio, A., and Zanda, M. (1998) *Eur. J. Org. Chem.*, **3**, 435–440; (e) Cotton, H., Elebring, T., Larsson, M., Li, L., Sörensen, H., and Von Unge, S. (2000) *Tetrahedron: Asymmetry*, **11**, 3819–3825; (f) Padmanabhan, S., Lavin, R.C., and Durant, G.J. (2000) *Tetrahedron: Asymmetry*, **11**, 3455–3457.

146 Bagherzadeh, M., Latifi, R., Tahsini, L., and Amini, M. (2008) *Catal. Commun.*, **10**, 196–200.

147 Palucki, M., Hanson, P., and Jacobsen, E.N. (1992) *Tetrahedron Lett.*, **33**, 7111–7114.

148 (a) Noda, K., Hosoya, N., Yanai, K., Irie, R., and Katsuki, T. (1994) *Tetrahedron Lett.*, **35**, 1887–1890; (b) Noda, K., Hosoya, N., Irie, R., Yamashita, Y., and Katsuki, T. (1994) *Tetrahedron*, **50**, 9609–9618; (c) Kokubo, C. and Katsuki, T. (1996) *Tetrahedron*, **52**, 13895–13900.

149 Saito, B. and Katsuki, T. (2001) *Tetrahedron Lett.*, **42**, 3873–3876.

150 Saito, B. and Katsuki, T. (2001) *Tetrahedron Lett.*, **42**, 8333–8336.

151 La Crois, R.M. (2000) Manganese Complexes as Catalysts in Epoxidation Reactions, a Ligand Approach, Ph.D. thesis, University of Groningen, Chapter 4.

152 Brinksma, J., La Crois, R., Feringa, B.L., Donnoli, M.I., and Rosini, C. (2001) *Tetrahedron Lett.*, **42**, 4049–4052.

153 Schoumacker, S., Hamelin, O., Pécaut, J., and Fontecave, M. (2003) *Inorg. Chem.*, **42**, 8110–8116.

154 (a) Bösing, M., Nöh, A., Loose, I., and Krebs, B. (1998) *J. Am. Chem. Soc.*, **120**, 7252–7259; (b) Mizuno, N., Nozaki, C., Kiyoto, I., and Misono, M. (1998) *J. Am. Chem. Soc.*, **120**, 9267–9272.

155 Herrmann, W.A., Fischer, R.W., Scherer, W., and Rauch, M.U. (1993) *Angew. Chem. Int. Ed.*, **32**, 1157–1160.

156 Sato, K., Aoki, M., Ogawa, M., Hashimoto, T., and Noyori, R. (1996) *J. Org. Chem.*, **61**, 8310–8311.

12
Biooxidation with Cytochrome P450 Monooxygenases
Marco Girhard and Vlada B. Urlacher

12.1
Introduction

The selective oxidation of organic molecules is not only fundamentally important for life, but also immensely useful in industry [1]. Among the myriad ways by which such oxidations may be performed, the biocatalytic oxyfunctionalization of nonactivated hydrocarbons is considered as 'potentially the most useful of all biotransformations' [2]. Cytochrome P450 monooxygenases (P450 monooxygenases, P450s or CYPs) that are able to catalyze such reactions are notable for many reasons: (i) they use molecular oxygen (O_2) as the primary oxidant, thereby operating at ambient conditions and thus presenting ideal systems for 'green' organic synthesis, (ii) they oxidize a vast range of substrates and are able to catalyze more than 20 different reaction types [3], and (iii) they often exhibit exquisite substrate specificity as well as regio- and/or stereoselectivity. Since applications utilizing P450s often yield compounds that are not easy to produce by traditional chemical synthetic processes, these enzymes have attracted considerable attention of chemists, biochemists, and biotechnologists, not only in academia, but also in industry.

P450s belong to the superfamily of heme *b*-containing monooxygenases found in all domains of life [4], where they play a central role in drug metabolism and have been shown to be involved in the biosynthesis of important natural compounds. The number of known P450s is constantly increasing. Currently there are more than 11 200 P450 sequences available in several online databases, for example in CYPED (http://www.cyped.uni-stuttgart.de, Institute of Technical Biochemistry, Universitaet Stuttgart, 2nd November 2009) [5, 6] or on the P450 homepage of Dr. Nelson (http://drnelson.utmem.edu/CytochromeP450.html, Molecular Sciences, University of Tennessee, 2nd November 2009) [4]. Extensive studies have revealed the key chemical principles that underlie the efficacy of P450s as biocatalysts for aerobic oxidations, and the large oxidative potential of these enzymes has been used for drug development, bioremediation, and synthesis of fine chemicals. Significant progress has been made over the past decade in engineering P450s with widely altered substrate specificities, substantially increased activity, and enhanced process stability. Since

Modern Oxidation Methods. Edited by Jan-Erling Bäckvall
Copyright © 2010 WILEY-VCH Verlag GmbH & Co. KGaA, Weinheim
ISBN: 978-3-527-32320-3

several comprehensive reviews on P450s have been written [7–14], we mainly focus on the recent progress in applying P450s as biocatalysts and discuss the engineering issue of P450s.

12.2
Properties of Cytochrome P450 Monooxygenases

12.2.1
Structure

P450s were recognized and defined as a distinct class of hemoproteins about 50 years ago [15, 16]. These enzymes got their name from their unusual property of forming reduced (ferrous) iron/carbon monoxide complexes in which the heme absorption Soret band shifts from 420 nm to ~450 nm [17, 18]. Essential for this spectral characteristic is the axial coordination of the heme iron by a cysteine thiolate which is common to all P450s [19, 20]. The phylogenetically conserved cysteinate is the ligand proximal to the heme iron, with the distal ligand generally assumed to be a weakly bound water molecule [21]. Despite relatively low sequence identity across the gene superfamily, crystal structures of P450s show the same structural organization [22] (Figure 12.1). The highest structural conservation is found in the core of the protein around the heme, reflecting a common mechanism of electron and proton transfer, and oxygen activation. The conserved core is formed by a four-helix bundle (named D,

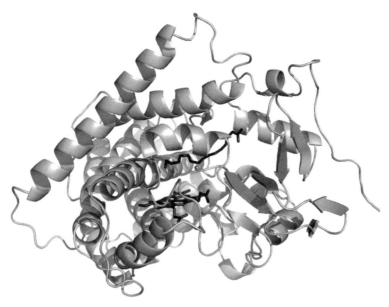

Figure 12.1 The crystal structure of the P450BM3 monooxygenase domain with palmitoleic acid bound (adapted from pdb 1SMJ): heme and palmitoleic acid in black.

E, L, and I), which is surrounded by the helices J and K and two sets of β-sheets. These regions comprise (a) the heme-binding loop, containing the most characteristic P450 consensus sequence (Phe-Gly/Ser-X-Gly-X-His/Arg-X-Cys-X-Gly-X-Ile/Leu/Phe-X) with the absolutely conserved cysteine that serves as a fifth ligand to the heme iron, (b) the conserved Glu-X-X-Arg motif, probably needed to stabilize the core structure through a salt bridge, and (c) the consensus sequence, considered as P450 signature (Ala/Gly-Gly-X-Asp/Glu-Thr), which is thought to play a role in oxygen activation through proton transfer [23]. Next to these organized elements, two further regions can be found: an unstructured region called 'meander' and the cysteine pocket [24].

Besides the highly conserved structural regions, there also exist some extremely variable ones. These constitute the substrate-binding site that causes the acceptance of a wide range of chemically different molecules. Other flexible regions are the B-C and F-G loops, which are located along the substrate access channel and therefore situated distal of the protoporphyrin system. Substrate recognition and binding is mainly arranged through six substrate recognition sites (SRS): the B' helix (SRS1), parts of helices F (SRS2), G (SRS3), and I (SRS4), as well as the β4-hairpin (SRS5) and the β2-loop (SRS6) [22]. Mutations in these regions have a high impact on substrate specificity. Crystal structures obtained from X-ray analysis of P450s with bound substrate indicate that the substrate-binding region is very flexible and often susceptible to structural reorganization upon substrate binding, encouraging an induced-fit model [25] accounting for the broad substrate spectra of many P450 monooxygenases, especially the microsomal ones.

The first structure of a P450 to be uncovered was that of $P450_{cam}$ from *Pseudomonas putida* (CYP101) in 1985 [26]. For a long time only the structures of soluble, microbial P450s were resolved, for example, those from $P450_{BM3}$ [27], $P450_{terp}$ [28], $P450_{eryF}$ [29], or $P450_{nor}$ [30]. Eukaryotic P450s are membrane-bound and therefore more difficult to crystallize. Nevertheless, in 2000 the first structure of a mammalian P450, CYP2C5 from *Oryctolagus cuniculus*, was uncovered [31], followed by the first structure of a human P450, CYP2C9 in 2003 [32]. This led to great developments in the crystallization and structure determination of eukaryotic P450s, for example that of CYP3A4 in 2004 [33], CYP2D6 in 2006 [34], CYP46A1 in 2008 [35], or recently CYP19A1 in 2009 [36]. At least thirty crystal structures of eight mammalian cytochrome P450s (CYP 2C5, 2C8, 2C9, 3A4, 2D6, 2B4, 2A6 and 1A2) have been published [37].

12.2.2
Enzymology

The vast majority of cytochrome P450 monooxygenases catalyze the reductive scission of dioxygen, which requires the consecutive delivery of two electrons to the heme iron. P450s utilize reducing equivalents (electrons in the form of hydride ions) ultimately derived from the pyridine cofactors NADH or NADPH and transferred to the heme via special redox proteins [38, 39].

The P450 catalytic cycle was recently revised by Sligar and colleagues [40] and is shown in Scheme 12.1. Substrate binding in the active site induces the dissociation of

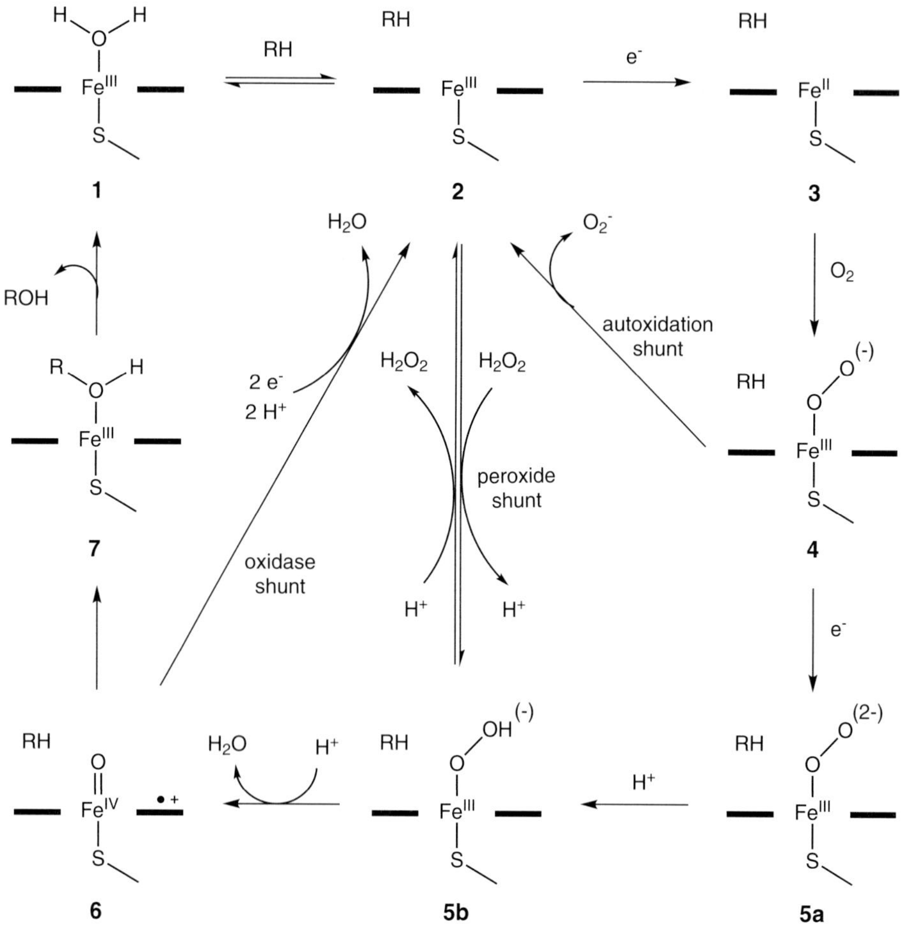

Scheme 12.1 The catalytic cycle of cytochrome P450 monooxygenases (reproduced from Ref. [44], with permission).

the water molecule that is bound as the sixth coordinating ligand to the heme iron (**1**). This, in turn, induces a shift of the heme iron spin state from low-spin to high-spin accompanied by a positive shift of the reduction potential in the order of 130–140 mV [41]. The increased potential allows the delivery of the first electron, which reduces the heme iron from the ferric Fe^{III} (**2**) to the ferrous Fe^{II} form (**3**). After the first electron transfer, the Fe^{II} (**3**) binds dioxygen, resulting in a ferrous dioxygen complex (**4**). The consecutive delivery of the second electron converts this species into a ferric peroxy anion (**5a**). Subsequent steps in the P450 cycle are considered to be relatively fast with respect to the electron transfer. The ferric peroxy species is protonated to a ferric hydroperoxy complex (**5b**; also referred to as compound 0). The following protonation of this complex results in a high-valent ferryl-oxo complex (**6**; also referred to as compound I) accompanied by the release of a water molecule through heterolytic scission of the dioxygen bond in the preceding intermediate (7). Compound I (**6**) is considered to be the intermediate catalyzing the majority of P450 reactions; however, compound 0 (**5b**) may also be important for some P450-dependent catalytic reactions [42], for example, the epoxidation of C=C double bonds [43].

Under certain conditions P450s can enter one of three so-called uncoupling pathways (Scheme 12.1). Autoxidation shunt occurs if the second electron is not delivered to reduce the ferrous dioxygen complex (**4**), which can decay forming superoxide. The inappropriate positioning of the substrate in the active site is often the molecular reason for the two other uncoupling cycles. The ferric hydroperoxy complex (**5b**) can collapse and release hydrogen peroxide (peroxide shunt), while decay of compound I (**6**) is accompanied by the release of water (oxidase shunt). For industrial applications, it is particularly important to note that the uncoupling pathways in all cases consume reducing equivalents from NAD(P)H without product formation.

Catalysis with cytochrome P450 monooxygenases requires two electrons to be transferred to the heme via redox proteins. Depending on the topology of the protein, components involved in the electron transfer to the heme, P450s can be grouped, for example by a classification system with ten different classes as has recently been suggested by Bernhardt and colleagues [39]. For industrial applications the fusion enzymes of class VIII are of particular interest, since they come along with their natural redox proteins. This group covers enzymes consisting of a P450 monooxygenase fused to a CPR-like reductase. The best studied representatives are P450$_{BM3}$ (CYP102A1) from *Bacillus megaterium* [45, 46] and its two homologs CYP102A2 [47] and CYP102A3 [48] from *Bacillus subtilis*, as well as their eukaryotic counterpart, CYP505A1 (P450$_{foxy}$) from the ascomycete fungus *Fusarium oxysporum* [49].

It is notable that there are other P450s not requiring electron transfer proteins, for example natural P450 peroxygenases, which employ the 'peroxide shunt' for catalysis [50]. Three enzymes with a potential for biocatalytic applications are the H_2O_2-utilizing fatty acid hydroxylases of the CYP152 family, namely CYP152B1 (SP$_\alpha$) from *Sphingomonas paucimobilis* [51], CYP152A1 (P450$_{Bs\beta}$) from *Bacillus subtilis* [52], and CYP152A2 (P450$_{CLA}$) from *Clostridium acetobutylicum* [53].

12.2.3
Reactions Catalyzed by P450s

The ability of P450s to catalyze a large variety of oxidative and also some reductive reactions – collectively involving thousands of substrates – has been discussed in a number of reviews [7, 11, 54–57], so here we will reflect only the most important oxidative reactions catalyzed by P450s. These reactions include hydroxylation of nonactivated sp^3-hybridized carbon atoms, epoxidation, aromatic hydroxylation, C–C bond cleavage, heteroatom oxygenation, heteroatom release (dealkylation), oxidative ester cleavage, oxidative phenol- and ring-coupling, isomerization via (abortive) oxidation, oxidative dehalogenation, and other complex reactions like dimer formation via Diels-Alder reactions of products or Baeyer-Villiger-type oxidations.

Hydroxylation of nonactivated sp^3-hybridized carbon atoms belongs to the classical oxidative reactions catalyzed by P450s. Examples of this reaction include the hydroxylation of saturated fatty acids (8) to hydroxy fatty acids (9) catalyzed, for example, by eukaryotic CYP4 and bacterial CYP102 enzymes [58], as well as the stereospecific hydroxylation of D-(+)-camphor (10) to 5-*exo*-hydroxycamphor (11) through P450$_{cam}$ [59] (Scheme 12.2).

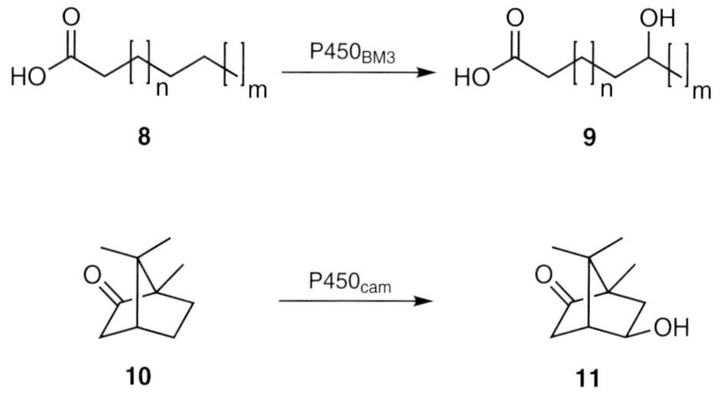

Scheme 12.2 Fatty acid hydroxylation catalyzed by P450$_{BM3}$ and camphor hydroxylation catalyzed by P450$_{cam}$.

Epoxidation of C=C double bonds is another major reaction type catalyzed by P450s. Particularly attractive in this regard is P450$_{EpoK}$ from *Sorangium cellulosum*, which catalyzes the epoxidation of the thiazole-containing macrolactone epothilone D (12) to epothilone B (13) [60] (Scheme 12.3). Epothilones are important anti-tumor polyketides with high microtubule stabilizing activity.

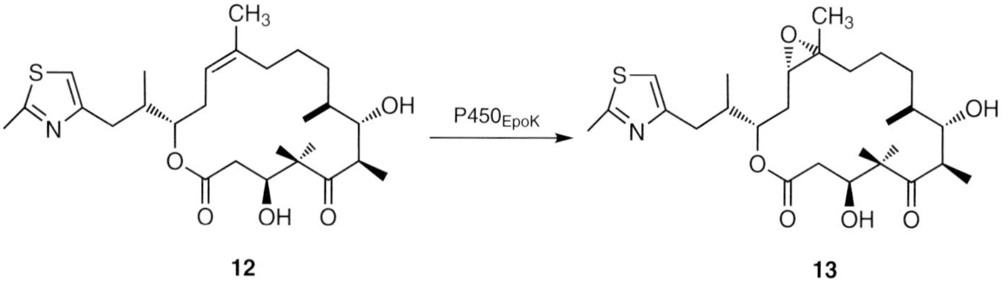

Scheme 12.3 Epoxidation of Epothilon D (12) to Epothilon B (13) by P450$_{EpoK}$.

Aromatic hydroxylation also belongs to the common P450 reactions. P450$_{NikF}$ from *Streptomyces tendae* Tü901 has been claimed to catalyze hydroxylation of pyridylhomothreonine (14) to form hydroxypyridylhomothreonine (15) in the biosynthesis of nikkomycin, an inhibitor of chitin synthase [61] (Scheme 12.4). Many P450s have been engineered toward aromatic hydroxylation, since this ability makes them attractive candidates for the production of fine chemicals (see Sections 12.4 and 12.5.3).

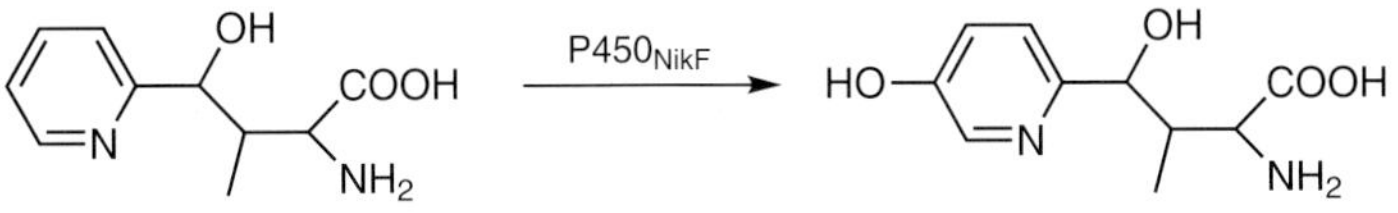

Scheme 12.4 Aromatic hydroxylation of pyridylhomothreonine (**14**) to hydroxypyridylhomothreonine (**15**) by P450$_{NikF}$ during nikkomycin biosynthesis.

Some P450s are able to catalyze the cleavage of C–C bonds via multiple substrate oxidations. For example, the demethylation of lanosterol (**16**) to a precursor of cholesterol, 4,4-dimethyl-5α-cholesta-8,14,24-diene-3β-ol (**19**), by a lanosterol 14α-demethylase (CYP51) [62] includes three steps and proceeds via initial hydroxylation of the C14 methyl group followed by further oxidation of the alcohol (**17**) to the aldehyde (**18**). Finally, acyl cleavage occurs leading to formation of a double bond in the steroid (Scheme 12.5). A similar cascade of reactions is assumed to be catalyzed by CYP107H1 (P450$_{Biol}$) from *Bacillus subtilis* during conversion of long-chain fatty acyl CoA esters to pimeloyl CoA in the biotin biosynthesis pathway [63].

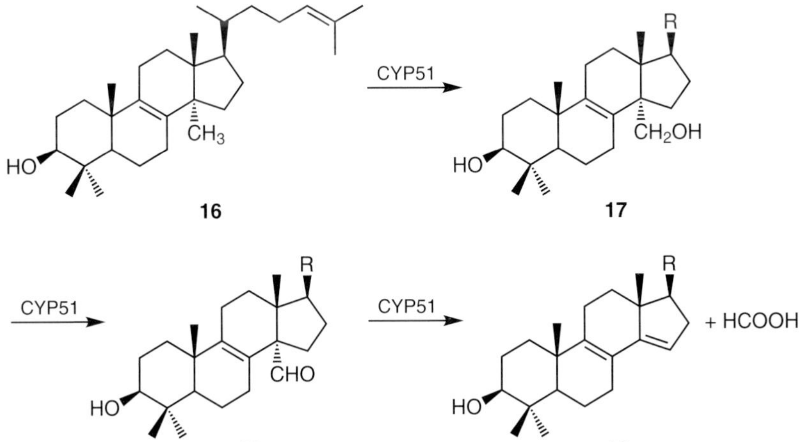

Scheme 12.5 Demethylation of lanosterol (**16**) to 4,4-dimethyl-5α-cholesta-8,14,24-diene-3β-ol (**19**) catalyzed by lanosterol 14α-demethylase (CYP51).

P450s catalyze oxidations, not only at carbon-atoms, but also at N-, S-, and O-atoms, and they also catalyze dealkylations, which are believed to be the next step in the hydroxylation of an α-carbon atom. Examples include the N-oxygenation of N,N-dimethylaniline (**20**) and N,N-dialkylarylamines by mammalian CYP2B1 and CYP2B4, respectively [64, 65], and O-demethylation of 5-methoxytryptamine (**21**) by CYP2D6 [66] (Scheme 12.6).

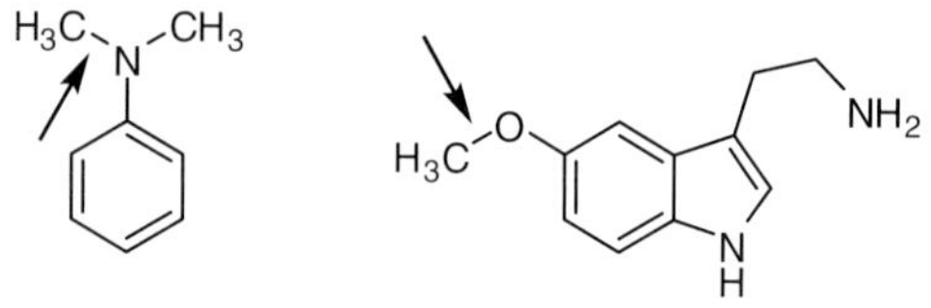

Scheme 12.6 N-oxygenation of N,N-dimethylaniline (**20**) and O-demethylation of 5-methoxytryptamine (**21**); reaction sides are indicated by arrows.

Some P450s are able to catalyze oxidative phenol coupling, a reaction usually carried out by peroxidases. Three independent P450 monooxygenases with such activity have been shown to be involved in the synthesis of vancomycin-type antibiotics in *Amycolatopsis balhimycina* [67].

Dimerization of thiophene S-oxide (**23**) via a Diels-Alder reaction was observed when ticlodipine (**22**) was oxidized by CYP2C19 or CYP2D6 [68] (Scheme 12.7).

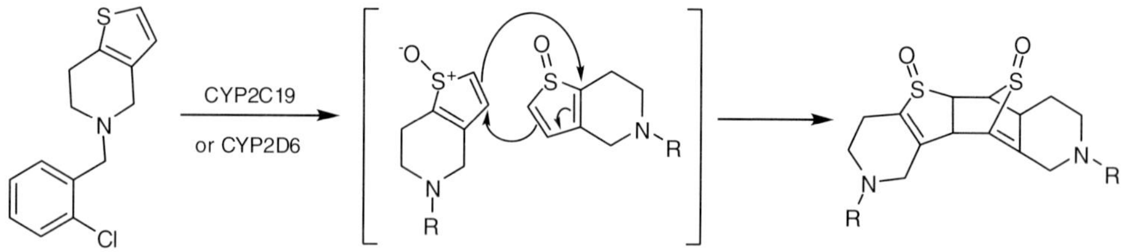

Scheme 12.7 Dimerization of thiophene S-oxide (**23**) via a Diels-Alder reaction through oxidation of ticlodipine (**22**).

Baeyer-Villiger-type oxidations can also be catalyzed by some P450s, for example CYP85A2 from *Arabidopsis thaliana*, which catalyzes the conversion of castasterone (**24**) to brassinolide (**25**) [69] (Scheme 12.8).

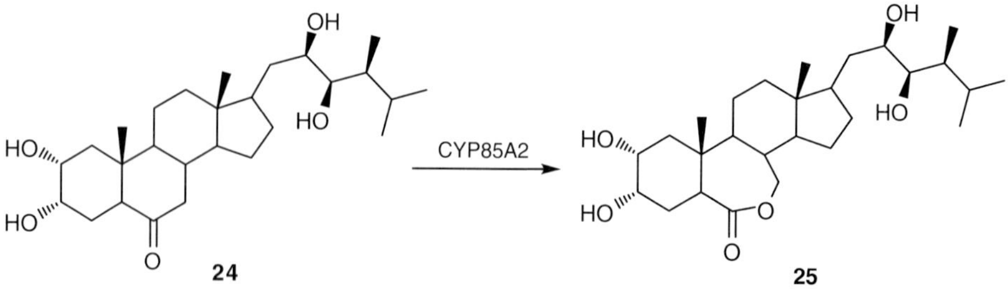

Scheme 12.8 Conversion of castasterone (**24**) to brassinolide (**25**) via Baeyer-Villiger-type oxidation.

Many other unusual types of oxidative and also some reductive reactions catalyzed by P450s have been described in the literature, including oxidative deamination, desulfurylation, oxidative dehalogenation, isomerization, dehydrogenation, dehydration, reductive dehalogenation, epoxide reduction, and others [54, 57, 70].

12.2.4
P450s as Industrial Biocatalysts

12.2.4.1 Advantages
P450 biocatalysts operate – like any enzyme applied for industrial biocatalysis – under ambient conditions, thereby often exhibiting exquisite substrate specificity as well as regio- and/or stereoselectivity. Compared with other biocatalysts, however, P450s potentially have additional advantages for industrial applications:

- Since their discovery, P450s have been studied in enormous detail because of their involvement in a plethora of crucial cellular roles – from carbon source assimilation, through biosynthesis of hormones, to carcinogenesis, drug activation, and degradation of xenobiotics.
- As mentioned above, P450s are able to catalyze more than 20 different reaction types and can oxidize a wide range of molecules [3]. Many of the compounds occur naturally and can be important industrial precursors.
- P450s can be produced industrially by fermentation. Considerable progress has been made during the last decade for recombinant expression of P450s in the well-studied hosts *Escherichia coli*, *Pseudomonas putida* and yeasts *Saccharomyces cerevisiae* and *Pichia pastoris*, which facilitates their use as industrial catalysts [71–77].
- The number of identified P450s is huge and constantly increasing as a result of microbial screenings and increasing information on sequenced genomes. The collection of P450s in (recombinant) libraries allows high-throughput screenings as well as functional characterization of new members of the P450 family and offers a route to diverse building blocks.

12.2.4.2 Challenges in the Development of Technical P450 Applications
Besides the questions concerning process stability and activity of P450s, which apply to all industrial biocatalysts, the development of technical applications for P450s faces specific problems. Probably the most important drawback restricting industrial applications is the fact that nearly all P450s require costly cofactors NADPH or NADH, which makes their application impossible if the cofactor has to be added in a stoichiometric amount. Closely linked to this cofactor dependency is the challenge (a) to find suitable redox proteins that can adequately deliver the electrons to the heme and (b) to construct auxiliary redox modules. However, many efforts have been made to overcome these hurdles by either minimizing or removing the need for NAD(P)H or by designing new strategies for simplified transmission of reducing power [11, 78, 79], some of these are discussed in the following sections.

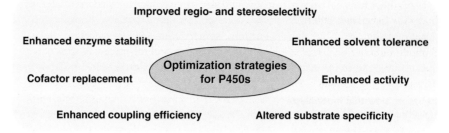

Figure 12.2 P450 optimization strategies for potential biotechnological application.

12.2.4.3 General Aspects of Industrial Application and Engineering of P450s

Because of the cofactor dependence of P450s, their industrial applications have so far been restricted to whole-cell systems, which take advantage of the host's endogenous cofactor regeneration systems. In such instances, however, physiological effects like limited substrate uptake, toxicity of substrate or product, product degradation, and elaborate downstream processing must be taken into account [80]. Moreover, when concentrations of recombinant P450 biocatalysts within the cell reach a certain level, the cofactor concentration may again become a bottleneck for the overall process.

Another factor important for efficient biocatalysis with P450s is the yield of product based on NAD(P)H consumed, or the coupling efficiency. Besides reducing the efficiency of cofactor usage, uncoupling between NADPH oxidation and product formation results in reactive oxygen species (such as superoxide anions and hydrogen peroxide) that cause oxidative destruction of the heme and oxidative damage of the protein.

Thus, optimization strategies for cytochrome P450 monooxygenases target diverse areas. These include identification of key residues involved in substrate binding, the extension of substrate spectra, the substitution or regeneration of the cofactor NAD(P)H, and the enhancement of enzyme stability, activity, selectivity, and coupling efficiency (Figure 12.2). To achieve these objectives, various techniques of protein engineering have been applied, for example site-directed and random mutagenesis, DNA recombination, or combinations of these methods. The applied strategies have their advantages and drawbacks, which are discussed in numerous excellent reviews and books [81–85].

12.3
Application and Engineering of P450s for the Pharmaceutical Industry

P450s have a central role in drug metabolism, where they catalyze a large proportion of the most complex and chemically challenging steps in the biosynthesis of many natural products used in medicine today. Thus, potential applications of P450s concern their involvement in drug biosynthesis, as well as strategies for developing new derivatives of drugs based on P450 engineering [86]. Given the diversity of reactions catalyzed by P450s, however, few of them have been exploited in industry so far.

12.3.1
Microbial Oxidations with P450s for Synthesis of Pharmaceuticals

Drug development is based on the detailed characterization of metabolic pathways and their relevance for drug safety. This type of analysis often requires milligram quantities of metabolites, which are difficult to synthesize by chemical routes, especially when the metabolites result from stereoselective oxidations. Microbial oxidations using fungi, yeast, archea, and bacteria can be performed either by using native P450-producing strains – which can be altered by metabolic engineering – or with the aid of recombinant whole-cells harboring microbial P450s. Microbial equivalents of human P450-catalyzed oxidations are an additional alternative, especially of interest when several hundred milligrams to many grams of metabolites are requested, both for identification purposes and for production of non-human metabolites with new biological properties.

Microbial oxidations of steroids represent very well-established large-scale commercial applications of P450 monooxygenases. The 11β-hydroxylation of 11-deoxycortisol (**26**) to hydrocortisone (**27**) using a P450 from *Curvularia* sp. [87] (Scheme 12.9) is applied by Schering AG (in 2006 acquired by Merck, Germany) at an industrial scale of approximately 100 tonnes per year [80]. Another example is the regioselective hydroxylation of progesterone to 11α-hydroxyprogesterone by *Rhizopus* sp. developed in the 1950s by Pharmacia & Upjohn (later acquired by Pfizer Inc., USA) [88, 89]. Both processes are one-step biotransformations, which cannot be achieved by chemical routes.

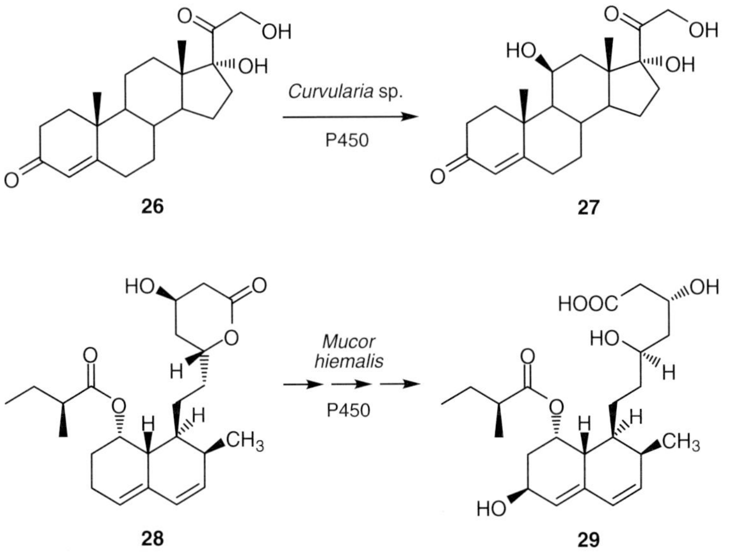

Scheme 12.9 Two examples of microbial oxidations of steroids: 11β-hydroxylation of 11-deoxycortisol (**26**) to hydrocortisone (**27**), and production of pravastatin (**29**) by oxidation of compactin (**28**).

Production of the cholesterol-reducing pravastatin (**29**) by oxidation of compactin (**28**) catalyzed by a P450 monooxygenase from *Mucor hiemalis* (Daiichi Sankyo Inc., USA, and Bristol-Myers Squibb, USA) is another example of a commercial application of microbial oxidations [90, 91] (Scheme 12.9). The same reaction can be catalyzed by *Streptomyces* sp. Y-110. In a batch culture with continuous feeding of compactin into the culture medium a conversion rate of 15 mg $L^{-1} h^{-1}$ pravastatin and a final concentration of 1 g L^{-1} pravastatin were achieved [92].

Diverse activities of microbial cytochrome P450 monooxygenases have potential applications in the synthesis of new antibiotics, especially in view of the widespread resistance to bacterial antibiotics. *Streptomyces* strains and other bacterial actinomycete species produce many important natural products including the majority of known antibiotics, and P450s catalyze important biosynthetic steps [93]. Particularly intriguing is the fact that *Streptomyces* has a large P450 complement reflecting the ecological niche that the organism finds itself in. The first complete *Streptomyces* genome (*Streptomyces coelicolor* A3(2)) was published in 2002, revealing the presence of 18 P450 genes [94]. Subsequently, genomes of *Streptomyces avermitilis* (33 P450 genes [95]) and *Streptomyces peucetius* (15 P450 genes [96]) have been reported. Many efforts have been undertaken to identify gene clusters involved in synthesis of pharmaceutically important compounds and to increase the product yield of these compounds by the use of engineered *Streptomyces* strains.

Recent examples include the elucidation of the biosynthetic gene cluster organization for pladienolide – a polyketide antitumor macrolide – in *Streptomyces platensis* Mer-11107, where a P450 of the CYP107 family acts as a 6-hydroxylase [97]. Pladienolide B (**30**) and its 16-hydroxylated derivative pladienolide D (**31**) show strong antitumor activity (Scheme 12.10). The original strain of *S. platensis* produces mainly pladienolide B, while pladienolide D is produced to a lesser extent. Consequently, to facilitate the production of pladienolide D, an engineered strain was constructed by over-expression of a pladienolide B 16-hydroxylase PsmA (a P450 from the CYP105 family) from *Streptomyces bungoensis* A-1544 in *S. platensis*. The recombinant strain produced pladienolide D at a production level comparable to that of pladienolide B [98].

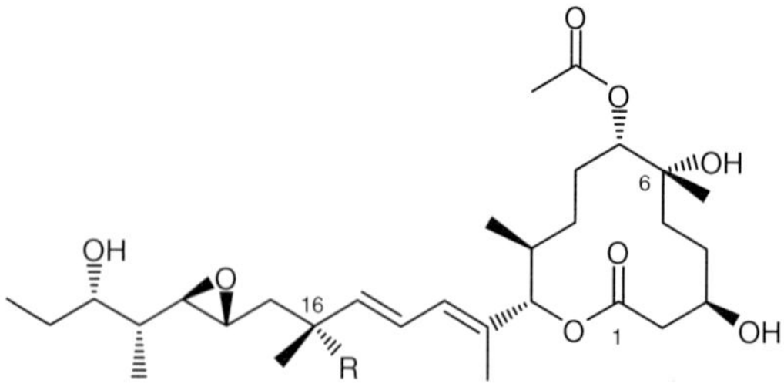

30: R = H; **31:** R = OH

Scheme 12.10 Structures of pladienolide B (**30**) and pladienolide D (**31**).

Pimaricin is an important antifungal agent used in human therapy for the treatment of fungal keratitis as well as in the food industry to prevent mold contamination. The enzymes and the gene cluster responsible for their production by *Streptomyces natalensis* was identified recently. The cluster contains a P450 that is responsible for 'ring decoration' [99].

Production of epothilones (potential anticancer agents) was carried out with recombinant *Streptomyces* strains. Therefore, the entire epothilone biosynthetic gene cluster from the myxobacterium *Sorangium cellulosum* was heterologously expressed in an engineered *Streptomyces venezuelae* strain. The resulting strains produced approximately $0.1\,\mu g\,L^{-1}$ epothilone B as a sole product after 4 days cultivation. Deletion of a gene encoding a cytochrome P450 epoxidase gave rise to a mutant that selectively produced $0.4\,\mu g\,L^{-1}$ epothilone D [100].

Biosynthesis of the sesquiterpene antibiotic albaflavenone in *Streptomyces coelicolor* A3(2) was studied in detail. The mechanism and stereochemistry of the enzymatic formation of epi-isozizaene by multistep cyclization of farnesyl diphosphate was investigated [101] and a P450 (CYP170A1) was identified to carry out two sequential allylic oxidations to convert epi-isozizaene to an epimeric mixture of albaflavenols and thence to albaflavenone [102].

While functions of a number of P450s from *Streptomyces* are known, in the vast majority of these species the function of P450s so far remains unknown. Efforts are being made to shed light on this issue. The long-term goal is to identify orphan P450 functions and to establish a strategy for the production of novel secondary metabolites that have new biomedical function [103, 104]. Some crystal structures of P450s from *Streptomyces* are already available, for example, that of $P450_{PikC}$ from *Streptomyces venezuelae* [105], $P450_{StaP}$ (CYP245A1) from *Streptomyces* sp. TP-A0274 [106], $P450_{SU1}$ (CYP105A1) from *Streptomyces griseolus* [107], or recently CYP105P1 from *Streptomyces avermitilis* [108]. The structural knowledge of these P450s opens up a way for enzyme optimization and engineering leading to new antibacterial properties.

While *Streptomyces* produce the majority of known antibiotics, there are also other bacterial species with interesting activities for pharmaceutical application. OxyA, OxyB, and OxyC (CYP165A1, CYP165B1, and CYP165C1) from *Amycolatopsis orientalis*, for example, are involved in the biosynthesis of the glycopeptide antibiotic vancomycin [67]. A similar set of P450s with high identity to the oxygenases in vancomycin biosynthesis (92% to 94%) is involved in the biosynthesis of balhimycin in *Amycolatopsis balhimycina* [109].

Another example is $P450_{eryF}$ (CYP107A1) from *Saccharopolyspora erythraea*, which catalyzes a hydroxylation step leading to the functional molecule in the biosynthesis of the macrolide antibiotic erythromycin [110]. The crystal structures of OxyB and OxyC [111, 112] as well as that of $P450_{eryF}$ [113] have been published.

Nonrecombinant wild-type strains are often used as biocatalysts for the production of pharmaceutically relevant metabolites, as in many of the examples described above. The application of wild-type strains, however, has some drawbacks. These include product degradation, by-product formation, and catabolite repression [75]. Further, there is little or no possibility of engineering the biocatalyst itself. The application of recombinant engineered strains, combined with bioprocess and biocatalyst engineering, provides an attractive alternative. For industrial biocatalysis *E. coli* is in general the organism of choice because of the well-developed molecular biology techniques and high levels of biocatalyst expression. Besides this, high cell densities can be achieved during fermentation of *E. coli* [114]. Recombinant *Pseudomonas* strains are of industrial interest as well, since they could be useful for biotransformations in the presence of toxic organic solvents. A number of solvent-tolerant *P. putida* strains have been isolated and used as hosts for the expression of P450 genes. An excellent review which includes many examples of the use of recombinant *E. coli* and *P. putida* whole-cell biocatalysts on a scale up to 30 L has been written by Park [75].

Recently construction and application of a novel P450 library, based on about 250 bacterial P450 genes (about 70% originating from actinomycetes), co-expressed with putidaredoxin and putidaredoxin reductase in *E. coli* has been reported [115]. A good example demonstrating the exquisite regioselectivity of P450s, which is impossible to reach by chemical oxidations, is presented by a screening of this library with testosterone. Within the screening 24 bacterial P450s were identified, which monohydroxylate testosterone regio- and stereoselectively at the 2α-, 2β-, 6β-, 7β-, 11β-, 12β -, 15β-, 16α- and 17-positions [116]. Most of these hydroxylations are common for both prokaryotic and human P450s. Thus, the identified bacterial candidates will be applied for the production of drug metabolites on a preparative scale.

A screening among 1800 bacterial strains identified the *Mycobacterium* sp. strain HXH-1500, catalyzing regio- and stereoselective hydroxylation of limonene at carbon-atom C7 to yield the anticancer drug (−)-perillyl alcohol. The biocatalytic production of (−)-perillyl alcohol from limonene was performed using the *Mycobacterium* sp. P450 alkane hydroxylase (CYP153 family) recombinantly expressed in *P. putida* cells [117]. The whole-cell process was performed in a two-phase system resulting in $6.8 \, \text{g L}^{-1}$ of product in the organic phase.

12.3.2
Application of Mammalian P450s for Drug Development

An ongoing field of research in drug discovery and drug development is the use of recombinant human P450s. Human P450s expressed in baculovirus-infected insect cell lines have been in use for quite some time now to identify human P450s involved in drug metabolism [118, 119]. They are also used to synthesize drug metabolites in order to assess the toxicity of potential pharmaceuticals [120–122]. Recent achievements in the expression of recombinant human P450s in *Saccharomyces cerevisiae* [76, 123, 124], *Yarrowia lipolytica* [125] and *Pichia pastoris* [126, 127] facilitate their use for the synthesis of drug metabolites.

Besides the achievements in the expression of mammalian P450s in eukaryotic organisms, *E. coli* is an attractive expression system and is of particular interest for industrial applications (see also Section 12.3.1). Efforts to express human P450s in *E. coli* have been made [128–130]; in this case, however, mammalian cDNAs have to be modified before they can be expressed [131].

12.3.2.1 Enhancement of Recombinant Expression in *E. coli*

Initial attempts to enhance the expression of human P450s in *E. coli* involved modification of the N-terminal sequence, since this region has been proposed to be important in determining protein yield [132, 133]. This strategy was applied to express a number of mammalian P450s including CYP 2E1 [134], 4A4 [135], 6A1 [136], 1A1 [137], as well as recently 20A1 [138] and 4X1 [139]. A different approach was pursued by truncating the N-terminal hydrophobic region – which is responsible for the interaction with the membrane – to a greater or lesser extent [140–142]. Three human P450s (3A4, 2C9, and 1A2) with truncated N-termini were each co-expressed with human CPR in *E. coli* using a bicistronic expression system. Intact *E. coli* cells and membranes containing P450 and CPR were used in the preparative synthesis of drug metabolites. The optimized biooxidation conducted on a 1-L scale yielded 59 mg 6β-hydroxytestosterone (**32**), 110 mg 4-hydroxydiclofenac (**33**), and 88 mg acetaminophen (**34**) [130] (Scheme 12.11).

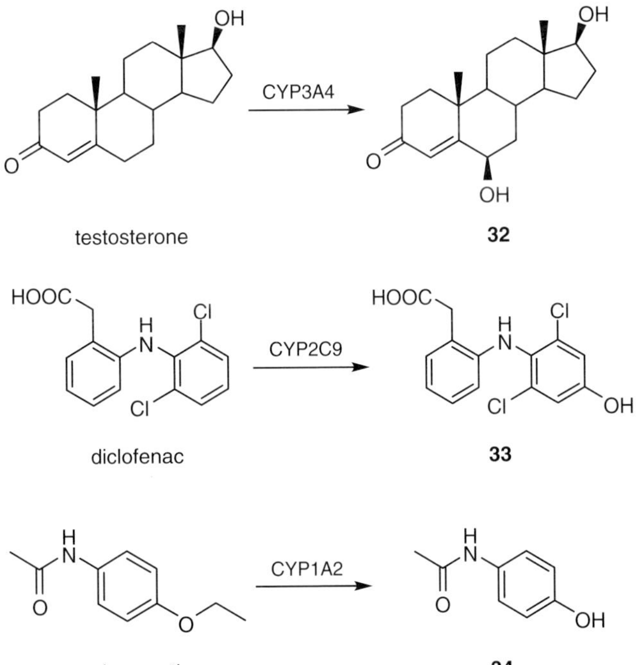

testosterone **32**

diclofenac **33**

phenacetin **34**

Scheme 12.11 Oxidation of testosterone, diclofenac, and phenacetin to yield 6β-hydroxytestosterone (**32**), 4-hydroxydiclofenac (**33**) and acetaminophen (**34**), respectively.

Further optimization could be achieved by fusions of a bacterial signal peptide (for example a modified ompA-leader sequence) to human P450 cDNAs [122]. This leader is removed during P450-processing, and the native P450 is released. Following this strategy, 14 recombinant human P450s co-expressed with NADPH P450 reductase in

E. coli have been used by several pharmaceutical companies, both as biocatalysts for the preparation of metabolites of drug candidates and for high-throughput P450 inhibition screenings. Up to 300 mg of different metabolites could be obtained using permeabilized recombinant *E. coli* cells expressing human P450s [143].

12.3.2.2 Enhancement of Activity and Selectivity and Engineering of Novel Activities

Since structure-function relationships for mammalian (and other eukaryotic) P450s are far less clear than for bacterial ones, most protein engineering efforts conducted on mammalian P450s so far focus on the identification of key residues involved in substrate binding. Nevertheless, as more crystal structures of mammalian P450s are becoming available, the structural knowledge can also be used to engineer novel enzyme properties.

Recent examples include the engineering of CYP2B1 by a combination of random and site-directed mutagenesis to accept the anti-cancer prodrugs cyclophosphamide and ifosfamide [144]. Homology modeling and rational design performed on CYP2B1 allowed the regioselectivity of progesterone 16-α-hydroxylation to switch to 21-hydroxylation [145].

The ability of human CYP1A2 to catalyze O-demethylation of 7-methoxyresorufin was improved by three rounds of mutagenesis. The triple-mutant E163K/V193M/K170Q exhibited turnover rates more than five times faster than wild-type CYP1A2 [146].

The Gillam group constructed a chimeric library of CYP1A1 and CYP1A2 employing restriction enzyme-mediated DNA family shuffling. They observed different activity profiles toward various luciferin derivatives among active clones, including improved specific activity, novel activities, and broadening of substrate range [147, 148].

The substrate spectrum of human CYP2D6 was expanded toward bulky indole derivatives, like 4- and 5-benzyloxyindoles, by random mutagenesis and site-directed mutagenesis [149].

12.3.2.3 Construction of Artificial Self-Sufficient Fusion Proteins

A very effective mode to eliminate the reconstitution procedure of redox chains is to tailor self-sufficient electron transport chains emulating natural fusion enzymes. A large variety of these artificial fusion enzymes have been reported. Fusions of P450s with a diflavin (FAD- and FMN-containing) reductase, P450s with flavodoxins, P450s with ferredoxins and ferredoxin reductases, P450s with dioxygenase reductase-like enzymes, and even semisynthetic heme-containing flavoproteins that can act as a monooxygenase have been described [70, 150]. Genes of various mammalian P450s and their redox partners were used to generate functional multi-domain proteins with designed properties ('molecular Lego') beyond the restrictions imposed by the naturally occurring protein domains [151].

For example, the N-terminal human CYP2D6 sequence was linked to the C-terminal human cytochrome reductase module, resulting in a self-sufficient enzyme exhibiting activity toward a wide range of pharmaceutical compounds such as bufuralol, metoprolol, or dextromethorphan [152]. Other examples are fusion proteins of human CYP1A1 with rat CPR [153], human CYP2C9 with BMR [154], and human CYP3A4 with rat or human CPR [155]. As cytochrome b$_5$ was detected to remarkably enhance the coupling efficiency if co-expressed with the CYP3A4-fusion systems [156], recombinant three-component systems with CYP3A4/CPR/cytochrome b$_5$ were tailored, maintaining even higher oxidation activity for testosterone and nifedipine [157].

12.4
Application of P450s for Synthesis of Fine Chemicals

Another valuable application for P450s, aside from the production of pharmaceutically relevant metabolites, is the fabrication of sought-after fine chemicals and flavors that are of interest for the food industry [158]. P450s are interesting biocatalysts for such syntheses, since chirality of the products is often very important.

Celik et al. have described an example of substrate hydroxylation by P450s with a high degree of regio- and diastereoselectivity [159]. The authors used recombinant E. coli whole-cell systems, in which cytochromes P450$_{SU1}$, P450$_{SU2}$, and P450$_{SOY}$ were over-expressed with their cognate ferrodoxins for hydroxylation of α-ionone (35) and β-ionone (38) to their corresponding mono-hydroxylated derivatives. For α-ionone the reaction was diastereoselective, yielding only the anti-isomers (3S,6S)- and (3R,6R)-3-hydroxy-α-ionone (36 and 37). For β-ionone an enantiomeric excess of 35% in favor of the (S)-enantiomer (39) was observed (Scheme 12.12).

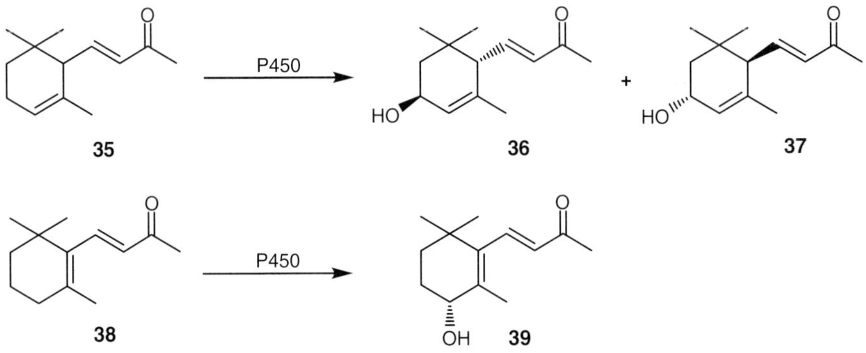

Scheme 12.12 Regio- and diastereoselective hydroxylation of α-ionone (35) and β-ionone (38).

In another screening of the recombinant P450-library described in Ref. [115] (see Section 12.3.1), CYP109B1 was identified to be capable of regioselective hydroxylation of (+)-valencene (48) at allylic C2-position, thereby producing (+)-nootkatone – a sought after fragrance for flavoring that has a strong grapefruit odor and bitter taste [160, 161] – via nootkatol. By product formation, which accounted for 35% of the total product during biooxidation with recombinant E. coli in an aqueous milieu, was significantly reduced when aqueous-organic two-liquid-phase systems were set up leading to accumulation of nootkatol and (+)-nootkatone of up to 97% of the total product. Up to 120 mg L^{-1} of nootkatol and (+)-nootkatone were produced within 8 h employing this system [162].

Functional expression of tricistronic constructs consisting of P450$_{cam}$ from *P. putida* and the auxiliary redox partners putidaredoxin reductase and putidaredoxin in *E. coli* has been reported. The transformed whole-cells efficiently oxidized (1R)-(+)-camphor (**10**) to 5-exo-hydroxycamphor (**11**) and limonene to (−)-perillyl alcohol [163]. Since these bioengineered e-cells possess a heterologous self-sufficient P450 catalytic system, they may have advantages in terms of low cost and high yield for the production of fine chemicals.

12.5
Engineering of P450s for Biocatalysis

12.5.1
Cofactor Substitution and Regeneration

The application of cytochrome P450 monooxygenases requires functional and efficient electron suppliers and is dependent on the availability of the costly cofactors NAD(P)H. Thus, various approaches have been made to substitute or regenerate NAD(P)H and to increase the coupling efficiency for both *in vitro* and *in vivo* applications of P450s. These attempts include direct chemical or electrochemical reduction of P450s, the use of cheap chemicals to directly replace NAD(P)H, the development of (enzymatic) cofactor recycling systems [164], and the engineering of artificial fusion proteins [150].

12.5.1.1 Cofactor Substitution *In Vitro*
Direct chemical reduction can be achieved by the use of organometallic complexes such as Co^{II} sepulchrate trichloride [165, 166] by addition of Ti^{III} citrate [167] or by the strong reducing agent sodium dithionite [168] – an inexpensive agent routinely used to produce the ferrous-carbonyl form of P450. The limitations of these approaches are, however, low system efficacy, low sensitivity of mediators to molecular oxygen, and the production of reactive oxygen species or mediator aggregation.

Peroxides that directly convert the heme iron of P450s to a ferric hydroperoxy complex by the 'peroxide shunt' (e.g., hydrogen peroxide, cumene peroxide, or *tert*-butyl oxide) could be useful for oxidation of various substrates. The essential problem in utilizing the 'peroxide shunt' for P450 biocatalysis seems to lie in the time-dependent degradation of the heme and in oxidation of the protein [169, 170]. Methods of directed evolution, like random and site-specific mutagenesis, were applied to evolve P450s to enhance the efficiency of the 'peroxide shunt' pathway [171].

Electrochemical reduction of P450s by using an electrode seems to be the most convenient way of fabricating bioreactors [172] and has been studied in detail for more than 15 years. However, heme reduction on a cathode is complicated and most often insufficient for biocatalysis, which can partly be attributed to the deeply buried heme iron and partly to the instability of the enzyme upon interaction with the electrode surface. Improvements to this approach include the modification of the electrode surface [150].

12.5.1.2 Cofactor Regeneration *In Vitro*
One of the most common approaches to overcome the stoichiometric need for NAD(P)H involves enzymatic regeneration systems. Strategies for the regeneration of NADPH-dependent P450s are based on D/L-isocitrate dehydrogenase [173], an engineered formate dehydrogenase from *Pseudomonas* sp. 101 [174], glucose-6-phosphate dehydrogenase [175], or alcohol dehydrogenase from *Thermoanaerobium brockii* [176]. Although these systems work reasonably well, further reduction of costs for enzymatic oxyfunctionalization can be achieved by engineering of NADPH-dependent P450s to accept NADH [177, 178]. NADH has the advantage that it is about ten times less expensive and more stable than NADPH. Furthermore, more NAD^+-dependent enzymes at lower prices are available for cofactor regeneration. Our group used homology modeling and site-directed mutagenesis to construct $P450_{BM3}$ mutants showing altered cofactor specificity from NADPH to NADH. The best mutant, W1046S/S965D, demonstrated high turnover rates ($k_{cat} = 14600\,min^{-1}$) during cytochrome c reduction and low K_M for NADH. It had equal affinity to both NADH and NADPH and displayed a 2.6-times higher activity when NADH was used [179].

12.5.1.3 Cofactor Regeneration in Whole-Cells
Cofactor concentration within the cell may become a bottleneck for the overall process if concentrations of the recombinant P450 biocatalyst and/or their activities reach a very high level [180]. Strategies to overcome this limitation employ the use of artificial cofactor recycling systems supporting the endogenous ones and the employment of native electron transfer systems to increase the coupling efficiency of P450s and reduce metabolic stress for the host organism.

The catalytic efficiency of $P450_{cam}$ expressed in *E. coli* could be increased up to 25-fold by the coupling of its native electron transfer system (putidaredoxin reductase and putidaredoxin) to enzymatic NADH regeneration catalyzed by a recombinantly expressed glycerol dehydrogenase. This whole-cell system was applied for camphor hydroxylation in aqueous solution, as well as in a biphasic system with isooctane [181].

A recent approach in this field is the construction of a novel *E. coli* whole-cell biocatalyst with improved intracellular cofactor regeneration driven by external glucose [182]. In this system, additional recombinant intracellular NADPH regeneration takes place through co-expression of a glucose facilitator from *Zymomonas mobilis* for uptake of unphosphorylated glucose and an $NADP^+$-dependent glucose dehydrogenase from *Bacillus megaterium* that oxidizes glucose to gluconolactone. When a quintuple mutant of $P450_{BM3}$ that oxyfunctionalizes α-pinene (**49**) – a cheap waste product of wood industry – to yield α-pinene oxide, verbenol and myrtenol [183] was expressed in this system, the engineered strain showed a 9-fold increased initial α-pinene oxide formation rate and a 7-fold increased α-pinene oxide yield in the presence of glucose compared to glucose-free conditions. Further bioprocess engineering addressed the low water solubility and the toxicity of α-pinene by setting up an aqueous-organic two-phase bioprocess with diisononyl phthalate as a biocompatible organic carrier solvent. With an aqueous-to-organic phase ratio of 3 : 2 and 30% (v/v) of α-pinene in the organic phase, a total product concentration of over $1 \, g \, L^{-1}$ was achieved [184]. This process marks a promising step toward a future application of recombinant microorganisms for the selective oxidation of terpenoids to value-added products.

12.5.2
Construction of Artificial Fusion Proteins

As in the case of human P450s (described in Section 12.3.2.3), a variety of artificial fusion enzymes have also been generated for bacterial P450s aiming to increase enzyme activity and coupling efficiency.

Concerning P450$_{cam}$, the hemoprotein was fused to its natural electron donors, putidaredoxin and putidaredoxin reductase in different orders of the individual parts and variable linkers between them. However, despite increased coupling, the catalytic activity of the engineered assembly was only 30% of that of the reconstituted wild-type mixture, reconstituted in a 1 : 1 : 1 ratio with the separate redox partners [185]. The catalytic activity of the fusion enzyme toward D-(+)-camphor (**10**) could be increased 10-fold by the development of an enzymatic cross-linking of putidaredoxin with a peptide fragment between putidaredoxin reductase and P450$_{cam}$, accommodating a reactive glutamyl residue prone to attack by transglutaminase [186].

The putidaredoxin reductase-like unit of the self-sufficient CYP116B2 from *Rhodococcus* sp. (P450RhF) [187] was ligated to the heme domain of P450$_{cam}$, CYP153A, or CYP203A. The chimeric proteins showed catalytic activity against a variety of compounds like *n*-alkanes, D-(+)-camphor (**10**), and 4-hydroxybenzoate [188].

12.5.3
Engineering of New Substrate Specificities

In this chapter focus exclusively on the two best characterized P450s: P450$_{cam}$ (CYP101A1) from *P. putida* [189] and P450$_{BM3}$ (CYP102A1) from *B. megaterium*. Their structure, catalytic mechanism, and biochemistry have been studied in detail extensively [10]. Mutagenesis studies on these P450s have led to significant improvements in our understanding of general aspects of P450 catalytic function, and numerous approaches have been undertaken to engineer new substrate specificities for these enzymes.

12.5.3.1 P450$_{cam}$ from *Pseudomonas putida*
For P450$_{cam}$ it was demonstrated that mutations of the active site residue Y96 to more hydrophobic ones considerably increased its activity toward the oxidation of hydrophobic molecules smaller than camphor (**10**), like styrene (**40**) and alkanes [190, 191],

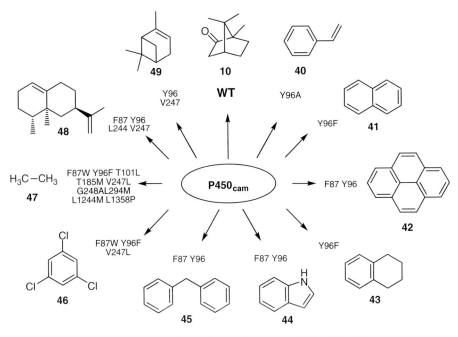

Figure 12.3 Mutations of P450$_{cam}$ leading to altered substrate specificities. WT: wild-type enzyme without mutations.

or larger than camphor, such as naphthalene (**41**) and pyrene (**42**) [192, 193]. The Y96F mutant was capable of regio- and stereoselective hydroxylation of tetralin (**43**) to the 1-(R)-alcohol [194]. Using saturation mutagenesis at positions Y96 and F87, several mutants were constructed with activity toward indole (**44**) and diphenylmethane (**45**) [195]. Some mutants have been engineered for biodegradation of environmentally harmful pollutants, such as di-, tri-, tetra-, penta-, and hexachlorobenzene (**46**) [196, 197] (Figure 12.3).

P450$_{cam}$ was further engineered to an alkane hydroxylase by step-by-step adaptation of the enzyme to smaller *n*-alkanes beginning with hexane [198], then proceeding to butane and propane [199], and finally to ethane (**47**) [200]. The best mutant, with eight substitutions, oxidized propane at 500 min^{-1} with 86% coupling, which was comparable with that of the wild-type enzyme toward camphor (**10**) [200].

Successful attempts to engineer mutants for selective oxidation of cheap terpenes to expensive oxidized derivatives have also been accomplished. Rational mutants of P450$_{cam}$ exhibited activity toward the sesquiterpene (+)-valencene (**48**) and produced > 85% of (+)-*trans*-nootkatol and (+)-nootkatone [201]. The monoterpene (+)-α-pinene (**49**) was oxidized by P450$_{cam}$ mutants to 86% (+)-*cis*-verbenol or to a mixture of (+)-verbenone and (+)-*cis*-verbenol (Figure 12.3) [202]. Verbenol and verbenone are active pheromones against various beetle species.

12.5.3.2 P450$_{BM3}$ from *Bacillus megaterium*

P450$_{BM3}$ is an obvious target enzyme for the development of biotechnological applications, since it is a self-sufficient single-component protein with a high catalytic activity. The turnover rates of P450$_{BM3}$ toward fatty acids are among the highest activity values reported for P450s [203].

The mutant A74G/F87V/L188Q, designed by saturation mutagenesis, was shown to oxidize indole, *n*-octane, highly branched fatty acids and fatty alcohols, polychlorinated dibenzo-*p*-dioxins, polyaromatic hydrocarbons, styrene, and many other chemical compounds [204–208]. The monoterpene geranylacetone was converted by P450$_{BM3}$ (R47L/Y51F/F87V) with high activity (2080 min^{-1}) and stereoselectivity (97% *ee*) to a single product, namely 9,10-epoxygeranylacetone [209]. The mutant A74E/F87V/P386S exhibited an 80-fold improved activity toward β-ionone (**42**) compared to the wild-type enzyme and produced the flavoring (*R*)-4-hydroxy-β-ionone as the only product [210].

Using a combination of directed evolution and site-directed mutagenesis Arnold and coworkers altered the selectivity of P450$_{BM3}$ from hydroxylation of dodecane (C12), first to octane (C8) and hexane (C6) and further on to gaseous propane (C3) and ethane (C2) [211–214]. Some mutants were found with high stereoselectivity, leading either to (*R*)- or to (*S*)-2-octanol [215].

In our group, a systematic analysis of the structures of 29 P450s and 6379 P450 sequences, with the aim of identifying selectivity- and specificity-determining residues, led to identification of a positively charged heme-interacting residue in the SRS5, which was present in about 98% of the sequences analyzed. This residue is located in close vicinity to the heme center and restricts the conformation of the SRS5. It is preferentially located at position 10 or 11 after the conserved ExxR motif (in about 95% of the sequences). Replacing this residue by hydrophobic residues of different size has been shown to change substrate specificity and regioselectivity for P450s of different superfamilies [216]. Based on this analysis, a minimal P450$_{BM3}$ mutant library of only 24 variants plus wild-type was constructed by combining five hydrophobic amino acids (alanine, valine, phenylalanine, leucine, and isoleucine) in positions 87 and 328. The library was screened with four terpene substrates geranylacetone, nerylacetone, (4*R*)-limonene, and (+)-valencene (**52**). Eleven variants demonstrated either a strong shift or improved regio- or stereoselectivity during oxidation of at least one substrate as compared to P450$_{BM3}$ wild-type [217].

Although wild-type P450$_{BM3}$ is not able to metabolize any drug-like compound tested so far, it has been turned by protein design and directed evolution into an enzyme that oxidizes human drugs [218]. The R47L/F87V/L188Q mutant was shown to metabolize testosterone, amodiaquine, dextromethorphan, acetaminophen, and 3,4-methylenedioxymethylamphetamine [219]. Several mutants were obtained by means of directed evolution which are able to convert propranolol, a multi-function beta-adrenergic blocker [220]. Another mutant was capable of stereo- and regioselective hydroxylation of the peptide group of buspirone to yield (*R*)-6-hydroxybuspirone – an anti-anxiety agent – with > 99.5% *ee* [221].

Recent approaches in this field reported mutants with applicability as biocatalysts in the production of reactive metabolites from the drugs clozapine, diclofenac, and acetaminophen [222]. Other mutants were reported that are capable of oxidizing 7-ethoxycoumarin [223] or of metabolizing trimethoprim – an antibacterial agent [224].

A new approach for DNA recombination (the so-called SCHEMA algorithm) was applied to obtain chimeras based on $P450_{BM3}$ and its homologs CYP102A2 and CYP102A3, sharing only 63% amino acid identity [225]. A survey of the activities of new chimeras demonstrated that this approach created completely new functions, absent in the wild-type enzymes, including the ability to accept drugs (e.g., propranolol, tolbutamide, zoxazolamine, verapamil, and astemizole) [226].

Although in most cases the yields of human metabolites obtained with $P450_{BM3}$ mutants were comparable to those produced by recombinant human P450s in bacterial or baculovirus systems, the catalytic activities, coupling efficiencies, and total turnover numbers achieved were much lower than those obtained with natural substrates of $P450_{BM3}$. This suggests that further mutagenesis is required to increase coupling and activity of these mutants.

12.6
Future Trends

Many interesting P450-mediated oxidations have been described in the literature so far. Examples of process implementation and scale-up to pilot or industrial scales are, however, comparatively rare. Obviously, plenty of time and excellent knowledge of the corresponding P450 systems are needed to develop industrial processes, bearing in mind the complexity of these enzymes.

The invention of new methods of protein engineering in the past decade has led to the construction of an abundance of P450 mutants with new tailored properties. Recent major achievements include significant increases in productivities, yields, and rates of catalytic turnover as well as modification of substrate specificity and efficient multistep reactions in whole-cell biocatalysts, coming one step closer to the technical application of cytochrome P450 monooxygenases.

The number of identified orphan P450s associated with unknown pathways of secondary metabolism is constantly increasing through the sequencing of genomes and microbial screenings. Functional characterization and utilization of unusual P450s and P450 mutants in drug development, agro- and fine chemical synthesis, bioremediation, biosensors, and plant improvement present new and exciting perspectives for future applications.

Abbreviations

CPR	NADPH-cytochrome P450 oxidoreductase
NAD(P)H	nicotinamide adenine dinucleotide (phosphate), reduced form
FAD	flavin adenine dinucleotide
FMN	flavin mononucleotide
DMSO	dimethyl sulfoxide

References

1. Que, L. Jr. and Tolman, W.B. (2008) *Nature*, **455**, 333–340.
2. Davies, H.G., Green, R.H., Kelly, D.R., and Roberts, S.M. (1989) *Biotransformations in Preparative Organic Chemistry: The Use of Isolated Enzymes and Whole-Cell Systems in Synthesis*, Academic Press, London.
3. Sono, M., Roach, M.P., Coulter, E.D., and Dawson, J.H. (1996) *Chem. Rev.*, **96**, 2841–2888.
4. Nelson, D.R. (2006) *Methods Mol. Biol.*, **320**, 1–10.
5. Fischer, M., Knoll, M., Sirim, D., Wagner, F., Funke, S., and Pleiss, J. (2007) *Bioinformatics*, **23**, 2015–2017.
6. Sirim, D., Wagner, F., Lisitsa, A., and Pleiss, J. (2009) *BMC Biochem.*, **10**, 27.
7. Bernhardt, R. (2006) *J. Biotechnol.*, **124**, 128–145.
8. Kelly, S.L., Lamb, D.C., and Kelly, D.E. (2006) *Biochem. Soc. Trans.*, **34**, 1159–1160.
9. Lewis, D.F. (1996) *Cytochromes P450 Structure, Function and Mechanism* (ed. D.F. Lewis), Taylor and Francis Ltd., London, pp. 115–167.
10. Ortiz de Montellano, P.R. (2005) *Cytochrome P450 Structure, Mechanism and Biochemistry* 3rd edn, Kluwer Academic/Plenum Press, New York.
11. Urlacher, V.B. and Eiben, S. (2006) *Trends Biotechnol.*, **24**, 324–330.
12. Urlacher, V.B., Lutz-Wahl, S., and Schmid, R.D. (2004) *Appl. Microbiol. Biotechnol.*, **64**, 317–325.
13. Werck-Reichhart, D. and Feyereisen, R. (2000) *Genome Biol.*, **1**, REVIEWS3003.
14. Wong, L.L. (1998) *Curr. Opin. Chem. Biol.*, **2**, 263–268.
15. Garfinkel, D. (1958) *Arch. Biochem. Biophys.*, **77**, 493–509.
16. Klingenberg, M. (1958) *Arch. Biochem. Biophys.*, **75**, 376–386.
17. Omura, T. and Sato, R. (1964) *J. Biol. Chem.*, **239**, 2370–2378.
18. Raag, R. and Poulos, T.L. (1989) *Biochemistry*, **28**, 7586–7592.
19. Bayer, E., Hill, H.O.A., Röder, A., and Williams, R.J.P. (1969) *Chem. Commun. (Camb.)*, 109.
20. Hill, H.A.O., Roder, A., and Williams, R.J. (1969) *Biochem. J.*, **115**, 59–60.
21. Poulos, T.L., Finzel, B.C., and Howard, A.J. (1986) *Biochemistry*, **25**, 5314–5322.
22. Graham, S.E. and Peterson, J.A. (1999) *Arch. Biochem. Biophys.*, **369**, 24–29.
23. Graham-Lorence, S. and Peterson, J.A. (1996) *FASEB J.*, **10**, 206–214.
24. Mestres, J. (2005) *Proteins*, **58**, 596–609.
25. Li, H. and Poulos, T.L. (2004) *Curr. Top. Med. Chem.*, **4**, 1789–1802.
26. Poulos, T.L., Finzel, B.C., Gunsalus, I.C., Wagner, G.C., and Kraut, J. (1985) *J. Biol. Chem.*, **260**, 16122–16130.
27. Ravichandran, K.G., Boddupalli, S.S., Hasemann, C.A., Peterson, J.A., and Deisenhofer, J. (1993) *Science*, **261**, 731–736.
28. Hasemann, C.A., Ravichandran, K.G., Peterson, J.A., and Deisenhofer, J. (1994) *J. Mol. Biol.*, **236**, 1169–1185.
29. Cupp-Vickery, J.R. and Poulos, T.L. (1995) *Nat. Struct. Biol.*, **2**, 144–153.
30. Park, S.Y., Shimizu, H., Adachi, S., Nakagawa, A., Tanaka, I., Nakahara, K., Shoun, H., Obayashi, E., Nakamura, H., Iizuka, T., and Shiro, Y. (1997) *Nat. Struct. Biol*, **4**, 827–832.
31. Williams, P.A., Cosme, J., Sridhar, V., Johnson, E.F., and McRee, D.E. (2000) *J. Inorg. Biochem*, **81**, 183–190.
32. Williams, P.A., Cosme, J., Ward, A., Angove, H.C., Matak Vinkovic, D., and Jhoti, H. (2003) *Nature*, **424**, 464–468.
33. Williams, P.A., Cosme, J., Vinkovic, D.M., Ward, A., Angove, H.C., Day, P.J., Vonrhein, C., Tickle, I.J., and Jhoti, H. (2004) *Science*, **305**, 683–686.
34. Rowland, P., Blaney, F.E., Smyth, M.G., Jones, J.J., Leydon, V.R., Oxbrow, A.K., Lewis, C.J., Tennant, M.G., Modi, S., Eggleston, D.S., Chenery, R.J., and Bridges, A.M. (2006) *J. Biol. Chem.*, **281**, 7614–7622.
35. Mast, N., White, M.A., Bjorkhem, I., Johnson, E.F., Stout, C.D., and Pikuleva, I.A. (2008) *Proc. Natl. Acad. Sci. USA*, **105**, 9546–9551.
36. Ghosh, D., Griswold, J., Erman, M., and Pangborn, W. (2009) *Nature*, **457**, 219–223.

37 Wang, J.F., Zhang, C.C., Chou, K.C., and Wei, D.Q. (2009) *Curr. Med. Chem.*, **16**, 232–244.
38 Bernhardt, R. (2004) *Chem. Biol.*, **11**, 287–288.
39 Hannemann, F., Bichet, A., Ewen, K.M., and Bernhardt, R. (2007) *Biochim. Biophys. Acta*, **1770**, 330–344.
40 Denisov, I.G., Makris, T.M., Sligar, S.G., and Schlichting, I. (2005) *Chem. Rev.*, **105**, 2253–2277.
41 Sligar, S.G. (1976) *Biochemistry*, **15**, 5399–5406.
42 Sligar, S.G., Makris, T.M., and Denisov, I.G. (2005) *Biochem. Biophys. Res. Commun.*, **338**, 346–354.
43 Jin, S., Bryson, T.A., and Dawson, J.H. (2004) *J. Biol. Inorg. Chem.*, **9**, 644–653.
44 Urlacher, V.B. (2009) *Green Catalysis, Vol. 3: Biocatalysis*, Vol. 3, 1st edn (ed. R.H. Crabtree), Wiley-VCH, Weinheim, pp. 1–25.
45 Fulco, A.J. (1991) *Annu. Rev. Pharmacol. Toxicol.*, **31**, 177–203.
46 Miura, Y. and Fulco, A.J. (1975) *Biochim. Biophys. Acta*, **388**, 305–317.
47 Budde, M., Maurer, S.C., Schmid, R.D., and Urlacher, V.B. (2004) *Appl. Microbiol. Biotechnol.*, **66**, 180–186.
48 Gustafsson, M.C., Roitel, O., Marshall, K.R., Noble, M.A., Chapman, S.K., Pessegueiro, A., Fulco, A.J., Cheesman, M.R., von Wachenfeldt, C., and Munro, A.W. (2004) *Biochemistry*, **43**, 5474–5487.
49 Nakayama, N., Takemae, A., and Shoun, H. (1996) *J. Biochem.*, **119**, 435–440.
50 Lee, D.S., Yamada, A., Sugimoto, H., Matsunaga, I., Ogura, H., Ichihara, K., Adachi, S., Park, S.Y., and Shiro, Y. (2003) *J. Biol. Chem.*, **278**, 9761–9767.
51 Matsunaga, I., Yokotani, N., Gotoh, O., Kusunose, E., Yamada, M., and Ichihara, K. (1997) *J. Biol. Chem.*, **272**, 23592–23596.
52 Matsunaga, I., Ueda, A., Fujiwara, N., Sumimoto, T., and Ichihara, K. (1999) *Lipids*, **34**, 841–846.
53 Girhard, M., Schuster, S., Dietrich, M., Durre, P., and Urlacher, V.B. (2007) *Biochem. Biophys. Res. Commun.*, **362**, 114–119.
54 Cryle, M.J., Stok, J.E., and De Voss, J.J. (2003) *Aust. J. Chem.*, **56**, 749–762.
55 Guengerich, F.P. (2001) *Curr. Drug. Metab.*, **2**, 93–115.
56 Guengerich, F.P. (2007) *J. Biochem. Mol. Toxicol.*, **21**, 163–168.
57 Isin, E.M. and Guengerich, F.P. (2007) *Biochim. Biophys. Acta*, **1770**, 314–329.
58 Hilker, B.L., Fukushige, H., Hou, C., and Hildebrand, D. (2008) *Prog. Lipid. Res.*, **47**, 1–14.
59 Rheinwald, J.G., Chakrabarty, A.M., and Gunsalus, I.C. (1973) *Proc. Natl. Acad. Sci. USA*, **70**, 885–889.
60 Tang, L., Shah, S., Chung, L., Carney, J., Katz, L., Khosla, C., and Julien, B. (2000) *Science*, **287**, 640–642.
61 Bruntner, C., Lauer, B., Schwarz, W., Mohrle, V., and Bormann, C. (1999) *Mol. Gen. Genet.*, **262**, 102–114.
62 Shyadehi, A.Z., Lamb, D.C., Kelly, S.L., Kelly, D.E., Schunck, W.H., Wright, J.N., Corina, D., and Akhtar, M. (1996) *J. Biol. Chem.*, **271**, 12445–12450.
63 Bower, S., Perkins, J.B., Yocum, R.R., Howitt, C.L., Rahaim, P., and Pero, J. (1996) *J. Bacteriol.*, **178**, 4122–4130.
64 Hlavica, P. and Kunzel-Mulas, U. (1993) *Biochim. Biophys. Acta*, **1158**, 83–90.
65 Seto, Y. and Guengerich, F.P. (1993) *J. Biol. Chem.*, **268**, 9986–9997.
66 Yu, A.M., Idle, J.R., Herraiz, T., Kupfer, A., and Gonzalez, F.J. (2003) *Pharmacogenetics*, **13**, 307–319.
67 Bischoff, D., Bister, B., Bertazzo, M., Pfeifer, V., Stegmann, E., Nicholson, G.J., Keller, S., Pelzer, S., Wohlleben, W., and Sussmuth, R.D. (2005) *Chembiochem*, **6**, 267–272.
68 Dalvie, D.K. and O'Connell, T.N. (2004) *Drug. Metab. Dispos.*, **32**, 49–57.
69 Kim, T.W., Hwang, J.Y., Kim, Y.S., Joo, S.H., Chang, S.C., Lee, J.S., Takatsuto, S., and Kim, S.K. (2005) *Plant. Cell*, **17**, 2397–2412.
70 Munro, A.W., Girvan, H.M., and McLean, K.J. (2007) *Biochim. Biophys. Acta*, **1770**, 345–359.
71 Arsenault, P.R., Wobbe, K.K., and Weathers, P.J. (2008) *Curr. Med. Chem.*, **15**, 2886–2896.
72 Barnes, H.J. (1996) *Methods Enzymol.*, **272**, 3–14.

73 Gonzalez, F.J. and Korzekwa, K.R. (1995) *Annu. Rev. Pharmacol. Toxicol.*, **35**, 369–390.

74 Guengerich, F.P., Gillam, E.M., and Shimada, T. (1996) *Crit. Rev. Toxicol.*, **26**, 551–583.

75 Park, J.B. (2007) *J. Microbiol. Biotechnol.*, **17**, 379–392.

76 Yabusaki, Y. (1998) *Methods Mol. Biol.*, **107**, 195–202.

77 Yun, C.H., Yim, S.K., Kim, D.H., and Ahn, T. (2006) *Curr. Drug. Metab.*, **7**, 411–429.

78 Gilardi, G. and Fantuzzi, A. (2001) *Trends Biotechnol.*, **19**, 468–476.

79 Hollmann, F., Hofstetter, K., and Schmid, A. (2006) *Trends Biotechnol.*, **24**, 163–171.

80 van Beilen, J.B., Duetz, W.A., Schmid, A., and Witholt, B. (2003) *Trends Biotechnol.*, **21**, 170–177.

81 Cirino, P.C. and Arnold, F.H. (2002) *Curr. Opin. Chem. Biol.*, **6**, 130–135.

82 Gillam, E.M. (2007) *Arch. Biochem. Biophys*, **464**, 176–186.

83 Gillam, E.M. (2008) *Chem. Res. Toxicol.*, **21**, 220–231.

84 Miles, C.S., Ost, T.W., Noble, M.A., Munro, A.W., and Chapman, S.K. (2000) *Biochim. Biophys. Acta*, **1543**, 383–407.

85 Tee, K.L., and Schwaneberg, U. (2007) *Comb. Chem. High Throughput Screen.*, **10**, 197–217.

86 Lamb, D.C., Waterman, M.R., Kelly, S.L., and Guengerich, F.P. (2007) *Curr. Opin. Biotechnol.*, **18**, 504–512.

87 Petzoldt, K., Annen, K., Laurent, H., and Wiechert, R. (1982) Schering Aktiengesellschaft (Berlin, Germany), Germany.

88 Peterson, D.H., Murray, H.C., Eppstein, S.H., Reineke, L.M., Weintraub, A., Meister, P.D., and Leigh, H.M. (1952) *J. Am. Chem. Soc.*, **74**, 5933–5936.

89 Hogg, J.A. (1992) *Steroids*, **57**, 593–616.

90 Serizawa, N. (1997) *J. Synth. Org. Chem. Jpn.*, **55**, 334–338.

91 Serizawa, N., Nakagawa, K., Hamano, K., Tsujita, Y., Terahara, A., and Kuwano, H. (1983) *J. Antibiot. (Tokyo)*, **36**, 604–607.

92 Park, J.W., Lee, J.K., Kwon, T.J., Yi, D.H., Kim, Y.J., Moon, S.H., Suh, H.H., Kang, S.M., and Park, Y.I. (2003) *Biotechnol. Lett.*, **25**, 1827–1831.

93 Chun, Y.J., Shimada, T., Sanchez-Ponce, R., Martin, M.V., Lei, L., Zhao, B., Kelly, S.L., Waterman, M.R., Lamb, D.C., and Guengerich, F.P. (2007) *J. Biol. Chem.*, **282**, 17486–17500.

94 Lamb, D.C., Skaug, T., Song, H.L., Jackson, C.J., Podust, L.M., Waterman, M.R., Kell, D.B., Kelly, D.E., and Kelly, S.L. (2002) *J. Biol. Chem.*, **277**, 24000–24005.

95 Lamb, D.C., Ikeda, H., Nelson, D.R., Ishikawa, J., Skaug, T., Jackson, C., Omura, S., Waterman, M.R., and Kelly, S.L. (2003) *Biochem. Biophys. Res. Commun.*, **307**, 610–619.

96 Parajuli, N., Basnet, D.B., Chan Lee, H., Sohng, J.K., and Liou, K. (2004) *Arch. Biochem. Biophys.*, **425**, 233–241.

97 Machida, K., Arisawa, A., Takeda, S., Tsuchida, T., Aritoku, Y., Yoshida, M., and Ikeda, H. (2008) *Biosci. Biotechnol. Biochem.*, **72**, 2946–2952.

98 Machida, K., Aritoku, Y., and Tsuchida, T. (2009) *J. Biosci. Bioeng.*, **107**, 596–598.

99 Martin, J.F. and Aparicio, J.F. (2009) *Methods Enzymol.*, **459**, 215–242.

100 Park, S.R., Park, J.W., Jung, W.S., Han, A.R., Ban, Y.H., Kim, E.J., Sohng, J.K., Sim, S.J., and Yoon, Y.J. (2008) *Appl. Microbiol. Biotechnol.*, **81**, 109–117.

101 Lin, X. and Cane, D.E. (2009) *J. Am. Chem. Soc.*, **131**, 6332–6333.

102 Zhao, B., Lin, X., Lei, L., Lamb, D.C., Kelly, S.L., Waterman, M.R., and Cane, D.E. (2008) *J. Biol. Chem.*, **283**, 8183–8189.

103 Lamb, D.C., Guengerich, F.P., Kelly, S.L., and Waterman, M.R. (2006) *Expert. Opin. Drug. Metab. Toxicol.*, **2**, 27–40.

104 Zhao, B. and Waterman, M.R. (2007) *Drug. Metab. Rev.*, **39**, 343–352.

105 Sherman, D.H., Li, S., Yermalitskaya, L.V., Kim, Y., Smith, J.A., Waterman, M.R., and Podust, L.M. (2006) *J. Biol. Chem.*, **281**, 26289–26297.

106 Makino, M., Sugimoto, H., Shiro, Y., Asamizu, S., Onaka, H., and Nagano, S. (2007) *Proc. Natl. Acad. Sci. USA*, **104**, 11591–11596.

107 Sugimoto, H., Shinkyo, R., Hayashi, K., Yoneda, S., Yamada, M., Kamakura, M., Ikushiro, S., Shiro, Y., and Sakaki, T. (2008) *Biochemistry*, **47**, 4017–4027.

108 Xu, L.H., Fushinobu, S., Ikeda, H., Wakagi, T., and Shoun, H. (2009) *J. Bacteriol.*, **191**, 1211–1219.

109 Stegmann, E., Pelzer, S., Bischoff, D., Puk, O., Stockert, S., Butz, D., Zerbe, K., Robinson, J., Sussmuth, R.D., and Wohlleben, W. (2006) *J. Biotechnol.*, **124**, 640–653.

110 Andersen, J.F., Tatsuta, K., Gunji, H., Ishiyama, T., and Hutchinson, C.R. (1993) *Biochemistry*, **32**, 1905–1913.

111 Pylypenko, O., Vitali, F., Zerbe, K., Robinson, J.A., and Schlichting, I. (2003) *J. Biol. Chem.*, **278**, 46727–46733.

112 Zerbe, K., Pylypenko, O., Vitali, F., Zhang, W., Rouset, S., Heck, M., Vrijbloed, J.W., Bischoff, D., Bister, B., Sussmuth, R.D., Pelzer, S., Wohlleben, W., Robinson, J.A., and Schlichting, I. (2002) *J. Biol. Chem.*, **277**, 47476–47485.

113 Nagano, S., Cupp-Vickery, J.R., and Poulos, T.L. (2005) *J. Biol. Chem.*, **280**, 22102–22107.

114 Julsing, M.K., Cornelissen, S., Buhler, B., and Schmid, A. (2008) *Curr. Opin. Chem. Biol.*, **12**, 177–186.

115 Arisawa, A. and Agematu, H. (2007) *Modern Biooxidation*, 1st edn (eds R.D. Schmid and V.B. Urlacher), Wiley-VCH, Weinheim, pp. 177–192.

116 Agematu, H., Matsumoto, N., Fujii, Y., Kabumoto, H., Doi, S., Machida, K., Ishikawa, J., and Arisawa, A. (2006) *Biosci. Biotechnol. Biochem.*, **70**, 307–311.

117 van Beilen, J.B., Holtackers, R., Luscher, D., Bauer, U., Witholt, B., and Duetz, W.A. (2005) *Appl. Environ. Microbiol.*, **71**, 1737–1744.

118 Peters, F.T., Meyer, M.R., Theobald, D.S., and Maurer, H.H. (2008) *Drug. Metab. Dispos.*, **36**, 163–168.

119 Yamanaka, H., Nakajima, M., Fukami, T., Sakai, H., Nakamura, A., Katoh, M., Takamiya, M., Aoki, Y., and Yokoi, T. (2005) *Drug. Metab. Dispos.*, **33**, 1811–1818.

120 Miners, J.O. (2002) *Clin. Exp. Pharmacol. Physiol.*, **29**, 1040–1044.

121 Gillam, E.M., Aguinaldo, A.M., Notley, L.M., Kim, D., Mundkowski, R.G., Volkov, A.A., Arnold, F.H., Soucek, P., DeVoss, J.J., and Guengerich, F.P. (1999) *Biochem. Biophys. Res. Commun.*, **265**, 469–472.

122 Pritchard, M.P., McLaughlin, L., and Friedberg, T. (2006) *Methods. Mol. Biol.*, **320**, 19–29.

123 Hanioka, N., Okumura, Y., Saito, Y., Hichiya, H., Soyama, A., Saito, K., Ueno, K., Sawada, J., and Narimatsu, S. (2006) *Biochem. Pharmacol.*, **71**, 1386–1395.

124 Hanioka, N., Tsuneto, Y., Saito, Y., Sumada, T., Maekawa, K., Saito, K., Sawada, J., and Narimatsu, S. (2007) *Xenobiotica*, **37**, 342–355.

125 Nthangeni, M.B., Urban, P., Pompon, D., Smit, M.S., and Nicaud, J.M. (2004) *Yeast*, **21**, 583–592.

126 Dietrich, M., Grundmann, L., Kurr, K., Valinotto, L., Saussele, T., Schmid, R.D., and Lange, S. (2005) *Chembiochem*, **6**, 2014–2022.

127 Kolar, N.W., Swart, A.C., Mason, J.I., and Swart, P. (2007) *J. Biotechnol.*, **129**, 635–644.

128 Breinholt, V.M., Rasmussen, S.E., Brosen, K., and Friedberg, T.H. (2003) *Pharmacol. Toxicol.*, **93**, 14–22.

129 Miyazaki, M., Nakamura, K., Fujita, Y., Guengerich, F.P., Horiuchi, R., and Yamamoto, K. (2008) *Drug. Metab. Dispos.*, **36**, 2287–2291.

130 Vail, R.B., Homann, M.J., Hanna, I., and Zaks, A. (2005) *J. Ind. Microbiol. Biotechnol.*, **32**, 67–74.

131 Gold, L. (1990) *Methods Enzymol.*, **185**, 11–37.

132 Barnes, H.J., Arlotto, M.P., and Waterman, M.R. (1991) *Proc. Natl. Acad. Sci. USA*, **88**, 5597–5601.

133 Chen, G.F. and Inouye, M. (1990) *Nucleic. Acids. Res.*, **18**, 1465–1473.

134 Winters, D.K. and Cederbaum, A.I. (1992) *Biochim. Biophys. Acta*, **1156**, 43–49.

135 Nishimoto, M., Clark, J.E., and Masters, B.S. (1993) *Biochemistry*, **32**, 8863–8870.

136 Andersen, J.F., Utermohlen, J.G., and Feyereisen, R. (1994) *Biochemistry*, **33**, 2171–2177.

137 Guo, Z., Gillam, E.M., Ohmori, S., Tukey, R.H., and Guengerich, F.P. (1994) *Arch. Biochem. Biophys*, **312**, 436–446.

138 Stark, K., Wu, Z.L., Bartleson, C.J., and Guengerich, F.P. (2008) *Drug. Metab. Dispos*, **36**, 1930–1937.

139 Stark, K., Dostalek, M., and Guengerich, F.P. (2008) *FEBS J.*, **275**, 3706–3717.
140 Gillam, E.M., Guo, Z., and Guengerich, F.P. (1994) *Arch. Biochem. Biophys.*, **312**, 59–66.
141 Gillam, E.M., Guo, Z., Martin, M.V., Jenkins, C.M., and Guengerich, F.P. (1995) *Arch. Biochem. Biophys.*, **319**, 540–550.
142 Sandhu, P., Baba, T., and Guengerich, F.P. (1993) *Arch. Biochem. Biophys.*, **306**, 443–450.
143 Hanlon, S.P., Friedberg, T., Wolf, C.R., Ghisalba, O., and Kittelmann, M. (2007) *Modern biooxidation* (eds R.D. Schmid and V.B. Urlacher), Wiley-VCH, Weinheim, pp. 233–252.
144 Kumar, S., Chen, C.S., Waxman, D.J., and Halpert, J.R. (2005) *J. Biol. Chem.*, **280**, 19569–19575.
145 Kumar, S., Scott, E.E., Liu, H., and Halpert, J.R. (2003) *J. Biol. Chem.*, **278**, 17178–17184.
146 Kim, D. and Guengerich, F.P. (2004) *Arch. Biochem. Biophys.*, **432**, 102–108.
147 Huang, W., Johnston, W.A., Hayes, M.A., De Voss, J.J., and Gillam, E.M. (2007) *Arch. Biochem. Biophys.*, **467**, 193–205.
148 Johnston, W.A., Huang, W., De Voss, J.J., Hayes, M.A., and Gillam, E.M. (2007) *Drug. Metab. Dispos.*, **35**, 2177–2185.
149 Wu, Z.L., Podust, L.M., and Guengerich, F.P. (2005) *J. Biol. Chem.*, **280**, 41090–41100.
150 Hlavica, P. (2009) *Biotechnol. Adv.*, **27**, 103–121.
151 Sadeghi, S.J., Meharenna, Y.T., Fantuzzi, A., Valetti, F., and Gilardi, G. (2000) *Faraday Discuss.*, **116**, 135–153; discussion 171–190.
152 Deeni, Y.Y., Paine, M.J., Ayrton, A.D., Clarke, S.E., Chenery, R., and Wolf, C.R. (2001) *Arch. Biochem. Biophys.*, **396**, 16–24.
153 Chun, Y.J., Shimada, T., and Guengerich, F.P. (1996) *Arch. Biochem. Biophys.*, **330**, 48–58.
154 Dodhia, V.R., Fantuzzi, A., and Gilardi, G. (2006) *J. Biol. Inorg. Chem.*, **11**, 903–916.
155 Shet, M.S., Fisher, C.W., Holmans, P.L., and Estabrook, R.W. (1993) *Proc. Natl. Acad. Sci. USA*, **90**, 11748–11752.
156 Hayashi, K., Sakaki, T., Kominami, S., Inouye, K., and Yabusaki, Y. (2000) *Arch. Biochem. Biophys.*, **381**, 164–170.
157 Inui, H., Maeda, A., and Ohkawa, H. (2007) *Biochemistry*, **46**, 10213–10221.
158 Wiseman, A. (2003) *Lett. Appl. Microbiol.*, **37**, 264–267.
159 Celik, A., Flitsch, S.L., and Turner, N.J. (2005) *Org. Biomol. Chem.*, **3**, 2930–2934.
160 Fraatz, M.A., Berger, R.G., and Zorn, H. (2009) *Appl. Microbiol. Biotechnol.*, **83**, 35–41.
161 Haring, H.G., Boelens, H., Vanderge, A., and Rijkens, F. (1972) *J. Agr. Food Chem.*, **20**, 1018–1021.
162 Girhard, M., Machida, K., Itoh, M., Schmid, R.D., Arisawa, A., and Urlacher, V.B. (2009) *Microb. Cell Fact.*, **8**, 36.
163 Kim, D. and Ortiz de Montellano, P.R. (2009) *Biotechnol. Lett.*, **9**, 1427–1431.
164 Faber, K. (2004) *Biotransformations in Organic Chemistry*, 5th edn, Springer-Verlag, Berlin.
165 Faulkner, K.M., Shet, M.S., Fisher, C.W., and Estabrook, R.W. (1995) *Proc. Natl. Acad. Sci. USA*, **92**, 7705–7709.
166 Schwaneberg, U., Appel, D., Schmitt, J., and Schmid, R.D. (2000) *J. Biotechnol.*, **84**, 249–257.
167 Li, S. and Wackett, L.P. (1993) *Biochemistry*, **32**, 9355–9361.
168 Fang, X. and Halpert, J.R. (1996) *Drug. Metab. Dispos.*, **24**, 1282–1285.
169 Gutteridge, J.M. (1989) *Acta Paediatr. Scand. Suppl.*, **361**, 78–85.
170 Gutteridge, J.M. (1990) *Free Radic. Res. Commun.*, **9**, 119–125.
171 Cirino, P.C. and Arnold, F.H. (2003) *Angew. Chem. Int. Ed.*, **42**, 3299–3301.
172 Shumyantseva, V.V., Bulko, T.V., and Archakov, A.I. (2005) *J. Inorg. Biochem.*, **99**, 1051–1063.
173 Schwaneberg, U., Otey, C., Cirino, P.C., Farinas, E., and Arnold, F.H. (2001) *J. Biomol. Screen.*, **6**, 111–117.
174 Maurer, S.C., Schulze, H., Schmid, R.D., and Urlacher, V. (2003) *Adv. Synth. Catal.*, **345**, 802–810.
175 Falck, J.R., Reddy, Y.K., Haines, D.C., Reddy, K.M., Krishna, U.M., Graham, S., Murry, B., and Peterson, J.A. (2001) *Tetrahedron Lett.*, **42**, 4131–4133.

176 Kubo, T., Peters, M.W., Meinhold, P., and Arnold, F.H. (2006) *Chemistry*, **12**, 1216–1220.

177 Dohr, O., Paine, M.J., Friedberg, T., Roberts, G.C., and Wolf, C.R. (2001) *Proc. Natl. Acad. Sci. USA*, **98**, 81–86.

178 Neeli, R., Roitel, O., Scrutton, N.S., and Munro, A.W. (2005) *J. Biol. Chem.*, **280**, 17634–17644.

179 Maurer, S.C., Kuhnel, K., Kaysser, L.A., Eiben, S., Schmid, R.D., and Urlacher, V.B. (2005) *Adv. Synth. Catal.*, **347**, 1090–1098.

180 Buhler, B., Park, J.B., Blank, L.M., and Schmid, A. (2008) *Appl. Environ. Microbiol.*, **74**, 1436–1446.

181 Mouri, T., Michizoe, J., Ichinose, H., Kamiya, N., and Goto, M. (2006) *Appl. Microbiol. Biotechnol.*, **2**, 514–520.

182 Schewe, H., Kaup, B.A., and Schrader, J. (2008) *Appl. Microbiol. Biotechnol.*, **78**, 55–65.

183 Lentz, O., Li, Q.S., Schwaneberg, U., Lutz-Wahl, S., Fischer, P., and Schmid, R.D. (2001) *J. Mol. Cat. B*, **15**, 123–133.

184 Schewe, H., Holtmann, D., and Schrader, J. (2009) *Appl. Microbiol. Biotechnol.*, **83**, 849–857.

185 Sibbesen, O., De Voss, J.J., and Montellano, P.R. (1996) *J. Biol. Chem.*, **271**, 22462–22469.

186 Hirakawa, H., Kamiya, N., Tanaka, T., and Nagamune, T. (2007) *Protein. Eng. Des. Sel.*, **20**, 453–459.

187 Roberts, G.A., Celik, A., Hunter, D.J., Ost, T.W., White, J.H., Chapman, S.K., Turner, N.J., and Flitsch, S.L. (2003) *J. Biol. Chem.*, **278**, 48914–48920.

188 Nodate, M., Kubota, M., and Misawa, N. (2005) *Appl. Microbiol. Biotechnol.*, **4**, 1–8.

189 Gunsalus, I.C. and Wagner, G.C. (1978) *Methods Enzymol.*, **52**, 166–188.

190 Nickerson, D.P., Harford-Cross, C.F., Fulcher, S.R., and Wong, L.L. (1997) *FEBS Lett.*, **405**, 153–156.

191 Stevenson, J.A., Bearpark, J.K., and Wong, L.L. (1998) *New. J. Chem.*, **22**, 551–552.

192 England, P.A., Harford-Cross, C.F., Stevenson, J.A., Rouch, D.A., and Wong, L.L. (1998) *FEBS Lett.*, **424**, 271–274.

193 Harford-Cross, C.F., Carmichael, A.B., Allan, F.K., England, P.A., Rouch, D.A., and Wong, L.L. (2000) *Protein Eng.*, **13**, 121–128.

194 Grayson, D.A., Tewari, Y.B., Mayhew, M.P., Vilker, V.L., and Goldberg, R.N. (1996) *Arch. Biochem. Biophys.*, **332**, 239–247.

195 Celik, A., Speight, R.E., and Turner, N.J. (2005) *Chem. Commun. (Camb.)*, 3652–3644.

196 Chen, X., Christopher, A., Jones, J.P., Bell, S.G., Guo, Q., Xu, F., Rao, Z., and Wong, L.L. (2002) *J. Biol. Chem.*, **277**, 37519–37526.

197 Jones, J.P., O'Hare, E.J., and Wong, L.L. (2001) *Eur. J. Biochem.*, **268**, 1460–1467.

198 Stevenson, J.A., Westlake, A.C.G., Whittock, C., and Wong, L.L. (1996) *J. Am. Chem. Soc.*, **118**, 12846–12847.

199 Bell, S.G., Stevenson, J.A., Boyd, H.D., Campbell, S., Riddle, A.D., Orton, E.L., and Wong, L.L. (2002) *Chem. Commun. (Camb.)*, 490–491.

200 Xu, F., Bell, S.G., Lednik, J., Insley, A., Rao, Z., and Wong, L.L. (2005) *Angew. Chem. Int. Ed.*, **44**, 4029–4032.

201 Sowden, R.J., Yasmin, S., Rees, N.H., Bell, S.G., and Wong, L.L. (2005) *Org. Biomol. Chem.*, **3**, 57–64.

202 Bell, S.G., Chen, X., Xu, F., Rao, Z., and Wong, L.L. (2003) *Biochem. Soc. Trans.*, **31**, 558–562.

203 Munro, A.W., Leys, D.G., McLean, K.J., Marshall, K.R., Ost, T.W., Daff, S., Miles, C.S., Chapman, S.K., Lysek, D.A., Moser, C.C., Page, C.C., and Dutton, P.L. (2002) *Trends Biochem. Sci.*, **27**, 250–257.

204 Appel, D., Lutz-Wahl, S., Fischer, P., Schwaneberg, U., and Schmid, R.D. (2001) *J. Biotechnol.*, **88**, 167–171.

205 Budde, M., Morr, M., Schmid, R.D., and Urlacher, V.B. (2006) *Chembiochem*, **7**, 789–794.

206 Li, Q.S., Ogawa, J., Schmid, R.D., and Shimizu, S. (2001) *Appl. Environ. Microbiol.*, **67**, 5735–5739.

207 Li, Q.S., Schwaneberg, U., Fischer, P., and Schmid, R.D. (2000) *Chemistry*, **6**, 1531–1536.

208 Sulistyaningdyah, W.T., Ogawa, J., Li, Q.S., Shinkyo, R., Sakaki, T., Inouye, K.,

Schmid, R.D., and Shimizu, S. (2004) *Biotechnol. Lett.*, **26**, 1857–1860.

209 Watanabe, Y., Laschat, S., Budde, M., Affolter, O., Shimada, Y., and Urlacher, V.B. (2007) *Tetrahedron*, **63**, 9413–9422.

210 Urlacher, V.B., Makhsumkhanov, A., and Schmid, R.D. (2005) *Appl. Microbiol. Biotechnol.*, **70**, 53–59.

211 Farinas, E.T., Schwaneberg, U., Glieder, A., and Arnold, F.H. (2001) *Adv. Synth. Catal.*, **343**, 601–606.

212 Fasan, R., Chen, M.M., Crook, N.C., and Arnold, F.H. (2007) *Angew. Chem. Int. Ed.*, **46**, 8414–8418.

213 Glieder, A., Farinas, E.T., and Arnold, F.H. (2002) *Nat. Biotechnol.*, **20**, 1135–1139.

214 Meinhold, P., Peters, M.W., Chen, M.M.Y., Takahashi, K., and Arnold, F.H. (2005) *Chembiochem*, **6**, 1765–1768.

215 Peters, M.W., Meinhold, P., Glieder, A., and Arnold, F.H. (2003) *J. Am. Chem. Soc.*, **125**, 13442–13450.

216 Seifert, A. and Pleiss, J. (2009) *Proteins*, **74**, 1028–1035.

217 Seifert, A., Vomund, S., Grohmann, K., Kriening, S., Urlacher, V.B., Laschat, S., and Pleiss, J. (2009) *Chembiochem*, **10**, 853–861.

218 Yun, C.H., Kim, K.H., Kim, D.H., Jung, H.C., and Pan, J.G. (2007) *Trends Biotechnol.*, **25**, 289–298.

219 van Vugt-Lussenburg, B.M., Damsten, M.C., Maasdijk, D.M., Vermeulen, N.P., and Commandeur, J.N. (2006) *Biochem. Biophys. Res. Commun.*, **346**, 810–818.

220 Otey, C.R., Bandara, G., Lalonde, J., Takahashi, K., and Arnold, F.H. (2006) *Biotechnol. Bioeng.*, **93**, 494–499.

221 Landwehr, M., Hochrein, L., Otey, C.R., Kasrayan, A., Backvall, J.E., and Arnold, F.H. (2006) *J. Am. Chem. Soc.*, **128**, 6058–6059.

222 Damsten, M.C., van Vugt-Lussenburg, B.M., Zeldenthuis, T., de Vlieger, J.S., Commandeur, J.N., and Vermeulen, N.P. (2008) *Chem. Biol. Interact.*, **171**, 96–107.

223 Kim, D.H., Kim, K.H., Kim, D.H., Liu, K.H., Jung, H.C., Pan, J.G., and Yun, C.H. (2008) *Drug. Metab. Dispos.*, **36**, 2166–2170.

224 Damsten, M.C., de Vlieger, J.S., Niessen, W.M., Irth, H., Vermeulen, N.P., and Commandeur, J.N. (2008) *Chem. Res. Toxicol.*, **21**, 2181–2187.

225 Otey, C.R., Landwehr, M., Endelman, J.B., Hiraga, K., Bloom, J.D., and Arnold, F.H. (2006) *PLoS Biol.*, **4**, e112.

226 Li, Y., Drummond, D.A., Sawayama, A.M., Snow, C.D., Bloom, J.D., and Arnold, F.H. (2007) *Nat. Biotechnol.*, **25**, 1051–1056.

Index

a

ent-abudinol B   103f.
acetaminophen   435
acetone   388
acetone monooxygenase (ACMO)   360
4-acetoxy-2-azetidinone   261
3-acetoxy-1-cyclohexene   247
N-acetoxyphthalimide (NAPI)   199
acetylenic ketone   204
acrylonitrile-butadiene-polystyrene (ABS) polymer   17
acyclic alkanone   136
N-acylproline   261
AD-mix   2
adamantane   188ff., 245
– carboxylation   211
2-adamantanecarboxylic acid   212
1,3-adamantanediol   193
1-adamantanol   193, 212
2-adamantanone   193, 212
additive
– MTO-catalyzed epoxidation   59
adriamycin   248
aggregation-induced hydrogen bonding enhancement   122
air   9
albaflavenone   433
alcohol
– amination   252
– primary   159ff., 242
– secondary   159ff., 242, 345
alcohol oxidation   204
– aerobic   337
– aldehyde   253
– aqueous biphasic reaction   344
– biocatalytic   179
– [Cu(II)BSP]-catalyzed   171
– environmentally benign oxidant   147ff.
– gold nanoparticle   169
– hydridometal pathway   152
– ketone   253
– metal-mediated   151
– Mn-catalyzed   405
– oxoammonium based   147
– oxometal pathway   152
– oxygen   159ff.
– Pd-bathophenanthroline catalyzed   167
– Pd-neocuproin   168
– peroxometal pathway   152
– perruthenate-catalyzed   159
– Ru catalysis   242ff.
– Ru(OH)$_3$-Al$_2$O$_3$ catalyzed   161
– TEMPO   147, 337
alcohol solvent
– fluorinated   117ff., 279ff.
aldehyde
– metal-free oxidation to carboxylic acid   354
– Mn-catalyzed oxidation   405
– oxidation   353
– perruthenate-catalyzed oxidation of primary and secondary alcohols   159
aliphatic ketone monooxygenase (AKMO)   360
Aliquat® 336   153, 177, 254
alkane
– carboxylation   212
– catalytic nitration with NO$_2$ 212
– functionalization by NHPI   212
– Mn-catalyzed oxidation   405
– oxidation   188, 265f.
– oxyalkylation of alkene   222
– sulfoxidation catalyzed by vanadium   214
alkene
– biomimetic dihydroxylation   307
– chiral ketone-catalyzed asymmetric epoxidation   85

Modern Oxidation Methods. Edited by Jan-Erling Bäckvall
Copyright © 2010 WILEY-VCH Verlag GmbH & Co. KGaA, Weinheim
ISBN: 978-3-527-32320-3

– dihydroxylation   13
– *cis*-dihydroxylation   375, 396ff.
– enantioselective epoxidation   393
– epoxidation   375ff., 392, 406
– epoxidation in fluorinated alcohol solvent   123
– epoxidation with chiral ketone   90
– epoxidation with dioxygen   208
– epoxidation with ketone   97
– epoxidation with H_2O_2 123ff.
– hydroacylation   224ff.
– hydroxyacylation   224f.
– osmylation   2
– oxidation   245, 379
– oxyalkylation with alkane   222
– radical-type pathway to epoxidation   406
– steroidal   266
– terminal   77, 334
– *trans* and trisubstituted   91
– *trans*-1,2-disubstituted   68
– transition metal-catalyzed epoxidation   37ff.
– Wacker reaction oxidation   334
alkene bearing support   19
alkyl hydroperoxide   295
– terminal oxidant   295
alkylarene
– aerobic oxidation   193
2-alkyl-2-aryl cyclopentanone   95
alkylated polyethyleneimine   345
alkylpyridine   199
alkyne
– direct oxidation   204
allenes   244
allyl acetate   248
allylic alcohol
– epoxidation   331
Amberlyst A-26   158
amide   252ff.
– $RuCl_2(PPh_3)_3$-catalyzed oxidation   260
– selective oxidation   260
amination
– alcohol   252
amine   252
– *N*-demethylation of tertiary amine   256
– α-methoxylation of tertiary amine   256
– oxidation   255, 277ff., 300ff.
– *N*-oxidation   300, 306
– secondary   258
– tertiary   257, 300ff.
amine *N*-oxide   300
– tertiary   257
3-amino-4-acetoxyazetidinone   262
aminodiol   100
(aminomethyl)phosphonic acid   43

(+)-2-aminomethylpyrrolidine   64
ammonium molybdate   177
annamycin   248
anthracene
– Mars–van Krevelen-type oxygenation   338
arsine   129
arsine oxide   129
arsonic acid   130ff.
aryl alkene
– asymmetric epoxidation with chiral ketone   89
aryl alkyl sulfide
– asymmetric oxidation   412
γ-aryl-γ-butyrolactone   95
2-aryl cyclopentanone   95
ascorbic acid   391
asymmetric dihydroxylation (AD)   403f.
– catalytic   2
– Mn-tmtacn   404
asymmetric epoxidation (AE)
– alkene   85ff.
– aryl alkene with chiral ketone   89
– Fe-catalyzed   72
– *trans*-2-heptene   31
– ketone-catalyzed   85ff.
– MTO   64
– Sharpless-Katsuki protocol   39
– titanium-catalyzed   40
asymmetric oxidation
– aryl alkyl sulfide   282ff., 412
– sulfide   411
(+)-aurilol   101f.
autooxidation   333
2-azetidinone   261
3-azide-1-cyclohexene   248
2,2′-azobisisobutyronitrile (AIBN)   210
α-azohydroperoxide   301

b

Baeyer-Villiger monooxygenase (BVMO)   358ff.
– biocatalytic property   359
– regeneration system   363
– type I and type II   358
Baeyer-Villiger oxidation   95
– acid-catalyzed   139
– cytochrome P450   428
– KA oil   209
– ketone in fluorinated alcohol solvent   136ff.
Baeyer-Villiger reaction   356
– biocatalytic   358
– catalytic asymmetric   356
– chemocatalytic   357
– enantioselective enzyme   360

benzenedicarbaldehyde 217
benzeneseleninic acid 132
benzoic acid 194, 205
benzyl alcohol
– [Cu(II)BSP]-catalyzed aerobic oxidation 171
benzyl phenyl selenide 9
benzylic compound 201
benzylidenecyclopropane 95
benzyltriphenyl phosphonium peroxymonosulfate 279
(R)-BINAP 285
biocatalytic oxidation 306
biocatalytic reaction 298
biooxidation
– cytochrome P450 monooxygenase 421ff.
biotin
– biosynthesis pathway 427
– vanadyl ion 297
2,2'-bipyridine (Bpy, bipy) 170ff.
bisacetoxyiodo benzene [PhI(OAc)$_2$] 76
(S,S)-1,2-bis-^{t}butyl-1,2-ethanediol 296
bis-(picolyl)(N-methylimidazol-2-yl)amine (bpia) 374
bis-pyridyl-cyclohexane-1,2-diamine based ligand
– chiral 412
1,4-bis(9-O-quininyl)phthalazine [(QN)$_2$PHAL] 23
bis-trimethylsilyl peroxide (BTSP) 60
bistetrahydrofuran C17-C32 fragment 100f.
bleaching
– Mn-catalyzed 375
[BMIm]PF$_6$ 291ff., 308
BPMEN (N,N'-dimethyl-N,N'-bis(2-pyridylmethyl)-diaminoethane) 65f.
– [(BPMEN)Fe(CH$_3$CN)$_2$](OTf)$_2$ 67
– (BPMEN)Fe(SbF$_6$)$_2$ 30
brassinolide 428
brevitoxin B 107f.
bromoperoxidase (VBrPO)
– V-containing 299
bromotriterpene polyether 101
BSP 170
BTX (benzene, toluene, xylene) aromatics 11
bufuralol 436
buspirone 442
α,β-butenolide 223
tert-butoxymethyltoluene 217
tert-butyl benzyl ether 217
(10R,3S)-3-[10-(tert-butyldimethylsilyloxy) ethyl]azetidin-2-one 262
α-(tert-butyldioxy)alkylamine 255

4-(tert-butyldioxy)-4-alkylcyclohexadienone 263
α-tert-butyldioxyamine 256
α-t-butyldioxypyrrolidine
– TiCl$_4$-promoted reaction 260
tert-butyl hydroperoxide (TBHP) 39, 325
tert-butyl nitrite (t-BuONO) 219

c
D-(+)-camphor 440
– stereospecific hydroxylation 426
(1R)-(+)-camphor 438
ε-caprolactam (CL) 210
– precursor 210ff.
ε-caprolactone 139, 209
carbapenem 262
carbon dioxide
– supercritical 343
carbon monoxide 212
carbon radical generation 222
carbon radical-producing catalyst (CRPC) 188, 229
carbon-carbon bond-forming reaction 222, 257
carbonyl compound
– oxidation 353ff.
4-carboxybenzaldehyde (4-CBA) 198
carboxylation 212
– alkane 212
carboxylic acid 353
– metal-free oxidation of aldehyde 354
2-(carboxy)phenyl methyl sulfide 299
2-(carboxy)vinyl methyl sulfide 299
castasterone 428
catalyst
– development 86
– heterogeneous 46
– homogeneous 42
catechol
– Ru-catalyzed oxygenation 264
cerium ammonium nitrate (CAN)
– Ritter-type reaction 220f.
cetylpyridinium chloride 178
chemocatalytic oxidation 302
chemocatalytic reaction 280
Chimassorb 944 149
S,S-Chiraphos 77
chlorite 8
m-chlorobenzoic acid (MCBA) 205
chlorohydrin process 37
m-chloroperbenzoic acid (m-CPBA) 382
chloroperoxidase (CPO) 298
4-chlorostyrene 75
chromenes

– epoxidation 93
chromium
– CrO$_3$(pyridine)$_2$ 204
– CrO$_3$/TBH 204
cobalt
– Co(II) 196
– Co(II) 2-ethylhexanoate 194
– Co(II) Schiff base complex 76
– Co(III) acetate 189
– [Co(III)W$_{12}$O$_{40}$]$^{5-}$ 334
– Co(acac)$_3$ 175
– [Co$_4$(H$_2$O)$_2$P$_2$W$_{18}$O$_{68}$]$^{10-}$ 319
– Co/Mn/Br system 198
– Co(OAc)$_2$ 191ff.
– Keggin-type polyoxometalate 333
– [PCo(II)Mo$_{11}$O$_{39}$]$^{5-}$ 334
– [PCo(H$_2$O)W$_{11}$O$_{39}$]$^{5-}$ 319
– Schiff-base complex 305
cofactor substitution and regeneration 438f.
compactin 432
copper
– Cu(I) 173
– [Cu^{II}(bipyridine ligand)] complex 174
– Cu^{II}Br$_2$(Bpy)–TEMPO 174
– [Cu(II)BSP] 170
– Cu(acac)$_2$ 204
– CuCl$_2$/phenanthroline catalyst 205
– Cu(OAc)$_2$ 191
– oxidation with O$_2$ 170
Coprinus cinerem peroxidase (Cip) 299
cortisone acetate 248
cryptophycin 52 99
cumene-phenol process 202
cyanation
– oxidative 257
cyanohydrin 250
N-cyanomethyl-*N*-methylaniline
– TiCl$_4$-promoted reaction 257
3-cyanopyridine 57
4-cyanopyridine 305
cycloalkane
– nitrosation with *t*-BuONO 219
cycloalkanone 209
cyclodecanone monooxygenase (CDMO) 360
cyclododecanone monooxygenase (CDMO) 299
cyclohexane 267
cyclohexanone 136
cyclohexanone monooxygenase (CHMO) 299ff., 359f.
– *Acinetobacter* sp. (CHMO$_{Acineto}$) 359
cyclohexylbenzene 202
cyclohexylbenzene-1-hydroperoxide (CHBH) 202

cyclooctane 267
cyclooctene 66
– epoxidation 27, 45
cis-cyclooctene
– Mn-porphyrin catalyzed epoxidation 48
Z-cyclooctene 123
cyclopentadecanone monooxygenase (CPDMO) 360
cyclopentanone monooxygenase (CPMO) 299, 360
cyclophosphamide 436
cytochrome P450 monooxygenase 332, 421ff.
– artificial fusion protein 440
– *Bacillus megaterium* 442
– biocatalysis 438
– biooxidation 421ff.
– catalytic cycle 424
– catalyzed reaction 425
– cofactor substitution and regeneration 438
– CYP1A2 435
– CYP102A1 425
– CYP102A2 425, 443
– CYP102A3 425, 443
– CYP107A1 433
– CYP107H1 427
– CYP170A1 433
– CYP2C9 435
– CYP3A4 435
– CYP51 427
– CYP505A1 425
– engineering 430ff.
– enzymology 423
– functional multi-domain protein 436
– fusion protein 436
– industrial application 430
– industrial biocatalyst 429
– mammalian 434
– pharmaceutical synthesis 431
– *Pseudomonas putida* 440
– recombinant expression in *Escherichia coli* 435
– selectivity 436
– structure 422f.
– substrate specificity 440
– synthesis of fine chemicals 437
– technical application 429

d

DBAD (dibutylazodicarboxylate) 171f.
DEAD-H$_2$ (diethylazo dicarboxylate) 172
2-decanone 136
1-decene 134
(*E*)-5-decene 241
trans-5-decene 66ff.

cis-dec-4-enoic acid
– epoxidation with ketone   96
4-demethoxyadriamycinone   248
demethylation   427
N-demethylation
– tertiary amine   256
O-demethylation
– 7-methoxyresorufin   436
– 5-methoxytryptamine   428
11-deoxycortisol   431
1-deoxy-5-hydroxy-sphingolipid analog   100
dextromethorphan   436
diacetoxyiodosylbenzene   249
3β, 21-diacetoxy-5α-pregn-17-ene   248
dialkoxyruthenium(IV) complex   161
dialkoxyruthenium(IV)-tmp complex   160
1,2-diaminocyclohexane   66
dibenzoyl peroxide (BPO)   227
1,3-di-tert-butoxymethylbenzene   217
di-tert-butyl hyponitrite (TBHN)   227
di-n-butylphenylarsine   129f.
2,6-dichloropyridine N-oxide (2,6-DCPNO)   74, 193
diclofenac   435
cis-dienes   96
– epoxidation with ketone   96
(R,R)-diethyl tartrate ((R,R)-DET)   296
(S,S)-diethyl tartrate ((S,S)-DET)   297
difluoro flavin   291
1,2-dihaloalkene   243
1,3-dihydro-2-benzofuran   217
2,3-dihydrobenzo[b]thiophene   299
1,2-dihydronaphthalene   384
1,1′-dihydroxydicyclohexyl peroxide   210
(R)-2,3-dihydroxy-2,3-dihydrosqualene   105
α,α-dihydroxyketone   244
dihydroxylation
– alkene   1ff., 307
– biomimetic   307
– metal-catalyzed   1ff.
– Os-catalyzed   2ff.
– Ru catalysis   24f.
cis-dihydroxylation
– alkene   375, 396
N,N-dihydroxypyromellitimide (NDHPI)   191
2,6-diisopropylnaphthalene   203
α-diketone   243
1,4-dimethoxymethylbenzene   218
1,3-dimethyladamantane   212ff.
N,N-dimethylamino derivative   303
N,N-dimethylaniline   428
2,2-dimethylchromene   403
dimethyldioxirane (DMD)   301

(S)-(N,N-dimethyl)-1-phenylethylamine   64
dimethyl tacn (dmtacn)   395
dimethyl terephthalate (DMT)   196
4,11-dimethyl-1,4,8,11-tetraazabicyclo[6.6.2]hexadecane (Me$_2$EBC)   386
1,3-dinitroadamantane   214
diol
– oxidation with dioxygen   206f.
dioxirane   86, 278
1,3-dioxolane   224
dioxygen
– hydroxyacylation   224
– oxidation   188, 206f.
– oxyalkylation of alkene   222
– terminal oxidant   293
dioxygenase mechanism   340
diphenylmethylamine   258
diselenide/seleninic acid   132
disulfide
– oxidation of thiol and thiophenol   138
1-dodecene   66
drug development
– mammalian cytochrome P450   434

e

early transition metal
– epoxidation of alkene   39
electron transfer mediator (ETM)   3f.
enantioselective catalyst   228
endo-oxacyclization   106
(+)-enshuol   102
cis-enyne
– epoxidation with ketone   96
epibatidine   225
epi-isozizaene   433
epothilone   99, 426
– B   426
– D   426ff.
epoxidation
– alkene   37ff., 91ff., 123ff., 208, 375ff., 392
– allylic alcohol   331
– asymmetric, see asymmetric epoxidation
– cyclooctene   27, 45
– dioxygen   208
– enantioselective   393
– ethene with H$_2$O$_2$ 128
– Fe-catalyzed   64ff.
– fluorinated alcohol solvent   123
– 1-hexene   389
– H$_2$O$_2$ 123ff., 387
– late transition metal   76
– mechanism of catalysis by fluorinated alcohol   123

- metalloporphyrin-promoted 247
- Mn-catalyzed 47
- Mn-tptn catalyst 405
- Mo-catalyzed 42
- MTO-catalyzed 56
- oxidant 38
- radical-type pathway 406
- Re-catalyzed 52
- Ru-catalyzed 74f., 243
- sandwich-type POM-catalyzed 331
- selective 38
- synthetic application 98f.
- transition metal-catalyzed 37ff., 80
- transition state 92ff.
- W-catalyzed 42
epoxide cyclization 107
epoxide rearrangement 94f.
cis- and trans-2,3-epoxyoctane 209
erythromycin 433
esomeprazol 296f.
ester
- oxidation of ketone 138
- unsaturated 223
ethane
- epoxidation with H_2O_2 128
1-ethyl-3-methylimidazolium tetrafluoroborate 62
5-ethyl-2-methylpyridine 200
2-ethylanthraquinone (EAQ)/2-ethylanthrahydroquinone (EAHQ) process 46
ethylbenzene hydroperoxide (EBHP) 39
o-ethyltoluene 194

f

FAD-containing monooxygenase (FADMO) 286ff.
flavin 4
- catalyst 287f.
- chiral 291f.
flavin adenine dinucleotide (FAD) 358
flavin hydroperoxide 3, 294
flavin mononucleotide (FMN) 358
flavoenzyme 295
flavoprotein monooxygenase 358
fluorinated alcohol 118f.
- Baeyer-Villiger oxidation of ketone 136ff.
- epoxidation of alkene 123
- mechanism of epoxidation catalysis 123
- sulfoxidation of thioether 136
fluoroketone
- catalyst 135
fluorous phase
- H_2O_2 117, 279

g

giant Pd cluster 163ff.
glabrescol 105
(+)-glisoprenin A 105f.
3-(glycidyloxy)propyltrimethoxysilane 395
glyoxylic acid methylester methyl hemiacetal (gmha) 397f.
gold
- $HAuCl_4/AgNO_3$ 294
- nanoparticle 169

h

H-bond enhanced H-bonding 119
haloperoxidase 298
heptanoic acid 207
heterocyclic additive 55
N-heterocyclic carbene-catalyzed (NHC-catalyzed) oxidative transformation 354
heterogenization
- homogeneous reaction 341
heterogeneous catalyst 46
heteropoly acid
- Keggin type 324
heteropolyoxometalate 189
hexafluoroacetone 135
hexafluoro-2-propanol 61
1,1,1,3,3,3-hexafluoro-2-propanol (HFIP) 117ff, 279
- aggregation-induced H bonding enhancement 122
- H bond donor feature 120
1-hexene
- epoxidation 389
Hock process 202
homogeneous catalyst 42
hydroacylation 224ff.
hydroaromatic compound 201
hydrocarbon
- oxidation 265f.
hydrocortisone 431
hydrogen bond
- low-barrier 126
hydrogen bond donor feature
- HFIP 120
hydrogen peroxide 3, 46
- alkene epoxidation 123
- catalytic oxidation in fluorinated alcohol solvent 117ff., 279
- catalytic oxidation of alcohol 176
- decomposition 373
- epoxidation 387
- epoxidation of ethene 128
- fluorous phase 279
- manganese catalase 373

– Mn-catalyzed oxidation 371ff.
– terminal oxidant 42, 280
α-hydroperoxyethylbenzene 201
2-hydroperoxy-2-hydroxypropane (hhpp) 388
hydroquinidine 1,4-phthalazinediyl diether (DHQD)$_2$PHAL 4ff.
hydrotalcite 169
– Ni-substituted 176
– Ru-exchanged (Ru-HT) 162
4-hydroxyacetophenone monooxygenase (HAPMO) 299, 360
hydroxyacylation 224f.
hydroxyapatite 168f.
4-hydroxybenzoate 440
ω-hydroxycaproic acid 139
α-hydroxy carbon radical 223
4-hydroxydiclofenac 435
α-hydroxy-γ,γ-dimethyl-γ-butyrolactone 223
1-hydroxy-1-hydroperoxycyclohexane 210
N-hydroxyimide 205
α-hydroxyketone 248
α-hydroxy-γ-lactone 223
hydroxylation
– D-(+)-camphor 426
– diastereoselective 437
– pyridylhomothreonine 426
– regioselective 437ff.
– stereoselective 441
N-hydroxyphthalimide (NHPI) 175, 187ff., 229, 294
– carbon-carbon bond-forming reaction 222
– chiral derivative 228
– fluorinated 190
– functionalization of alkane 212
– NHPI/CAN system 220f.
– NHPI-catalyzed aerobic oxidation 188
– NHPI/Co(II) system 205, 222
– NHPI/Co(acac)$_2$ system 193
– NHPI/Co(OAc)$_2$ system 189ff., 205
– NHPI/Cu(II) system 204
– NHPI/NO$_2$ system 214
– polarity reversal catalyst 226
α-hydroxy-γ-spirolactone 224
6β-hydroxytestosterone 435
hypochlorite 5, 148, 382
hypochlorite-procedure 7

i
idarubicin 248
ifosfamide 436
indole 260
inorganic metalloporphyrin 318
o-iodooxybenzoic acid (IBX) 298
iodosyl benzene 76, 249, 382

ionic interaction
– immobilization 21
ionic liquid 22, 254, 299ff.
– aerobic Ru-catalyzed oxidation 153
– H$_2$O$_2$-based sulfoxidation 291
– heterogeneous catalytic oxidation of sulfide 292
α-ionone 437
β-ionone 437ff.
iron
– Fe(III)-OOH intermediate 324
– FeCl$_3$ 189
– [Fe$_2$Ni(OAc)$_3$PW$_9$O$_{37}$]$^{10-}$ 333
– Fe(NO$_3$)$_3$-FeBr$_3$ 294
– (MEP)Fe(SbF$_6$)$_2$ 30
– polyfluorinated Fe(TPP)-catalyst 65
– [(TPA)Fe(CH$_3$CN)$_2$](OTf)$_2$ (TPA= trispicolylamine) 67
iron catalyst 26
– asymmetric epoxidation 72
– epoxidation 64ff.
iron complex
– chiral 31
iron porphyrin 64
iron-oxo species 321
isophthalic acid 194
isoquinoline 260

j
Jacobsen-catalyst 50
Julia-Kocienski alkenation 100

k
K, *see* potassium
KA oil 187ff., 209ff.
– Baeyer-Villiger oxidation 209
Keggin-type compound 318ff.
Keggin-type silicodecatungstate [γ-SiW$_{10}$O$_{34}$(H$_2$O)$_2$]$^{4-}$ 44
ketone 86, 267
– acetylenic 204
– ammonium 87
– N-aryl-substituted 93
– Baeyer-Villiger oxidation in fluorinated alcohol solvent 136ff.
– bicyclo[3.2.1]octan-3-ones 88
– bis(ammonium) 87
– C2-symmetric binaphthyl-based 87
– carbocyclic 86
– carbohydrate-based 88
– chiral 89
– epoxidation of alkene 97
– epoxidation of *trans* and trisubstituted alkenes 91

– fructose-derived   87
– oxidation   356
– oxidation to lactone and ester   138
ketone monooxygenase   299
ketone-catalyzed asymmetric epoxidation
– alkene   85ff.

l

β-lactam   260f.
– acyloxylation   261
– aerobic oxidation   262
lactone   108, 209
– oxidation of ketone   138
lanosterol
– demethylation   427
lanthanide
– catalyst   286
lauric acid   206
layered double hydroxide (LDH)   21, 46, 281
– W-exchanged Mg-Al   303
limonene   434ff.
(4R)-limonene   442
lipophiloselective oxidation   345
liquid-liquid reaction system   341ff.
liquid phase catalysis
– solvent-anchored supported   343
liquid phase oxidation
– polyoxometalate-catalyzed   315ff.
lithium
– LiNbMoO$_6$ 285
(−)-longilene peroxide   100f.
low-barrier hydrogen bond   126

m

magnesium
– Mg$_6$Al$_2$Ru$_{0.5}$(OH)$_{16}$CO$_3$ 162
manganese
– alkene oxidation   405
– bleaching   375
– bio-inspired oxidation catalyst   372
– manganese sulfate/H$_2$O$_2$ system   51
– [Mn^{III}$_2$O(μ-O)(μ-RCO$_2$)$_2$(tmtacn)$_2$]$^{2+}$ complex   402f.
– Mn(III) porphyrin/Pt/H$_2$ system   188
– Mn(III) tetra-2,6-dichlorophenylporphyrin   318
– Mn(III)(salen) complex   285, 410f.
– [Mn^{IV}$_2$ (μ-O)$_3$(tmtacn)$_2$]$^{2+}$ 399ff.
– Mn(acac)$_2$ 189
– Mn(Me$_2$EBC)Cl$_2$ 386
– Mn(OAc)$_2$ 200
– Mn(OAc)$_3$ 407
– Mn(OAc)$_3$·2H$_2$O   282
– [Mn$_2$O(OAc)(tmtacn)$_2$](PF$_6$)$_2$ 387
– [(Mn$_2$O(OAc)$_2$tptn)]$^{2+}$ 405f.
– [Mn$_2$O$_3$(TMTACN)]$^{2+}$ 50
– [Mn$_2$O$_3$(tmtacn)$_2$](PF$_6$)$_2$ 375, 397ff.
– Mn-porphyrin complex   379
– Mn(salen) complex   298, 382ff.
– Mn-tmtacn complex   395ff., 407ff.
– Mn-tmtacn derivative on solid suport   395
– Mn-tmtacn/oxalate system   389f.
– Mn(TPP)Cl   49
– Mn-tptn catalyst   405
– oxidation of alkane, alcohol and aldehyde   407
– ozonide complex   320
– [PMn(H$_2$O)W$_{11}$O$_{39}$]$^{5-}$ 319
– [P$_2$Mn(H$_2$O)W$_{17}$O$_{61}$]$^{7-}$ 319
– salt   376
manganese-catalyzed epoxidation   47
– porphyrin   48
manganese-catalyzed oxidation
– H$_2$O$_2$ 371ff.
manganese-substituted polyoxometalate   320
– sandwich structure   321
Mars–van Krevelen-type mechanism   338
maytenine   250
MCM-41   149, 159f.
– catalyst   46
(MEP)Fe(SbF$_6$)$_2$ 30
(−)-mesembrine   100f.
metal-mediated oxidation of alcohol   151, 174
metalloenzyme   297
metalloporphyrin   378
metalloporphyrin-promoted epoxidation   247
metallosalen compound   339
methane
– aerobic oxidation   339
methane monooxygenase (MMO)   30, 65, 332
1-(methoxycarbonyl)pyrrolidine
– Ru-catalyzed oxidation   260
α-methoxylation
– tertiary amine   256
4-methoxymethyltoluene   217
1-(4-methoxyphenyl)-1-phenylethane   203
7-methoxyresorufin   436
5-methoxytryptamine   428
N-methylamine   256
– oxidation   256
3-methyl-2-butanone   252
1-methylcyclohexane   247
methyl 3-(3,3'-dimethyladamantyl)-2-hydroxypropionate   223
methyl 3-(3,3'-dimethyladamantyl)-2-oxopropionate   223
N-methyl-N-(3-heptenyl)aniline   256
N-methylhomoallylamines   256

N-methylmorpholine (NMM) 3f., 306f.
N-methylmorpholine N-oxide (NMO) 2f., 157, 249ff., 306, 384
2-methyl-1,4-naphthaquinone 337
2-methyl-1-naphthol 337
methylpyridine 199
3-methylpyrazole 58
methylquinoline 199
3-methylquinoline 200
α-methylstyrene 10ff.
– Os-catalyzed dihydroxylation 4ff.
– Ru-catalyzed dihydroxylation 24
cis-β-methylstyrene 64, 90, 406
– epoxidation 93
trans-β-methylstyrene 72
methyl thioglycolate 227
methyltrioctylammonium chloride 177
methyltrioxorhenium (methylrhenium trioxide, MTO) 3f., 52ff., 79, 133, 178, 305, 307
– asymmetric epoxidation 64
– epoxidation catalyst 54
– oxidation catalyst 133
– solvents/media 61
metoprolol 436
microbial oxidation
– P450 for pharmaceutical synthesis 431
microencapsulated osmium tetroxide (MC OsO_4) 17ff.
molybdenum
– epoxidation 42
– $H_5PV_2Mo_{10}O_{40}$ 335ff.
– Mo(VI) peroxo complex 177
– $Ph_2PO_2[MoO(O_2)_2]_2^-$ 281
– $[PPh_4][Mo(O)(O_2)_2(QO)]$ 45
– $[PV_2Mo_{10}O_{40}]^{5-}$ 336ff.
– $[PZnMo_2W_9O_{39}]^{5-}$ polyoxometalate 328
– $Q_3PMo_{12}O_{40}$ 281, 322
– $Q_5[PV_2Mo_{10}O_{40}]$ (Q=quaternary ammonium cation) 328
mono-oxygen donor 317
– oxidation 317
MTO, see methyltrioxorhenium
myrtenol 439

n
NADH 439
ent-nakorone 103f.
nano-iron 178
nano-cluster
– Pd 168
naphthalene 441
naphthalene derivative
– oxidation 264
2,6-naphthalenediol 203

naphthoquinone 264
NADPH 299f., 359, 438
NAPI, see N-acetoxyphthalimide
NDHPI, see N,N-dihydroxypyromellitimide
Nexium® 296
NHPI, see N-hydroxyphthalimide
NHPI/CAN system 220f.
NHPI-catalyzed aerobic oxidation 188
NHPI/Co(II) system 205, 222
NHPI/Co(acac)$_2$ system 193
NHPI/Co(OAc)$_2$ system 189ff., 205
NHPI/Cu(II) system 204
NHPI/NO$_2$ system 214
nickel-substituted hydrotalcite 176
nicotinic acid 200
(+)-nigellamine A$_2$ 102
nikkomycin 426
nitration
– alkane with NO$_2$ 212
nitric oxide
– reaction with organic compound 217
nitrile
– Ru-catalyzed oxidation 266
nitroadamantane 214
nitroalkane 213
nitrogen-group donating support 16
nitrone 258
nitrosation
– cycloalkane with t-BuONO 219
NMO, see N-methylmorpholine N-oxide
2-nonanone 136
nootkatol 437f.
(+)-nootkatone 437f.
norbornane
– fused 245

o
1,2-octanediol 207
2-octanone 136
1-octene 134
cis-2-octene 209
trans-2-octene 68, 209, 406
trans-4-octene 406
4-octyn-3-ol 204
4-octyn-3-one 204
4-octyne 204
(+)-omaezakianol 102
omeprazol 297
organometallic polymer 63
osmium
– dihydroxylation 2ff.
– microencapsulated osmium tetroxide (MC OsO$_4$) 17ff.
– Os EnCat® 19

– osmium tetroxide 20
– osmium(VI) glycolate 2
– osmium(VI) mono(glycolate) 8
– osmium(VIII) mono(glycolate) 8
osmium catalyst 16ff.
– polyaniline 16
– supported 16
osmylation
– alkene 2
endo-oxacyclization 106
oxazolidine 228
– kinetic resolution 228f.
oxidant
– environmentally friendly terminal 3
– epoxidation 38
– hypochlorite 148
– terminal, *see* terminal oxidant
oxidation
– aerobic 151ff., 187ff., 253, 294
– alcohol 147ff., 204, 249, 407
– aldehyde 353, 407
– alkane 188, 265, 407
– alkene 245
– alkylarene 193, 322
– alkyne 204
– amide 260
– amine 277ff.
– Baeyer-Villiger oxidation of ketone in fluorinated alcohol solvent 136ff.
– benzylic compound 201
– biocatalytic 179, 306
– biooxidation, *see* biooxidation
– carbonyl compound 353ff.
– catalytic 117ff., 265
– chemocatalytic 302
– chiral *N*-hydroxyimide 205
– copper-catalyzed 170
– diol 206f.
– dioxygen 188, 206f., 331
– environmentally benign oxidant 147
– hydroaromatic compound 201
– hydrocarbon 265
– H_2O_2 176
– ionic liquid 153, 291
– ketone 356
– ketone to lactone and ester 138
– β-lactam 260
– lipophiloselective 345
– liquid phase 315ff.
– metal-free oxidation of aldehyde to carboxylic acid 354
– metal-mediated 151
– methane 339
– *N*-methylamine 256

– methylpyridine 199
– methylquinoline 199
– microbial 431
– Mn-catalyzed 371ff., 407
– mono-oxygen donor 317
– organocatalytic 85ff.
– oxoammonium based 147
– oxygen 159, 170ff., 188, 331
– Pd-catalyzed 163
– Pd-bathophenanthroline catalyzed 167
– peroxygen compound 323
– perruthenate-catalyzed 159
– primary amine 259
– primary and secondary alcohol to aldehyde 159ff.
– Ru-catalyzed 153, 241ff., 260
– $RuCl_3$-catalyzed 247ff.
– $RuCl_2(PPh_3)_3$-catalyzed 260
– $Ru(OH)_3$-Al_2O_3 catalyzed 161
– $Ru(TPP)(O)_2$ (TPP=5,10,15,20-tetraphenylporphyrinato) 264
– secondary amine 258f.
– sulfide 277ff., 294, 410
– tertiary amine 256f., 300ff.
– thiol and thiophenol to disulfide 138
– Ti-catalyzed oxidation 282
oxidative cyanation
– pyrrolidine derivative 257
N-oxide 300
oxidized electron transfer mediator (ETM_{ox}) 3
oxirane oxidation 278
oxoammonium based oxidation of alcohol 147
oxodiperoxomolybdenum(VI) complex 45
α-oxo-ene-diol 249
oxone® 86ff., 135, 279, 306, 324
– derivative 279
oxo-ruthenium complex 263
oxoruthenium(VI) complex 161
oxydehydrogenation 337
– vinylcylohexane 340
– vinylcylohexene 340
oxygen, *see also* dioxygen 9ff.
– α-oxygen 321
– alkane oxidation 188, 266
– carboxylation of alkane 212
– Cu-catalyzed oxidation 170ff.
– hydroxyacylation 224
– metal-calalyzed oxidation 174
– oxidation 331
– oxidation of alcohol 159ff.
– oxidation of diol 206f.
– oxyalkylation of alkene 222

- Pd-catalyzed oxidation 163
- Ru catalyzed oxidation 153
- $RuCl_3$-catalyzed oxidation 257ff.
- terminal oxidant 293
oxygenation
- anthracene 338
- catechol 264
- Mars–van Krevelen-type 338
- Ru-catalyzed 264
- xanthene 338
N-oxygenation
- N,N-dimethylaniline 428

p

P450, *see* cytochrome P450
$P450_{BioI}$ 427
$P450_{BM3}$ 425ff., 443
$P450_{cam}$ 439f.
$P450_{EpoK}$ 426
$P450_{foxy}$ 425
$P450_{NikF}$ 426
palladium
- catalyzed oxidation with O_2 163, 206
- giant cluster 163ff.
- nanocluster 168
- Pd(II)-bathophenanthroline-complex 165ff.
- Pd(II) neocuproin complex 167f.
- $PdCl_2(PhCN)_2$ 169
- Pd/DMSO system 164
- $Pd(OAc)_2$ 163, 206
- $Pd(OAc)_2[PV_xMo_{12-x}O_{40}]^{(3+x)-}$ 335
- $Pd_{561}phen_{60}(OAc)_{180}$ 163
- Pd/pyridine system 164
PEM-MC $OsO4$ 19
4-pentynoic acid 107
peptide
- oxidative modification 261
peptide Schiff-base library 284
peracetic acid 209
peracid 278, 301
perfluoroheptadecan-9-one 135
(−)-perillyl alcohol 434
peroxidase 298
- V-containing 298
peroxide shunt 438
peroxo complex
- Venturello-type 281
peroxo-cobalt(III) complex 196
peroxo-molybdate 281
peroxo-tungstate 281
1,1′-peroxydicyclohexylamine (PDHA) 210f.
peroxygen compound
- oxidation 323

perrhenic acid 134
phase transfer agent 178
phenacetin 435
phenol
- oxidation 262
phenylacetone monooxygenase (PAMO) 360
1-phenylcyclohexene 391
N-phenyl-2-(dicyclohexylphosphanyl)pyrrole 252
(R)-1-phenylethylamine 64
phenylmethylsulfoxide 322
trans-1-phenyl-3-propyl-4-chloropiperidine 256
phosphovanadomolybdate 335ff.
photosulfoxidation 215
photosystem II (PSII) 372
phthalimide N-oxyl (PNO) 193ff., 213ff.
- radical 188
β-picoline 200
γ-picoline 200
pinane 245
α-pinene 439
(+)-α-pinene 441
α-pinene oxide 439
piperidine 256
piperidinyloxyl copper(II) complex 173
PIPO (polymerimmobilized piperidinyl oxyl) 149
pladienolide
- B 100, 432
- D 432
platinum
- asymmetric sulfoxidation 285
- Pt(II) catalyst 357
- Pt(II) complex 77
- Pt(II)methylpyrimidinium-$H_5PV_2Mo_{10}O_{40}$ catalyst 339
PNN (2-(di-*t*-butylphosphinomethyl)-6-(diethylaminomethyl)pyridine) 252
polarity reversal catalyst
- NHPI 226
polyene-polyepoxide-polycyclization 107
polyepoxide 107
polyepoxide cyclization 106
polyfluorooxometalate 329
polyketide antitumor macrolide 432
polymer-supported perruthenate (PSP) 158, 254
polymethoxybenzene
- oxidation 264
polyoxometalate (POM) 281, 316ff., 327ff.
- epoxidation 331
- Keggin-type 44, 327
- lanthanide-containing 327

- lipophiloselective reaction 346
- liquid phase oxidation 315ff.
- Ru-containing 247, 267
- sandwich-type 318ff., 330f.
- sulfoxide complex 323
- V-substituted sandwich type 325
- Wells-Dawson-type 318
polyoxomolybdate 324
- Keggin type 323
- V-substituted 321
polyoxotungstate 189, 329
polyurea microencapsulated OsO4 19
poly(4-vinylpyridine) 63
poly(4-vinylpyridine) N-oxide 63
porphyrin
- catalysis 378
- epoxidation of cis-cyclooctene 48
potassium
- K_2CO_3 171f.
- $K_3[Fe(CN)_6]$ 2ff.
- $K_2[OsO_2(OH)_4]$ 2
- K_2RuO_4 249ff.
pravastatin 432
progesterone 436
L-prolineamide 64
proposed glabrescol 103
propyl 3,3-bis(1-methylimidazol-1-yl) propionate 67
pyrene 441
pyridazinones 225
pyridine-2,6-bisoxazoline 75
pyridine-2,6-dicarboxylic acid 68ff.
pyridinecarbonitrile 199
pyridinecarboxamide 199
pyridinecarboxylic acid 199
pyridyl ligand 405
pyridylhomothreonine
- hydroxylation 426
pyrrolidine 68

q

$Q_3[PMo_{12}O_{40}]$ 281, 322
quinoline 257
quinolinecarboxylic acid 200
8-quinolinol (QOH) 45

r

revised glabrescol 105
rhenium
- compound 133
- epoxidation 52
- $[Ho_{0.5}(CH_3)_{0.92}ReO_3]$ 63
rhenium oxide 61
ring decoration 433

Ritter-type reaction 220f.
- cerium ammonium nitrate (CAN) 220f.
ruthenium
- amide 260
- catalyst 23ff.
- $[Cn^*Ru(CF_3CO_2)_3(H_2O)]$ ($Cn^* =N,N',N''$-trimethyl-1,4,7-triazacyclononane) 250
- dihydroxylation of α-methylstyrene 24
- dioxoruthenium(IV) complex 243
- epoxidation 74, 243
- K_2RuO_4 249ff.
- low-valent 245, 266
- mechanism for Ru/TEMPO-catalyzed oxidation of alcohol 158
- cis-$[(Me_3tacn)-(CF_3CO_2)Ru^{VI}O_2]$ CF_3CO_2 26
- $Mg_6Al_2Ru_{0.5}(OH)_{16}CO_3$ 162
- oxidation for organic synthesis 241ff.
- oxidation with O_2 153ff.
- perfluorinated 267
- polyoxometalate 247, 267
- $(n-Pr_4N)(RuO_4)$ (TPAP) 249ff.
- $Ru^{III}(H_2O)SiW_{11}O_{39}]^{5-}$ 320
- Ru-Al-Mg-hydrotalcite 254
- $Ru-Al_2O_3$ 254
- Ru/AlO(OH) 254
- Ru/benzoquinone system 156
- cis-$[Ru(6,6-Cl_2bpy)_2(OH_2)_2]^{2+}$/t-BuOOH 265
- $RuCl_3$ with H_2O_2 249
- $RuCl_3$-catalyzed oxidation 247
- $[RuCl(dpp)_2]^+$ 265
- $RuCl_3·3H_2O$/didecyldimethylammonium bromide 177
- $[RuCl(Me_2SO)(TPA)]^+$ 265
- $RuCl_2(PPh_3)_3$ 249ff., 260ff.
- $RuCl_2(PPh_3)_3$-TEMPO 254
- $[RuCl_2(TPA)]^+$ 265
- $Ru_3(CO)_{12}$ 252
- $Ru(CO)_3(\eta^4$-tetracyclone) 251
- Ru-Co-Al-hydrotalcite 254
- Ru-Co-HT 162
- $[Ru(p-cymene)Cl_2]_2$ 252
- cis-$[Ru(dmp)_2(MeCN)_2]^{2+}$/H_2O_2 265
- $[Ru(dmp)_2(CH_3CN)_2](PF6)$ (dmp=2,9-dimethyl-1,10-phenanthroline) 247
- $RuH(PPh_3)_4$ 252
- RuHCl(PNN)(CO) (PNN=2-(di-t-butylphosphinomethyl)-6-(diethylaminomethyl)pyridine) 251
- Ru-hydroxyapatite (RuHAP) 254ff.
- cis-$[Ru(Me_3tacn)(O)_2(CF_3CO_2)]^+$/t-BuOOH ($Me_3tacn=N,N',N''$-1,4,7-trimethyl-1,4,7-triazacyclononane) 265

- $[Ru_3O(OCOCF_2CF_2CF_3)_6(Et_2O)_3]^+$ 267
- RuO_4-promoted oxidation 241
- $[Ru(OAc)_3(CO_3)]$ 254
- $Ru_2(OAc)_4Cl$ 259
- $Ru(OEP)(PPh_3)_3$ 265
- $Ru(OH)_2/Al_2O_3$ 264
- $Ru(OH)_3$-Al_2O_3 161
- $Ru(OH)_x/Al_2O_3$ 267
- $Ru(PMe_3)_2Cl_2(eda)$ (eda=ethylenediamine) 251
- $Ru(PPh_3)_3(CO)H_2$ 251
- Ru-polyoxometalate 340
- Ru-porphyrin complex 208, 245f., 259ff.
- Ru(pybox)(Pydic) complex 249
- Ru/quinone system 156
- Ru-substituted polyoxometalate 208
- Ru/TEMPO system 156
- $Ru(TMP)Cl_2$-catalyzed oxidation 246
- $Ru(TMP)(O)_2$ 259ff.
- Ru(TPFPP)(CO) (TPFPP=tetrakis (pentafluorophenyl)porphyrinato) 265ff.
- $Ru(TPP)(O)_2$ (TPP=5,10,15,20-tetraphenylporphyrinato) 264
- Shvo Ru-catalyst 154
- TEMPO 155
ruthenium glycolate 24
ruthenium oxide 74
ruthenium-exchanged hydrotalcite (Ru-HT) 162
- $Mg_6Al_2Ru_{0.5}(OH)_{16}CO_3$ 162

s

salen-based system 381ff.
sandwich-type polyoxometalate 318
scandium triflate $(Sc(OTf)_3)$ 286
seleninic acid 132ff.
selenium
- SeO_2/TBHP system 204
sesquiterpene 433
sexipyridine ligand 73
Sharpless-Katsuki protocol
- asymmetric epoxidation (AE) 39
Shvo Ru-catalyst 154
silica gelsupported cinchona alkaloid [SGS-$(DHQD)_2PHAL$]-OsO_4 22
silica–PEG/PPG–$H_5PV_2Mo_{10}O_{40}$ 343
silicodecatungstate $[\gamma\text{-}SiW_{10}O_{34}(H_2O)_2]^{4-}$ 44
sodium
- Na_2CrO_4/acetic anhydride 204
sodium percarbonate (SPC) 60
solid support 63
solid-liquid reaction 341
solvent 61

solvent-anchored supported liquid phase catalysis 343
steroid monooxygenase (SMO) 299, 360
stilbene 391
trans-stilbene 24
stoichiometric reaction 278, 301
streptavidin
- vanadyl-loaded 297
styrene 25, 393
- epoxidation 93
- oxidation 388
sulfide 409ff.
- aerobic oxidation 294
- oxidation to sulfone 280
- oxidation to sulfoxide 277ff., 410
sulfone 280, 409
2-sulfonyloxaziridine 301
sulfoxidation 137, 277ff.
- alkane catalyzed by vanadium 214f.
- asymmetric 285
- enantioselective 285
- H_2O_2-based 287ff.
- ionic liquid 291
- Pt-catalyzed 285
- thioether in fluorinated alcohol solvent 136
- V-catalyzed 214f., 285
sulfoxide 283, 322, 409f.
supercritical carbon dioxide 343
superoxide dismutase (SOD) enzyme 372
superoxocobalt(III) complex 196
support
- bearing alkene 19
- nitrogen-group donating 16
- solid 63

t

TADOOH 326
TBHP 76
TEMPH 157
TEMPO (2,2′,6,6′-tetramethylpiperidine-N-oxyl radical) 147ff., 165ff., 337, 354
TEMPOH (2,2′,6,6′-tetramethylpiperidine-N-oxyl hydroxylamine) 156, 173
terephthalic acid (TPA) 194ff.
terminal oxidant 3, 41f., 62, 80
- alkyl hydroperoxide 295
- dioxygen 208
- environmentally friendly 3
- H_2O_2 42, 280
- molecular oxygen 293
- sodium percarbonate (SPC) 60
terpyridine 74
testosterone 435
tetraazacycloalkane derivative 385

α,α,α′,α′-tetrabromoxylene 218
tetra-*n*-butylammoniumperruthenate
  (TBAP) 157
tetrahydrofuran lactone 103
tetrahydroquinoline 258
meso-tetrakismesitylporphyrin dianion (tmp)
  160
N,N,N′,N″-tetrakis(2-pyridylmethyl)-1,2-
  ethylenediamine (tpen) 406
N,N,N′,N″-tetrakis(2-pyridylmethyl)-1,3-
  propanediamine (tptn) 405
tetralin 441
tetramethylammoniumhydroxide 153
7,8,15,16-tetraoxadispiro[5.2.5.2]
  hexadecane 139
tetraphenylporphyrin (TPP) 49, 264
tetra-*n*-propylammoniumperruthenate
  (TPAP) 157ff.
thioether 285
– sulfoxidation in fluorinated alcohol
  solvent 136
thiol
– oxidation to disulfide 138
thiophene S-oxide 428
thiophenol
– oxidation to disulfide 138
ticlodipine 428
titanium
– asymmetric epoxidation 40
– di-μ-oxo titanium complex 410
– oxidation 282
– oxo complex 282
– TiCl$_4$-promoted reaction of N-cyanomethyl-
  N-methylaniline 257
– Ti(O^{i}Pr)$_4$ 296f.
titanium silicalite TS-1 178
titanium(IV)-silicate catalyst (TS-1) 39, 79
TMSOOTMS 249
para-toluenesulfonic acid 136
p-toluic acid (PTA) 194ff.
o-toluic acid 194
TPA
– *see* terephthalic acid
– *see* trispicolylamine
– *see* tris(2-pyridylmethyl)amine
[(TPA)Fe(CH$_3$CN)$_2$](OTf)$_2$ (TPA=
  trispicolylamine) 67
TPAP ((*n*-Pr$_4$N)(RuO$_4$)) 249ff.
TPP (5,10,15,20-
  tetraphenylporphyrinato) 49, 264
transition metal
– catalyst 280
transition metal-catalyzed epoxidation
– alkene 37ff.

transition state
– epoxidation 98
– spiro and planar 92
N,N,N-1,3,5-trialkylated flavin 289ff., 302
N,N,N,-3,5,10-trialkylated flavin 286ff., 304
tri-azacycloalkane derivative 385
1,4,7-triazacyclononane (TACN) 50, 374
– N-alkylated (TMTACN) 50
tricaprylylmethylammonium chloride 153
endo-tricyclo[5.2.1.0]decane 212
2,2,2-trifluoroethanol (TFE) 117ff., 134
trifluorotoluene (TFT) 190
1,4,7-trimethyl-1,4,7-triazacyclononane
  (tmtacn) 374f., 405
– chiral tmtacn based ligand 394
N,N′,N″-trimethyl-1,4,7-
  triazacyclononane 387
trispicolylamine (TPA) ligand 67
tris(2-pyridylmethyl)amine (TPA) 265
tris-pyrrolidine-1,4,7-triazacyclononane 394
tungstate-exchanged Mg-Al layered double
  hydroxide (LDH) 303
tungsten
– epoxidation 42
– (*n*-hexyl$_4$N)$_3$[PO$_4$(W(O)(O$_2$)$_2$)$_4$] 42
– γ-[H$_2$SiV$_2$W$_{10}$O$_{40}$]$^{4-}$ 329, 342
– [{γ-H$_2$SiTi$_2$W$_{10}$O$_{38}$}$_2$(μ-O)$_2$]$^{8-}$
  compound 329
– Keggin-type silicodecatungstate
  [γ-SiW$_{10}$O$_{34}$(H$_2$O)$_2$]$^{4-}$ 44, 328f.
– [M$_3$(H$_2$O)$_3$PW$_9$O$_{37}$]$^{6-}$ 334
– [M$_4$(H$_2$O)$_2$P$_2$W$_{18}$O$_{68}$]$^{10-}$ 319
– Ph$_2$PO$_2$[WO(O$_2$)$_2$]$_2^-$ 281
– {PO$_4$[WO(O$_2$)$_2$]}$^{3-}$ 327
– {PO$_4$[W(O)(O$_2$)$_2$]$_4$}$^{3-}$ 42
– PW$_9$-Fe$_2$Ni heteropolyanion 193
– [PW$_{12}$O$_{40}$]$^{3-}$ 327
– Q$_3$PW$_{12}$O$_{40}$ 281
– Se-contaning peroxotungstate [SeO$_4${WO
  (O$_2$)$_2$}]$^{2-}$
– [TM(H$_2$O)H$_2$W$_{17}$O$_{55}$F$_6$]$^{q-}$ 329
– Venturello-type peroxo complex Q$_3${(PO$_4$)[W
  (O)(O$_2$)$_2$]$_4$} Q=N-(*n*-C$_{16}$H$_{33}$) pyridinium
  281
– W(VI) peroxo complex 177
– {[(WZn$_3$(H$_2$O)$_2$][(ZnW$_9$O$_{34}$)$_2$]}$^{12-}$ 320f.,
  331ff.
– {[(WZnMn(II)$_2$(H$_2$O)$_2$][(ZnW$_9$O$_{34}$)$_2$]}$^{12-}$
  polyoxometalate 330
– {[(WZnMn$_2$(H$_2$O)$_2$][(ZnW$_9$O$_{34}$)$_2$]}$^{12-}$
  342f.
– WZnMn$_2$(ZnW$_9$O$_{34}$)$_2^{2-}$ 281
– [WZnRu$_2$(OH)(H$_2$O)(ZnW$_9$O$_{34}$)$_2$]$^{11-}$ 247,
  267

– $\{[(WZnTM_2(H_2O)_2][(ZnW_9O_{34})_2]\}^{q-}$ 318, 330f.

u
urea/hydrogen peroxide (UHP) 55ff., 409

v
(+)-valencene 437ff.
vanadium 175
– alkane sulfoxidation 214ff.
– enantioselective sulfoxidation 285
– $VO(acac)_2$ 215f.
vanadyl acetylacetonate 3f.
vanadium-containing Keggin-type polyoxometalate 333
vanadium-substituted polyoxomolybdate 321
vancomycin-type antibiotic 428
Venturello anion 46
Venturello catalyst 42f.
Venturello-type peroxo complex 281
verbenol 439ff.
(+)-cis-verbenol 441
(+)-verbenone 441

vinylcylohexane 340
vinylcylohexene 340
4-vinylcyclohexene 390

w
Wacker reaction oxidation
– terminal alkene 334
water oxidation center (WOC) 372
Wells-Dawson-type polyoxometalate 318

x
xanthene
– Mars–van Krevelen-type oxygenation 338
m-xylene 194
o-xylene 194
p-xylene (PX) 194ff.

y
ynone 204

z
zeolite
– NaY 63